PHYSICAL SCIENCE

Fifth Edition

R. Terrance Egolf
Donald Congdon

Greenville, South Carolina

PHYSICAL SCIENCE
Fifth Edition

R. Terrance Egolf, CDR, USN (Retired)
Donald Congdon, MA

Consultants
Cheryl Batdorf, MEd
Eugene Chaffin, PhD
Ron Samec, PhD
Linda Shumate
Brian Vogt, PhD

Bible Integration
Brian Collins, PhD
Bryan Smith, PhD

Project Editor
Rick Vasso, MDiv

Concept Design
Drew Fields

Cover Design
Drew Fields

Page Layout
Peggy Hargis
Northstar Creative

Designer
Sarah Ensminger

Permissions
Sylvia Gass
Lilia Kielmeyer

Illustration
R. Terrance Egolf
Sarah Ensminger
Preston Gravely
Brian Johnson
Ethan Mongin
Kathy Pflug
John Roberts
David Schuppert
Lynda Slattery
Heather Propst Stanley
Del Thompson
Courtney Godbey Wise

Project Coordinator
Donald Simmons, MBA

Third Edition published as *The Physical World*, An Introduction to Physical Science by Donovan Hadaway, David Hurd, PhD, John Jenkins, MS, and George Mulfinger, Jr., MS. Second Edition published as *Basic Science* by John Jenkins, MS, and George Mulfinger, Jr., MS. First Edition originally published as *Physical Science* by Emmett Williams, PhD, and George Mulfinger, Jr., MS.

Photograph credits appear on pages 585–87

Front and Back Covers
The photo on the covers is of a ferrofluid exposed to a strong magnetic field. A ferrofluid is a colloidal suspension of a ferromagnetic material in an organic solvent. The pinnacles are oriented in the direction of the magnetic lines of force emanating from the magnet beneath.

Produced in cooperation with the Bob Jones University Division of Natural Science of the College of Arts and Science, Bob Jones Academy, and BJU Press Distance Learning.

ISBN 978-1-60682-464-1

15 14 13 12 11 10 9 8 7 6 5 4 3

CONTENTS

Read Me First!

Every person is curious about the world. That's the way God designed us. Do you wonder how airplanes fly, why lightning strikes, or how those *E-ZPass* tags work at interstate tollbooths? What factors affect the rhythm of a pendulum? Why does it require a *lot* more braking force to stop a fast car compared to a slow car?

Well, OK, so maybe you haven't asked these particular questions. But you probably know that people are concerned about air pollution. In some cities, this *real* problem seriously affects the health of people. And what about car safety? Thousands of people die in car accidents every year in the United States. Why can car accidents be so lethal? We need cars, so is there a way to make them safer?

Smog is a serious health hazard in many large cities of developing nations.

Solving Problems Through Physical Science

This physical science course isn't just a "hoop you have to jump through" to graduate from high school on your way to college. *You need to learn how the world works first.* You might be able to get by without understanding forces or chemical reactions, but to really enjoy the world you need to know about these things—and much more. Even if you don't plan to be a scientist, knowing what causes color in paint and harmonics in sound allows you to more quickly develop your talents in art, photography, or music.

But there is something even more important. As you mature, you are learning that there are lots of challenges out there in the world. In the news, you read about climate change, growing human populations, devastating diseases, energy shortages, and so on. Christians need to be a part of the search for solutions to these concerns. In Matthew 22:37–40, Jesus proclaimed that the two greatest commandments are to love God with all our being and to love others as ourselves. Just as Jesus healed others during His ministry on Earth, so too do Christians have a similar responsibility to show the love of Christ to others as we work to find solutions to problems in our fallen world.

Solar thermal power plants are just one of the advanced renewable energy sources engineers and scientists are developing to meet the world's growing energy needs.

There are many ways to obey these great commandments, and physical science provides opportunities for both. By doing science, we model the laws or ordinances that God has established to govern the operation of the world (Job 38:33; Jer. 31:35). These laws make the world understandable so that we can study it. As we uncover the secrets of creation, a Christian is moved to glorify God and praise Him for His goodness and wisdom. At the same time, we can use the knowledge we gain from physical science to help other people. Physical science has contributed to improving medicine, developing tools and machines that are more efficient, and protecting and saving people's lives. Throughout this course you will encounter more ways that physical science has solved human problems in the past and is contributing to solutions to problems that remain to be solved.

Organization of This Textbook

Just as you must build the walls of a house before you can put on the roof, learning scientific knowledge must be done in an orderly way so that you can understand it. We have tried to arrange the sequence of topics in *Physical Science* so that you will have the necessary foundation on which to build more-advanced concepts as you work through the textbook.

The first unit (Chapters 1–3), titled *Foundations*, lays the groundwork. In these chapters you'll learn of three key concepts—you'll discover the nature of the world in which you live, and that what a scientist does depends on his beliefs. You'll learn the fundamentals of matter that underlie all areas of physical science. And you'll discover the principles of making scientific measurements.

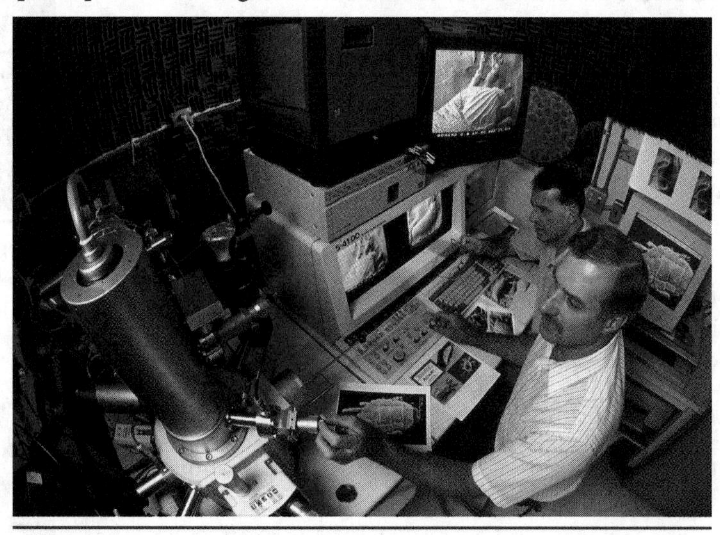

Observing and measuring are important elements of science.

In the second unit (Chapters 4–9), *Mechanics*, you'll explore how objects move and what causes them to move. You may also discover that more forms of energy exist and interact with matter than you realized. These energies can change into other useful or not-so-useful forms of energy. With this background you will be able to understand how liquids and gases move, transmit energy, and do work. Finally, you'll connect these principles of mechanics with their effects on matter at the particle level. These effects are measurable as thermal energy.

Magnetism—or in this case the lack of it—can result in some unexpected observations.

The third unit (Chapters 10–11), *Electromagnetism*, is a brief introduction to essential principles of electricity and magnetism. You'll discover that modern digital technology relies on these basic principles. They also account for the structure and properties of matter.

The fourth unit (Chapters 12–15), *Periodic Phenomena*, covers how energy and matter undergo periodic changes that allow energy to move over very short to very great distances. You'll begin with the principles of periodic motion and then apply these ideas to both sound and electromagnetic waves. This background will allow you to study the most familiar form of electromagnetic energy—visible light.

All this knowledge leads to the fifth unit (Chapters 16–17), *The Structure of Matter*. You'll learn about the discovery and study of the basic particles of matter. Then you will discover the history of the periodic table that displays the orderliness in the chemical elements.

In the final unit (Chapters 18–21), *Introduction to Chemistry*, you'll put all the preceding knowledge to use discovering how the atoms of elements join together and understanding the various kinds of changes chemical substances can experience. You will see that many forms of matter are really mixtures. Mixtures and solutions are important both to physical science and life itself.

A theme runs throughout the text—the development of scientific models. These models, constructed and modified many times through centuries of history, organize and give meaning to scientific research today. The change in scientific knowledge isn't a progression from ignorance to ultimate truth. The history of science records major changes in how we understood the world around us.

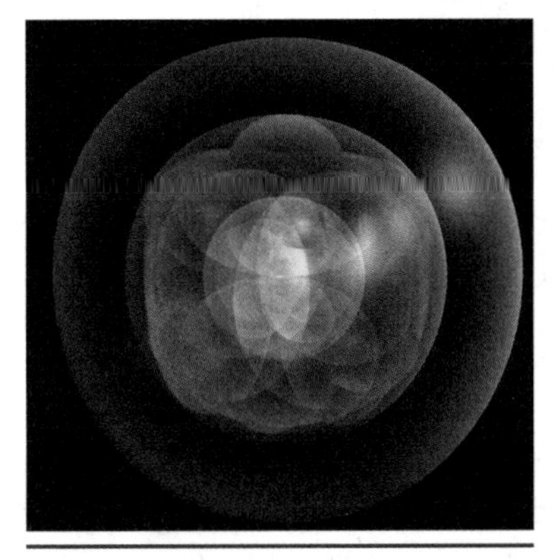

The quantum atomic model is one of the most complex and detailed scientific models today.

The more we know about the world, the more complex, interesting, and perplexing it seems. This is a testimony to the infinite complexity of our Creator. For the curious believer, physical science can be a life-changing journey.

The particles of many forms of matter are arranged in orderly patterns as shown in this electron microscope image.

How to Use This Textbook

This textbook has many features to help you get excited about physical science. Each unit begins with a unit opener, which introduces the topics covered in the unit. Take time to read these descriptions.

Dominion Science Problems

Every chapter begins with a **dominion science** (DS) problem. It describes a real-life problem that people face now or have faced in the past. The information that you will learn in the chapter has helped or is helping to solve this problem. During your studies, see if you can figure out the problem's solution before you come to it later in the chapter.

Dominion science is what we call any scientific work that helps us obey the Creation Mandate that God gave Adam in Genesis 1:28. You'll learn more about this in Chapter 1.

Chapter Divisions and Reviews

Each chapter is broken down into lettered sections (e.g., 1A, 1B, 1C) that cover major topics. These sections are further divided into subsections that are numbered consecutively through the chapter (e.g., 1.1, 1.2, 1.3). A list of objectives for the section, similar to the one in the adjacent margin, appears at the beginning of each section. If you can meet each of the learning objectives, then you have the minimum knowledge expected for that section. At the end of each section is a set of review questions. When reviewing for a chapter test, review the section objectives again to refresh your memory of the key ideas in each section.

Objectives

After completing this section, you will be able to

- ✓ state the purpose for studying physical science.
- ✓ describe the topical content of each unit in the textbook.
- ✓ identify the various features of Physical Science that aid learning.
- ✓ use the phonetic guide found in the introduction as an aid to pronouncing unfamiliar terms.

14B Section Review

1. Why is EM energy also called radiant energy?
2. How do the frequencies of radio waves compare to all other bands of EM energy? How do radio waves form?
3. How are microwaves similar to radio waves? How are they different?
4. What defines the upper limit for the IR band? How is IR radiation formed?
5. What is the range of frequencies for visible light in terahertz? How are the photons of visible light formed?
6. How is UV light produced? What are the main hazards from frequent exposure to UV light, such as from the sun and in tanning salons?
7. How are x-rays produced? Why do you sometimes wear a lead-lined apron when you receive a dental x-ray?
8. How are gamma rays formed? Can certain frequencies of EM waves include both gamma rays and x-rays? Explain.

DS ✪ 9. How is EM radiation used to treat cancer? What is this type of treatment called?

DS ✪ 10. Though radiation treatment can kill cancer, it can hurt healthy tissue too. Suggest some ways that doctors can avoid damaging healthy tissue during radiation treatment.

Many section and chapter reviews include questions with blue (✪) or red (✪) bullets. The blue bullets indicate that some extra effort is required, such as a math problem solution. Red bullets indicate that outside research or work should be done. Your teacher may assign these as interesting learning projects. Questions flagged by DS apply to the dominion science problem discussed in the chapter.

Special Words

As you read through each chapter, you will notice that some terms are in **boldface**. These are the vocabulary terms, which appear at the end of the chapter in the chapter review in a box labeled "Scientifically Speaking." Some words are *italicized*. These words help communicate an important idea by emphasizing a particular point. Others are terms that are generally important to scientists. You should be familiar with them to better understand scientific articles in news media or on the Internet. Pay close attention to these italicized terms.

Also, throughout the text you will see words or phrases colored blue. These words are linked directly to a box in the margin of the page on which they appear. The margin box contains information related to the word(s) in blue. We use these margin boxes to provide a guide for pronouncing the term, an etymology to help you better understand the term's meaning (also see Appendix A), a more complete definition, or a brief biography of a scientist whose name is linked to the margin box.

etymology: etymo- (Gk. *etymon*—true meaning) + -logy (Gk. *logos*—speak of). An etymology is the derivation of a word's meaning by describing the meanings of the parts of the word in their original languages.

Facets and Information Boxes

Most chapters include one or more facets or boxes that contain extra information set off from the text. *Facets* are articles that provide extra information that relates to the main text. Normally, review and test questions covering facet content are optional, unless your teacher says otherwise. Read them anyway—they're pretty cool! Large text boxes, such as *Math Help* boxes, contain important algebraic information that you should study. You may be able to use their contents in other courses you are taking.

What Is the "Actual Value"?

A lot of people believe that if we just had a good enough instrument, then we could obtain an exact measurement. But reality doesn't support this idea. Recall from Chapter 1 that we can't explain or describe natural phenomena exactly. If we could, this would mean we know something to be absolutely true. You could look at measuring as another kind of modelmaking. Each measurement *represents* a dimension approximately; it's not the dimension itself.

Imagine zooming in with a supermicroscope on the edge of a razor blade. At normal magnification, it looks sharp and straight. But as you zoom in, the blade starts looking rough. Closer still, the edge begins to look like mountains. At the atomic level, the atoms are jiggling around in a vast, heaving landscape of caverns. So how can you exactly measure something that is so irregular? All we can do is make measurements to an exactness that we define as acceptable.

Even though every measurement includes err
good reasons for trying to minimize the error. Meas
pect of exercising dominio
the error of measurements
able to develop really usefu
and technology. There wo
that could lead us to incorr
things that we build, like
wouldn't fit together right i
ent parts used different m
ment having different allow

We try to minimize t
urement by using instrum
curate as needed and that
required for our work. L
often-confused terms.

The Stuff in the Back That You Never Read

Physical Science also has three important sections that you should skim over right at the beginning of this course: eight interesting appendixes, an extensive glossary of definitions, and a complete index.

The Appendixes are important references to help you better understand and use information in this textbook. Take time to read Appendix H, *Worldview, Apologetics, and Evidence,* as you study Chapter 1.

The Glossary contains definitions for vocabulary terms and many other words found in this textbook. These definitions are generally more complete than those given in the running text. Entries in **bold** text are definitions of vocabulary terms listed at the end of each chapter. Entries in ***bold italic*** text are important scientific concepts and terms that will help you better understand the subject. Cross-references listed at the end of entries link to other terms in the Glossary.

The Index helps you quickly locate information in the textbook. Page references in standard typeface signal that the reference is located within the body of the text. Page numbers for terms mainly discussed in margin boxes are *italicized.* **Boldface** page references indicate that we have provided an image of the entry. Page ranges may include margin text and images of the entry.

Remember that a textbook is just one of many tools at hand in the adventure of science. As the authors, we hope and pray that Physical Science 5th Edition will strengthen your faith in the authority of Scripture. The Bible is the Word of God and is as applicable today as the day that its individual books were written thousands of years ago. We hope this course will inspire a lifelong love of learning and of loving service to others.

GLOSSARY

Glossary entries in **bold** text are definitions of vocabulary terms listed a
text are important scientific concepts and terms that will help you better
at the end of entries link to other terms in the Glossary.

A

absolute pressure **(155)** A pressure reading referenced to a vacuum. In contrast, see *gauge pressure.*

absolute temperature **(*T*) (140)** The temperature of a system referenced to absolute zero. The scale's degree is the kelvin (K). See also *Kelvin scale.* See *temperature (t); absolute zero; degree (°).*

absolute zero **(187)** The lowest theoretical temperature; defined as 0 K (about −273.15 °C or −459.67 °F). Physicists believe that all motion of matter ceases at this point. Laboratory temperatures have been attained to within billionths of a degree of this point. See *temperature (t).*

The Appendixes, Glossary (shown), and Index are important tools for better understanding physical science. Get to know them!

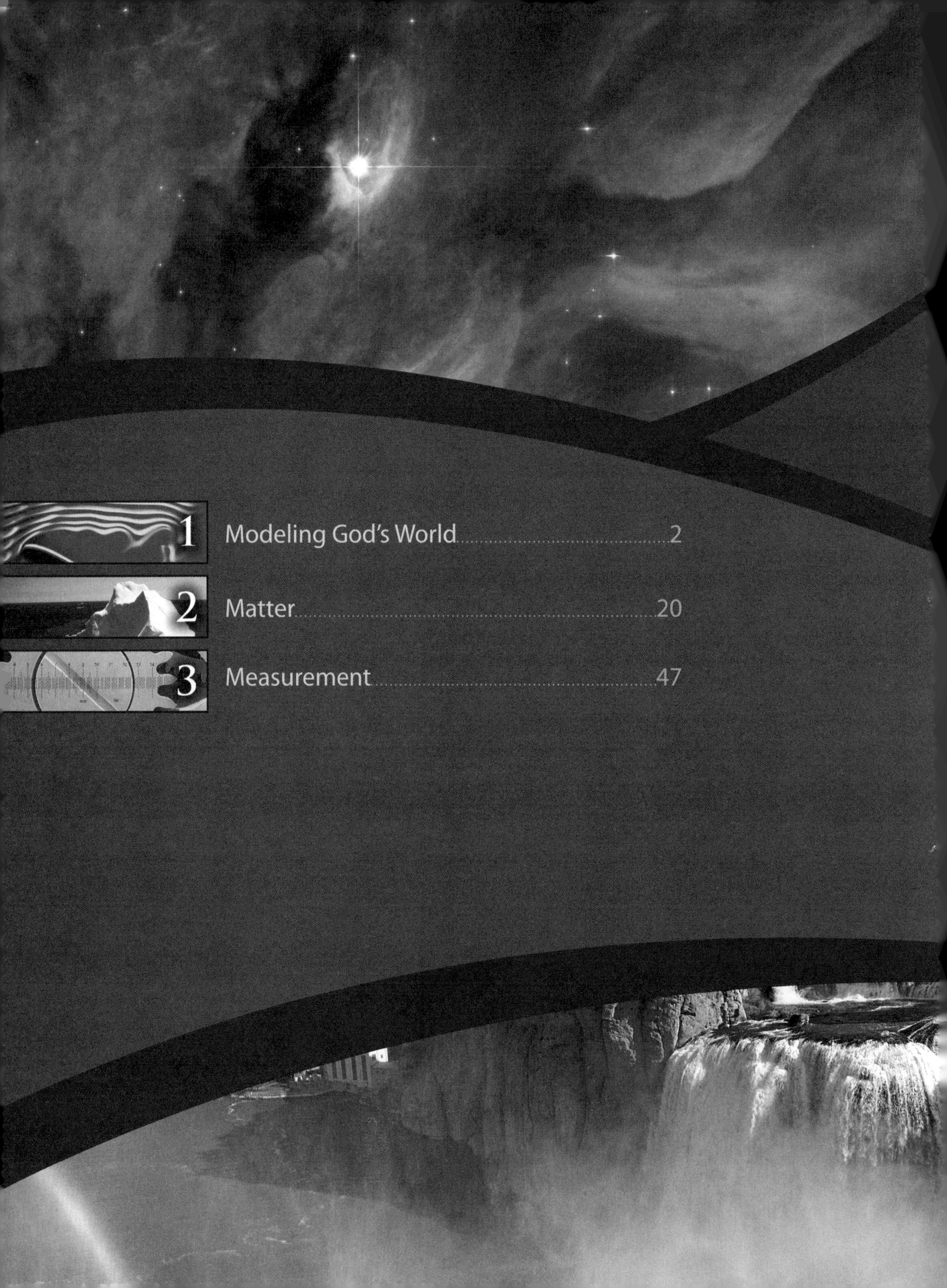

Science isn't what it used to be.

Generations of people have come and gone believing that science was the path to a true understanding of the natural world, but this perception is changing. People are realizing that different groups of scientists hold different assumptions about how the world works and that they come up with very different explanations for some observations based on their assumptions. The origin of matter is the most obvious issue. Most scientists assume that the entire universe came into being through natural processes over billions of years. Scientists who believe the Bible's historical narrative assume that the earth and all living things were specially created by God about 7000 years ago. Thus, each group interprets the same data quite differently. In the midst of such controversy, the general public is slowly realizing that scientific data must be interpreted within a philosophical framework.

In Unit 1 you will learn about the basis for this controversy—that real science is primarily about making workable models of the observable world, while truth is found only in God's Word. With this foundation you can develop a solid understanding of matter and scientific measurements, the framework of physical science.

UNIT 1

Modeling God's World

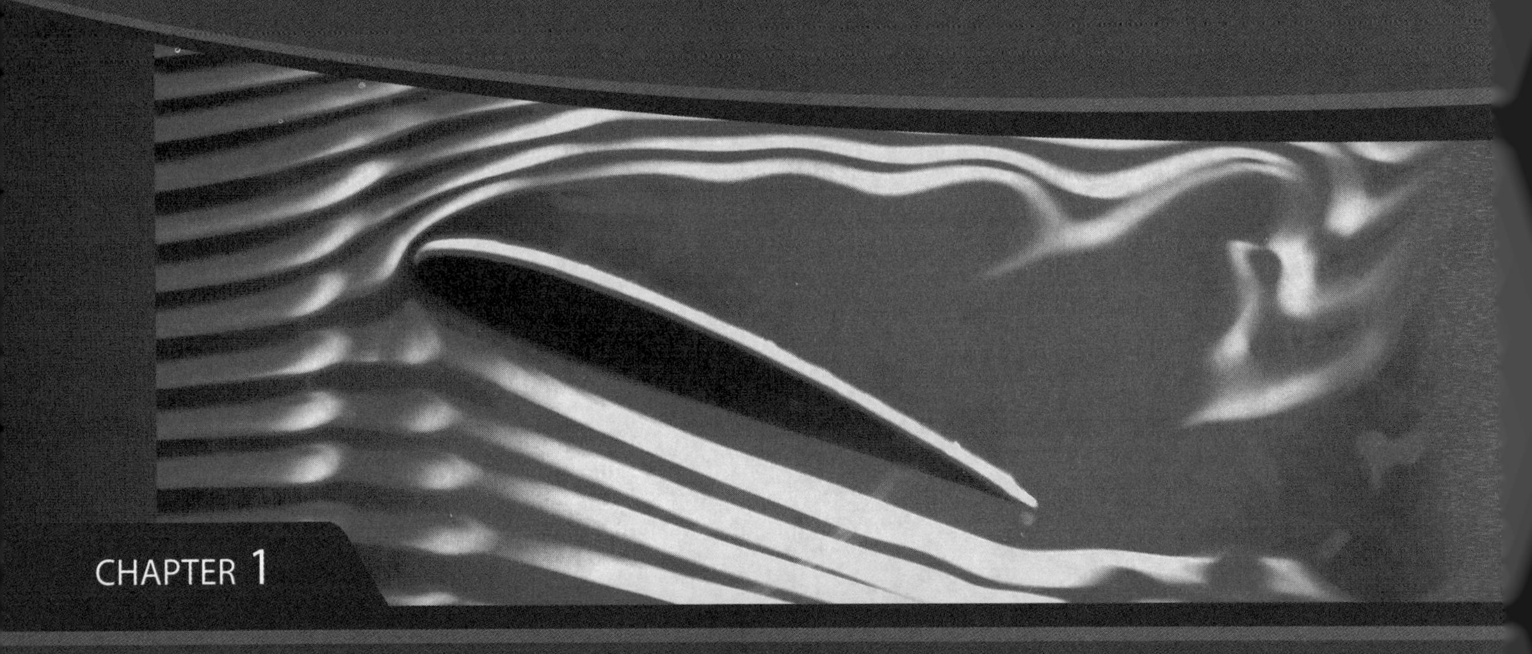

CHAPTER 1

1-1 An Airbus A330-203 similar to the Air France Flight 447 aircraft

Lost in the Dark

Everyone aboard Air France Flight 447 (AF 447) was expecting an uneventful trip. The passengers came from many countries around the world. They included newlyweds, doctors, students, lawyers, artists, scientists, and business persons. At 7:03 PM on 31 May 2009, the Airbus A330-203 jumbo jet carrying 216 passengers and 12 crewmembers took off from Rio de Janeiro International Airport in Brazil heading for Paris, France. Powered by two large turbofan engines, it quickly climbed to an altitude of 35,000 feet and settled in to a cruising speed of about 537 mph.

The flight's path took it northeast of Brazil into the stormy Equatorial Intertropical Convergence Zone above the Atlantic Ocean. For the first 3 hours of the 11-hour flight, everything went normally. Air traffic controllers spoke routinely with the pilots. But after that time, controllers couldn't regain radio or radar contact with the aircraft. Increasingly desperate, they tried over and over again to call the plane. But it was no use. After 8 hours of silence from AF 447, they realized that the aircraft was missing, along with all 228 people onboard. AF 447 was lost somewhere in the dark and stormy Atlantic—one of the worst disasters in commercial aviation history.

1A What in the World … ?

1.1 The World as It Is

Where was Air France Flight 447? What happened? Were there injured people floating, stranded in the dark, cold waters of the Atlantic? If so, where were they?

Rescue teams from countries on both sides of the Atlantic leaped into action. Brazilian and French aircraft began searching miles of ocean along the plane's expected flight path. Families of the passengers and crew gathered and began their long, frightened vigil, waiting for news. Experts collected what little data there was from the radar controllers, other aircraft, and ships near the plane's flight path.

That accidents and natural disasters kill and injure people every day makes them no less devastating. But why does God allow these things? Why is there suffering in a world created by a good, loving, all-powerful, all-knowing God? The Bible has the answer.

God didn't create things to be this way. The fault is ours. In Genesis 3 we learn that Adam and Eve disobeyed God, and there was a punishment—death. Their sin plunged the whole world into its broken, fallen state. People began to struggle for survival. Nature seemed set on fighting man's attempts to use it. Food was no longer free for the picking; man had to toil for it. Life became dangerous and hard. This is the world we inherited, the world where disasters happen.

But the effects of the Fall don't stop there. We are Adam and Eve's children, and we follow in their disobedience. Not only is everything outside of us broken, but everything inside us is too. Our thinking is fallen. Our reasoning, observing, and planning are clouded by pride, envy, fear, selfishness, and a desire to live independently of God and His Word. We don't think about God and our place in His world as we should.

Our fallen minds affect our ability to do physical science. Modern science largely encourages fallen people to see only natural processes and their effects. It views the earth as unimaginably old. Continents drift slowly as they have for billions of years. Life arose from nonlife long ago, requiring huge amounts of time to evolve. So every part of life appears different because of our fallen state. Even something like looking for crash survivors and trying to figure out what happened is affected by our fall into sin. We prefer not to think about how such disasters remind us of our own sinfulness and God's judgment on us for our sin. In our natural state, if we think deeply at all, we prefer to think that either God doesn't exist or He is unjust, and so life has no meaning.

1.2 People Matter

Right after the crash of AF 447, no one even had a clear idea of where the plane went down. As hours stretched into days, hope of finding any survivors was lost. Five days after the crash, searchers found and recovered the first bodies and bits of wreckage. Many ships, aircraft,

1A Objective

After completing this section, you will be able to

✓ justify from the Bible why a Christian should understand and use science.

1-2 Frightened, grieving people waiting for news

Theologians call the important moment when man sinned early in our history the *Fall.* All the struggles, disasters, pain, and death we suffer began with this event.

1-3 People often try to make sense of the world without knowing God and His Word.

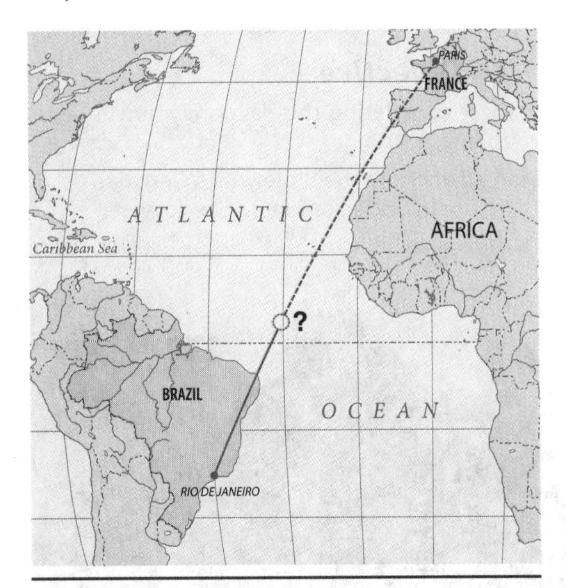

1-4 This is a map showing the approximate location where contact with Air France Flight 447 was lost. Searchers had no really good idea where to begin.

satellites, and even a French nuclear submarine searched the area for information. Why do people expend so much effort looking for the bodies of complete strangers?

People from around the world spent hours searching because people's lives mattered to them. People are valuable because they are created in the image of God (Genesis 1:26–27). It's not that we look like God. God is Spirit and doesn't have a physical head, body, and limbs as we do. Rather, we bear a likeness to the qualities of His person in almost every other way—intelligence, will, and emotion. God equipped us with these so that we can relate to Him and each other. Each one of us is valuable to God, now and for all eternity, because He created us. God's image in us makes us want to help, protect, and comfort others. When we live in ways that reflect His image, we live for God's glory.

Let's Find Out Why.

Scientists, pilots, ship and submarine crews, and politicians rallied from around the world to solve the mystery of AF 447 for another important reason. They felt the burden to keep something like this from happening again by finding the answer to the question, "What caused this plane crash?" Answering this question involved using science and technology. Experts pored over the very limited evidence they had in order to develop an explanation for the accident. The evidence they needed included the pieces of the aircraft, the kinds of injuries on the bodies of the victims, and—most importantly—the plane's automatic flight data and voice recorders. Investigators struggled to work out the events leading up to the crash as these pieces of evidence began to trickle in.

Exercising good and wise **dominion** means taking full responsibility for the care and management of every aspect of our world, and using it under God's authority.

We are driven to understand the world. The longing to answer questions and solve problems is woven into our being. God made humans to subdue the earth and to have **dominion** over it. Genesis 1:28 records the command called the **Creation Mandate**. God made people His representatives and caretakers of His world. If we live out this command in obedience to His Word, we are glorifying God.

1-5 Each of us bears God's image.

So how can we declare God's glory in physical science? We do this through *exploring*, *explaining*, and *describing*. Physical science opens up amazing opportunities for these activities. Through physical science we can explore how to best use the earth and its resources to live in a fallen world. To protect and improve people's lives—to love our neighbors—we need to describe the earth, its oceans, atmosphere, energy, forces, and chemistry. To manage the living things God created, we have to explore what they eat, and describe how they relate to each other and the physical world. Clearly, science is key to exercising dominion.

The work of explaining the mystery of the AF 447 crash is similar to other kinds of investigations that scientists do. They were limited in their ability to search vast tracts of ocean or the ocean floor so they made use of technology such as satellites, aircraft, submersibles, radio communications, and radar. Our ability to create technological things is evidence of the image of God in man. Creating is one way to reflect God's character, helping us do a better job obeying the Creation Mandate and declaring God's glory. Technology like that developed and used to unravel the mystery of AF 447 involved many areas of physical science.

1.3 *Living for Redemption*

Probably most of the people who worked to recover the victim's remains and piece together the evidence from AF 447 weren't believers. Though they did good and useful things, they didn't have the perspective that God's Word gives to Christians.

Christians, of all people, should seize opportunities to lessen the consequences of the Fall and maximize the usefulness of God's world. Jesus set the example for us. He taught the gospel to rescue people from fallen thinking and He healed people's physical hurts to rescue them from the consequences of living in a fallen world. His acts of mercy were proof that He was who He claimed to be, God's promised Redeemer.

Christ has redeemed His people from a guilty conscience (Heb. 10:22), a sinful life (1 Pet. 1:18), and wrong thinking (2 Cor. 10:3–5). Believers can follow Christ's example by doing science in ways that showcase biblical thinking and help people who live in a fallen world. God's Word gives us the reason and the way to do this. We can use science to heal diseases, provide people with fresh, clean water, warn them of natural disasters, and investigate the causes of accidents like AF 447. This is science that declares God's glory.

1-6 Aircraft searching for the crash site covered lots of ocean quickly, but seeing small pieces of debris was difficult.

1-7 The people who were involved in recovery of aircraft wreckage and victims probably weren't all believers. But they wanted to return to families the remains of the victims as well as to find evidence that would help prevent similar disasters in the future. This is evidence for the image of God in man.

1A Section Review

1. *Why* (not *how*) is the world today not the same one that God created in the beginning?
2. What kinds of character flaws resulting from our fallen condition make doing science especially difficult?
3. What are two possible conclusions that people in their natural state may come to when thinking about disasters, God, and the purpose of life?
4. Why are people generally concerned about the physical welfare of complete strangers?
5. How does the Creation Mandate shape our relationships with God, other people, and the world?
6. Why should Christians, of all people, try to lessen consequences of the Fall, such as plane crashes?
7. (True or False) The way a scientist does his work is unaffected by the Fall.

1B Objectives

After completing this section, you will be able to

✓ summarize the distinctives of a Christian worldview.

✓ contrast a Christian with a non-Christian view of science.

✓ explain the role of models in science.

✓ contrast the main kinds of scientific knowledge.

✓ compare operational and historical science.

Data is any kind of information or observation used in an investigation. Most scientific data is either measured or described.

pitot (PEA toe); from Henri Pitot (1695–1771), a French scientist who suggested a device for measuring liquid flow. A pitot tube is a tube bent at a right-angle mounted on an aircraft. It measures the pressure of the air flowing past it; in the cockpit, this calculation is displayed as the air speed of the plane.

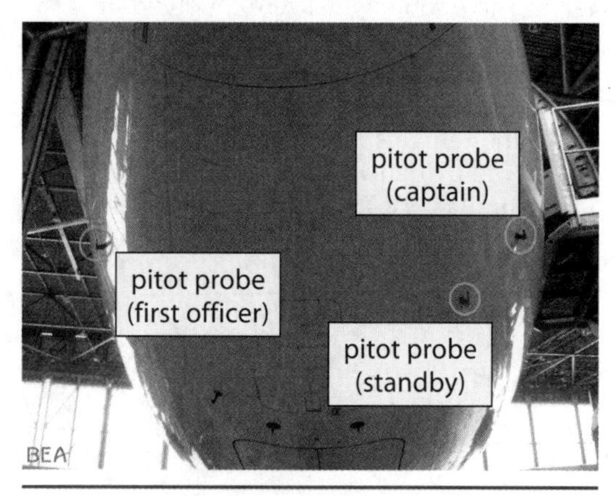

1-8 These are the three pitot tubes that investigators suspected may have frozen up, providing incorrect airspeed indications during AF 447's flight.

1B Science with a View!

1.4 The Search for Information

Crash investigators needed **data** to solve the mystery of AF 447's crash. Following the tragedy, information began to trickle in. Reports from other planes near AF 447's flight path of flashes and debris on the water provided some clues. Back in Paris, Air France technicians analyzed equipment error messages that the aircraft computer had automatically transmitted shortly after the plane flew out of radar contact. These messages indicated that some very strange things had happened.

Authorities from several countries began to piece together this data. Were terrorists involved? Did the plane blow up? Was the aircraft struck by lightning? Was there a major equipment failure? Other aircraft with the same type of air speed detectors, called ***pitot tubes***, had experienced problems when the tubes had frozen up in cold, moist air. This problem could account for many of the equipment error messages that the plane's computer sent.

What made all this so mystifying was that the Airbus A330 aircraft was a technological marvel. Things like this weren't supposed to happen to aircraft with state-of-the-art digital control and safety systems. There were backup systems for backup systems! And furthermore, three experienced pilots were flying it. It was difficult for investigators to believe that anything short of a major equipment failure could have caused such a crash.

Without more data, investigators realized that they had little hope of solving this mystery. However, because of crashes in the past, nearly all passenger planes now carry two important pieces of equipment to help with this problem. They are the *flight data* and the *cockpit voice recorders*. These devices automatically and continuously record information about the aircraft as well as the conversations of the pilots in the cockpit. Armored containers that can withstand nearly any kind of crash protect them. But to get their hands on this data, investigators had to find the wreckage of the plane at the bottom of the Atlantic.

1.5 Assumptions and Problem Solving

The work of the crash investigators trying to solve the mystery of AF 447 is the same kind of work scientists do. Both are trying to answer questions and solve problems. Both groups acquire a set of assumptions from their education, training, and professional experiences. These assumptions are an important but unconscious part of their work.

You operate the same way. You live with basic assumptions about life too. You assume that the world is a real place—not an illusion—and that it behaves predictably. When you sit on a chair, you automatically assume it will hold you up. You make plans for the weekend because you assume that you will be alive then. You have assumptions about right, wrong, and the nature of truth. These basic assumptions are called **presuppositions**.

The presuppositions of scientists affect how they solve problems and answer questions. What kinds of presuppositions do they have? For starters, they assume that the world they study is understandable and works in a predictable, reliable way. They assume that for every process there must be an adequate cause (the "cause and effect" principle). There is no way to scientifically prove these assumptions. But since scientists can't do their work without them, no professional scientist doubts their truth.

The concept that natural processes work in generally reliable, predictable ways is called the *principle of uniformity*.

1.6 Evidence and Worldview

Scientists rely on their presuppositions to make sense of a scientific problem. For example, in 1993, a scientist was studying rock-like fossilized bones from a *Tyrannosaurus rex* dinosaur that she believed were about 68 million years old. There was no question in her mind about this age because evolutionary and geologic science had dated the strata of these fossils authoritatively. Yet when she accidently broke open one of the fossil bones, she discovered what appeared to be soft, stretchy tissues and red, blood-like matter inside! Other scientists disputed her findings, suggesting that the substances were modern bacterial residue. But further testing in 2010 confirmed that the materials were actually from a dinosaur. What did the scientists conclude? Instead of questioning their assumption that the fossils were inconceivably old, they decided that fossilized tissues could stay soft far longer than anyone could have imagined—for millions of years!

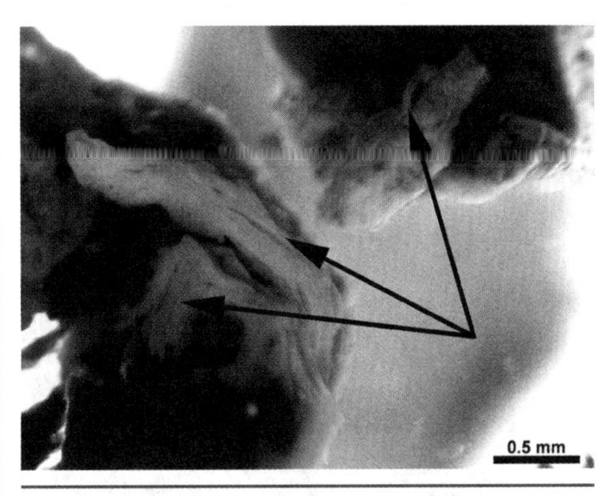

1-9 So is this stretchy dinosaur tissue 68 million years old or only about 5500 years old?

But there's another way to interpret this evidence. Christian biologists and geologists who believe that the earth is only a few thousand years old, as the Bible indicates, have a much different explanation. Standard scientific studies have shown that biological tissues in dead things do not last very long. Even in very unusual cases, like ancient, artificially mummified humans and animals, tissues break down and turn to dust after a few thousand years. So how do these believer scientists interpret this report? The dinosaur bones with soft tissues must be far younger than 68 million years. Not only does this interpretation better fit the standard scientific evidence; it supports the Bible's narrative suggesting a young earth.

How can qualified scientists looking at the same evidence come to such different conclusions? It's as if they were looking at these fossils through different kinds of glasses! Clearly, their presuppositions affected their interpretations of the evidence. In the case of the researcher, she changed what she believed about the decay of tissues rather than her presuppositions about the ages of dinosaur fossils. You can see how powerfully presuppositions can affect our thinking. Non-Christian or ***secular*** scientists build their ideas on top of their assumptions about long ages and biological evolution. Believing scientists make their framework the Bible's history—Creation, the Fall, the Flood, and God's redemption of the world.

The presuppositions that we rely on to solve a problem come from a much larger, deeper, and more basic part of who we are—our **worldview**. Everyone has a worldview. Your education, life experiences, family, training, and religious beliefs influence your worldview. Your ***presuppositions***, ***biases***, and ***prejudices*** are the way you express your worldview. Some presuppositions are practically "hard-wired" into you. Others change when you learn something new or have an experience that makes you change your mind about them.

We will occasionally refer to non-Christian professionals or ideas as **secular** ones. "Secular" comes from the Latin word *sæculāris*, which in philosophical or religious matters means "of the world," or "not sacred."

A **presupposition** is an assumption that a person makes about the world that helps him to investigate the world and understand it from a certain point of view.

A **bias** is an inclination one has about an idea after thinking about its pros and cons.

A **prejudice** is like a bias, but a person holds to it without thinking about it, and it is often unreasonable.

1.7 A Believer's Worldview

A person can have one of two basic worldviews. Either his worldview is built on the truths of the Christian Bible—or it isn't. A secular worldview starts with the assumption that some or all of the Bible is *not* true. People who have this worldview usually assume that mostly slow, natural processes have shaped the earth we see today. They also believe that biological evolution is responsible for producing all living things. If they hold to a belief in a god, as most of the world's religions teach them to do, they confine the god to certain areas of life or believe that he/she/it is mostly irrelevant. Everyone has faith in something. Even a secular view of life requires a leap of faith to believe in a purely material universe understood solely by human reasoning.

Since every scientist has a worldview, the ways he views data and the explanations that he creates to solve scientific problems are all affected by it. As we saw with the soft dinosaur fossil tissue, a scientist may think in ways radically different from someone who has a Christian worldview. However, a scientist with a secular worldview can still accomplish great things. He can work hard and successfully solve problems because he is fulfilling the Creation Mandate, though he may not know it. His worldview may likely even include, unconsciously, parts of a Christian worldview. Many of the scientists working to solve the AF 447 crash were probably operating this way.

1-10 These men are fulfilling the Creation Mandate as they recover debris from AF 447. It's possible that such data will help identify the cause of the crash so that similar accidents can be avoided in the future.

Nevertheless, when a person tries to understand the world without a Christian worldview, major problems appear. A secular scientist can't account for the obvious design and purpose in nature. To him, these are accidental results of blind chance. Also, though he wants to help people, he may not be able to explain why people are important. Or he may even suggest that people are no more important than "other animals."

A scientist working within a Christian worldview observes everything through the lens of Scripture. He sees evidence of design and of the Designer in nature. He should have an attitude of humility about his work as he discovers God's ordinances and explanations for things in God's creation. He values people and tries to help them because they bear God's image. Discoveries that he makes about Creation lead him to honor and praise his Creator (Matt. 22:37, 39; Rom. 11:34–36).

1.8 Physical Science and Modeling

Scientists work within their worldview to do science as they exercise dominion in either conscious or unconscious fulfillment of the Creation Mandate. But what *is* **science**? Many people, including philosophers, court justices, and scientists themselves, have tried to define science in simple terms. But science isn't so easily defined. It's more useful to *describe* what science is. Science is about people creating workable explanations or descriptions for the things and processes that we see in the world around us. An observable "thing or process" is called a ***phenomenon***. By *workable*, we mean that the explanations and descriptions are useful for understanding the world. They work.

phenomenon: (Gk. *phainomenon*—something that appears); pl. phenomena. A phenomenon is anything observable or measureable.

A **model** is a workable explanation or description. It's not complete, accurate, or even necessarily true. The best models are those that are most *useful* and help us to work toward new knowledge.

When a scientist tries to explain or describe a phenomenon, he constructs a scientific **model**. Models *represent* phenomena—they can never *be* the phenomena. For this reason, they are never perfect duplicates of what they represent. To say that an explanation or description of observed relationships is perfect is to say that it is true. Scientists do science, and scientists are humans with limited knowledge as well as fallen minds and hearts. It isn't possible for people to arrive at absolutely true knowledge about such matters apart from God's Word. So science is not actually about truth; it is about workability. Moreover, science can't even be called a gradual increase of knowledge building toward some undiscovered truth. The history of science is full of many examples of scientists replacing well-accepted scientific explanations by completely different explanations. See the facet on the next page to learn more about the nature and history of science.

For an example of a workable description as a model, study Figure 1-11. It is a model of a complicated molecule made of carbon atoms—a *fullerene*. It shows us the general arrangements of the atoms. But is it completely accurate in every detail? No. Real atoms don't have sticks connecting them. Carbon atoms aren't black. As far as we know, individual atoms don't have color. And real atoms and molecules aren't solid or rigid. We could build a *physical model* of a fullerene molecule to place on our desk, but it will be just a simplified representation to help us understand its overall shape.

When you think about models, you probably think about model airplanes, cars, or ships. These are also physical models. But there are other types of models. You may have played video games or seen movies featuring 3D digital images. These are *virtual models* of things that may or may not exist in the real world. Scientists can use both of these types of models, but they also use *conceptual models*. These models are explanations or descriptions that exist only in the form of math equations or in a computer program. Other models exist as relationships among points or lines in graphs that we can't view any other way.

Should this view of science bother us? No! Even though scientific models are limited, that often makes them more useful for their intended purposes. Figure 1-13 shows two different models of the iron atom. The first model is useful for understanding the energy of electrons within the atom; the second is far more complicated and represents the probable regions that those electrons may occupy inside an iron atom. Just like tools you might use to build a birdhouse or to make a sculpture, science and the models it produces are tools to help us live in, understand, and use God's world. Each tool is useful for its intended purpose.

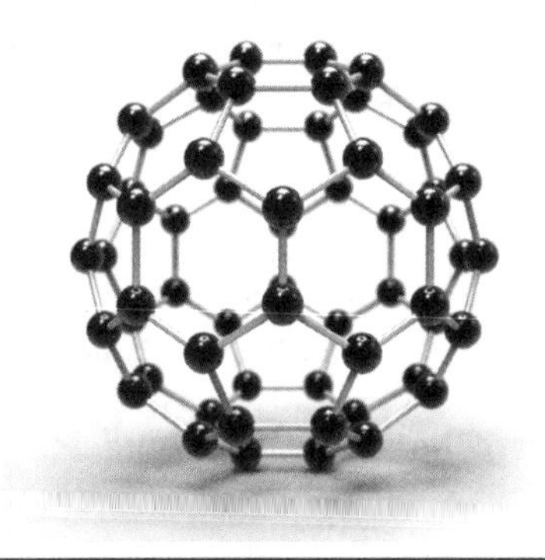

1-11 The geodesic structure of some organic molecules, like this fullerene made of 60 carbon atoms, demonstrates the unusual designs we can find in nature.

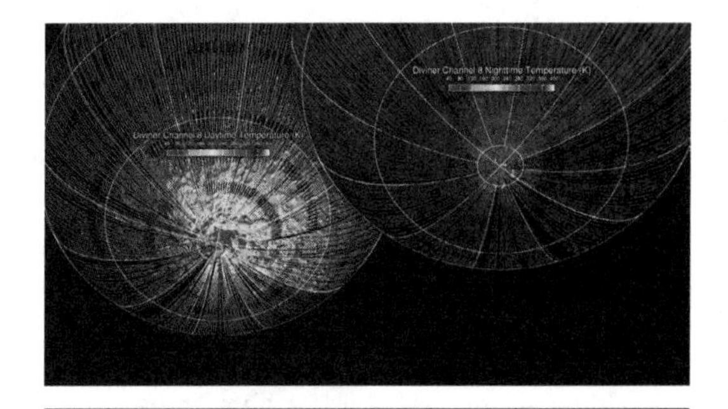

1-12 These virtual models represent the moon's surface temperature during daytime and nighttime.

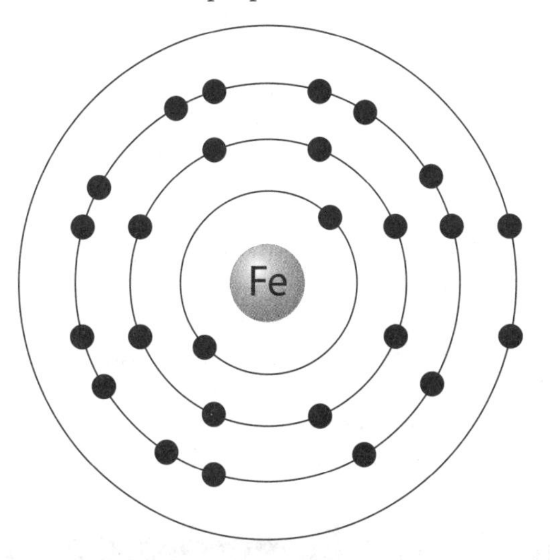

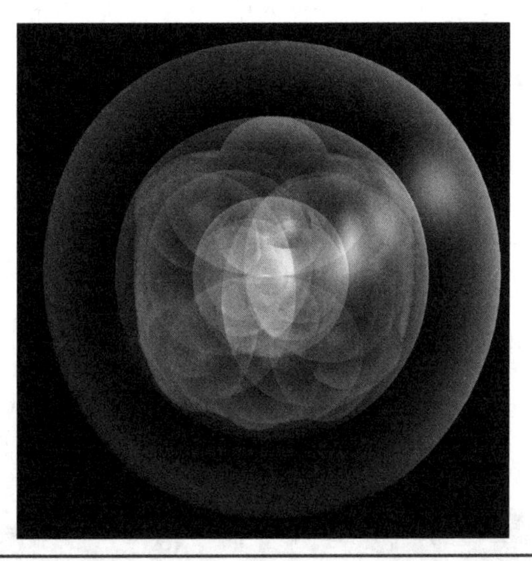

1-13 The diagram on the left is called a *Bohr model* of an atom of iron. Each ring contains electrons at similar energies. The right image shows a 3D model that represents the quantum mechanical description of an iron atom.

FACETS OF SCIENCE

Science and Truth

Can science determine what is true? Most people think that's the ultimate goal of science. Somewhere "out there" is a settled set of facts that will explain all we need to know about the universe and everything that is in it. Each scientific discovery brings us closer to the truth.

This has been an accepted view of science since the mid-1700s, when it came out of the European Enlightenment period. Philosophers call this view of science *positivism*. This is the idea that all matters of fact can be understood only by experience through the senses. Beyond sense-determined facts there are only pure logic and mathematics. Thus, reality can only be known through the formal sciences.

Since positivism denies any supernatural beings or processes, it became the logical philosophy behind science and society as Western cultures abandoned biblical authority. If an idea isn't supported by science, it can't be true. Many Christians have fallen into this way of thinking, even when the scientific "truth" contradicts what the Bible says.

However, not all philosophers agree with this view of science. One of the most influential philosophers of the twentieth century, the American physicist Thomas Kuhn, broke down these claims very effectively. First, he made popular the idea of the *scientific paradigm* (PAIR uh DIME). A scientific paradigm governs the way that a certain field of science is done. In general, a paradigm dictates the *working assumptions (presuppositions)* used in a branch of science. There are different paradigms for physics, chemistry, and biology, and some overlap all areas of science. You may have heard about the particle model of matter, the heliocentric theory, evolutionary theory, and Newtonian physics. These are all paradigms for their associated areas of studies.

All scientists must work within a paradigm to get science done. Kuhn notes that scientists often become so committed to their paradigm that they believe that they are dealing with fundamental truths of nature. This is why science is unlike many other branches of knowledge. Science allows only one paradigm to control an area of science at a time.

From day to day, scientists working within their paradigm solve interesting problems and make new discoveries. This activity is what Kuhn calls *normal science*. They develop better instruments to make measurements. They measure things more accurately and compute important number values to greater precision. And they try to solve current problems and challenges. Normal science helps to confirm the paradigm and to extend its workability.

The history of science, however, shows that scientific truths don't just keep piling up over time. This is where Kuhn took another jab at positivism. Scientists sometimes make observations that just don't fit into the paradigm that governs their work. Or the models that they must create in order to make the observation fit become really complicated. This can begin to spell trouble for the paradigm. If a scientific fact is true, then there shouldn't be any trouble finding a place for it within the body of scientific knowledge.

Consider a real example of this. Several centuries before Christ, the Greeks had decided that the earth was the center of the universe, and the sun and everything else revolved around it in circular orbits. This was obvious by just looking at the sky! This *geocentric model* was the paradigm for early astronomy and astrology. But to make the predicted positions of celestial objects agree with observations, the second century AD Greek mathematician Ptolemy had to increase the complexity of the model of the solar system. He added little circular orbits (called *epicycles*) to the main orbit circles (*deferents*). He moved the earth slightly off-center inside the circular deferents and added *equant* points. Others added more circles and revolving lines to keep the planets in their proper places. But the model was still inaccurate and had become almost impossibly complex to use.

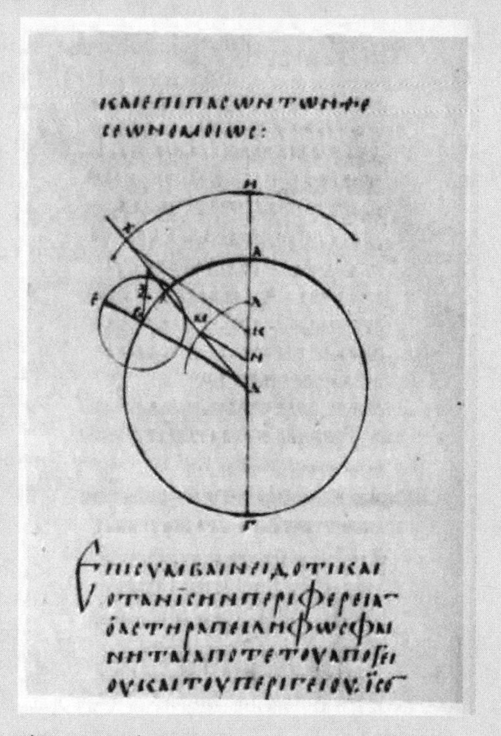

This page from Ptolemy's *Almagest* shows the main features of his geocentric model, including a *deferent*, an *epicycle*, and various *equants*.

In the sixteenth century, the Polish cleric Copernicus suggested a system with the sun in the center and all the planets, including the earth, orbiting around it in circular orbits. His *heliocentric model* wasn't much more accurate or easier to use than the geocentric one, but it was somewhat simpler. Galileo, Kepler, and others soon provided new evidence that this model was a better one for explaining the arrangement of the universe.

So which paradigm was the truth? Neither! Kuhn calls such changes in paradigms *scientific revolutions*. These have happened many times in all areas of science through history. So the idea that science builds toward an ultimate truth is simply not true. Kuhn doesn't see a movement toward truth, just movement. So normal science isn't about truth, but rather about solving problems .

1.9 Organization of Science

So how do scientists make models of phenomena they observe in the world? Models in science fall into one of two main categories—they either explain or describe phenomena. An accepted model that *explains* something is called a **theory**. A good theory must agree with the presuppositions of the scientific worldview that produced it. It must do a thorough job explaining all observations relating to it. And it should make predictions that can be tested, leading to further discoveries. Familiar scientific theories include the atomic theory, the theory of relativity, and evolutionary theory. Theories are necessary for doing science. Broader theories can include many more-specific theories under them.

Notice that an accepted scientific theory doesn't mean that it's *true* or even *plausible*. We can't determine truth through science. Plausibility rests on the presuppositions that produced the theory. To a secular scientist, evolution is both true and plausible. To a Christian, it is neither.

A model that *describes* a phenomenon but doesn't or can't explain it is called a scientific **law**. A law describes things that seem to hold true under certain conditions. For example, all objects that orbit the sun or a planet follow an elliptical path. This behavior is described by Kepler's first law. This law states that the shape of every natural orbit is an ellipse. It doesn't attempt to explain why this is so. In fact, Johannes Kepler himself didn't know why. A law may also describe relationships among very different quantities. The law of gravity states that the attraction between two objects is related in a special way to the distance between them. Gravitational attraction, distance, and the matter in objects are completely different quantities but the law of gravitation makes their relationship predictable.

Theories explain—laws describe.

Scientists approach their work in two main ways, and they often overlap. **Operational science** examines phenomena that exist or occur in the present. Such science observes and tests phenomena repeatedly to confirm a model. This is possible only for present-day phenomena or for evidence that observers documented well in the past. Operational science includes activities such as placing things into categories, researching the structure of matter, developing weather models, experimenting with new materials, analyzing the chemical processes in living things, and inventing the microprocessors for smaller and faster computers. Most science done today involves operational science. Since operational science allows us to use the world's resources and create things for our benefit, we can say that operational science is *dominion science*.

1-14 Engineers are testing the design of this rocket engine. This is an example of operational science. They can run the test as many times as necessary in order to obtain the data they need.

There's another kind of science that studies phenomena that happened in the unobserved past and that are *not* happening today (to our knowledge). We call this **historical science**. Models that scientists develop in historical science can't rely on direct observations of key events or processes. Instead, they rely much more on the scientists' presuppositions about the conditions they *believe* existed when the things they study happened. Historical science includes all areas of origins (the origin of the universe, the origin of the earth and its features, and the origin of life). Most aspects of ancient archaeology and anthropology fall into the historical science category if there is no credible written record. The fact that historical models can't be tested using the same methods as operational science is their greatest weakness.

1-15 These scientists are recovering the fossil bones of a large dinosaur. They can classify the animal and attempt to reconstruct its skeleton as part of operational science. But when they try to describe its living habits or how it came to be fossilized, they begin to work in historical science since none of these ideas can be confirmed in the present.

FACETS OF SCIENCE

Forensic Science

Forensic science, or *forensics*, brings to mind popular TV crime scene investigation (CSI) shows. Investigators try to piece together all kinds of evidence, with the goal of finding the guilty perpetrator (the "perp") and bringing him or her to justice.

Forensic science tries to identify the method of the crime, the time of the crime, and then link the crime to one or more specific people, when applicable. Forensics is most important when the crime was not witnessed as it happened.

When dealing with murders, forensic science can include just about any discipline within biology. Familiar forms of evidence include DNA testing to identify the criminal, dental records for identification of the victim, and pathology for determining the cause of death. Biological evidence can also include studying insects and plants present at the crime scene, and even the state of decay of the victim, to better understand the conditions and time of the murder.

Chemical forensics looks for poisons, explosive or gunpowder residues, or the source of chemical substances involved in a crime.

Forensics also relies on the physical sciences. Examples include ballistics tests on bullets to identify the weapon used in a crime. Impact evidence from blunt-force attacks indicates the type of weapon and the height and strength of the attacker. Skid mark lengths, tire impressions, bullet trajectories, and fabric analysis all involve branches of various physical sciences. Vehicle crash investigations (like the investigation of Air France Flight 447) rely on many applications of physics and chemistry to reconstruct the conditions and effects of the crash.

In today's technological age, many crimes are committed with a computer, or the evidence of a crime resides in computer memories. A whole division of forensics investigates computer data and usage and often involves cracking firewalls and passcodes to get at encrypted evidence. Authorized investigators can access electronic bank accounts, e-mail, cell phone calls and locations, and social media sites as necessary to gather evidence of a crime.

So how does forensic science relate to the work of science in general? You've learned that *operational science* deals with observable phenomena. On the other hand, *historical science* develops explanations for things we see in the present but that happened in the past. Historical scientists have to make important and sometimes far-reaching assumptions about the conditions that existed to produce the evidence that we see today.

Forensic science includes features of both operational and historical science. In most cases, forensic science is working with leftover evidence of an unwitnessed crime. In this way, it is very much like historical science. The crime has to be reconstructed from the evidence, leading to an explanation of what *most likely happened*. Worldviews also play a part in solving unwitnessed crimes. The investigator's biases or even prejudices can often influence the way he interprets the suspect's motivations and the evidence at the crime scene.

The word *forensics* comes from the Latin word *forensis*, which means "before the forum." A forum in Roman times was a theater-like structure seating many people and often surrounding a stage. For legal proceedings, the accuser and the accused would present their cases before the forum. The audience would declare guilt or innocence on the basis of the evidence presented and the quality of the speeches by the individuals. So forensics applies both to developing evidence in a legal case as well as to presenting an argument before an audience.

Sometimes the forensic model can be quite complete and very detailed. In other cases, a lack of evidence or conflicting evidence can make resolving a case very difficult. If the information that *is* known doesn't meet the standards for legally admissible evidence, the case goes unsolved.

A key difference between forensic and historical science is the ability to experiment with some forms of observable forensic evidence to identify a plausible cause. Since most crimes dealt with today occurred in the recent past, we can make better assumptions about conditions and processes that resulted in the evidence. This is very difficult to do for most kinds of historical science, such as the study of fossils, rock strata, or very old anthropological finds.

There is an ever-increasing need for people who specialize in some branch of forensic science. Besides knowing scientific principles, these people must be well educated in culture and history, and must have excellent reasoning skills. Forensics is just one more way a Christian can lead a life serving others while fulfilling the Creation Mandate.

The modelmaking aspect of science was crucial for solving the mystery of the AF 447 crash. Experts in aircraft design, flight, meteorology, and oceanography had to create useful models that could explain the crash based on the evidence they had. Since they didn't observe the crash, their investigation was very much like those of historical scientists or crime scene investigators (see the facet on the facing page). They had to draw from their assumptions about the plane's location, how high it was, its speed before the crash, and the way it crashed in order to create models predicting its location on the ocean bottom. Though these models weren't true and actually were quite imperfect, they still helped guide the search. They increased the probability of finding the plane, along with its much-needed data recorders, on the ocean floor.

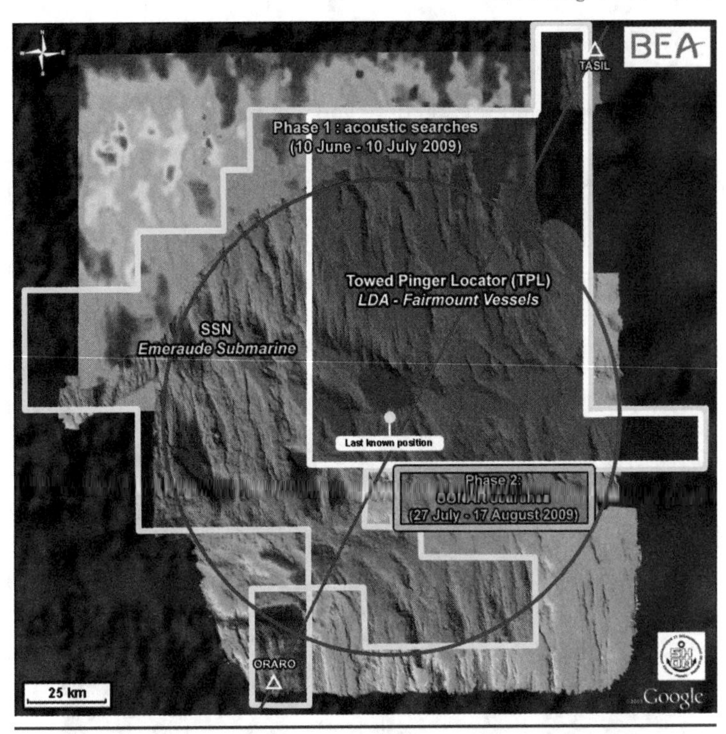

1-16 This map shows the initial circular search area determined by investigators as the most probable location of the crash. The center is the last known location of the aircraft. The rectangular boundaries are the zones searched by the French submarine *Émeraude* and surface ship-towed sonar. BEA is the acronym for the French crash investigation bureau.

1B Section Review

1. Why did investigators initially assume that the crash of Air France Flight 447 must have resulted from a major equipment malfunction?
2. What do we call the basic assumptions that we rely on for understanding life, making important decisions, and doing our work?
3. When a piece of scientific evidence contradicts a scientist's deeply held presuppositions, what usually happens when he attempts to explain the evidence?
4. ✪ Based on your answer to Question 3, would you say that the practice of science is completely objective and without bias? Explain.
5. What does every worldview require regarding faith?
6. How is a scientific phenomenon different from a thought in one's mind?
7. What is a scientific model? What kinds of models do scientists use?
8. How does a scientific law differ from a scientific theory?
9. What category of science would attempt to explain the purpose of the Stonehenge megalithic structure found in England?
10. ✪ (True or False) Biological evolution and the special creation of life followed by rapid speciation after the Flood are both valid scientific theories that account for the diversity of present life on Earth.

1C Objectives

After completing this section, you will be able to

- ✓ summarize how the scientific process works.
- ✓ give examples of how professionals use physical science.
- ✓ consider how you might prepare for a career in science.

1C The Work of Physical Science

1.10 The Search Continues

Things weren't going well in the crash investigation of AF 447. Searchers found the bodies of only 50 crash victims floating at the ocean's surface during the four weeks after the accident. Following this first search phase, which ended in June 2009, the French conducted three additional search phases with American assistance. The first was in July and August 2009, the second in April and May 2010, and the last began in March 2011. Each new search zone included a different section of the rugged ocean bottom. These were arranged around the best-guess position that the investigators made on the basis of the plane's assumed flight path.

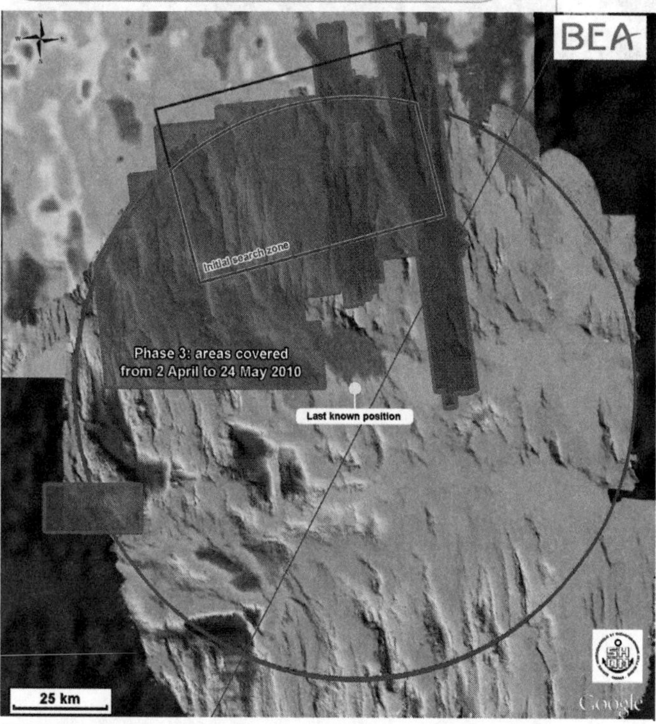

1-17 The later phases of the search for the wreckage of Air France Flight 447 used AUVs and a towed active sonar to map the bottom. The rectangular areas show their search areas.

Complex technology, the best that French and American science could produce, was used in the search. During the first search phase, the French nuclear submarine *Émeraude* (*Emerald*) listened in vain for the loud, acoustic locator pingers on the data recorders. Ships towed sensitive underwater hydrophones listening as well. But after about five weeks, the batteries powering the pingers had run down, and they went silent.

France next brought in ships equipped with autonomous undersea vehicles (AUV). The AUVs, hi-tech programmable submersible robots, sent out sonar beams to paint the bottom with swaths of high-frequency sound. As each search phase ended, the ships returned to port and a new, wider search area was mapped. This process went on for more than 18 months.

Finally, on 3 April 2011, they found something. An AUV discovered the wreckage of the plane at depths ranging from 3800–4000 m (12,000–13,000 ft) down. The small size of the debris field indicated that the plane had probably crashed all in one piece. An American remotely operated vehicle (ROV) surveyed the area and found many bodies still strapped in their seats and inside sections of the wreckage. On 5 May 2011, the searchers began recovering the bodies with the ROV. Other vehicles continued looking for the data recorders, which were finally found on 7 May. Searchers brought the last body to the surface on 3 June, two years after the crash. Altogether, 154 bodies were recovered from the sea, leaving 74 unaccounted for. But they had the data recorders, the key to unlocking this mystery, in their hands.

1-18 A Remus AUV discovered the wreckage of AF 447 in 2011. This graphic shows the invisible active sonar beams that the AUV uses as well as its photographic light source.

1.11 The Scientific Process

With the recovery of the flight recorder data, accident investigators could begin to test their models for the cause of the crash. This approach is very similar to the way scientists do their work in science. A scientist calls his initial model to explain the phenomenon that he is studying a **hypothesis**. It is his best-guess explanation, based on the information he has, along with the professional knowledge he brings to the problem. The crash investigators had several hypotheses, and they used them to guide the data-gathering process. Hypotheses that didn't explain the data were changed or thrown out.

A scientific process for solving a problem is often called the *scientific method*. The important thing to remember is that there is no single scientific method.

The scientific process that a scientist actually uses depends on what he wants to study. Every kind of scientific problem involves a different approach. Generally, a scientist must gather data related to the problem. He compares the data to the hypothesis to see if it can explain the data. In an operational science study, a scientist may set up a *controlled experiment* to test the hypothesis. During the experiment, he tries to identify cause-and-effect relationships that will confirm or disprove his initial model (the hypothesis). If the experiment's results don't make sense, or even if they do, he can repeat the experiment as many times as necessary so that he can confirm the results. In a historical science problem like a crash or crime scene investigation, the data helps us understand *what most likely happened.* But there is no way to prove beyond any doubt what actually happened since there were no eyewitnesses.

A scientist must eventually weigh his observations and decide whether his hypothesis is valid. If he believes it is, he will usually report his findings in a research paper or article so that other scientists can become aware of his idea and test the hypothesis using their own data. If the hypothesis holds up after much testing, it may become a theory generally accepted by scientists in the field (see Subsection 1.9).

The AF 447 investigators were able to recover all the data in the data recorder memory units even though they had been under two miles of water for two years. The data confirmed what the investigators suspected—the pitot tubes *had* frozen over, cutting off airspeed information. But the rest of the story stunned them and made the accident even more tragic.

1-19 This remotely operated vehicle (ROV) is a tethered submersible controlled from a surface ship. Its cameras and mechanical hands can do complicated work at extreme depths.

1.12 *Working in Physical Science*

The methods and technology made possible by physical science were crucial to the crash investigation of AF 447. Physical science includes two main branches of science—*physics* and *chemistry.* You will be learning about these two areas in this course this year.

Physics deals with how matter and energy interact. Crash investigators were experienced in aerodynamics—the study of methods and forces involved in flight. They also hired scientist consultants who were experts in the strength of materials and how they deform (bend or crumple) under stress. Their findings helped determine that the plane crashed by doing a "pancake dive" into the ocean with little forward motion. Experts in electronics and aircraft control systems helped investigators understand the events during the final minutes of the flight based on the information from the data recorders. Oceanographers modeled deep-water currents above the rugged bottom terrain to help searchers better understand where the wreckage might have drifted as it fell to the bottom. Other physical scientists and engineers were not directly connected with the investigation but had contributed the knowledge needed to develop the ships, aircraft, and submersibles that took part in the search and recovery. Physics helped them design the vehicles, their sensors, and their propulsion systems.

Chemistry is the study of the composition of substances and how they act in the presence of other substances. It also played a part in the AF 447 story. Early in the criminal investigation phase of the accident, chemists ruled out a bomb as the possible cause. They found no explosive chemical residue on the wreckage recovered during the first month after the crash. All aircraft wreckage taken from the ocean had to be soaked in special treatment fluids to remove the salts from them. Chemical experts supervised these steps. And just

1-20 The memory modules for the flight data recorder (top) and the cockpit voice recorder (bottom) as they were picked up from the ocean bottom

1-21 Scientists often provide expert testimony in courts of law when needed and where required by governments and businesses.

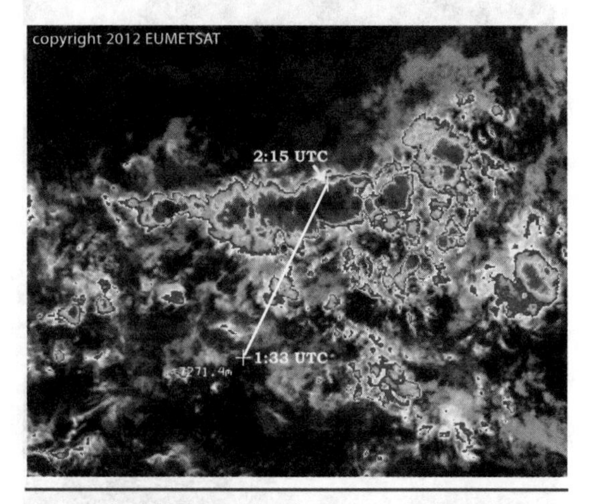

1-22 The flight path of AF 447 took it directly through some strong thunderstorms towering above the Equator.

as with physicists, chemists indirectly contributed to the search in many ways, including developing the metals and materials used in the search planes, their fuels, and even the chemical batteries that powered the data recorder locator pingers.

Working in physical science can involve more than these things. Scientists may act as paid expert consultants for industry, for government, or even in court cases. Some physical scientists educate the next generation of scientists in universities. These young scientists become the scientists and engineers of the future, whose work continues to save and improve people's lives.

1.13 Your Place in Physical Science

Investigators released the final AF 447 crash report in 2012, more than three years after the tragedy.

Here's what investigators think happened. After the plane left Brazilian airspace, it entered a region of huge thunderstorms near the Equator. Most airlines had rerouted their planes to avoid the worst storms, but the crew of AF 447 decided to go through them, flying around only the worst thunderheads. As the cold aircraft flew through a warm, moist updraft, ice formed on the pitot tubes. The pilots and the automated flight control system lost indication of airspeed, but everything else about the plane was working normally. Without airspeed input, the autopilot disengaged, sounding a warning alarm in the cockpit. The sudden shift to manual flight surprised the junior copilot, who was flying the plane (the senior pilot had left the cockpit for a scheduled rest period, an action which is normal on long flights). Thinking the plane had somehow actually slowed down, the pilot increased engine power to full and tilted the plane up to gain altitude. The aircraft climbed to 38,000 ft and slowed until its wings lost all lift—the plane stopped flying. Physicists call this condition an *aerodynamic stall.*

With no lift from its wings, the plane fell out of the sky, slowly spiraling downward with its nose slightly raised. The two copilots didn't understand what was happening. Everything they did kept the plane in a stalled condition. Only after the senior pilot returned to the cockpit several minutes later did they finally correctly analyze the situation. But it was too late. By then, the plane was less than 2000 ft above the water. It crashed nearly straight down on its belly with its nose slightly up. The time from the first stall alarm in the cockpit until the crash was just 4½ minutes. During that time, the plane fell nearly 7 miles.

1-23 Air France Flight 447 literally fell out of the sky. It crashed nearly straight down as this artist's conception shows.

The aviation industry learned a lot from this accident. The pitot tubes on these planes had to be replaced with a design that didn't freeze up as easily. But existing pilot training programs seemed to be the main problem. Some industry analysts have found that most passenger aircraft pilots spend only a small fraction of their time actually flying a plane. They are trained to let the computer autopilots do most of the flying and only control the aircraft during landings and takeoffs. And even these are done automatically in many cases. Pilots do practice emergency procedures to maintain flight readiness, but these sessions may be too infrequent to keep them ready to handle unusual situations. This crash showed that pilots need to be trained better to handle emergency events like these. The highly automated joystick-style flight controls in the A330 aircraft prevented the pilot from "feeling" the condition of the plane or allowing the other pilot to intervene. This design interfered with their ability to analyze and react properly. So the extremely advanced design of the plane may have also contributed to the crash.

How do these factors apply to you? You are entering a phase in your life when you should be preparing yourself for higher education and the life work God will lead you into. The world wants you to believe that the most important goal in life is to get a good job so you can earn plenty of money, live comfortably, and have good health insurance. But education for a believer—especially science education—should guide your thinking toward how you might serve God in fulfilling the Creation Mandate. In a fallen world that denies God, what can you do to direct people's hearts back to Him by declaring the glories of His creation? Of equal importance, how can you show love for other people through good works in science or technology that will help them lead safer, more fulfilled lives? You can achieve both of these goals through an interesting and satisfying career in science.

We used the Air France Flight 447 story to show you that there is a wide-open world of opportunities to glorify God and to serve other people. The work is not easy. The fallen creation as well as unbelievers will oppose your efforts. But your work through dominion science and your Christian testimony will declare the truth of the gospel in a dark and lost world. Let's get started learning about physical science!

1-24 Now is the time to begin preparing for a life working in science.

1C Section Review

1. How long after the crash did it take investigators to find the wreckage of Air France Flight 447 on the ocean bottom? What does this say about human desire to find answers to such disasters?
2. Before the investigators had recovered the information contained by the data recorders, what would we call their best guesses for the cause of the Air France Flight 447 crash?
3. What kinds of conclusions are the best that we can hope for in a historical science investigation? What do these conclusions heavily rely on, besides the data that we can collect in the present?
4. What are the two main branches of science that make up physical science?
5. Give one example for each branch of physical science, showing how it contributed to solving the mystery of the crash of Air France Flight 447.
6. What factor was the actual cause of the Air France Flight 447 crash? Explain.
7. Why is knowing the causes of man-made disasters important for exercising biblical dominion?
8. What should be a Christian's main motivation when seeking a lifelong vocation? When should preparation for that vocation begin?
9. (True or False) Improving technology has been shown to be the best way to exercise biblical dominion in fulfilling the Creation Mandate.

Scientifically Speaking

Creation Mandate	4
presupposition	6
worldview	7
science	8
model	8
theory	11
law	11
operational science	11
historical science	11
hypothesis	14

Chapter Review

1. What were three consequences of Adam's sin mentioned in Genesis 3?
2. How do the consequences in Question 1 affect the way we do science?
3. Does a person have to be a believer in order to express the image of God within? Explain how the Air France crash story supports your answer.
4. In your own words, what does the Creation Mandate mean?
5. How does developing new technology, such as aircraft, demonstrate God's image in us?
6. How does science done by Christians contribute in noticeable ways to God's redemptive plan for the world?
7. Why were the aircraft pitot tubes the focus of attention early in the Air France Flight 447 investigation?
8. What presuppositions do you have about your plans for the weekend?
9. What is the strongest *scientific* evidence against soft dinosaur tissue that is 68 million years old?
10. What are the two basic worldviews, and how are they different?
11. What do we mean by the idea of "workability?"
12. What is the key difference between operational science and historical science?
13. Every scientist uses some kind of scientific process in his work. What is usually the final thing he does after he has finished his research?
14. Besides doing research, what other kinds of things may a scientist do during his scientific career?
15. When is the best time in life to begin preparing to work in science?

True or False

16. God is unjust to allow disasters like the crash of Air France Flight 447.
17. People have essential value that is far above any other thing on Earth.
18. Scientists display the image of God within by exploring, explaining, and describing, even in their imperfect ways.
19. When a person's mind and heart are rescued from fallen thinking, he is restored to his original purpose of exercising biblical dominion.
20. The flight data recorders served as witnesses to the crash of Air France Flight 447.
21. Presuppositions lead to absolute truths.
22. Every person is born with his or her worldview.

23. A scientist with a secular worldview may still discover or accomplish great things in this life.
24. Every model is an imperfect representation of something.
25. For a scientific theory to be acceptable, it must be both true and plausible.
26. A chemist who is developing a new light-sensitive adhesive is doing operational science.
27. When proposing a hypothesis for the cause of the crash of Air France Flight 447, investigators were offering their best guess for the cause.
28. After examining the floating pieces of the Air France Flight 447 wreckage, materials scientists decided that the aircraft must have crashed into the ocean at high speed nose first.
29. The results of the Air France Flight 447 crash investigation suggested that even the best of well-designed technology can produce unintended consequences.

MATTER

CHAPTER 2

DOMINION SCIENCE PROBLEM

Protecting Those Who Protect Us

Men and women in police forces serve us by protecting our communities. While doing their job, they frequently put themselves at risk. For example, on 4 June 2012, Police Officer Kevin Ambrose of the Springfield Police Department, Massachusetts, was killed by gunfire while answering a call for a domestic disturbance. He was survived by his wife, two children, and a grandchild. Almost 14,000 American police officers have died in the line of duty since the first death was recorded in 1794. How can we protect those who protect us?

2-1 Police officers often risk their lives when apprehending those who break the law.

2A The Particle Model of Matter

2.1 Describing Matter

Criminals will continue to shoot, stab, and strike police officers because these actions come from their fallen, evil hearts. Scientific knowledge can't do anything about the heart. But by *using* scientific knowledge, we can try to protect those who go into harm's way to protect us. Placing some kind of material barrier between a weapon and a person could protect him from an assault and save his life. What properties does matter have that we can use for such an important purpose?

2A Objectives

After completing this section, you will be able to

- ✓ describe what matter is.
- ✓ give evidences that led to the acceptance of the particle model of matter.
- ✓ summarize the kinetic-molecular model of matter.
- ✓ describe the basic structure of an atom.
- ✓ distinguish between atoms, molecules, and ions.

Operational science relies on observation, so let's do some observing. Look around the room. What do you see? There are probably things made of wood, metal, fiber, plastic, and many other solids. There may be a glass or bottle containing a liquid beverage. If you blow on your hand, you can feel the air moving across your skin even though you can't see it.

What are all these different kinds of physical materials made of? That would be **matter**. All substances in the universe are forms of matter.

Carefully look around you again. Using all your senses, note those things that don't seem to be physical substances. What about the light shining through a window? What is the signal that a TV turns into images and sound? How about the warmth given off by a light bulb? While you can sense these things directly or indirectly, they aren't matter.

2-2 What forms of matter and nonmatter can you see in this room?

So what is matter? People have long defined **matter** as anything that takes up space and has mass. But this statement isn't a definition—it's a description. Matter, like many other things that scientists study, can't be defined using more basic ideas.

Matter is anything that occupies a volume of space and has mass. Nonmatter doesn't have these properties.

Matter can exist as immense objects, such as a supergiant star 500 times the size of our sun. Or it can exist as extremely small objects, such as atoms or subatomic particles. As you will learn, all matter is made of these tiniest of particles.

Try to group the forms of matter that you see in the room into a few main categories. Most of the visible matter is solid. The beverage in a glass is a liquid. The air you are breathing is a gas. These kinds of matter are probably familiar to you. If there is a fluorescent light in the room, a *plasma* exists inside the lamp. This is a less familiar but very common form of matter in the universe. Matter occurs in these different *states* because of the arrangement and motion of its particles.

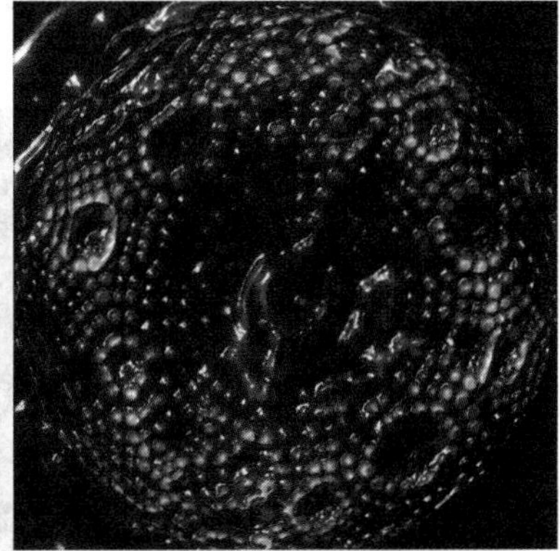

2-3 Matter can occupy great volumes, such as stars (left), and tiny volumes, such as atoms on the tip of a tungsten needle (right).

2.2 The Particle Model

About 400 years before Christ, Greek philosophers debated about the nature of matter. Several of these men believed that all matter was continuous—that is, it could be subdivided repeatedly into infinitely small particles. Other philosophers were convinced that matter wasn't continuous. After subdividing a substance enough, you would eventually reach a particle that couldn't be divided further. The philosophers called these basic particles *atoma*, meaning "indivisible," and those who held to this idea were called *atomists*. Remember, these people had no knowledge of atoms or molecules, so these ideas about material things that couldn't be seen were pretty amazing.

The debate in ancient times ended when the teachings of the influential Greek philosopher Aristotle, who opposed the atomists, eventually won out over his rivals. He objected to atomism because it required a void (or vacuum) to exist between atoms, something Aristotle rejected on philosophical grounds. How could "nothing" exist? Instead of atoms, Aristotle saw matter as continuous and full of the potential to become anything. When matter takes on form, it becomes the objects that we see. Aristotle's writings about nature influenced the models of European science for nearly 2000 years, including a large part of the Christian era. Sadly, Christian acceptance of Aristotle's ideas came about because church teachers tried to combine Greek and Christian worldviews together, even though they were largely incompatible.

Interestingly, scholars in India had developed an atomistic philosophy at least a century earlier than the Greeks. We don't know if these two nations shared this information or if they developed their models of atomism separately.

Worldviews and Science

Science deals with the natural world. We tend to think of worldviews as dealing with questions about God, morals, and life after death. Many people believe that worldviews have very little to do with science. But is this idea correct? What does the history of the particle theory of matter indicate about the relationship between worldviews and science?

During the Renaissance, European scholars studied ancient literature and poetry from Greek and Roman times. Even though most of the original atomistic writings had been lost, the writings of Aristotle survived, including his critiques of atomist philosophies. As early as the 1300s, European scholars were aware of a form of atomism. But a useful *particle theory* of matter didn't gain a foothold in European science until the 1600s. (See the facet on the facing page.) Scientists generally accepted the model by the mid-nineteenth century, but not until the twentieth century did scientists understand how small and complex the particles of matter are.

Physicists measure the diameter of a typical particle of matter in nanometers. (The nanometer is the metric unit used for measuring such small lengths.) If you could lay one million particles of water side by side, their combined lengths would be equal to the thickness of a page of this textbook!

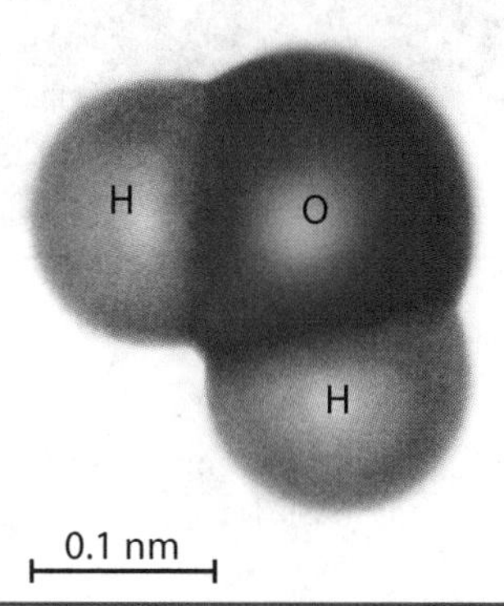

2-4 A water molecule is nearly one-third of a nanometer long.

There are different kinds of particles of matter, but the most common particle that scientists refer to is the *atom*. Atoms are the "building blocks" of matter. They can combine to form larger particles or masses of atoms.

Can we observe anything so small? A microscope known as the *scanning tunneling electron microscope (STM)* allows scientists to detect atoms on the surface of specially prepared materials. The machine does not see atoms directly. Instead, it detects the electrical influences surrounding the atoms and converts them into artificial images that represent the atoms (see Figure 2-5). Even before the STM, however, physicists had gathered enough indirect evidence to convince them that atoms do exist.

STM can refer to either the actual microscope or the techniques used by the microscope (scanning tunneling electron microscopy).

2-5 STM image of palladium on a graphite surface

FACETS OF SCIENCE

History of Atomism

In order to understand nature, you need to understand that atoms make up matter. Early in our education, we learn that these tiny particles are the "building blocks" of everything. For scientists, the particle model of matter has proven a very useful tool to describe the natural world and the materials in it.

The origin of *atomism*, the concept that matter is made of tiny unseen particles, is unknown. Long ago, people who supported this view were called *atomists*. The earliest writings on this theory appear in India in the sixth century BC. Certain sects of the Hindu culture developed a fairly complex theory of atomism that persists to this day.

The Chinese and other Asian cultures often made important scientific discoveries long before Europeans, but they seem to have never developed a distinct theory of atomism. Some researchers say this is because of their early adoption of *pictographic writing*. Western languages build many words from a relatively small set of letters, but Asian languages based on Chinese use thousands of complex, picture-like symbols to represent whole things and ideas. Some researchers suggest that this form of writing resulted in a sort of mental block against thinking of physical objects as being made of smaller particles.

The Greeks were the earliest Europeans to include the concept of atomism in their writings. Philosophers such as Leucippus and his student Democritus (c. 400 BC) developed a theory of atomism as part of a naturalistic worldview. They taught that atoms and the void (an infinite vacuum) in which the atoms moved were eternal. Everything else was temporary. The universe came about from the random collisions and connections of atoms. Humans came into existence at birth and dissolved into atoms at death. There was no soul or existence after death. The Greek atomists viewed life as totally purposeless.

Not all Greek philosophers agreed with the atomists. Plato and his student Aristotle (fifth and fourth centuries BC) could not accept the natural and materialistic origin of everything when there was so much beauty and order evident in nature. Plato believed that nature was two separate realities. Here on Earth, reality consisted of the imperfection and decay that we experience through our natural, uneducated, and closed minds. The other reality existed in celestial perfection. The heavens contained the perfect forms of things and ideas. They were the patterns that were imperfectly copied by the things on Earth. Only those people who were educated well enough to be aware of them could experience this reality.

Aristotle built on Plato's ideas and developed his own. He made many observations about the natural world and culture. He also proposed that the material world and even the immaterial world, such as thoughts and light, consisted of infinitely small particles, or "atoms," in constant motion. There was no void between the atoms. Atoms could change, so they were not eternal. An important difference between Aristotle's view of matter and that of Democritus is that Aristotle's "atoms" were essentially points having no dimensions. This resulted in the structure of matter being *continuous*, with no spaces between atoms.

Greek philosophers never tested any of these thoughts because they considered reasoning to be the only source of truth. They believed that experiments to test ideas in the material world could not produce truth because of the imperfect nature of matter.

Most people accepted Aristotle's teachings far more readily because they gave an impression that the world and human life had purpose. It's important to understand that the purpose implied in Aristotle's teaching did not come from knowledge of biblical truth, but from a pagan worldview. Indeed, the "God" described by Aristotle in his writings (the "unmoved mover") is an impersonal deity with no interest in anything but himself, a far cry from the personal God of the Bible.

During the following centuries, the philosophy of atomism faded to a dim memory. Aristotle's teachings were adopted throughout the Roman Empire. Eventually, long after the rise of Christianity in Europe, Aristotle's ideas were accepted by the Catholic Church as religious doctrine.

It wasn't until the fifteenth and sixteenth centuries, nearly 2000 years after Democritus, that interest was rekindled in an atomic theory. Men of learning began studying the ancient Greek and Roman writings that had been preserved by Islamic scholars during the Middle Ages. But the continuous nature of matter was firmly rooted in scholastic thought. The idea of atomism was considered interesting but unimportant, and possibly heretical.

By the time of Copernicus's death in 1543, however, the acceptance of his model of the world changed people's view of nature. The earth was no longer the center of the universe. It was merely a planet orbiting a sun similar to other stars in space. This sun-centered model of nature shook the faith man had in the philosophy of Aristotle, which held to the geocentric view. It became politically safe for scholars to question Aristotle's writings. They could now develop a particle theory of matter that would not be immediately rejected as a God-denying philosophy. Scientists

such as Kepler, Galileo, Boyle, Hooke, Newton, and ultimately Albert Einstein firmly established the scientific basis for a particle theory that better describes the way matter behaves. This is the goal of every scientific model.

In the 1700s, chemists, such as Lavoisier, showed how matter seemed to combine in specific, predictable ratios. This was not likely to occur if matter were continuous. In 1803 John Dalton published his famous theory of the atom and it changed the course of chemistry. His particle model was not without problems, but it provided a starting point from which the classification of known chemical elements and the discovery of new ones could proceed.

From the early 1800s on, a description of the particles of matter gained more detail. By 1860 nearly all scientists were convinced of the existence of atoms and molecules. Even if they were not convinced, they saw how useful such a model was in solving scientific problems. By 1900 the electron had been discovered, demonstrating that the indivisible atom of John Dalton could be subdivided. By 1910 physicists knew that the atom was mostly empty space but had a dense nucleus containing positively charged particles called protons. In 1935 scientists discovered another nuclear particle—the neutron.

Many physicists believed they had finally laid the mystery of the atom to rest. However, there were some minor but persistent problems with the atomic and nuclear models. In 1931 physicist Wolfgang Pauli suggested that some nuclear changes emitted an unidentified particle. In 1956 that particle—the neutrino—was detected. Since then, physicists have discovered hundreds of so-called *elementary particles*. These particles combine to make up the familiar electrons, protons, and neutrons, as well as other particles that inhabit the nucleus.

Another discovery during the early decades of the twentieth century did serious damage to the relatively simple particle model of matter. Electrons could act as waves! Scientists then realized that all matter of any size acts as a wave—even you! But only for the smallest particles was this property observable with current technology. These findings demonstrated that scientific models could only describe some things in workable ways under certain conditions. Sometimes, no one model can explain all observations.

The question that faces physicists today is whether even more fundamental particles of matter remain to be discovered. Some scientists believe that the structures of such particles can be understood only with a model having more than three physical dimensions. Such questions are profitable for a new generation of scientists to pursue. Who better to consider these questions than a Christian—someone who has a personal relationship with the Creator of matter!

We don't normally observe single particles of matter all by themselves. Matter that we see with our unaided eyes contains *lots* of particles. The properties of matter that we can sense or measure, such as color, hardness, weight, and flammability, are due to the way the uncountable particles work together in the substance. We aren't observing the properties of the individual particles. For example, a single atom of silver doesn't have a silvery luster, and a tiny crystal of silicon dioxide sand in a brick doesn't have the same density as the brick itself. A material's properties are results of the kinds of particles that make up the material and the way those particles connect to each other.

Atomic Bonding

In solid matter, atoms are normally connected to other atoms to form molecules and other structures. If they were not, all matter would be more like liquids or gases, having no rigidity and no definite shape. Life and even the earth could not exist.

The links that atoms make with each other are called *chemical bonds,* or just *bonds*. Bonds between atoms cause them to form relatively stable arrangements with other atoms. They give structure to matter and are the source of its properties.

We discuss how bonds determine matter's properties later in this text. For now, you need to know only that atoms form connections with other atoms through chemical bonds.

2.3 Evidence for the Particle Model

A good scientific model explains all laboratory observations. If the particle model of matter is workable, then it should be able to explain how matter behaves in everyday situations.

The particle model of matter assumes two important things about matter: it is made up of tiny particles (atoms or molecules), and the particles are in constant motion. The way sugar dissolves in water—so that every part of a sugary drink is sweetened—agrees with this model. You may have noticed how quickly a perfume or the smell of ammonia can spread through the air in a room. A crystal of a colored chemical compound (see Figure 2-6) that disperses into a beaker of water, even without stirring, supports the particle model. Particles of the compound break away from the crystal and slowly move off into the water. Eventually, the water is evenly colored. The dispersion of a substance through another by particle motion is called *diffusion*. These examples seem to support the particle model but they support the continuous model of matter as well. Remember that Aristotle believed that matter could be subdivided indefinitely, so taste, smell, or color should easily diffuse throughout liquids or gases. We need to look further for stronger evidence of particles in matter.

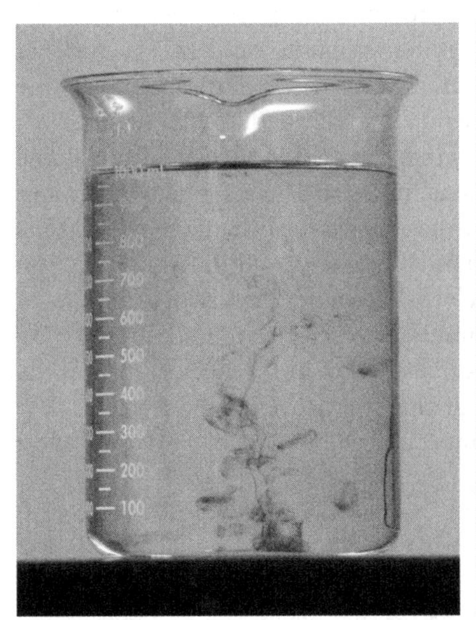

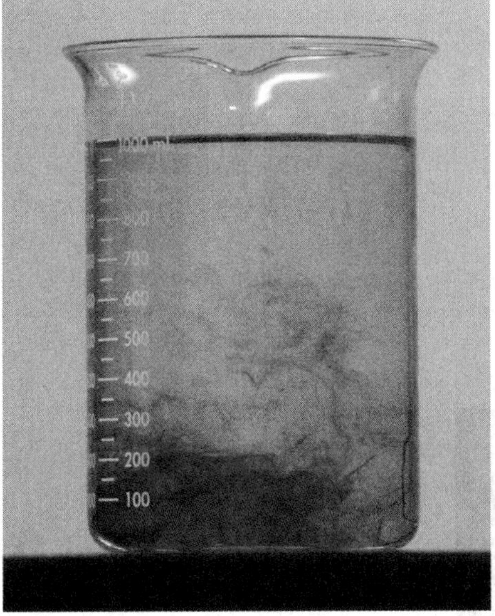

2-6 These three photos illustrate diffusion of a substance in a liquid.

Chemists discovered the first indirect evidence for particles of matter in the late 1700s. They noticed that when they weighed certain chemicals and then burned them in air, the weight of their residue after burning was different from the weight before. The ratios of the before and after weights were often small whole numbers like 1:2 or 3:5. Other kinds of chemical changes produced similar ratios. Chemists began to suspect that they were observing the evidence of individual particles chemically combining in definite ratios.

Scientists first recognized the direct effects of particles of matter in motion in 1827. Robert Brown, an English botanist, was examining the interior of tiny plant spores with a light microscope. He saw cellular parts within the spores jostle back and forth as if something

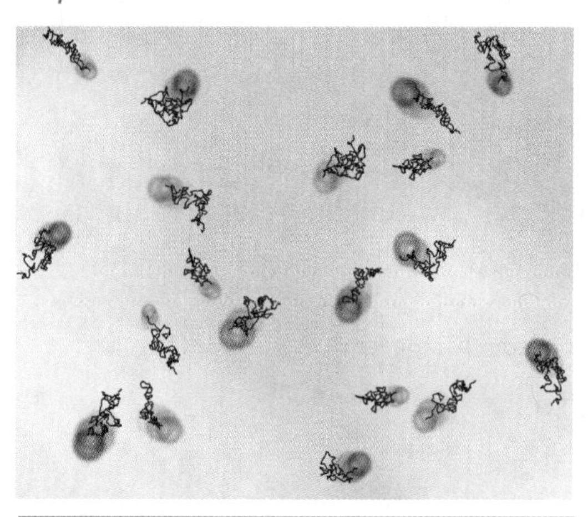

2-7 Brownian motion as observed through a microscope. The black squiggly lines show the paths of these particles over time.

A **subatomic** particle is any particle of matter smaller than an atom. Nearly all the hundreds of subatomic particles known can help build the structure of an atom.

Fundamental Electrical Charge

The electrical charges on protons and electrons are equal in size but opposite in sign. The size or *magnitude* of either of these charges is the *fundamental electrical charge*. All measureable amounts of electrical charge, such as that from a battery, are simply huge multiples of the fundamental charge.

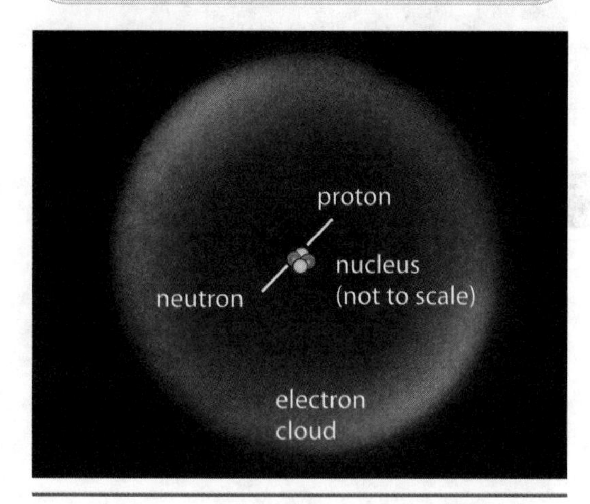

2-8 A highly simplified model of the atom showing the nucleus, which contains protons and neutrons, and a cloud of electrons, which forms the volume of the atom

were striking them repeatedly from different sides. Yet nothing was touching them except the cellular fluid in which they were floating. Although other scientists had observed this motion before him, he was the first to report his findings in a widely read paper. Others after Brown suggested that the jiggling motion of the cell parts was due to the impacts of fluid particles striking them randomly from different directions. Today, we call this movement of bits of matter in liquids and gases **Brownian motion** in his honor.

These are only a few of the evidences that have led scientists to conclude that tiny particles in random, constant motion make up all matter. This fundamental model of matter is often called the **kinetic-molecular theory** of matter.

2.4 The Atom

Matter that we can detect with our senses is made of atoms, molecules, or large structures of one or the other. The *atom* is probably the most familiar to you. It's the basic particle that makes up all matter. Atoms contain protons, electrons, and usually neutrons. These ***subatomic*** particles exist within a tiny spherical region in space that is the atom.

At the center of the atom is the *nucleus*. The nucleus contains one or more *protons*. The number of protons in the nucleus determines the kind of atom. Each proton carries a single positive *charge*. The nuclei of most atoms also contain *neutrons*. These particles have about the same mass as a proton, but they have no electrical charge—they are *neutral*.

Electrons are strange things that can exist in the spherical volume of space around the nucleus or as tiny particles flying between atoms. Each electron carries a single negative electrical charge. The electron's mass is only a tiny fraction of a proton's or neutron's mass. The electron "cloud" that surrounds the nucleus determines the size of the atom.

In an ideal atom, the numbers of electrons and protons are equal—so the number of positive charges in the nucleus equals the number of negative charges in the space around the nucleus. The opposite electrical charges cancel each other so we say that an ideal atom is *neutral*.

2.5 The Molecule

A molecule is a distinct particle formed when two or more atoms bond together and become separate from other atoms. A molecule can be very simple, like an atmospheric oxygen molecule, which is made of two identical oxygen atoms. Molecules can also be very large and complex, containing hundreds of millions of atoms, like DNA. Physicists usually treat individual molecules as if they were BB-like particles, since the details of what they are made of aren't as important as how they move and affect other particles. Chemists, though, are more concerned with the kinds and arrangements of the individual atoms in a molecule, and how they determine the molecule's properties.

Molecules are made of atoms, so they also contain the protons and electrons that form their atoms. The number of protons within an ideal molecule equals the number of electrons in the molecule, so it is also electrically neutral.

2.6 Ions

Under certain conditions, atoms and molecules can gain or lose electrons. If they have more or fewer electrons than protons, then there is an electrical charge imbalance. Particles with unbalanced electrical charges are called *ions*. ***Anions*** have more electrons than protons and have a negative electrical charge. The opposite condition results in ***cations***, which have a positive charge. Ions form important and common forms of matter like rocks and salts. Electric fields can also attract and repel ions. Later in this course, you will see how atoms, molecules, and ions build the matter from which the universe is made.

Anions (AN eye un) are negatively charged atoms—atoms with more electrons than protons.

Cations (CAT eye un) are positively charged atoms—atoms with fewer electrons than protons.

sodium ion
(atomic ion)
Na
+1 charge
11 protons (+11)
10 electrons (−10)

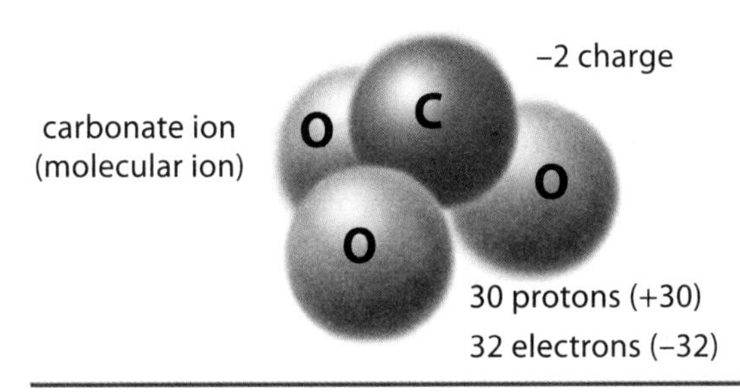

2-9 Ions may be single atoms or several atoms bonded together acting as a single charged particle.

2A Section Review

DS 1. Why is the problem of bodily assaults by criminals not one that can be solved by science?
2. What is matter?
3. Compare matter and nonmatter.
4. What question about the structure of matter took more than 2000 years to answer?
5. What is the name of the current model that explains the structure of matter?
*6. Why are chemical bonds between atoms important?
*7. What was Aristotle's strongest argument against the existence of indivisible atoms?
8. What particles make up an atom?
9. What is a molecule?
10. What is an ion?
11. (True or False) The spreading of odors is evidence for the particle model of matter and proof against the continuous model of matter.

2B Objectives

After completing this section, you will be able to

- ✓ classify examples of matter.
- ✓ list the common states of matter and give examples of each.
- ✓ compare and contrast the three common states of matter using the kinetic-molecular theory.
- ✓ explain the causes of viscosity and gas pressure using the kinetic-molecular theory.

2-10 Adam classified the kinds of animals when he named them.

2B Classification of Matter

2.7 A Classification Model

Classification is one of the first things that humans did in obedience to the Creation Mandate (Gen. 2:19–20). As Adam named the animals in the Garden of Eden, he likely separated them into their kinds based on their appearance (see Figure 2-10). Classification helps us understand our world. We organize what we know into familiar patterns in a way that helps us remember and make sense of what we learn.

In Section 2A, you classified all the matter you could see in the room around you. Beyond simply grouping the substances into states of matter—solids, liquids, and gases—the task seems almost hopeless. There are millions of substances. And many materials, like concrete or granite, are made of more than a single substance. Where do you begin? We need a classification model to help us categorize matter. As with any model, the system won't be perfect, but it will be workable for the purposes of this course. Figure 2-11 illustrates the model we will use for the classification of matter.

2.8 Pure Substances

We classify substances by the kinds of particles they are made of and how they are combined. The first category we will use includes all the **pure substances**—those materials made of only one kind of substance. An *element* is the simplest kind of pure substance and is made of only one kind of atom. Elements can exist as single atoms, as molecules of one kind of atom, or as masses of uncountable atoms of the same kind. Compounds are the other category of pure substances. A *compound* contains more than one kind of atom chemically bonded together. The atoms of

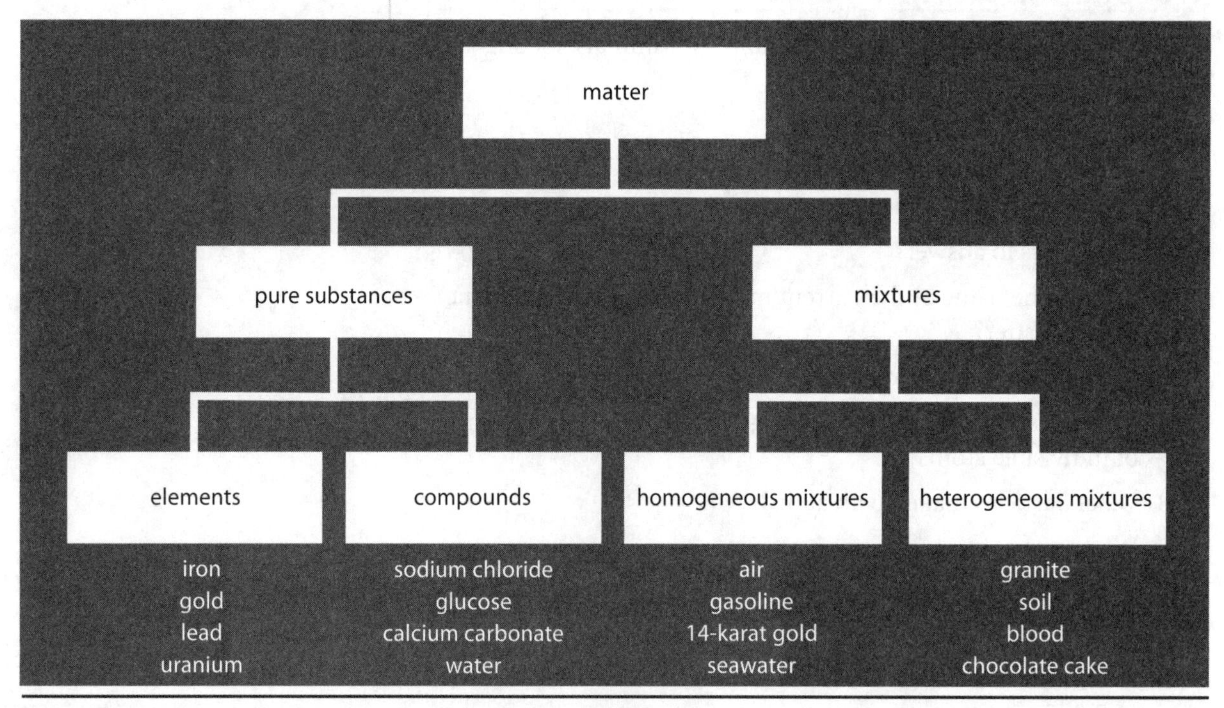

2-11 Matter generally falls into one of four major categories.

compounds may link up in molecules or in great masses of ions arranged in a repeating geometric pattern called a *lattice*. But for a given compound, they always combine in the same proportions.

Chemists sometimes subdivide compounds into two major groups—organic and inorganic compounds. *Organic compounds* always contain the element carbon, and many of these are found in or are produced by living things. Other elements, such as hydrogen, oxygen, nitrogen, and phosphorus, can be present in organic compounds. There are over 50 million known organic compounds, and most new compounds developed each year are organic compounds. The other compounds, including a few that *do* contain carbon (like carbon dioxide), as well as all that don't, are *inorganic compounds*.

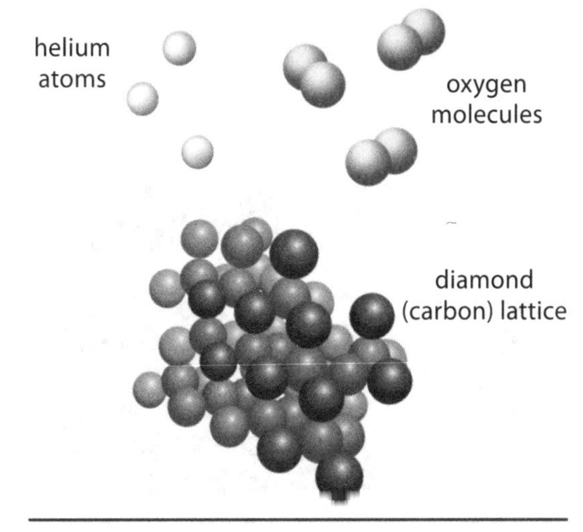

2-12 Elements may occur as single atoms (helium), as molecules (oxygen), or as lattices of atoms (diamond).

2.9 Mixtures

Mixing two or more substances together forms—you guessed it—a **mixture**. These substances may be elements, compounds, or even other mixtures. The substances in a mixture are *not* chemically combined. This is the main difference between a compound and a mixture. The proportions of the substances that make up mixtures can also vary widely. The parts can be mixed in equal amounts, or only a few molecules of one substance could be mixed into another substance that forms most of the mixture.

Mixtures are very common in our world. Rock, soil, air, blood, and seawater are just a few examples of mixtures. We can sort mixtures into two main groups based on how the particles of their substances mix.

In ***homogeneous** mixtures*, the particles of the different substances are roughly the same size and evenly mix. We say that the mixture of substances is *uniform*—by sight and at the particle level. If these mixtures are gases or liquids, we can also call them *solutions*. The air that you breathe, tap water, gasoline, and many metal alloys such as those used in gold jewelry are common examples of homogeneous mixtures.

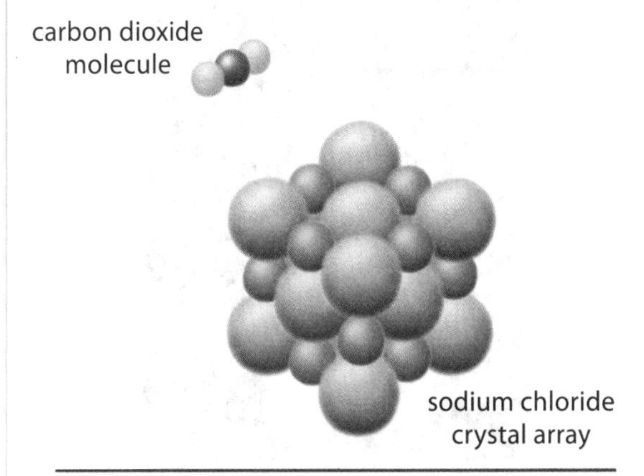

2-13 Carbon dioxide molecule (top) and a small piece of sodium chloride (bottom)

homogeneous (HOE muh GEE nee us or huh MAHJ e nus)

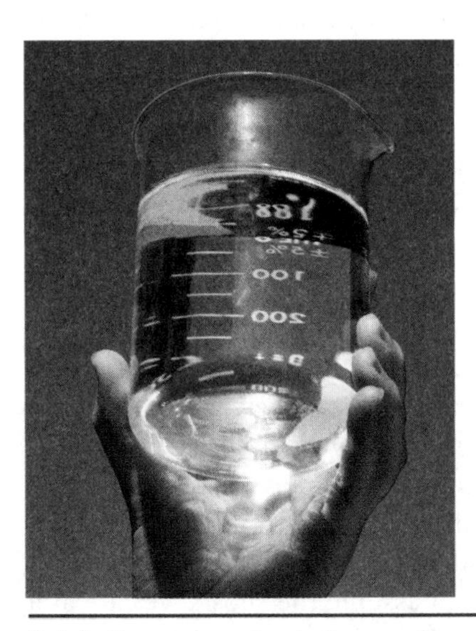

2-14 Examples of solutions or homogeneous mixtures

heterogeneous (HET ur uh GEE nee us; more rarely, HET ur AHJ e nus)

2-15 The different mineral phases are clearly visible in this granite sample.

2-16 This heterogeneous mixture contains three states (or phases) of the same substance—water.

The observation that the speeds of particle random motion and the temperature of the substance are directly related is an important one to remember.

Atmospheric Pressure

There are many kinds of units for atmospheric pressure, including the *atmosphere* (atm). The unit we use depends on the purpose. Meteorologists commonly use the *millibar* for weather reports.

Mixtures made of particles that differ greatly in size are ***heterogeneous mixtures***. These mixtures are *not* uniform. You can often see the different substances with the unaided eye or under a low-power microscope. A familiar example of a heterogeneous mixture is the rock called *granite*. It usually contains three visible types of minerals—dark biotite, light pinkish feldspar, and white or clear quartz. Scientists call visually distinct parts of a mixture *phases*. Thus, there are three mineral phases in granite (see Figure 2-15). Other examples of heterogeneous mixtures include whole blood, soil, whole milk, home-cooked stew, and gravel.

2.10 States of Matter

We normally find matter on Earth in one of three physical forms, or **states**: solid, liquid, or gas. Because many materials can exist together in more than one physical state, scientists refer to the different states as phases, just as if they were a heterogeneous mixture. The glass of water in Figure 2-16 contains three phases of water. Ice is the solid phase. The water itself is the normal liquid phase. The bubbles in the water contain water vapor—the gaseous phase. How can one material exist in different physical states?

In our kinetic-molecular model of matter, extremely tiny particles are in constant, random motion. Long ago, scientists found that a substance's physical state depends on the strength of the attraction between its particles and how fast they're moving. Atoms and molecules that are similar have some attraction for each other. The strength of attraction is related to the distance between particles. Close together, the particles strongly attract. The attraction drops off rapidly as the particles draw apart. At a distance, the particles have no attraction. The speed of the particles' random vibrations and the temperature of the substance are also related. The slower the particles move, the lower the temperature; the faster they move, the higher the temperature. You will learn why this is so later in the course.

Standard Conditions

The state of a substance depends on the kind of substance and the conditions surrounding it. The state of matter also relates in some way to temperature. Many properties of substances can change as temperature, atmospheric pressure, and other conditions change.

To accurately classify and compare different substances, scientists need to compare their properties when measured under identical conditions. For example, an oceanographer who wants to compare the density of a sample of seawater with that of pure water would need to make sure that they were the same temperature, since temperature affects density. Otherwise, the comparison would be meaningless.

Most measureable properties of many substances have been recorded in reference lists used by scientists and educators. Researchers (mostly *materials engineers* and *mineralogists*) likely measured these properties at a *standard temperature and pressure (STP)* when necessary. These conditions are typically 0 °C and average atmospheric pressure at sea level (e.g., 1013 millibars). When discussing the properties of the different states of matter and the properties of matter in general later in this chapter, we will assume the substances are at a standard temperature or pressure.

Solids

Solid materials make up most of the earth's matter and most of the physical objects familiar to you. Your pencil, your desk, a rubber band, the fabric in your clothing, a computer, and even the room you are in are solids. What makes a **solid** different from the other states of matter? Solid objects don't easily change shape, they don't flow, and their volume is basically constant. Most are hard and rigid.

The range of particle random motion in solids is very limited. This is what determines its main properties. Particles of a solid are close together and occupy definite positions. They randomly vibrate in those locations, continuously rattling against their neighboring particles. But otherwise, they don't move from their fixed locations. This explains why the solid state of a substance is usually the densest—the particles are normally much closer together in a solid than in its liquid or gaseous states. The vibrations of the particles of a solid substance also have the slowest motion of any of the three states of that substance—it is cooler than when in its liquid or gaseous states.

The cozy arrangement of particles in a solid explains another important property of most solids—low *compressibility*. Since particles are already practically touching, it is difficult to squeeze them closer. Atoms strongly *repel* other atoms if they get too close together.

The arrangement of particles in a solid varies depending on its chemical composition and the way the solid formed. The particles in a *crystalline solid* are arranged in geometric patterns that can repeat without limit. Solids of this kind often form *crystals*. Atoms and molecules in solids can also be randomly arranged. These are *amorphous solids.*

The particle structure of real solids may lie anywhere between these extremes or can be a combination of them. Material scientists have grouped solids into at least eleven types, though some may overlap with others. Most *metals* are clearly crystalline, with closely packed, repeating patterns of atoms. Crystalline earth *minerals* often produce beautiful crystals. Most glasses are common examples of amorphous solids. In between crystalline and amorphous structures are materials like ceramics and ceramic glasses. Advanced ceramics are composed of tiny crystalline particles that are compressed together under high temperature and pressure, producing a ***poly****crystalline* structure. Pure *organic solids* like wax are generally amorphous. However, some natural organic solids have definite structures, like *wood* or *bone*, which consist of organic fibers embedded in amorphous or crystalline compounds. Artificial and natural organic compounds can form useful solids called *polymers*, which are long chains or nets of identical,

Characteristics of a Solid

1. Definite volume
2. Definite shape
3. Dense
4. Low compressibility
5. Particles are close together and vibrate in fixed positions.
6. *Crystalline solids:* Particles are in a fixed, repeating structure that forms a large array.
7. *Amorphous solids:* Particle arrangement is fixed but random.

Completely solid matter (that which is without air spaces) is usually dense and has low compressibility.

The array of particles in a crystalline solid is called a *crystal lattice*. The lattice structure is unique for each kind of substance.

The prefix *poly-* means "many" or a "great many" of whatever the root word is that follows.

an iron pyrite crystal

milled wood

carbon-resin bike frame

2-17 Which of the above is a composite solid? a crystalline solid? an organic solid?

linked molecules. Latex, silk, plastics, and synthetic fabrics are all made from polymers. Advanced materials that human creativity has developed within the last century include *composite solids* like carbon fiber resin, which is both strong and somewhat flexible. The computer chips that are the hearts and brains of all digital electronics are composed of artificial crystalline *semiconductor* compounds. We continue to develop even stranger but very useful solids, including *aerogels* and *aerographites*, as we exercise dominion, shaping the resources of the world in ways that best serve our needs.

Liquids

A **liquid** has two main physical properties—it has a definite volume like a solid but it flows—it isn't rigid. This is why scientists call liquids ***fluids***. On Earth, a liquid takes on the shape of its container and fills the container from the bottom up. Because of this, a liquid normally has a *free surface* if it doesn't completely fill its container. An undisturbed free surface lies flat but waves can move through it. In a microgravity environment, like that inside the International Space Station, a liquid naturally forms a wobbly sphere. Its free surface is the outside of the sphere.

fluid: (L. *fluidus*—to flow); any substance that flows; a liquid or gas

Characteristics of a Liquid

1. Definite volume
2. Shape determined by container
3. Low compressibility
4. A fluid
5. Particles are completely mobile but still close together.
6. A liquid forms a free surface if it doesn't completely fill its container.
7. Liquids have viscosity.

The particles in a liquid are much freer to move around than those in a solid because they aren't trapped in fixed positions by strong attractions to their neighbors. They vibrate and freely spin. But the still-strong attraction between particles keeps them close together, so liquids have low compressibility like solids. The particles in a liquid move easily among each other, colliding almost continuously with other particles. The mobility of particles in a liquid allows them to flow easily from one place to another. This fluid behavior allows liquids to pour, to flow through pipes, and to form rivers and streams.

The ability to flow varies from one liquid to another. For example, rubbing alcohol pours easily and seems to flow everywhere when it spills. On the other hand, corn syrup is very thick and takes a long time to pour out of a bottle. A liquid's resistance to flow is called *viscosity*. A liquid's viscosity depends on the strength of attraction between the liquid's particles, how tangled up the particles become due to their shape, and how rapidly the particles randomly move (again, related to the liquid's temperature). Liquids generally become less viscous as their temperature increases and more viscous as they get colder. Viscosity is especially important in liquids called *lubricants*. These substances reduce friction by forming a thin film of liquid between smooth sliding surfaces, which improves slipping and reduces wear.

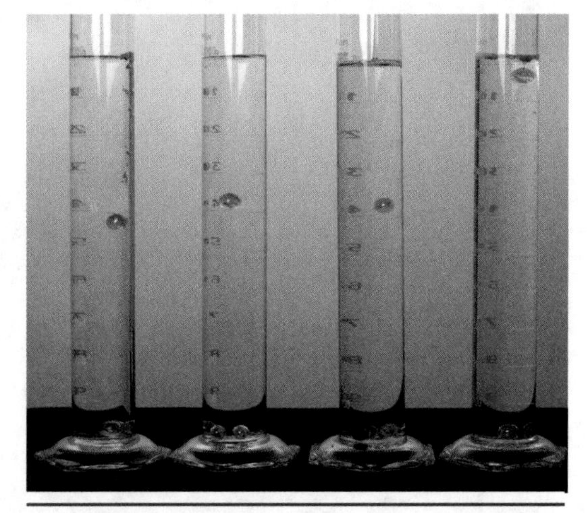

2-18 These liquids are arranged in order of least viscous on the left to most viscous on the right. The falling marbles indicate the relative viscosities of the four liquids.

Gases

The molecules in a **gas** are far apart and move at high speeds compared to the particles in liquids or solids. They randomly vibrate just like the particles in solids and spin as in liquids. But particles in the gaseous state also move very fast in straight lines until they collide with something. A typical particle in a gas at room temperature moves at about 490 meters per second, or about 1.5 times the speed of sound! Gas molecules near the surface of the earth collide with each other billions of times a second and act like tiny billiard balls bouncing off one another when they do collide.

Characteristics of a Gas

1. Volume and shape determined by container
2. Highly compressible
3. A fluid
4. Particles are far apart and move at high speeds.

A quantity of gas doesn't have a definite shape like a solid. Neither does it have a definite volume like a liquid. When a gas flows into an empty, sealed container, its molecules will quickly move to every part of the container. Very shortly, the gas uniformly fills the available volume. If a sealed container has one kind of gas in it and another kind of gas enters the container, the second gas may take a long time to diffuse into the original gas, especially if their densities are very different. They will first separate into layers.

Gases, like liquids, can flow from one place to another. You can pour pure carbon dioxide into a beaker because it is denser than air. Cold, dense, nighttime air can flow down a mountain into a valley, displacing a blanket of warmer air. Compressed air can flow through pipes. Thus, just like liquids, gases are fluids.

Some containers, such as the space inside a bicycle tire pump, can change shape and volume. If you push in on the pump handle while holding your thumb over the nozzle, the air in the pump is squeezed into a much smaller volume. You can see that gases are highly compressible.

Molecules of a gas in a sealed container collide not only with one another but also with the walls of the container. Imagine how the gas molecules bombard the walls of the container—billions of them hit each square centimeter every second! You would never notice the force of a single molecule, but the impacts of billions of molecules can produce a significant push spread out over the area of the container's walls. The force a gas exerts on an area is *gas pressure*. Gas pressure exists even if a gas isn't trapped in a container. The air molecules in the atmosphere exert a pressure that we can measure with special instruments.

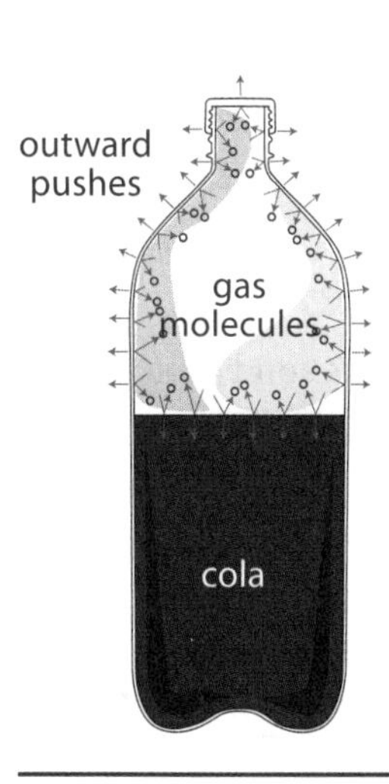

2-20 As billions of gas molecules collide with the boundary of a container, the tiny pushes combine to produce gas pressure.

2-19 You can pour carbon dioxide gas just like a liquid.

Table 2-1 summarizes the properties and effects of the three common states of matter.

2-1 *Comparison of Three Physical States of Matter* (at standard pressure)

Water as an Example		
ice (solid)	water (liquid)	steam (gas)
Particle Arrangements		
Temperature Ranges		
0 °C (32 °F) or below	0–100 °C (32–212 °F)	100 °C (212 °F) or above
Characteristics		
definite volume	definite volume	indefinite volume
definite shape	shape determined by container	shape determined by container
Attraction between molecules dominates because they are very close together (resulting in low compressibility). Randomly vibrating particles occupy fixed positions. Frequent collisions with neighboring molecules occur.	Attraction between molecules still dominates because they are close together (resulting in low compressibility). Molecules have no fixed positions and move randomly around each other. Frequent collisions with neighboring molecules occur.	Attraction between molecules is almost nonexistent because they are far apart (resulting in high compressibility). Molecules mainly move in straight lines at high speeds. Collisions are relatively rare.

2B Section Review

1. What is the first instance of classifying in human history?
2. How would you classify the following substances according to the model discussed in Subsections 2.8–9?
   a. oxygen
   b. air
   c. carbon dioxide
   d. peanut-roll candy bar
   e. oil and vinegar dressing
   f. stainless steel alloy
3. What kinds of pure substances are made of different kinds of atoms chemically bonded together?
4. ✪ Briefly discuss the difference between an element and a compound.
5. What kinds of matter are combinations of substances that are *not* chemically bonded together?
6. ✪ What do scientists mean when they speak of a phase of matter?
7. What are common temperature and pressure conditions used when documenting the properties of matter?
8. The one property besides hardness that sets solids apart from the other states of matter is that they are ______.
9. Like solids, liquids are nearly ______.
10. Gases and liquids are both ______ (a kind of matter).
11. A state of matter depends on what two factors?
12. ✪ Identify one physical property each of a solid, liquid, and gas that is *not* shared with the other states of matter. More than one choice may be possible.
13. ✪ After reading this section, how do you think the speed of the particles in a substance and the temperature of the substance are related?
14. (True or False) There is one physical property that clearly distinguishes the solid, liquid, and gaseous states.

2C Objectives

After completing this section, you will be able to

- ✓ make a list of physical and chemical properties that apply to a specific type of matter.
- ✓ classify changes in matter as physical or chemical.
- ✓ paraphrase the law of the conservation of matter.
- ✓ compare and contrast physical, chemical, and nuclear changes in matter.

2C Changes in Matter

2.11 *Physical Properties and Changes*

Physical Properties

A *physical property* is anything about a substance or object that we can observe or measure *without changing its chemical composition*. These properties include color, density, hardness, crystal form, texture, and physical state. We use physical properties to help us identify substances, classify them, and most importantly, determine the best material to use for a given task. Some properties are useful for classifying only certain substances in a specified state. For example, hardness applies only to solids.

2-21 Diamonds (left) and quartz crystals (right) have many similar properties. However, diamond is much harder than quartz.

2-22 Industrial diamonds make the teeth of this rock drill wear resistant.

Consider the diamond teeth on a well drill bit. Diamonds are extremely hard and don't quickly wear down when boring through rock. If we were to replace the diamonds with quartz crystals, the drill would quickly grind to a halt! Why do diamonds work better than quartz? They share several of the same physical properties. Both minerals are transparent. Both can have beautifully faceted shapes. Yet the industrial diamonds are between 8.5 and 16 times as hard as quartz, depending on which measurement scale you use. Diamond's physical property of hardness makes it ideal for drilling through rock.

> You usually can't identify an unknown substance based only on the measurement of one physical property. For example, dozens of minerals may have the same hardness number. Only when we compare the observations and measurements of several physical properties do the results point to one substance.

Physical Changes

When we study the world around us, we see that matter is continuously changing. Some of the changes affect the shape or state of the material. Any change of the arrangement of the matter in a material or object is a *physical change*. A physical change alone *can't* result in a change of the material's chemical composition.

> The chemical identity of materials made of molecules comes from the molecules themselves. So a physical change of a molecular substance can be any rearrangement of its molecules as long as the molecules themselves don't change.

Most kinds of matter on Earth occur in the solid, liquid, or gaseous state. Matter can change states, but it remains the same kind of matter. Water freezes into solid ice, but it is still water. Alcohol may evaporate into a gas, but it is still alcohol. You know that sugar can dissolve in water, but it is still sugar (the sugar molecules have not been broken down into atoms). Even though the ions that make up table salt separate in water as a salt solution, when the water evaporates (a physical change), the salt ions come together again to form the crystalline solid. And you can grind wood into sawdust, but the sawdust is still wood. These are all examples of physical changes.

People use physical changes in all kinds of ways. For example, we produce much of our electrical energy by huge steam turbine-generators located in electrical power plants. Physical changes in water power these huge machines. Heated water produces steam under high pressure. The steam flows through the turbine blades and the turbines spin electrical generators. As thermal energy in the steam changes into mechanical energy to rotate the turbine, the steam cools and turns back into water. Pumps return the liquid water to the boiler to repeat the process. See Figure 2-23 on the next page.

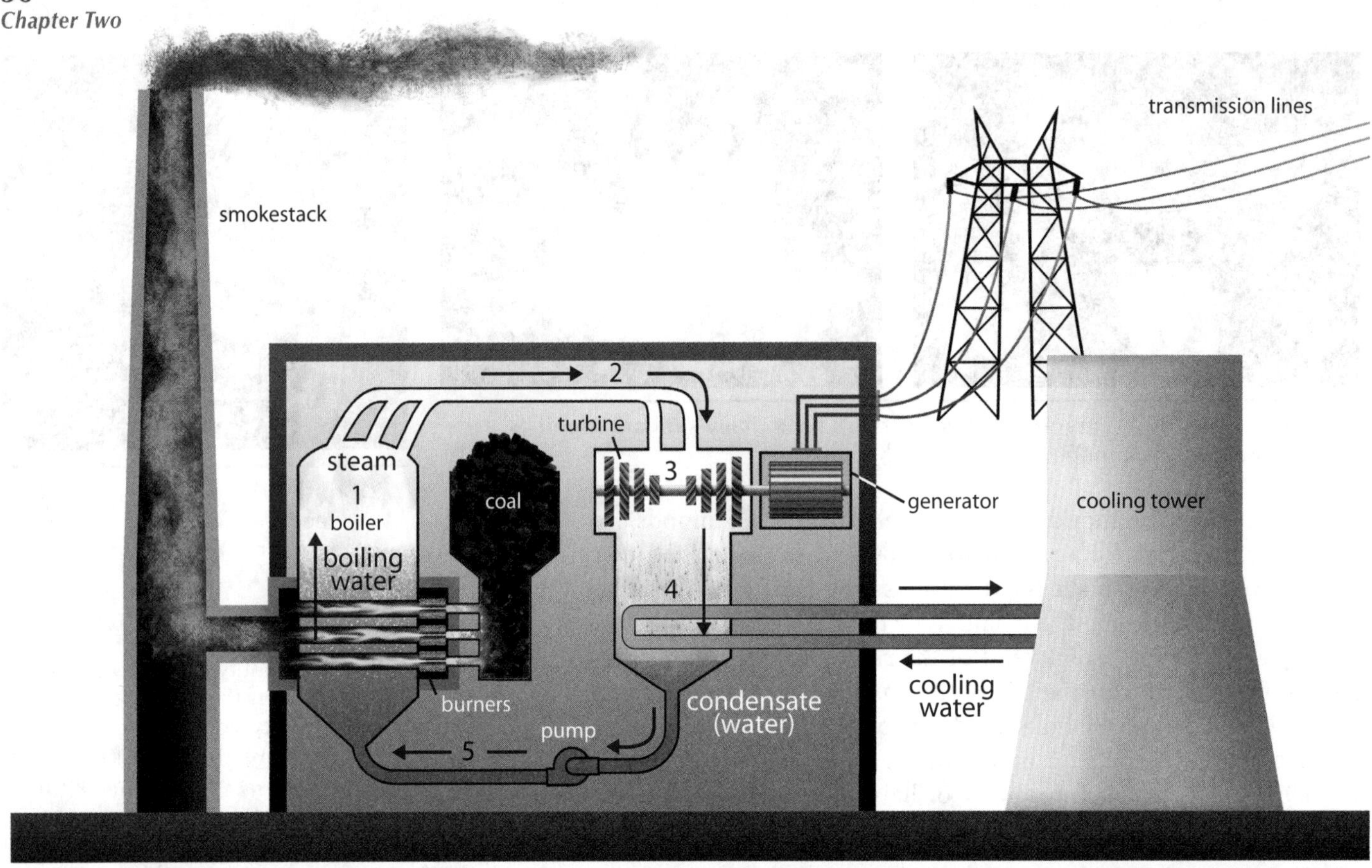

2-23 Most electrical power plants use a physical change to produce electricity. A heat source (nuclear, coal, or gas) boils water into steam (1); steam under high pressure flows through pipes to steam turbines (2); the steam strikes the turbine blades and rotates the turbine, which rotates the generator (3); after transferring energy to the turbine, the cooler steam condenses back into water (4) and returns to be heated again (5).

In another example, we make many of our metal hand tools using a die-casting process. A furnace heats the solid metal until it melts, and then the liquid metal flows into a mold. As the metal cools, another physical change occurs: the metal solidifies. The worker breaks away the mold to release the solid metal tool.

2.12 Chemical Properties and Changes

Chemical Properties

We can use other characteristics besides physical properties to identify or classify matter. For instance, why does pouring liquid water on a fire extinguish it, but pouring liquid gasoline on the fire causes an explosion? We say that water is *nonflammable* but gasoline is highly ***flammable***. Liquid oil doesn't corrode metals, but protects them from corrosion. On the other hand, a liquid acid can eat metals away. Iron rusts in the presence of water vapor and oxygen, but stainless steel is rustproof. We can learn of a material's *chemical properties* by observing how it acts and changes in the presence of other substances.

Note that in use, *flammable* and *inflammable* mean the same thing—"easily ignited, burns rapidly." However, some people think the prefix *in-* means "not" and have been seriously burned as a result. It is best to use only the term *flammable* to avoid confusion.

Chemical Changes

Materials can change from one kind of chemical substance into other kinds of chemical substances. We use this property of matter all the time. Such changes in matter are ***chemical changes.*** Any change in a substance that affects the kinds of atoms and/or the way they are bonded together is a chemical change. We often speak of a chemical change as a *chemical reaction*, since in many chemical changes the

Chemical changes occur when atoms of different substances break apart and combine in different ways.

substances react in obvious and energetic ways. Common chemical changes include burning of any kind of fuel, photosynthesis in plants, digestion in your stomach, and rusting or corrosion. The physical changes that occur in a conventional power plant begin with a chemical change as the coal, gas, or oil burns in the boiler.

2.13 Using Properties of Matter to Solve Problems

Can you think of a way to prevent the serious injury or death of a police officer by gunfire? You could enclose him in a military-style tank. But it would be hard to climb a stairway to a third floor apartment in one! How about an armored suit? The military is working on these, but a usable one is still several decades in the future, and they are pretty bulky. One of the most obvious ways is to use body armor, like bulletproof vests. Police Officer Ambrose wasn't wearing a bulletproof vest when he was killed.

Bulletproof vests are often made of a substance called *Kevlar*. Kevlar is ideal for protective vests because of its physical properties of strength and flexibility. Strength means resistance to penetration by a bullet or knife. It is five times stronger than steel and is almost as strong as spider's silk, one of the strongest natural substances known. Its strength, a physical property, comes from its molecular structure. Kevlar is made of polymerized molecular chains—a chemical property. These molecules tangle together and are very hard to pull apart.

Strands of Kevlar are spun together into flat yarns and then woven into a fabric. Typically, seven layers of this fabric make up a bulletproof vest. Under normal circumstances, Kevlar is extremely strong, but it loses its strength when it gets wet or is exposed to ultraviolet light. Manufacturers coat Kevlar fabric with a protective substance as it is made into a vest. The lives of more than 3000 law enforcement officers and an unknown number of soldiers have been saved by personal body armor since it was first introduced in the 1970s.

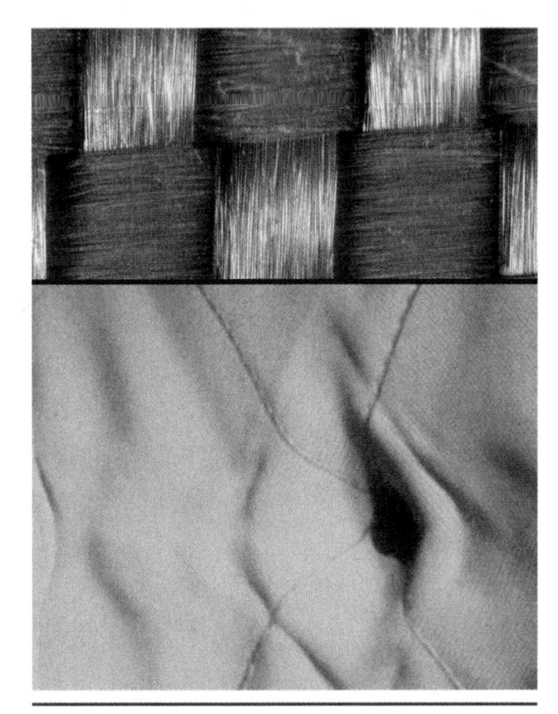

2-24 The strength of Kevlar can stop bullets and jagged pieces of shrapnel.

2.14 Conservation of Matter

Physical and chemical changes occur around us all the time. In many changes, matter seems to disappear. Water evaporates and wood burns. In other changes, matter seems to appear out of nowhere. Frost forms on cold surfaces. Plants grow more leaves. Your little brother or sister gets bigger. The models of matter that the ancient philosophers held were unable to explain these changes. The pseudoscience of *alchemy* included the idea that mystical chants and the influence of talismans could result in changing matter or bringing new matter into existence.

After many experiments during the 1700s and 1800s, scientists concluded that the quantity of matter remaining *after* a physical or chemical change was equal to the quantity of matter present *before* the change, even if their forms were greatly different. The particle model of matter helped scientists to realize that the total number of particles never changed. Scientists described this important observation in the *law of the conservation of matter*. Scientists often state this law:

> *Matter can be neither created nor destroyed, but only changed from one form to another.*

Conservation of Matter and the Bible

If the law of the conservation of matter is an established scientific law, how did Christ feed the multitude with just five loaves and two fish (John 6:5–14)?

Most historians credit the famous French scientist Antoine Lavoisier (1743–94) with being the first to confirm the law of the conservation of matter through experiments conducted around 1785.

2.15 Nuclear Changes

In physical and chemical changes, the kinds of atoms involved always remain the same. These kinds of changes involve whole atoms or molecules and their arrangements. But the atoms themselves remain unchanged. What kind of change, if any, can alter the identity of an atom? Since the number of protons in a nucleus determines the kind of atom, to change an atom would mean changing the number of protons. Other changes in the nucleus involve changing the number of neutrons or emitting or absorbing high-energy rays or particles. Physicists call such events ***nuclear changes***.

A **nuclear change** occurs when an atom's nucleus emits or absorbs a nuclear particle or ray. Splitting the nucleus into smaller nuclei is also a nuclear change.

Nuclear changes also include processes where large nuclei split into smaller nuclei or very small nuclei combine to form a larger nucleus. Any of these changes can result in totally different kinds of atoms. These reactions take place in nuclear power plants and nuclear bombs. Physicists believe that a continuous nuclear reaction produces the sun's energy.

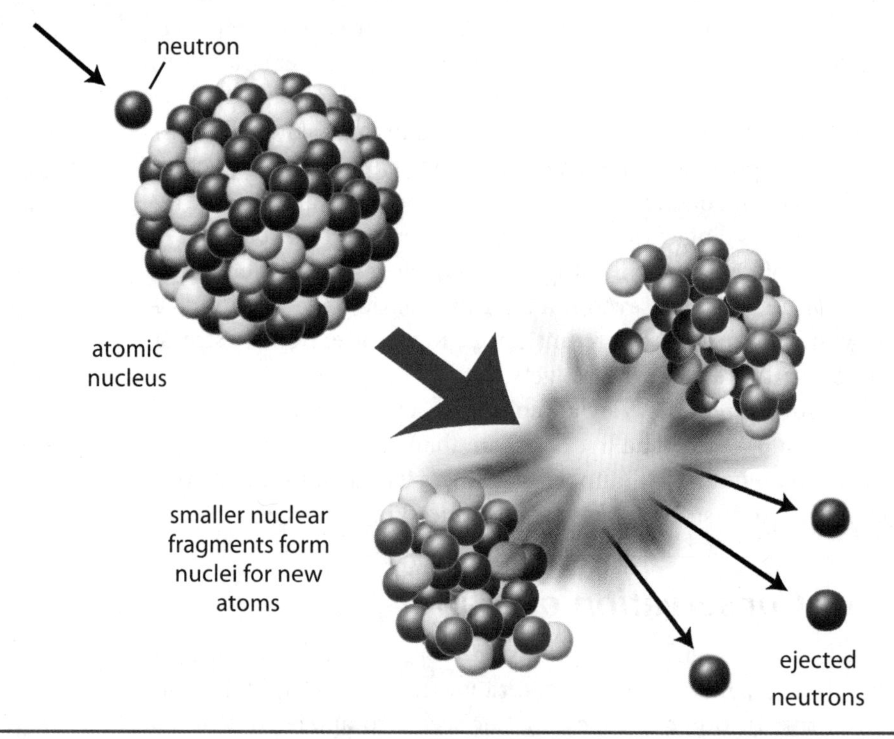

2-25 When a heavy nucleus absorbs a high-speed neutron, it can split into smaller pieces that form lighter atoms.

2C Section Review

1. Pure quartz and the gemstone amethyst are nearly alike chemically. What physical property could be used to tell them apart?
2. What *doesn't* change in a substance during a physical change?
3. Physical and chemical changes happen when the particles of the substances involved become rearranged. How is the particle rearrangement of a chemical change different from a physical change?

4. Which kind of property of matter helps us distinguish between pure water and a strong liquid acid like nitric acid?
DS 5. Is the Kevlar fiber's strength a physical property or a chemical property?
6. What doesn't change in a substance during a chemical change?
✪ 7. If a scientist could prove by experiments that the total amount of matter after a physical or chemical change was different from the amount of matter before the change, what would be the effect on our assumptions about the nature of matter?
8. What happens in a nuclear change that makes it different from a chemical or physical change? What is often the result of a nuclear change?
9. (True or False) Physical changes include grinding, dissolving, burning, bending, and boiling.

2D Changes of State

2.16 *Temperature and Changes of State*

2D Objectives

After completing this section, you will be able to

✓ list the types of changes that matter can undergo.
✓ relate temperature to change of state.
✓ relate pressure to change of state.
✓ explain each change of state using the kinetic-molecular model of matter.

If you drop several ice cubes onto a hot frying pan, the spitting, popping cubes will dance about the pan as they melt into water (a liquid) and then sizzle into steam (a gas). The water changes physical states but remains the same substance. When matter changes from one state to another, we call this result a *phase change* or *change of state.* Ice, liquid water, and steam are all composed of water molecules. What happened as the ice melted into liquid water and the water boiled into steam?

As a substance changes state, something happens to the position and movement of its particles. The faster the particles move or vibrate, the higher the temperature. You can also say that the higher the temperature, the faster the material's particles are moving. Since the speed of particle motion and their distance apart determine the state of a substance, temperature is an important factor in the state of a material. For the remainder of this section, let's assume that, except for temperature, air pressure and other conditions that can affect states of matter are constant.

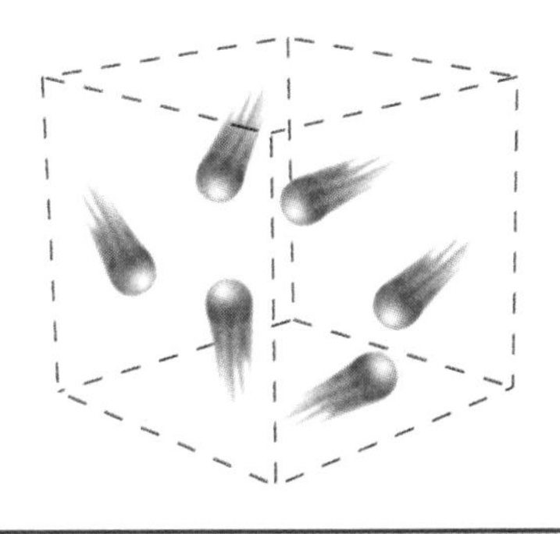
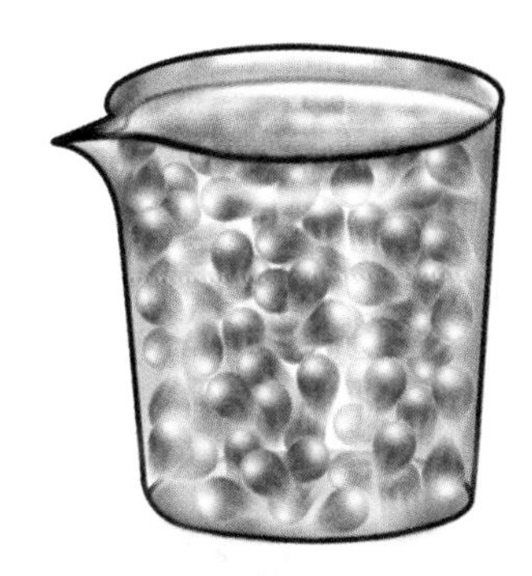

2-26 The particles of a hot gas (left) are moving faster and are farther apart than those of a cooler liquid (middle). The particles in the cold solid (right) have the least amount of motion. Notice that as the temperature decreases, the amount of particle motion also decreases.

An object that gains thermal energy becomes hotter—its temperature increases. When an object loses thermal energy, it cools—its temperature decreases.

The temperature of a substance is an indicator but not a direct measure of the thermal energy present in the substance. *Thermal energy* is related to the microscopic random vibrations and motions of the particles of matter. For a given object, the greater the amount of thermal energy it contains, the hotter it is. Thermal energy flows into an object by *heating* it. A loss of thermal energy results in *cooling*.

2.17 Melting and Freezing

Melting

As a solid, pure substance—an element or compound—gains thermal energy, its temperature increases. If the solid absorbs enough thermal energy, its particles vibrate so rapidly that the attractions between particles can't hold them in fixed locations any longer. They break free and begin moving around in the liquid state. We call this physical change **melting**. The temperature at which a pure solid turns into its liquid phase is its melting point. The melting point of pure ice, for example, is 0 °C (32 °F). Every *pure* solid has a distinct melting point. This is just one of many physical properties that can help identify pure substances.

Solid mixtures can melt too. However, the melting point for a mixture of substances isn't a unique temperature. This is because a mixture's melting point depends on the proportions of the materials in the mixture. The ratios can differ from one mixture to the next, and so does the melting point. For example, assume a metal mixture (an alloy) contains only gold and silver. The alloy will melt at different temperatures depending on the percentages of gold and silver (see Figure 2-27).

We usually say that water and water-like substances *freeze* when they change from the liquid to the solid state. For other substances, we say they *solidify*.

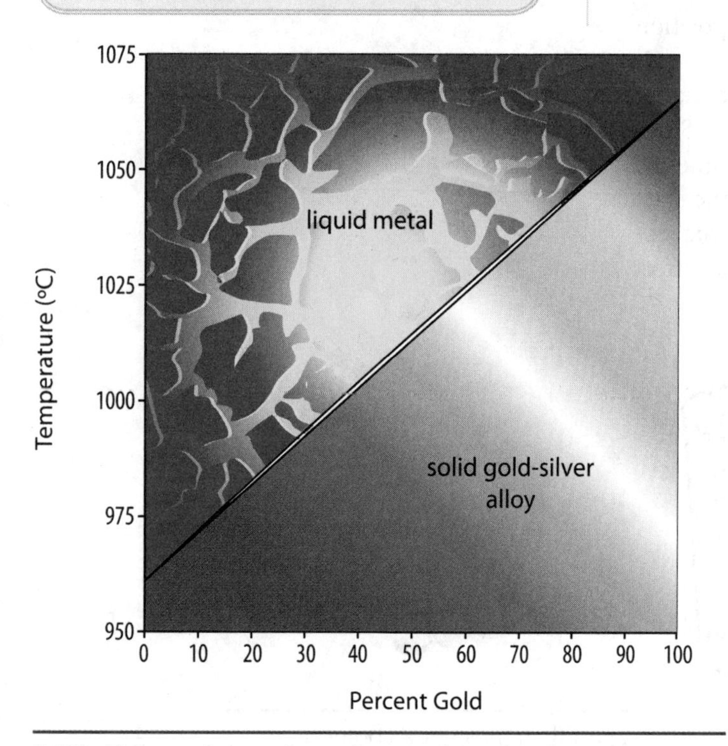

2-27 This graph is a phase diagram for gold-silver mixtures. The melting point rises with more gold in the mix. The thin middle region shows where both solid crystals of alloy and liquid exist together in the molten state.

Freezing

When a liquid cools, its particles slow down. Eventually the motion is slow enough that the attraction between the particles can hold them in fixed locations. As particles clump together, they enter the solid state. When all the particles in the liquid are trapped and can no longer move about, the change to a solid is complete. We call this process **freezing**. This occurs for pure substances at a unique temperature called the *freezing point*. The freezing and melting points for any pure substance are the same temperature under the same conditions.

The freezing point of a mixture depends on the ratios of the different substances in the mixture. The ratio of the parts in the liquid mixture changes as freezing takes place because the different parts usually have different freezing points. The substances with the highest freezing points freeze first as the mixture cools, increasing the ratio of the others in the remaining liquid phase.

2.18 Vaporization and Condensation

The Particle Model and Vaporization

Vaporization is *any* process resulting in a solid or liquid substance changing directly to a gas. Let's look at water to see what's happening. Molecules in liquid water are moving constantly in random motion. Collisions with other molecules jostle them. They roll, slide around each other, and vibrate. Most molecules move at a moderate speed. In some collisions, a molecule's motion may slow when another molecule strikes it, or it may pick up speed. A few molecules are moving faster than the average speed of the others. When one of these collides with another fast-moving molecule, the second may gain a lot of extra speed. If this molecule is near the surface of the liquid, it may be able to escape the attractions of the other molecules in the liquid and fly free. The molecule then enters the gaseous phase above the liquid.

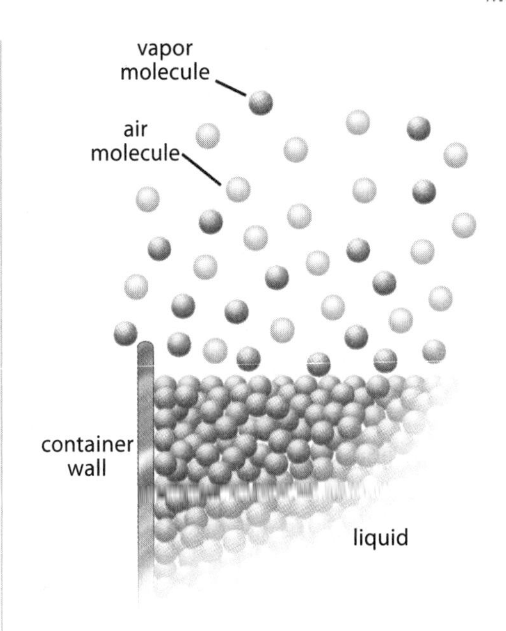

2-28 The vapor of a substance is the gaseous state of the liquid. In this diagram, the vapor mixes with the atmospheric gases above the liquid.

Temperature, Pressure, and Vaporization

The *rate of vaporization* refers to how fast liquid water changes to a gas. It depends on two factors: the temperature of the water and the pressure the water's vapor exerts on its liquid or solid phase. Let's look at the liquid-to-vapor phase change to understand what happens.

As the water's temperature rises, molecules in the water are moving faster, so it's easier for them to gain enough speed to escape the liquid phase. Warmer water causes more molecules to leave the liquid phase, increasing the vaporization rate. As a liquid cools, the opposite happens and vaporization slows down.

At the surface of any liquid, molecules in the liquid and vapor phases continually exchange places. Some of the vapor molecules near the water's surface slow down enough to be captured by the liquid phase. Other molecules from the liquid enter the vapor phase.

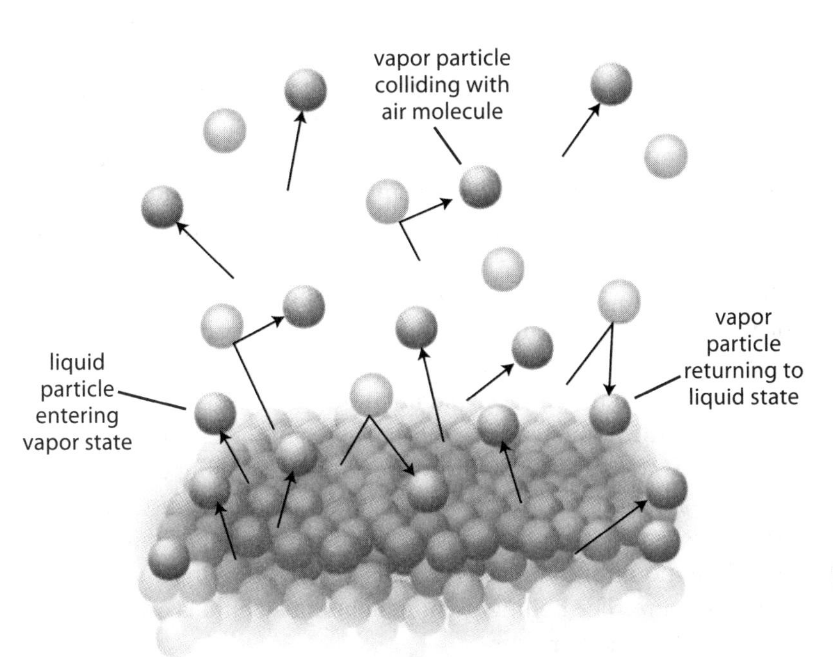

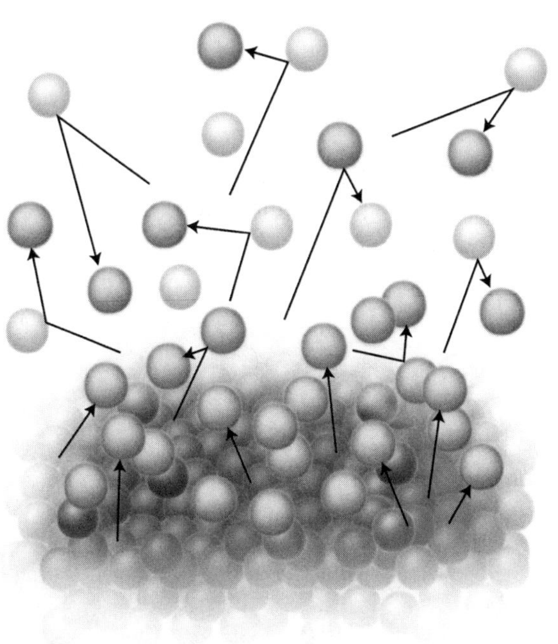

2-29 The vapor pressure above the liquid at a given temperature, not the atmospheric pressure, determines how fast the liquid vaporizes.

For liquid water open to the atmosphere, more molecules are lost to the air than return. Over time, the number of water molecules in the liquid grows smaller, and the liquid water will eventually disappear.

In a *closed* container, such as a capped soft drink bottle, water molecules escaping the liquid phase collect in the vapor space above the liquid. At first, the number of molecules becoming vapor is much greater than that returning to the liquid. But as the number of water vapor molecules increases, more and more are likely to return to the liquid phase. Eventually, the number of water molecules vaporizing equals the number of water vapor molecules returning to the liquid. As long as the temperature doesn't change, this can go on indefinitely in a closed container. The amount of liquid water doesn't appear to change because both processes are occurring at equal rates—they are in *equilibrium*.

Scientists call the condition of equal but opposing processes a *dynamic equilibrium*.

Vapor Pressure

The water's vapor molecules exert a pressure on the liquid water. This is in addition to any other gas pressure in the container. When the gas and liquid phases of any pure substance are in equilibrium as we just described, scientists call the pressure that the gaseous phase exerts ***vapor pressure***. Vapor pressure is a property of the *liquid* substance and is directly related to temperature—the higher the temperature of the liquid is, the higher its vapor pressure. For two substances under the same conditions, the one with a higher vapor pressure will vaporize faster. For example, rubbing alcohol has a higher vapor pressure than water, so it evaporates off your skin faster. A pure substance's equilibrium vapor pressure *measured at standard conditions* is an identifying property of the substance.

Vapor pressure is the gas pressure exerted by just the vapor molecules, ignoring the pressure caused by the rest of the air molecules. The vapor pressure of a liquid is *not* related to air pressure above it.

Kinds of Vaporization

Boiling is the fastest kind of liquid vaporization. The liquid's temperature is high enough that vapor bubbles form *inside* the liquid and rise to the surface. Bubbles appear wherever the liquid's vapor pressure becomes greater than the surrounding liquid pressure. Air pressure above a boiling liquid doesn't affect the vaporization rate. It only determines at what temperature bubbles can form.

The temperature at which boiling begins is the *boiling point* of the liquid. Vapor bubbles can form anywhere in the liquid where the temperature is at the boiling point. In a pot of water on the stove, the hottest point is nearest the heating source. So bubbles form in the bottom layer of water since that reaches the boiling point first. As the hot water circulates in the pot, the whole mass reaches the boiling point. Hot water vapor bubbles (steam) form nearly everywhere in the liquid, rise to the surface, and burst.

Boiling occurs
- at the boiling point temperature, and
- throughout the liquid.

The boiling point of a pure substance is a unique property of that substance. Since atmospheric pressure affects its boiling point, scientists must measure this temperature at standard conditions so that they can compare it to other substances' boiling points. For example, the boiling point of pure water is 100 °C (212 °F) at normal atmospheric pressure. The boiling point of rubbing alcohol is only 82.5 °C (181 °F) at the same pressure.

We cover different units of air pressure in Chapter 8.

Evaporation is the kind of vaporization that happens when a liquid's temperature is anywhere between its melting and boiling points. Unlike boiling, evaporation can occur only at the surface of a liquid. We described the process of evaporation in open-container vaporization earlier in this section. The rate of evaporation of a liquid in an open container depends not only on the liquid's temperature but also on the surface area of the liquid that is evaporating. A liter of water in a wide pan will evaporate much faster than a liter of water in an open soft drink bottle at the same temperature.

Evaporation occurs
- at any temperature between the melting and boiling points, and
- only at the liquid's surface.

Sublimation happens when certain kinds of solids seem to "evaporate." The solid's molecules enter the vapor state without becoming a liquid first. Sublimation occurs at temperatures below the solid's melting point. Solids that sublimate into a vapor appear to gradually shrink over time and disappear. Ice cubes shrinking in your freezer is probably the most common example of sublimation. Mothballs used to protect clothing against moth larva damage also sublimate and disappear over time. Just as with evaporation, sublimation occurs only at the surface of a solid.

Sublimation occurs
- between the solid and vapor phases;
- at temperatures below the melting point of the solid; and
- at the solid's surface.

Condensation and Deposition

Freezing is the opposite of melting, and **condensation** is the opposite of vaporization. Condensation occurs when a vapor cools and changes to the liquid state. Vapor temperature and pressure determine when condensation occurs.

Dew is a good example of condensation. In the cool of the evening, water vapor molecules slow down when they collide with slower moving air molecules in the cool atmosphere. As the water vapor temperature drops, the molecules move more and more slowly. Eventually their attraction for each other causes them to clump together, forming microscopic water droplets. The exchange of water molecules between these droplets and the air still occurs, but at a lower rate than the clumping, so the droplets grow larger. These microscopic droplets come together and form visible drops of dew on cool surfaces, such as blades of grass, spider webs, and car windshields. Drops of water on cold beverage glasses result when the water vapor molecules slow as they collide with the cold glass surface.

2-30 Dew results from the condensation of water vapor on cool surfaces.

The opposite process to sublimation is **deposition**. Solids that sublimate to a vapor also readily deposit as a solid under certain conditions. Deposition occurs when the rate of vapor molecules returning to the solid exceeds the number of molecules sublimating away. Perhaps you have noticed the feathery crystals of ice that form on packages stored in a freezer, especially if you live in a humid locality. These ice crystals build up due to the relatively large amount of water vapor in the air contacting a surface that is below the freezing point of the vapor's liquid state. This is also the reason frost appears on outside objects in chilly weather.

2-31 Frost is a common example of the deposition of water vapor to ice crystals.

2D Section Review

1. What do we call changes between the solid, liquid, and gaseous forms of a substance? What remains unchanged in these processes?
2. How does the speed of atoms or molecules in a substance relate to its temperature?
3. What does a substance gain or lose when its temperature changes?
4. What do we call the change from the liquid to the solid state? from a solid to a liquid? For a pure substance, what temperature do these two processes have in common?
5. Where in a liquid does evaporation occur?
6. What do we call conditions such as the constant exchange of particles between the liquid and vapor phases within a closed container?
7. What does a liquid's vapor exert on the liquid, especially in a closed container?
8. ✪ Compare and contrast the three types of vaporization.
9. ✪ Why does water evaporate in an open container?
10. ✪ Why must we state the atmospheric pressure at which we measure the boiling points for pure substances?
11. ✪ Explain the difference between condensation and deposition.
12. (True or False) Thermal energy affects the motion of particles in matter, and temperature is an indicator of a material's particle motion.

Chapter Review

1. Read Hebrews 11:3 and 2 Peter 3:5. What do these verses say about the origin of matter?
2. How does matter differ from nonmatter?
3. Why does matter exist in different states?
4. What is the smallest building block of matter we can sense?
*5. What is a connection between two atoms called? Why are these connections important?
6. What are the two main ideas of the kinetic-molecular theory of matter?
7. As a Christian and creationist, analyze the problems with the origin and structure of matter proposed by the Greek philosophers Leucippus and Democritus in the third century BC.
8. Describe the basic structure of an atom, including the locations of the three main subatomic particles.
9. What do we call a molecule-like particle whose atoms all together have a greater number of electrons than protons?
10. What is the most important difference between a compound of three elements and a mixture of three elements?
11. What do we call the different substances that you can see in a heterogeneous mixture?
12. Why must scientists identify standard conditions when comparing similar properties among different substances?
13. What properties of solids distinguish them from liquids and gases?
14. Every chemical reaction demonstrates what important scientific principle?
15. The temperature of an object usually changes with the addition or removal of what quantity?
16. For a given substance under standard conditions, how do the temperatures of its solid, liquid, and gaseous states compare to each other?
17. What property of a liquid affects its rate of vaporization at a given temperature?
18. What vaporization process occurs at temperatures below the freezing point of a solid?

Scientifically Speaking

Term	Page
matter	21
Brownian motion	26
kinetic-molecular theory	26
pure substance	28
mixture	29
state	30
solid	31
liquid	32
gas	32
melting	40
freezing	40
vaporization	41
boiling	42
evaporation	43
sublimation	43
condensation	43
deposition	43

True or False

19. Both matter and nonmatter can be sensed, measured, or both.
20. Brownian motion and diffusion are both evidences for the particle model of matter.
21. The proton is the smallest particle of matter that makes up all other matter.
22. In a heterogeneous mixture, the distinctly different phases of matter are uniformly mixed together.
23. Wood is a good example of a crystalline solid because the wood fibers form an orderly structure shown in the grain.
24. All liquids have a free surface if they don't completely fill their containers.
25. The melting and retreat of a glacier is a physical change.
26. Iron turning to rust in warm, humid air is an example of a physical change.
27. Only a nuclear change is capable of producing atoms that are different from the atoms originally present in matter before the change.
28. The high speed of gas molecules compared to their speed in the liquid state implies that they have much greater thermal energy.
29. The air pressure above a boiling liquid affects the rate of vaporization of the liquid.

DS ✪ 30. What kind of injuries do you think a police officer might receive when hit by a bullet in a place protected by Kevlar? Explain.

DS ✪ 31. What does the need for bulletproof vests reveal about the state of humanity?

DS ✪ 32. Why don't law enforcement officials wear vests made of a material stronger than Kevlar, such as diamond?

✪ 33. Research the definition of *organic compound* in at least two sources other than this textbook. Compare the definitions and note which seems the clearest.

✪ 34. Physicists have discovered that electrons, protons, and neutrons are made of even smaller particles, called *elementary particles*. Research and write a one-page paper discussing these particles. Discuss the major groups that exist. State how many particles have been discovered so far. Identify the particles that make up the electron, proton, and neutron.

✪ 35. What do many astronomers believe is the most common form of matter in the universe? Give the source for your answer.

MEASUREMENT

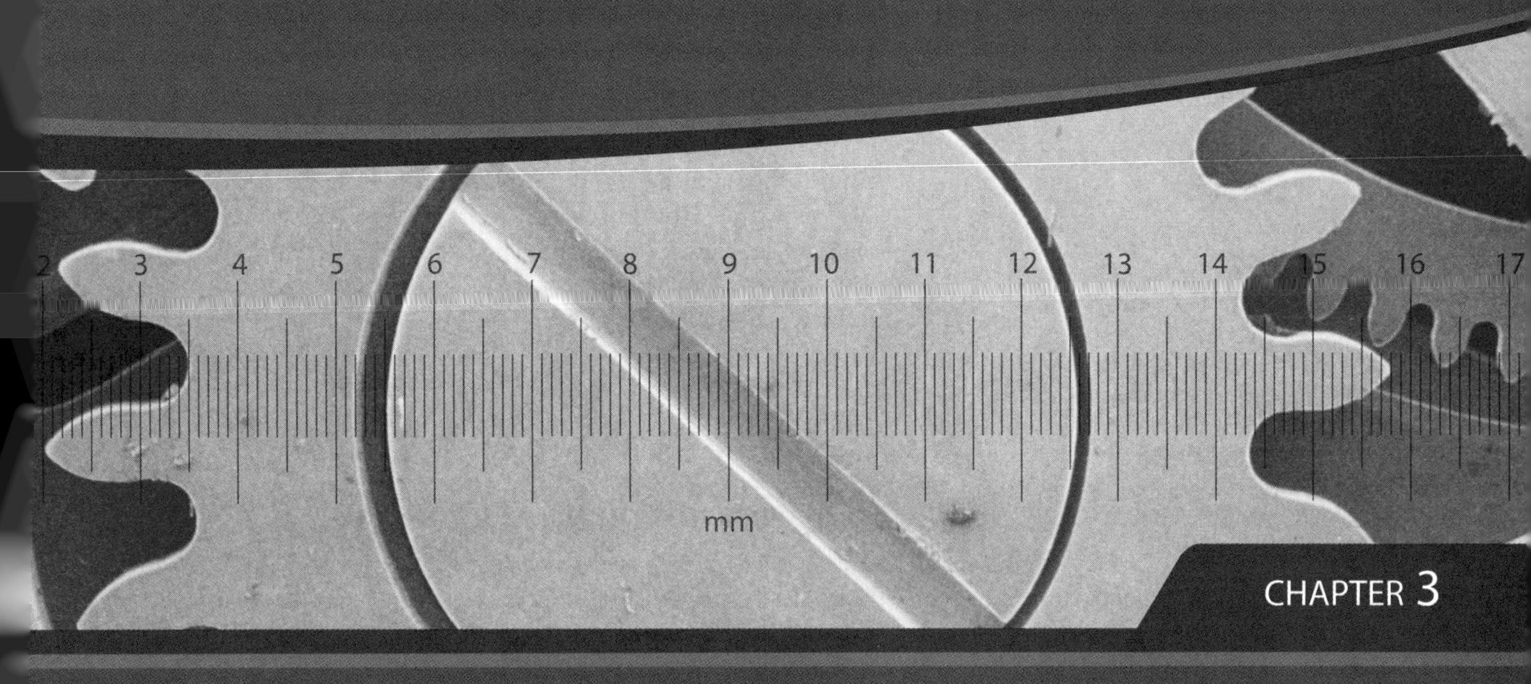

CHAPTER 3

3-1 *Curiosity*, the newest rover on Mars, is using the most advanced technology ever developed in its exploration of the Martian surface.

DOMINION SCIENCE PROBLEM

Satisfying Curiosity

In August 2012, Mars's surface became accessible to mankind in a whole new way. NASA's Mars Science Laboratory mission safely landed a nuclear-powered rover named *Curiosity* on the Martian surface.

But it wasn't easy. To do this, *Curiosity* had to endure what NASA engineers call "seven minutes of terror" to land there. *Curiosity* was enclosed in a capsule protected by a heat shield that resisted scorching temperatures as it entered the atmosphere. A supersonic parachute slowed it down to about 200 mph. Just a few miles above the surface, the probe assembly released the parachute and rockets fired to further slow it and direct it to the landing site at the bottom of a crater. At 20 m above the surface, a hovering "sky crane" gently lowered the rover to the surface. The rocket pack then flew off to the side and safely crashed some distance away.

Curiosity is equipped with the most advanced science instruments yet to explore the Martian surface. It can drive around on the surface collecting, analyzing, and even vaporizing rock samples with its built-in laser. Scientists are hoping this will give them valuable new data about Mars's surface and history. But all this relies on a robotic arm and wheels to function flawlessly in a very unforgiving and dirty environment. And these parts have to last a whole Martian year in extreme winds and temperatures. How could NASA build parts that meet the needed tight standards to ensure that the rover can fulfill its mission exploring Mars?

3A Objectives

After completing this section, you will be able to
- ✓ describe how scientists make measurements.
- ✓ compare and contrast quantitative and qualitative data.
- ✓ recall the decimal equivalents for metric prefixes.
- ✓ convert data reported in one unit of measurement to another unit.

Life on Mars?

One of the reasons scientists are so curious about Mars is because they see it as the planet most likely to support life. From the Viking missions of 1976 up to the present day, Mars missions have searched for signs of past or present life, or at least the conditions that could support life.

Most scientists believe that finding life on Mars would confirm their evolutionary worldview. Scientists holding to a Christian worldview would reject that motivation behind the search for life on Mars. But they would still see exploration of the solar system as a means to learn more about God's creation and to glorify Him.

In science, we call any measurable aspect of something a **dimension**. Common dimensions include length, time, temperature, volume, and speed.

3-2 Measuring involves comparing a quantity against a graduated scale.

unit: (L. *unus*—one)

Observational **bias** occurs when an observer imposes his presuppositions on the observational results. This is more likely to happen when collecting qualitative data.

3A Scientific Measurements

3.1 Measuring and the Creation Mandate

The Mars Science Laboratory mission and *Curiosity* are all about measuring. The rover itself is a marvel of design and measurements. What size must a gear on *Curiosity's* robotic arm be to make it move properly? How hot must its laser be in order to vaporize a rock? But these are just a few of the questions of space science. Here on Earth, what's the average volume of air in a person's fully inflated lungs? What's the distance between carbon atoms in the diamond of an engagement ring? How low is the air pressure inside a tornado? These and similar scientific questions can be answered only by measuring.

Gaining useful knowledge about God's creation—the purpose of true science—demands that we make measurements of every facet of nature. Measured information allows us to use what we know more effectively, to return glory to God, and to help our fellow human beings.

3.2 Measurement—The Basics

What exactly *is* a measurement anyway? When you measure something, you compare some aspect of the thing that you are measuring to an appropriate measuring standard. Such a standard is usually an instrument with a graduated scale. We need to define some key words in this description so that we understand fully what a measurement is.

Once you have identified the ***dimension*** you want to measure, you must compare it against an appropriate standard. A *standard* is a known quantity of a dimension that everyone agrees to use for comparison when measuring. Later in this chapter, you will learn about the important standards scientists refer to when they make measurements.

An *instrument* is any man-made device a scientist uses that includes the standard for measuring. Human senses aren't reliable and can't produce numerical measurements. We can't even sense some phenomena, such as gamma radiation or bacteria. So an instrument extends, refines, or substitutes for human senses.

Mechanical instruments include a *graduated scale,* and electronic ones display a numerical value. The spacing of the markings on a scale shows the units of the dimension that the mechanical instrument measures, while a digital display usually shows the abbreviation for the unit. A ***unit*** is the name and the size of a portion of a dimension that has a value of 1. The dimension standard defines the size of the unit.

3.3 Measured Data

Data is information collected from scientific observations. For scientific data to be useful, scientists must be able to collect and report data clearly. For this reason, scientists report observations using either numerical measurements (*quantitative data*) or precise word descriptions (*qualitative data*). Observers are far less likely to misinterpret (due to **bias**) instrument-measured quantitative data compared to qualitative data. Thus, it is the preferred kind of scientific data. Scientists collect qualitative data when they can only describe an observation, not measure it.

Measured data usually has two parts—a number and a unit. The unit accomplishes three things: it signals that the preceding number is a measurement; it indicates the kind of dimension; and it identifies the size of the dimensional unit.

3.4 The Metric System

As international shipping expanded during the seventeenth and eighteenth centuries, trading in foreign ports became more difficult. Weights of grain, lengths of cloth, and volumes of oil all had different units in different countries. Many merchants limited their trades to ports that used the same units. At the same time, scientists all over Europe were communicating more and more as they shared scientific theories. Although the language of science was standardized (Latin), the units of measurements were not. Increasingly, the European scientific community was finding it difficult to communicate important discoveries so that fellow scientists could understand. Many scientists realized that they needed to settle on a standard set of measures. But which one? Every nation wanted to make its own set of units the basis for the new common measuring system. International bickering about this matter went on for decades.

Finally, in 1799, France established a decimal **metric system**. The "decimal" designation means that the units of each dimension and any other units based on these are related by powers of 10. The French metric system first contained two standards, the *meter (m)* for measuring length and the *kilogram (kg)* for measuring the quantity of matter. The French originally defined the meter as one ten-millionth of the distance from the earth's North Pole to the Equator along the line of longitude that passed through Paris. They manufactured a bar made of a corrosion-proof metal alloy to show the length of 1 meter between two carefully scribed marks. The French defined a kilogram as the mass of one **cubic decimeter** of water. They made the standard kilogram in the form of a cylinder of the same alloy as the meter bar. Scientists all over Europe eventually adopted the system because its standards came from physical quantities and it was easy to use.

New standard units had to be included in the metric system as the existing fields of science progressed and new branches appeared. One of the roles of normal science is refining the standards by which it operates (see the Science and Truth facet on page 10), so scientists developed more exact ways of defining the existing unit standards in ways that did not rely on physical objects. For example, in 1983 scientists agreed to define the meter as exactly the length of the path that light travels in a vacuum during 1/299 792 458 second. The second itself is also a metric unit, exactly defined as the time required for a specified number of oscillations of a certain kind of light wave.

Physicists required these demanding standards because of the increase in accuracy and precision of scientific measurement. However, the kilogram is still based on a physical object because scientists have not yet been able to develop a standard of mass based on other exactly defined units. The **standard kilogram** resides in the International Bureau of Weights and Measures (BIPM) at Sèvres, France, under carefully controlled environmental conditions. There are six

Examples of quantitative data:
- water temperature
- aircraft speed
- time of an accident

Examples of qualitative data:
- audibility of a sound
- pattern of veins in a butterfly's wing
- shape of a mineral crystal

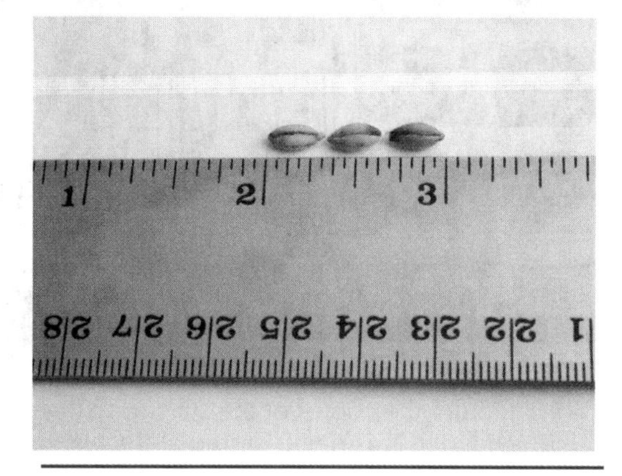

3-3 During the Middle Ages, the inch was equal to a thumb's width in one place and the length of three barleycorns in another. A standardized system of measurements was needed.

metric: (Fr. *mètre* from Gk. *metron*—to measure)

A **cubic decimeter** (dm^3) is equal to one one-thousandth of the volume of a cubic meter (m^3). Today, a cubic decimeter is equal to a volume of 1 liter (L).

The **standard kilogram** is officially called the *International Prototype Kilogram (IPK)*. BIPM is the abbreviation formed by the initials of the French words for International Bureau of Weights and Measures.

3-4 One of the copies of the standard kilogram platinum-iridium cylinder. This one is stored at the National Institute of Standards and Technology (NIST) in the United States.

copies of the standard kilogram stored with the original. National institutes around the world hold about 100 other copies.

Even with a workable metric system, many countries continued to use local traditional systems for trade and government work well into the twentieth century. In 1960, the countries of the world finally agreed to adopt a single global metric system. An international conference renamed the system the *Système International d'Unités* in French. The more common abbreviation **SI** comes from this name. The French metric system began in 1799 with two base units. Today, the SI has seven. A *base unit* is one defined by a set of measurable conditions that can be recreated in any properly equipped laboratory anywhere in the world (true of all but the kilogram). Appendix B lists the SI base units and the conditions that define each dimensional standard. Table 3-1 lists the seven SI base units along with their dimensions and unit symbols.

SI Base Units and Derived Units

SI base units have their own unit symbols and can't be expressed as combinations of any other SI units. The SI also includes many other units scientists require for their work that are combinations of the SI base units. These other SI units are called *derived units*. In this course, you will become familiar with many SI derived units, such as the *liter (L)*, which is a unit of volume, the *newton (N)*, which is a unit of force, and the *joule (J)*, which is a unit of energy and work.

3-1 **SI Base Units**

Unit	Dimension	Symbol
meter	length	m
kilogram	mass	kg
second	time	s
ampere	electric current	A
kelvin	absolute temperature	K
mole	amount of substance	mol
candela	radiant intensity	cd

Metric Prefixes

The metric base units alone are useful for a very limited range of measurements. How would you like to measure the thickness of aluminum foil in meters? Or the volume of an ocean in cubic meters? Since the SI is a decimal system, we can define units larger or smaller than a base unit by multiplying the base by a power of 10 (see the Math Help box below).

Math Help: Powers of 10

Understanding powers of 10 is easy and useful for work in science. A "power" of a number is just an exponent that shows how many times that number is multiplied by itself. The exponent hangs out in the upper right corner of the base number. *Exponential notation* helps you to write very large or very small numbers compactly. For example, 10^2 means 10×10, or 100, and 10^6 means $10 \times 10 \times 10 \times 10 \times 10 \times 10$, or 1,000,000. Study Table 3-2 on the facing page to see the usefulness of this form of notation.

There are a few special power rules to be aware of. The exponent "1" simply means the base number itself. This is why you almost never see 10^1, because that is the same as 10. Also, any base number with the exponent "0" is equal to 1. So $10^0 = 1$. (You will learn why in a math class.) Finally, any exponent beginning with a negative sign signals that the number's value is a fraction equal to 1 over the exponential value of the number. So $10^{-1} = 1/10$; $10^{-2} = 1/10^2$, and $10^{-6} = 1/10^6$. Keep these rules in mind as you learn about the metric system.

You can know the size of a metric unit by its *unit prefix*. A prefix indicates the exponential factor needed to give the size unit you want. For example, the prefix *kilo-* tells you that the metric base unit is multiplied by the factor 1000, or 10^3. A *kilo*meter (km) is 1000 meters (m). You can see by this that the relation 1 m = 1/1000 km is also true. The prefix *milli-* stands for 1/1000 (or 10^{-3}) times the base unit. A *milli*liter (mL) is 1/1000 of a liter. Similarly, it takes 1000 mL to equal 1 L.

Each prefix has its own abbreviated symbol that combines with the symbol of the base unit. (See Table 3-2 below.) We can adapt the base unit for any metric measurement. For example, we measure distances between cities in kilometers (km or 10^3 m), while we measure the thickness of aluminum foil in millimeters (mm or 10^{-3} m) or micrometers (μm or 10^{-6} m).

Nearly every country has an organization that standardizes the rules for using the SI prefixes since these are established by international agreement. In the United States, the National Institute of Standards and Technology (NIST) enforces these rules for scientific work.

3-5 We can add metric prefixes to SI derived units just as with SI base units. For example, the jaws of a trap jaw ant (*Odontomachus sp.*) can exert a force of about 11 *milli*newtons (mN). This ant also has the fastest moving jaws of any animal in the world!

3-2 Table of Metric Prefixes

Prefix	Pronunciation	Abbreviated Symbol	Meaning	Factor	Exponential Form
peta-	PEAT uh	P-	quadrillion	1,000,000,000,000,000	10^{15}
tera-	TEHR uh	T-	trillion	1,000,000,000,000	10^{12}
giga-	GIG uh	G-	billion	1,000,000,000	10^{9}
mega-	MEG uh	M-	million	1,000,000	10^{6}
kilo-	KILL oh	k-	thousand	1000	10^{3}
hect-/hecto-	HEK toh	h-	hundred	100	10^{2}
deka-	DEK uh	da-	ten	10	10^{1}
—	—	(none)	(base)	1	10^{0}
deci-	DESS ee	d-	tenth	0.1	10^{-1}
centi-	SEN tih	c-	hundredth	0.01	10^{-2}
milli-	MIL lih	m-	thousandth	0.001	10^{-3}
micro-	MY kroh	*μ-	millionth	0.000 001	10^{-6}
nano-	NAN oh	n-	billionth	0.000 000 001	10^{-9}
pico-	PEE koh	p-	trillionth	0.000 000 000 001	10^{-12}
femto-	FEM toh	f-	quadrillionth	0.000 000 000 000 001	10^{-15}

Note: You should memorize the prefixes marked in blue.
*The symbol "μ" is the Greek letter *mu*.

3.5 Unit Conversions

Metric Conversions

A teacher gave three groups of students the assignment of measuring the length of a table in their classroom. Each group received a different ruler. One ruler was graduated in whole meters, the second in just centimeters, and the third in millimeters. The teacher recorded their results on the board:

Group A	Group B	Group C
2 meters	200 centimeters	2000 millimeters

Do their measurements agree? They measured the same quantity, but came up with different numbers. We can check the results by

> Recall that a **factor** is one of the numbers in a multiplication operation. In the math expression 5 × 7, the numbers 5 and 7 are factors.

converting all the measurements into the same unit, say, the meter, by using a conversion factor. A *conversion factor* is any **factor** that multiplies a measurement to convert its unit to a different but equivalent unit. A conversion factor has the following properties:

- It is a fraction.
- Its overall value is equal to 1, since same-sized quantities are in both the numerator and denominator.
- The value in the numerator has the unit you are converting *to*; the value in the denominator has the unit you are converting *from*.
- After multiplying the original measurement with the factor and cancelling like units, the result is the same quantity but is a new number with the desired unit.

Math Help: Conversion Factors

While you are rummaging around in your refrigerator for a cold soft drink, your mother asks you to look and see how many dozen eggs are left on the egg shelf. You see 6 eggs and report back that she has ½ dozen eggs left. How did you come up with this conversion from "single-egg" units to "dozen" units? You were aware of two things: there are 12 items in one dozen, and 6 is half of 12.

If you summarized your thoughts in a mathematical equation, you could say:

$$12 \text{ eggs} = 1 \text{ dozen}.$$

Dividing both sides of the equation by 12 eggs,

$$1 = \frac{\textbf{1 dozen}}{\textbf{12 eggs}} \quad \textbf{(conversion factor)},$$

because dividing a number by itself always equals 1. Any factor equal to 1 that is a ratio of two units can be used as a conversion factor. Study the following expression of your mental "egg" calculation to see how a conversion factor is used.

Notice that units in division divide out, or cancel, just like numbers.

$$6 \cancel{\text{eggs}} \times \frac{\textbf{1 dozen}}{\textbf{12} \cancel{\textbf{eggs}}} = \frac{6 \text{ dozen}}{12}$$

$$= \frac{1}{2} \text{ dozen}$$

In general, you do unit conversions as follows:

- When finding a conversion factor, first set up an equation with equal quantities. These values will always come from definitions of unit relationships: 60 s always equal 1 min; 12 in. always equal 1 ft; 10 mm always equal 1 cm.
- Note which side of the equation contains the unit you are converting *from* and divide both sides by that quantity. This gives a fraction equal to 1, with the new unit (the one you are converting *to*) in the numerator and the unit to cancel in the denominator. This is the conversion factor.
- Multiply the measurement you want to convert by the conversion factor to yield the same measurement in the new unit.

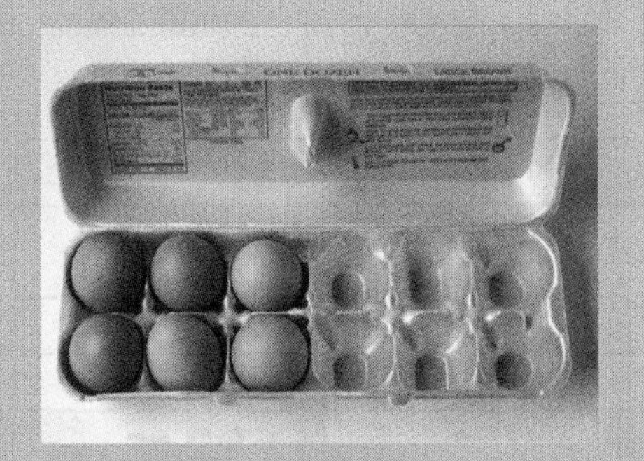

A half-dozen eggs

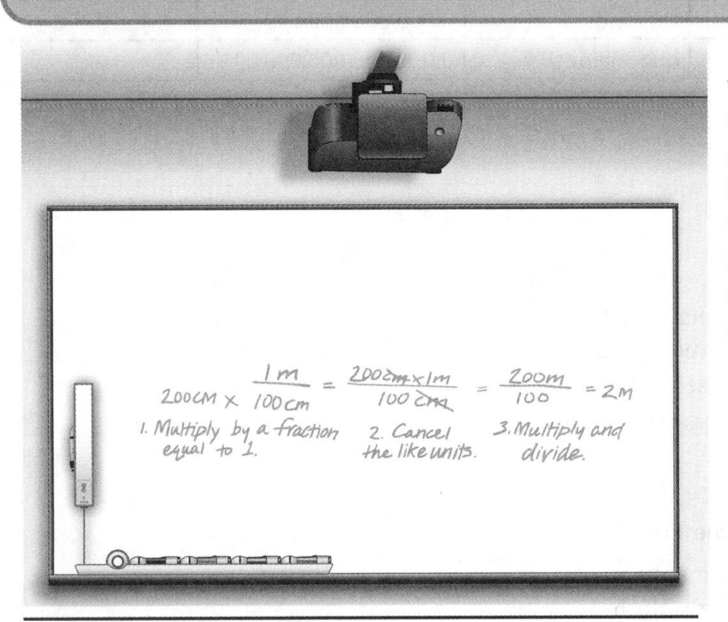

3-6 Thinking through the cancellation of metric units

Let's see how a conversion factor works when converting between metric units. Group A's measurement is already in meters (2 m). You can convert Group B's measurement, 200 centimeters (cm), to meters by knowing that there are 100 cm in 1 m (see Table 3-2). If 100 cm is the same as 1 m, then 200 cm must equal 2 m. However, most unit conversions aren't this simple. The following steps show how to find the conversion factor and then make the unit conversion.

$$100 \text{ cm} = 1 \text{ m} \qquad \text{(equal quantities)}$$

$$\frac{\cancel{100 \text{ cm}}}{\cancel{100 \text{ cm}}} = \frac{1 \text{ m}}{100 \text{ cm}} \qquad \text{(divide by 100 cm)}$$

$$1 = \frac{1 \text{ m}}{100 \text{ cm}} \qquad \text{(conversion factor)}$$

$$200 \cancel{\text{cm}} \times \frac{1 \text{ m}}{100 \cancel{\text{cm}}} = \frac{20\cancel{0} \text{ m}}{10\cancel{0}} = 2 \text{ m}$$

The slash marks show cancellation that occurs during division. Recall that like factors cancel in the numerator and denominator of a fraction. If there are like units in both numerator and denominator, they cancel just like numbers. Remember that multiplying by a value equal to 1 does not change the size of the original measurement. The goal here is simply to change the measurement into a value with different units.

Check the measurements made by Group C. Is 2000 millimeters (mm) the same as 2 m? We can find the answer by starting with the relationship 1000 mm = 1 m.

$$1000 \text{ mm} = 1 \text{ m}$$

$$1 = \frac{1 \text{ m}}{1000 \text{ mm}} \quad \text{(conversion factor)}$$

$$2000 \cancel{\text{mm}} \times \frac{1 \text{ m}}{1000 \cancel{\text{mm}}} = \frac{2\cancel{000} \text{ m}}{1\cancel{000}} = 2 \text{ m}$$

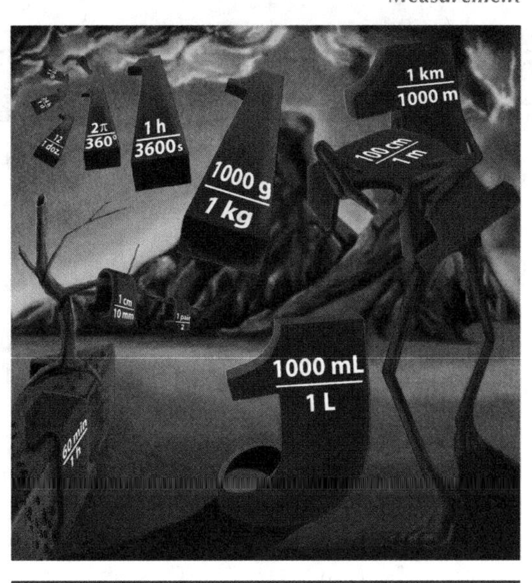

3-7 Ratios created from unit equalities always have an overall value equal to 1.

Examine the conversion factors in the following examples.

Example Problem 3-1

Practicing Unit Conversion

How many milliliters are there in a 2-liter (L) soft drink bottle?

Relationship: 1 L = 1000 mL
(given unit: liters; desired unit: milliliters)

Conversion factor: $1 = \frac{1000 \text{ mL}}{1 \text{ L}}$ (divided both sides by 1 L)

Unit conversion: $2 \cancel{\text{L}} \times \frac{1000 \text{ mL}}{1 \cancel{\text{L}}} = 2000 \text{ mL}$ (like units cancel)

Answer: There are 2000 mL in a 2-liter bottle.

Which contains more mass, a 350 g box of chocolates or a 3.5 kg box of chocolates? Let's convert kilograms to grams.

Relationship: 1 kg = 1000 g
(given unit: kilograms; desired unit: grams)

Conversion factor: $1 = \frac{1000 \text{ g}}{1 \text{ kg}}$ (divided both sides by 1 kg)

Unit conversion: $3.5 \cancel{\text{kg}} \times \frac{1000 \text{ g}}{1 \cancel{\text{kg}}} = 3500 \text{ g}$

Answer: The 3.5 kg box contains 10 times more chocolate!

3-8 Comparing 350 g of chocolate (left) to 3.5 kg of chocolate (right)

You may occasionally need to convert between two units whose direct relationship you do not know. For example, how many seconds are in a week? Simply use more than one conversion factor in a series of steps to convert from the given unit to the desired unit. You probably do not know the number of seconds in a week (wk), but you do know the number of days (d) in a week, the number of hours (h) in a day, the number of minutes (min) in an hour, and the number of seconds (s) in a minute. The conversion from weeks to seconds requires four factors.

$$\underset{\text{wk}}{1} \times \underset{\text{d/wk}}{7} \times \underset{\text{h/d}}{24} \times \underset{\text{min/h}}{60} \times \underset{\text{s/min}}{60} = 604{,}800 \text{ s}$$

FACET OF SCIENCE

How Long Was the Cubit?

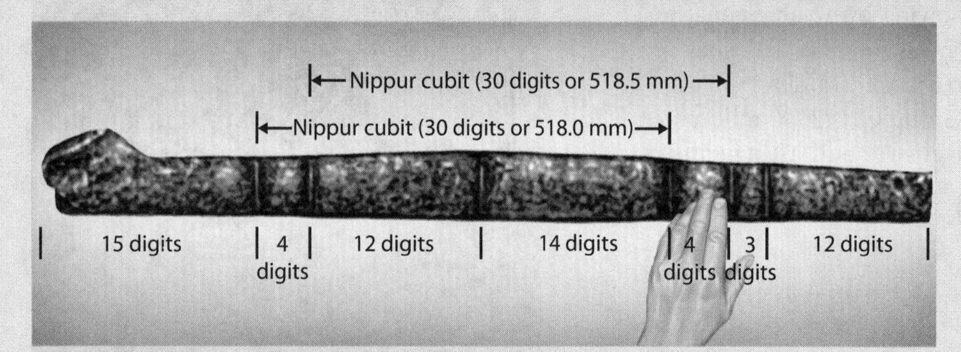

The cubit is the world's oldest known unit of measure. It is the first unit mentioned in Scripture (Gen. 6). The cubit defines the dimensions of Noah's Ark, placing its use before the Flood. Tim Lovett, a Christian engineer who is an authority on the vessel, has done extensive research on the cubit that may have been used in building it. This facet discusses some of his findings.

A cubit was given as the distance from elbow to fingertip, usually of some prominent person in a kingdom. The word *cubit* comes from the Latin word for forearm, *cubitum*. Scholars traditionally assume that the biblical cubit was about 457 mm (18 in.) long. However, archeologists discovered long ago that early important civilizations like Egypt and Babylon used two different cubits. One was the "common" cubit, about 450 mm (17.7 in.), and the other was a longer "royal" cubit of around 520 mm (20.5 in.).

The royal cubits were about one handbreadth (3–4 in.) longer than common cubits. For example, researchers established that the Royal Egyptian cubit used to construct the pyramids at Giza was 524 mm (20.6 in.) by comparing construction records to direct measurements of the pyramids.

A 2.2 m copper bar found near ancient Nippur in former Mesopotamia is known as the *Nippur Cubit* (illustrated above). This irregular piece of metal contains six markings. Four marks describe two overlapping cubits of about 518 mm (20.4 in.) each. Historians believe the Nippur cubit is one of the oldest existing measuring instruments known.

The lengths for the common cubit have been documented from numerous locations throughout the Middle East and Europe. These cubits were clearly based on human anatomical measures. Most of them were about 6 handbreadths long. The common cubits seem to have come into use after the royal cubit, at least for the Hebrews.

Lovett observes that measurement units change very little over time in cultures, so it would not be surprising that the only cubit known in the world after the Flood—Noah's cubit—would have been the basis of the royal cubit used to build the great civilizations that appeared just a few hundred years later. This idea is also supported by the fact that royal cubits have been found in ancient English, Chinese, Indian, and Aztec cultures that fit within the range of 518–31 mm (20.4–20.9 in.). This range is only a 6% variation—important evidence that the royal cubit was held in high regard as the language groups dispersed after the confusion of tongues at the Tower of Babel.

The Bible also indicates that the different cubits were used for different purposes. For example, the length of King Og's bed was measured "after the cubit of a man" (Deut. 3:11). This reference seems to indicate a common anatomical standard. The common cubit was also used in the construction of the Siloam water tunnel by King Hezekiah in Jerusalem (2 Kgs 20:20). The tunnel was rediscovered with its length in cubits engraved at the entrance.

In contrast, a number of scriptural references clearly indicate that the longer royal cubit was used for temples and other special purposes. King Solomon used a cubit "after the first measure" to build the temple (2 Chron. 3:3). The cubit of the "first measure" was probably a royal cubit, which was older and held in higher regard.

The strongest biblical evidence for the association of the royal cubit with religious structures occurs in Ezekiel 40:5 and 43:13. In Ezekiel's vision of the New Jerusalem, an angel measured the temple with a reed that was 6 cubits long. This cubit is specified as a "cubit and an hand breadth," which is how royal cubits were defined—even in Egypt and Babylon.

So how long was the first cubit, the one that was used by Noah to build the Ark? Lovett reasons that the longer cubit may have come from an important leader living before the Flood. It is reasonable to assume that he was taller than the average for modern humans. Biblical genealogies indicate that humans lived longer before the Flood than afterward, and this longevity may have included a taller stature, since there is fossil evidence that many kinds of living things were larger before the Flood compared to afterward. Studies have shown that the length of an average human's forearm is approximately 27% of his height. A forearm length of 20.4 in. (the Nippur Cubit) would mean that humans were on the average 6 ft 3 in. tall before the Flood.

Lovett also suggests that if the ancients traditionally used the royal cubit for measuring religious structures and government buildings, and God even specified it for the future temple, Noah probably used the royal cubit for constructing the Ark just a few hundred years earlier. He believes the best existing artifact for this cubit is the ancient copper Nippur Cubit, discovered near the center of the post-Flood civilization.

This facet is a summary of Tim Lovett's article "Which Cubit for Noah's Ark?" www.worldwideflood.com/ark/noahs_cubit/which_cubit.pdf, 2006.

Unit Conversions between Different Systems

Conversion between different units within the SI is nearly always a matter of finding the right power of 10 to relate the two units. But how do you convert measurements between SI units and English units, such as those used in the United States? For example, while vacationing in the Canadian Rockies, you see a road sign that says, "Banff, Alberta, 35 km." How far is that in miles?

Unit conversion between different measurement systems follows the same process as above. However, conversion factors from English to metric units are *almost never* equal to exactly 1. Because of this, the unit "equations" involve one rounded number and are only approximately equal. To convert the distance to Banff to miles, use the following relationship:

$$1.6\text{ km} \approx 1\text{ mile (mi)}$$

$$1 \approx \frac{1\text{ mi}}{1.6\text{ km}}$$

$$35\,\cancel{\text{km}} \times \frac{1\text{ mi}}{1.6\,\cancel{\text{km}}} \approx 21.9\text{ mi}$$

If you use the relationship 1 mi ≈ 1.609 km, which is a better approximation, the calculated distance is about 21.8 mi. For these kinds of conversions, refer to Appendix D to find the appropriate unit relationships.

3-9 How many miles to Banff?

The ≈ symbol means "approximately equal to."

3A Section Review

1. What is a dimension? Give three examples.
2. Provide a definition of a measuring unit. Give an example of a unit of length, mass, and volume.
3. Why do most scientists prefer quantitative data over qualitative data?
4. What was the main reason that the French developed the metric system?
5. What is generally true about unit conversion within the metric system that is not generally true for unit conversion between the metric and the English systems?
6. (True or False) Measuring involves comparing a dimension with a graduated scale of an appropriate instrument or other standard.
7. (True or False) The liter is an SI base unit.
8. *Which unit do we know was used both before and after the Flood?
9. Convert each measurement as indicated. Show your work.
   a. 16.2 km to meters
   b. 12 mg to grams
   c. 103 μm to meters
   d. 23.5 mcd to candela (cd)
   e. 386 μK to kelvins (K)
   f. 1.5 MA to amps (A)
10. Convert the following measurements to the units indicated. See Appendix D for unit relationships. Show your work.
    a. 0.48 L to milliliters
    b. 34 g to kilograms
    c. 75 m to feet
    d. 21 mi to kilometers
    e. 5.6 m to millimeters
    f. 455 g to ounces (dry)

3B Objectives

After completing this section, you will be able to

- ✓ list some common sources of measurement error.
- ✓ compare and contrast accuracy and precision in measurement.
- ✓ identify the significant digits in a scientific measurement.
- ✓ explain the importance of significant digits as an aspect of scientific modeling.

3B Accuracy and Precision in Measuring

3.6 *The Limits of Measuring*

Take a ruler graduated in centimeters and measure the length of this textbook's spine. It should be about 27.6 cm long. You may have noticed that the edge of the book fell between 27.5 cm and 27.7 cm. But how long is it exactly? This is impossible to answer. If your ruler is metal, it can expand and contract with temperature differences; if your ruler is wooden, it changes length with humidity. Most plastic rulers came from molds that vary in accuracy. Also, if you measure the book's spine a second time, you might notice that the reading is slightly different from the first because you placed the ruler zero mark in a slightly different place.

These thoughts show many of the problems of measuring. Measuring can *never* be exact for one or more of the following reasons:

- The instrument markings can't be fine enough to align *exactly* with the dimension you are measuring.
- Environmental conditions such as temperature, pressure, or humidity may alter the instrument's scale.
- You may use or read the instrument incorrectly.
- The instrument may be defective from manufacturing, damage, or wear.
- Conditions during measurement can interfere with obtaining a good measurement.

3-10 No instrument has markings that are fine enough to measure things exactly.

Any of these problems will produce a measurement that is not the "actual" or, more correctly, the *acceptable value*. The difference between a measurement and the dimension's acceptable value is the ***error*** of the measurement. Error in measurement is an ever-present fact of science. Every measurement contains error, no matter how good the instrument and the observer are. For that reason, much of the work scientists do in their research goes into figuring how much error is present in their measurements.

The measurement **error** is the difference between the measured value and the acceptable value.

Even though every measurement includes error, Christians have good reasons for trying to minimize the error. Measuring is just one aspect of exercising dominion. Without limiting the error of measurements, we would not be able to develop really useful models in science and technology. There would be inaccuracies that could lead us to incorrect conclusions. Or things that we build, like the *Curiosity* rover, wouldn't fit together right if builders of different parts used different methods of measurement having different allowable errors.

We try to minimize the error in measurement by using instruments that are as *accurate* as needed and that have the *precision* required for our work. Let's examine these often-confused terms.

What Is the "Actual Value"?

A lot of people believe that if we just had a good enough instrument, then we could obtain an exact measurement. But reality doesn't support this idea. Recall from Chapter 1 that we can't explain or describe natural phenomena exactly. If we could, this would mean we know something to be absolutely true. You could look at measuring as another kind of modelmaking. Each measurement *represents* a dimension approximately; it's not the dimension itself.

Imagine zooming in with a super-microscope on the edge of a razor blade. At normal magnification, it looks sharp and straight. But as you zoom in, the blade starts looking rough. Closer still, the edge begins to look like mountains. At the atomic level, the atoms are jiggling around in a vast, heaving landscape of caverns. So how can you exactly measure something that is so irregular? All we can do is make measurements to an exactness that we define as acceptable.

3.7 Accuracy

The **accuracy** of a measurement is an assessment of the measurement error. A scientist would state that measurement is more accurate, less accurate, or just "accurate" or "inaccurate." A smaller error means a *more* accurate measurement while a larger error indicates a *less* accurate measurement.

An instrument's accuracy depends on how well it is constructed and maintained. The scale graduations must be accurately printed or engraved. The instrument's accuracy must not change when environmental conditions change. Most instruments rarely meet these ideals. However, many scientific instruments have built-in features that allow the user to adjust the instrument's accuracy back to within an acceptable range of error. To do this, the user first measures an accurately known object or quantity, called a *prime standard*, with the instrument. Then the user adjusts the instrument to display the value of the prime standard.

The instrument user also affects the accuracy of a measurement. Because of the limitations of human observation, there is always some human error associated with making any direct measurement. Scientists try to eliminate these kinds of errors by averaging several measurements of the same quantity by the same or different persons. This leads us to two related but different concepts—*precision* and *repeatability*.

Accuracy evaluates how close a measurement is to the acceptable value of the measurement.

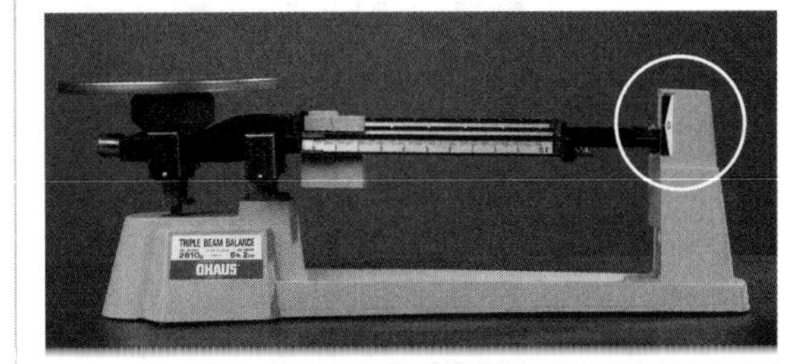

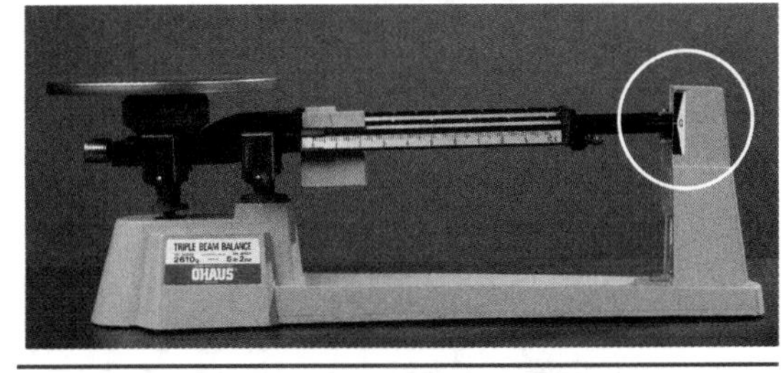

3-11 The prime standard for aligning a lab balance is the zero mark on the instrument. The top photo shows that the empty balance is not zeroed. In the bottom photo, the balance is zeroed.

3.8 Precision

Suppose you want to find the density of a sample of quartz in your mineral collection. Density is found by dividing the mass of the sample by its volume. The measured quantities are mass, 27.55 g, and volume, 10.4 cm³. When you perform the division on a calculator, the answer is

$$2.649\,038\,462 \text{ g/cm}^3.$$

As you can see, calculators can compute answers to many decimal places—in this case, nine. But is it proper to report your result to that many decimal places? Look at the data. You measured the mass to two decimal places and the volume to only one. Any digits in decimal places beyond the given digits are unknown. Do you think the answer to a calculation involving measured data should be more exact than the original numbers?

Precision is an assessment of the exactness of a measurement. A more precise measurement has more certain digits than a less precise measurement of the same quantity. The number of digits in a scientific measurement and the location of the decimal point indicate something of the instrument's precision too. The finer the spacing of the graduation marks on the instrument scale, the more digits and decimal places are possible in the measurement, and the more exact the measurement.

You determine the exactness of measured data by reading the major and minor divisions on the instrument scale. Most of the digits in a measurement come directly from the scale marks; you are certain about these digits. But measurements almost never fall

The Glory of God and Measurement

Our ability to measure things with precision and accuracy testifies to the glory of being intelligent, creative beings. The Bible indicates that God knows about measurements too. Isaiah 40:12–18 states in poetic language God's extensive measuring ability. According to this passage, what sort of God is the God of the Bible? What does verse 18 state about how we should respond to this God?

Precision evaluates how exactly a measurement is made.

If a quantity is defined or is an exact count, it is *not* measured data. It is exact and the concept of precision does not apply.

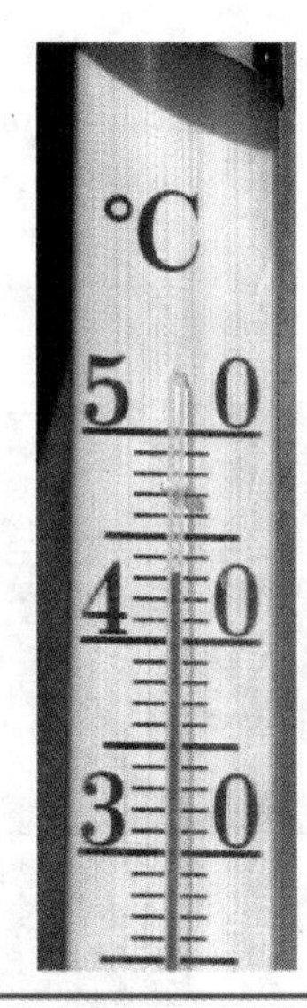

3-12 This thermometer is indicating 43.3 °C. The reading is estimated to tenths of a degree because the smallest scale division is 1 °C.

The half-degree subdivisions of some Celsius thermometers are not *decimal* subdivisions of the scale. These result from dividing a degree by 2, not 10. They *do* help you estimate tenths of a degree, so you should report temperatures to the nearest 0.1 °C.

exactly on a scale mark. You must estimate where the measurement falls in between the marks and include that estimation as part of the measurement. This leads to our first rule of measuring.

When measuring with a *metric* instrument, estimate the measurement to 1/10 of the smallest *decimal* subdivision on the instrument scale.

Metric instruments have scales subdivided into tenths, so estimating the fraction of the smallest division in tenths just adds a decimal place to the measurement. This is not true for most instruments measuring in true English units. A foot ruler, for example, commonly has 16th inch graduations. Estimating a tenth of a 16th is not very useful. Users will estimate to a 32nd or 64th of an inch in such cases.

Most measurements made in a high school laboratory come from metric instruments. The precision of a measurement is limited by the size of the smallest decimal subdivision on the instrument scale.

3.9 Repeatability

Precision applies to a given instrument, but the data an instrument produces is usually part of a larger, often complex measuring process. Making a measurement involves the instrument, to be sure, but it also includes the operator, the thing being measured itself, and the surrounding environmental conditions. How do you know that any single measurement is actually useful?

Scientists evaluate the usefulness of their measurements by repeating them as many times as needed to see how they vary. These variations represent measuring errors from some average value. With enough measurements, scientists will calculate this average and *it* becomes the measurement. The differences between each measurement and the average are called *random errors*. Random errors are more easily detected with instruments that are more precise, but they tend to cancel out when enough data is collected to compute an average. Small random errors mean the measurement has good **repeatability**.

Having a good idea of what the result of an experiment should be is normal for scientific work. If the results of a series of measurements continue to be different from the expected or accepted value, then the scientist begins to look at the measurement process. Similar differences indicate a *systematic error*, one that shifts the results of measurements the same amount and same direction each time. Systematic errors come from inaccurate instruments, improper measuring methods, or environmental factors affecting the process. These errors are harder to detect in school laboratories. Measurement results may be repeatable, but they could be inaccurate.

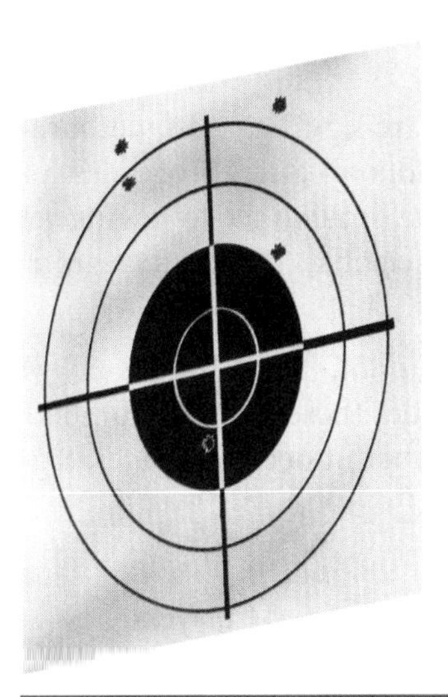
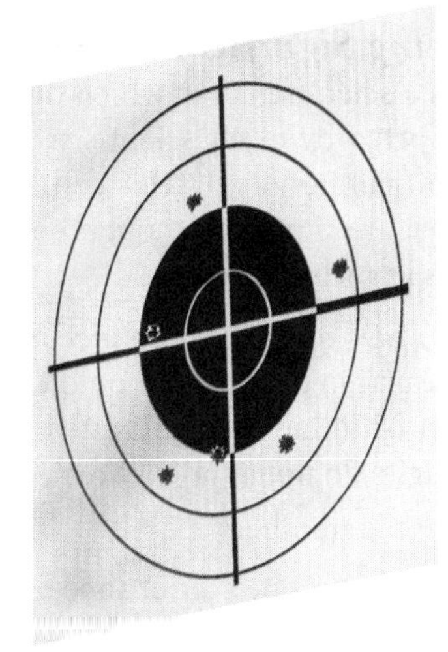
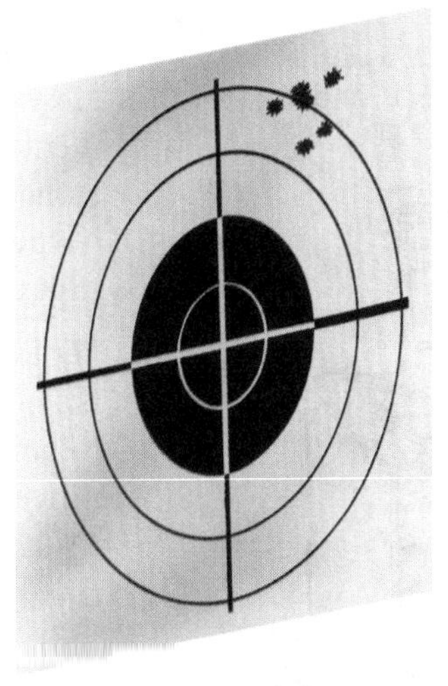
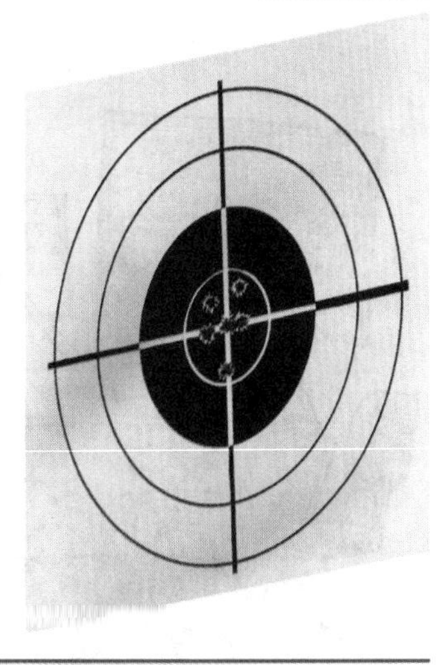

3-13 These targets illustrate repeatability and accuracy. The two on the left show poor repeatability due to larger random errors (larger distances from the *center of each group*).The third target shows good repeatability but poor accuracy due to a large systematic error. The fourth target shows good repeatability and good accuracy

3.10 Significant Digits

Suppose you use a metric ruler to measure the width of a piece of notebook paper. The smallest divisions on the ruler are millimeters. You carefully position the ruler across the paper and find that the paper is between 215 and 216 mm wide. What should you report as the paper's width? Reporting 215 mm wouldn't be correct since you can see that the paper extends a little bit beyond the 215 mm mark. Likewise, 216 mm would also be incorrect since the paper falls short of that mark. Should you split the difference and report 215.5 mm? Suppose the paper ends a little closer to the 216 mm mark. Should you report 215.7 mm? How about 215.75 mm? Or even 215.759 84 mm?

Scientists have a way to indicate the precision of their measurements. They use the concept of **significant digits (SD)**. You learned in Subsection 3.8 that in any measurement, you are certain of all the digits obtained from the scale markings, but you can only estimate one additional digit. The significant digits of a measurement include the certain digits plus this estimated digit. The only other digits that might be required when recording the measurement are "placeholder zeros" for locating the decimal point.

A measurement of 215.7 mm, for example, has four significant digits. You read the 2, 1, and 5 directly from the ruler scale. You estimated the 7/10 of a millimeter and added it to the certain number—215 mm. Because you estimated the last digit, any value between 215.0 mm and 216.0 mm would be a valid measurement. Any reading in this range would indicate the correct significant digits, but the accuracy of the measurement might suffer! Also, reporting more digits, like 215.76 mm with the ruler used in Figure 3-14, would be wrong since there is no way to estimate the value of a digit in the second decimal place.

Suppose that your ruler indicated a reading of "exactly" 216 mm. Since a scientific measurement *must* include an estimated digit, you should report the reading as 216.0 mm, which includes the estimated 0/10 of a millimeter. Adding a zero in the first decimal place of the number does not make the measured paper any wider, but it does indicate the precision of your measuring instrument. *Not* including the zero would indicate your ruler had only centimeter (10 mm) graduations, and the 6/10 cm was the estimated digit.

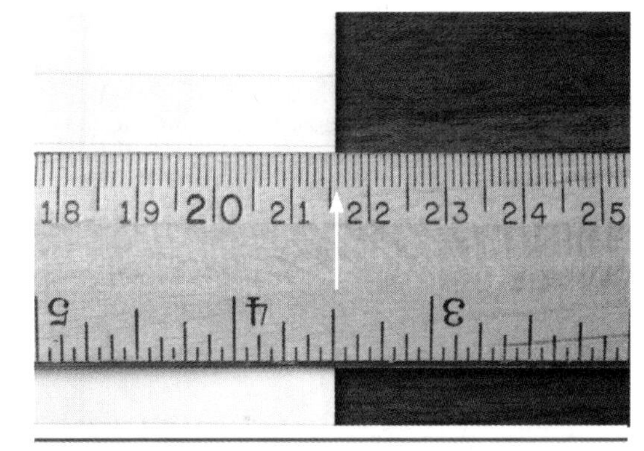

3-14 How many digits should you report for this measurement?

The purpose of observing significant digits is to establish the precision of the measurement, not improve its accuracy.

The significant digits in a measurement are all the digits known from the instrument scale plus one estimated digit.

On many instrument scales, only the major graduations have numbers. But you can identify the values of the minor markings by their relationship to the numbered ones. For example, on a ruler graduated in millimeters, the centimeter marks have numbers and the millimeters don't. But you can count the millimeter marks to find each one's value.

3-15 Scientists must be able to identify the significant digits in a measurement.

The symbol NM is used to stand for the *international nautical mile*. Compare with nm (nanometers).

Identifying Significant Digits

How does a scientist know which digits are significant in measurements reported by other scientists? He follows a series of rules for using significant digits. In this course, we provide the following set of rules that are similar to those used by scientists to identify significant digits in measurements.

SD Rule 1: SDs apply only to measured data. The numbers below are *not* measured data. You should consider these types of numbers to be exact or to have an unlimited number of decimal places. Significant digits do not apply to them.

- Counted numbers

  5 students, one pair of shoes, a dozen eggs, 11 players

- Pure numbers

  $-9, 2, \pi\ (= 3.141\,59\ \ldots), \sqrt{5}\ (=2.23606\ldots)$

- Fractions that are exactly 1 by definition

  $$\frac{60\text{ sec}}{1\text{ min}}, \frac{100\text{ cm}}{1\text{ m}}, \frac{1852\text{ m}}{1\text{ NM}}$$

SD Rule 2: All nonzero digits in measured data are significant.

112.54 m	(5 SDs; estimated to the nearest 0.01 m)
34 °C	(2 SDs; estimated to the nearest 1 °C)

Measurements that are whole numbers where the last digit is *not* a zero do not have decimal points. SD Rules 3, 4, and 5 describe how to deal with zeros in measurements with and without decimal points.

SD Rule 3: All zeros between nonzero digits are significant.

10.6 mL	(3 SDs; estimated to the nearest 0.1 mL)
15.06 m	(4 SDs; estimated to the nearest 0.01 m)
103 K	(3 SDs; estimated to the nearest 1 K)

SD Rule 4: Decimal points define significant zeros.

A. *If a decimal point is present*, all zeros to the right of the last nonzero digit are significant.

45.0 s	(3 SDs; estimated to the nearest 0.1 s)
8.500 L	(4 SDs; estimated to the nearest 0.001 L)

Zeros following the last nonzero digit are called *trailing zeros.*

B. *If a decimal point is* not *present*, no trailing zeros are significant. They are placeholders to locate the assumed decimal point.

10 cm	(1 SD; estimated to the nearest 10 cm)
120 °C	(2 SDs; estimated to the nearest 10 °C)
1800 g	(2 SDs; estimated to the nearest 100 g)

Zeros to the left of the first nonzero digit in a decimal fraction are called *leading zeros.*

C. *If a decimal point is present*, none of the zeros to the left of the first nonzero digit are significant. They are placeholders to locate the decimal point.

0.050 m	(2 SDs; estimated to the nearest 0.001 m)

SD Rule 5: A decimal point must follow an estimated zero in the one's place. If the estimated digit of a measurement is a zero and falls in the one's place, you cannot tell for sure what the precision of the measurement should be. In these cases, we include a decimal point to indicate that the zero in the one's place (and any zero to the left of it) is significant.

110 m ⇒ 110. m (3 SDs; estimated to the nearest 1 m)

1000 K ⇒ 1000. K (4 SDs; estimated to the nearest 1 K)

3.11 Scientific Notation

Measurements in science often deal with very large or very small numbers. Geophysicists have calculated the mass of the earth to be 5,972,200,000,000,000,000,000,000 kg. Nuclear physicists have determined the mass of a hydrogen atom to be 0.000 000 000 000 000 000 000 000 001 674 kg! Such large and small numbers are difficult to read and write, and even more difficult to use in a mathematical calculation.

You can express such numbers much more conveniently in a form called **scientific notation**. In scientific notation, you convert very large or small numbers into a product of a small number and a power of ten. The symbolic format for scientific notation is

$$M \times 10^n,$$

where M is a number greater than or equal to 1 and less than 10, and n is an integer exponent (positive or negative) of the factor 10. Using scientific notation, the mass of the earth is 5.9722×10^{24} kg and the mass of a hydrogen atom is 1.674×10^{-27} kg. You can see that this is a much more compact way of writing these kinds of numbers.

3-16 A scale would require a display having 25 digits to indicate the mass of the earth to the nearest kilogram.

In some situations, the number in standard format may be negative. In these cases, when converting to scientific notation, M will be a number between −1 and −10. The exponent is not affected.

Scientific Notation and Significant Digits

According to *SD Rule 4B* the measurement

3800 mm

appears to be estimated to the nearest 100 mm. But what if the estimated digit in this measurement is actually in the 10s place? This means that the zero in the tens place is significant, and the zero in the ones place is not. So then the measurement properly has 3 SDs. You can't indicate the precision of this measurement by just inserting a decimal point at the end of the number.

3800. mm (WRONG!)

In a scientific measurement, the only way to correctly indicate the precision of the measurement is to use scientific notation. Write the significant digits as the factor M in scientific notation ($1 \leq M < 10$) and drop the remaining zero.

3.80×10^3 mm (RIGHT!)

This method preserves the precision of the measurement in a way that standard decimal notation can't.

For numbers in scientific notation of the form $M \times 10^n$, the factor M contains *only* significant digits.

In this section, so far we have introduced you to the big concepts of accuracy, precision, repeatability, and significant digits. More than anything else, these concepts should show you that even operational science is not about absolute truth. The best we can do is define workable standards of accuracy and precision and then use tools to help us stay within those limits.

3-17 The calculated density of a quartz sample

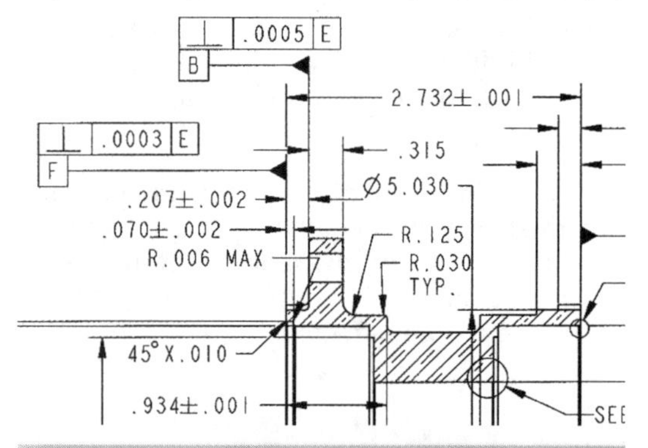

3-18 The bottom photo shows the machined component that is a part of several major joints in *Curiosity's* main instrument arm. The photo on top is a close-up of a manufacturing blueprint for the part, showing milling tolerances in thousandths of an inch.

3.12 Math with Measurements

In Subsection 3.8, we figured the density of a sample of quartz by dividing its mass by its volume using a calculator. The calculator displayed the answer to nine decimal places, which was far more precise than the measured mass and volume (see Figure 3-17). How do you know how many decimal places to report when performing calculations with measurements?

Calculators can give results with many decimal places. However, reporting an answer with greater precision than any of the original data is misleading and not scientifically valid. To avoid this problem, scientists have developed sophisticated methods for identifying the significant digits in calculated results. You can imitate these methods by using some simple rules we have provided in Appendix E. We recommend that you become familiar with the rules by practicing them since they will be important in more advanced science courses. Solutions to Example Problems in this textbook will show you how the rules apply. However, you shouldn't get "hung up" on using the SD rules. Think about how these ideas affect your results in your homework, and especially in your lab activities. Don't get bogged down in the details of determining SDs in your answers and miss the big science principles.

3.13 Using Measurements to Solve Problems

Imagine that you work at Andrew Tool and Machining, the machine shop that NASA chose to make some of the most complex parts for the Mars rover *Curiosity.* How would you make gears with a microscopic precision of two ten-thousandths of an inch (which is much less than the thickness of a human hair)?

The "wiggle room" in parts that are made to fit together, like a hole and shaft or interlocking gears, is called *mechanical tolerance.* Engineers express tolerance as the desired dimension plus or minus (±) an allowable variation. For example, the diameter of the earbud jack on your MP3 player is specified as 3.60 ± 0.05 mm in diameter. That means that it could be as small as 3.55 mm or as large as 3.65 mm. Any smaller, and the earbuds won't work. Any larger and the earbud plug won't fit in the jack.

Being able to measure and do math with significant digits is key to designing parts with small acceptable tolerances. But stating small tolerances in design drawings is useless without good machines and skilled operators that can reliably cut, drill, and shape parts. The tolerances that Andrew Tool and Machining had to work with were pretty tiny. But they specialize in extremely accurate machining and rigorous quality testing. Another strength that set them apart as a company was the extra human care they demonstrated to get the job done with microscopic precision.

And the result? In the weeks after *Curiosity's* landing, its mission control team was overjoyed that all of its mechanical systems tested perfectly—a tribute to their designers and builders.

3B Section Review

1. What are three sources of measurement error?
2. Discuss the difference between accuracy and precision.
3. How can you tell that an instrument has a decimal scale?
4. Why do scientists repeat a measurement when possible? What does good repeatability with the same instrument *not* tell you about the measurement?
5. Why do scientists use significant digits when reporting scientific measurements?
6. How many estimated digits do you include in the significant digits of a scientific measurement?
7. What should you do to indicate that all digits in a measurement ending with a zero in the one's place are significant?
8. What is the main reason for using scientific notation? How can you use scientific notation to identify the significant digits in certain numbers?

DS 9. A piece of PVC drain pipe for a sink has an outside diameter of 1.90 in. (48.26 mm) and a tolerance of ±0.01 in. (0.25 mm). What is the largest and smallest diameter the pipe could be? Express your answers in both inches and millimeters.

DS 10. How is developing mechanical tolerance and making parts that fit an example of good and wise dominion?

11. (True or False) Engineers can measure the diameter of a steel ball bearing to the nearest 0.000 01 cm. This number of decimal places ensures that it is a very accurate measurement.

✪ 12. State the number of significant digits and the place value of the estimated digit for each of the following measurements.
   a. 3.53 g
   b. 0.640 A
   c. 10.05 mL
   d. 10,000 L
   e. 1050 cm
   f. 6.380×10^{-2} kg

3C The Science of Measuring

3.14 How Much Matter?

Chapter 2 discusses matter, its properties, and the kinds of changes it can undergo. Matter has two basic dimensions—mass and volume. This section discusses some ways to measure the amount of matter in an object. Let's look at your ordinary, everyday brick of gold such as you would find at the Fort Knox bullion depository. What are some ways to measure the amount of matter (gold) in the brick? You could measure its *weight*, *mass*, *volume*, or *density*. Let's look at each of these properties more closely and discuss how you can measure them.

3-19 Gold bullion

3.15 Weight

Weight can indicate the amount of matter in an object, but it does not measure the amount of matter directly. This is because weight is the force of the earth's gravity acting on the matter of the object. We measure weight in newtons (N), which is also the metric unit of force. One newton equals about 0.224 **pound-force (lbf)**. The average ninth grader weighs 602 N (135 lbf).

Gravity at the earth's surface depends on a number of factors, including the distance of the earth's surface from its center and the amount and density of rock beneath the earth's surface. Therefore, the weight of an object can change slightly from place to place and at different heights above sea level. The way objects move can also affect their measured weight. For example, an astronaut in orbit around the earth seems to have no weight at all (a condition caused by orbital motion).

We use the terms pound-force (lbf) and pound-mass (lbm) to emphasize the difference between the two kinds of quantities.

The *spring scale* is a familiar laboratory instrument that measures weight. It balances the pull of the earth's gravity on the object being weighed against the pull of a stretched spring inside the instrument. A pointer connected to the spring moves with the spring to indicate the object's weight on a scale.

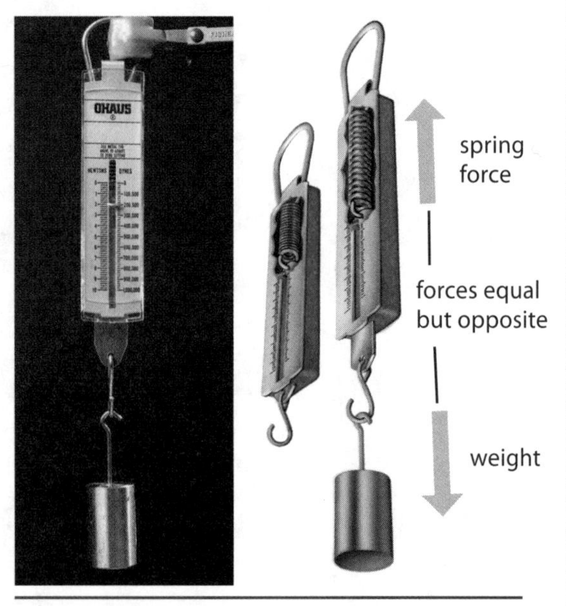

3-20 A spring scale works by balancing the weight of the object with the force exerted by a stretched spring.

An *electronic scale* is another common laboratory instrument used to weigh things. You place the object on a plate on top of the instrument. The scale converts the pull of gravity on the object into an electronic signal that a small computer processes. The electronic scale displays the weight on a digital screen. Most modern digital scales can display weight in a variety of units, including metric, English, and others.

3-3 Common Weight Units

Unit	Symbol	Relation
newton	N	1 N = 0.224 lbf
pound-force	lbf	1 lbf = 4.46 N
ton ("short")	t	1 t = 2000 lbf
ton (metric)	T	1 T ≈ 2205 lbf

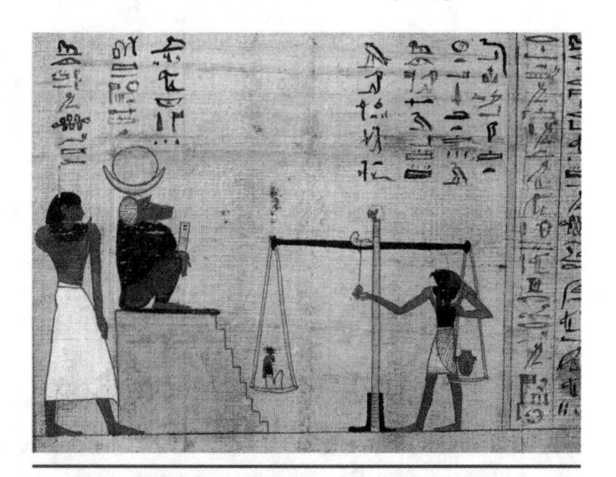

3-21 An ancient Egyptian balance illustrated on a papyrus scroll

3.16 Mass

Mass is the measure of the amount of matter in an object. There is no way to directly measure mass, so we must use one of several indirect methods. One way is to compare an unknown mass with a known mass using a *mass balance*, or simply a *balance*. This method has probably been in use since the earliest ancient times. A balance in its simplest form is an arm that has two pans hung from its ends and is supported in the center by a pivot. There is usually a pointer indicating when the arm is level and balanced. The object whose mass is unknown is placed in one pan, and known masses are placed in the other pan one at a time until the arm balances.

When balanced, the unknown weight of the object equals the weight of the known masses. Since gravity affects all of the masses in the same way, we can assume that equal weight means equal mass. Thus, a balance requires gravity to measure mass. A balance would not work in earth orbit or interstellar space where there is no apparent gravity. The object, the balance, and the masses would just float apart! On earth, however, variations in gravity do not affect a balance because gravity affects all the weights in the same way, canceling out any variation.

Measuring Honestly

A common instrument for measuring weight in Bible times was the balance. Its accuracy depended on the honesty of its owner. What does Leviticus 19:36 have to say about measuring? How does that apply to us?

Though the standard SI unit for mass is the kilogram (kg), most measurements in a school laboratory will likely use the gram (g), which is a more convenient unit size. Unlike weight, which depends on the pull of gravity, mass is a constant for a given object.

3-4 Common Mass Units

Unit	Symbol	Relation
kilogram	kg	1 kg = 1000 g
gram	g	—
milligram	mg	1 mg = 1/1000 g
pound-mass	lbm	1 lbm ≈ 453.6 g
ton (metric)	T	1 T = 1000 kg

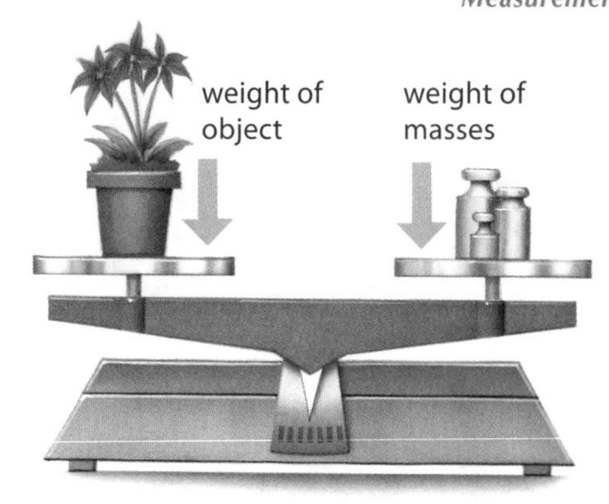

3-22 The weight of a known mass balances the weight of an unknown mass to zero the balance arm.

3.17 Volume

All matter occupies some amount of space. The space enclosed or occupied by an object is its **volume**.

Calculating Volume

A solid object is rigid, so its size and shape are stable. If it has a simple geometric shape, you can measure its dimensions and compute its volume using a simple math equation called a ***formula***.

Let's consider a rectangular solid. One way to measure its volume is by using its three linear dimensions, length, width, and height, arranged at right angles to each other. You can calculate its volume using the *word formula*:

$$\text{volume} = \text{length} \times \text{width} \times \text{height}.$$

We show most formula symbols in italics. The volume *symbol formula* looks like this:

$$V = l \times w \times h.$$

Length, width, and height are all dimensions of length. Using this formula, you can derive the SI unit for volume—the cubic meter (m^3)—from the SI base unit for length—the meter (m). In a cube, if $l = 1$ m, $w = 1$ m, and $h = 1$ m, then

$$V = 1\ \text{m} \times 1\ \text{m} \times 1\ \text{m}$$
$$= 1\ \text{m}^3.$$

Solving Formulas

A **formula** is a special kind of math equation that relates different measurable quantities using symbols. When you solve a formula, you replace the symbol with the number value of the quantity it represents in the problem, and then you do the arithmetic to find the value of the equation.

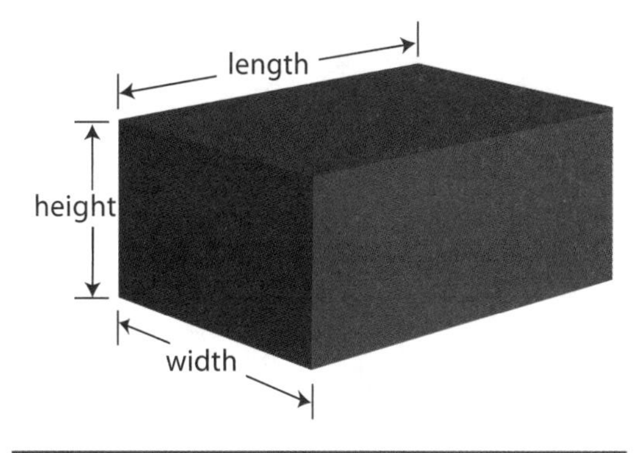

3-23 A rectangular solid is also called a *parallelepiped* (PAHR uh LEL uh PIE pid).

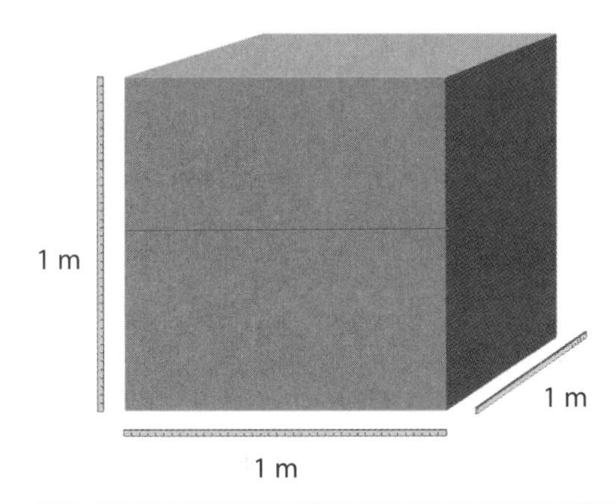

3-24 The volume of a cubic meter comes from the product of its three dimensions.

The cubic meter is the SI derived unit of volume. However, a cubic meter is too large for typical laboratory volume measurements—it is equivalent to about 264 gallons! One solution is to use a volume based on the centimeter (cm). A volume 1 cm × 1 cm × 1 cm is 1 cubic

The derived SI unit for volume is the *cubic meter (m^3)*.

centimeter (1 cm^3). Scientists usually report the volumes of solids in cubic centimeters or cubic meters.

Scientists use the liter (L) for measuring volumes of liquids and gases (fluids). A liter is equal to one cubic decimeter (dm^3), which is one-thousandth of a cubic meter. For even smaller fluid volumes, the milliliter (mL) is a more convenient unit. ***The milliliter and the cubic centimeter are equal volumetric units.***

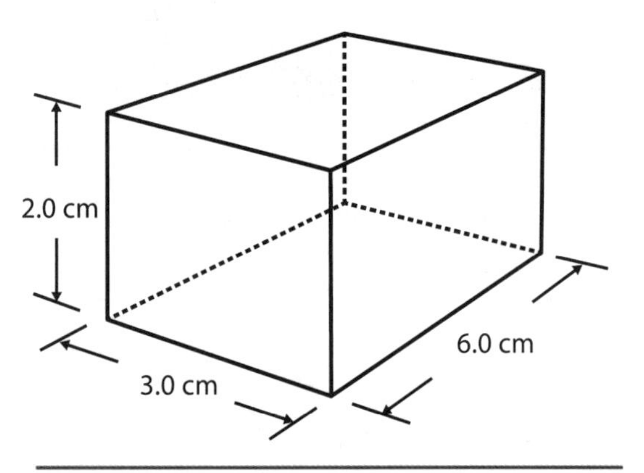

3-25 Diagram of the rectangular solid for Example Problem 3-2

Example Problem 3-2

Calculate the Volume of a Geometric Solid

What is the volume of a rectangular solid 6.0 cm long, 3.0 cm wide, and 2.0 cm high?

Known: length (l) = 6.0 cm
width (w) = 3.0 cm
height (h) = 2.0 cm

Unknown: volume (V)

Required formula: $V = l \times w \times h$

Substitution: V = 6.0 cm × 3.0 cm × 2.0 cm

Solution: V = 36 cm^3 (2 SDs allowed)

As long as a solid or hollow object has a standard geometric shape, such as a cube, cylinder, or sphere, we can calculate its volume using a formula. Review the common solid shapes and their volume formulas in Table 3-5.

3-5	***Volume Formulas***	
Shape	**Relationship**	**Formula**
cube	volume = (edge × edge × edge)	$V = e^3$
solid rectangle	volume = length × width × height	$V = l \times w \times h$
sphere	volume = 4 × π × (radius × radius × radius) ÷ 3	$V = \frac{4\pi \times r^3}{3}$
cylinder	volume = π × (radius × radius) × height	$V = \pi \times r^2 \times h$
cone	volume = π × (radius × radius) × height ÷ 3	$V = \frac{\pi \times r^2 \times h}{3}$

Measuring Volume Directly

Since a liquid takes the shape of its container, you can use any container to measure the volume of a liquid directly if it has a scale marked in volumetric units. Similarly, you can estimate the volumes of granular solids, such as sand or flour. These solid materials can fill in the shapes of their containers just like a liquid does.

Scientists have long used special containers graduated to measure liquid volume. The graduated cylinder is the most familiar. Volume capacity can range from 10 mL to several liters. To read the graduated cylinder, note the level of the liquid against the graduated markings on the side of the cylinder. Other instruments to measure volume have various shapes depending on their purpose. Pipettes (pie PET) are like long, thin eyedroppers; they can measure and deliver tiny volumes of liquid drop by drop if needed.

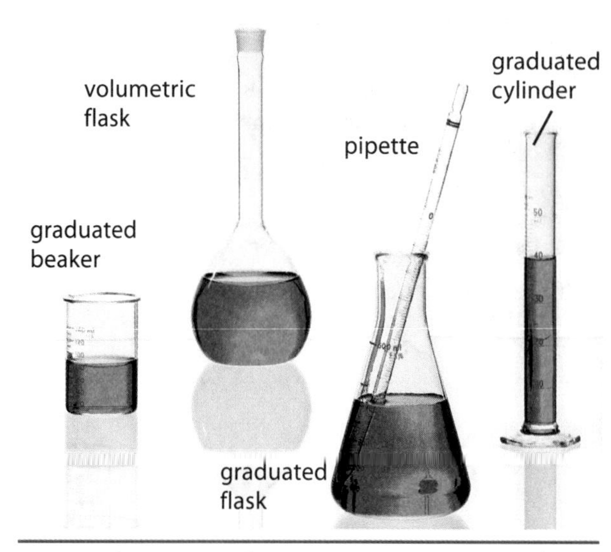

3-26 Laboratory technicians use many types of volumetric instruments to measure liquids. Volumetric flasks, graduated cylinders, and pipettes are marked for precise volume measurement.

Measuring Volume by Water Displacement

You can measure the volume of relatively small, irregularly shaped, solid objects by the *water displacement method.* You completely immerse the object in a graduated cylinder or an overflow container filled with water to the verge of spilling over. (See Figures 3-27 and 3-28.) Read the change in volume of water in the graduated cylinder directly to find the object's volume. Or, using an overflow container, capture the overflow water from the container in a graduated cylinder and measure it. You can practice these methods in lab activities.

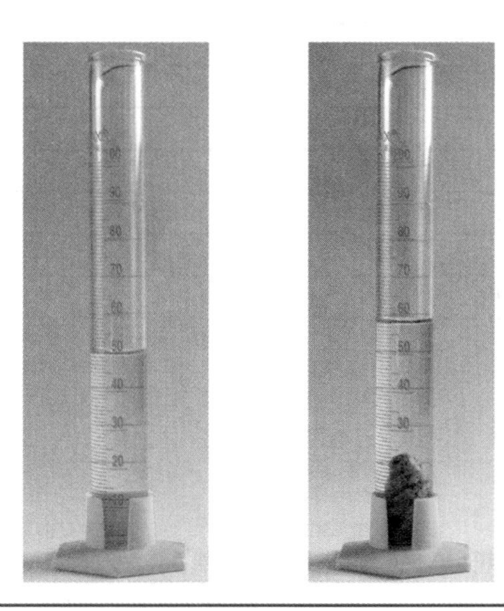

3-27 To find the volume of this irregular mineral sample, use the displacement method.

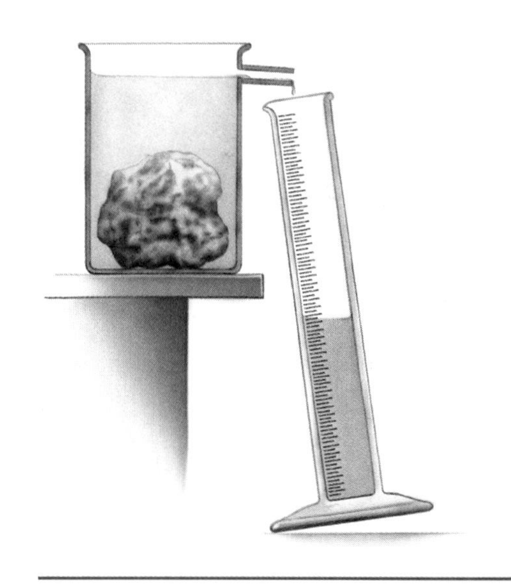

3-28 The volume of larger irregular objects can be measured by using an overflow container.

3.18 Density

Density is a property of matter that you can use to identify certain substances when you measure them under standard conditions. The density of something is the amount of matter (mass) it contains compared to its volume. The SI density unit is kilograms per cubic meter (kg/m^3), but in a lab setting, smaller units, such as grams per cubic centimeter (g/cm^3), are more useful. The standard formula symbol for density is the lower-case Greek letter *rho* (ρ).

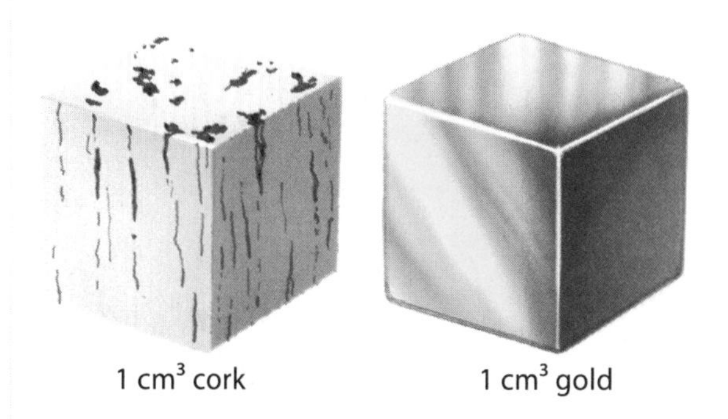

3-29 Substances with greater densities have more matter packed into a given volume.

Recall that a milliliter (mL) is the same volume as a cubic centimeter (cm^3).

A Weighty Thought

If the weight of a brick of iridium is about half that of an average ninth grader, how does its mass compare to the student's?

You can easily calculate density using the following formula:

$$\text{density} = \frac{\text{mass}}{\text{volume}},$$

$$\rho = \frac{m}{V}.$$

The density of the metal lead is about 11 g/cm^3. The densest natural material on Earth is the metallic element iridium, with a density of 22.65 g/cm^3. A chunk of iridium the size of a masonry brick would weigh about half as much as an average ninth grader! Scientists usually state the density of liquids in grams per milliliter (g/mL). The density of pure water is 1.00 g/mL at 4 °C.

3-6 *Densities of Some Elements and Common Materials (in g/cm^3)*

iridium	22.65	opal	2.10
osmium	22.61	clay	1.8–2.6
platinum	21.46	magnesium	1.74
gold	19.28	sugar	1.59
uranium	18.95	brick	1.43–1.95
mercury	13.53	gelatin	1.27
lead	11.34	bone	1.2–1.9
copper	8.96	hard rubber	1.19
iron	7.87	seawater	1.02
diamond	3.52	pure water	1.00
aluminum	2.70	ice	0.92
chalk	2.50	butter	0.87
concrete (set)	2.3	ethyl alcohol	0.79
glass	2.2–7.2	seasoned oak	0.74–0.77
rock salt	2.17	cardboard	0.69

The following example demonstrates a density calculation.

Example Problem 3-3

Calculate Density

If a solid sample of lead displaces 36.0 mL of water and has a mass of 408.2 g, what is its density?

Known: lead's volume (V) = 36.0 mL = 36.0 cm^3

lead's mass (m) = 408.2 g

Unknown: lead's density (ρ)

Required formula: $\rho = \frac{m}{V}$

Substitution: $\rho = \frac{408.2 \text{ g}}{36.0 \text{ cm}^3}$

Solution: $\rho \approx 11.3 \text{ g/cm}^3$ (3 SDs allowed)

Note that when units combine in mathematical operations they can form products or fractions, just like numbers do.

Engineers and machinists like those at NASA and Andrew Tool and Machining use the science you've learned in this chapter to make things that we use every day. Others help farmers to feed families, doctors to save and improve lives, and engineers to design trucks that move products from the factory to you.

So how does measuring apply to you? How do things like units, scientific notation, significant digits, and density make a difference in your life and in the lives of others? God could call you to be an engineer or machinist in the future. One young machinist said, "I hated high school. I was a fish out of water. But then I was introduced to a machine shop through a local foundation." Today he's helping to fabricate parts for prosthetic limbs. That's using measurement and technology to help others!

3C Section Review

1. State four ways that we can measure the amount of matter in an object. Which methods, if any, measure matter directly?
2. Under what conditions might measuring the mass of an object with an extremely precise portable electronic scale *not* provide the same reading every time?
3. How does an inanimate object's actual mass vary from place to place on Earth?
4. List the two methods discussed in the text that are used for measuring volume. Which method would require only a ruler to find the volume of an object?
5. ✪ You find what appears to be a flattened American Civil War musket ball while weeding your garden. Which method would you use to determine the volume of the lead ball? Explain why you chose the method.
6. Explain how the density of an object is different from its mass.
7. (True or False) The mass of an object at rest is constant but its weight may vary depending on where it is located and how it is moving.
8. ✪ A spherical tank for storing liquefied natural gas (LNG), an important heating fuel, is 60.0 m in diameter. What is the volume of the tank in cubic meters? Express your answer to the nearest 1000 m^3.
9. ✪ A student measures the volume of a cat's-eye marble using the displacement method. She fills a graduated cylinder to the 10.3 mL mark. Then, she carefully places the marble in the cylinder. The final volume is 11.9 mL. What is the volume of the marble? Report your answer to the nearest 0.1 cm^3.
10. ✪ A 1.00 L sample of water from the Dead Sea in Israel has a mass of 1.24 kg. What is its density? Report your answer to the nearest 0.01 g/mL.

Scientifically Speaking

Chapter Review

1. Describe the process of measuring. Discuss finding the mass of a lead fishing sinker as an example to illustrate your answer.
2. Which kind of data (quantitative or qualitative) is more likely to be affected by the bias of the observer? Why is this true?
3. What two pieces of information does measured data usually include?
4. What property makes the metric system a decimal system?
5. Which SI base unit is the only one that has a physical object for its standard?
6. When converting a measurement from one unit to another, what does not change in the conversion?
7. Why is it impossible to make an exact measurement, no matter how precise the instrument?
8. What kind of error does averaging many measurements of the same quantity minimize?
9. What kind of error might you detect by making a measurement of the same quantity with different instruments?
10. State the number of significant digits and the place value of the estimated digit for each of the following measurements.
    a. 30.5 m
    b. 0.04 g
    c. 6080. ft
    d. 6400 g
    e. 9.83×10^2 km
    f. 0.006 10 m^3
11. What is the main goal of identifying significant digits in the results of a calculation involving measured data?
12. Why might the weight of the same object vary slightly from one place to another?
13. Basically, how does a laboratory mass balance work?
14. Which dimensions do you need to measure to calculate the volume of a cylinder?
15. How would you measure the volume of a granular solid like sand?

True or False

16. According to the definition given in this chapter, counting things is a measurement.
17. Quantitative data is more objective than qualitative data for scientific purposes because it is less likely to be affected by a scientist's bias.
18. You may use metric prefixes to modify only the seven SI base units.
19. Unit conversion factors must always be exactly equal to 1.
20. Measurement always involves some error.
21. It's possible for measured data to be extremely precise but not accurate.

22. There is no need to repeat a measurement if you know an instrument is both accurate and precise.
23. Professional engineers don't concern themselves with measurement tolerances because tolerances do not affect the accuracy of their measurements.
24. We assume that the value of *pi* (π) is exact, so significant digits don't apply to it in calculations with measurements.
25. When you determine the mass of an object using a lab balance, you are actually comparing the weight of the object to the weight of a known amount of matter.
26. When using the water displacement method, you assume that the object does *not* soak up any of the water.

✪ 27. State the number of significant digits and the *actual value* of the estimated digit in each of the following quantities.
   a. 3 cm
   b. 0.107 g
   c. 100. mL
   d. 4.8×10^{-3} J
   e. π
   f. 30,010 m
   g. 101,300 Pa
   h. 60 min

✪ 28. The smallest divisions on a certain 100 mL graduated cylinder are 2 mL. What would be the problem with estimating the volume to 1/10 of the smallest divisions using this instrument?

DS ✪ 29. How does an engineer, using correct mechanical tolerances when developing a design for a product, aid the manufacturer and ultimately the customer?

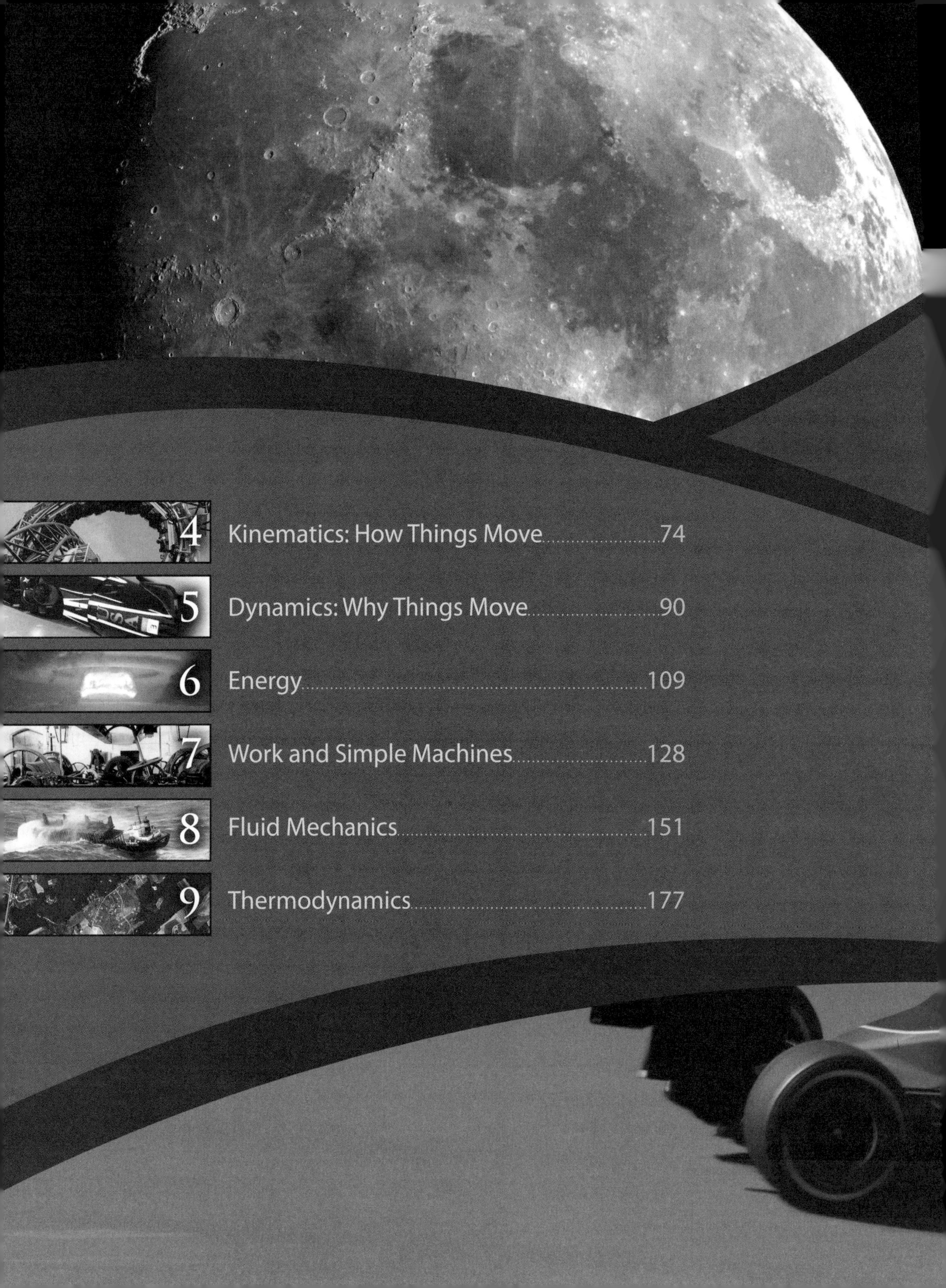

MECHANICS

Our universe is filled with motion.

Electrons whiz around the nuclei of atoms. Atoms vibrate within molecules and crystals. Cells coast through our blood vessels. We rack up miles on motorcycles, cars, buses, trains, and airplanes. Our earth revolves around the sun, which speeds around our galaxy, the Milky Way. And our Lord controls it all, "upholding all things by the word of his power" (Heb. 1:3).

Where does all this motion come from? How does it happen? In Unit 2 you will learn how motion occurs and how it is measured. The motion of matter depends ultimately on the application of forces. Newton's three laws of motion relate the action of forces to all familiar forms of motion—linear, rotational, fluid, and even the thermal motion of particles of matter.

UNIT 2

KINEMATICS: HOW THINGS MOVE

CHAPTER 4

DOMINION SCIENCE PROBLEM

Children Are Not Crash-Test Dummies

With the widespread use of automobiles in the mid-1900s came vehicle-related accidents, injuries, and deaths. As accidents peaked in the 1980s, safety, especially child safety, became a major concern. Even today, car accidents are the leading cause of child deaths, with more than 1300 children 14 years old and younger killed and 180,000 injured each year. How can we make car travel safer for children?

4-1 A car seat from the 1940s. How much protection would this seat provide in an auto accident?

4A Introduction to Mechanics

4.1 *Historical Development of Mechanics*

Several centuries before the birth of Christ, Greek philosophers believed that matter was motivated to behave in certain ways. Solid matter "desired" to move toward the center of the world, so solid objects fell toward the ground. Light matter "desired" to ascend to the heavens, so smoke and vapor tended to rise. According to Greek philosophers, all natural phenomena and matter acted according to mysterious internal tendencies. Because of the presuppositions inherent in their model of the natural world, most Greeks, including Aristotle, could not conceive that inanimate physical principles cause natural phenomena.

As Chapter 2 discusses, Aristotle's teachings became the basis of European science for more than 2000 years. One of the first to successfully challenge the Aristotelian worldview was the Italian scientist Galileo Galilei. In the seventeenth century, Galileo argued that rather than assume that mystical motivations animate matter, scientists should examine how things move and then describe their motions in mathematical ways. This method was a radical departure from how nature had been studied in the past. Many natural philosophers gladly accepted his ideas and began to make real scientific progress. Others were less enthusiastic because many of Galileo's ideas seemed to contradict the doctrine of the Catholic Church. Some Catholic leaders accused Galileo of heresy.

The scientific revolution begun by Copernicus a century earlier rapidly moved forward as Galileo described motion and provided observational evidence for a mechanical, sun-centered model of the universe. Sir Isaac Newton continued this revolution in the eighteenth century. However, a more complete model of the causes and effects of motion and other processes wasn't introduced until the early twentieth century by Albert Einstein.

The modern study of motion, called **mechanics**, is based largely on Newton's three laws of motion and his law of universal gravitation. Today, physicists split mechanics into three parts: *kinematics* (the description of how things move), *dynamics* (the description of what causes things to move), and *statics* (the description of how stationary things react to pushes and pulls).

As you study mechanics, you will encounter many terms and concepts that are familiar to you. However, you may not have a sound understanding of what they mean. Before exploring the fascinating world of kinematics, you need to

4-3 The design of a modern roller coaster considers kinematics, dynamics, and statics—all parts of the physics of mechanics.

4A Objectives

After completing this section, you will be able to

✓ summarize the development of modern mechanics.

✓ give examples of physical systems.

✓ explain the purpose of reference frames in observing motion.

✓ identify different kinds of reference frames.

✓ express time intervals using correct symbols.

Natural philosophers were early scientists who studied all aspects of the physical world. Later, as physics and chemistry became distinct from one another, the name went out of use.

4-2 Galileo Galilei is credited with beginning the modern scientific era.

To early scientists, using mechanical explanations for natural phenomena meant that God had created underlying principles to govern the behavior of nonliving matter in His Creation. The Bible calls these principles *ordinances*, or *laws* (Job 38:33; Jer. 31:35). Once accepted, this worldview was completely compatible with historical and scientific statements in the Bible.

understand the key terms of mechanics. You will also learn how engineers use the laws of motion to improve vehicle safety, especially for children.

4.2 Physical Systems

4-4 The boundary around each of the subjects in this picture identifies a system.

In mechanics, as in nearly all of science, scientists must first clearly define what they are going to study. This helps them decide what factors they must take into account, both in their experiments and in any scientific models they develop. They mentally construct a boundary that clearly defines the limits of the objects or processes they are studying. The boundary may be an actual surface, such as the walls of a container, or it may be an imaginary box. Everything inside this boundary is called the **system**. All processes, forces, and measurable properties originating within the boundary belong to the system. Everything outside the boundary is the system's *surroundings*. In most cases, scientists are interested in how the surroundings affect the system.

A system can be anything you are studying. Some examples include a baseball, a pot of water, a volume of air, an elevator cab, an airplane, a light bulb, or even your body. For our dominion science problem, the system under study is a child traveling in a car. In mechanics, we can identify the pushes and pulls on a system and then observe how the system responds to these influences.

In the discussion of kinematics later in this chapter, the system and its boundaries are clearly identified so that you can learn how to identify them for yourself. After studying dynamics in Chapter 5, you will be able to recognize the system of interest within its boundaries and the causes of motion imposed on it.

4.3 Frames of Reference

4-5 Pioneers took great risks crossing the western deserts of the United States—there were few landmarks.

American pioneers often crossed stretches of deserts on their westward journey. Sometimes a group of travelers lost the trail and wandered until they ran out of water and died in the arid wasteland. A water hole might have been just a few miles away, but without landmarks to guide them, people were often misled by the vast, trackless desert.

In kinematics, we need landmarks to anchor our observations. Without them, our studies are as futile as the lost pioneers' wanderings. Physicists call such landmarks *points of reference*. A point of reference is nothing more than the stationary location from which we choose to make our observations. The geometric space containing the point of reference is the **frame of reference**, or *reference frame*.

In the study of motion, the frame of reference determines how motion will be described. For example, how fast are you moving right now? Suppose that you are sitting at a desk in a classroom in Baltimore, Maryland. A person in the same room would say that you're stationary because his reference frame is the classroom. However, if he were somehow suspended high above the earth and could observe your school from that location as the earth rotates in space, he would say that you were moving at 360 m/s! In that reference frame, the positions of the earth and sun are fixed, but the earth is free to rotate about its axis.

If our observer were moved to a point in space far above the sun's position relative to the plane of the earth's orbit, he would observe the earth whirling around the sun at almost 30,000 m/s! Your motion would be a complicated snaking motion superimposed on the earth's orbit. That point of view places the sun in a frame of reference at a location that is stationary in relation to the background stars.

The plane of the earth's orbit is called the *ecliptic plane*.

When describing motion in mechanics, no single reference frame is more correct than another. The way motion is observed depends on what you choose as the frame of reference. The choice is determined somewhat by the kind of motion you are studying. You pick a reference frame that makes the study of motion easiest.

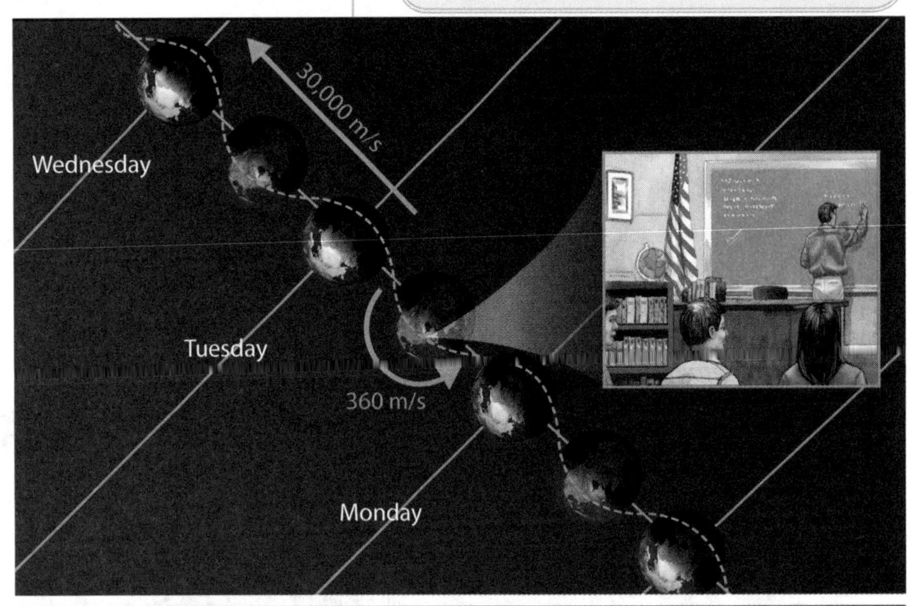

4-6 The kind of motion you observe depends on your point of view—your frame of reference.

Kinds of Reference Frames

Scientists generally choose a reference frame where they know the causes of motion and where moving systems follow the laws of motion (which you will study in Chapter 5). Systems move in straight lines or in simple curved paths caused by forces acting on the systems from within the frame of reference. We call such a reference frame a fixed or *inertial reference frame*. We assume such a point of view is stationary, even though it may actually be moving at a constant speed in one direction. For instance, as a car passenger, you watch cars passing your car on a highway from an inertial reference frame as long as your car's driver maintains a constant speed. The motions of systems that you will study in this course will use an inertial reference frame.

When the observational point of view is *not* moving at a constant speed or direction, it is *accelerating*. We are then observing systems within an *accelerated reference frame*. Continuing the above example, when your car's driver slams on his brakes, everyone in the car jerks forward. The passengers' motion seems to be the result of a force suddenly acting on them. But the force isn't real—it results from the deceleration of the car, which is the reference frame in which you are viewing their motion.

Scientists often observe systems from a reference frame that is not stationary and in which the speed or direction is changing. This is the case when meteorologists observe the motion of weather systems from a point on the rotating earth's surface. Air masses should flow directly from regions of high pressure to lower pressure. But an apparent influence seems to turn them from a straight path. In the Northern Hemisphere, they twist to the right. In the Southern Hemisphere, they deflect to the left. This is called the *Coriolis effect*, and it affects everything moving on the earth. Every point on the earth's surface lies within a *rotational reference frame*, which is a special case of an accelerated reference frame. The apparent forces you feel while riding a spinning merry-go-round aren't real, but you sense the effect within the rotating reference frame of the playground ride.

4-7 Hobbyists control their model planes from an inertial, or fixed, frame of reference.

4-8 A rider on a merry-go-round views the world from a rotational frame of reference.

The Coordinate Axis

We measure straight-line motion of a system in relation to a coordinate axis. A *coordinate axis* is an imaginary line marked off in distance units. We often choose its direction to point in the direction of motion of the system to simplify our study. A *coordinate* is a value of the system's position along the axis. The *origin* of the axis is at the zero position on the axis and is the point of reference for determining position. We usually choose to have coordinates increase to the right for horizontal axes or in the upward direction for vertical axes. Coordinate values increase in the positive and negative directions from the axis origin.

This introductory course covers only straight-line, one-dimensional motion. Studies of the other kinds of motion and their specialized reference frames require more advanced mathematics than you are likely to know at this point in your education. As you learn each new concept, relate it to one-dimensional motion.

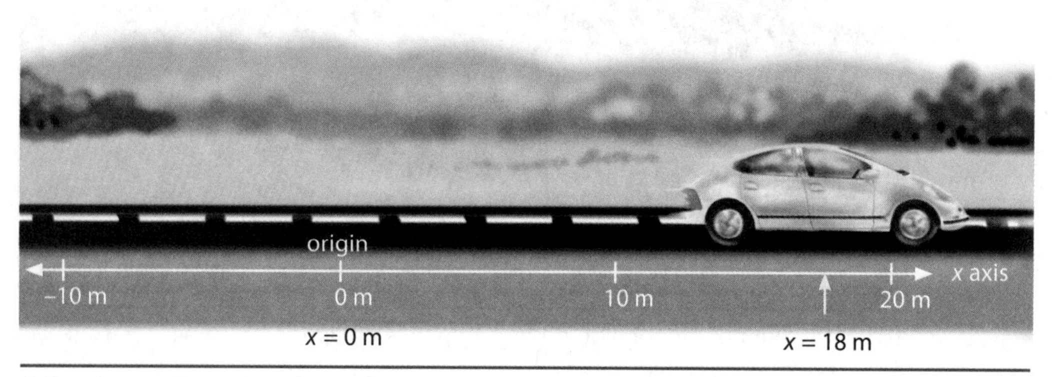

4-9 A one-dimensional coordinate system of position

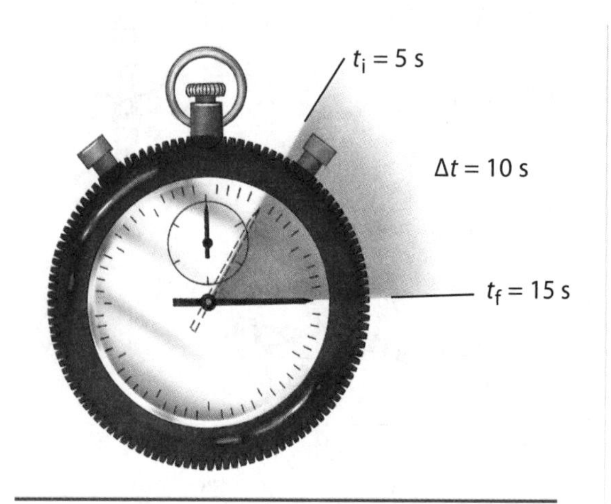

4-10 A time interval is the span of time between an initial time and a final time.

A time **interval** is represented by

$$\Delta t = t_f - t_i.$$

In science, the Greek capital letter delta (Δ) means a "change in" whatever dimension follows it.

4.4 *Time Intervals*

We can define *time* as "a nonphysical continuum that orders the sequence of events and phenomena." Let's examine this technical description of time. *Nonphysical* simply means that time can't be touched. It's not matter. A *continuum* is a scientific word for an unbroken expanse or series. Time is continuous. Without the existence of time, we couldn't experience the world as a series of events. Every instant would be "now." We would have to be like God, who is outside of time, to experience the world.

We can infer from the Bible that time began the instant God began to create the heavens and the earth. The Bible even starts with a time-related phrase—"in the beginning." Time will likely continue into future eternity, since as created material beings, we will exist, think, and move forever. Therefore, time is a continuum that began at Creation and will continue into the future without end.

Any natural phenomenon must occur within a span of time. A time span has an initial time (t_i) and a final time (t_f). The length of a time span is an **interval**. A time interval is represented by the symbol Δt (spoken "delta tee") and is calculated by subtracting the initial time from the final time. A time interval is always a positive number because a later time always has a larger positive value than an earlier time. Correctly identifying time intervals is an important part of understanding how things move. For example, time intervals in car accidents can make the difference between a minor injury and death. An adult human skull can bear a load close to a ton and a half for a fraction of a second, but if the time interval is longer, the skull can be crushed.

4.5 Scalars and Vectors

In physics, certain dimensions require just a single piece of information to be complete. These are *scalar* quantities. You are already familiar with many examples of scalars, such as the volume of a soft drink can (355 mL), the temperature outside (–12 °C), the time of day (10:55 AM), and the mass of a brick (1.5 kg). A scalar dimension can be positive, negative, or zero.

Other dimensions that physicists use require two pieces of information to describe them completely. These quantities are *vectors*. A vector is something measureable that includes both a scalar value and a direction. For example, the weight of a 20 N box is a force vector of 20 N directed straight down. A car moves down Main Street with a velocity vector of 15 m/s east. The scalar value of the weight or velocity alone cannot completely describe these quantities. The direction of each is also necessary.

A vector has a direction measured relative to each coordinate axis in the observer's frame of reference. In straight-line (one-dimensional) motion, a vector may point only in the positive or negative directions. So we will indicate vector quantities by their positive or negative scalar values to simplify concepts in this textbook. When we show a vector quantity in a diagram, such as Figure 4-11, we use an arrow pointing in the direction the vector affects the system.

We write the symbols for most scalar quantities in formulas with italic letters. We may use boldfaced symbols for vectors.

Scalar: temperature, t

Vector: force, $\mathbf{F}$

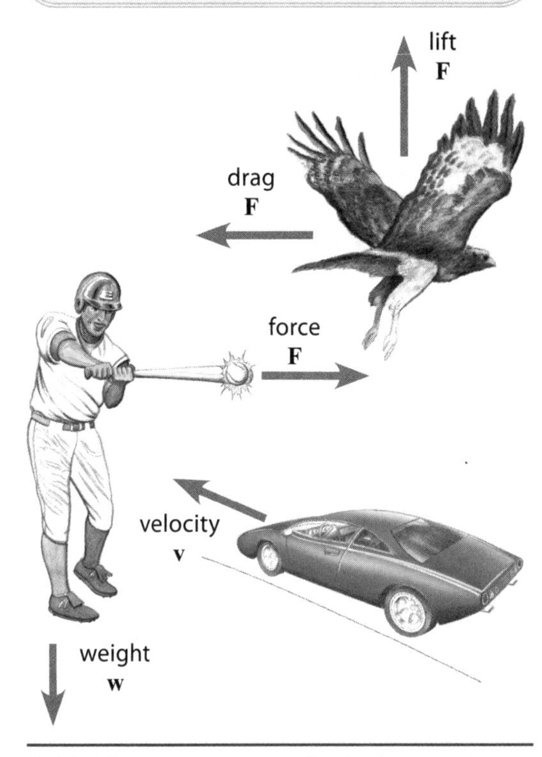

4-11 Vectors represent physical quantities that have both a scalar value and a direction.

4A Section Review

1. Define *mechanics* and describe its three major divisions.
2. In physics, what is a system and how is it used?
3. Why are frames of reference important to the study of motion in physics? What is the correct reference frame for a given situation? the best reference frame?
4. You are riding in a car on a four-lane highway. The driver of your car decides to pull ahead of the car in the right lane to get ready to exit. Your car speeds up and you observe the car to the right fall behind. What type of reference frame produces this kind of observed motion?
5. How is a time interval different from a specific time?
6. Discuss the properties of a scalar quantity. Give an example of a scalar.
7. How are vectors illustrated in diagrams?
8. (True or False) The fact that one's point of view affects one's observations applies to many aspects of scientific investigation, including motion.
9. A certain digital camera in "movie mode" can take 30 pictures per second. If a photographer presses the shutter button when his watch indicates 36 s and releases it at 39 s, what is the time interval of this series of exposures? How many photos did he take?

4B Objectives

After completing this section, you will be able to

✓ diagram a system's position in one-dimensional motion and assign symbols to each position.

✓ contrast a system's displacement and distance traveled.

✓ calculate speed and velocity from position data.

✓ define acceleration and identify the factors that can change during acceleration.

✓ calculate average acceleration given speed or velocity data.

✓ compare and contrast motion in two and three dimensions with motion in one dimension.

Axes (AK sees), plural of *axis*, are reference lines along which dimensional units are measured.

4B Kinematics: Describing Motion

4.6 *Introduction*

Kinematics does not deal with the properties of moving objects themselves, only with the way they move. **Kinematics** measures the *positions, speed* or *velocity,* and *acceleration* of objects. As Galileo believed, it is necessary for a scientist to describe how things move before he can explain the causes of their motion.

4.7 *What Is Motion?*

Motion occurs when a system's position changes during an interval of time. Depending on the number of coordinate axes in one's frame of reference, a system may require one, two, or three coordinates to identify its position. For one-dimensional motion, only a single position coordinate is needed to identify a system's position, symbolized as x. Subscript numbers are used to indicate a sequence of positions over time, for example, x_1, x_2, and so on. If only the initial and final positions are important, you can use the symbols x_i and x_f respectively.

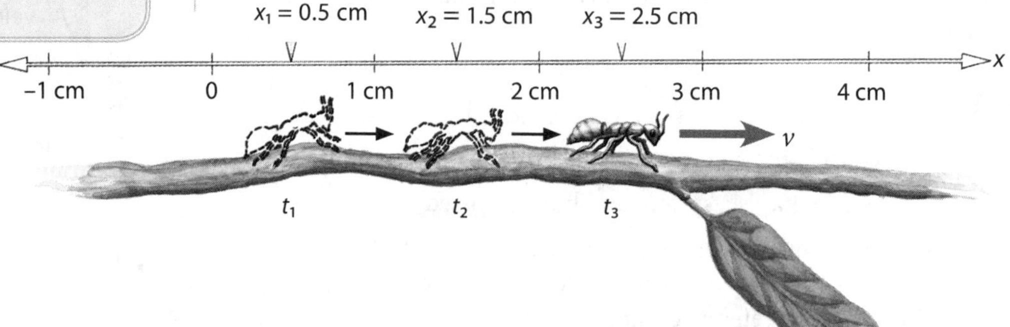

4-12 Simple, one-dimensional motion can be represented by positions along a single coordinate axis.

4.8 *Distance and Displacement*

Imagine that just before dinner you are reading a book in your living room. Your mother calls you to set the table for supper. You walk to the kitchen, make several trips between the table and the kitchen carrying utensils, food, and condiments, and then sit down at the table. What was the total distance you walked between your living room seat and taking your place at the table? If you measured the length of every step between the chair, the kitchen, and the table for every trip you made and added up the individual lengths, you could arrive at the total distance you traveled. Distance is a positive scalar quantity because there is no direction associated with it. Thus, for example, you would report the total distance walked as 20 m.

Distance is represented by *d* in formulas. Note that distance is always positive.

Now, what was your overall change in position between your living room seat and the table? In most cases, this is a very different quantity from the distance you traveled. The change in position involves a distance, to be sure, but it also includes a direction. What kind of quantity is this? It is a vector quantity physicists call displacement. The distance part of your displacement vector is the

The vector quantity displacement is represented by **d**.

shortest distance between the couch in the living room and the chair at the table. The direction portion of the vector from the living room to the table could be measured relative to some directional reference, like geographic north. The complete displacement vector might be stated as 3.4 m northeast, for example. Notice that two pieces of information are required to completely describe a vector—size (a scalar value) and direction.

To simplify the analysis of this example, you could choose a coordinate axis that lies on a line connecting your living room couch and your seat at the table. You could place the origin at the couch. Then, your displacement would simply be the distance between the two seats in the direction of the table (see Figure 4-13). The distance in this case is the *magnitude* of the displacement vector, or just its length. The magnitude of a number or vector is its absolute value and is always positive. The magnitude of a displacement vector and the distance traveled are equal only when the object has traveled a straight line in one direction between the beginning and ending points. This is always true of one-dimensional motion.

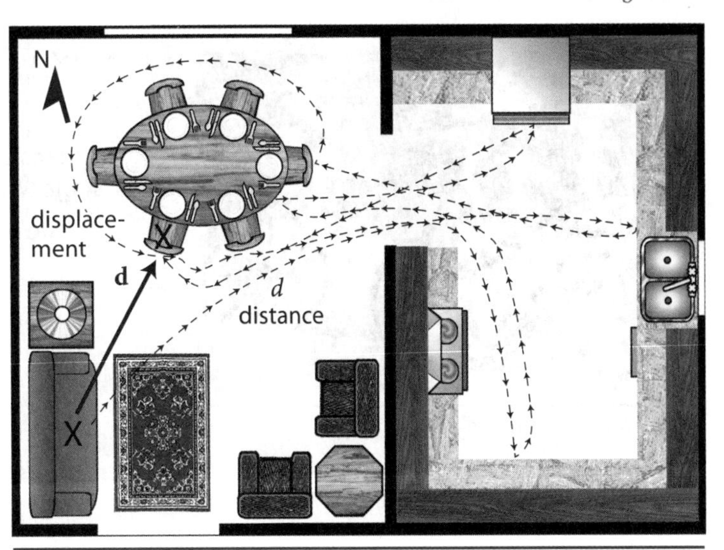

4-13 The distance you have walked and your displacement may not be the same thing.

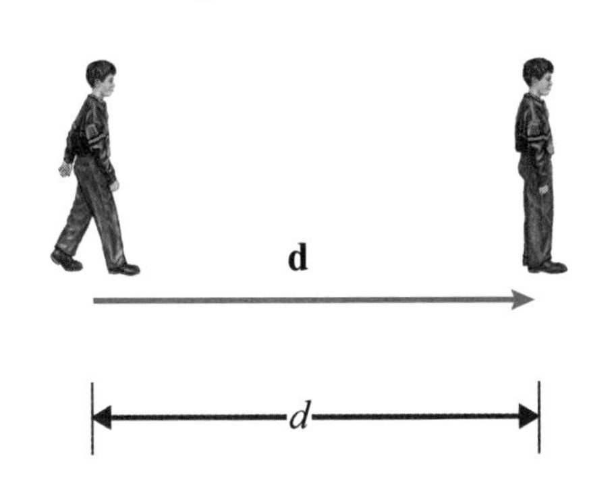

4-14 The magnitude of a vector is simply its size expressed as a positive number.

> The scalar quantity speed is represented by *v* in any formula.

4.9 *Speed and Velocity*

A honeybee (*Apis mellifera*) may fly 1 m in 0.1 s, while a glacier may take a year to move the same distance. They've moved an equal distance, but the motion occurred over very different time intervals. How fast motion occurred is the quantity speed. The **speed** of a system is simply the *rate of motion.* A *rate* compares the change in the value of a quantity to a certain span of time. Mathematically, speed is equal to the distance traveled in an interval of time divided by the time interval. In a formula, speed is written as

$$\text{speed } (v) = \frac{\text{distance}}{\text{time interval}} = \frac{d}{\Delta t}$$

$$\boxed{v = \frac{d}{\Delta t}}$$

Since distance and time are always positive, speed is too. You don't concern yourself with the direction of motion because speed is a scalar quantity.

The *average speed* (v_{avg}) of a system during an interval is equal to the total distance traveled divided by the time interval. In the real world, the actual speed at any moment, called *instantaneous speed,* is probably lower or higher than the average speed of the system for the entire interval. For example, the speed indicated on a car's speedometer is instantaneous speed. When you calculate your speed from the total distance of your trip and the time it took, you are calculating average speed. Average speed is the constant speed you would have to travel with no slowing down or speeding up to make the trip in the same time interval.

4-15 Racers in the Tour de France

Example Problem 4-1

Calculating Average Speed

It takes a Tour de France cyclist 0.86 h (51.6 min) to cover a 41.5 km stage. What is his average speed in km/h?

Known: time interval (Δt) = 0.86 h

distance (d) = 41.5 km

Unknown: average speed (v_{avg})

Required formula: $v_{avg} = \frac{d}{\Delta t}$

Substitution: $v_{avg} = \frac{41.5 \text{ km}}{0.86 \text{ h}}$

Solution: $v_{avg} = 48 \text{ km/h}$ (2 SDs allowed)

When both speed and direction are important, you use velocity. **Velocity** is the rate of completing a displacement. Since displacement is a vector quantity, velocity is also a vector. You calculate the average velocity of a system by dividing its displacement by the displacement's time interval:

$$\text{velocity } (\mathbf{v}) = \frac{\text{displacement}}{\text{time interval}} = \frac{\mathbf{d}}{\Delta t}$$

$$\boxed{\mathbf{v} = \frac{\mathbf{d}}{\Delta t}}$$

The bold characters stand for vector quantities that might point in any of three dimensions. Since you will be studying only one-dimensional (straight-line) motion in this course, we will use the scalar symbols in these formulas to keep them simple. As long as the system's direction of motion doesn't change during a time interval, distance and displacement will be identical.

In a simple description of one-dimensional motion, you can use a + or − to designate the direction of the velocity vector. For instance, consider a hybrid-energy car traveling east at 30 m/s and a hydrogen fuel-cell car traveling west at 45 m/s. The velocity of the east-bound car is $v = +30$ m/s, and the velocity of the west-bound car is $v = -45$ m/s. The selection of which direction to label positive is completely arbitrary, but the sign of the opposite direction must be negative.

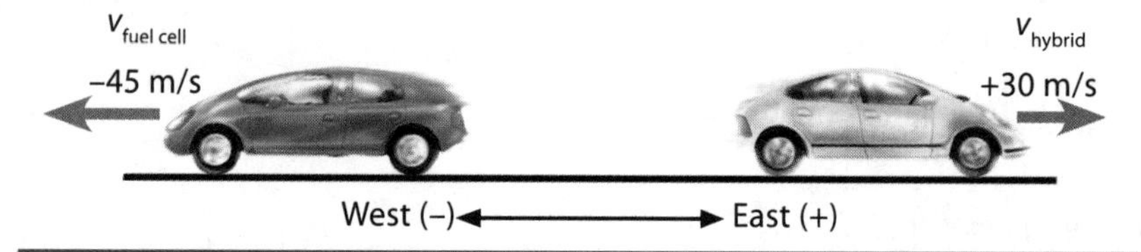

4-16 Velocities of the hybrid-energy and hydrogen fuel-cell cars. If both vehicles are on the same road heading away from each other, how fast are they moving apart?

Physicists find velocity much more meaningful than speed for calculating momentum, centripetal force, and torque, as well as other quantities. Several of these are discussed in other chapters.

Calculations Involving Speed and Velocity (Optional)

The formulas for finding speed and velocity contain three variable quantities: distance (or displacement), time interval, and speed (or velocity). Such formulas are useful because knowing any two quantities enables you to find the third. The formulas can be rearranged using simple algebra. The following example problems illustrate just a few real-world applications.

Example Problem 4-2

Calculating the Time Interval Given Speed and Distance

A mechanical engineer in a large manufacturing plant is asked to bring a set of plans as soon as possible to his manager's office in a building 1.2 km away. The engineer decides to use one of the company's Segway Personal Transporters to carry him on his errand. If the Segway can move at 16 km/h, how long (in minutes) will it take the engineer to transport the plans to his manager?

Known: distance (d) = 1.2 km
transporter speed (v) = 16 km/h

Unknown: time interval (Δt)

Required formula:

$$v = \frac{d}{\Delta t}$$

$$v \times \Delta t = \frac{d}{\cancel{\Delta t}} \times \cancel{\Delta t} \quad \text{(multiply by time interval)}$$

$$\left(\frac{1}{\cancel{v}}\right) \times \cancel{v} \times \Delta t = d \times \left(\frac{1}{v}\right) \quad \text{(divide by speed)}$$

$$\Delta t = \frac{d}{v}$$

Multiply both sides by $\frac{1}{v}$ and cancel.

Substitution:

$$\Delta t = \frac{1.2\ \cancel{\text{km}}}{16\ \cancel{\text{km}}/\text{h}} = 0.075\ \text{h}$$

Solution:

$$\Delta t = 0.075\ \text{h} \times 60\ \text{min/h} = 4.5\ \text{min}$$

$$\Delta t = 4.5\ \text{min} \quad \text{(2 SDs allowed)}$$

Multiply by 60 min/h to convert from hours to minutes. (The conversion factor does not affect the number of SDs since it is an exact definition.)

4-17 The Segway® Personal Transporter (PT) is a useful technological development that moves people quickly and safely in situations in which more conventional vehicles cannot be used.

Example Problem 4-3

Calculating Distance Given Average Speed and the Time Interval

The 500 Series Japanese bullet trains are designed to travel at a top speed of 320 km/h (200 mi/h). If a bullet train averages 206 km/h and takes 2.5 h to travel between Tokyo and Osaka, Japan, what is the approximate distance between those two cities by rail?

Known: speed (v) = 206 km/h
time interval (Δt) = 2.5 h

Unknown: distance (d)

Required formula:

$$v = \frac{d}{\Delta t}$$

$$v \times \Delta t = \frac{d}{\cancel{\Delta t}} \times \cancel{\Delta t} \quad \text{(multiply both sides by } \Delta t\text{)}$$

$$d = v \times \Delta t$$

Transpose the equation to place the unknown on the left side of the equal sign.

Substitution: $d = (206\ \text{km/h})(2.5\ \text{h})$

Solution: $d = 5\underline{1}5\ \text{km}$

$d \approx 520\ \text{km}$ (2 SDs allowed)

4-18 A Japanese 500 Series bullet train

4.10 Acceleration

When a speed skater hears the starting gun, he changes his speed from zero to a high value in a short amount of time. The change of his speed is his acceleration. **Acceleration** is the rate of change of the speed or velocity during a given time interval. Note that acceleration may apply to a change in a scalar or a vector, so acceleration may itself be either a scalar or a vector.

To calculate average acceleration involving a change of speed, subtract the initial speed from the final speed and then divide by the time interval of the speed change:

$$\text{average acceleration } (a_{avg}) = \frac{\text{change in speed}}{\text{time interval}} = \frac{\Delta v}{\Delta t}$$

$$\boxed{a_{avg} = \frac{v_f - v_i}{\Delta t}}$$

Vector acceleration involving changes in velocity can result from a change in speed, direction, or both.

The formula used for calculating acceleration in which velocity changes is similar to this formula, except that velocities are substituted for speeds. This difference is important. Remember that speed is simply the rate of change of position—direction is not considered. With velocity, both speed and direction must be considered. Therefore, a change in velocity might involve a change in speed, a change in direction, or both. An acceleration occurs in any of these three circumstances. For this course, you will solve problems involving only one-dimensional speed changes. Just be aware that acceleration includes changes in both speed and direction.

4-19 Driving on this street involves one-dimensional accelerations.

Example Problem 4-4

Calculating Vector Acceleration

A car moving at +5.0 m/s smoothly accelerates to +20.0 m/s in 5.0 s. Calculate the car's acceleration. North is positive.

Known: car's initial velocity (v_i) = +5.0 m/s
final velocity (v_f) = +20.0 m/s
time interval (Δt) = 5.0 s

Unknown: average acceleration (a_{avg})

Required formula: $a_{avg} = \frac{v_f - v_i}{\Delta t}$ (velocity symbols are in scalar form)

Substitution: $a_{avg} = \frac{(+20.0 \text{ m/s}) - (+5.0 \text{ m/s})}{5.0 \text{ s}}$

$a_{avg} = \frac{+15.0 \text{ m/s}}{5.0 \text{ s}}$ (include sign for vectors)

Solution: $a_{avg} = +3.0 \text{ m/s/s} = 3.0 \text{ m/s}^2$ north

There are several things to note in the above example. First, the car is traveling in one dimension. Second, when substituting velocities into the formula, parentheses are used to keep the signs of the velocities separate from the subtraction operation sign. Also, observe that the sign of the acceleration is positive. This means that velocity is increasing during the time interval in the positive direction of the coordinate system (north). Finally, the units of acceleration are meters per second per second (m/s/s). This expression is usually shortened to meters per second squared (m/s^2).

Example Problem 4-5

Negative Acceleration

A car moving at +20.0 m/s smoothly slows to a stop (0 m/s) in 6.0 s. Calculate the acceleration of the car. East is positive.

Known: car's initial velocity (v_i) = +20.0 m/s
final velocity (v_f) = 0.0 m/s
time interval (Δt) = 6.0 s

Unknown: average acceleration (a_{avg})

Required formula: $a_{avg} = \frac{v_f - v_i}{\Delta t}$ (acceleration and velocities are in scalar form)

Substitution:
$$a_{avg} = \frac{(0.0\text{ m/s}) - (+20.0\text{ m/s})}{6.0\text{ s}}$$
$$a_{avg} = \frac{-20.0\text{ m/s}}{6.0\text{ s}}$$

Solution: $a_{avg} = -3.\overline{3}$ m/s/s = $3.\overline{3}$ m/s^2 West (2 SDs allowed)

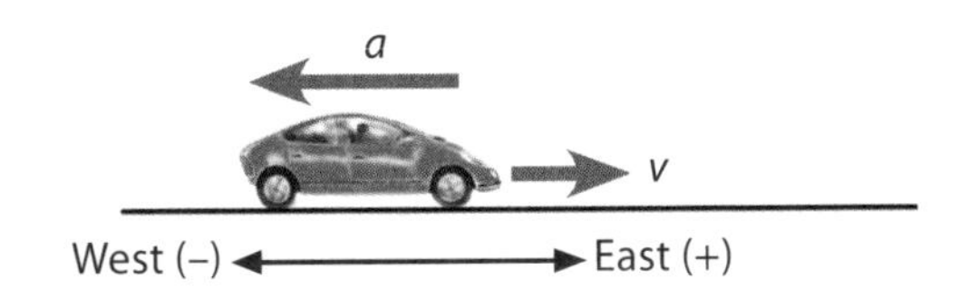

4-20 A car experiencing negative acceleration

This problem illustrates something important about vectors. A negative vector points in the direction opposite the positive direction of the coordinate axis. In the above example, the acceleration vector is pointing west, even though the velocity of the car was toward the east. In a one-dimensional coordinate system, negative acceleration (or *deceleration*) acts to slow, stop, or reverse the original positive velocity. Negative acceleration can also *increase* velocity in the negative direction.

4-21 Navy aircraft are rapidly accelerated to attain flight speed in a short distance.

4.11 Motion in Two and Three Dimensions

Although your study of motion in this course concerns mainly one dimension, most real motion occurs in two or three dimensions. Therefore, you should be familiar with two- and three-dimensional motion.

Motion in two dimensions involves changes of position on a coordinate *plane*. Two-dimensional motion may occur in a horizontal plane. For example, players on a soccer field, cars on the road, or ships on the surface of the ocean move essentially in two dimensions. It may also occur in a vertical plane. For example, the path of a thrown ball or a falling aerial bomb lies in a vertical plane. Position in a plane is determined using two coordinate axes perpendicular to each other. The origin is located where the axes intersect. Each position in the system must be described by two coordinates (e.g., x and y).

4-22 Two-dimensional motion occurs in a plane, such as the ocean's surface.

The term spatial (SPAY shul) describes something relating to or having the properties of space.

Three-dimensional motion occurs in all three spatial dimensions—length, width, and height. Roller coasters, acrobatic aircraft, thunderstorm thermal updrafts, and geologic fault motion are examples of motion in three dimensions. Position in three dimensions requires reference to three mutually perpendicular coordinate axes. Their common intersection is the origin for the spatial reference system. Positions are described by three coordinates (e.g., x, y, and z).

4-23 A hurricane represents a complex system of motion in three dimensions.

The definitions of distance, displacement, speed, velocity, and acceleration are the same in two- and three-dimensional motion as they are in one-dimensional motion. Calculating their values can be more complicated because you must understand a kind of geometric math called *trigonometry*. You will have an opportunity to study two- and three-dimensional motion in more advanced courses in physics.

4.12 Using Kinematics to Solve Problems

In a car accident, two collisions take place. The car decelerates as it collides with something, and the occupants decelerate when they collide with the interior of the car or when something restrains them. Car accident decelerations take place quite rapidly, causing injuries. As the acceleration equation in Subsection 4.10 predicts, if the same change in speed happens over a longer time interval, then the rate of acceleration decreases. Padding and restraint systems used in car seats increase the likelihood that a child will survive a car accident by increasing the time interval for the same change in speed.

According to studies by the Washington School of Medicine, car seats drastically decrease the risk of death—by 71% for infants and 55% for toddlers. This is why every state in the United States requires that small children travel in car seats.

One of the child-sized crash-test dummies recently developed for testing car seats and airbags goes by the acronym *CRABI*. Using this dummy, scientists reconstruct the conditions of actual accidents that have resulted in child deaths to come up with possible solutions that will help keep children safe.

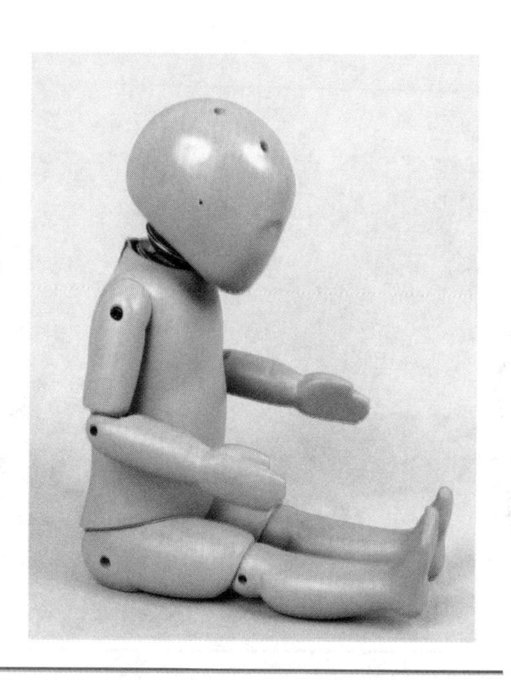

4-24 The Child Restraint Air Bag Interaction (CRABI) child-sized test dummy and similar devices are helping engineers design safer car seats and car interiors.

4B Section Review

1. How many position coordinates are required to identify the position of a system undergoing one-dimensional motion? What symbol(s) should you use to identify a system's position in one-dimensional motion?
2. A common black ant (*Formica fusca*) discovers a piece of bread 85 cm east of the entrance to her nest. If the ant carries 10 bits of bread back to her nest on separate trips before moving on to other chores, what distance did she travel after discovering the bread? What was her displacement from the nest after dropping off the last bit of bread?
3. An archer shoots an arrow into a target 53 m north of his position. What do you call the magnitude of the arrow's displacement? How large is that magnitude?
4. Define *speed* and give its formula. Identify each quantity in the formula. What quantity does this formula give in real-world problems?
5. State two ways of defining *velocity*. What is another name for the magnitude of the velocity vector?
6. Define *acceleration* in your own words. Write the formula for average acceleration and identify each term.
7. In what ways can a change in velocity be an acceleration?
8. Give an example of two-dimensional motion and an example of three-dimensional motion.
9. (True or False) Negative acceleration always means that a system is slowing down.
10. a. Convert the speed formula to a formula that solves for distance when you are given average speed and the time interval. Show each algebraic step.

    b. Convert the speed formula to a formula that solves for the time interval when you are given the average speed and the distance traveled. Show each algebraic step.
11. A freight train must approach a road crossing at 16 km/h. After passing through the intersection, the train speeds up to 65 km/h, taking 10. min to do so. What is the train's acceleration during this time in meters per second squared?
12. DS Describe the kinds of conditions scientists should investigate when using child-sized dummies to reconstruct accidents that have resulted in child deaths. What kind of data can the test dummies provide?

Scientifically Speaking

mechanics	75
system	76
frame of reference	76
interval	78
kinematics	80
distance	80
displacement	80
speed	81
velocity	82
acceleration	84

Chapter Review

1. Contrast the views of the influential Greek philosophers and those of scientists after Galileo and Newton regarding why objects move.
2. How does kinematics differ from dynamics?
3. A skier is schussing down a snowy slope. Describe everything that is part of the skier-system.
4. When playing a flight-simulator computer game (a game where you act and see the world as the pilot of an aircraft), what is the reference frame for the computer image in the game?
5. What evidence do we have that time began at Creation and will continue forever into the future?
6. A vector quantity can be represented in a diagram by a stylized arrow. Discuss how this symbol represents a vector quantity.
7. When would actual distance traveled equal the magnitude of a system's displacement?
8. Explain why speed is a rate. Explain why acceleration is also a rate.
9. The average speed during a car trip is calculated by dividing the trip odometer reading at the end of the trip by the total time the trip took. What car instrument displays the *instantaneous speed* of the car, or the speed the car is moving at any instant in time?
10. What is the significance of the sign of velocity in one-dimensional motion?
11. The following is a radio exchange between a large passenger aircraft (Heavy 315) and an air traffic control tower in Washington DC:

    Controller: "Heavy 315, this is National tower. Turn left to heading 180 degrees and slow to 160 knots."

    Pilot: "National tower, this is Heavy 315. Roger, turning left to heading 180, slowing to 160."

    Using kinematics terms you learned in this chapter, describe what is happening to the passenger plane.
12. Which kind of coordinate system is required to describe most real-world motion? Why is this so?

True or False

13. In the science of mechanics, it is vital to understand how things move before you try to understand the causes of motion.
14. When studying the motion of a thrown ball, the system is the ball and the hand that throws it.
15. The weathervane on top of a barn measures the wind vector.
16. Position in one-dimensional motion requires only a single number.
17. The magnitude of −15.0 m/s is 15.0 m/s.

✪ 18. The instantaneous speed of a car must equal its average speed at least once during a trip.

19. In most cases of motion, scientists try to avoid measuring velocities because it complicates measuring the data.

20. Increasing speed always indicates positive acceleration, and decreasing speed always indicates negative acceleration.

21. Acceleration is the rate of change of the rate of change of position.

22. The definitions of distance, displacement, speed, velocity, and acceleration are the same in one-, two-, and three-dimensional coordinate systems.

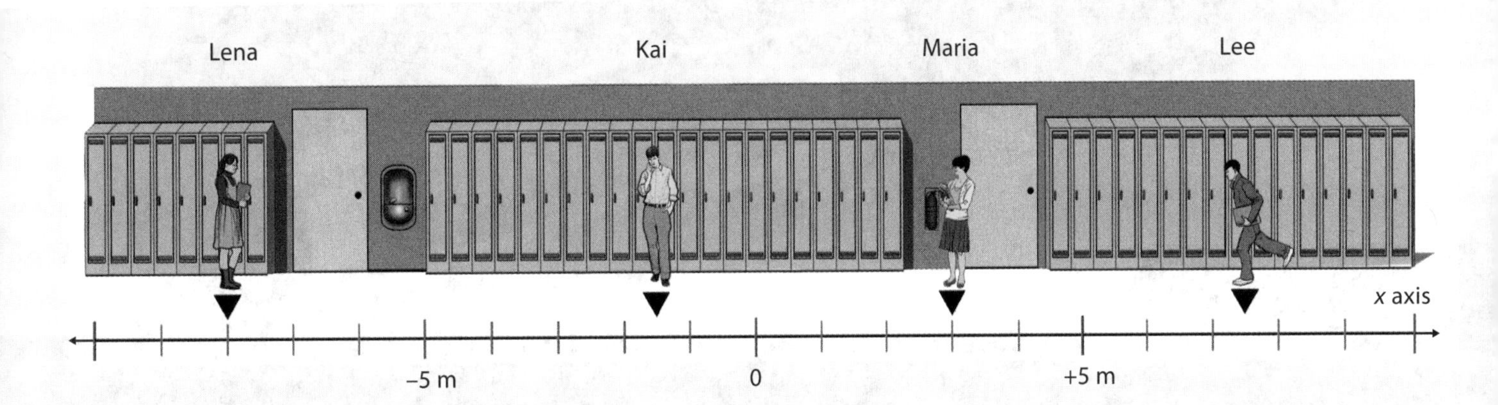

Questions 23–25 refer to the diagram above.

✪ 23. State the positions of the four students in the diagram.

✪ 24. If Lena walks to where Maria is standing, what distance has she covered?

✪ 25. If Lee takes 3.0 s to jog to where Kai is standing (no running in the hall, please!), what is his velocity?

✪ 26. A nitro-fueled dragster can accelerate from 0 to 440 km/h in 5.0 s. Assuming that the vehicle's acceleration is constant, calculate the vehicle's acceleration in meters per second squared.

DYNAMICS: WHY THINGS MOVE

CHAPTER 5

DOMINION SCIENCE PROBLEM

Giving People a Lift

Since ancient times people have lived together in cities. Throughout history, cities became more convenient, more prosperous—and more crowded. Sprawling cities taking up more and more land often meant longer distances between home and work. As the world's population grew, land near cities became extremely valuable and more expensive to buy and own. The solution to crowding was obvious: build the cities vertically. But taller buildings required strong building materials and a convenient way to access the upper floors. The Industrial Revolution and the development of cheap steel solved the first problem. Buildings could be constructed dozens of stories tall. But who has the endurance to climb many flights of stairs to go to work every day? How can the physically challenged access floors above ground level? How can you move heavy equipment and products from floor to floor?

5-1 As cities built upward, the ability to access the upper parts of the buildings became a problem.

5A Forces

5.1 Forces Weren't Always with Us

"Captain! The ship's shields are down to less than 50 percent! We'll never hold off the next attack!" Entertaining science fiction stories of improbable futures (and pasts) have made much use of the basic principles of immense forces and force fields. However, the real world of forces is no less fascinating.

In ancient Greece, Aristotle believed that all matter possessed an urge to move in certain ways. Matter naturally moved up or down depending on its nature. Once things were set in motion, their natural inclination took over and guided their behavior. He believed that all natural changes of motion came from within material things.

During the third century BC, Archimedes, a Greek mathematician, engineer, and astronomer, was the first to write about the causes of motion. About 1800 years later, Galileo realized in a moment of inspiration that the motions of objects change because of a push or a pull on them, not because of some impulse from within themselves.

Isaac Newton was born around the time Galileo died. He called Galileo's pushes and pulls *forces*. He also developed mathematical equations describing how forces affect motion. Perhaps his most important contribution to science was his description of the gravitational force, which explained why the moon and planets stayed in their orbits. In the 1700s, the British physicist Henry Cavendish (1731–1810) performed an experiment that allowed later scientists to calculate the gravitational forces between objects. Since the early 1800s, physicists have devoted much effort to developing models to describe all the forces they observe in nature.

Scientists aren't entirely sure what ultimately causes forces, though they have developed complex models that work for this purpose. They have described how forces cause phenomena in nature through **dynamics**. Some of the most important aspects of dynamics are

- identifying the kinds of forces,
- explaining how forces interact with matter, and
- describing how they affect the motions of objects.

In this section, you will learn about the different kinds of forces found in nature.

5.2 Classifying Forces

A **force** is a push or a pull on a system. This description fits well with your experience and observations of forces in nature. You have probably pushed a grocery cart or pulled on the leash of an ornery pet. You know that powerful engines push rockets into space and that cables pull elevators upward.

When a force acts on a system, it pushes or pulls—at least at any given moment—in a single direction and with a certain magnitude. So forces are vectors. Sketches that show how forces act on systems are *force diagrams*. These use an arrow to show which object a force acts on, how large the force is, and the direction of the pull or push on the object (see Figure 5-3).

5A Objectives

After completing this section, you will be able to

✓ describe the characteristics of a force.

✓ categorize a force as a contact or field force and describe it.

✓ explain the existence and effects of balanced and unbalanced forces on an object.

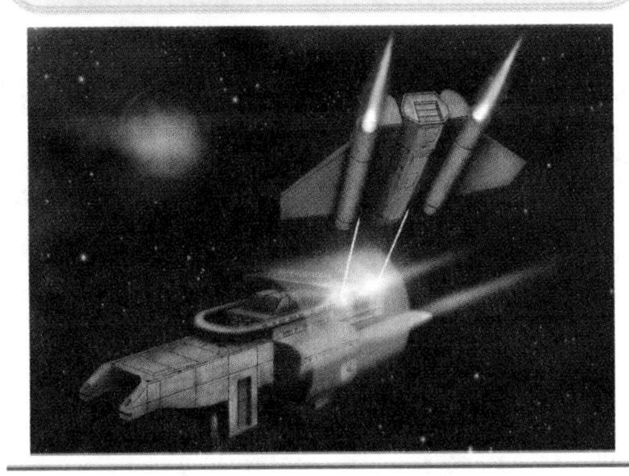

5-2 Science fiction force fields stretch physics beyond the limits of what we currently know. But real forces and the fields that create them are extremely important to us.

In the early 1900s, physics entered a new realm of knowledge called *modern physics*. Modern physics studies the effects of extreme gravity and the unusual behavior of things that are extremely fast (near the speed of light) or extremely small (at the subatomic level).

The area of mechanics called **dynamics** investigates the causes of motion and changes in motion.

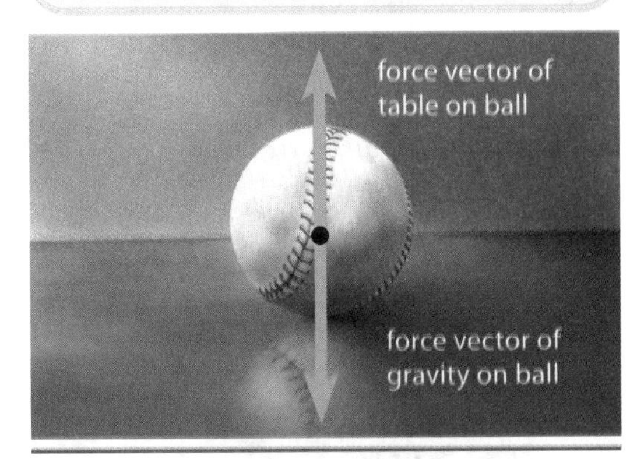

5-3 A labeled force diagram

The symbol "*F*" in diagrams and formulas stands for the magnitude of a force. The unit of force in the SI is the newton (N), named in honor of the English physicist.

We can classify forces in our everyday experience into two broad categories: *contact forces* and *field forces*. Physicists believe that they are the visible results of four *fundamental forces* that are present in all matter (see the facet below). Let's take a closer look at these two kinds of forces.

Contact Forces

As their name implies, **contact forces** act only when one system touches another. Contact forces originate at the atomic level of matter. Atoms resist the intrusion of other particles into their space, so when the atoms of another object push on them, they push back. You

FACETS OF SCIENCE — FUNDAMENTAL FORCES

Physicists have identified four forces that seem to underlie all other known forces in nature. These forces are responsible for holding together the structure of all matter in the universe, from the subnuclear to the intergalactic. We list these forces from the strongest to the weakest.

The *strong nuclear force* strongly attracts at distances equal to the diameter of a proton. It holds the protons and neutrons together inside atomic nuclei. However, at much shorter distances it repels strongly. The strong nuclear force is the most powerful of all fundamental forces.

The *electromagnetic force* can also be attractive or repulsive. These forces can act over distances ranging from the subatomic to the very large. The electromagnetic force originates in static (unmoving) and moving electrical charges. It affects electrical charges and magnetic materials. The electromagnetic force is 1/137 of the strength of the strong nuclear force.

The *weak nuclear force* is an extremely short-range force, attracting at distances smaller than the radius of a proton. It holds together the elementary particles of matter that make up protons and neutrons. It is also involved in some forms of nuclear decay. The weak nuclear force is about one-millionth the strength of the strong force.

All matter exerts the *gravitational force* on other matter. Although gravitational force can act over intergalactic distances, it is the weakest of all fundamental forces. Its relative strength is 10^{-39} of the strong nuclear force.

Theoretical physicists have developed a model of forces that shows that the strong nuclear, weak nuclear, and electromagnetic forces are mathematically related. Many physicists believe that they will eventually find that the gravitational force is related to the other fundamental forces as well, making them all just different aspects of one unified force.

Christians need to consider becoming physicists who can contribute to solving this great question. Who better to study the mysteries of God's universe than a person who has a personal relationship with the Creator? And who better to interpret what this means for how we think about God and how we live our lives?

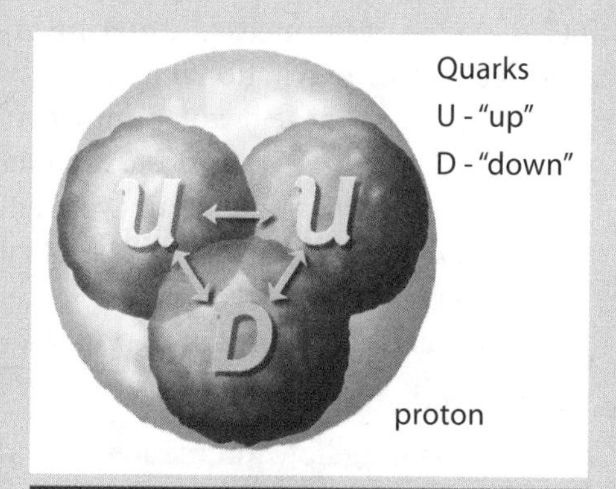

The weak nuclear force holds the particles inside protons and neutrons together.

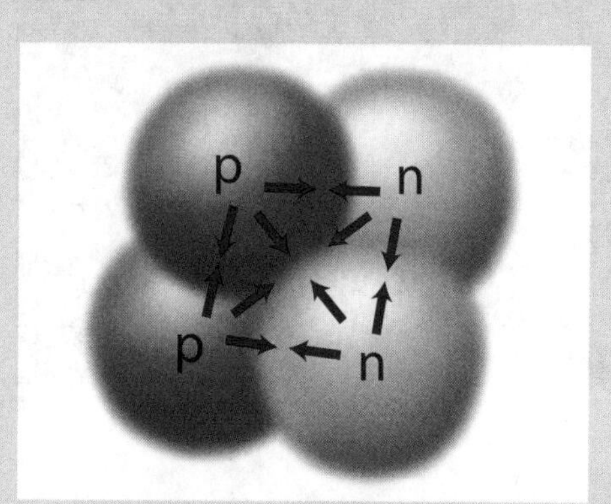

The strong nuclear force acts between protons and neutrons.

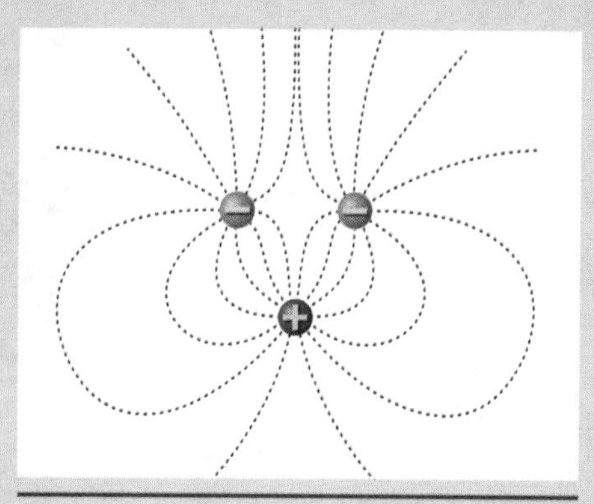
The electromagnetic force is exerted between electrical charges.

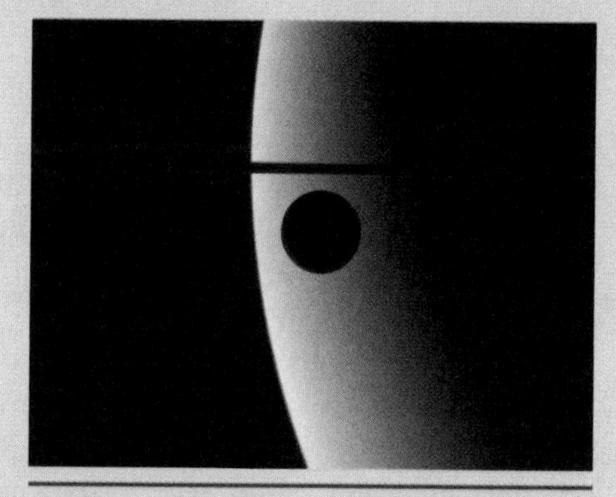
Only large celestial bodies exert significant gravitational force. In this photo, the moon Rhea orbits Saturn in the background.

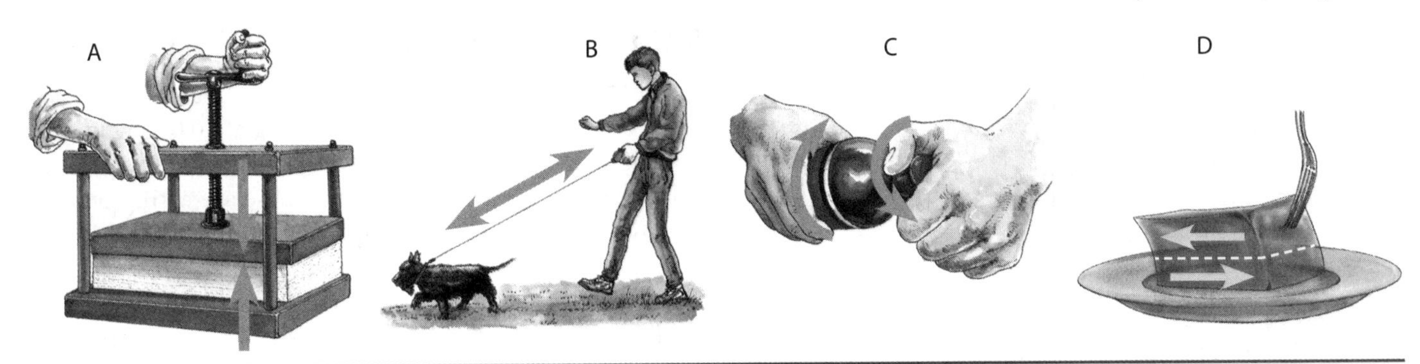

5-4 Contact or mechanical forces: compression (A), tension (B), torsion (C), shear (D)

exert a contact force when you open a drawer, step on the pavement, open a jar, or apply a strip of adhesive tape.

Scientists and engineers properly call contact forces *mechanical forces*. These include *compression*, which pushes things together; *tension*, which pulls things apart; *torsion*, which twists objects; and *shear*, which can cause one part of a material to distort or move parallel to the direction of the force relative to other parts.

Field Forces

Forces also act between objects that are *not* touching. Though such forces originate in matter, they do not need to act *through* matter. Long ago, people observing pushes and pulls between objects that weren't touching considered such phenomena abnormal, even magical. Newton called them *action-at-a-distance forces*.

Today, physicists explain that such forces act through a *force field*. So what is a field? In physics, a *field* is a volume of space in which every point can have a different value of a dimension of interest. For instance, imagine a room with a fireplace containing a roaring fire. Nearer the fire, the temperature is higher. Farther away, the temperature is cooler. Every point in the room, ceiling to floor, has its own temperature value. This model of the room's temperature is a temperature field. An object surrounded by a space in which every point can exert a push or pull produces a **field force**. In other words, each point in the field has a force vector associated with it. Field forces produced by matter affect susceptible objects by attracting or repelling matter.

The principal field forces are the gravitational, magnetic, and electric fields. A *gravitational field* surrounds anything made of matter, although only planet-sized masses or larger exert significant gravitational force. Gravity affects everything, including light. The sources of *magnetic fields* are magnets and moving electrical charges. Magnetic fields affect susceptible materials such as iron, magnets, or moving charges. *Electric fields* surround and affect electrical charges. All fields extend outward from their sources without limit but their influence weakens rapidly with distance (see Figure 5-5).

Susceptible objects are those that have properties that respond to the influence of a field force.

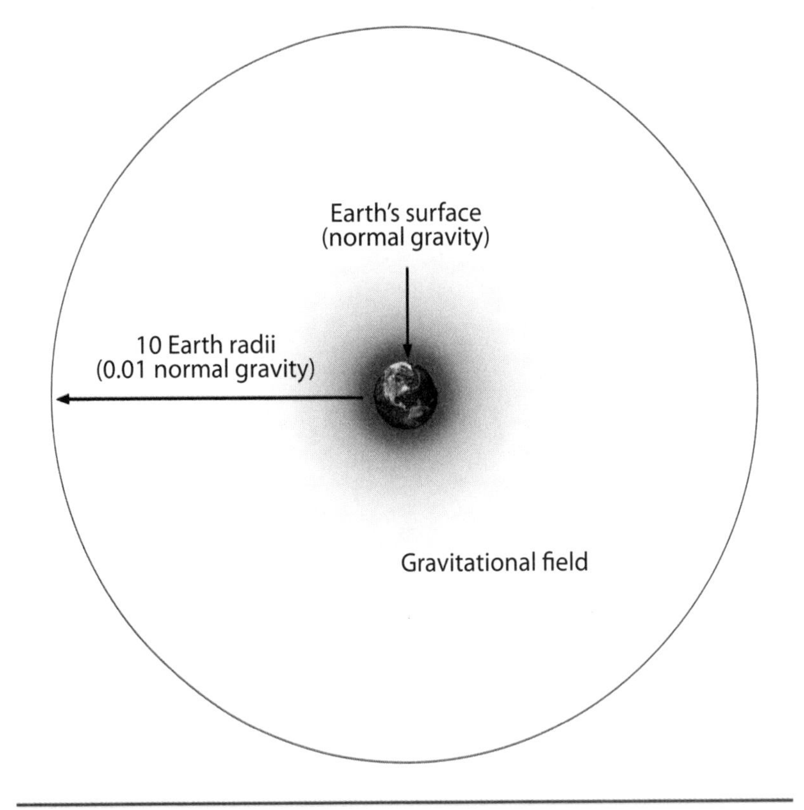

5-5 A model of Earth's gravitational field. The darker the blue color, the stronger the field.

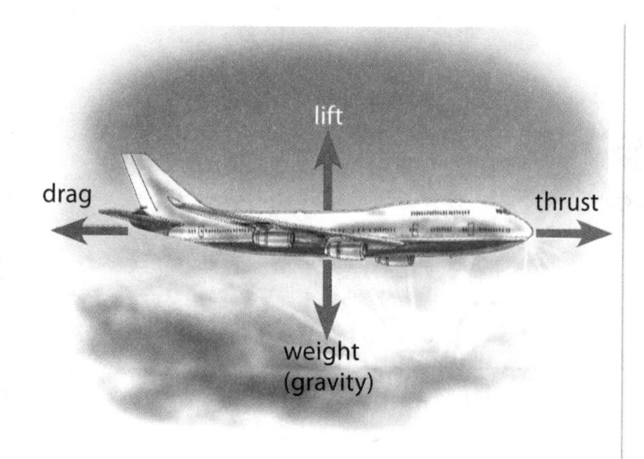

5-6 Even when a plane is flying straight, level, and at a constant speed, many forces are acting on it. But they are balanced and cancel each other out. This shows that a system can be moving even though balanced forces act on it.

5.3 Balanced and Unbalanced Forces

Every object in the universe has more than one force acting on it. For example, the forces acting on an airliner flying at 10,000 ft include the thrust of its engines, the lift on its wings, air resistance, and, of course, Earth's gravity. So, if forces like these are acting on each object all the time, why isn't everything in continual motion? It's because most objects experience *balanced forces*—multiple, simultaneous forces whose pushes and pulls cancel each other out.

Let's look at a simple example of this. Assume that two groups of students are pulling on a tug-of-war rope, as in Figure 5-7. The *external forces* that the teams apply acting on the rope-system come from outside the system's boundary. If the teams pull equally hard in opposite directions, the two forces are balanced and the rope doesn't move. As far as the rope is concerned, there is no ***net force*** acting on it. The rope-system acts as though there were no forces applied to it at all.

On the other hand, if one team pulls harder than the other does, then an *unbalanced force* acts on the rope. The net force equals the difference of the two forces. The net force acts in the direction of the larger force. The stronger team will eventually drag the other team over the line in the direction of their pull.

Balanced and Unbalanced Forces

In this discussion, you may assume that the forces are acting in the same line, directly opposite each other. If not, then they couldn't balance each other.

Net or Difference

When comparing two related but opposite things, their difference is the *net* quantity. For example, if your weekly allowance is $10 and you spend $7 on a lunch out, your net allowance savings for the week is $3.

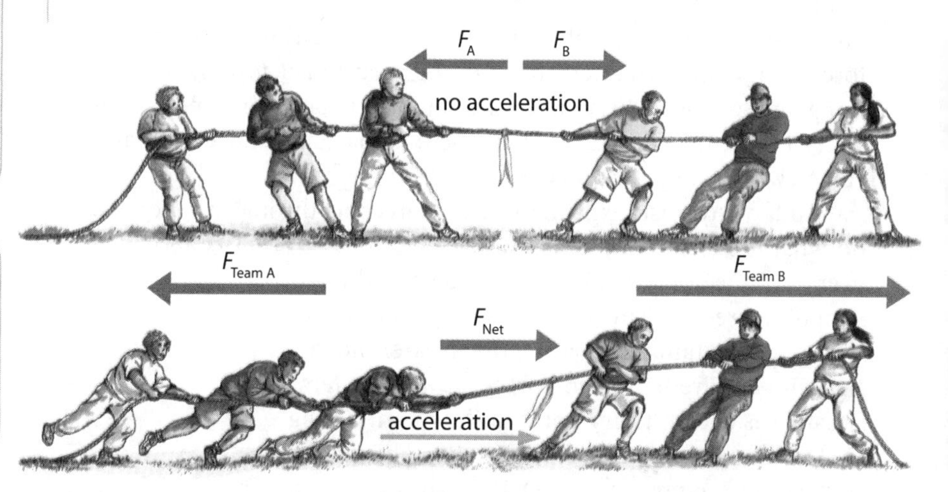

5-7 When forces are balanced, there is no net force (top). When forces are unbalanced, there is a single net force acting on the system (bottom).

In the real world, there are usually more than two forces acting on a system, often in multiple directions. If these forces cancel each other out, they are balanced and no net force exists. If they don't cancel, there is a single net force exerted in a specific direction. The result of an unbalanced force on a system is a change in motion. This will be the topic of Section 5B.

5A Section Review

1. How does dynamics differ from kinematics?
2. What is a force? How do physicists represent forces in diagrams?
3. What is the SI unit for force?

*4. Which fundamental force is responsible for holding the universe's matter in recognizable structures such as planets, stars, and galaxies?

5. What must occur for an object to exert a contact force on another object?
6. In physics, what is a field?
7. What is susceptible to the gravitational field?
8. If two external forces acting on a system are equal in magnitude but opposite in direction, what is the size of the net force acting on the system?
9. (True or False) Compressive and tensional forces act in opposite directions on a system.

5B Newton's Laws of Motion

5.4 *Imagining Inertia*

Scientists in the Middle Ages, relying on Aristotle's explanations for natural phenomena, believed that all things would slow to a stop if left to themselves. This idea seems logical enough. If the engine in your car stops, your car rolls to a standstill. If you shove a book across the table, it slides to a stop. But experiments proved that models relying only on reason failed to explain everything about the causes of motion.

The physicist Galileo is the first person known to have investigated these ideas. He spent a great deal of time experimenting by rolling balls down inclined planes, measuring projectile motion, and timing pendulums. From his experiments, Galileo concluded that objects continue moving indefinitely unless some external influence affects their motion. He was able to infer this even though he couldn't directly observe such a situation on Earth. Johannes Kepler (1571–1630) first gave this property of matter a name: **inertia**.

Early scientists, including Galileo, never settled on an explanation for how forces caused changes in motion. One reason for this was the lack of necessary technology. To study motion, you must be able to measure time intervals accurately. Although Galileo developed several methods to subdivide time for his work, it wasn't until a few years after Galileo's death that the invention of accurate and precise pendulum clocks solved this problem. This new technology allowed Isaac Newton to build on Galileo's concept of inertia. The results of this important work were his three *laws of motion.*

5B Objectives

After completing this section, you will be able to

- ✓ summarize the historical development of the concept of inertia.
- ✓ state Newton's first law of motion in your own words and explain its significance.
- ✓ state Newton's second law in your own words and express it as an equation.
- ✓ state Newton's third law in your own words and explain why identifying the system of interest is necessary.
- ✓ describe friction and the factors that generally affect its magnitude.

5.5 *Newton's First Law—The Law of Inertia*

Newton merged Galileo's principle of inertia into a more general statement of physics that included his own discoveries about the actions of forces. Newton's first law or **law of inertia** states:

> *Objects at rest remain at rest, and objects in motion continue in a straight line at a constant velocity (speed and direction) unless acted upon by a net external force.*

We can't measure inertia directly and so it has no units associated with it. Physicists quantify inertia by mass. In fact, the concept of mass was "invented" by physicists as a measure of inertia.

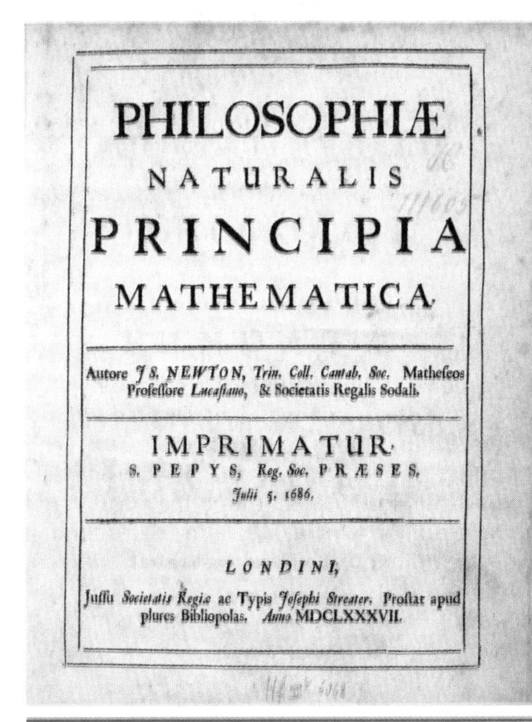
PHILOSOPHIÆ
NATURALIS
PRINCIPIA
MATHEMATICA.

Autore *J S. NEWTON, Trin. Coll. Cantab. Soc.* Matheseos Professore *Lucasiano*, & Societatis Regalis Sodali.

IMPRIMATUR.
S. PEPYS, *Reg. Soc.* PRÆSES.
Julii 5. 1686.

LONDINI,
Jussu *Societatis Regiæ* ac Typis *Josephi Streater*. Prostat apud plures Bibliopolas. *Anno* MDCLXXXVII.

5-8 The title page of Sir Isaac Newton's great work, *Principia,* in which he reports his three laws of motion

5-9 The *Voyager 1* probe is crossing over into interstellar space. Only extremely weak gravitational forces affect this vehicle.

Newton's first law of motion defines the conditions for zero net force on a system. Recall from Section 5A that a zero net force means that all forces on the system are balanced. As long as this condition exists, no change to a system's motion will occur. The system is in *mechanical equilibrium.* For example, a car driver maintains a constant speed by balancing the slowing forces of friction, air resistance, and gravity (on hills) with the driving force of the engine. The driver adjusts engine force by changing the position of the gas pedal.

Newton's genius was his ability to conceive an idea that is totally outside of common experience. This law is not easy to prove here on Earth. All kinds of familiar motions are subject to two ever-present forces—friction and gravity. When you kick a soccer ball, it doesn't sail away in a straight line. It curves gracefully toward the ground because of gravity. Only deep-space probes seem to be nearly free of external forces. But even at interstellar distances, the sun's weak gravity and even the gravity of the galaxy itself can affect the motion of these distant objects.

Thus, any combination of unbalanced contact or field forces acting on a system can result in a change in motion. You will learn more about friction and gravity later in this chapter.

5.6 Newton's Second Law—The Law of Accelerated Motion

5-10 An SUV requires much more force to accelerate than a smaller vehicle.

A small, lightweight car takes relatively little force to accelerate. Such cars have small engines that use very little fuel. Large, heavy sport-utility vehicles need much bigger and more powerful engines. The larger vehicles have more mass, so they have more inertia.

Newton's most significant accomplishment in developing his three laws of motion was relating the quantities force, mass, and acceleration in an equation that always works, at least in normal, everyday experience. His **law of accelerated motion** says:

> *The acceleration of a system is directly proportional to the net force acting on the system and is inversely proportional to its mass.*

Thus, a net force on an object of a given mass results in a certain acceleration of the object. If you double the net force on the object, you double its acceleration. On the other hand, if the original net force acts on another object with twice the mass, it results in only half the acceleration.

Newton originally thought of his law in the form of this equation:

$$\text{acceleration of system} = \frac{\text{net force on system}}{\text{mass of system}}$$

Physicists usually write Newton's second law in symbols as

$$\boxed{F = m \times a},$$

which is simpler to remember. In this equation

- F is the net external force on a system, in newtons,
- m is the mass of the system, in kilograms, and
- a is the acceleration of the system, in meters per second per second.

Remember that the unit of force, the newton (N), is an SI derived unit. Derived units are combinations of two or more SI base units. You can express the newton in its base SI units by substituting the mass and acceleration unit values into the second law equation.

$$F = m \times a$$

$$\text{N} = \text{kg} \times \frac{\text{m}}{\text{s}^2}$$

$$\text{N} = \frac{\text{kg}\cdot\text{m}}{\text{s}^2}$$

The dot (·) means multiplication in a unit symbol.

Newton's first law defines mechanical equilibrium by showing what happens when there is no net force on a system—no change in motion occurs. Newton's second law shows how the motion of a system changes when the net force acting on it is *not* zero.

5-11 A sports car has a much higher acceleration compared to a truck for the same force delivered by their engines.

Example Problem 5-1

Acceleration and Newton's Second Law

Calculate the force exerted by the engine of a flashy sports car as it accelerates down an empty stretch of highway. Assume that it accelerates at 3.3 m/s² and that its mass is 1500 kg (3300 lb).

Known: acceleration (a) = 3.3 m/s²
mass (m) = 1500 kg

Unknown: force (F)

Required formula: $F = m \times a$

Substitution: $F = (1500\ \text{kg})(3.3\ \text{m/s}^2)$

Solution: $F \approx 4\underline{9}50\ \text{kg·m/s}^2$
$F \approx 5000\ \text{N}$ or $5.0 \times 10^3\ \text{N}$ (2 SDs allowed)

5.7 *Newton's Third Law—The Law of Action-Reaction*

Unlike Newton's second law, we can't easily express the **law of action-reaction** as an equation. We can state Newton's third law of motion in these words:

> *For every force exerted on a system by its surroundings, the system exerts an equal but opposite force on its surroundings.*

If you push a box across the floor, it moves because the box exerts a force on your hands. If it didn't, your hands would pass right through the box! The resistance of the box on your hands is actually the box pushing back on you. If the third law weren't true, changes of motion would be impossible because no force could act on a system. If things didn't exert forces on other things, the material universe could not exist; particles of matter must exert forces on each other to hold matter together.

Newton's third law of motion is often called the *action-reaction principle*. However, this title suggests that some change of motion (a reaction) always results from a force (an action). But this isn't the case every time. For example, when you push against a wall, the wall pushes against you. No motion results unless you push so hard that you fling yourself backward—or you punch a hole in the wall!

Let's look at the earth and the moon. The earth-system exerts a gravitational force on the moon-system. This external force keeps the moon in orbit around the earth. You now know that, because of Newton's third law, the moon also exerts a gravitational force on the earth. It has the same magnitude as the earth's force, but it's in the opposite direction (see Figure 5-12). This force causes ocean tides and movements in the earth's crust.

5-12 The earth exerts a gravitational force on the moon. The moon exerts an equal but opposite force on the earth.

5-13 The weight of this climber is exactly balanced by the reaction force in the rope.

The symbol for friction in force diagrams and equations is *f*.

Newton's third law also applies in the case of a rope attached to a rock climber. The downward weight of the climber is exactly supported by the upward tension in the rope, which is the third-law reaction force to the climber's weight (Figure 5-13).

5.8 Friction

Friction is everywhere and affects all motion here on Earth, slowing things down and wearing parts out. So what is friction? Friction is a contact force that works against the movement of systems past each other. You can think of "systems" more broadly here. Not only are they objects touching or rubbing along surfaces (such as a pencil on paper) but they are also fluids flowing around or against a rigid object (such as a current flowing around a buoy in a river channel).

The direction of friction force acting on a system of interest is opposite to its direction of motion through its surroundings. If you slide a box down a ramp off the back of a truck, the ramp exerts a friction force acting *up* the ramp. The magnitude of friction between solid objects is proportional to the force holding the two objects together. In most familiar cases, this force depends on the weight of the object and the angle of the surface it rests on. For instance, the friction on the box just mentioned decreases if the angle of the ramp is increased. The box slides more easily. Analyzing the factors that affect friction and drag can be complicated, so we will only summarize them here. These factors generally apply to friction between solids. Some also apply to fluid drag, which is affected by other factors applicable only to fluids.

- Friction always occurs between systems in contact.
- Friction depends on the kinds of materials in contact but *not* on the size of the areas in contact.
- Friction acts on the system of interest opposite to the direction of motion relative to the system causing the friction.
- The amount of friction is proportional to the force holding the two systems in contact together.
- The friction between two stationary objects (e.g., a box resting on the floor) is greater than the friction between the same two objects when they are sliding past each other.

Physicists recognize several different kinds of friction. They include *static friction*, or friction between stationary objects; *kinetic friction*, or friction between sliding objects; *rolling friction*, or friction between a rolling object and a stationary surface; and *fluid friction*, or the drag of a liquid or gas exerted on a moving object immersed in the fluid.

kinetic (Gk. *kinetikos*—moving); in physics, an adjective that identifies any phenomenon involving motion

5B Section Review

1. What did Galileo discover about matter that contradicted Aristotle's writings? What do we call this property of matter?
2. What condition in dynamics does Newton's first law of motion define?
3. What do we really mean by "force" in Newton's second law of motion?

4. When you stand on the sidewalk, your weight acts downward on the sidewalk. What law of motion assures you that you won't fall to the center of the earth? Why?
5. Since friction is a force, how do you determine its magnitude and direction? Use a box sliding across a carpet to illustrate your answer.
6. How does the static friction between the box and the carpet in Question 5 compare to the kinetic friction as the box is sliding across the carpet?
7. (True or False) An object will accelerate whenever a force acts on it.
8. A student is helping her teacher move a 9.5 kg box of books. What net sideways force must she exert on the box to slide it across the floor so that it accelerates at 1.0 m/s²?
9. If the kinetic friction acting on the sliding box in Question 8 was 23.5 N, what was the total force the student had to exert to slide the box?

5C Gravity and Free Fall

5.9 *The Gravitational Field*

Of all forces, we are most familiar with gravity. We live with gravity from the time we are born until we die. Most of the time, we don't even think about it. It's just always there. It may surprise you to learn that every object exerts a gravitational force on every other object in the world. If this is true, then why doesn't everything just clump together into one gigantic ball of matter? Mainly because common objects such as chairs and lamps—or even jumbo jet airliners and large ships—can't exert enough gravitational force to affect similarly sized objects.

Gravity is a basic property of all matter. After studying the moon's motions, Newton concluded that the earth's gravity acting on the moon depends on just three quantities: the mass of the earth, the mass of the moon, and the distance between their centers. By inference, he concluded that this principle must be applicable to *any* two objects.

In words, Newton's **law of universal gravitation** says that the force of gravity between two objects is proportional to the product of their masses. Thus, if you replace one of the objects with one having twice the mass, the gravitational force acting between the two objects doubles. For example, the earth's gravity attracts a 10 kg concrete block twice as strongly as a 5 kg brick. Gravity also depends on the distance (d) between the centers of mass of two objects. The factor is $1/d^2$.

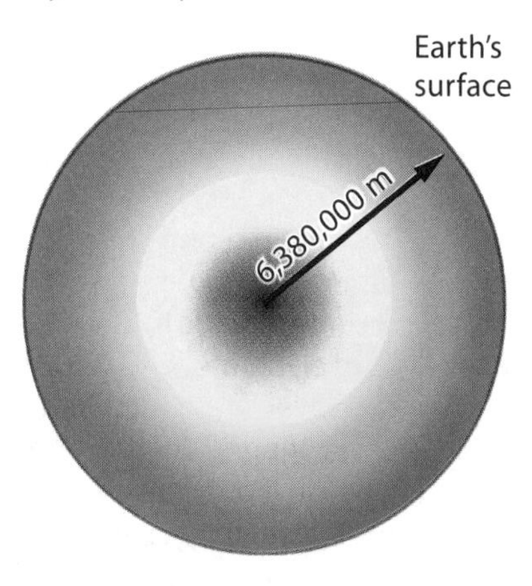

5-14 The earth's surface is about 6.38×10^6 m from the earth's center of mass.

5C Objectives

After completing this section, you will be able to

- ✓ summarize the principle of universal gravitation in your own words.
- ✓ write the formula for the law of universal gravitation and identify each variable and constant in the formula.
- ✓ define free fall and describe the conditions under which it exists.
- ✓ show how weight is related to Newton's second law.
- ✓ calculate velocity and distance of free-falling bodies, given the necessary data.
- ✓ explain how drag affects falling objects and how it relates to terminal velocity.
- ✓ give examples of how air resistance can be a help or a hindrance.

5-15 A Nimitz-class aircraft carrier is the largest warship ever built. Even so, it generates only a 0.000 32 N (0.000 072 lb) gravitational force on a typical 85 kg (190 lb) sailor.

FACETS OF SCIENCE

Johannes Kepler

Is it possible to serve God in politically troubled times? You may be asking that question right now as you observe developing events in your own country. Mathematician and astronomer Johannes Kepler probably asked it many times during his lifetime, since he lived in one of the most tumultuous periods of European history.

Kepler was born in 1571 near modern Stuttgart, Germany, 54 years after the start of the Protestant Reformation. At that time Europe was divided into many small nation-states whose rulers decided whether their subjects would be Protestant or Roman Catholic. The Thirty Years' War (1618–48), a bloody religious conflict, overshadowed the last 12 years of Kepler's life.

A genuine believer, Kepler held to the Lutheran persuasion throughout his life and frequently experienced persecution from Catholic rulers. He also came into conflict with Lutheran leaders over doctrine, a situation that frustrated him and further complicated his life.

Kepler lived in a time when superstition and unbiblical ideas distorted the worldviews even of genuine Christians. Greek philosophy and mysticism influenced the thinking of both educated and uneducated alike. Many practiced *astrology*, the art of predicting the future from the positions of the sun, moon, stars, and planets. Throughout his life, Kepler supplemented his income by casting horoscopes for political leaders, a practice that Christians later came to recognize as forbidden by the Bible.

At the same time, Kepler held the strong conviction that God revealed Himself through the beauty and order of creation. Kepler spent much of his life trying to explore and express this beauty through his mathematical and astronomical work. Biographers speculate that Kepler sought in mathematics the peace that was missing from his world.

Much of Kepler's inspiration for his view of the universe came from one of his professors at the University of Tübingen, Michael Maestlin. Though Maestlin taught the geocentric theory as fact, he also quietly introduced Copernicus's controversial heliocentric theory to his students. Kepler immediately appreciated the simplicity and elegance of the Copernican model and became an enthusiastic supporter.

In 1591, Kepler completed his master's degree and began theological studies, intending to become a Lutheran minister. Before he completed his training and final examinations, however, he was unexpectedly offered a teaching position in Graz, Austria.

Kepler's years at Graz were both challenging and rewarding. There he produced his first serious scientific book, *The Cosmic Mystery*, an ambitious (and faulty) geometric model of the solar system based on ancient Greek philosophical ideas. This book got the attention of several prominent astronomers and paved the way to Kepler's next step in life.

As the sixteenth century drew to a close, the political and religious conditions in Graz grew increasingly turbulent. Faced with severe religious persecution, Kepler accepted a position in Prague as an assistant to the great astronomer and imperial mathematician, Tycho Brahe.

Tycho had been making remarkably accurate observations of the stars and planets for 20 years. He wished to compile his data into tables that could be used for astronomical study and prediction. Impressed with Kepler's mathematical ability, Tycho was eager to add Kepler to his staff. While Kepler and Tycho had great difficulty getting along, each respected the other's ability. When Tycho died in 1601, Kepler continued Tycho's work and eventually published the *Rudolphine Tables* (an important astronomical reference used mainly by astrologers) in 1627.

Thanks to the quality of Tycho's data, Kepler was also able to improve the Copernican model of the solar system. After exploring numerous blind alleys, he finally arrived at the three laws of planetary motion familiar to every student of astronomy:

1. Planets move in ellipses with the sun at one focus.
2. An imaginary line from the sun to a planet sweeps over an equal area in equal time.
3. The square of a planet's orbital period is directly proportional to the cube of the planet's average distance from the sun.

Johannes Kepler died in 1630, but his reputation as a mathematician lives on in his three laws of planetary motion. Indeed, his third law provided the foundation for Newton's law of universal gravitation half a century later. A prominent crater on the moon has been named in his honor, and his native Germany has erected elaborate monuments in his memory. Kepler always remained a humble man, writing in a letter, "Let also my name perish if only the name of God the Father . . . is thereby elevated."

So if you double the distance between the center of the earth and the brick, the resulting gravitational force on the brick is only one-fourth of the original. For this reason, gravitational force drops off very rapidly with distance from the earth. A typical ninth grader weighs 602 N (135 lbf) at the earth's surface but less than 0.2 N (0.7 oz) at the distance of the moon's orbit!

Since gravity is a field force, it extends outward indefinitely. It is theoretically possible to calculate the weight of an object due to the earth's gravity at any distance from the earth. However, the gravitational force diminishes so rapidly with distance that human-sized objects beyond the orbit of the moon are essentially weightless with respect to the earth. At even farther distances, other gravitational forces, such as that from the sun, become more significant than the earth's gravity.

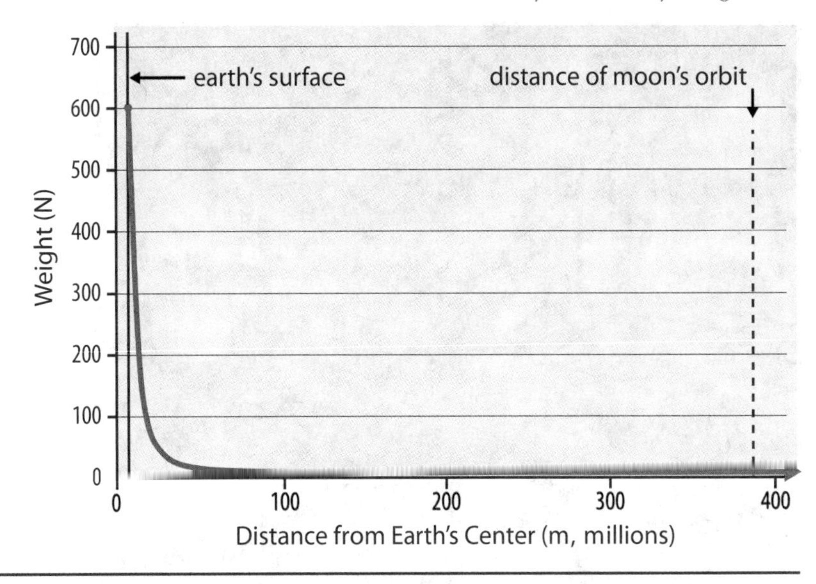

5-16 This graph shows how the weight of a 61 kg (600 N) student changes with distance from the earth.

5.10 *Free Fall*

For nearly 2000 years after Aristotle, teachers and students of science believed that the speed at which an object fell to the earth was proportional to its weight. They held Aristotle in such high regard that it wasn't until the sixteenth century that anyone actually tested his idea. Galileo was a professor of natural philosophy at the University of Pisa in the state of Tuscany, Italy, when he began to question this ancient view.

According to a popular story, Galileo dropped different-sized cannonballs off Pisa's famous leaning tower to disprove Aristotle's theory. It's more likely that this story was based on a "thought experiment" that Galileo described in his papers. He imagined two small equal weights dropped side by side from the same height at the same instant. According to Aristotle, they should hit the ground at the same time. Next, Galileo imagined connecting the two weights with a very light loose chain and then dropping the weights again, with the same results. In his mind, he repeatedly dropped the weights after shortening the chain for each drop. Eventually, the chain was so short that the two weights were essentially a single mass weighing twice as much as either original weight. But the doubled weight would fall in the same time as a single weight. This would disprove Aristotle's principle, which predicted that the doubled weight would fall a given distance twice as fast as the smaller weights.

In his actual experiments, Galileo rolled balls down graduated wooden ramps to "slow down" the rate at which balls fell. He observed that metal balls of different weights rolled down the ramp in the same amount of time. He also discovered that the balls accelerated as they rolled. The acceleration was the same, within his ability to measure, no matter what the size or weight of the ball. Following Aristotle's reasoning, the larger, heavier balls should have rolled to the bottom of the ramp first since they were more "motivated" to reach the earth's center than lighter balls.

5-17 Galileo's "thought experiment"

5-18 Galileo used long ramps to "slow down" the falling motion of objects. He correctly realized that rolling objects on a smooth ramp behaved the same way as objects falling freely did.

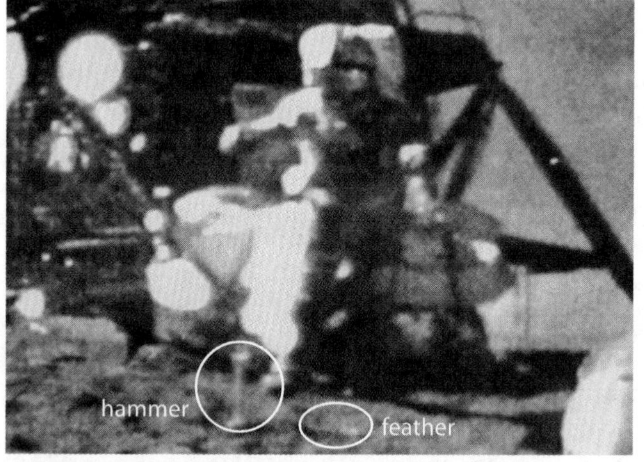

5-19 In August 1971, astronaut David Scott, the commander of the Apollo 15 moon mission, demonstrated ideal free fall in a vacuum using a hammer and a feather from a falcon, the US Air Force's mascot. He dropped the feather and the hammer before the watchful eye of a television camera broadcasting to an audience of millions. In exactly 1.33 s, the feather and hammer hit the surface of the moon together, proving Galileo's hypothesis.

Extraterrestrial g's

The value of g we provide here is the surface gravity for Earth. Other planets and moons have quite different surface gravities. Their surface gravity symbols often include a subscript to identify the body to which the value applies. For instance, the average surface gravity of Mars (g_{Mars} or $g_{♂}$) is 3.71 m/s² or 0.38 g.

Galileo also observed that the larger an object's surface area was in comparison to its weight, the more air resistance affected its motion. You can demonstrate this yourself by using two sheets of paper in a drop test. Crumple one of the sheets into a compact wad. Drop the two pieces of paper at the same time with the uncrumpled sheet horizontal to the floor. The crumpled wad will land first because its surface area compared to its weight is much smaller than the surface area of the flat sheet having the same weight. *Air resistance*—friction on an object moving through the air—significantly affects the rate of fall of bulky, low-density objects.

Objects that fall due to gravity alone with no other forces acting on them are in **free fall**. An object can experience this condition only in a complete vacuum where there is no air resistance. Objects in free fall change speed with a constant acceleration, regardless of their masses. Objects that have more mass have greater gravitational force acting on them. But because they have more mass, they also have more inertia. A larger force acting on an object with a larger inertia produces the same acceleration as a smaller force acting on an object with a smaller inertia. The two effects exactly cancel each other out.

5.11 Gravitational Acceleration and Weight

Careful experimentation at many locations all over the earth has demonstrated that the average *gravitational acceleration* (g) at sea level is about 9.8 m/s² (32 ft/s²). This value is acceptable for any place on the surface of the earth. The discovery that all dense objects accelerate at the same rate regardless of their masses has important practical applications in physics.

Let's look again at Newton's second law to see how this information can be useful. We know that the earth exerts a force on all objects at its surface. We call this force *weight* (w). We know that all objects have mass (m). And we know that all objects have the same gravitational acceleration (g) on earth. When you substitute these quantities for their equivalent symbols in the formula for Newton's second law, $F = m \times a$, you get

$$\boxed{w = m \times g}$$

where

- w is the weight of an object, in newtons,
- m is the mass of the object, in kilograms, and
- g is gravitational acceleration (a constant), in meters per second per second.

The product $m \times g$ results in the SI unit kg·m/s², or newtons.

EXAMPLE PROBLEM 5-2

Calculating Weight Using Newton's Second Law

The Golden Jubilee Diamond is currently the world's largest cut diamond. Its mass is 545.67 carats, or 109.13 g. What is the weight of the diamond in newtons?

Known:	mass (m) = 109.13 g = 0.109 13 kg
	gravitational acceleration (g) = 9.8 m/s²
Unknown:	weight (w)
Required formula:	$w = m \times g$
Substitution:	$w = (0.109\ 13\ \text{kg})(9.8\ \text{m/s}^2)$
Solution:	$w \approx 1.\underline{0}6\ \text{kg}\cdot\text{m/s}^2$
	$w \approx 1.1\ \text{N}$ (2 SDs allowed)

5-20 The Golden Jubilee Diamond

The weight formula derived from Newton's second law is a convenient way to calculate the weight of any object quickly, given its mass. Recall that a law simply describes relationships among measurable quantities; it doesn't try to explain them. So don't make the mistake of assuming that an object must be accelerating in order to have weight.

5.12 Free Fall and Distance

Imagine that you are a missionary radio technician making repairs to the top of a 54.9 m radio tower on the island of Carriacou in the Caribbean. You accidentally drop a bolt. How far does the bolt move each second it falls? How fast is it falling when it hits the ground? Answers to these questions involve the dynamics of free-falling objects.

Gravitational acceleration near the earth's surface is 9.8 m/s²; that is, the bolt's downward speed increases 9.8 m/s for every second of free fall. After the first second, the bolt would be falling 9.8 m/s. After two seconds, it would be falling 19.6 m/s (2 s × 9.8 m/s²); after the third second, it would be falling 29.4 m/s (3 s × 9.8 m/s²). The increase of speed follows the formula for free fall,

$$\boxed{v = g \times \Delta t},$$

where

- v is the free-fall speed, in meters per second,
- g is the gravitational acceleration, in meters per second per second, and
- Δt is the total elapsed time of free fall, in seconds.

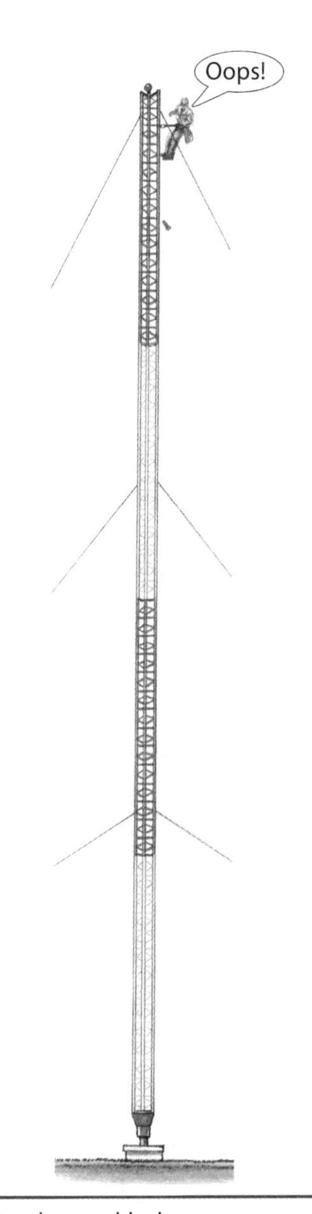

5-21 The dropped bolt

Knowing how fast the bolt is falling allows you to calculate how far it has fallen in a given amount of time. You can calculate the distance traveled by a moving object using the familiar equation, distance equals rate multiplied by the time traveled, or

$$d = v_{\text{avg}} \times \Delta t,$$

where

- d is the distance, in meters,
- v_{avg} is the average speed, in meters per second, and
- Δt is the elapsed time interval, in seconds.

Average Speed

$$v_{\text{avg}} = \frac{v_i + v_f}{2}$$

(valid only for constant acceleration)

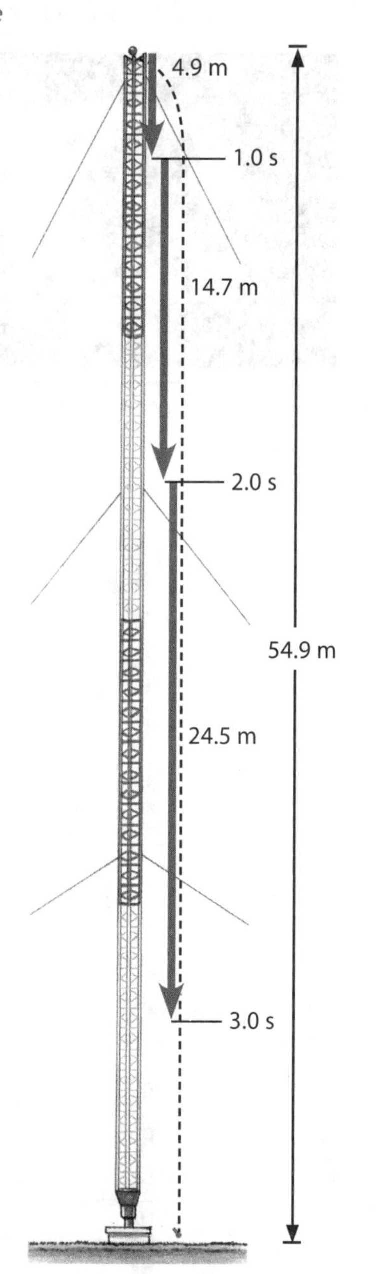

5-22 The bolt falls a greater distance with each second of drop.

When an object has constant acceleration, as in free fall, its average speed during an interval is just the sum of the initial and final speeds in the interval divided by 2.

During the first second of free fall, the bolt's initial speed is 0 m/s and it accelerates to 9.8 m/s. Its average speed during the first second is (0 m/s + 9.8 m/s)/2, or 4.9 m/s. The bolt falls 4.9 m/s × 1 s, or about 4.9 m in the first second. Between one and two seconds, the bolt's average speed is (9.8 m/s + 19.6 m/s)/2, or 14.7 m/s. The bolt falls 14.7 m during the interval 1 s to 2 s, for a total drop of 19.6 m. You can do the same analysis for every 1 s interval during the fall of the bolt, ignoring air resistance. The total distance the bolt travels is the sum of the distances covered in all the 1 s intervals. The table below summarizes this problem-solving approach. Since the radio tower is only 54.9 m tall, the bolt will hit the ground sometime between 3 s and 4 s after the technician drops it.

When objects accelerate at a constant rate, as they do in free fall, they experience *uniform acceleration*. Galileo observed that the distance an object moves during uniform acceleration is proportional to the square of the time interval. Physicists since his time have confirmed his observations and have developed a formula that relates distance, acceleration, and time:

$$d = \tfrac{1}{2}a \times (\Delta t)^2.$$

This formula is valid only for cases of uniform acceleration such as gravity. For an object in gravitational free fall, the formula becomes

$$\boxed{d = \tfrac{1}{2}g \times (\Delta t)^2},$$

where g is the constant acceleration 9.8 m/s². This formula is valid for any object starting from rest, dropping vertically in free fall, if air resistance is not significant.

5-1 Free Fall Data

Time (t_n) n is the time interval number	Time interval (Δt)	Speed ($v = g \times \Delta t_n$)	Average speed during interval (v_{avg})	Interval distance ($d = v_{avg} \times \Delta t$)	Total distance (d_T)
$t_0 = 0$ s	—	0.0 m/s	—	—	—
$t_1 = 1$ s	1 s	9.8 m/s	4.9 m/s	4.9 m	4.9 m
$t_2 = 2$ s	1 s	19.6 m/s	14.7 m/s	14.7 m	19.6 m
$t_3 = 3$ s	1 s	29.4 m/s	24.5 m/s	24.5 m	44.1 m
$t_4 = 4$ s	1 s	39.2 m/s	34.3 m/s	34.3 m	78.4 m

Example Problem 5-3

Calculating Distance in Free Fall

How far will an object starting from rest drop in free fall in 3.346 s?

Known: time interval (Δt) = 3.346 s
$g = 9.8\ \text{m/s}^2$

Unknown: distance (d)

Required formula: $d = ½g \times (\Delta t)^2$

Substitution: $d = ½(9.8\ \text{m/s}^2)(3.346\ \text{s})^2$

Solution: $d = ½(9.8\ \text{m/s}^2)(11.1\underline{9}5\ \text{s}^2)$
$d \approx 5\underline{4}.9\ \text{m} \approx 55\ \text{m}$ (2 SDs allowed)

This distance just happens to be the height of the radio tower with which we began this subsection. The time interval 3.346 s is how long the machine bolt would take to fall to the ground from the top of the missionary radio tower.

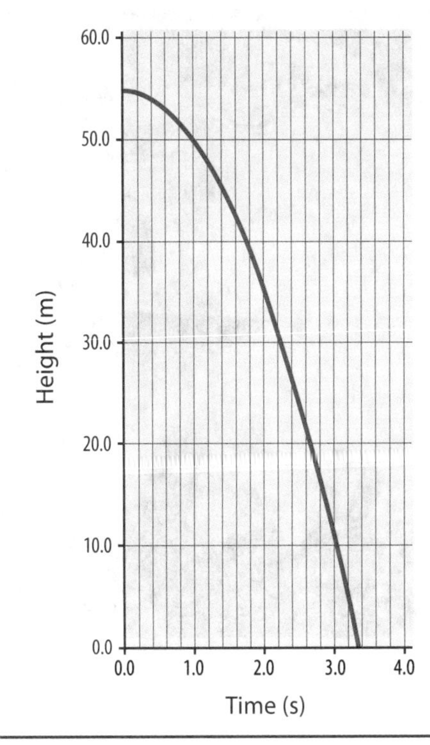

5-23 The distance an object in free fall travels is proportional to the square of the time it has fallen.

5.13 Free Fall and Air Resistance

In our discussion of free fall, we have neglected air resistance to simplify learning the concept of free fall. But in the real world, falling objects *do* encounter air resistance. Though air resistance complicates our scientific formulas, it isn't all bad. In fact, *no* air resistance would generally cause problems.

Air resistance is a form of friction called **drag**. As a falling object's speed increases, drag on the object increases, exerting a force opposite to the direction of its fall. This force opposes the force of gravity on the object (its weight). This reduces the net force accelerating it toward the ground. The falling object still accelerates but at a decreasing rate. Eventually, if the object can fall far enough, the drag increases to the point that it equals the object's weight—the net force on the falling object is zero. According to Newton's first law, when the net force is zero, its acceleration is also zero. With zero acceleration, the object's speed is constant for the remainder of its fall. Physicists call this constant falling speed *terminal velocity*.

An object falling at its terminal velocity is in mechanical equilibrium.

The value of terminal velocity depends mainly on the object's cross-sectional area compared to its mass. A small, compact, dense object has a much higher terminal velocity than a broad, light, low-density object. These factors greatly affect things designed to fall through the air, like parachutes and ballistic missile warheads. The purpose of a parachute is to slow a falling human or space probe to a speed that will avoid injury or damage on landing. The terminal velocity of a parachutist needs to be very low. The chute is made of light but strong materials to provide the necessary amount of drag. On the other hand, engineers design a ballistic missile warhead for exactly the opposite purpose. Heavy, extremely dense, and having a narrow conical shape, it produces the smallest drag possible. It has to fall through the atmosphere from the edge of space quickly to minimize the effects of wind and to avoid interception by antiballistic missile defenses.

5-24 Parachutists (left) depend on air resistance; warhead reentry (right) is opposed by it.

5-25 Elisha Otis demonstrating his safety elevator in 1854 at the New York World's Fair. Don't try this at home!

5.14 Using Newton's Laws to Solve Problems

How can we use man-made forces to transport people and things between floors of tall buildings against the force of gravity? The obvious answer is the elevator. Elevators have been in use since ancient times, including at the Coliseum and other amphitheaters in Roman times. Animals, water, or even people powered these elevators. With buildings growing taller in the 1800s, people needed a reliable elevator to move loads between floors safely. Early elevators were dangerous because they had no safety mechanism in case the supporting cable broke.

Elisha Otis solved this problem in 1852. He worked for a New York furniture company that manufactured beds. His company needed an elevator to hoist heavy equipment safely to the upper floors of its factory. He invented a device that stopped the downward motion of an elevator in the event that its cable or lifting machinery broke. His first device consisted of mechanical latches that engaged rails on both sides of the car whenever the cable failed. This invention prevented loss of equipment and, even more importantly, loss of human life.

He showcased his invention at the 1853 New York World's Fair by staging a demonstration at the New York Crystal Palace. He rode his elevator to a dizzying height and then had a helper cut the supporting cable with an axe! The gathered crowd gasped and then applauded as the safety mechanism locked into place, stopping the elevator's fall. He installed his first passenger elevator a few years later in a New York City department store.

5C Section Review

1. What three quantities determine the force of gravity between two objects?
2. How does the gravitational force between two objects relate to each of the quantities given in Question 1?
3. Under what conditions is an object in theoretical free fall? Can a stone falling from a cliff in the Grand Canyon meet these conditions? Explain.
4. ✪ Compare the weight formula $w = m \times g$ with the formula for Newton's second law, $F = m \times a$. What arc the similarities and the differences between them?
5. How does the speed of a dropped object in free fall near the surface of the earth relate to the time it has fallen?
6. Discuss how air drag reduces the acceleration of a falling object.
7. (True or False) A dropped object will fall half as fast as an object twice as heavy dropped at the same time.
8. ✪ What is the weight, in newtons, of a 442 g World Cup soccer ball? Convert this weight to pounds-force. (The conversion factor is about 0.22 lbf/N.)

Chapter Review

1. Completely describe a force, including its definition, SI unit, its symbol, and how it is illustrated in a diagram.
2. List the four mechanical (contact) forces and give an example of each.
3. Do all field forces visibly affect all matter? Explain.
4. Identify the external forces acting on a dinner plate resting on a table. Why does the plate not move?
5. What is the tendency for matter to continue in its current state of motion? What quantity is a measure of this tendency?
6. Use Newton's second law of motion to prove his first law.
7. According to Newton's third law of motion, if you exert 20 N of force on a lump of bread dough while kneading it, what else must be true?
8. A farmer pulls on his obstinate mule with 250 N of force to the right. The ground exerts a reaction force to the mule's resistance of 250 N to the left. What is the net force on the mule system? What is the mule's acceleration?
9. Under what condition of motion is the friction between two objects or materials that are touching generally greatest?
10. How does distance between the centers of mass of two objects affect the gravitational force between the objects?
11. Why did Galileo roll balls down a ramp rather than simply dropping them as he investigated inertia?
12. In order for you to use the distance formula $d = \frac{1}{2}a \times (\Delta t)^2$, what assumption must you make? How does this assumption apply to free-falling objects?
13. Explain the main function of a parachute from a physics point of view.

True or False

14. Physics has explained the why and how of all natural forces.
*15. Gravity is the strongest of the four fundamental forces.
16. A field force can produce a mechanical force in a system.
17. You wind up the rubber band propeller of a toy airplane and let it go. The rubber band exerts an external force on the airplane system.
18. No one has actually observed that an object will continue to move indefinitely unless acted on by an outside force.
19. In Newton's second law, the net force on a system is the cause and the acceleration of the system is the effect.
20. Friction is proportional to the surface area of contact between two objects.
21. According to the law of universal gravitation, the weight of a brick held at 6 ft above the ground will be one-fourth its weight when it is held only 3 ft above the ground.

Scientifically Speaking

22. After being dropped, an object falling for 4 s will have traveled 16 times the distance of the same object after having fallen for only 1 s.

Questions 23–26 concern the motion of an elevator. Use the information given in the earlier questions to help answer the later ones.

DS ✪ 23. An elevator rises at a constant speed of 2 m/s.

a. Is the elevator accelerating?

b. What can you say about the net force acting on the elevator? What is the justification for your answer?

DS ✪ 24. The elevator and passengers have a mass of 5500 kg. After a stop at the fifth floor, the elevator accelerates upward uniformly at 1.0 m/s^2 until it is rising at 2.0 m/s. What is the magnitude of the net force acting on the elevator as it accelerates?

DS ✪ 25. Calculate the weight of the elevator. In which direction does this force vector act on the elevator?

DS ✪ 26. Just as the last passenger steps out of the elevator at the thirteenth floor, a malfunction occurs and it begins dropping toward the ground floor in free fall. Ignoring friction (and the fact that elevators are designed to avoid this type of accident), what is its downward acceleration? How fast is it moving after 2.5 s?

ENERGY

CHAPTER 6

DOMINION SCIENCE PROBLEM

The Urban Heat Island Effect

Blazing summer sun, the exhaust of many cars, scorching hot asphalt and concrete—these are the summertime conditions of American cities. But step into a nearby alfalfa field and things become much cooler, as much as 6 °C (10 °F) cooler. Climate scientists call this difference in temperature between urban areas and the rural areas that surround them the *urban heat island effect*. The graph below illustrates how temperatures in a city and its surrounding rural areas can be so different. In an average year, more people die from extreme heat than from any other weather-related condition, including lightning, tornadoes, hurricanes, and floods. How can we make our cities cooler?

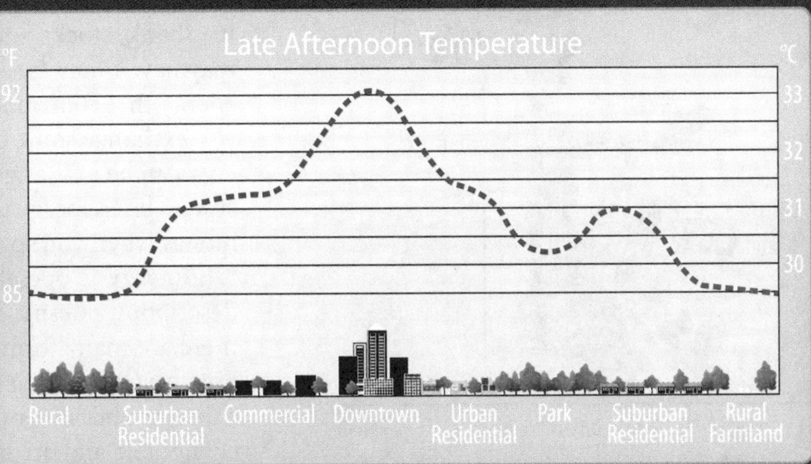

6-1 Cities are islands of extreme temperatures in the summer.

6A Objectives

After completing this section, you will be able to

✓ discuss the importance of energy.

✓ state what energy can do and the unit in which energy is measured.

✓ compare and contrast potential and kinetic energy.

6A The Nature of Energy

6.1 *Energy's Importance*

People talk a lot about energy. Ever since the great gas shortage of the early 1970s, people have realized that there is no cheap, convenient, inexhaustible source of energy. Technologically advanced nations, including the United States, gobble up energy at ever-increasing rates, while developing countries in Asia, Africa, and South America have their own increasing energy needs. Oil-rich Middle Eastern nations use their resources for political purposes by threatening to cut off exports to more powerful enemies. Nations make "energy independence" an economic and political goal. So why is energy so important?

6-2 Control of energy sources such as oil (left) and nuclear power (right) is a matter of national and international politics.

Energy moves everything. Humans, animals, and plants require energy to live. Energy drives all the processes in nature that recycle water and other forms of matter. Wind and ocean currents circulate because of energy. Internal and external sources of energy heat the earth. Humans use many forms of energy for building, communicating, exploring, and conducting warfare. Without usable energy, life would not be possible and the universe would be a vast, dark, cold, empty place.

The universe would cease to exist without energy. Just as space, matter, and time could not exist without each other, so matter could not exist without energy. Energy holds matter together. In fact, matter and energy are different forms of that which makes up the universe!

6.2 *What Is Energy?*

Energy cannot be defined any more easily than the other fundamental properties of creation. No one knows what energy is—we know only what it does or has the potential to do. It is a property of matter that can sometimes be measured, although measuring some forms of energy is very difficult or impossible with our current level of technology.

Energy is a positive scalar quantity measured in joules (J).

$$1\ \text{J} = 1\frac{\text{kg}\cdot\text{m}^2}{\text{s}^2}$$

To understand some aspects of the nature of energy, you can compare it to wealth. Having a lot of energy is like having money invested in a bank, stocks and bonds, property, and possessions. Just as you may have many forms of wealth, a system may have many forms of energy that contribute to its total energy. When you buy something, you exchange some of your money for something else. Similarly, energy can be exchanged for work, matter, or other forms of energy. A system loses energy to its surroundings in the same way that you lose money when you spend it, give it away, or it is stolen.

Energy is operationally defined as the ability to do work. This description doesn't *define* energy; it just says what it can do. Just as there are many forms of energy, there are many forms of work. For now, we will say that *work* is done whenever an object moves through a distance because of a force acting on it. Work and energy are interchangeable and are measured in the same unit, i.e., the joule. When the surroundings do work on a system, the system gains energy.

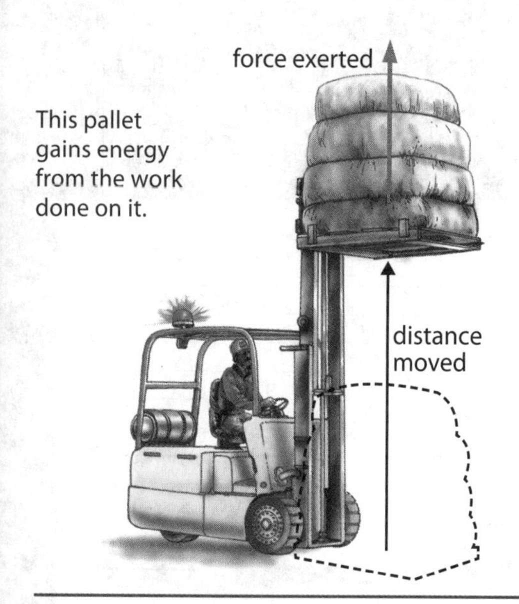

6-3 Work is done when a force acts on an object as it moves through a distance. Work and energy are interchangeable.

When a system does work on its surroundings, the system loses energy. Work is discussed in detail in Chapter 7.

6.3 Potential and Kinetic Energy

Energy can be broadly divided into two main kinds—potential energy and kinetic energy.

Potential Energy

Potential energy (PE) is sometimes called the *energy of position* or the *energy of condition*. A system's potential energy can depend on its position and a force that has the potential to move the system from its current position. Some contact forces (except friction) and all field forces can generate potential energy in susceptible materials and objects. For example, a magnet can produce magnetic potential energy in a steel paper clip holding papers together. The paper clip can't move, but it has the potential to.

Potential energy can also exist as a consequence of the composition or condition of the system. This principle is particularly true at the particle level. Forces in chemical bonds store energy that can do work under the right conditions. Gasoline molecules contain chemical potential energy that is released during combustion—a change of chemical condition.

6-4 This rock has the potential to fall under the force of gravity. It has a lot of potential energy.

Kinetic Energy

The **kinetic energy (KE)** of a system is its *energy of motion*. All moving matter, regardless of size, has kinetic energy. Kinetic energy depends only on the system's mass and speed. The more mass an object has and the greater its speed, the greater its kinetic energy.

Systems may have both potential and kinetic energy at the same time. In some situations, a system's total energy can be converted back and forth between potential and kinetic energy. The following sections discuss various forms of energy, how energy is converted, and how total energy is conserved.

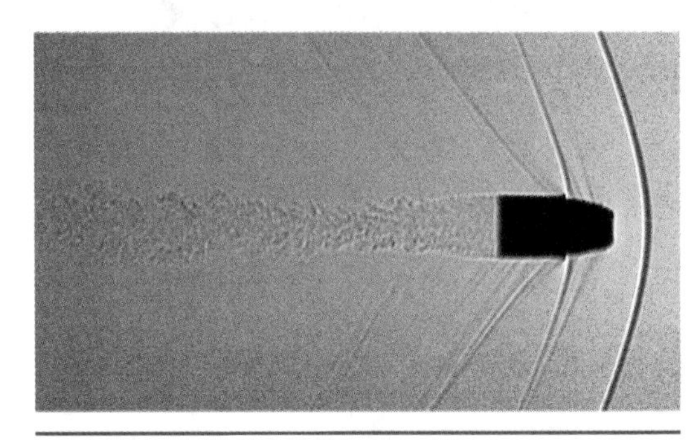

6-5 A traveling bullet has a high speed and thus much kinetic energy.

6A Section Review

1. Why is energy important?
2. How do we describe energy? What is the SI derived unit for energy?
3. What three things can we obtain from a given form of energy?
4. What is the potential energy of a system? Give an example of two kinds of potential energy.
5. What is the kinetic energy of a system? On what factors does a system's kinetic energy depend?
6. (True or False) Energy is present wherever there is matter.

6B Objectives

After completing this section, you will be able to

- ✓ list the nine forms of energy given in the text.
- ✓ briefly discuss the nature of each kind of energy and state man-made and natural sources for each.
- ✓ compute an object's gravitational potential energy (GPE) and kinetic energy (KE).
- ✓ explain the relationship between matter and energy.

6-6 Richard Feynman (1918–88) was one of the greatest American physicists of the twentieth century. He expanded our understanding of quantum physics and worked on the first atomic bomb, receiving the Nobel Prize in Physics in 1965. Feynman was a member of the presidential committee that investigated the loss of the space shuttle *Challenger*, which occurred in 1986.

mechanical: (Gk. *mekhane*—machine). Mechanical in this sense references mechanics, the modern study of motion (see Chapters 4 and 5).

6B Classification of Energy

6.4 *Identifying Forms of Energy*

We use energy in many different forms. Some scientists and engineers devote their lives to discovering how to produce, exchange, and use energy more efficiently. We can classify energy by how we sense it or by its principal source. Many forms of energy come from the different ways that matter possesses kinetic and potential energy.

The famous physicist Richard Feynman categorized energy into the following divisions: gravitational (a form of mechanical potential energy), kinetic, heat (a form of kinetic energy), elastic (a form of mechanical potential energy), electrical, chemical, radiant, nuclear, and mass energy. However, there are many other ways to classify energy. For the purposes of this text, we combine similar types of energy and identify others by their sources. They are mechanical, thermal, acoustic, electrical, magnetic, radiant, chemical, nuclear, and mass energy.

6.5 *Mechanical Energy*

Systems that are physical objects can have **mechanical energy**. Mechanical energy consists of mechanical potential energy and kinetic energy.

Mechanical Potential Energy

Mechanical potential energy is the energy of an object due to its physical position. Specifically, potential energy exists when a force acting on an object has the potential to move it. Potential energy is proportional to the size of the force and to the distance that the object could possibly move when pushed or pulled by the force. We call the position from which we measure this distance the *zero reference position*. The farther the object is from this position toward which the force tries to move it, the greater the object's mechanical potential energy.

A common example of mechanical potential energy is *gravitational potential energy (GPE)*. The gravitational force acts on objects vertically by pulling them downward toward the center of the earth. The zero reference position is below the object, and we usually choose it to be the ground or some other horizontal surface. Every object subject to gravity has the potential to fall to a lower height unless it is already resting at its zero reference height.

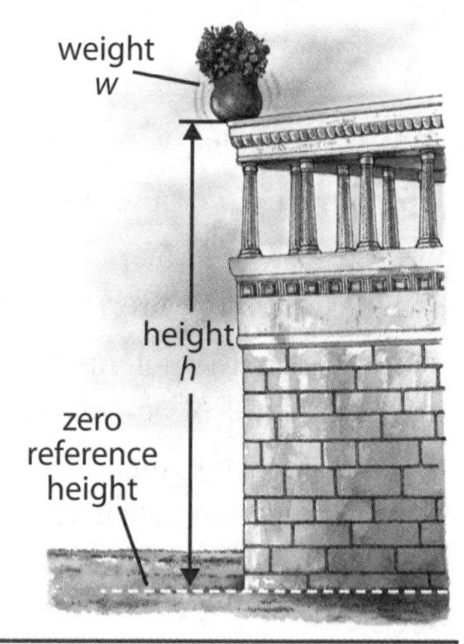

6-7 This vase has GPE when its height is measured from the ground. Its GPE is zero when its height is measured from the railing.

Gravitational potential energy can be easily calculated if one knows the object's weight and its height above the zero reference height. The formula is

$$\boxed{\text{GPE} = w \times h},$$

where

- w is the object's weight ($w = m \times g$), in newtons (1 N = 1 kg·m/s^2), and
- h is the object's height above the reference height, in meters.

Gravitational potential energy may also be calculated by the formula

$$GPE = m \times g \times h.$$

EXAMPLE PROBLEM 6-1

Gravitational Potential Energy (GPE)

What is the potential energy of Isaac Newton's 0.45 kg apple hanging from a tree at a height of 5.3 m above the ground?

Known: apple's mass (m) = 0.45 kg
apple's height (h) = 5.3 m
gravitational acceleration (g) = 9.8 m/s^2

Unknown: apple's GPE

Required formula: $GPE = w \times h = m \times g \times h$

Substitution: GPE = (0.45 kg)(9.8 m/s^2)(5.3 m)

Solution: GPE ≈ 2$\underline{3}$.3 kg·m^2/s^2
GPE ≈ 23 J (2 SDs allowed)

6-8 Tradition says that Sir Isaac Newton gained inspiration for his law of gravity from observing an apple drop from a tree.

The GPE of an object changes as its height changes. A falling object rapidly loses GPE as gravity does work on the object. A rocket gains GPE as its engines do work to lift it against gravity.

Elastic potential energy (EPE) is another example of mechanical potential energy. Elastic potential energy is produced in a system or material affected by an elastic force. In physics, *elastic* means that a system returns to its original form or state after its shape is stretched or compressed by an outside force. Springs, rubber bands, and similar materials have elastic properties. The contact force that an elastic material exerts varies in relation to the change of its shape. The more a rubber band is stretched, the stronger it will snap back. The force a spring produces is related to how much the spring is stretched or compressed. A spring's zero reference position is the spring's length in its completely relaxed condition. The farther from its relaxed position it is stretched, the greater the force it exerts.

If a spring is attached to a system, such as a screen door, it generates EPE in the door as the door is opened and the spring is stretched. When the door is released, the spring closes the screen door, converting the EPE to work by moving the door.

6-9 These huge springs contain elastic potential energy. They support the former Cheyenne Mountain Command Center to protect the structure against nuclear weapon blasts.

Kinetic Energy

Kinetic energy is mechanical energy by definition because it involves the physical movement of matter. The formula for kinetic energy is

$$\boxed{KE = \tfrac{1}{2}m \times v^2},$$

where

- m is the mass of the system or object, in kilograms, and
- v is its speed, in meters per second.

Note that in the formula the speed is squared; therefore, changing speed has a much greater effect on kinetic energy than changing mass does.

6-10 Kiriani James, the world's fastest 400-meter sprinter in the 2012 Olympics

EXAMPLE PROBLEM 6-2

Calculating Kinetic Energy

What is the kinetic energy of a 75 kg sprinter running at 8.7 m/s?

Known: sprinter's mass (m) = 75 kg
sprinter's speed (v) = 8.7 m/s

Unknown: sprinter's kinetic energy (KE)

Required formula: $KE = \frac{1}{2}m \times v^2$

Substitution: $KE = \frac{1}{2}(75\ kg)(8.7\ m/s)^2$

Solution: $KE \approx 2\underline{8}30\ kg \cdot m^2/s^2$
$KE \approx 2800\ J$ (2 SDs allowed)

External forces that increase or decrease an object's speed do work on the entire object and thus change its kinetic energy. The kinetic energies of large objects and individual isolated particles are easily calculated using the kinetic energy formula. The kinetic energies of atoms or molecules of objects and substances at rest are difficult to measure for several reasons: these particles are extremely small, are very numerous, and vibrate and move at different speeds. Therefore, scientists must use a different method to measure their kinetic energies.

6.6 Thermal Energy

The motion of particles in matter due to their thermal energy is called *thermal motion.*

thermal: therm- (Gk. *thermē*—heat) + -al

The kinetic-molecular model of matter states that matter consists of innumerable tiny particles in constant, random motion. Each particle, no matter its state of matter, moves at some speed at any given moment. Thus, every particle of matter has kinetic energy. The average sum of the kinetic energies of all the particles in an object is its **thermal energy**.

Theoretically, all particle motion ceases at absolute zero (0 K or about −273 °C). However, absolute zero cannot be reached, so all matter has some particle motion and therefore thermal energy. It is impossible with our current technology to measure the average kinetic energy of particles in common objects at any given moment. Since thermal energy can move across a system's boundaries, it is much easier for us to measure the *changes* of thermal energy in a system, which are indicated by changes in its temperature. The movement of thermal energy into or out of a system is called *heat*. The process of gaining thermal energy is called *heating*, and the process of losing thermal energy is called *cooling*. Thermal energy, heat, and temperature are discussed further in Chapter 9.

6-11 A glowing strip of metal contains a large amount of thermal energy.

The principal sources of thermal energy on our planet are the sun and the earth itself. Thermal energy from the sun maintains the earth's surface within habitable temperature limits. The earth's internal heat source keeps the planet from being an icy cold cinder like many of the asteroids. Humans have discovered ways of converting other forms of energy found on and in the planet into thermal energy for warmth and industrial purposes.

6.7 Acoustic Energy

Unlike thermal energy, which involves the random vibration and motion of particles of matter, acoustic **energy** is the transmission of energy through matter by particle oscillations that occur in specific directions. Acoustic energy differs from thermal energy in several important respects:

- Acoustic energy, when present, must *always* move through matter; it cannot move through a vacuum (a region lacking matter). In contrast, thermal energy is always present in matter, but it moves or is exchanged only between points of different temperatures.
- Acoustic energy causes particles to move in *periodic motion* consisting of back-and-forth vibrations. The periodic motion of particles occurs as their kinetic and potential energies are repeatedly exchanged. Thermal particle motion is random in direction and duration.
- When acoustic energy is transferred, the distance that the particles move is usually much larger than the size of the particles. Thermal particle vibrations, in contrast, are approximately the size of the particles.

The way acoustic energy moves is often described as waves, but acoustic waves are different from water waves. (See Chapter 12 for a comparison of various kinds of waves.) Sources of acoustic energy vary in many ways. Familiar sources include earthquakes, explosions, bass speakers, submarine sonar equipment, human vocal cords, piccolos, jet engines, dog whistles, bats, and medical ultrasound diagnostic equipment. Acoustic energy and sound waves are discussed further in Chapter 13.

acoustic (uh KOO stik): (Gk. *akoustikos*—related to hearing, sound)

oscillation, *from* oscillate (AHS suh LATE): oscill- (L. *oscillum*—swing) + ate. Oscillation is a repetitive, back-and-forth motion, as in a swing.

Periodic refers to the *period* of time needed to complete one full cycle of a repeated motion or process.

6-12 You can hear bells only because acoustic energy moves through air and matter.

6.8 Electrical Energy

One of the single most important discoveries in physics was the discovery of electrical charges in matter. All atoms consist of even smaller particles that either are neutral or carry a positive or negative charge. Electrical charges exert field forces on their surroundings. It is these forces acting on other electrical charges that can do work as **electrical energy**. Two objects carrying the same charge repel each other. Objects with opposite charges attract. The most common everyday source of electrical energy is found in the excess, the lack, or the movement of electrons.

Useful sources of electrical energy are almost entirely man-made. Electrical batteries, generators, photovoltaic (solar) cells, and radiothermal generators are a few of the devices that store and produce electrical energy. Natural sources include lightning, electric rays and eels, and the immense electrical currents circulating in the earth's magnetic fields in space. The properties and use of electrical energy are discussed further in Chapter 10.

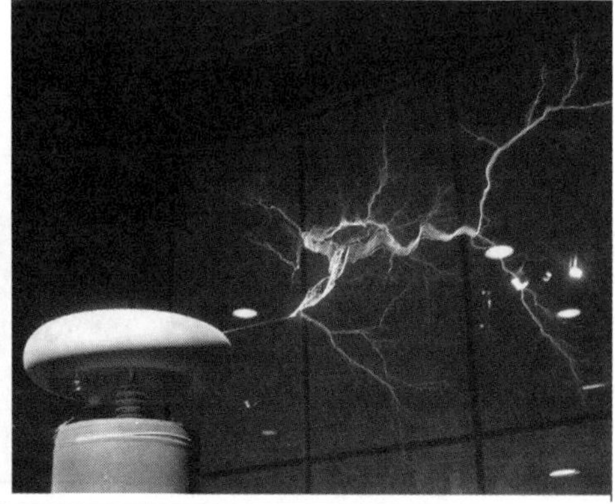

6-13 Electricity is an essential part of modern life. An electrical discharge is a dramatic demonstration of its potential to do work.

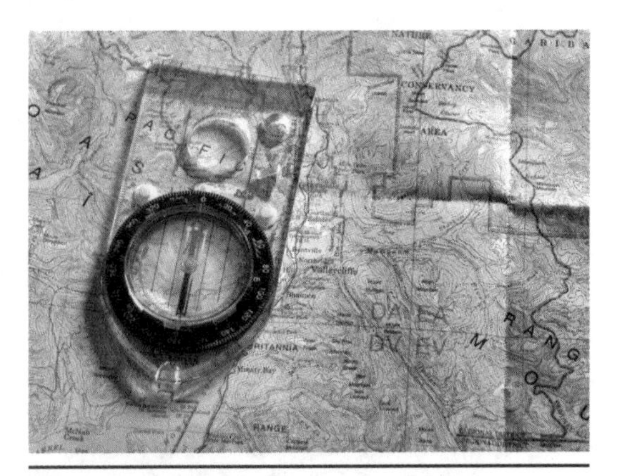

6-14 Magnetic energy has long been used for navigation.

6.9 Magnetic Energy

Magnetism was originally thought to be a mysterious and occult (hidden or magical) property of matter because seemingly ordinary matter can exert a force over a distance to attract or repel objects. For instance, magnetic compass needles unerringly point north. As with electrical energy, magnetism exerts a field of force. **Magnetic energy** is the ability to do work through the influence of a magnetic field. Magnets can exert an attractive or repulsive force on a magnetic object depending on its material and orientation. While an object may have either a positive or a negative electrical charge, all magnetic objects have both a north and a south pole.

Natural sources of magnetism include certain kinds of rocks, the earth itself, and many of the largest bodies in our solar system. Man-made sources of magnetic energy include magnets and magnetic materials such as computer hard drives. Any object, such as a wire, that is conducting electricity also generates a magnetic field. Magnetism and magnetic energy are discussed further in Chapter 11.

6-15 Intense laser light represents just a small portion of the entire electromagnetic spectrum.

6.10 Radiant Energy

Magnetism in matter ultimately comes from the order and motions of electrons within atoms. In fact, the motion of electrons is the source of all magnetic fields. A rapid change in a magnetic field around a wire can cause the acceleration of electrons. This discovery led physicists to conclude that electricity and magnetism are two forms of the same energy, today called **electromagnetic energy**. This energy, in various forms, can be radiated by matter and transmitted through a vacuum; therefore it is also called **radiant energy**.

Visible light is the most familiar form of radiant energy. Other types of this energy include ultraviolet light, radar, radio waves, x-rays, microwaves, infrared light, and gamma rays. These types of radiant energy are naturally emitted from astronomical objects such as stars, black holes, pulsars, and nebulas. Artificial sources of radiant energy are used for communication, illumination, imaging, and medical purposes. Chapter 14 discusses all forms of electromagnetic energy.

6-16 High-energy chemical bonds are formed by the process of photosynthesis.

6.11 Chemical Energy

Atoms in matter are usually bonded together as elements or compounds. The electromagnetic forces that hold atoms together are the source of the potential energies shared among the atoms. Bonds break and are remade as atoms rearrange themselves during chemical changes. These rearrangements release or absorb energy in various forms. Potential energy stored in chemical bonds is called **chemical energy**.

The amount of energy in chemical bonds depends on the kinds of atoms and the way they are bonded together. In most chemical changes, chemical energy is released as thermal, radiant, or acoustic energy. An important exception to this rule is the chemical change of photosynthesis. Radiant energy from the sun is stored in the chemical bonds of sugar molecules and other materials in green plants. When an animal eats and digests the plant matter, the chemical bonds are broken and the energy is released for the animal's life processes. Chemical bonds and chemical reactions are discussed in Chapters 18 and 19.

6.12 Nuclear Energy

As its name implies, **nuclear energy** is associated with the nucleus of the atom. Physicists discovered early in the twentieth century that the mass of an atom's nucleus was less than the sum of the individual masses of its protons and neutrons. The difference in mass, called the *mass defect*, was exactly equal to the energy required to hold the nucleus together (i.e., *binding energy*) against the extremely strong repulsion between the protons (remember that like charges repel). How is mass equal to energy? Recall Einstein's famous equation, $E = m \times c^2$, where m is the mass of matter, c^2 is the speed of light squared, and E is the total energy released if the matter were converted entirely to energy. He showed that energy and matter are equivalent to each other.

6-17 Nuclear energy is released when protons and neutrons in nuclei are separated or combined.

Nuclear energy is released in two different ways. Large nuclei with many protons and neutrons can be split into two or more smaller nuclei in a process called *fission*. The sum of the masses of the smaller nuclei is less than the mass of the original nucleus. The difference in mass is converted to different forms of energy (thermal, radiant, etc.) according to Einstein's equation. Nuclear energy is also released when small nuclei are smashed together to form larger ones in a process called *fusion*. The binding energy for the particles in the larger nucleus is released as huge quantities of energy in various forms.

Nuclear energy is liberated in man-made nuclear fission reactors for electrical power generation and research. It is also explosively generated in fission and fusion bombs, which are extremely destructive. A natural site of fission was found in Africa, where concentrated deposits of uranium experienced nuclear fission long ago because of or perhaps following the Flood. Natural nuclear fusion occurs in every sun-like star, where hydrogen nuclei combine to form helium. A discussion of nuclear changes and the structure of the nucleus is provided in Chapter 16.

Evidence for natural fission reactions was discovered in the Oklo uranium mines in Gabon, Africa. The uranium ore vein is sandwiched within thick layers of sandstone rock, which suggests that it was deposited during the Flood.

6.13 Mass Energy

The largest potential source of energy in the universe is **mass energy**, the energy equivalent to all matter itself. Just as nuclear energy is produced when fission or fusion takes place, mass energy is the energy you would obtain if you could convert all of an atom's mass into energy according to $E = m \times c^2$. This is a general application of Einstein's equation from his special theory of relativity. However, since matter is inherently stable, it is unlikely that we will ever be able to completely convert a significant amount of matter into megatrillions of joules of useful energy.

6-18 The matter in a compact media player completely converted into energy would release more energy than the largest military nuclear warheads.

Example Problem 6-3

Mass Energy

How much mass energy, in joules, could be obtained from the complete conversion of a compact 138 g (4.9 oz) media player?

Known: mass of player (m) = 138 g = 0.138 kg
speed of light (c) = 3.00×10^8 m/s

Unknown: mass energy released (E)

Required formula: $E = m \times c^2$

Substitution: $E = (0.138 \text{ kg})(3.00 \times 10^8 \text{ m/s})^2$

Solution: $E = 1.2\underline{4}2 \times 10^{16} \text{ kg·m}^2/\text{s}^2$
$E \approx 1.24 \times 10^{16}$ J

This amount of energy is equivalent to that released by a blast of 3 *million* tons of TNT—a 3-megaton bomb!

6B Section Review

1. List the nine forms of energy that you learned about in this section. Explain why these are not the only possible categories of energy.
2. How do you identify the zero reference position for calculating mechanical potential energy? What is the most common zero reference position for a system that has gravitational potential energy (GPE)?
3. Compare thermal energy and acoustic energy at the particle level of matter.
4. What is electrical energy? On which subatomic particle does electrical energy relate to in everyday situations?
5. What form of energy is produced when both magnetic and electrical energy act together? What is the significance of this form of energy?
6. Where is chemical energy stored in matter? What can release chemical energy and what may you observe that indicates that this has happened?
7. Why does nuclear fission release nuclear energy?
8. Why is it not possible to completely convert the mass energy of a piece of lead into electrical energy for the purpose of powering a community?
9. (True or False) Electromagnetic energy is the kind created by horseshoe magnets.
10. ✪ What is the GPE of a 0.443 kg World Cup soccer ball that has been kicked to a height of 3.5 m?
11. ✪ What is the kinetic energy of a 14.26 g rifle bullet moving at 600. m/s? (*Hint*: Remember to convert grams to kilograms first.)
12. ✪ The average air gun BB has a mass of 0.345 g. How much mass energy does a BB possess?

6C Conservation Laws

6.14 Energy Transformation

Energy often changes from one type to another. We use energy transformations every day to light our homes, cook our food, and travel to school. For example, a lamp transforms electrical energy into radiant energy (light) and thermal energy. An electric oven transforms electrical energy into thermal energy. A car transforms chemical energy (from gasoline, diesel fuel, or ethanol) into mechanical energy. Our bodies transform chemical energy from food into thermal energy in our cells, electrical energy in our nerves, and mechanical energy in our muscles.

Every form of useful energy can change into at least one other form of useful energy. However, the conversion is never 100%—some energy that we can't use always escapes the process. We measure the **efficiency** of the process by comparing the amount of energy we can actually use from the process to the original amount of energy that went into the process.

Most man-made energy transformations are not more than 20%–40% efficient. Many familiar ones are much less. A standard incandescent light bulb is about 5% efficicent in converting electrical energy into light. The other 95% is radiated away as unusable heat. In contrast, a fluorescent bulb is around 20% efficient in the same energy conversion.

6C Objectives

After completing this section, you will be able to

- ✓ give examples of energy conversions.
- ✓ define the efficiency of an energy conversion process.
- ✓ state the law of conservation of energy and the conditions under which it is true.
- ✓ state the first law of thermodynamics.
- ✓ define momentum and state the conditions under which it is conserved.
- ✓ classify collisions as elastic, partially elastic, or inelastic.
- ✓ compare and contrast momentum and kinetic energy.

6-19 An incandescent lamp is very inefficient when converting electrical energy to light.

6.15 Using Energy Transformations to Solve Problems

According to the US Environmental Protection Agency, the energy usage and costs to cool homes and businesses in large cities increase approximately 2% with every 1 °F rise in temperature. The urban heat island problem discussed at the beginning of the chapter results from unwanted energy transformations in the confined spaces of a city. Concrete, black asphalt roofs and streets, and the many nooks and crannies all trap incoming radiant energy from the sun, which turns into thermal energy.

The thermal energy in the city's structural materials radiates back into the air above the city as infrared rays. Certain gases, particularly carbon dioxide and water vapor, absorb the infrared, warming the air. The large quantities of glass reflect the sun's energy onto more absorbent materials, making the problem even worse. In addition, the presence of many thousands of people inside buildings creates a source of thermal energy that must be pumped from inside buildings into the environment by air conditioning. Finally, vehicles add their hot exhausts to the already scorching air.

Solutions to prevent these energy transformations are just now being put into action. The most successful solutions focus on eliminating the radiant-energy absorption in the first place. What do you think would be some good and workable ways of doing this? Engineers have determined that heat island effects can be reduced significantly by painting roofs white or surfacing them with highly reflective coatings. Increasing the amount of green spaces in cities and roofing buildings with grass and shrubs also reduces the amount of thermal energy absorbed in building structures.

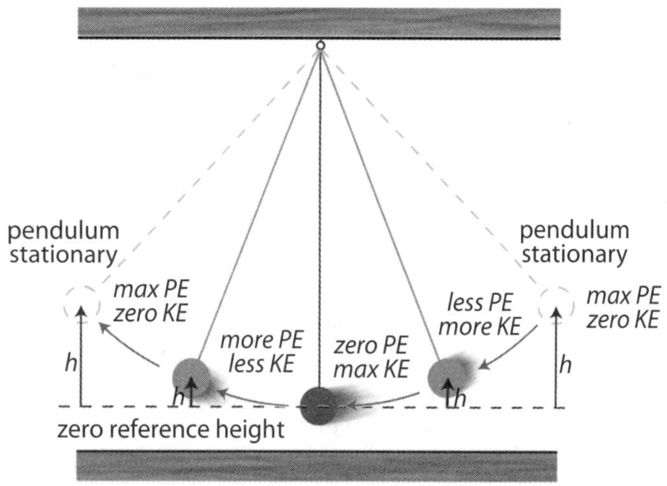

6-20 A pendulum converts energy from potential to kinetic and back again.

The motion of a pendulum and other forms of periodic motion are discussed in Chapter 12.

Because of Einstein's discovery that matter and energy are interchangeable, many scientists prefer a combined *mass-energy conservation law*. This law states that matter and energy can never be created or destroyed, only changed from one form to another. Thus, the total amount of matter and energy in the universe is constant.

6.16 Energy Conservation—First Law of Thermodynamics

A grandfather clock keeps time using a pendulum to mark regular time intervals. A *pendulum* is a heavy weight on the end of an arm that swings back and forth on a pivot point at its upper end. Pendulums have been used for centuries to study various aspects of motion and energy.

When the pendulum is pulled to the side, the pendulum mass is higher than when in its lowest position. It has gained gravitational potential energy (GPE = $m \times g \times h$) relative to this zero reference position. When it is released, it gains speed and kinetic energy as the mass drops toward the bottom of its swing. At that point its GPE is zero and its kinetic energy is maximum. As the pendulum begins to rise to the other side of the swing, it gains potential energy but loses kinetic energy. Eventually it comes to a momentary stop at the same height as where it was released. At this point potential energy is at a maximum, and kinetic energy is zero. The pendulum begins to fall to repeat the cycle again. If friction were not present, a pendulum would continue its periodic swings forever.

A pendulum is a clear, simple example of the law of the **conservation of energy**, a thoroughly tested scientific law stating that energy can never be created or destroyed, only changed from one form to another. This law is directly related to the law of the conservation of matter (see Chapter 2). In every process where energy is transformed into another form, the total amount of energy entering the process equals the total of all the forms of energy that exist at the end of the process. This principle is also known as the **first law of thermodynamics**. No exceptions to the law of energy conservation have ever been observed.

The first law of thermodynamics accounts for all forms of energy used and released in a process within a system's boundaries. For example, consider your car as you begin a road trip. Its fuel tank is full of gasoline, a compound that contains a large amount of chemical potential energy in its chemical bonds. As the fuel burns in the engine, it turns the engine to produce mechanical energy. Light and acoustic energy are emitted when the explosion in the engine cylinder occurs, and thermal energy is transferred to the engine metal. The engine also turns an alternator that generates electrical energy to charge the battery, storing energy in the chemical bonds of the battery materials. Thermal energy is also generated in this process. Electricity can produce radiant energy in the headlights and sound in the radio and horn. Mechanical energy is converted into thermal energy in the brakes when they are applied to slow the car. In a single vehicle, there are dozens of energy transformations that take place. If all the forms of energy are accounted for that came from within the system bound-

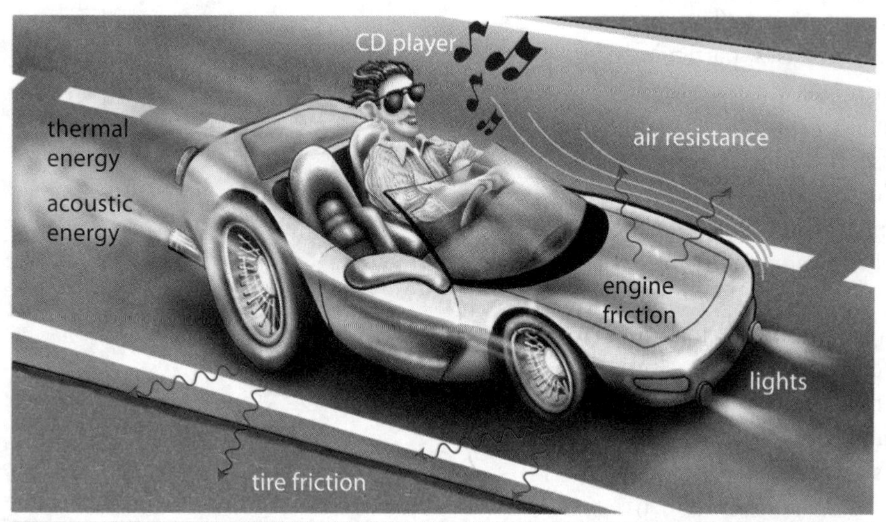

6-21 The energy leaving the car is equal to the energy input from the gasoline in its tank.

ary, they equal the energy initially contained in the chemical bonds of the gasoline. Most of the energy seems to have disappeared, but it still exists as thermal energy distributed throughout the landscape along the route you took while traveling.

6.17 Newton's "Quantity of Motion"

Newton's second law tells us that a net force is needed to start, stop, or change the motion of an object. Two factors determine the size of the force needed and how fast the object's motion will change in a certain amount of time—the speed and mass of the object. It takes a much larger force to stop a speeding car than to stop the same car when it is barely moving. It also takes more force to stop a large truck than a motorcycle moving at the same speed.

As Isaac Newton was developing his laws of motion, he noticed that moving objects have a property that determines how much force is required to change their motion. He called this property the *quantity of motion*. Today, we call it **momentum**.

6-22 The large truck has much more momentum than the motorcycle.

6.18 Motion and Momentum

Experiments revealed that the momentum of an object moving in a straight line is directly proportional only to its mass (m) and to its velocity (v). We can express this relationship mathematically in a single formula,

$$\boxed{p = m \times v},$$

where

- p is the symbol for momentum,
- m is the mass of the system, in kilograms, and
- v is the velocity magnitude (speed) of the system, in meters per second.

This formula describes the momentum of a system moving in a straight path, i.e., *linear momentum*.

In vector form, the momentum equation is

$$\mathbf{p} = m \times \mathbf{v}.$$

You can see from the momentum formula that the unit for momentum comes from the product of mass and velocity, or kilogram-meters per second (kg·m/s). Velocity is a vector quantity, so momentum is also a vector, since it has both magnitude and direction. For the purpose of this course, you will work with only the magnitude of momentum vectors unless otherwise instructed.

The unit for momentum is the kilogram-meters per second (kg·m/s). There is no SI derived unit for momentum.

Rotating systems, such as a Ferris wheel or a helicopter rotor, also have momentum. But because these systems do not move in a straight line, it is more complicated to calculate their momentums. Even so, the *angular momentum* of a rotating system still depends on the mass of the system and how fast it's rotating. You will learn about angular momentum in more advanced science courses.

6-23 The total momentum of this billiard-ball system after the break is equal to the original momentum of the cue ball alone (since it was the only object moving in the system before the break).

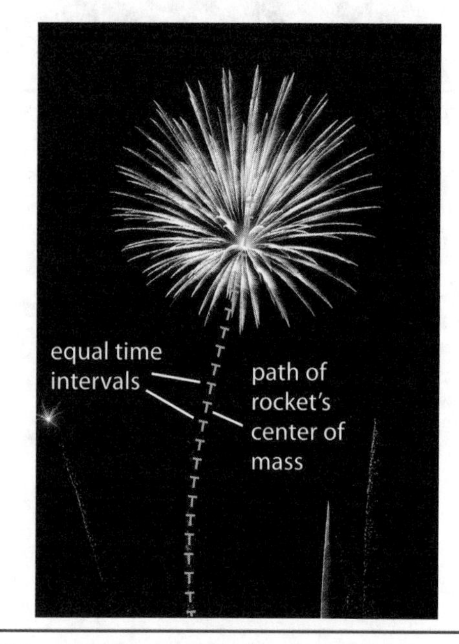

6-24 At the instant the fireworks rocket system explodes, the sum of the momentums of all its particles equals the momentum vector of the rocket just before it bursts. This is because the explosion comes from *within* the system. Momentum is conserved, ignoring external forces like air resistance.

6.19 Conservation of Momentum

A collision occurs when two objects smash into each other, such as with billiard balls or cars in an auto accident. At least one of the colliding objects must be moving for a collision to occur. When an object collides with another, its momentum changes because its speed, direction, or both change. The opposite process in physics is an explosion. Imagine a firecracker or a 4th of July fireworks rocket exploding. In one moment, the pieces of the fireworks are all together, intact; the next, they are flung outward, each with its own speed, direction, and momentum.

When physicists analyze a collision or explosion, the "system" includes all the objects inside the boundaries of the event. The momentum of the system is the sum of the momentums of each of the objects that are part of the system. So what happens to the total momentum of the system when a collision occurs inside the system? Physicists discovered that as long as no *external* net force acts on the system of objects to change their motions, the system's total momentum remains unchanged, no matter how violent the collision or the number of objects involved in the collisions. For an explosion, the force comes from *inside* the system boundaries, so the sum of all the explosion fragment momentums equals the momentum of the system before the explosion.

This principle is another important universal conservation law called the **conservation of momentum**. It is true for all kinds of collisions, explosions, and similar phenomena. We can demonstrate the conservation of momentum when firing a gun. Assume that the bullet and gun are parts of the system of interest. Just before firing, the gun and bullet have zero total momentum. The instant after firing, the small bullet has a high speed in one direction, giving it a large positive momentum. The force of the exploding cartridge accelerates the bullet. Again, this force comes from *inside* the system boundary. The explosion creates an equal but opposite force acting on the gun that kicks it backward (Newton's third law). The bullet's change of momentum involves a high speed and a low mass; the gun's equal but opposite change of momentum involves a low speed and a large mass. If we add the two momentums together, taking into account their equal magnitudes and opposite directions, their sum is zero, which agrees with the law of the conservation of momentum.

6.20 Momentum, Energy, and Collisions

Both momentum and kinetic energy are properties of moving systems. In fact, they are very similar except that momentum is a vector defining the system's quantity of motion, while kinetic energy is a scalar describing the mechanical energy of the moving system. Is kinetic energy conserved in collisions as momentum is? The answer depends on the kind of collision. There are three kinds of collisions recognized by physicists—elastic, partially elastic, and inelastic.

Elastic Collisions

An *elastic collision* occurs when two objects collide and rebound so that the sum of their momentums and the sum of their kinetic energies are the same before and after the collision. No frictional forces from inside or outside the system slow the colliding parts of the system down. No real-world collision is perfectly elastic. There is always some deformation of a colliding object, no matter how slight, that converts a bit of the kinetic energies to some other form of energy (sound, light, thermal, etc.).

An approximate example of an elastic collision is when ball bearings collide as they roll on a hard surface. Air hockey pucks that collide as they glide along on a cushion of air also experience near-elastic collisions.

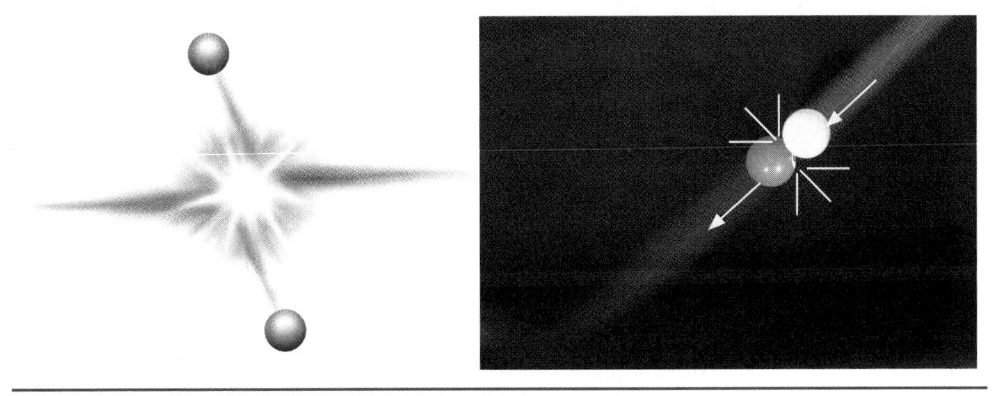

6-25 Elastic collisions conserve both momentum and kinetic energy.

Partially Elastic Collisions

When one or both objects in a collision system deform, some of the kinetic energy in the original objects is converted into heat. Since total energy is conserved, the kinetic energy of the system decreases. A *partially elastic collision* has occurred when after the collision the rebounding objects are temporarily or permanently deformed in some way. Most real-world collisions are of this kind. Examples of partially elastic collisions include two soccer players colliding on the field or two cars colliding at an intersection. The greater the deformation, the more the system's total kinetic energy after the collision is reduced.

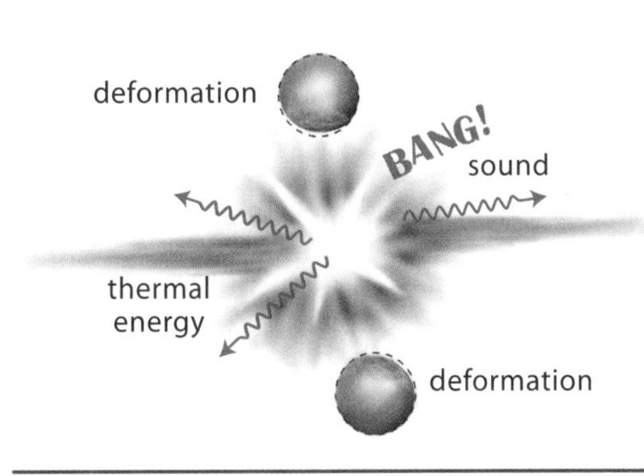

6-26 Partially elastic collisions may conserve momentum but do not conserve kinetic energy.

Inelastic Collisions

An *inelastic collision* occurs when the colliding objects stick together after the collision. The colliding objects may or may not be in motion after the collision, depending on their orientations and momentums before the collision. The kinetic energy of the two objects together after the collision is significantly different from the sum of their separate, initial kinetic energies. Two hockey players colliding and holding on to each other as they fall to the ice is an example of an inelastic collision. Objects in an inelastic collision are greatly and usually permanently deformed.

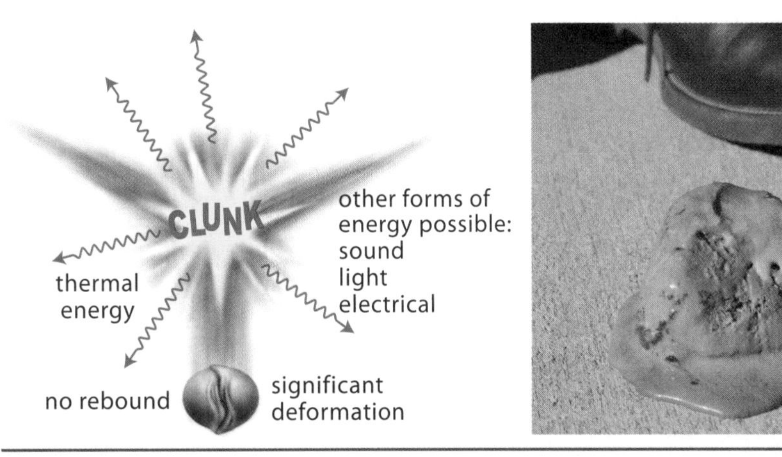

6-27 These images show cases where neither kinetic energy nor momentum is conserved in an inelastic collision. The illustration (left) shows that kinetic energy wasn't conserved because most of the energy resulted in deforming the objects. Momentum would have been conserved, except that gravity (an external net force) pulled the objects downward after the collision. In the photo (right), both the momentum and the kinetic energy of the falling ice cream scoop were lost when the external force exerted by the ground deformed the scoop and stopped its motion.

Momentum is always conserved in collisions as long as no external forces act on the parts of the system. Kinetic energy is conserved only in elastic collisions. In any other kind of collision, kinetic energy is not conserved because forces internal to the system change the speeds of the parts of the system, reducing the final total kinetic energy.

6C Section Review

1. What do we call changes such as nuclear energy into thermal energy or thermal energy into electrical energy?
2. How complete is the transformation of one usable form of energy into another usable form? What term describes the effectiveness of an energy transformation?
3. DS Suggest at least two ways that you could change a black asphalt roof to decrease the amount of heat added to the air above the building.
4. What do we mean by *conservation of energy*?
5. Since both mass and energy are conserved, and Einstein showed that they are both equivalent, how do physicists usually restate the first law of thermodynamics?
6. What are the two properties of a moving object that determine its momentum?
7. List at least two important differences between linear momentum and kinetic energy.
8. Where does a physicist set the boundaries of a system when studying the changes of momentum due to collisions between two or more objects or to an explosion producing moving objects?
9. During a collision or an explosion within a system, what condition must hold true for the momentum of the system to be conserved?
10. What are the three kinds of collisions? Which kind is the most common?
11. (True or False) Two billiard balls colliding in outer space would be an example of a true elastic collision.

FACETS OF SCIENCE

STOP!

The kinetic energy of a car must "go somewhere" when the car stops. This energy is normally converted to thermal energy in the brakes. If a car comes to a skidding halt, some of the energy is also converted to thermal energy in the tires and on the road.

The US Department of Transportation has conducted extensive tests to determine the distance it takes standard passenger cars to stop on dry, clean, level pavement. The results of these tests are given in the table.

The driver-reaction distance is based on a reaction time of 0.75 s. Though the reaction time stays the same at all speeds, the driver-reaction distance increases in direct proportion to the speed. In other words, the faster you go, the more ground you cover during your 0.75 s reaction time.

The braking distance is the distance needed to stop once the brakes are applied. Researchers found that when the speed is doubled, the braking distance is more than quadrupled. This dramatic increase is caused by two facts: doubling the speed quadruples the kinetic energy, and automobile brakes work less efficiently at higher speeds.

The total stopping distance is the sum of the driver-reaction distance and the braking distance. Stopping a car requires more distance than you might think. The figures given here are for ideal conditions. The stopping distance can be much greater when the car is going downhill, when the weather is bad, or when the road is dusty, wet, or covered in gravel.

Speed (mi/h)	Driver-reaction distance (ft)	Braking distance (ft)	Total stopping distance (ft)
20	22	18–22	40–44
25	28	25–31	53–59
30	33	36–45	69–78
35	39	47–58	86–97
40	44	64–80	108–124
45	50	82–103	132–153
50	55	105–131	160–186
55	61	132–165	193–226
60	66	162–202	228–268
65	72	196–245	268–317
70	77	237–295	314–372
75	83	283–353	366–436
80	88	334–418	422–506

Scientifically Speaking

Chapter Review

1. Is energy something that an object *has* or something that it *does*? Explain.
2. What is *work* as a physicist would define the term?
3. Compare potential energy and kinetic energy. Does a stationary car have kinetic energy? Does a flying bird have potential energy?
4. Which contact force cannot produce potential energy in an object? How can a contact force generate potential energy?
5. What conditions must exist for an object to have mechanical potential energy?
6. What are some factors that make measuring the individual kinetic energies of atoms and molecules in clumps of matter essentially impossible?
7. Compare the kinetic energy represented by thermal energy to kinetic energy in the sense of mechanical energy.
8. What is the everyday source of electrical energy? Name one natural source of electrical energy.
9. Why were magnets considered to be magical in ancient times?
10. What form of energy can move through the vacuum of space without needing matter?
11. When a log burns on a campfire, what is the energy source for the combustion? Into what other kinds of energy is this energy converted?
12. ✪ Suggest why a fission or fusion energy source would be safer than one that instantly converts the entire mass of an object into energy.
13. Give the most complete statement of the first law of thermodynamics. What are the main consequences of this law relating to matter and energy?
14. When you strike a softball with a bat, what kind of collision occurs? Is the total amount of kinetic energy in the bat-ball system conserved? Are their momentums conserved? Explain.
15. DS ✪ What methods are being used to reduce the amount of thermal energy trapped in buildings and the air around cities?
16. ✪ Why are there no perfectly elastic collisions?

True or False

17. The energy of an object is one of its properties.
18. A falling snowflake can have both kinetic and potential energies at the same time.
19. All physicists agree that there are nine forms of energy.
20. Thermal energy is a form of kinetic energy at the particle level.
21. Acoustic energy is transferred through small, random, repeated particle displacements.

22. Electrical and magnetic energies are so closely related that they are often combined in a single form called *electromagnetic energy*.
23. Chemical energy is a form of potential energy bound up in the bonds between atoms.
24. The most common source of fusion nuclear energy is found in man-made fusion bombs.
25. In the most efficient energy transformations, more useful energy is produced than was put in.
26. Both total momentum and total kinetic energy are conserved during a perfectly elastic collision, if no external forces act on the system of colliding objects.

Questions 27–31 relate to the following problem: A high-school freshman is enjoying an ice cream cone. The mass of ice cream in the cone is 100. g. As he tries to lick the soft scoop, it falls out of the cone from a height of 1.75 m to the ground!

✪ 27. What was the ice cream's GPE before it fell?

✪ 28. What was the ice cream's GPE on the ground?

✪ 29. If all the ice cream's potential energy was converted into kinetic energy just before it hit the ground, how much kinetic energy did the ice cream have?

✪ 30. Which of the three kinds of collision did the scoop of ice cream experience when it hit the ground?

✪ 31. How fast was the scoop of ice cream falling just as it struck the ground?

WORK AND SIMPLE MACHINES

CHAPTER 7

DOMINION SCIENCE PROBLEM

Time for the People

Britain was the birthplace of steam technology, and in the early 1800s steam locomotion transformed the lifestyles of the British people. A day-long trip by a horse-drawn coach now took only hours by passenger train. People loved the convenience, and it was great for business, causing the British economy to flourish in the early decades of the Industrial Revolution.

There was, however, a problem for the more than one million London residents—everyone's clock told a different time. Before trains and other events were run on a tight schedule, the difference didn't matter as much. As England industrialized, time became important to the minute. How could city planners synchronize the time for all the people of London?

7-1 Trains quickly became a preferred mode of transportation for the British public in the 1800s.

7A Work and Mechanical Advantage

7.1 *Introduction*

Imagine what life may have been like in the early years after Creation. The entire world lies before Adam and Eve, waiting to be explored. Our first parents face many new challenges to meet their daily needs and to satisfy their curiosity. To "have dominion over . . . all the earth" (Gen. 1:26), they must do things that their bodies are not able to do—or not able to do well.

Out of a desire to explore a river that runs out of Eden (Gen. 2:10), Adam cuts down a tree and they shape it into a canoe. They soon discover that the canoe is difficult to steer. And no matter what they do, they cannot make it go upstream.

But then Eve notices a group of ducks. "*They* don't have any trouble getting around on the river," she says. The couple approaches the ducks and carefully observes their movements. Adam then picks one up and studies its webbed feet. Placing it back in the water, he notices that by repeatedly moving only one foot, the duck can steer. Eve later concludes that the large surface area of the webbed feet enables it to get traction in the water and move upstream.

Within hours, the world's first canoe paddles are made. Like a duck's leg, they are thin at the top and broad at the bottom. And, like a duck's leg, they make it possible to move through the water with ease.

This piece of historical fiction describes the invention of a simple machine, and it highlights the importance of machines in obeying the Creation Mandate. Exercising good and wise dominion often requires more than the human body can do. And the need for machines has only increased because of the Fall and the Curse. For example, it is not just calm rivers that must be traveled but also stormy oceans. Adam and Eve finished their work that day with a very simple machine, but we often need something far more complicated.

Our little story also introduces an important fact: *some of our best machines are imitations of God's creation.* From Genesis 1:26–28, we learn that humans declare God's glory by being like Him. He has made the world with His infinite intelligence and strength. We are to manage and nurture that world with our own limited intelligence and strength. It should not surprise us, then, that one of the best ways to enhance our work is to imitate His. God delights in using levers, planes, and wedges. So should we.

7.2 *Calculating Mechanical Work*

You probably associate work with heavy physical or mental labor. Mowing the grass is work. Shoveling snow is work. Solving math problems is work. However, scientists view work differently. Work involves the exchange of energy for some change in a physical system. In fact, work and energy are so closely related that they have the same unit, the joule (J).

The most familiar form of work is mechanical work. **Mechanical work (W)** occurs when a force acts on an object to move it in a direction parallel to the force vector. Two factors determine work—the

7A Objectives

After completing this section, you will be able to

- ✓ explain how simple machines reflect the glory of God and assist in exercising dominion over creation.
- ✓ describe mechanical work and relate work and energy.
- ✓ compute work, given appropriate quantities and conditions.
- ✓ define and calculate power.
- ✓ list the three general categories of simple machines.
- ✓ show how simple machines exhibit the distance principle when doing work.
- ✓ explain why no machine can be 100% efficient.
- ✓ define and calculate MA in various ways.

7-2 Simple machines have long been used to solve problems. Around 200 BC, the military defenses of Syracuse held off the Roman fleet for over three years. Archimedes designed the weapons to use levers and pulleys.

7-3 We often associate work with heavy physical labor.

7-4 Mowing the grass is work from a physicist's point of view.

The product of two vectors (force and displacement) in the work formula equals a scalar (work). To simplify the formula, we will use the scalar values of force and displacement and restrict how we use the formula.

application of a specific force and the occurrence of a displacement. The formula for mechanical work is

$$\boxed{W = F \times d},$$

where

- W is the work done on a system, in joules,
- F is the force applied to the system parallel to the system's displacement, in newtons, and
- d is the displacement of the system, in meters.

Work is a scalar quantity. It can be positive, negative, or zero. In this discussion, we will consider only positive work. *Positive mechanical work* is done when the force acts on a system in the same direction as the system's motion. For example, if a student lifts a book vertically, his hand applies an upward force on the book and it moves upward. A system gains energy when positive work is done on it. The book has more gravitational potential energy after it is raised.

If either the force or displacement of an object is zero, or the force acts perpendicularly to the object's displacement, then no work is done by the force on the object. You can push and push against a wall, but as long as the wall remains stationary ($d = 0$ m), you have done no mechanical work. Similarly, a person in-line skating along a level street has gravity acting on him the entire time, but gravity does no work on the skater because its force vector is perpendicular to his motion.

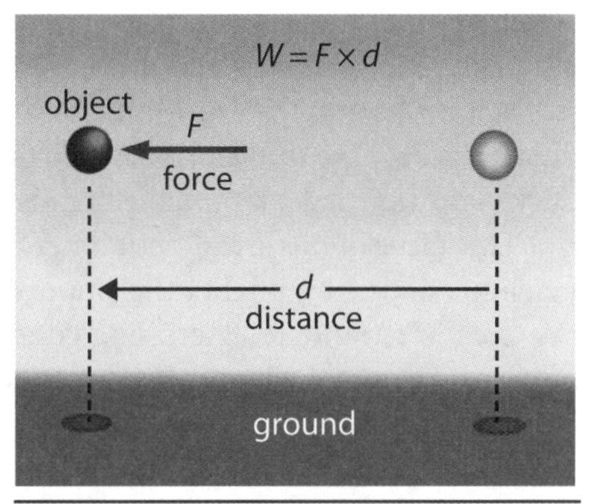

7-5 Work is done when a force is applied to an object parallel to its displacement.

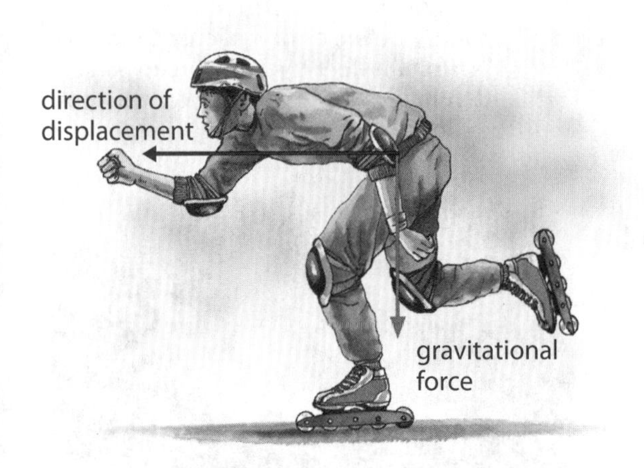

7-6 A force exerted perpendicular to an object's displacement does no work on the object.

7.3 Using Simple Machines to Solve Problems

To provide a timepiece to the residents of London, Parliament ordered the construction of a huge clock tower. The Great Clock of Westminster strikes the hour with an accuracy of within 1 s. The clock's mechanism must do work on the hands, sometimes against the added force of snow accumulation or wind—all the while maintaining this accuracy. The clock, built in 1854, uses gears, wheels, pendulums, and weights to keep time on each of its four faces, each with a minute hand weighing 980 N (220 lb) and an hour hand weighing 2940 N (660 lb)! The clock's mechanism uses a falling weight to provide the potential energy to power the clock. The Great Clock houses a 13.8-ton bell known as Big Ben, which sounds on the hour.

EXAMPLE PROBLEM 7-1

Winding the Great Clock

London's Great Clock mechanism is run by a falling weight, which turns the many gears that position the hands. The 2490 N (560 lb) clock weight has to be periodically wound up by an electric winch. Assuming that the weight is raised 53 m and the winch exerts a force equal to the clock weight, how much work does the winch do on the weight?

Known: applied force (F) = 2490 N
weight displacement (d) = 53 m

Unknown: work on the weight (W)

Required formula: $W = F \times d$

Substitution: $W = (2490\ \text{N})(53\ \text{m})$

Solution: $W \approx 1\underline{3}1{,}000\ \text{N}\cdot\text{m}$

$W \approx 130{,}000\ \text{J}$ (2 SDs allowed)

7-7 The Great Clock of Westminster in London

After studying Example Problem 7-1, can you tell what other combination of SI units the joule is equal to besides kg·m²/s²?

7.4 Power

What is the practical difference between a motorbike with a 49 cc engine and a motorcycle with a 1200 cc engine? The larger engine can deliver a lot more force to move its vehicle than the smaller engine. According to Newton's second law of motion, a larger force results in a greater acceleration for a given mass. Greater acceleration means a higher speed in less time, resulting in a higher rate of displacement. In other words, the larger engine can do more work in a shorter time interval—it can produce more power. **Power (*P*)** is the rate of doing work.

We compute mechanical power by dividing the work done by the time it takes to do the work:

$$\boxed{P = \frac{W}{\Delta t}},$$

where

- P is power, in joules per second (J/s),
- W is work, in joules, and
- Δt is the time interval, in seconds.

The unit of power in the SI is the *watt (W)*, named in honor of James Watt, the inventor who vastly improved the efficiency of the steam engine. One watt is 1 J of work done in 1 s.

7-8 An aircraft carrier catapult does a large amount of work in a short interval—it produces a lot of power.

The watt (W) is the SI unit of power.

$$1\ \text{W} = 1\ \frac{\text{N}\cdot\text{m}}{\text{s}} = 1\ \frac{\text{J}}{\text{s}}$$

7-9 A clock tower weight

When working with answers from previous calculations, use the unrounded values to minimize rounding errors in later calculations.

7-10 Noah must have used all the simple machines to build a structure as large as the Ark.

Example Problem 7-2

Lifting Power

The electric winding mechanism in the Great Clock takes about 40. min. to rewind the clock weight cable. Refer back to Example Problem 7-1 on the previous page, and then calculate the power of the electric winch in watts.

Known: work on clock weight (from Example Problem 7-1)
$W = 1\underline{3}1{,}000\ \text{J}$
winching time
$\Delta t = 40.\ \text{min} \times 60\ \text{s/min} = 2400\ \text{s}$

Unknown: winch power output (P)

Required formula: $P = \dfrac{W}{\Delta t}$

Substitution: $P = \dfrac{131{,}000\ \text{J}}{2400\ \text{s}}$

Solution: $P \approx 5\underline{4}.5\ \text{J/s} \approx 55\ \text{W}$ (2 SDs allowed)

7.5 Simple Machines and Efficiency

Example Problem 7-2 illustrates the usefulness of man-made tools to do work faster and easier. They often reduce the toil of work that is part of the curse on Adam's sin. Machines also allow us to perform tasks that would be difficult or impossible to complete without them. These tools, whether complex or simple, are composed of basic mechanical parts called **simple machines**. Most simple machines can be classified as levers, wheels and axles, or inclined planes. You've probably used all these simple machines in one form or another.

As our opening illustration about Adam and Eve suggests, simple machines were likely discovered very early in human history. Their use seems to be almost intuitive to people and even to some animals, such as chimpanzees and several species of birds. Wheel-like machines and inclined planes were likely in use before the Flood. It is difficult to imagine Noah and his sons dragging logs long distances to the building site of the Ark and erecting its hull without the use of levers, wheels, pulleys, and ramps. There was certainly ample time and intelligence to invent these kinds of tools before the Flood.

When you do work using a simple machine, you exert a force on one part of it. The machine then exerts a force to do work on something else—the task you want to accomplish. Using the machine allows you to do the work more easily or faster than if you did it directly. Ideally, the machine produces only as much work as you put into it. This is another example of the conservation of energy. Since a simple machine does not create energy, *you cannot get more work out of a machine than you put into it.*

As discussed in Chapter 6, no process or machine is capable of converting all the energy put into it into the same amount of usable energy or work. No machine is 100% efficient. The ratio of work obtained from a machine (work output, or W_{out}) is always less than the energy or work put into a machine (work input, or W_{in}). A machine's efficiency is calculated by dividing the work output by the work input.

$$\text{efficiency} = \frac{W_{out}}{W_{in}} \times 100\%$$

Machine efficiency is usually expressed as a percentage.

The lost energy or work is expended overcoming friction or is released in unusable forms of energy, such as thermal, radiant, or acoustic energy. However, the *total* energy or work produced by a simple machine is always equal to the energy or work put into the machine.

7.6 Mechanical Advantage and the Distance Principle

Most machines are designed to make work easier by reducing the force needed to do a given amount of mechanical work. The system to be moved is called the *load* or *resistance*. The least amount of force that a simple machine needs to exert on the load to begin moving it equals the force that the load exerts on the machine. We will call this the *resistance force* (F_r). The force actually exerted on the simple machine is called the *effort force* (F_e), or simply the *effort*.

To illustrate these relationships, consider lifting a 3300 N (740 lb) grand piano 2 m vertically to a stage. This task requires 6600 J of work (3300 N × 2 m). Since you can't lift the piano unassisted, you must resort to a machine.

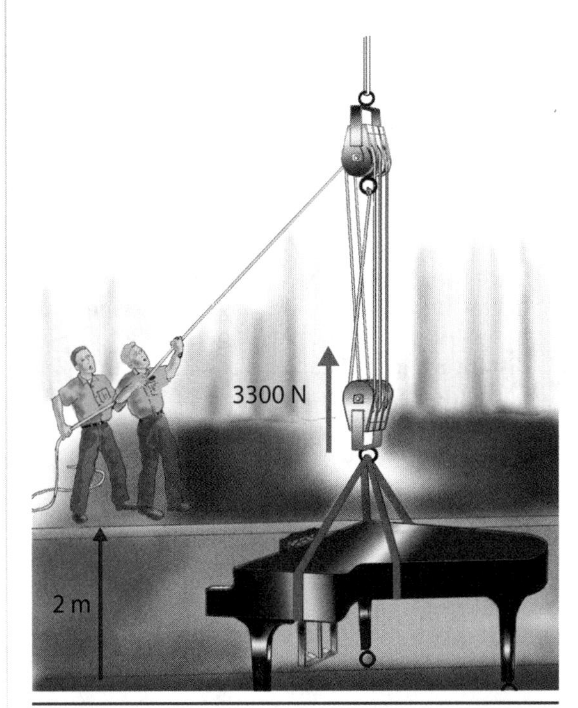

7-11 One method to raise a grand piano to an auditorium stage

Let's say you decide to use a block-and-tackle system of six pulleys and connecting rope (see Section 7C). You exert an effort of only 550 N (123 lb) on the end of the rope, and the piano rises slowly and smoothly to the stage height. The block and tackle gives you a **mechanical advantage (MA)** when doing work.

One way to calculate MA is to compare the force needed to directly do the work to the force applied when using a machine. The MA of the block-and-tackle system is

$$\text{MA} = \frac{\text{weight } (F_r)}{\text{effort } (F_e)} = \frac{3300 \not{\text{N}}}{550 \not{\text{N}}}$$

$$\text{MA} = 6$$

In other words, your effort force is 6 times as effective in moving the load's weight when using the block and tackle. The MA of most machines is greater than 1, meaning that machines reduce the effort you must exert to do a given amount of work.

Actual MA is determined by actual measurements of work input and work output. Ideal MA is determined just from the dimensions of the simple machine.

Actual MA is always less than ideal MA for a given simple machine.

It is important to understand that the piano example is idealized. We're assuming that the block and tackle is 100% efficient. In reality, friction and other factors sap energy in the process of using a machine so that no simple machine is 100% efficient. When calculating MA by using actual effort and resistance forces, you obtain the *actual mechanical advantage* (AMA). For simplicity, we will ignore the effects of friction. In this textbook, you will normally compute the *ideal mechanical advantage* (IMA). It's important to keep this distinction in mind.

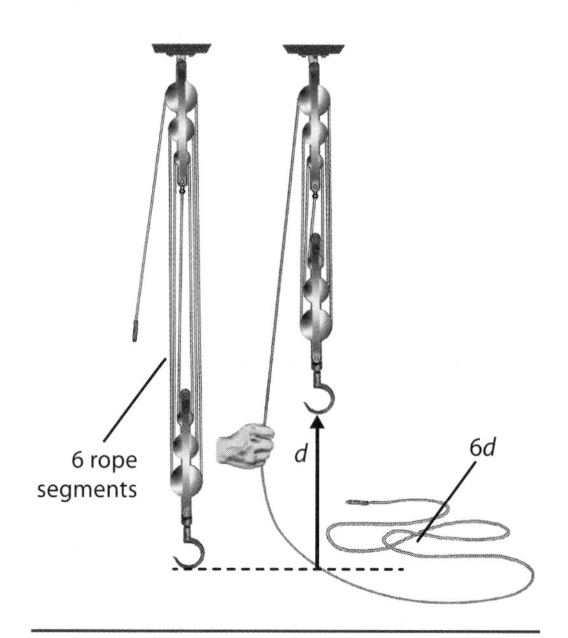

7-12 The real-world effect of the distance principle

Once the MA of a simple machine is known, it can be used to find any relationship between effort and resistance or effort distance and resistance distance for that particular machine.

But wait! Work input must equal work output, because energy must be conserved. In our example, you need to do 6600 J of work on the piano. The 3300 N piano moves only 2 m, but you find that you had to pull 12 m of rope through the block-and-tackle system while exerting 550 N of effort.

$$\text{work input} = \text{work output}$$

$$12 \text{ m} \times 550 \text{ N} = 2 \text{ m} \times 3300 \text{ N}$$

$$6600 \text{ N·m} = 6600 \text{ N·m}$$

To conserve work and energy, simple machines compensate for a smaller effort force by exerting that force over a longer distance. This property is called the **distance principle**.

Another way to compute the IMA of a simple machine is to divide the displacement of the effort by the displacement of the resistance.

$$\text{IMA} = \frac{d_{\text{effort}}}{d_{\text{resistance}}} = \frac{12 \cancel{\text{m}}}{2 \cancel{\text{m}}}$$

$$\text{IMA} = 6$$

For this pulley system, the distance over which you exert effort is 6 times as great as the distance the piano moves.

7A Section Review

1. What is one of the best sources of good ideas for developing technologies such as simple machines?
2. What is the SI unit for work? How do you determine the mechanical work done on a system?
3. State the SI unit for power and its equivalent in more basic SI units. What two factors determine the power applied to a system?
4. What are the three general categories of simple machines? Why are they called simple machines?
5. What is always true in the real world about the value of the ratio
$$\frac{W_{\text{out}}}{W_{\text{in}}}?$$
6. How does a simple machine provide a mechanical advantage? What does a simple machine *not* do?
7. How does the user "pay" for obtaining a mechanical advantage from a simple machine? What is this property of simple machines called?
8. (True or False) A simple machine allows you to accomplish a task with less work than if you did the work without the machine.
9. ✪ A warehouse worker exerts 420 N horizontally on a 540 N crate to shove it 3.5 m across a concrete floor. How much work, in joules, did he do on the crate?
10. DS ✪ Before the installation of an electric winding winch in the Great Clock of Westminster, winding the clock required 5.0 hours of intense physical labor for two men. How much total work did the men do, and how much power did they generate in watts? Refer to the example problems in this section for the necessary information.

7B Levers and the Law of Torques

7.7 Introduction

Have you ever marveled at the quick motion of a hummingbird's wing? What allows it to suspend its body in midair? How does a hammerhead shark change direction to circle its prey, or a humpback whale rocket out of the water? It may not be apparent, but God made all these animals to use their wings, fins, or tails as levers to help them move.

A **lever** is simply a rigid bar that rests on a pivot point called a **fulcrum**. However, a lever can be any rigid object that can pivot or rock back and forth. The lever has been used since ancient times as a pry bar and as the rigid arm in a balance. Archimedes was the first to record the geometric principle by which levers work.

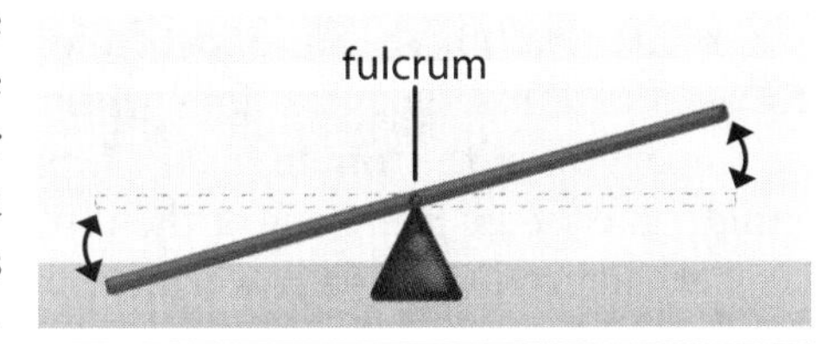

7-13 An ideal lever consists of a rigid bar and a fulcrum.

7B Objectives

After completing this section, you will be able to

- ✓ classify a lever and name its parts.
- ✓ describe how a torque is exerted around a point of rotation.
- ✓ state the law of torques and the conditions under which it can be used.
- ✓ solve lever problems using the law of torques.
- ✓ calculate the MA of a lever system.
- ✓ give real-world examples of the three types of levers.
- ✓ state the possible values of MA for each class of lever and explain where these values come from.

7.8 The Law of Torques

On a seesaw, it's difficult to get yourself and the person at the other end to balance just right. Either the other person is heavier than you and you end up high in the air with your legs dangling helplessly, or you're sitting on the ground and the other person is hoisted high. Ideally both of you could find a seat position where the seesaw balances perfectly and the teetering motion is equal.

Movement of an object about a central axis or pivot point is rotational motion. Levers and wheels rotate as they move. When the seesaw is balanced and has no tendency to teeter about the fulcrum, it is in a state of **rotational equilibrium**. This means that there are no unbalanced forces acting on the seesaw, and therefore, according to Newton's first law of motion, no accelerations occur. One major difference between linear kinetics and the kinetics of rotating systems is that in rotating systems, not only are the magnitudes and directions of the balanced forces important, but their distances from the fulcrum must also be considered. This is the solution to our seesaw problem.

7-14 Balancing a seesaw can be a problem when the riders have significantly different weights.

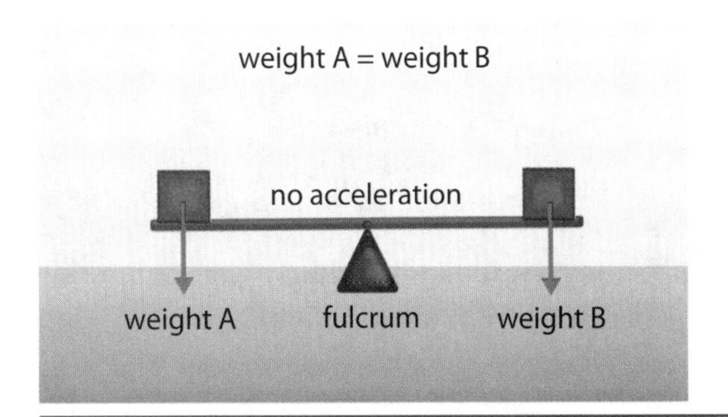

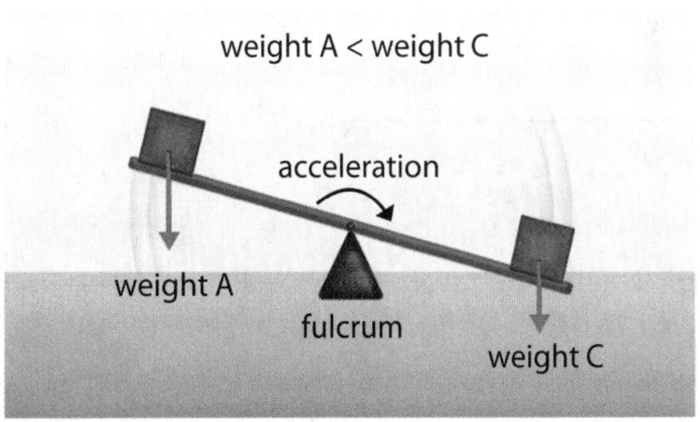

7-15 No motion when rotational equilibrium exists (left); rotational acceleration when equilibrium does not exist (right)

Torques have direction, so they are a vector quantity. Torques on a lever must be equal in magnitude and opposite in direction for rotational equilibrium to exist.

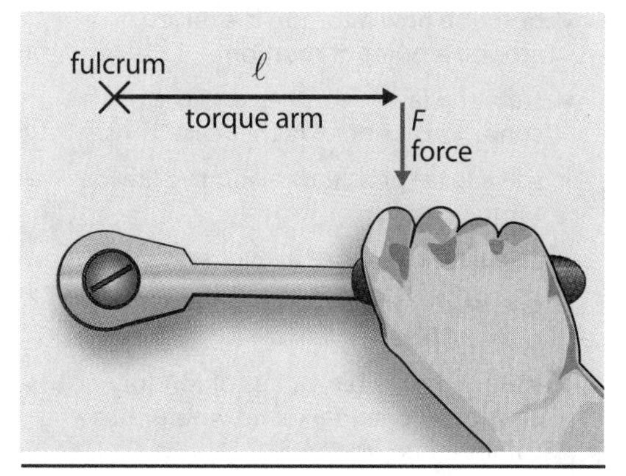

7-16 A torque applied to a bolt

When a force acts perpendicularly on a lever rotating around a pivot point, it produces a quantity called a *moment,* or **torque**. If you've ever tried to remove a nut from a bolt using a wrench, you generated a torque to turn the nut. Similarly, when you sit on one end of a seesaw, your distance from the seesaw's fulcrum and your weight combine to create a torque around the fulcrum. Since the seesaw is rigid, the person at the other end also creates a torque that acts opposite to your torque. If the torques are equal, the seesaw balances and is in rotational equilibrium. If the torques are not equal, the net torque will accelerate the seesaw in the direction of the larger torque.

A torque is simply the product of a force and the distance from the pivot point where the force is applied. This fixed distance is called the *torque arm* (ℓ). The fulcrum of a seesaw divides the rigid bar into two torque arms. The positions where the riders sit determine the actual lengths of the torque arms. To make the torques equal in magnitude, the riders must adjust the distances they sit from the pivot so that the product of their weights and the distances are equal. In a formula, this relationship becomes

$$\boxed{w_1 \times \ell_1 = w_2 \times \ell_2},$$

where

- w_1 is the weight of the first person, in newtons,
- ℓ_1 is the first person's torque arm, in meters,
- w_2 is the weight of the second person, in newtons, and
- ℓ_2 is the second person's torque arm, in meters.

This formula expresses the **law of torques**, which applies to all lever systems, such as seesaws, but only when the lever system is in rotational equilibrium.

A more general form of the law of torques is

$$F_1 \times \ell_1 = F_2 \times \ell_2,$$

because any force, not just weight, can produce a torque.

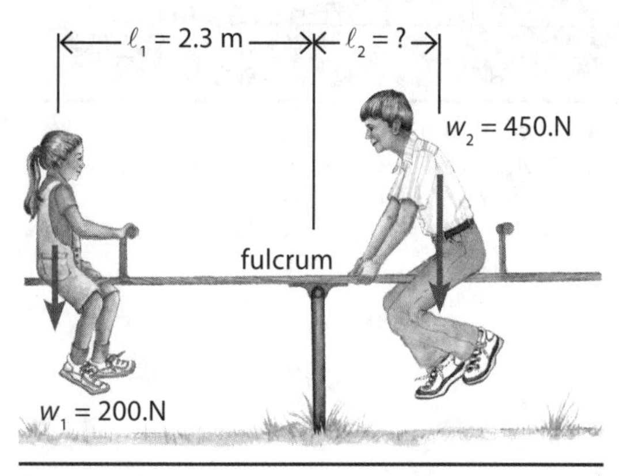

7-17 Balancing a seesaw

EXAMPLE PROBLEM 7-3

Seesaw Torques

A 200. N girl asks her 450. N brother to ride a seesaw with her. She sits 2.3 m from the fulcrum of the seesaw. How far from the fulcrum must her brother sit to balance the seesaw?

Known: girl's weight (w_1) = 200. N
brother's weight (w_2) = 450. N
girl's torque arm (ℓ_1) = 2.3 m

Unknown: brother's torque arm (ℓ_2)

Required formula: $w_1 \times \ell_1 = w_2 \times \ell_2$

Substitution: $(200.\ \text{N})(2.3\ \text{m}) = (450.\ \text{N})(\ell_2)$

Solution:

$$\ell_2 = \frac{(200.\ \text{N})(2.3\ \text{m})}{450.\ \text{N}} = 1.0\underline{2}\ \text{m}$$

$\ell_2 \approx 1.0$ m (2 SDs allowed)

7.9 First-Class Levers

In every lever system designed to do work, the lever arm is subdivided into the *effort arm* (ℓ_e), which is the torque arm of the effort force, and the *resistance arm* (ℓ_r), which is the torque arm of the force exerted by the resistance, or load. In a *first-class lever,* the fulcrum is located between the effort and the resistance. For example, in human anatomy, the foot acts as a first-class lever when flexing. The seesaw, pry bar, and equal-arm balance are also examples of first-class lever systems.

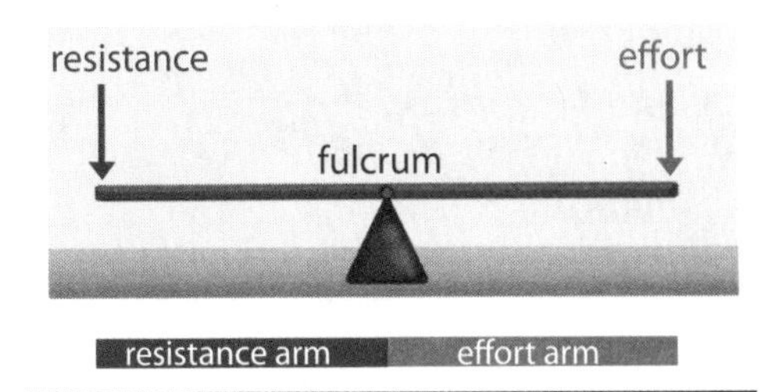

7-18 A first-class lever

The law of torques can be used to estimate the effort needed to move a given resistance using a first-class lever. Suppose that you live in New England and you discover one spring day that your front lawn is "growing" a rock as frost heaves it into view. You decide to dig around the 2450 N (550 lb) rock and pry it out with a 2.00 m spud bar, a heavy metal bar used for digging holes and breaking up rock. If you place a brick 15 cm from the base of the rock to act as the fulcrum, how much effort will you have to exert just to begin moving the rock? The law of torques is written as

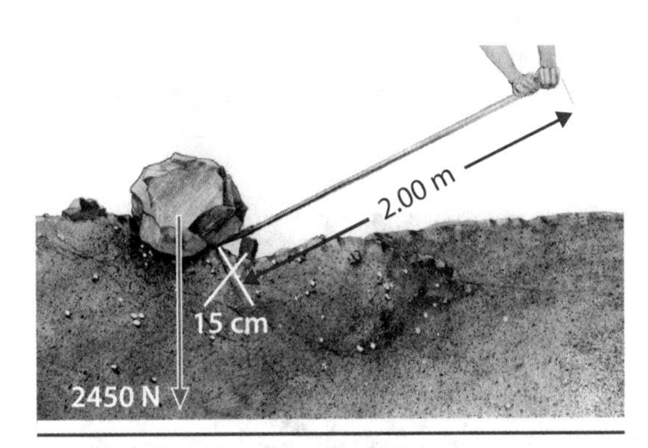

7-19 A first-class lever used for prying

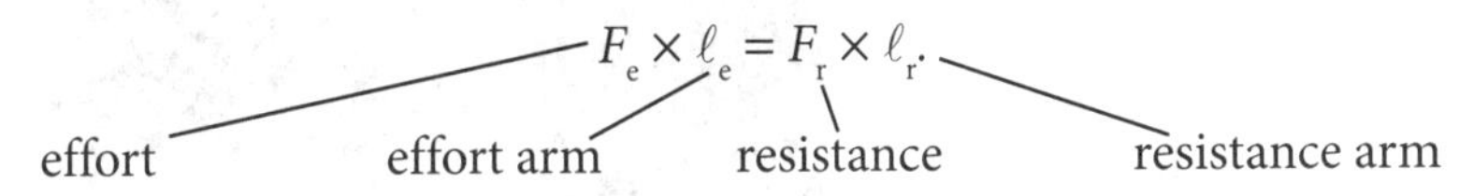

$$F_e \times \ell_e = F_r \times \ell_r.$$

Since we know that the resistance arm is included in the length of the bar, the effort arm must be the difference of 2.00 m and 0.15 m (15 cm), or 1.85 m. Substitute the known information into the law of torques formula.

$$F_e \times (1.85 \text{ m}) = (2450 \text{ N}) \times (0.15 \text{ m})$$

Next, solve for the effort force by dividing both sides of the equation by the effort arm.

$$F_e = \frac{(2450 \text{ N})(0.15 \cancel{\text{m}})}{1.85 \cancel{\text{m}}}$$

$$F_e \approx 1\underline{9}8 \text{ N or } 2.0\times10^2 \text{ N} \quad \text{(2 SDs allowed)}$$

Thus, you will have to exert only 200 N (45 lb) of force just to budge the rock.

Mechanical Advantage

What is the actual MA of the spud lever system discussed above? Recall from Subsection 7.6 that AMA is the ratio of the resistance to the effort.

$$\text{AMA} = \frac{\text{resistance}}{\text{effort}} = \frac{2\underline{4}50 \cancel{\text{N}}}{1\underline{9}8 \cancel{\text{N}}} \approx 1\underline{2}.3$$

$$\text{AMA} \approx 12$$

You can also use the ratio of the effort arm to the resistance arm to calculate the IMA for this system. This is an application of the distance principle, since you will move the effort arm much farther than the resistance will move. The distance the effort must move is directly proportional to the length of the effort arm. Similarly, the distance the resistance moves is directly proportional to the length of the resistance arm.

$$\text{IMA} = \frac{\ell_e}{\ell_r} = \frac{1.85 \cancel{\text{m}}}{0.15 \cancel{\text{m}}} \approx 1\underline{2}.3$$

$$\text{IMA} \approx 12$$

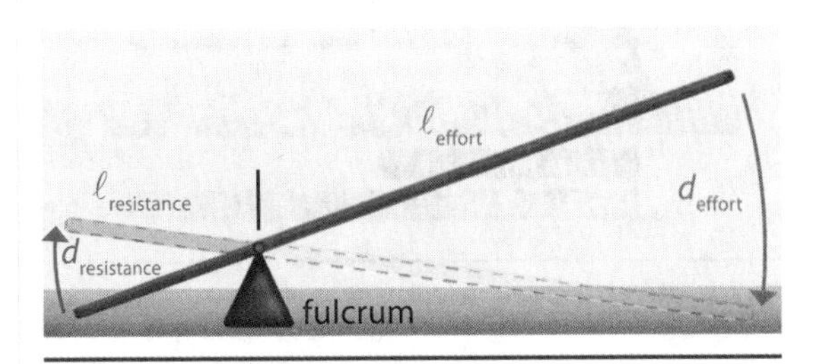

7-20 The distances that the effort and resistance move determine the amount of work done.

First-class levers do not always have a mechanical advantage greater than 1. If the fulcrum is placed exactly in the middle of the lever, then the IMA = 1 because $\ell_e = \ell_r$. If the fulcrum is placed so that the effort arm is shorter than the resistance arm, then the IMA is less than 1. However, there is little benefit to using a first-class lever whose MA is less than 1 because it makes the job more difficult rather than easier.

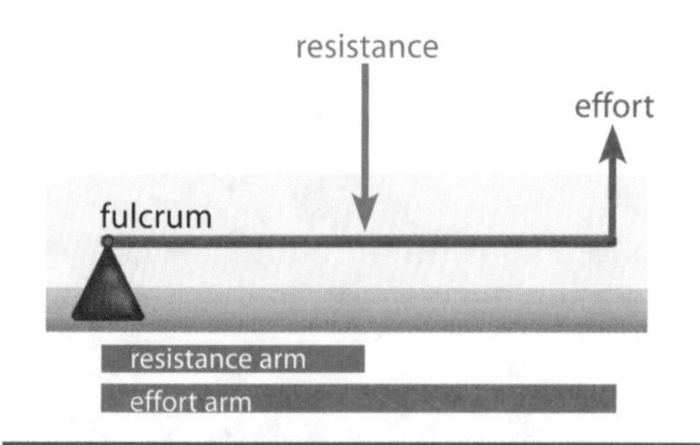

7-21 A second-class lever

7.10 Second-Class Levers

Under some circumstances, a compact lever system is needed or a compressive force must be employed against the resistance. In such cases, the locations of the fulcrum and the effort and resistance forces must be rearranged. A *second-class lever* has its fulcrum at the end of the lever arm and the resistance between the fulcrum and the effort. With this arrangement, the effort arm is always longer than the resistance arm, so the IMA of a second-class lever is always greater than 1.

The human foot acts as a second-class lever when rising up on the ball of the foot (see Figure 7-22). Other second-class lever systems include wheelbarrows, nutcrackers, an office paper cutter, and a door-hinge system. Can you see the action of the second-class lever in each case?

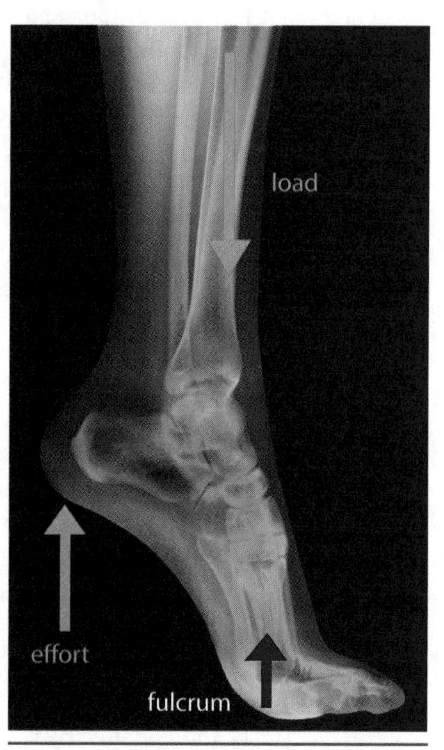

7-22 The foot acts as a second-class lever when you rise up on the ball of your foot. The bones at the base of your toes act as the fulcrum, the weight of your body is the load, and the effort force is supplied by the muscles in the calf of your leg.

7-23 A hand truck uses the second-class lever principle.

7.11 Third-Class Levers

The *third-class lever* seems at first to be something of an oddity. The fulcrum is at the end of the lever arm as in a second-class lever, but the effort is between the resistance and the fulcrum. Therefore, the effort arm is *always* shorter than the resistance arm, and the IMA is *always less than 1.* Why even bother with a lever system that takes more effort to use than the resistance itself?

Raise your right arm to shoulder height. Now bend your arm to bring your hand toward your shoulder. You just operated a third-class lever. In fact, nearly every bone in your body that is involved in motion is part of a third-class lever system! Two major exceptions are the skull and the foot, both of which form first-class levers. Arms, legs, hands, fingers, toes, backbones, and jaw are all third-class levers.

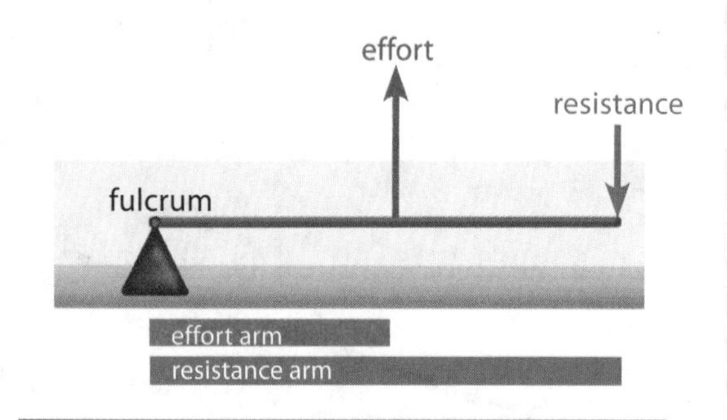

7-24 A third-class lever

Because the third-class lever has an IMA of less than 1, the resistance arm is longer than the effort arm, and thus the resistance force must travel farther than the effort force moves. In a third-class lever system, you can obtain a large amount of motion of the resistance for a small motion of the effort. This is exactly why God designed humans and animals with third-class levers for moving their skeletons.

Muscular force is generated in organisms at the cellular level. Even with millions of muscle cells acting together, they can produce only a few centimeters of motion with each muscle contraction. However, the cells' combined force can be hundreds or even thousands of newtons. When a muscle contracts (the effort), it pulls on an insertion point in a bone very close to a joint (the fulcrum). The free end of the bone (the resistance) moves through a large distance.

The advantage of the third-class lever is that it multiplies effort motion, as opposed to the other lever classes, which multiply the effort force. Just imagine how your body would appear and the limited range of motion you would have if God had designed your body with first- or second-class levers in your limbs! Other applications of third-class levers include fingernail clippers, baseball bats, sweep brooms, and rakes.

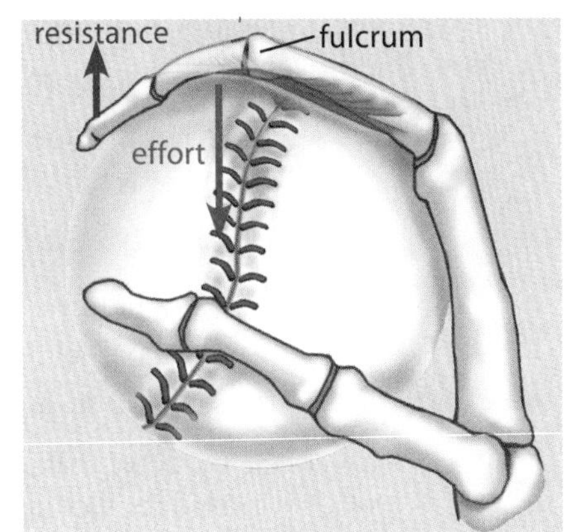

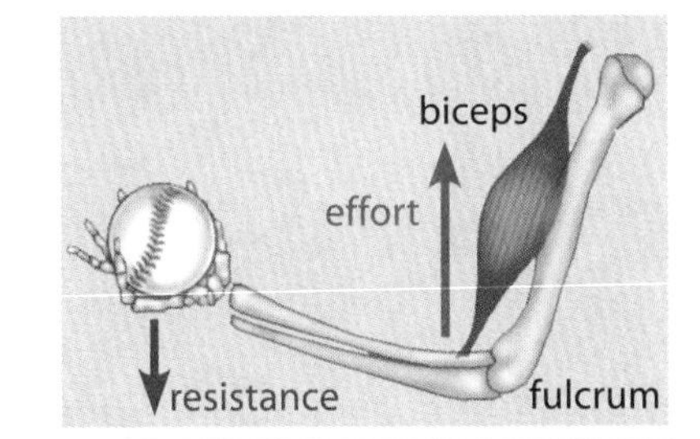

7-25 Most movable bones in the human body form third-class levers.

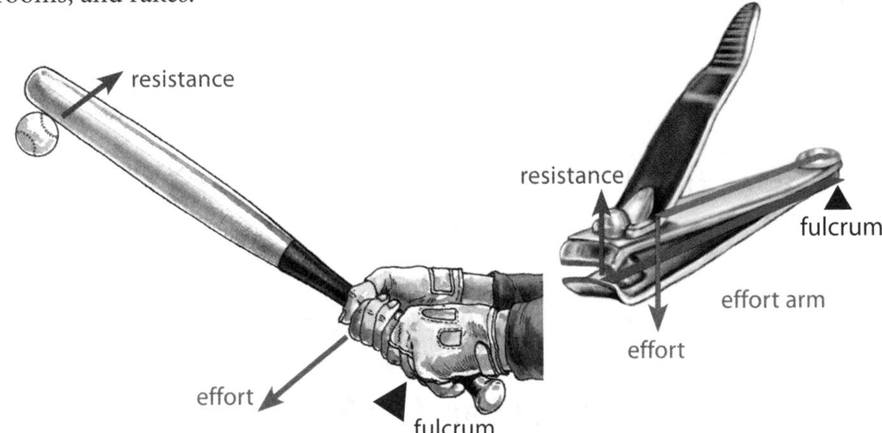

7-26 Applications of the third-class lever

7-27 Human movement would be grossly restricted without third-class levers!

7B Section Review

1. Describe the two essential parts of an ideal lever system.
2. What condition exists when a lever system is not accelerating around its pivot point? What two quantities must cancel for this condition to occur?
3. Write the general formula for the law of torques and identify each variable.
4. What are the potential values of IMA for a first-class lever? What is the relationship between the effort and resistance arms in each case?

(continued)

5. What are the potential values of IMA for a second-class lever? What is the relationship between the effort and resistance arms?
6. What are the potential values of IMA for a third-class lever? What is the relationship between the effort and resistance arms?
7. Classify each of the following levers:

   a. a hammer pulling nails
   b. a baseball bat
   c. a three-hole punch
   d. a nutcracker
   e. a leg bending at the knee
   f. scissors cutting paper

8. (True or False) The law of torques applies to a lever system under all conditions.
9. ✪ A public-works technician wants to pry up a large manhole cover to open an access to the sewer system below the street. The worker has only a 1.00 m crowbar with a resistance arm of 3.0 cm (0.030 m). How much effort will he have to exert at the end of the crowbar to lift the manhole cover if the cover exerts 445 N on the crowbar?
10. ✪ What is the IMA of the crowbar in Question 9?

7C Wheels, Gears, and Pulleys

7.12 *Introduction*

Levers are simple and effective tools for doing work, but they have one inherent restriction—they're limited in how far the effort force or resistance can move. If a lever's IMA is 5, the effort has to move five times the distance that the resistance moves according to the distance principle. Work input must be at least equal to work output. Real-world levers can accomplish only so much work before motion is stopped by some obstruction.

Other simple machines can overcome this limitation. As with levers, their designs are found in the relics of the earliest civilizations. All were probably known from the time before the Flood.

7C Objectives

After completing this section, you will be able to

- ✓ summarize how a wheel-axle system works.
- ✓ determine the IMA of a wheel-axle system.
- ✓ give examples of various kinds of wheel-axle systems.
- ✓ relate gear systems to wheel-axle systems.
- ✓ show how gear ratio relates to IMA.
- ✓ compute IMA for gear systems.
- ✓ explain how a pulley is related to a wheel and axle.
- ✓ determine the IMA for various pulley systems.

7.13 *The Wheel and Axle*

One way to continuously apply a force to a load is to use a **wheel and axle**. A wheel is usually a disk that rotates around an axis, but, functionally, it can be any rotating object in which a force is applied at some distance from the axis, creating a torque. An axle is a smaller disk, cylinder, or shaft that attaches to the wheel at its rotational axis so that the wheel and axle rotate together as a unit. A force applied to the axle is converted to a torque around the rotational axis, turning both the axle and the wheel. The wheel's radius is longer than the axle's, and their ratio, along with the locations of the effort and resistance forces, determines the IMA of the wheel-axle system.

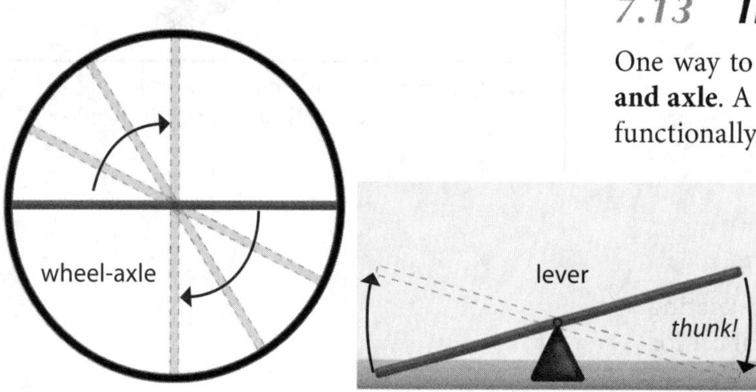

7-28 A wheel and axle is a modification of the lever principle that eliminates the restriction of motion.

A bicycle's pedal-and-sprocket-wheel system is an example of a wheel-axle system. The effort force is exerted on the pedals at a relatively large distance from the rotational axis of the front sprocket. The pedal arms transmit the force to the sprocket's axle. The resistance force exerted by the chain on the sprocket is exerted at a smaller distance from the axis. The system acts like a continuously rotating second-class lever. The IMA is the ratio obtained by dividing the length of the pedal arm (the effort arm) by the sprocket wheel's radius (the resistance arm).

$$\text{IMA} = \frac{\text{effort arm}}{\text{resistance arm}} = \frac{\text{pedal arm}}{\text{sprocket radius}}$$

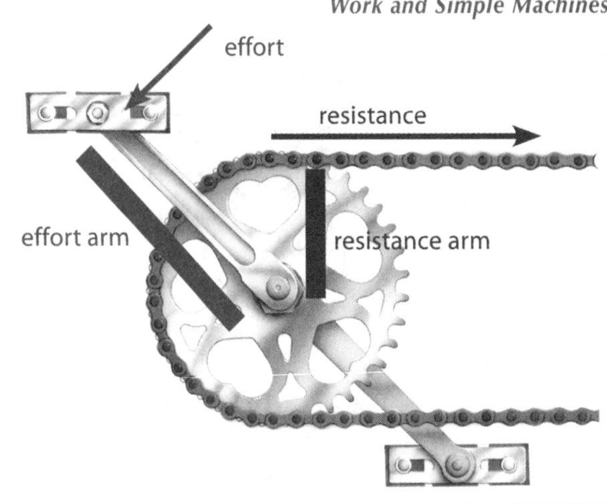

7-29 The pedals and sprocket of a bicycle act as a wheel-axle system.

A more familiar example of a wheel and axle is the drive wheel of a car. In this case, the effort is applied to the axle and the resistance is the road acting against the wheel. The effort arm (the radius of the axle) is very short—just a few centimeters. The resistance arm is the radius of the wheel and tire. The IMA of this wheel-axle system is very small, probably around 1/10 or less. Thus, to turn the wheel of a car requires a great deal of force. What advantage does it provide? It converts the small distance of the axle's rotation to a large distance of the rotating tires, yielding higher speeds. This kind of wheel-axle system is like a rotating third-class lever. The engine provides much more power than is needed to turn the wheel directly in order to obtain greater motion.

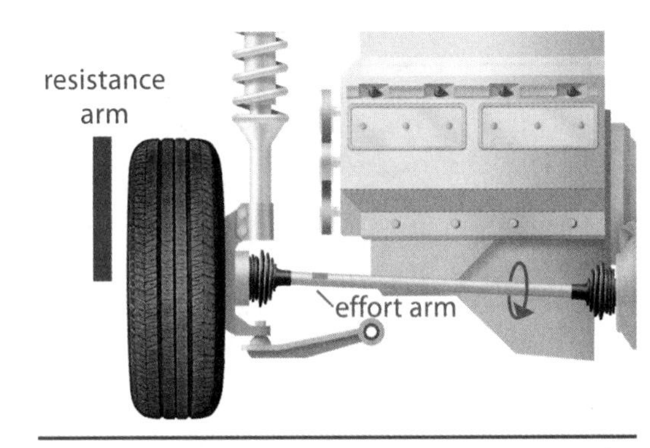

7-30 The drive wheel of a car is an example of a wheel-axle system with a mechanical advantage of less than 1.

Wheel-axle systems may not always look like a disk attached to an axle. Any device that includes a crank handle employs the wheel-axle principle. This includes can openers, fishing reels, and manual pencil sharpeners. Almost any machinery, appliance, or tool turned by a motor depends on a wheel-axle system. On a microscopic level, a natural "motor" powers the flagellum of an *E. coli* bacterium, having an efficiency of almost 100%!

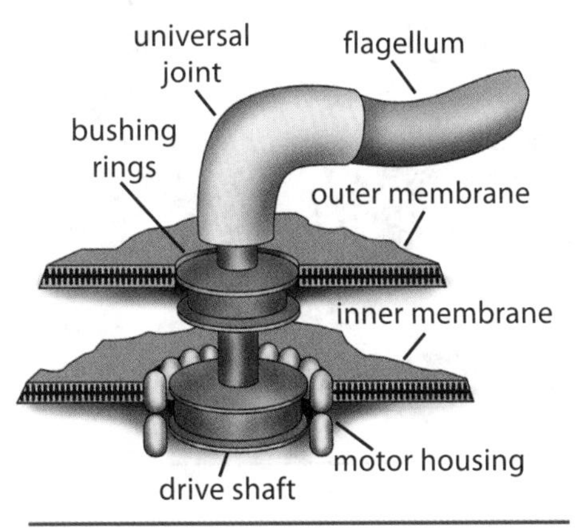

7-31 The wheel-axle principle can be found in the most unusual places—the "motor" of a bacterium!

7.14 Gears

A **gear** is a modified form of wheel and axle. It consists of a wheel with notches or teeth around its circumference. These teeth mesh with the teeth of one or more other gears to provide a reliable, nonslip way to transmit energy and do work in a wheel-axle system. Gears usually work in pairs—an *effort gear* and a *resistance gear.* The effort gear drives the resistance gear, and they rotate in opposite directions when in operation.

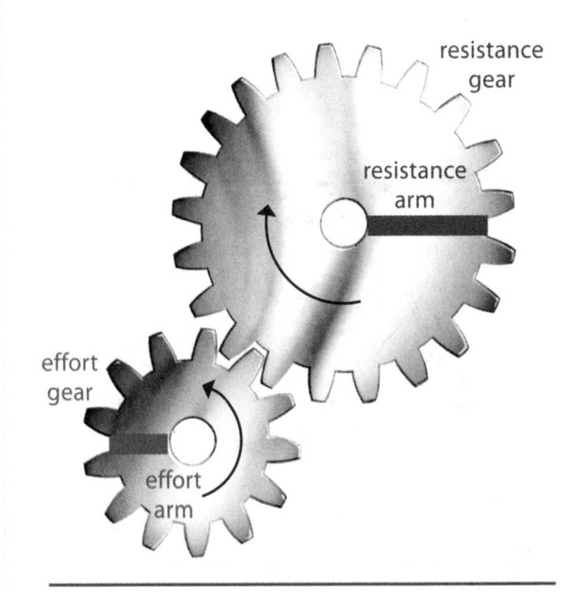

7-32 Gears are a modification of the wheel-axle principle.

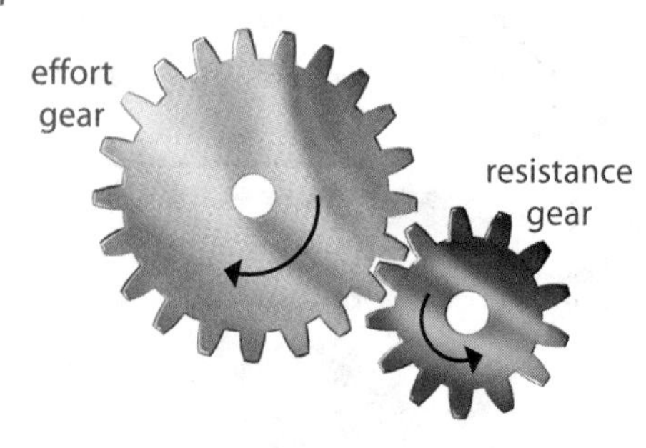

7-33 The gear ratio of the upper pair is 13:21, or 0.62; for the lower pair it is 21:13, or 1.62.

The IMA of a gear system equals the number of teeth in the resistance gear divided by the number of teeth in the effort gear.

Usually gears of different sizes work together. Pairs of gears must have the same size teeth, but the larger gear has more. If the smaller gear has a diameter that is half the size of the larger, then its circumference is half as long and it will have half as many teeth. The IMA of a pair of gears can be determined by the ratio of the number of teeth on the resistance gear to the teeth on the effort gear. Let's assume that the effort gear has 12 teeth and the resistance gear has 36 teeth. The IMA of this pair of gears is 3. The IMA also indicates how many times the effort gear will turn for every rotation of the resistance gear. This is an application of the distance principle to gears. The higher the MA, the more times the effort gear must rotate per turn of the resistance gear.

Different arrangements of gears can be used to change the speed and direction of motion. Gears can connect axles that are parallel or perpendicular to each other. Other arrangements piggyback gears on the same axle with other gears to make gear systems more compact. The transmission of a car is an example of a multipurpose, complex gear system. Examine the photograph at the beginning of this chapter. It shows the complex system of gears that operates the Great Clock of Westminster.

7-34 The manual transmission of a modern truck contains many gears.

7.15 *Pulleys and Pulley Systems*

Another application of the wheel and axle is the pulley. A **pulley** is a wheel with a groove around its outer circumference mounted on an axle. A rope, cable, or belt rests in the groove and moves with the pulley as it rotates. As with other wheel-axle systems, the pulley is a modification of a lever. The effort arm is the radius of the pulley from its center to the point where the rope leading to the effort force first touches the pulley's rim. The resistance arm is the radius from its center to the point on the rim where the rope leaves the pulley as it acts on the resistance. Since the radii are equal in a circular pulley, the effort and resistance arms are equal. The *minimum* IMA of a single pulley is 1. A pulley's IMA can be modified by how it is used.

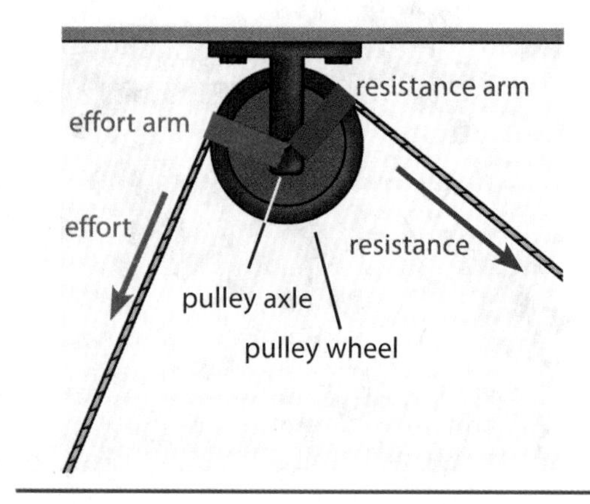

7-35 A pulley system

Single Pulleys

Pulleys must be fastened to something so that they can function. Single pulleys are mounted in a rigid frame that can be attached to a stationary support or to an object that is being moved. A pulley mounted on a fixed support is called a *single fixed pulley.* The force of the effort is transmitted through tension in the rope around the pulley to the point where the rope attaches to the resistance. When a pulley is in rotational equilibrium (the load is off the ground but not moving), the effort force and the resistance force are equal, so the MA of a single fixed pulley is 1. This type of pulley does not reduce effort, but it *is* useful for changing the direction of the effort. For most people, it is much easier to pull downward on a rope using one's weight than it is to lift an object directly from above it.

If a pulley is attached to the load being moved, it is called a *single movable pulley.* One end of the rope passing through the pulley is attached to a rigid location. The effort pulls on the free end. Notice in Figure 7-36 that the resistance load is supported entirely by the pulley, but the pulley rope is supported at two points. The resistance force is divided evenly between the rigid support and the effort, so the effort is only half of the resistance. Thus, the IMA of a single movable pulley is 2. Because of the distance principle, you must pull 2 m of rope through a single movable pulley to lift a load 1 m. Generally, *the IMA of a pulley system equals the number of rope segments supporting the load.*

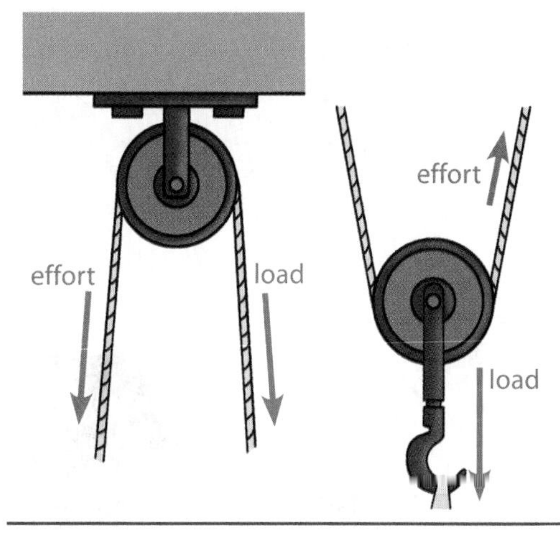

7-36 A single fixed pulley (left); a single movable pulley (right)

Pulley Gangs

Since the MA of a single pulley depends on how it is used, assembling multiple pulleys in "gangs" can combine their mechanical advantages. Figure 7-37 shows two fixed pulleys and two movable pulleys connected together in a gang by a single rope. Such a system of pulleys and cable is called a *block and tackle*, a well-established sailor's term for this equipment. How many rope segments support the load hooked to the lower set of pulleys? There are four. The effort is applied to only one of the four supporting rope segments. The rest of the load is transferred to the rigid support by the upper pulleys. You can see that the effort is only 1/4 of the resistance. The IMA for this pulley system is 4. The IMA for a block and tackle is the same as the number of rope segments or cables supporting the load.

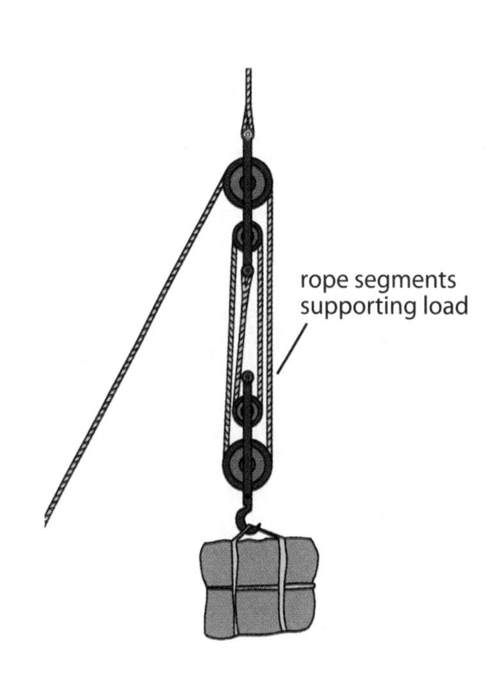

7-37 This block and tackle has four ropes supporting the load. Its IMA is 4.

Pulley gangs are used in many applications. Heavy-lift cranes commonly use a block-and-tackle system. Some weight-training machines use pulleys that allow people to exercise muscles in a variety of motions on a single machine. Engineers installing Big Ben, the 13.8-ton bell in London's Great Clock Tower, made use of many block-and-tackle systems to get it to the top of the tower.

7-38 What is the MA of this pulley system?

7-39 Rescuing the cow with a pulley system

Example Problem 7-4

Working with Pulleys

After a period of heavy rain, a rancher's 5280 N cow has become stuck in mud up to her belly. He rigs up a framework of long poles over the unhappy beast and fastens a block and tackle at the top of the frame consisting of three fixed and three movable pulleys. After fitting a belly band around the cow's middle, he hooks the block and tackle to the belly band, and the cow is lifted free from the muck.

What is the IMA of the block and tackle? Ignoring the suction force of the mud and any friction in the pulley system, how much force had to be exerted to lift the cow? If the cow was hoisted 1.3 m vertically, how much rope had to be pulled through the block and tackle?

IMA

Known: 6 supporting ropes

Solution: IMA = 6

Effort

Known: cow's weight (w_r) = 5280 N
IMA = 6

Unknown: effort (F_e)

Required formula: $\text{IMA} = \frac{w_r}{F_e} \Rightarrow F_e = \frac{w_r}{\text{IMA}}$

Substitution: $F_e = \frac{5280 \text{ N}}{6}$

Solution: $F_e = 880. \text{ N}$ (3 SDs allowed)

Effort Distance

Known: distance cow was lifted (d_r) = 1.3 m
IMA = 6

Unknown: rope pulled, or effort distance (d_e)

Required formula: $\text{IMA} = \frac{d_e}{d_r}$

Substitution: $6 = \frac{d_e}{1.3 \text{ m}} \Rightarrow d_e = 6 \times 1.3 \text{ m}$

Solution: $d_e = 7.8 \text{ m}$ (2 SDs allowed)

7C Section Review

1. What is a major mechanical limitation of lever systems? How can this be overcome?
2. How is the IMA of a wheel-axle system determined?
3. Give an example of a device that uses the wheel-axle principle, though it does not appear to *be* a wheel.
4. What is a gear? How is the IMA of a pair of gears determined?
5. In a bicycle, which has both a pedal sprocket wheel and a rear sprocket wheel, which is the effort gear and which is the resistance gear?
6. Why is a pulley's minimum IMA equal to 1?

7. If a block and tackle consists of a single fixed pulley and a single movable pulley, what is the effect of each on the pulley system's IMA and ease of use?
8. (True or False) The front-wheel drive of a modern car is an example of a wheel-axle system.
9. The hand wheel of an outdoor faucet is 7.50 cm in diameter. The faucet stem is only 0.65 cm in diameter. What is the IMA of the faucet-hand-wheel system?
10. A block and tackle has three fixed pulleys and three movable pulleys assembled together. What is the IMA of this pulley system?

7D Inclined Planes, Wedges, and Screws

7.16 *Inclined Planes*

An **inclined plane** is a two-dimensional tilted surface that allows a resistance load to be moved from a lower position to one that is higher with less effort than lifting it vertically. Inclined planes and structures based on them can be found everywhere—stairways, escalators, loading ramps, and grain conveyors. Using an inclined plane reduces the required effort force by lengthening the distance over which the force is applied, according to the distance principle.

Like the other simple machines, the origin of inclined planes in ancient history is obscure. The earliest civilizations document the use of inclined ramps for the purpose of constructing great structures such as the Egyptian pyramids. Marine engineers believe that the door built into the Ark probably provided access to its second floor, more than 10 cubits (18 ft) above the ground. If Noah used a ramp to load and unload the animals, food, and passengers, then the inclined plane would have been known as early as the Flood, and likely much earlier.

7-41 An inclined ramp might have provided access to the Ark.

7D Objectives

After completing this section, you will be able to

- ✓ give real-world examples of inclined planes.
- ✓ calculate the IMA of inclined planes.
- ✓ give real-world examples of wedges.
- ✓ give real-world examples of screws.
- ✓ compare and contrast inclined planes, wedges, and screws.
- ✓ explain how screw TPI/pitch relates to the IMA and ease of use of a screw.

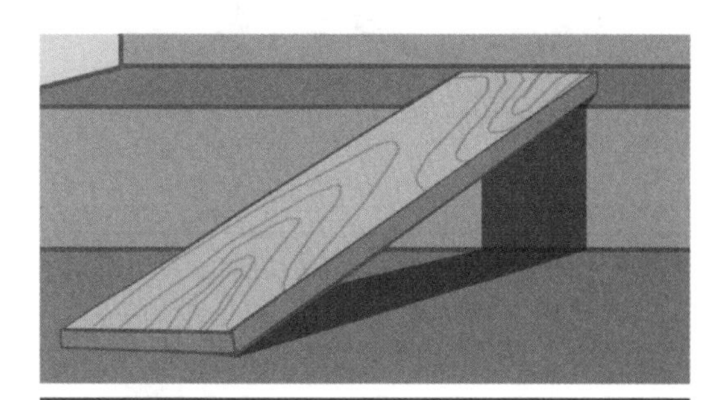

7-40 An inclined plane

How do inclined planes provide MA? Recall from Subsection 7.5 that one way to raise a grand piano 2.0 m to an auditorium stage is to use a block and tackle. What if the ceiling is too high or too weak to use a block and tackle to support the piano? One solution is to use a ramp, say 5.0 m long. How much force is required to roll the piano up the ramp?

We assume that the piano's acceleration while being lifted is negligible, so the lifting force is essentially equal to the piano's weight.

Begin by calculating the total work to be done. The 3300 N piano must be raised 2.0 m to the stage, no matter what method is used. The total work required is 6600 J ($W = F \times d = 3300 \text{ N} \times 2.0 \text{ m}$). Pushing the piano up the 5.0 m ramp still accomplishes 6600 J of work. We can use the work formula to calculate the force required to roll the piano up the ramp, neglecting friction.

$$W = F \times d$$

$$6600 \text{ J} = F \times 5.0 \text{ m}$$

$$F = \frac{6600 \text{ J}}{5.0 \text{ m}} = 1\underline{3}20 \frac{\text{N}\cdot\cancel{\text{m}}}{\cancel{\text{m}}}$$

$$F \approx 1300 \text{ N}$$

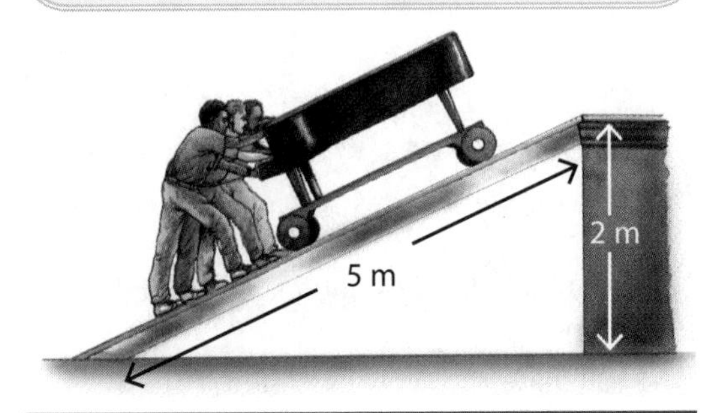

7-42 An inclined plane can be used to move heavy objects from one level to another.

Though this value (nearly 300 lb) is still a lot of force to exert, several strong men could probably manage it. A block and tackle could even be attached horizontally to something on the stage and used to pull the piano up the ramp.

What is the MA of the inclined ramp system? We can divide the resistance by the effort to find the AMA, ignoring friction. In this case, the resistance is the piano's weight, 3300 N, and the effort using the ramp is 1$\underline{3}$20 N.

$$\text{AMA} = \frac{\text{weight}}{\text{effort}} = \frac{3300 \cancel{\text{N}}}{1\underline{3}20 \cancel{\text{N}}}$$

$$\text{AMA} \approx 2.5$$

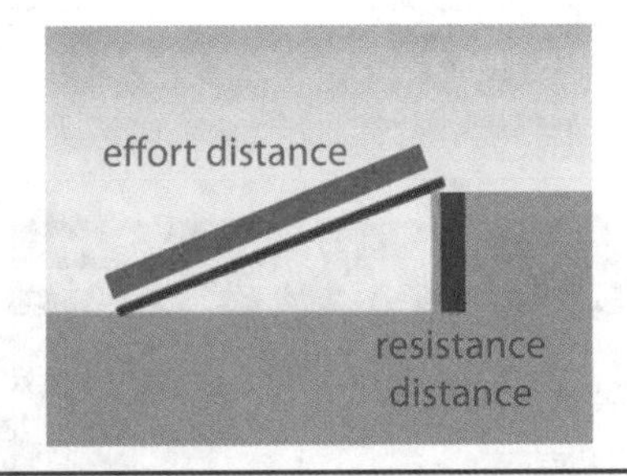

7-43 The IMA of an inclined plane is equal to its length (effort distance) divided by its height (resistance distance).

Another, more direct method to calculate the IMA for an inclined plane is to divide the ramp's length by its height. This is equivalent to dividing the effort distance by the resistance distance.

$$\text{IMA} = \frac{\text{incline length}}{\text{height}} = \frac{5.0 \cancel{\text{m}}}{2.0 \cancel{\text{m}}}$$

$$\text{IMA} = 2.5$$

The purpose of an inclined plane is to reduce the effort needed to move from a lower to a higher elevation. Highway engineers design roads in mountainous terrain to have small inclines that keep forces low as vehicles climb a hill. This reduces the power required and saves gas. Similarly, the steepness of stairs in public buildings must be held below a certain value to make climbing easier for people of all ages and physical conditions.

7-44 Stairs use the inclined plane principle to allow people to move from one level to another.

7.17 Wedges

One application of the inclined plane is the *wedge*. A wedge is similar in shape to an inclined plane but is a three-dimensional object that exerts a force to spread material apart as it is forced into the material. The magnitude of the spreading force depends on the IMA of the wedge. The IMA for a wedge is calculated just like that of an inclined plane—the length of the inclined surface divided by the width of the wedge. Thus, the narrower the wedge, the more effective it is in splitting the material.

Wedges come in a variety of shapes depending on their use. A woodpecker's beak and the teeth of the *Tyrannosaurus rex* are both examples of wedges in nature. Mangrove seeds are shaped like wedges so that they can penetrate swampy soil when they fall. Artificial wedges include door stops, shovel blades, scissors, the claws of a carpentry hammer, wind deflectors on trucks, and some sports car profiles.

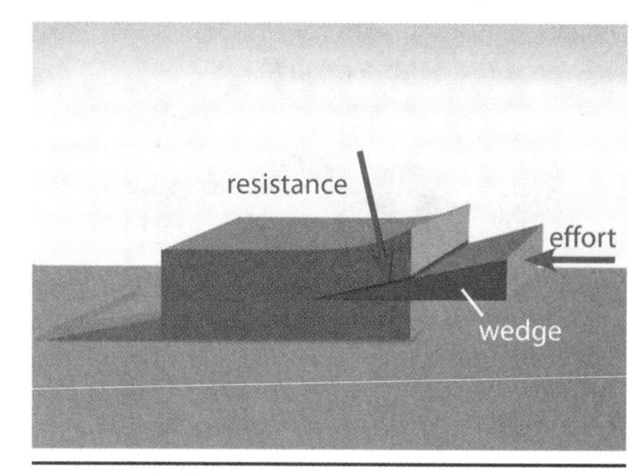

7-45 A wedge in action

Some wedges, such as the blades of kitchen knives, wood carving tools, and carpentry planes and chisels, are used to cut relatively soft materials. They contain wedge-shaped surfaces that are very narrow, having sharp edges for rapid cutting. This type of tool has a large MA. A tool for cutting denser or harder materials has a wider wedge-shaped cutting edge with a smaller IMA. Some applications require a wedge that looks like two wedges placed base-to-base, called a *double wedge*. This application is necessary for tools where the tip needs more support close to the edge to strengthen it and keep it from dulling or breaking. In this case, tool designers trade off cutting efficiency for tool durability. Examples of double wedges are wood and metal lathe-cutting tools, chopping axes, and masonry cold chisels. The bows of ships and boats also employ a double wedge to cut efficiently through the water.

7-46 Applications of the wedge principle

7.18 Screws

The inclined-plane concept finds several additional applications in the use of screws. A *screw* is a fastener that exerts a large amount of force to hold objects together. It consists of a long, thin wedge wrapped around a shaft in a spiral or **helix**. This forms ridges on the shaft called *threads* (see Figure 7-47). Threads anchor the screw in place and provide a means to pull it into the material when it is turned.

helix: (Gk. *helix*—spiral-like)

A screw's purpose will determine its shape and the space between its threads. Cylindrical screws with closely spaced threads are called *bolts*. Bolts may be screwed directly into pre-threaded holes in metal objects or inserted through two pieces and fastened by a threaded nut. In the United States, bolts are classified by their *thread count*, the number of *threads per inch (TPI)*. The more threads per inch, the more turns must be used to screw the bolt. This is another application of the distance principle. More turns per inch

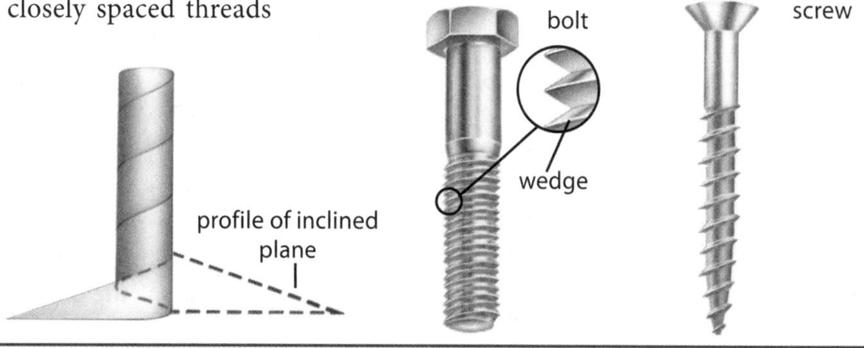

7-47 Threaded fasteners use both the inclined plane and the wedge.

A large TPI yields a large MA.

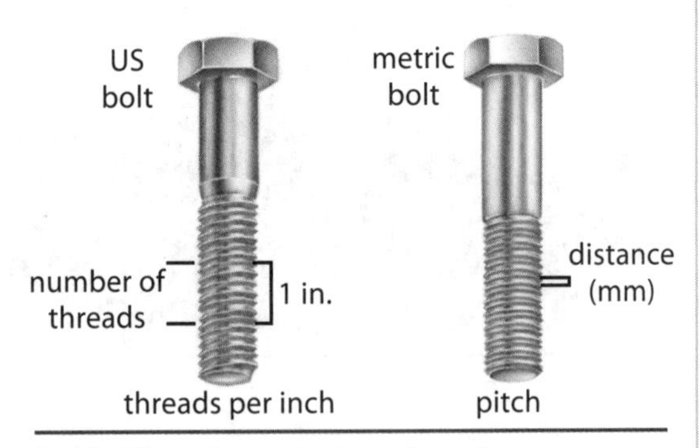

7-48 Threads per inch and pitch

7-49 This huge screw helps manage storm runoff water in Texas.

means that the angle of the thread is smaller, like a long inclined plane with a small slope, which produces a larger IMA. Metric fasteners are classified by the distance between adjacent threads in millimeters. This is the *pitch* of a bolt's thread. A smaller pitch implies a larger MA. Bolts and screws with large TPI or small pitch are used to grip tightly in hard, dense materials, such as metals. These bolts are generally easy to turn when fastening them into place.

In addition to use as fasteners, screw systems are used for lifting heavy objects safely and simply. *Jackscrews* are tools used by construction workers to lift building beams into place or lift a house off its foundation. They are also used in some repair garages for lifting vehicles. In large theaters, jackscrews may be used to elevate or lower parts of a stage. Since most jackscrews are operated by hand or by relatively small motors, they use a high TPI to gain the most MA possible. Using a manual jackscrew can be laborious even though the required force is small, because it has to be turned many times to obtain a small amount of lift.

Some things in nature move like a screw. Time-lapse photography shows that a wild oat seed, driven by the wind, burrows into the soil like a screw. The flagella of some bacteria, such as the spirochete that causes Lyme disease, spin like corkscrews to propel the cells along.

7D Section Review

1. What is the general purpose of an inclined plane? How is an inclined plane's IMA determined?
2. Ideally, how does moving an object up a ramp compare to lifting the object the same distance vertically, in terms of work done? What is the advantage of using an inclined plane?
3. Describe a wedge. How does the height of a wedge determine how much force it can exert to cut or split?
4. Why is a double wedge sometimes necessary?
5. Describe a screw. Note the ways an inclined plane relates to the design of a screw's thread.
6. How does a screw's thread count or pitch determine how strongly it fastens and how hard it is to screw into a material?
7. Give an application of a screw that does not involve fastening two things together.
8. ✪ How could the corkscrew flagellum of a spirochete be redesigned to move the bacterium at a higher speed? Assume that the flagellum's molecular motor spins at a constant rate.
9. (True or False) The larger the pitch, the greater the MA of the screw.
10. ✪ The vertical distance between floors in a house is 3.0 m. How much work must be done by a 602 N student to walk up a 4.25 m stairway?
11. ✪ The threaded shaft of a bolt is 33 mm long and its pitch is 1.1 mm. How many turns would be required to completely screw the bolt into a hole?

Scientifically Speaking

Chapter Review

1. Write the work formula and completely describe each of the terms in the formula.
2. Gravity is a force that acts continuously on all objects on Earth. Under what condition does gravity do *no* work on a moving object?
3. Define *power*. On the basis of the power formula, what change in conditions can *increase* the amount of power?
4. Explain why you can't obtain as much useful work out of a simple machine as you put into it. What property of the machine is indicated by the percentage of work input that is converted into usable work?
5. What property of a simple machine tells how much effort is decreased proportionately by using that machine compared to doing the work directly? What principle ensures that energy is conserved when using a simple machine?
6. How does a balanced seesaw relate to Newton's laws of motion? What kind of equilibrium exists when a lever system is balanced?
7. For a simple lever to be balanced, what two conditions must be true about the torques acting on the lever?
8. Describe a first-class lever. Is it always possible to obtain an IMA of more than 1 with a first-class lever? Explain.
9. Describe a second-class lever. Why is the IMA of this class of lever always greater than 1?
10. Describe a third-class lever. Why is this kind of lever useful?
11. Under what condition would the IMA of a wheel-axle system be greater than 1?
12. How is a gear system's IMA determined?
13. In the real world, how will the AMA of a pulley (or any simple machine) compare to its IMA?
14. What is the most direct method for calculating the IMA of an inclined plane?
15. What design feature of a wedge improves its cutting or wedging ability? Why would a different shape be necessary for some uses?
16. How does TPI relate to how easily a screw can be turned into a material? Explain.

True or False

17. Work is done on an object to increase its kinetic or potential energy.
18. In lever systems, the IMA is found by dividing the length of the resistance arm by the length of the effort arm.
19. Simple machines must have mechanical advantages greater than 1 to be useful.

20. One important purpose of gears is to change the direction of rotary motion.
21. Unlike other wheel-axle systems, pulleys and pulley systems always have an IMA of 1 or more.
22. Inclined planes decrease the amount of work that must be done when raising an object from a lower level to a higher one.
23. A thicker wedge takes more effort to use than a thinner one.
24. The main purpose of the helical thread on a bolt or screw is to allow insertion using a twisting motion.
25. The best jackscrew to use for lifting a house off its foundation is one with a large pitch.

✪ 26. The wreck of the Confederate submarine CSS *H. L. Hunley*, the first military submarine to sink an enemy warship, was discovered 9.5 m below the surface of the water. If the recovery crane exerted 62,000 N to slowly lift the sub's remains to the ocean's surface, how much work did the crane do?

✪ 27. American aircraft carriers have huge elevators to lift aircraft from the hangar deck inside the ship to the flight deck. Lifting a fully ready Navy fighter-bomber requires 540,000 J of work. If an elevator can complete the lift in 10. s, how much power, in kilowatts, must the elevator use just to lift the plane?

✪ 28. A homeowner is dismantling his family-room wall to do some renovations. His pry bar is 0.65 m long. When prying two studs apart using the bar as a first-class lever, the fulcrum point is 3 cm from the prying end. How much resistance do the nails holding the boards together exert if he applies 130 N to the bar to remove them? (*Hint*: Sketch out the pry bar and label the different lever arms.)

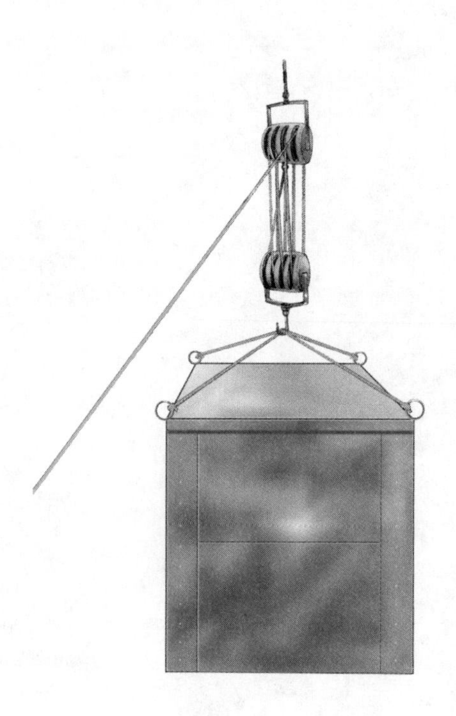

✪ 29. Assume that a man's bicep connects to the major forearm bone at a point 3 cm from the elbow joint. If the palm of his hand is 30 cm from the elbow, what is the IMA of the forearm lever? What class lever is the forearm?

✪ 30. The block and tackle shown in the diagram is used to lift a 33,000 N load. What is the MA of the pulley system? How much effort will be required to move the load, ignoring friction?

✪ 31. The 3.5 m loading ramp to a moving van spans a distance of 3.3 m between the door of the house and the truck. The high end of the ramp is 1.2 m above the low end. What is the ramp's IMA?

✪ 32. The thread of a 4.50 cm-long machine bolt wraps 36 complete turns around the bolt. What is the pitch of this bolt?

FLUID MECHANICS

CHAPTER 8

DOMINION SCIENCE PROBLEM

Tsunami Detection and Warning

On 11 March 2011, an earthquake measuring 9.0 occurred in the seabed 72 km (45 mi) east of the main island of Honshu, Japan. This was the fifth largest earthquake ever recorded and the largest to ever strike Japan. While the earthquake itself caused significant damage, even at that distance, a powerful tsunami generated by the sea-floor movement soon swamped the Japanese islands and was detectable as far away as Chile in South America. At one location in Japan, the wave measured nearly 39 m (128 ft) high! As a direct result of this event, nearly 16,000 people died, more than 6000 were injured, and 2000 are still missing. Hundreds of thousands of buildings collapsed or were severely damaged. The tsunami wave also disabled the emergency power generators for many nuclear power plants located near the coast, resulting in several damaged reactors.

Unlike the countries surrounding the Indian Ocean at the time of the December 2004 tsunami caused by the powerful Sumatra earthquake, Japan had some warning of the incoming wave. In the 2004 event, more than 200,000 people died because they were caught unawares near the seacoast. Even so, entire Japanese villages disappeared under the wave. What can we do in an attempt to avoid such tragic loss of life?

8-1 This tsunami wave wiped out the village of Natori, Miyagi Prefecture, Japan, following a major earthquake in 2011.

8A Objectives

After completing this section, you will be able to

- ✓ identify what is studied in fluid mechanics.
- ✓ define pressure.
- ✓ show how different physical properties affect pressure.
- ✓ calculate pressure when given applied force and area.
- ✓ recognize units of pressure.
- ✓ describe how instruments can measure pressure.
- ✓ state Archimedes's principle in your own words.
- ✓ calculate specific gravity.

8A Properties of Fluids

8.1 Introduction

Tsunamis can be explained by the mechanical forces exerted by air and water. Chapter 6 describes how moving objects possess energy, and Chapter 7 explains how forces acting on objects can do work. Fluids can exert forces and do work too. Because of the special physical properties of fluids, such as their lack of rigidity, physicists have established a distinct area of study within mechanics called fluid mechanics. **Fluid mechanics** is the study of how fluids flow and how forces and energy are transmitted through fluids.

Both liquids and gases flow and assume the shape of their containers. Matter that has these properties is called a *fluid*. Fluids exert pressure, which allows them to transmit forces. They exert an upward force that acts against the force of gravity on any immersed part of an object. In this chapter you will learn how motion, forces, work, and energy apply to liquids and gases.

8-2 Flowing fluids can do work, such as turning the turbines that generate electrical energy or running an air tool.

8.2 Pressure

While riding in a car on a trip through mountainous terrain, you perhaps have experienced a sharp pain in your ears that is relieved only by swallowing. This condition is caused by a rapid change in air pressure against your eardrum. These abnormal forces on the delicate eardrum membrane can cause great discomfort. So what exactly is pressure?

The **pressure (*P*)** on a surface is a ratio of force to area—the net force acting perpendicular to the surface and the area over which it acts. Pressure is calculated using the formula

$$\boxed{P = \frac{F}{A}},$$

where

- P is pressure, in newtons per square meter,
- F is the net force acting perpendicular to the surface, in newtons, and
- A is the total surface area over which the force is applied, in square meters.

The larger the area over which a given force is distributed, the lower the pressure. For example, suppose you place a book flat on a table. The book exerts 10. N on the table. The book's cover has an area of 580 cm^2 (0.058 m^2), and the pressure the book exerts on the table is distributed over this area.

$$P = \frac{F}{A} = \frac{10.\ \text{N}}{0.058\ \text{m}^2}$$
$$P \approx 170\ \text{N/m}^2$$

Now take the same book and balance it on top of a short piece of dowel rod 2.5 cm in diameter. The dowel's cross-sectional area is calculated as $A = \pi r^2 \approx 4.9\ \text{cm}^2$, or $4.9 \times 10^{-4}\ \text{m}^2$. The pressure exerted on the table by the book through the dowel is now distributed over this smaller area.

$$P = \frac{F}{A} = \frac{10.\ \text{N}}{4.9 \times 10^{-4}\ \text{m}^2}$$
$$P \approx 20{,}400\ \text{N/m}^2$$

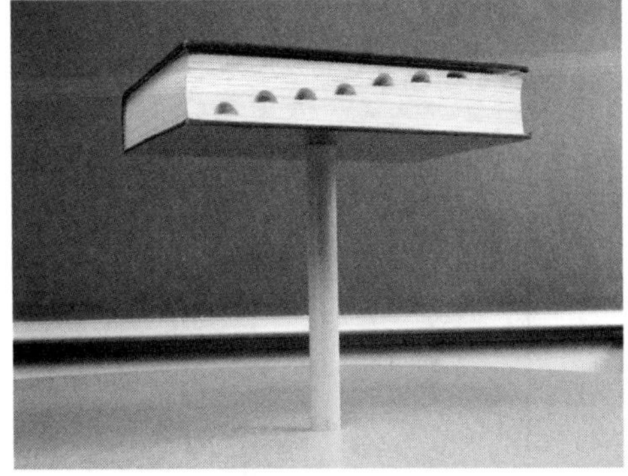

8-3 Pressure exerted by a force is determined by the area over which the force is applied.

Fluid Pressure

In the book-dowel example, the only pressure exerted was the vertical weight of the book on the surface area in contact with the table. Pressure was not exerted in any other direction because solid, rigid objects can transmit a net force in only one direction.

With liquids and gases, the situation is somewhat different. Imagine that you could isolate an exceedingly thin horizontal slab of water with an area of 1 m^2 in the center of a still swimming pool (see Figure 8-4). We know, by Newton's first law of motion, that if the water in the slab is not accelerating, all forces acting on it must be balanced. In particular, the downward force above the water slab must equal the upward force below the slab. Since both forces act on equal surface areas, the pressures above and below the slab must be equal. Now, rotate the imaginary slab 90°. The same conditions exist. The horizontal forces, and thus the pressures, on both sides of the vertical slab are equal because it does not accelerate. We can conclude that at a given point in a body of water, the pressure is exerted equally in all directions. This property of fluids is called **fluid pressure**.

Fluid pressure exists because liquid and gas particles are not held rigidly in place. The kinetic-molecular model of matter says that the particles of matter, especially in fluids, are in constant, random motion. When a force is exerted on a fluid particle, it transmits the force in the direction the particle is moving. Because of the innumerable randomly moving particles, the force is transmitted equally in all directions.

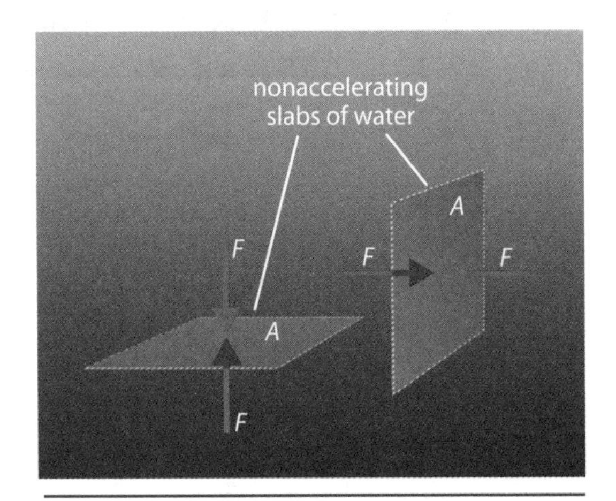

8-4 Thin slabs of water in a still swimming pool do not accelerate. Therefore, the forces on them are balanced.

Blaise Pascal (1623–62) was a French mathematician, physicist, and philosopher known for his theory of probability and his work with pressure.

Units of Pressure

1 bar = 1×10^5 Pa

1 bar = 1000 mb

1 atm = 1013 mb

1 atm = 14.7 lb/in.^{2} (English unit, psi)

1 atm = 760 mm Hg (Hg barometer)

Units of Pressure

In the SI, pressure is measured in newtons per square meter (N/m^2) or with the *pascal (Pa)*, the derived SI unit for the same quantity. This unit is named after the prominent seventeenth-century philosopher and physicist Blaise Pascal.

$$1\ \text{N/m}^2 = 1\ \text{Pa}$$

There are also other units of pressure. The average atmospheric pressure at sea level is one atmosphere (atm). One atmosphere is 101,325 Pa. Scientists defined the bar (b) as being equal to 100,000 Pa to make studying the atmosphere easier. Average atmospheric pressure (1 atm) is slightly higher, 1.013 bar. Meteorologists today report air pressure in millibars (mb). The bar is no longer used in most scientific measurements.

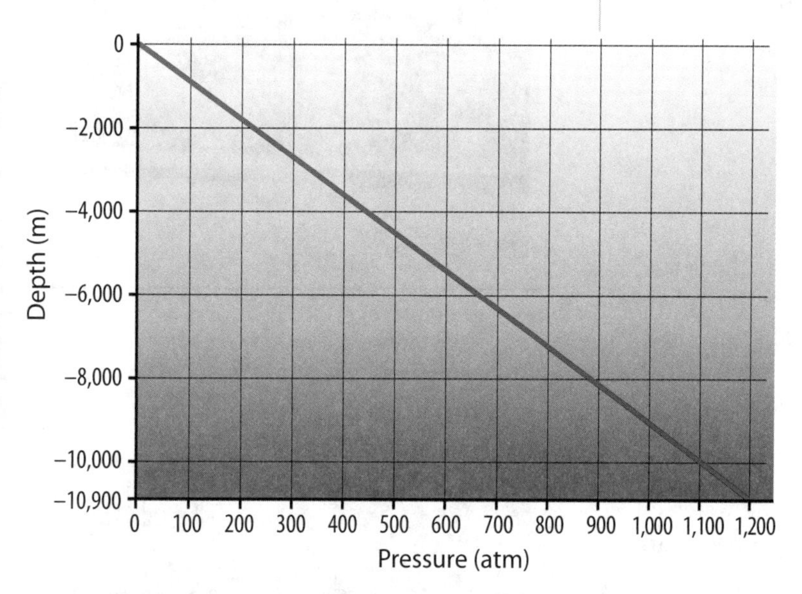

8-5 Water pressure increases with depth.

Factors Affecting Fluid Pressure

Fluid pressure may exert itself equally in all directions at any specific point in a fluid, but that doesn't mean that the pressure is equal throughout the fluid. Pressure can change greatly within a fluid. That is why you feel pressure in your ears when changing altitude in a plane or changing depth in the water. Under natural conditions, gravity and fluid properties determine the pressure at a given point in the fluid.

Fluids have weight because every fluid particle has a mass that experiences the pull of gravity ($w = m \times g$). Each fluid particle exerts its weight on surrounding particles, as Newton's third law predicts. That force, or weight, is transmitted downward through the fluid. Thus, the deeper the volume of a fluid, the more weight is experienced at the bottom of the fluid and the higher the pressure. For example, the earth's surface is at the bottom of its atmosphere, so the maximum pressure in the atmosphere is experienced there. As one goes higher in elevation, there is less air above, so air pressure decreases. Similarly, water pressure is zero at the surface of a body of water. As one goes deeper, the amount of water above increases, and so pressure increases with depth. The rate of pressure increase in water is approximately 1 atm (14.7 lb/in.^{2}) for every 10 m (33 ft) of depth. Deep-water instruments and submarine vehicles must be designed to withstand the high pressures of deep depths.

The volume or shape of a fluid's container does not affect the pressure at a given depth or height in the fluid. For example, the

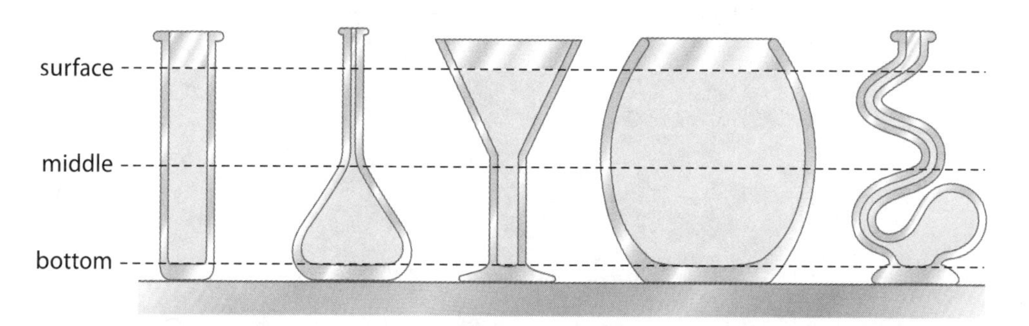

8-6 Fluid pressure does not depend on the shape or volume of the fluid.

water pressure at a depth of 10 cm is essentially the same in a glass of water as in the Pacific Ocean. Water pressure depends almost entirely on the water depth. Water pressure due to depth alone is called **hydrostatic pressure**.

Fluid density is another factor that determines fluid pressure. When particles are closer together, the weight they exert for a given change in depth or height is greater. The effect of gravity on *liquid* density is fairly insignificant because liquids are nearly incompressible. Therefore, the increase of density with depth in liquids is insignificant. Water temperature and dissolved matter are more important factors in determining water density. For example, cold water is denser than warm water, and salty water is denser than fresh water. Overall, depth is the most important factor controlling hydrostatic pressure in liquids.

Recall that density is the amount of mass per unit of volume. Density increases if the number or mass of the particles present in a volume increases.

Atmospheric gases are highly compressible. Air molecules closer to the earth are more strongly attracted by the earth's gravity than those higher up. As a result, the earth's gravity hugs the blanket of air closely to the surface. Air density and atmospheric pressure decrease rapidly with increasing altitude. Half the matter in the atmosphere is below 5.5 km (3.5 mi) and 99% of the atmosphere is below 32 km (20 mi). Nearly all the atmosphere is below 47 km (29 mi). Because temperature affects the volume of a gas much more than it affects the volume of liquids, temperature also has a much greater effect on atmospheric density than on water density. The pressure differences between cold and warm air masses can vary up to 10% of total atmospheric pressure.

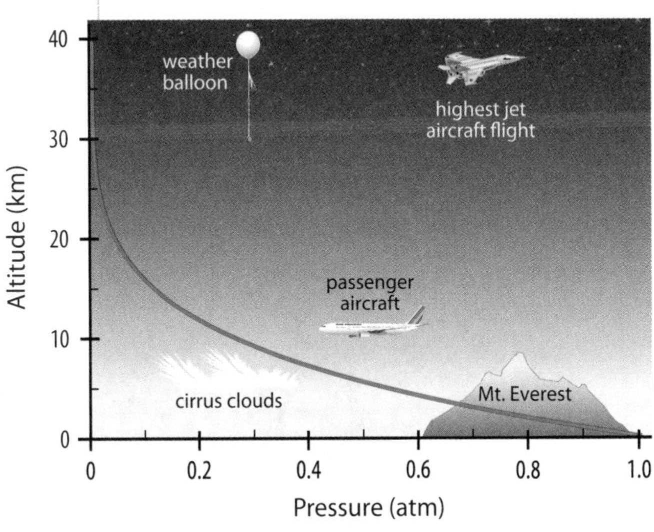

8-7 Atmospheric pressure decreases rapidly with altitude because air is highly compressible. Air is much denser near the earth's surface than higher up.

Pressure Instruments

Nearly every pressure instrument is designed to compare a measured pressure to a known pressure. The earliest accurate pressure instrument, built by Evangelista Torricelli in 1643, measured atmospheric pressure. Torricelli's **mercury barometer** consisted of a thick-walled glass tube filled with liquid mercury and sealed on one end. When the tube was upended in a bowl of mercury so that air wouldn't enter the tube, the level of mercury in the column dropped a short distance. At that point the atmospheric pressure on the mercury in the bowl equaled the weight of the mercury in the column. The height of the mercury column, about 760 mm, was proportional to atmospheric pressure. The vacuum in the sealed end of the tube was the standard to which atmospheric pressure was compared. The pressure indicated by a mercury barometer is sometimes called *absolute pressure* because its standard is a vacuum, a condition of zero pressure.

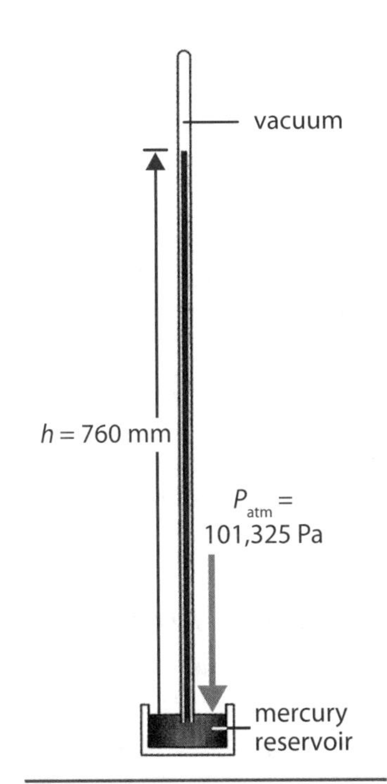

8-8 Schematic of a mercury barometer

Evangelista Torricelli (1608–47) was an Italian physicist who served as Galileo's secretary. He is best known for creating the first true vacuum and inventing the mercury barometer.

aneroid (AN uh ROYD): a- (Gk. *a*—without) + -ner- (Gk. *neros*—wet or dampness) + -oid (Gk. *-oeides*—a likeness to)

Modern ***aneroid barometers*** are more compact than the meter-long mercury barometer. Aneroid barometers contain a sealed accordion-like can filled with a low-pressure gas. The can, which flexes with changing atmospheric pressure, is linked to a pointer through a gear mechanism. It uses the gas pressure within the flexible can as its standard. Digital barometers rely on an aneroid mechanism to sense the air pressure and convert the flexing motion into a digital signal.

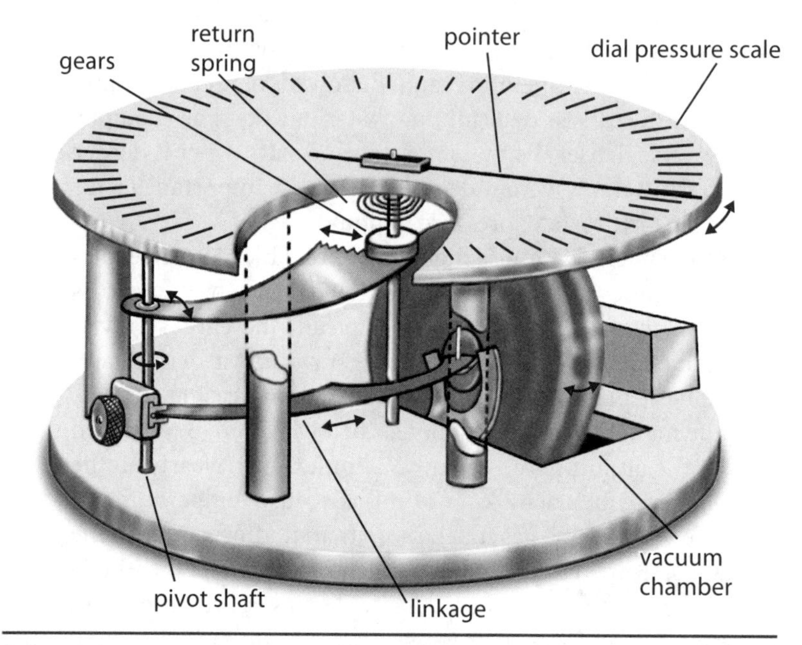

8-9 Schematic of an aneroid barometer

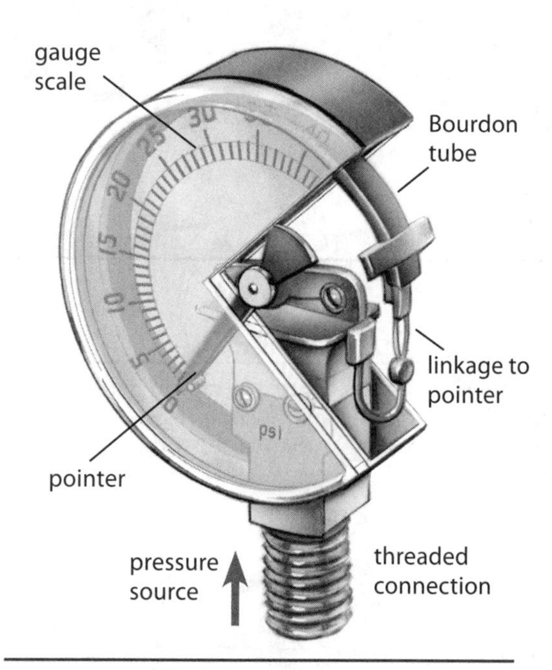

8-10 As pressure increases in the tube, it tries to straighten, actuating a needle or digital position sensor.

Fluid pressures in mechanical systems and pipes are measured with mechanical *gauges.* These instruments are built around a flexible sealed can or a sealed tube that is bent into a U or a coil. As fluid pressure increases, the tube unbends or the can expands, moving a mechanism attached to an indicator. The reference (zero) pressure for these instruments is set at atmospheric pressure. So when the gauge indicates exactly zero pressure, it is actually measuring atmospheric pressure. *Gauge pressure* in a system equals the absolute pressure in the system minus atmospheric pressure.

Some pressure-sensing instruments use a C-shaped sealed tube detector called a ***Bourdon tube.*** A Bourdon tube senses changes in pressure by flexing as fluid pressure changes inside the tubing. Unlike other pressure instruments, Bourdon tube instruments are very rugged and suitable for extreme environments.

The Bourdon (BOOR dn) tube pressure detector was named in honor of Eugène Bourdon (1808–84), a French engineer who invented a reliable steam pressure gauge.

Did you know . . .

Fluids stored under artificial or abnormal conditions can exert extreme pressures. Gas bottling plants store some gases at millions of pascals in strong cylindrical tanks. Heated ground water can be trapped deep underground at pressures of hundreds of millions of pascals.

8.3 Using Fluid Pressure to Solve Problems

Nine years before the Indian Ocean tsunami in 2004, the United States National Oceanic and Atmospheric Administration (NOAA) began installing special instruments in the Pacific Ocean. These instruments are designed to detect passing tsunamis that could threaten the United States and other countries around the Pacific basin. The sensors, called *tsunameters,* are placed on the bottom of the ocean with anchors at depths of 1000–6000 m.

As a tsunami passes overhead, the sea level changes less than a few meters in deep water. Tsunameters use a Bourdon tube pressure detector to measure the tiny change in sea pressure. Its C-tube straightens slightly as it expands in response to the pressure rising inside it. The free end of the C-tube moves a digital sensor, whose data is analyzed by a computer. When a tsunami is detected, an acoustic signal is transmitted like sonar through the water to a buoy on the surface. The buoy sends an alerting radio message to an overhead satellite, which relays the alert to a tsunami warning station.

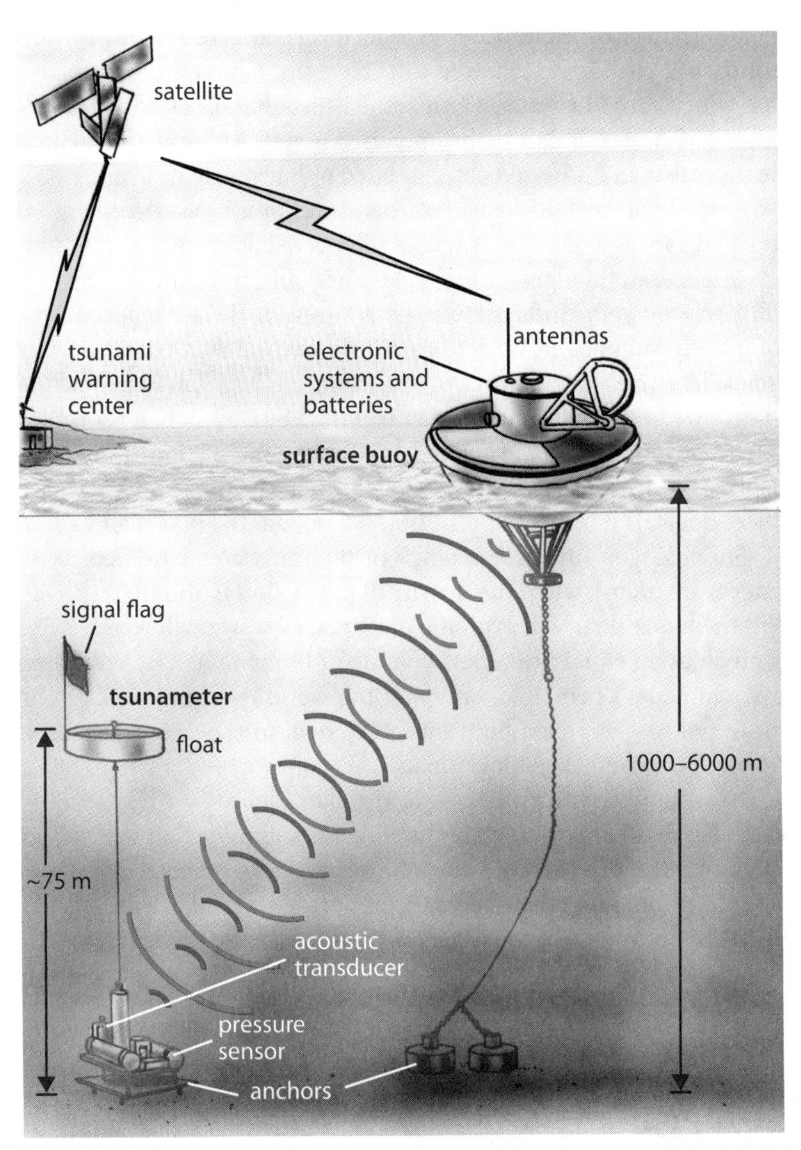

8-11 NOAA's Deep-ocean Assessment and Reporting of Tsunami (DART II) buoy system

Because of the 2004 Indian Ocean tsunami, the United States moved quickly to complete its Pacific Ocean system of 32 buoys and to expand tsunami warning coverage to the Atlantic Ocean and the Caribbean Sea. Japan received alerts from the Pacific system in 2011, but the earthquake was so close to shore that little time was available to move people away from the danger areas. Southeast Asian nations are now seeking to install a similar buoy warning system in the seismically active waters surrounding them to provide needed alerts.

8-12 Archimedes discovered the principle of buoyancy when he tested the purity of his king's gold crown.

8-13 The upward buoyant force equals the weight of the water displaced by the object. However, the object does not actually have to move that quantity of fluid aside to float.

8.4 Buoyancy

While playing in the family pool, ten-year-old Beth asked her dad to pick her up and throw her back into the water. He quickly discovered that it was easy to pick her up partway, but as he lifted her out of the water she became heavier and heavier. His plan of flinging her halfway across the pool was quickly reduced to merely tossing her a few feet. At first she was so light—but she apparently gained weight as she came out of the water! What happened?

Objects weigh less in water or other fluids than they do when they are not immersed. Archimedes explained this observation in the third century BC. He learned that when an object sinks into water, it displaces a certain volume of water. He reasoned that the displaced water has weight and exerts a force on the immersed object. **Archimedes's principle** states that an immersed object is lifted or buoyed up by a force equal to the weight of the fluid displaced. This lifting force is called the *buoyant force* and applies to floating as well as submerged objects.

In general, the densities of a fluid and an object immersed in it will likely be very different. So the amount of fluid displaced by a completely submerged object has a different weight than the object itself. Thus, the buoyant force of the displaced fluid will not equal the object's weight. Whether an object will float or sink depends on how much buoyant force is generated by the volume of fluid the object displaces. If the buoyant force is greater than the object's weight, the object floats. If it is less than the object's weight, the object sinks.

Suppose you immerse a block of pine in water. The density of water is 1.0 g/cm^3, while the density of pine wood is about 0.5 g/cm^3. Will the block float? The volume of displaced water will weigh twice as much as an equal immersed volume of the pine block. According to Archimedes's principle, only half the wood's volume needs to be immersed to generate a buoyant force equal to the entire weight of the block of wood. The block floats!

Now suppose you immerse a lead sinker (density = 11.3 g/cm^3) in water. No matter how much of the sinker is submerged, it will always weigh much more than the buoyant force of its displaced volume of water. Without any other forces to support it, the sinker will accelerate to the bottom (Newton's second law). Hence the name "sinker."

Objects that float are *buoyant* or *positively buoyant*.

Objects that sink are *negatively buoyant*.

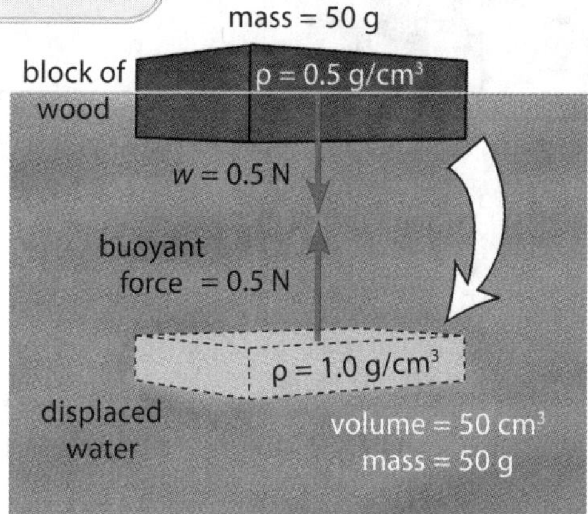

8-14 A floating object's density is less than that of the supporting fluid.

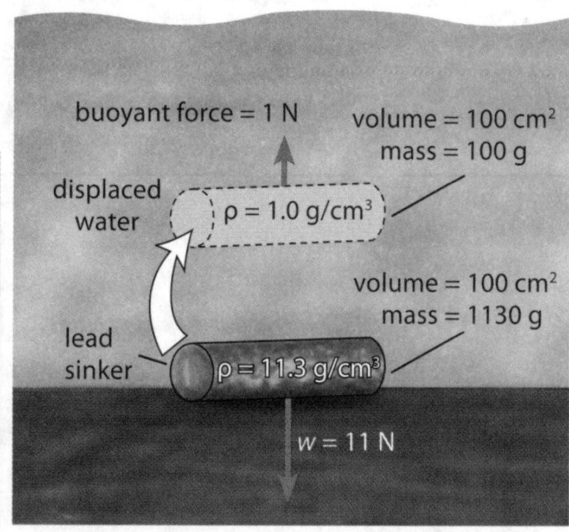

8-15 An object that sinks has a higher density than the fluid in which it is immersed.

You may be wondering—since steel has a density much greater than water—why a steel ship floats. The density of an entire ship is not equal to the density of the steel used to construct it. Its density is determined by dividing the total mass of the materials in the ship by the structural volume of its hull, which is mostly air. A ship is not simply a block of steel. The naval architect's goal is to design a ship whose overall density is much less than that of water so that the ship floats.

Gases are governed by the principles of buoyancy too. Helium-filled balloons float upward because the helium in them is less dense than the surrounding air. A helium balloon displaces air, and the resulting buoyant force of the displaced air is greater than the weight of helium and the balloon. Hot-air balloons operate similarly because the internal hot air is less dense than the cooler surrounding air.

Sometimes the weight of an object exactly equals the weight of fluid it displaces. This happens when the object and the fluid have the same density. The object's weight and the buoyant force cancel each other exactly, and the object is suspended in the fluid, neither rising nor sinking (Newton's first law). Dirigibles, hot-air balloons, submarines, and many aquatic organisms are able to adjust their densities so that they nearly match those of the surrounding fluids. These objects and organisms seem to float effortlessly within their media.

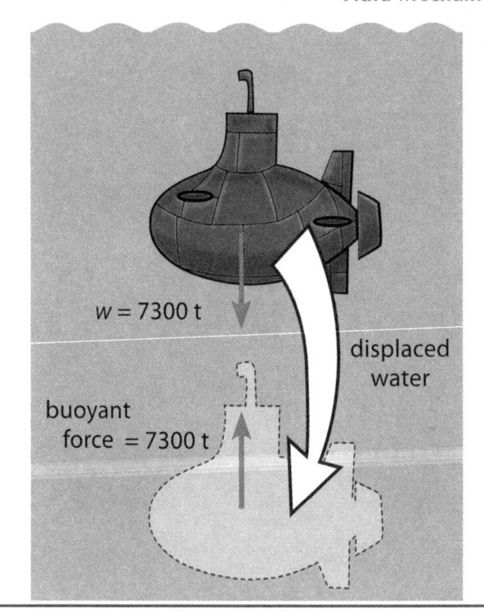

8-16 An object that neither sinks nor floats has a density equal to that of the fluid.

Submerged objects in liquids or gases that do not rise or sink in the fluid are *neutrally buoyant.*

8.5 Specific Gravity

Scientists and engineers have established a way to compare the density of a substance with the density of water. This comparison is called *relative density,* or **specific gravity (s.g.)**. A substance's specific gravity is a unitless number obtained by dividing the density of the substance by the density of water. Since water's density at standard conditions is 1.0 g/cm³, the specific gravity of a substance is simply the numeric value of its density without the units. For example, the specific gravity of pine wood is calculated as follows:

$$\text{s.g.} = \frac{\text{density of pine}}{\text{density of water}} = \frac{0.5\ \cancel{\text{g/cm}^3}}{1.0\ \cancel{\text{g/cm}^3}}$$

$$\text{s.g.} = 0.5$$

Similarly, the specific gravity of lead is 11.3 and of steel is 7.8.

Specific gravity is useful as an identifying property of substances, especially solutions and other mixtures. For example, the condition of a lead-acid battery, such as those used in cars, is determined by the specific gravity of the battery's acid solution. A specific gravity of 1.27 indicates that the battery is fully charged, while a value of 1.12 indicates that the battery is dead. Geologic minerals can also be classified by their specific gravities, especially when their visual appearance is not distinctive.

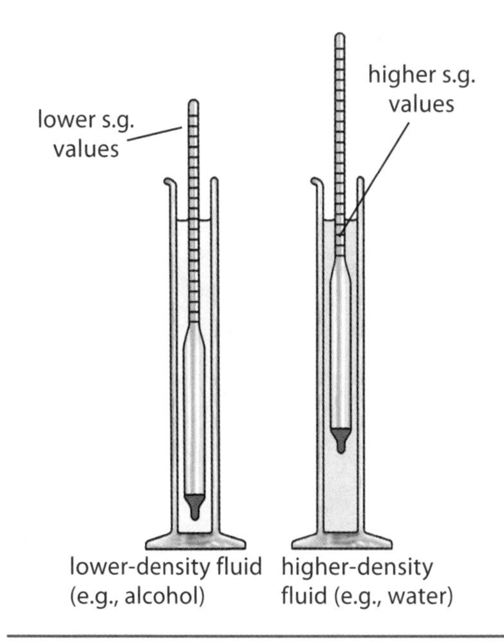

8-17 Special instruments called hydrometers measure the specific gravity of liquids. The higher the hydrometer floats, the denser the liquid is and the greater its specific gravity.

8A Section Review

1. What important property of fluids necessitated the development of fluid mechanics as an area of science separate from standard mechanics?
2. How is pressure calculated? How does pressure change when each of its contributing factors is increased?
3. How does pressure exerted by a solid object on a surface differ from pressure exerted within a liquid or a gas?
4. What is the derived unit of pressure in the SI? To what combination of SI units is it equal? What pressure unit is used in weather prediction?
5. Discuss the difference between gauge pressure and absolute pressure. What is a gauge pressure of zero equivalent to?
6. What two factors affect the pressure in a fluid? How does gravity affect each of these? What two aspects of a fluid container do *not* affect fluid pressure?
7. State Archimedes's principle in your own words.
8. List the three kinds of buoyancy by comparing the densities of the fluid and the object in the fluid. Describe what happens to the object in each case.
9. Why is specific gravity useful?
10. (True or False) If a mercury barometer were taken into a deep mine on a typical day, the mercury column would probably be taller than 760 mm.

DS ✪ 11. How does hydrostatic pressure change at a tsunameter when the crest of a tsunami wave passes overhead? when the trough of the wave passes over it? Describe what happens to the Bourdon tube in each case.

DS ✪ 12. If ocean pressure increases at a rate of 1 atm every 10 m, what is the water pressure (in atm) on a tsunameter anchored at a depth of 6000 m?

DS ✪ 13. Why is the tsunami-detection buoy network worth the expense, constant maintenance, and support staff?

8B Hydraulics and Fluid Flow

8.6 *Pascal's Principle*

The octopus is among the most intelligent invertebrate animals. It also has excellent eyesight and sophisticated defense mechanisms. One of these mechanisms is the ability to move rapidly away from harm, using a process similar to that of a rocket engine. The octopus draws water into an internal cavity, and then rapidly shoots the water out through its *siphon*, a tubelike structure. The muscular force exerted on the water in the cavity is used to accelerate the octopus. This action is similar to squirting water out of a full sports bottle. The force applied to a contained fluid in one location is transmitted through the fluid to another location.

8B Objectives

After completing this section, you will be able to

- ✓ state Pascal's principle and discuss the conditions under which it applies.
- ✓ describe a simple hydraulic machine and how it relates to other simple machines.
- ✓ discuss the causes of fluid flow and explain how they apply in familiar examples.
- ✓ summarize Bernoulli's principle and identify the three quantities whose sum must be conserved in a fluid flowing through a closed system.
- ✓ describe the Coandă effect and explain how it is responsible for exerting forces in fluids.

8-18 The Common Octopus (*Octopus vulgaris*) uses its "jet propulsion" to escape danger.

This observation was first explained by Blaise Pascal. In a famous experiment, he attached a long narrow pipe to the top of a tightly sealed, water-filled cask. A relatively small amount of water that filled the long tube above the cask provided enough pressure to burst the cask's staves. From his experiment Pascal reasoned that *changes* of pressure on the surface of a confined fluid are exerted equally throughout the fluid and at all points on the fluid's container. This statement is known as **Pascal's principle**.

Several conditions must exist for Pascal's principle to be true. First, the fluid must be confined in a completely enclosed container. If not, any change of pressure would cause it to move out the opening, similar to the sports water bottle. Second, the fluid must also be incompressible. In other words, it cannot absorb pressure changes by allowing its particles to be squeezed closer together. For practical purposes, the only real fluids that approach this condition are liquids. Gases are far too compressible to completely obey Pascal's principle.

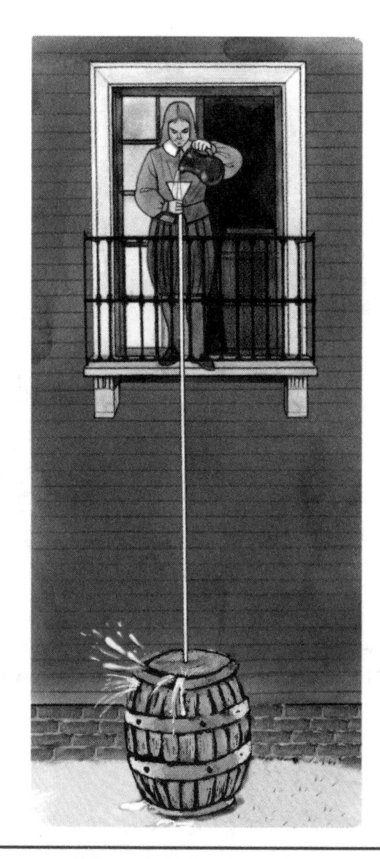

8-19 Pascal bursting the barrel with a column of water

8.7 Simple Hydraulic Machines

The most common application of Pascal's principle is found in machinery such as vehicle lifts in auto shops and brake systems in cars and trucks. These devices produce a mechanical advantage by magnifying a relatively small effort force on a confined liquid to produce a much larger force that does the work on the resistance. These *hydraulic machines* serve the same function as the simple machines discussed in Chapter 7. Since the active medium transmitting the force is a liquid, these machines use the principle of **hydraulics**.

To illustrate how hydraulics works, let's examine the basic function of a hydraulic lift found in an auto shop. The support on which the vehicle rests is connected by a vertical rod to a large-diameter cylindrical *piston* (see Figure 8-21). The rod penetrates a hole in the end of a smooth cylinder in which the piston slides. This is the *resistance*, or *load*, *piston* and its cylinder. A strong pipe connects the load cylinder to a smaller-diameter cylinder containing another piston. This is the *effort piston* and its cylinder. A rod connected to the effort piston penetrates the end of its cylinder. The entire system is filled with *hydraulic fluid*, a kind of low-viscosity oil. The effort pressure is delivered by a *hydraulic pump*, which may be hand powered or part of a motorized pumping system that includes the effort piston and cylinder. An electrical hydraulic pump is standard for powering lifts that involve large distances.

hydraulics: hydr- (Gk. *hudro*—water) + -aul- (Gk. *aulos*—pipe) + -ics. Hydraulics is the science of the transmission of forces and energy by liquids.

8-20 A common application of Pascal's principle

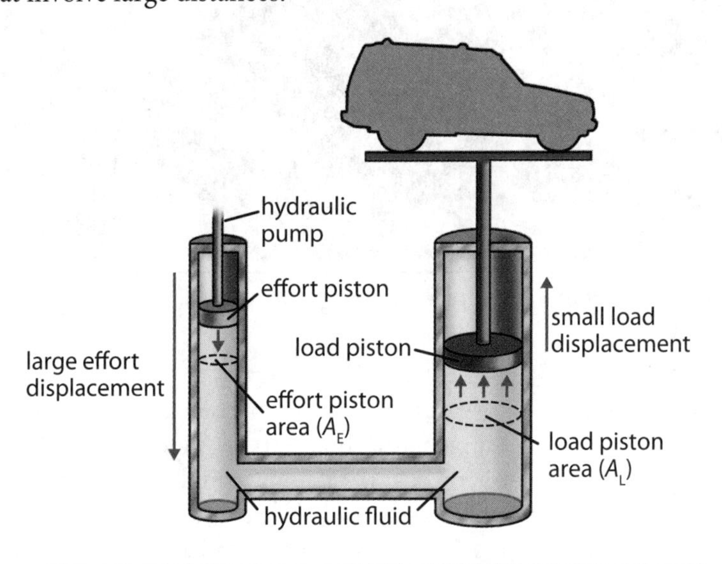

8-21 A simple hydraulic machine

According to Pascal's principle, the hydraulic pressure is distributed uniformly throughout the lift system by the hydraulic fluid. Therefore, the pressure generated by the hydraulic pump (the force it exerts on its piston area) acts on the face of the larger load piston. Through hydraulic pressure, a larger piston face generates a larger force to lift the load ($F = P \times A$). The ratio of the load-piston area to the effort-piston area is the IMA of a hydraulic machine. This calculation is very similar to the way IMA is calculated for simple machines. By increasing the ratio of the load-piston to effort-piston areas, we can reduce the effort force needed to do a certain amount of work.

8-22 A hand-operated hydraulic pump

The IMA of a simple hydraulic machine can be found using the following formula:

$$\text{IMA} = \frac{\text{load piston area}}{\text{effort piston area}} = \frac{A_L}{A_E}$$

8.8 Flowing Fluids

You know that liquids and gases have the ability to flow. But what *causes* them to flow? According to Pascal's principle, when an external force is exerted on a contained liquid, the resulting pressure is transmitted equally throughout the liquid. A water pipe in your house illustrates this principle. As long as the pipe's boundary is intact, the pipe's wall exerts the same amount of pressure on the liquid as the liquid exerts on the pipe (Newton's third law). When the water is not running, all forces are in equilibrium at every point in the liquid; thus, there is no motion of liquid inside or across the boundary of the pipe (Newton's first law).

When you turn on the faucet at your sink, you create an opening in the pipe system, which results in an unbalanced force. The water molecules accelerate in the direction of the unbalanced force (Newton's second law of motion). Because water is a fluid, it easily moves around and through the bends and turns of the pipe and faucet valve parts.

After the faucet is opened, the water flows, due to the pressure difference between the water inside and the water outside the faucet. The unconfined water flows from a region of high pressure to one of lower pressure. The rate of flow is proportional to the difference in pressure, so the larger the difference in pressures, the greater the rate of water flow. Liquid flow rate is related to many other factors as well, such as pipe size, liquid density, liquid viscosity, and how smoothly the liquid flows, but these conditions are generally secondary to pressure difference. When the water leaves the faucet, it no longer experiences the confining forces of the pipe, and it enters free fall under the force of gravity.

Gases flow because of a difference in pressure as well, but their behavior is influenced much more by the way they diffuse and are compressed. You've observed the flow of atmospheric gases as wind. Wind blows from high-pressure centers (its source) toward low-pressure centers. The kinds of pressure systems present determine the kind of weather a specific location experiences.

The physics of flowing and stationary gases is called *pneumatics*. The study of the physics of objects moving through gases is *aerodynamics*. The study of flowing and stationary liquids is called *hydrodynamics*.

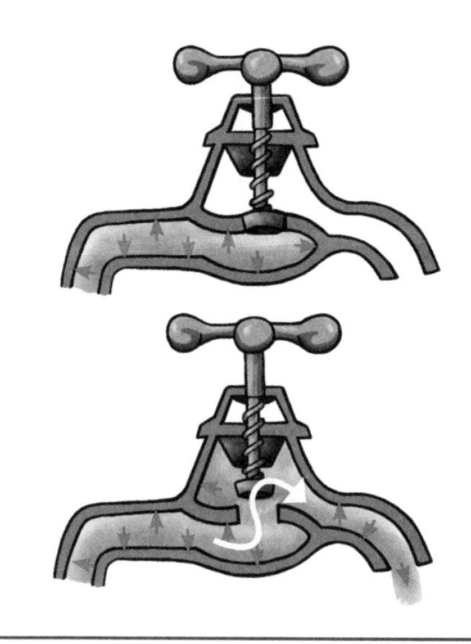

8-23 A pressure difference in an unconfined fluid causes flow.

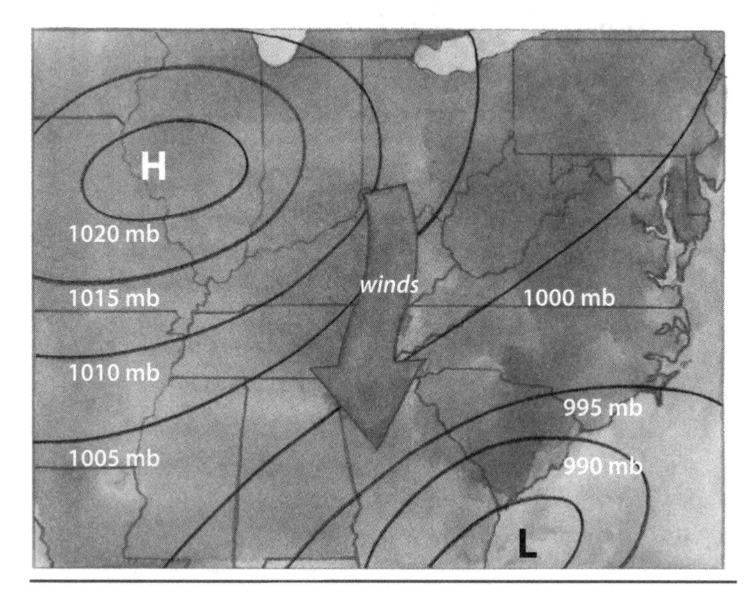

8-24 Wind blows from high-pressure zones to regions of low pressure.

There are many other familiar examples of fluid flow caused by pressure differences. A vacuum cleaner fan creates a low-pressure region inside the vacuum case, which causes high-velocity air to flow into the vacuum, picking up and carrying dirt with it. When you inhale, your rib cage expands and your diaphragm contracts, allowing your lungs to occupy a larger volume. The larger lung volume reduces the density and thus the inside air pressure, allowing higher-pressure air to flow into your lungs until the pressures equalize. Sucking on a straw placed in your favorite carbonated beverage reduces the pressure in the straw above the liquid. Atmospheric pressure pushes the drink up the straw and into your mouth.

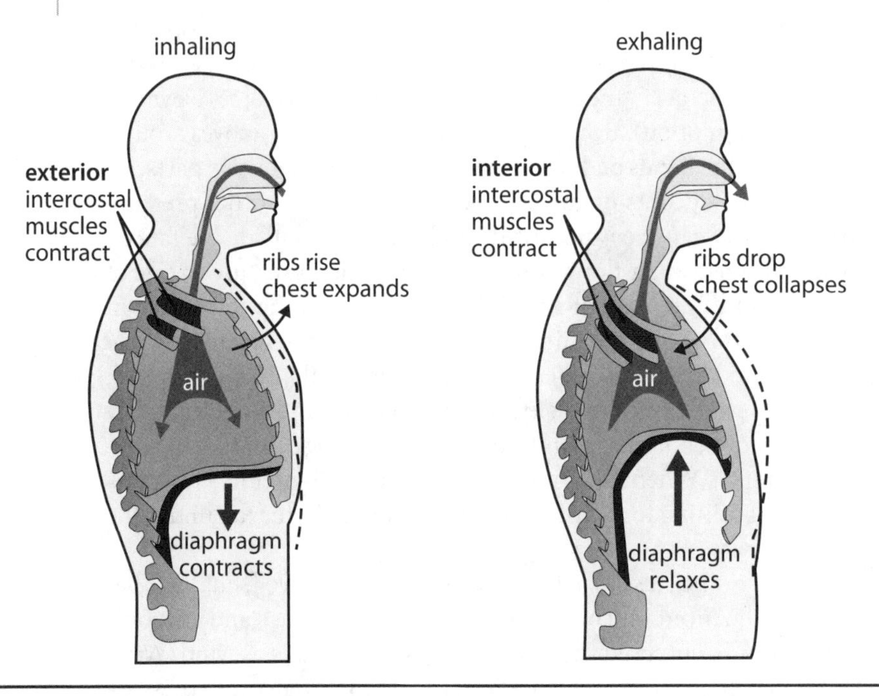

8-25 Breathing involves creating pressure differences so that air flows into and out of your lungs.

8.9 *Bernoulli's Principle*

Daniel Bernoulli (1700–1782) was born in Holland into a famous Swiss family of mathematicians. He is best known for his studies of fluids and their properties.

Daniel Bernoulli studied the flow of fluids through pipes of different diameters that ran horizontally and vertically. He recognized that liquids like water were incompressible but gases were highly compressible, though both could flow through pipes. He knew from Pascal's work that the pressure of nonmoving water in a vertical pipe was related to the height of the water. Water higher in a pipe had lower (static) pressure than water at the bottom of the pipe. He verified that water in straight, smooth pipes flows from higher to lower pressure. But he also discovered that the total mechanical energy of flowing water was conserved. In other words, the total kinetic and potential energies of a small volume of water, say a cubic centimeter, would change in predictable ways with pressure as the water flowed through a pipe. After experimenting, Bernoulli constructed the following model for an ideal fluid:

$$P_1 + \mathrm{KE}_1 + \mathrm{PE}_1 = P_2 + \mathrm{KE}_2 + \mathrm{PE}_2$$

where the subscripts 1 and 2 indicate two different points within a flowing fluid confined inside a pipe.

A flowing ideal fluid in Bernoulli's model has the following properties:

- It is incompressible; its density is constant.
- It flows smoothly with no turbulence.
- The speed of the fluid at a point in the pipe does not change with time, though its speed at different points in the pipe can be different.
- Friction has no effect on the fluid.

The equation is a simplified form of what physicists call **Bernoulli's principle**. Most liquids and even slow-moving confined gases obey these conditions. The model helps scientists and engineers make predictions about the properties of flowing fluids.

Let's look at a few special cases. When an ideal liquid flows through a horizontal section of a pipe, its potential energy remains the same and the two PE terms in Bernoulli's equation cancel out. If we ignore friction, then the fluid's speed is constant and the two KE terms cancel. Thus, we can predict that the pressure in a flowing ideal fluid in a horizontal pipe is everywhere constant. However, if we raise the downstream end of the pipe, the potential energy of the fluid increases as it rises in the pipe. To conserve energy, both the static pressure and the rate of flow in the entire pipe decrease. The greater the rise of the pipe, the lower the pressure and the lower the flow rate. You may be able to observe this if you have a multiple-story house. Water pressure in the second floor shower is lower than in the first story sink. Many water companies use fluid potential energy to supply water at useful pressure to communities by building large water storage tanks on nearby hills. This is Bernoulli's principle at work.

You may have noticed that when you hold your thumb over the open end of a garden hose and turn on the water, the water squirts out far faster than when it flows out of the hose unhindered. Bernoulli's principle explains this as well. We can assume that the water's potential energy inside the hose just behind your thumb is the same as it is after it escapes, so the only factors that change are pressure and water kinetic energy. Water pressure inside the hose is much higher than outside (atmospheric pressure) when you hold it back with your thumb. According to the law of conservation of mass, the mass flow rate of the water inside the hose must be equal to the flow rate of water outside. To conserve kinetic energy as the water passes by the constriction of your thumb, the water has to speed up, squirting out

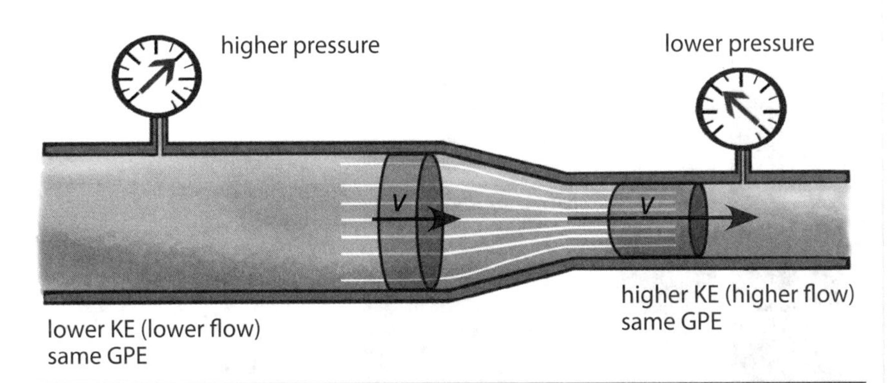

8-28 When a flowing fluid's height is constant (constant potential energy), but it encounters a constriction in the pipe diameter, it gains kinetic energy and loses pressure.

greater height
higher potential energy
lower pressure
rising pipe
increasing
potential
energy
narrower pipe
higher speed
lower pressure
lower height
lower potential energy
higher pressure

8-26 Bernoulli's principle describes the conservation of matter and energy of an ideal fluid flowing inside a pipe.

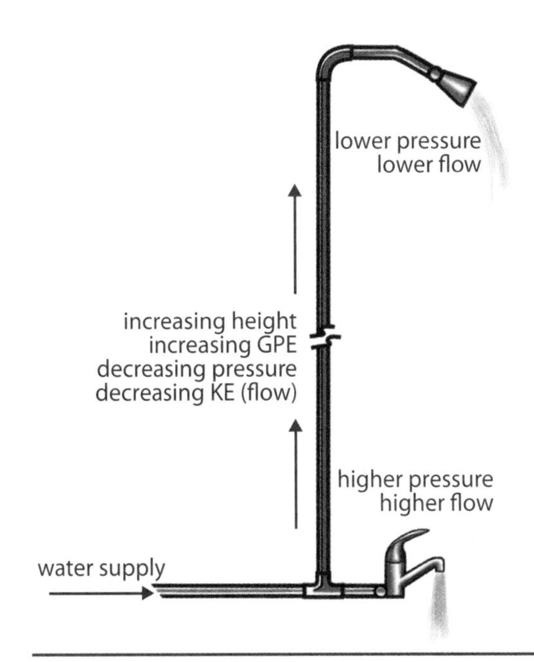

8-27 When an ideal flowing fluid changes height, fluid pressure and kinetic energy drop with increasing potential energy.

An ideal fluid flows through a horizontal pipe at constant pressure because it has inertia and obeys Newton's first law of motion. However, in a real pipe there is friction, so pressure has to decrease with flow downstream in order to overcome flow resistance.

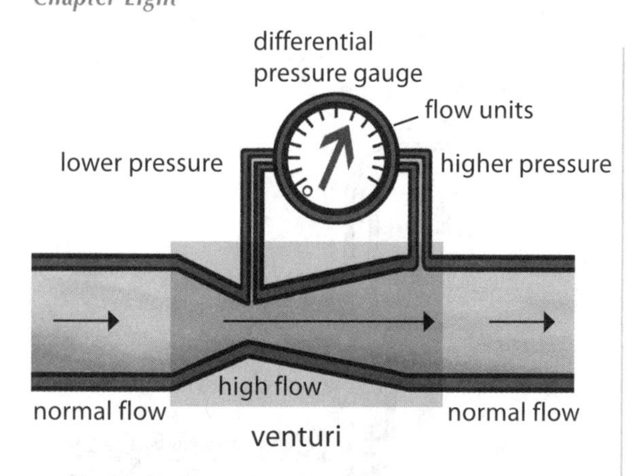

8-29 A flow meter measures pressure difference in the fluid stream using a venturi.

venturi (ven TOOR ee), from Giovanni Battista Venturi (1746–1822), an Italian physicist and historian of science

of the hose. The arrows in the equation below show the changes to the pressure and KE terms.

$$P_{\text{inside}} + \text{KE}_{\text{inside}} = \downarrow P_{\text{outside}} + \uparrow \text{KE}_{\text{outside}}$$

This happens in any constriction in a pipe containing a flowing ideal fluid. To conserve mass flow rate, the fluid speeds up in the smaller sized pipe (see Figure 8-28 on the previous page). As the flow increases through the constriction, liquid pressure drops because pressure is exchanged for kinetic energy to conserve matter and energy throughout the system. Some liquid flow meters use this principle. A *flow meter* is a pressure gauge connected to a specially designed constriction in the pipe called a ***venturi***. The gauge measures the change in liquid pressure between a normal flow point of the venturi and its greatest constriction. The pressure difference is proportional to the flow rate, which is indicated on a graduated scale.

When we imagine fluids flowing in pipes, we usually think of liquids. However, gases also can flow in pipes as well as around objects. Early physicists studying fluid flow noticed that gas pressure drops as velocity increases. The effect, however, is not as dramatic as that observed in liquids, because gases are compressible and tend to be far less dense than liquids. Even so, in the 1800s several researchers with the objective of building human gliders began investigating how air pressure differences around a bird's wing contribute to flight.

8.10 The Coandă Effect

Henri Marie Coandă (1885–1972) was a Romanian aerodynamics engineer who invented the first jet engine and discovered the principle of aerodynamic lift that bears his name.

Many people believe that the only forces responsible for generating the supporting force, or *lift*, on an airplane wing are the pressure differences predicted by Bernoulli's principle. But this idea ignores Newton's laws of motion! In 1934 the Romanian physicist Henri Coandă discovered that fluids flowing close to a curved surface follow the shape of the surface rather than a straight path. This effect is caused by the viscosity within a flowing fluid, which tends to drag adjacent portions of the fluid with it. Figure 8-30 illustrates how the **Coandă effect** can divert the path of water flowing around a cylinder. The same principle applies to leaf-shedding roof gutters, which allow rainwater to flow into them without trapping leaves. The water flows around the curved surface and drips into the gutter. The leaves are too stiff and heavy to follow the curve with the water and fall to the ground.

An airfoil is any surface that is designed to deflect air or create a force as it moves through the air.

The Coandă effect is responsible for most of the lift in an airplane wing and other ***airfoils***. The wing's motion through the air creates a downward force in the air above the wing. This force diverts large quantities of air from above the wing downward. What does Newton's third law require? The air must exert an upward reaction force on the airfoil—lift. This is the force that supports an airplane.

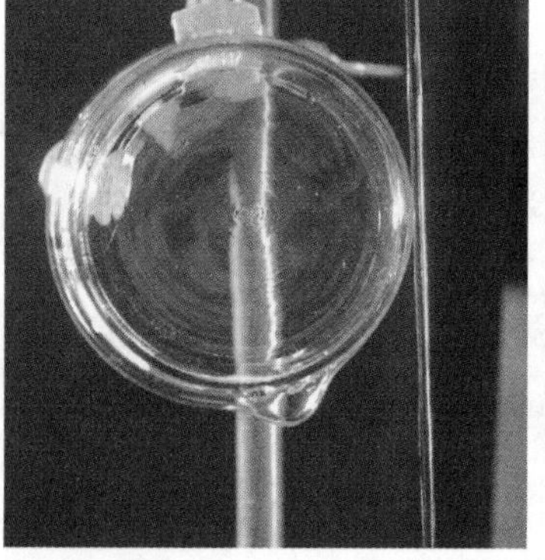
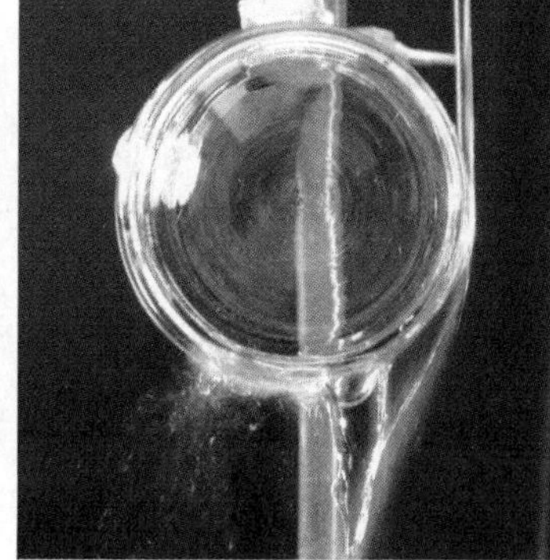

8-30 The Coandă effect in water flowing past a beaker. Leaf-shedding gutters use the Coandă effect.

FACETS OF SCIENCE

How Do Airplanes Really Fly?

Many books have been written on how airplanes fly. Most do a good job of explaining drag (or air resistance); thrust (what pushes the airplane through the air); and gravity (what tries to pull the airplane down or hold it on the ground). But very few of them accurately describe *lift*, that which enables the airplane to rise off the ground in the first place.

Most people learn that lift is the result of the Bernoulli principle. This idea is misleading. One common way of applying the Bernoulli principle to airplane wings is to assume that the air divides at the front of the wing and comes together at the back edge. In most diagrams illustrating lift, the top of the wing is curved and the bottom is flat. It is said that this shape requires the air passing over the wing to travel farther and therefore faster than the air passing under the wing if the two air streams are to meet at the trailing edge of the wing at the same time. This faster speed results in a lower pressure above the wing according to Bernoulli's principle. The higher pressure beneath the wing pushes upward, creating lift.

This simple yet inaccurate theory has been taught for generations from elementary school through advanced pilot training, but it ignores some significant facts. The amount of lift generated in this way is only 2%–5% of the total. The idea that air divides at the front of the wing and joins together at the back edge is a misconception that has been proven false in wind tunnel tests and by advanced computer models of wings. The air under the wing actually slows due to drag and arrives at the rear edge much later, if ever.

A better explanation of how airplanes fly involves two important factors: the Coandă effect and Newton's third law of motion. These two principles account for 60%–90% of the total lift generated by an airplane's wings.

The Coandă effect is an unfamiliar but very important property of flowing fluids. Fluids moving by a gently curving surface will tend to follow the surface rather than move in a straight path. The force that draws the fluid to the surface is related to the viscosity of the liquid or gas. Molecules of the fluid right next to a moving surface are motionless because they adhere to the material. Fluid particles just a short distance away move very slowly relative to the surface; with greater distance from the surface, they move more rapidly. This is called the *boundary layer effect*. A surface moving relative to a fluid drags the fluid along in a path that parallels the curvature of the surface.

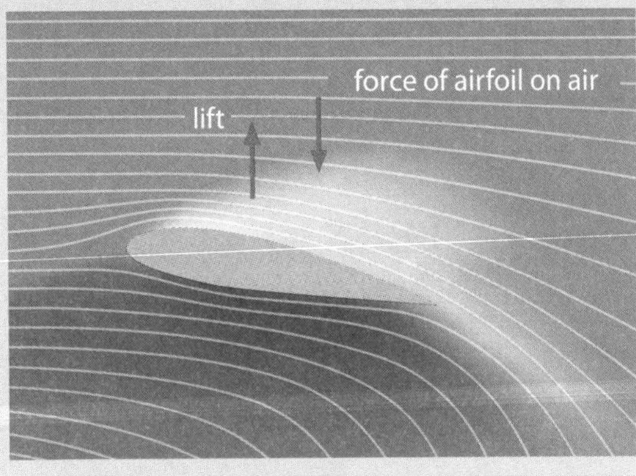

The Coandă effect draws air from above the wing and directs it downward. The reaction force is lift on the wing.

An aircraft wing generating lift is not parallel to the path of the wing through the air but is tilted up somewhat. The air under the wing compresses slightly, causing the air in front of the wing to rise up and flow over the wing. The Coandă effect causes the air to follow the wing surface. The air molecules spread out farther above the wing due to inertia and low air viscosity. This action strongly reduces the density of the air and thus the air pressure above the forward half of the wing. Air farther above the wing rushes into the low-pressure region.

This is where Newton's third law comes in. In the Coandă effect the wing exerts a downward force on the air above the wing that accelerates the air downward toward the top of the wing. According to Newton's third law, the air must exert an upward reaction force to this force on the wing. This upward force is lift. In this model the lifting force is generated by conditions above the wing rather than below it.

Unlike the Bernoulli flow model, this model accounts for upside-down flight in aerobatic and military aircraft and the use of wings that are streamlined on both upper and lower surfaces.

What is truly astounding is that the large quantities of downward flowing air pour nearly straight down off the back of the wing. Aeronautical engineers estimate that a plane must redirect nearly five times its weight in air downward to stay airborne! See the photo below.

Whether the Bernoulli principle or the Coandă-Newton concept is a better model for describing the lift phenomenon is hotly debated by physicists. Several other models that involve vortex mathematics and streamline analysis have been proposed. NASA engineers note that lift is an extremely complex state that involves aspects of both Bernoulli's principle and Newton's laws as well as energy and mass conservation principles. This is a good lesson to those who think that there is only one explanation for a particular scientific question, even something many authorities believe is as "obvious" as lift.

Garden hose chemical sprayers, artist's airbrushes, and some perfume aspirators use the Coandă effect in conjunction with Bernoulli's principle. In these cases, air or water is directed across the open end of a tube inserted in a liquid. This rapidly moving fluid flows over the tube end, creating a low-pressure region compared to the air pressure on the liquid. The liquid is forced up the tube and mixes with the fluid flowing by it. The chemical, paint, or perfume mixture sprays from the nozzle of the applicator.

8B Section Review

1. How did Pascal explain the bursting of the water cask's staves in his experiment?
2. What conditions must exist for Pascal's principle to be true? What real fluids seem to obey Pascal's principle fairly well?
3. What do we call a mechanical device that uses a fluid to make work easier? On what principle do these devices rely?
4. A simple hydraulic machine uses pistons to apply a force to a fluid and to exert a force to do work. What is one way to compute the IMA of a simple hydraulic machine?
5. What is the most important condition that must exist for a fluid to flow? In a piping system, what other factors affect the flow of a liquid?
6. ✪ Identify an example of a flowing fluid that was not mentioned in Section 8B. Identify the cause for the pressure difference that causes the fluid to flow.
7. According to Bernoulli's principle, the combination of what three properties of a liquid flowing in a pipe must be the same at all points in the pipe? What factors prevent this principle from being exactly true in real-world flowing liquids?
8. ✪ Identify one application of the Coandă effect other than those given in this chapter.
9. (True or False) A manually operated hydraulic jack applies the Bernoulli principle to amplify effort to do work.

8C Gas Laws

8C Objectives

After completing this section, you will be able to

- ✓ state the gas laws.
- ✓ show how the gas laws are predicted by the particle theory of matter.
- ✓ perform calculations using the gas laws.

8.11 History of the Gas Laws

Scientists began experimenting with gases almost as soon as Galileo demonstrated the value of scientific observation. His secretary, Evangelista Torricelli, invented a barometer in 1643. Earlier, Gasparo Berti proved that gases have weight. Shortly after, Robert Boyle investigated the "springiness" of air. In 1702 the French scientist **Guillaume Amontons** built a thermometer that worked on the principle that the pressure of an isolated quantity of gas was proportional to its temperature. Using his thermometer, he was the first to make an approximation of absolute zero. In 1787 Jacques Charles, another French scientist, built and flew in the first hydrogen balloon. His research revealed the relationship between the temperature and the volume of an isolated quantity of gas.

Guillaume Amontons (1663–1705) was an inventor of scientific instruments and a physicist. He invented a long-distance flag signaling system (a semaphore), possibly because he was deaf from his youth.

In the last decades of the 1700s, much evidence was accumulating to discredit the Aristotelian theory of continuous matter. **John Dalton**, who is best remembered for his atomic theory, set the stage for that revolutionary theory when he observed in 1801 that atmospheric pressure is just the sum of the *partial pressures* of individual atmospheric gases. In other words, each gas exerts its own pressure as if it were the only gas present. The partial pressure of each gas depends only on its temperature and the density and molecular mass of its particles.

John Dalton's *law of partial pressures* governs the diffusion of gases across membranes, such as the exchange of oxygen and carbon dioxide in our lungs.

In 1808 **Joseph Gay-Lussac** discovered through a series of experiments that when gases chemically react at a constant pressure and temperature, their volumes combine in ratios of small whole numbers. For example, two volumes of hydrogen combine with one volume of oxygen to produce one volume of water vapor under the same conditions. This discovery offered significant support of Dalton's atomic theory. Just three years later, **Amedeo Avogadro** explained Gay-Lussac's results by showing that equal volumes of gas contain equal numbers of particles if they are at the same temperature and pressure. These breakthroughs contributed to the acceptance of the particle theory of matter.

Joseph Louis Gay-Lussac (1778–1850) was a French chemist and physicist. He published the law bearing his name after studying atmospheric gases extensively.

Amedeo Avogadro (1776–1856) was an Italian lawyer, mathematician, and physicist. He was the first to distinguish between an atom and a molecule. His studies were the basis for determining the number of particles in a certain amount of matter.

As you can see, scientists made great progress in understanding fluids, especially gases, in the eighteenth and nineteenth centuries. The rest of Section C discusses two of the most important gas laws—Boyle's law and Charles's law.

8.12 Boyle's Law

You can properly understand gas pressure only by understanding what occurs at the particle level. At room temperature, a single gas molecule speeds along at about 1000 m/s. It has momentum because of its speed and mass ($p = m \times v$). When it collides with a solid surface, it exerts a force on the surface and rebounds in an elastic collision. The force of this collision is minuscule for a single molecule, but with many trillions of molecules colliding with a surface, this force is multiplied trillions of times per second, resulting in a measurable pressure.

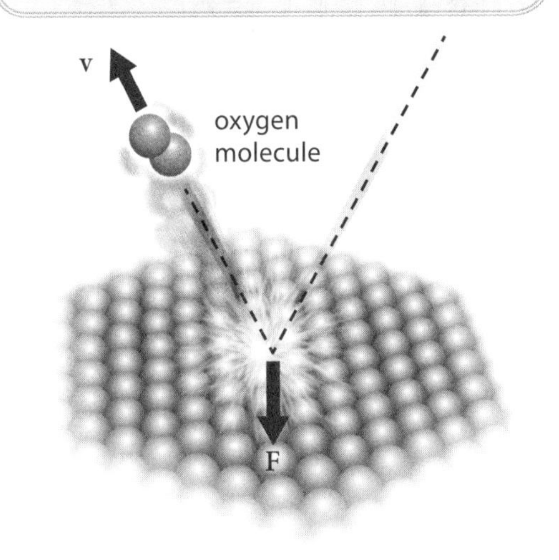

8-31 A single collision of an oxygen molecule

Gas pressure depends on the number of molecules present in a volume of gas, their masses, and how fast they are moving. The average speed of all gas molecules in a sample of gas determines the temperature of the gas. The faster the gas molecules move, the higher the temperature. A gas enclosed in a container has a fixed number of molecules in its volume. The average number of collisions per second of gas molecules with each other and the container's surfaces is constant for a constant temperature.

What would happen to gas pressure if the container's size were reduced without changing the number of gas molecules or their temperature? Reducing the container's size decreases its volume and the distance between molecules. Thus, the number of collisions per second increases because the molecules have less distance to travel between collisions. Keeping temperature constant means that the molecules' average speed is constant. Therefore, the average molecular momentum, and thus the average force per collision, is constant. Combining these factors, you can see that the total pressure on the container's walls will increase as its volume decreases. If the container's volume

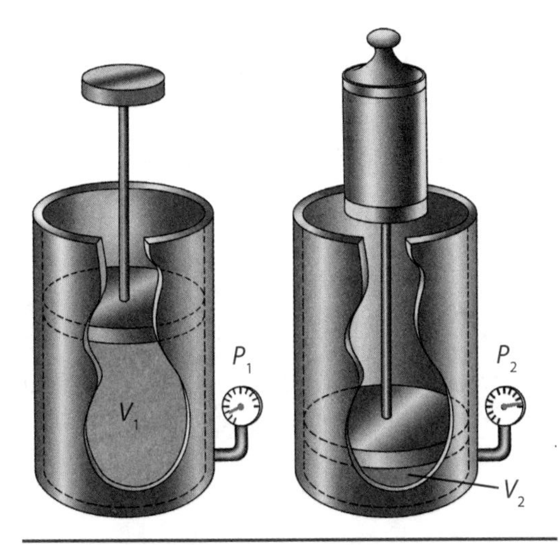

8-32 Compressing a gas increases the number of gas-molecule collisions per second and thus the gas pressure.

Robert Boyle (1627–91) was an English scientist who worked in many areas of science, especially chemistry and physics. He is generally considered to be the first modern chemist. He is best known for discovering Boyle's law, which relates the pressure of an enclosed gas to its volume.

is enlarged, gas pressure decreases because molecules must travel farther between collisions, reducing the rate of molecule collisions.

Robert Boyle discovered this relationship between volume and pressure in experiments during 1659 and 1660. He trapped a small amount of air in the sealed short end of a glass J-tube by pouring liquid mercury into the open long end. When he added more mercury, the volume of trapped air decreased as the height and weight of the mercury column increased. He discovered that the volume of the trapped air was inversely proportional to the pressure exerted by the mercury. In other words, if he doubled the pressure on the gas, its volume decreased by half. He found this relationship to be true for any volume or pressure of air as long as he measured a fixed amount of air at a constant temperature. The inverse relationship between the volume and pressure of a confined gas is called **Boyle's law**. This law can be expressed by the formula

$$P = \frac{k}{V} \text{ or, equivalently, } P \times V = k,$$

where

- P is the pressure of the gas, in pressure units such as newtons per square meter or pascals;
- V is the volume of the confined gas, in volume units such as cubic meters; and
- k is a constant (its value is not important).

For a fixed amount of gas confined at a constant temperature, the product of any pressure and its corresponding volume will equal the same constant:

$$P_1 \times V_1 = k \text{ and } k = P_2 \times V_2$$

Since $k = k$, we can substitute the two products of PV for k,

$$\boxed{P_1 \times V_1 = P_2 \times V_2},$$

which is the standard formula for Boyle's law.

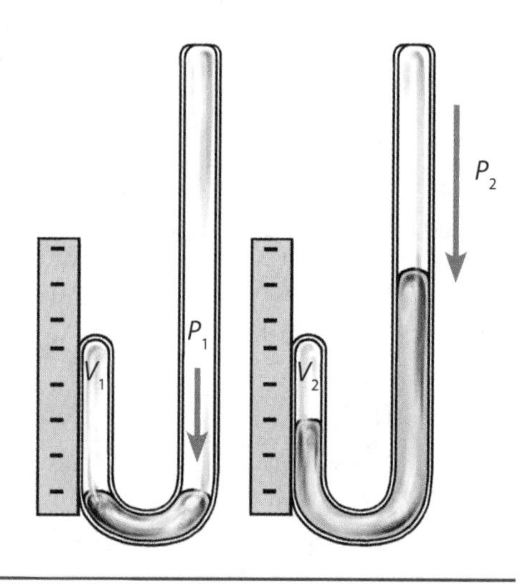

8-33 Boyle's J-tube apparatus. The mercury column height is the vertical distance between the two meniscuses.

Math Help: Inverse Proportionality

Suppose you're running late for school, but you've got to be there on time because you have a test first hour that you still need to study for! How can you make up for lost time? Obviously, the distance from school to your house can't be shortened. However, your mom can drive faster—within legal limits, that is—so that you can get to school earlier.

Speed, distance, and time are related by the equation

$$v = \frac{d}{\Delta t},$$

where

- v is the speed at which you travel,
- d is the distance you travel to school, and
- Δt is the time it takes you to get to school.

If you want to get to school earlier, you need to decrease Δt. Decreasing Δt means that you need to increase your speed (v). In this case, speed and the time interval are *inversely proportional*. If the value of one increases, the other decreases in proportion.

In equation form, an inverse proportion looks like this:

$$A = \frac{k}{B} \quad \text{or} \quad A = k \times \frac{1}{B},$$

where k is the proportionality constant. In our example, distance is the proportionality constant, A is the speed, and B is the time interval.

The quantities with the subscript 1 are the pressure and volume at one set of conditions. The subscript 2 indicates the pressure and volume of the same amount of gas at another set of conditions. If you know any three of these quantities, you can calculate the fourth using simple algebra.

EXAMPLE PROBLEM 8-1

Boyle's Law

Normal atmospheric pressure supports a column of mercury 760. mm tall in a barometer. A confined sample of gas occupies 3.00 L at 760. mm Hg (1 atm) of pressure. If the volume of the gas sample is changed so that its pressure increases to 2280 mm Hg, what is its volume?

Known: initial pressure (P_1) = 760. mm Hg
initial volume (V_1) = 3.00 L
final pressure (P_2) = 2280 mm Hg

Unknown: final volume (V_2)

Required formula: $P_1 \times V_1 = P_2 \times V_2$

Substitution: (760. mm Hg)(3.00 L) = (2280 mm Hg)(V_2)

Solution:

$$V_2 = \frac{(760.\ \cancel{\text{mm Hg}})(3.00\text{ L})}{2280\ \cancel{\text{mm Hg}}}$$

$$V_2 = \frac{2280\text{ L}}{2280} = 1.00\text{ L} \quad \text{(3 SDs allowed)}$$

Note that when pressure tripled, volume was reduced to one-third what it was originally.

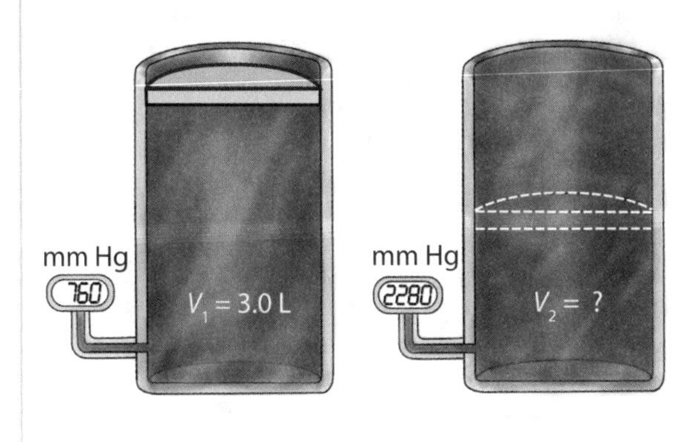

8-34 Solving for final volume using Boyle's law

EXAMPLE PROBLEM 8-2

Pistons and Boyle's Law

The piston in a pneumatic shock absorber compresses the air inside when a car wheel hits a bump, then expands when the bump has passed. If the pressure of the fully compressed gas is 4940 mm Hg, what is the volume of the gas in the compressed shock absorber? The shock absorber then relaxes to a normal volume of 1.30 L at one atmosphere of pressure (760. mm Hg).

Known: initial pressure (P_1) = 4940 mm Hg
final volume (V_2) = 1.30 L
final pressure (P_2) = 760. mm Hg

Unknown: initial volume (V_1)

Required formula: $P_1 \times V_1 = P_2 \times V_2$

Substitution: (4940 mm Hg)(V_1) = (760. mm Hg)(1.30 L)

Solution:

$$V_1 = \frac{(760.\ \cancel{\text{mm Hg}})(1.30\text{ L})}{4940\ \cancel{\text{mm Hg}}}$$

$$V_1 = \frac{988\text{ L}}{4940.} = 0.200\text{ L} \quad \text{(3 SDs allowed)}$$

When pressure decreased as the shock absorber relaxed, the gas volume increased.

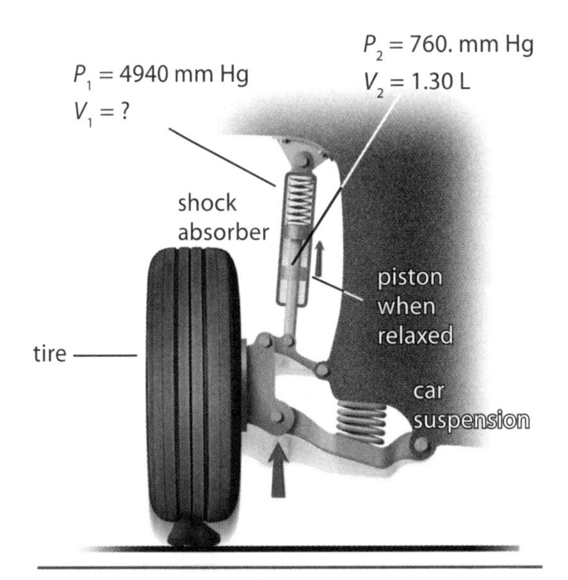

8-35 A shock absorber

8-36 A high-altitude weather balloon is only partially filled before launch.

What are some practical applications of Boyle's law? People use compressed air for all kinds of purposes, such as filling sports balls, operating powerful shop tools, and even painting with airbrushes. Manufacturers of compressed air tanks know that a small tank containing the same amount of air as a larger tank has a greater pressure exerted on its walls. Such considerations are crucial. If the smaller tank were not constructed with thicker walls, it could explode! Boyle's law also explains why meteorologists only partially fill their high-altitude balloons with helium before they release them. As the balloon rises into the atmosphere, the surrounding air pressure decreases and the volume of the lifting gas increases to fill the balloon. If the balloon were full when released, it would burst before reaching the desired altitude.

8.13 Charles's Law

Tightly cover a glass bowl with a sheet of polyethylene food wrap and place the bowl in a microwave oven. After heating the sealed bowl for a short time, you'll notice that the food wrap has bulged upward like an inflated balloon. What has happened, and how can we explain it?

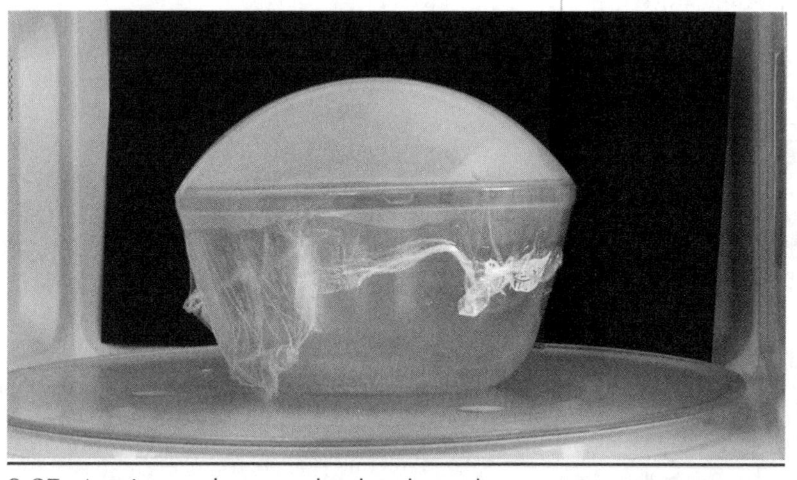

8-37 An air sample expands when heated.

The food wrap traps a fixed amount of air in the bowl. When microwave energy is added to the trapped air, its molecules vibrate faster and gain kinetic energy. When they collide with the food wrap and the sides of the bowl, they exert a greater force with each collision and there are more collisions per second. Both of these effects increase the pressure of the trapped air. The elastic food wrap stretches because of the trapped air's higher pressure. (The force exerted by the warmed food wrap on the trapped air is negligible and can be ignored for this illustration. It merely forms a boundary between the hot trapped air and the cooler outside air.) The trapped air expands because its pressure exceeds the outside air pressure exerted on the food wrap. An equilibrium occurs only when the microwave is turned off and stops heating the air. All forces on the food wrap are balanced (Newton's first law). Though the trapped air particles are moving at higher speeds, they are now spread out because of thermal expansion. Thus the rate of molecule collisions drops. The net effect is a lowering of inside air pressure so that it equals outside air pressure. The original amount of air now occupies a larger volume at a higher temperature.

We can observe from this demonstration that, for gases, a direct relationship exists between volume and temperature. For a given quantity of pure gas, as temperature increases, its volume increases if pressure is held constant. This principle is known as **Charles's law**, in honor of Jacques Charles, a prominent eighteenth-century scientist known for his exploits as a balloonist.

Jacques Alexandre César Charles (1746–1823) was a French chemist, physicist, and balloonist. He invented a number of scientific instruments and formally identified the gas law that bears his name. He was better known in his day for inventing the hydrogen balloon.

Charles's law says that the volume of a fixed amount of gas changes in the same way as its absolute temperature, at constant pressure. This relationship can be expressed in the formula

$$V = k \times T,$$

where

- V is the volume of the gas, in a volumetric unit such as liters;
- k is a proportionality constant (its value is not important); and
- T is the absolute temperature of the gas, in kelvins.

The formula for Charles's law can be rearranged to isolate the constant factor on one side of the equation, just as with Boyle's law. The gas sample's volume and temperature at different conditions can be predicted because their ratio always equals the same constant:

$$\frac{V_1}{T_1} = k \quad \text{and} \quad k = \frac{V_2}{T_2}$$

Since $k = k$, we can say

$$\boxed{\frac{V_1}{T_1} = \frac{V_2}{T_2}},$$

which is the standard formula for Charles's law.

As with Boyle's law, if you know any three of the quantities, you can easily solve for the fourth. Remember that this relationship holds only for a fixed amount of a gas at constant pressure. Also remember that the absolute temperature (in kelvins) must be used as the unit for temperature. Gas-molecule kinetic energy decreases continuously as temperature decreases toward absolute zero. The only temperature scale that can correctly relate properties that depend on particle energies is the Kelvin temperature scale.

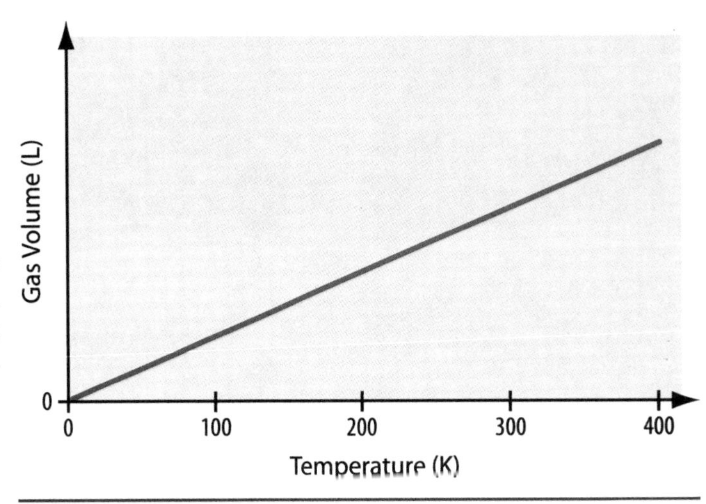

8-38 If one assumes that gas molecules have no volume, Charles's law can be used to predict that gas volume vanishes when there is no temperature.

Guillaume Amontons estimated absolute zero using the principle of Charles's law years before Charles studied the relationship between volume and temperature.

EXAMPLE PROBLEM 8-3

Charles's Law

A weather balloon has an uninflated volume of 0.66 m³. This volume of helium is used to inflate the balloon at 30. °C (303 K) at 1 atm of pressure. After the balloon sits in the sunlight for several hours, the helium warms to 39 °C (312 K). What volume does the helium occupy at the higher temperature, assuming that the gas is still at 1 atm? Assume that the balloon exerts no forces on the confined gas.

Known: initial helium volume (V_1) = 0.66 m³
initial helium temperature (T_1) = 303 K
final helium temperature (T_2) = 312 K

Unknown: final helium volume (V_2)

Required formula: $\frac{V_1}{T_1} = \frac{V_2}{T_2}$

Substitution: $\frac{0.66\ \text{m}^3}{303\ \text{K}} = \frac{V_2}{312\ \text{K}}$

Solution: $(312\ \cancel{\text{K}}) \times \frac{0.66\ \text{m}^3}{303\ \cancel{\text{K}}} = \frac{V_2}{\cancel{312\ \text{K}}} \times (\cancel{312\ \text{K}})$

$V_2 = \frac{(312)(0.66\ \text{m}^3)}{303}$

$V_2 \approx 0.6\underline{7}9\ \text{m}^3$

$V_2 \approx 0.68\ \text{m}^3$ (2 SDs allowed)

Just as Charles's law predicts, the warmer the gas is, the larger the volume it occupies.

Kelvin temperature (T) is equal to Celsius temperature (t_C) plus 273°.

$$T = t_C + 273°$$

8-39 A weather balloon's volume increases as it is warmed by the sun.

Multiply both sides by 312 K.

8C Section Review

1. Eighteenth- and nineteenth-century research in gases led to the acceptance of what principal theory of science?
2. What three factors contribute to gas pressure? If you reduce the amount of gas in a given volume and keep the other factors the same, what happens to the pressure of the gas?
3. If the volume of a confined gas triples, what happens to the pressure of the gas (assuming that its temperature doesn't change)?
4. When a confined gas experiences a temperature change, which of the three factors from Question 2 could change the pressure of the gas? According to Charles's law, what would have to happen to keep gas pressure constant?
5. (True or False) An oxygen molecule (O_2) has eight times as much mass as a hydrogen molecule (H_2). For equal masses of oxygen and hydrogen at the same volume and the same temperature, the oxygen sample will exert more pressure than the hydrogen sample.
6. A weather balloon is inflated with 0.80 m^3 of helium (He) at ground level (pressure is 1.0 atm). The balloon is released and rises into the air. What is the atmospheric pressure when the gas in the balloon expands to a volume of 3.32 m^3?
7. A sealed, empty 1.0 L plastic sports bottle is sitting on a porch in the hot sunlight. The temperature of the air inside the bottle is 39 °C (312 K). When the sun goes down, the air in the bottle cools to 20.°C (293 K). Assuming that the bottle is completely flexible, what is the volume of air at the cooler temperature?

Chapter Review

1. Why is fluid mechanics distinct from other areas of mechanics?
2. What is the difference between pressure exerted by a solid and by a fluid?
3. Pressure instruments indicate fluid pressure relative to some standard. What is the standard pressure for a mercury barometer?
4. Why does a ship made of steel, which has a density nearly eight times that of water, float?
5. How can you determine the IMA of a simple hydraulic machine from its piston dimensions? from the movement of its pistons?
6. Does a fluid need to be enclosed to experience pressure differences? Give at least one example to support your answer.
7. Through a supply pipe, water flows horizontally into a house, rises vertically to the second floor, and then flows horizontally to the bathtub to fill it. Assuming that the pipe diameter is constant, discuss the changes in pressure, kinetic energy, and potential energy of the flowing water in each section of its path, ignoring the effects of friction.

*8. Why do fluids flow along gently curved surfaces? Why does this model explain aerodynamic lift better than Bernoulli's principle alone?

9. Describe what happens to a gas molecule's velocity, kinetic energy, and momentum when it collides with a surface or another gas molecule?
10. What assumptions apply to Boyle's law?
11. According to Charles's law, how does the volume of a gas in a flexible container change when its temperature is doubled?
12. What assumptions apply to Charles's law?

True or False

13. Fluids cannot exert forces because they are not rigid.
14. Pressure can be increased by decreasing the area over which a given force is applied or by increasing the applied force.
15. The height of a column of mercury in a glass tube with one sealed end upended in a bowl will always be 760 mm tall (at 1 atm), no matter how long the glass tube is or the amount of mercury used to fill it.
16. In the atmosphere, air pressure decreases at a constant rate as altitude increases.
17. The amount of water that supports a floating object has to be actually pushed out of the way (displaced) for the object to float.

*18. Bernoulli's principle is the most important factor in generating lift in aircraft wings.

19. Experimentation with gases in the seventeenth through nineteenth centuries significantly contributed to the development of the particle theory of matter.

Scientifically Speaking

20. Gas pressure depends partly on the number of gas molecule collisions per second.
21. Charles's law is applicable when inflating a party balloon.

✪ 22. What is the pressure, in pascals, exerted on the ground by a 64.6 kg student whose shoes contact the ground over an area of 280. cm^2 (0.0280 m^2)?

✪ 23. Convert 990. mb to millimeters Hg.

✪ 24. What is the specific gravity of hard rubber? (Refer to Table 3-6 on page 68.)

✪ 25. A snorkeling vest is an inflatable bag draped around the diver's neck and strapped to his chest. It is used to adjust the diver's buoyancy to be neutral so that he doesn't automatically rise or sink. If the vest contains 9100 ml of air at 1.0 atm when he is at the surface, what volume will the air in his vest occupy when he is swimming near the bottom (where pressure on the vest is 2.3 atm)? Assume that air temperature in the vest doesn't change.

✪ 26. World-class balloonists, who try to circle the earth without stopping, notice that they tend to rise higher in the daytime when the gas in their balloon is warm and that they sink to a lower altitude at night when the gas cools off in the frigid air. One such balloon contained 15,600 m^3 of helium during the daytime when gas temperature was −21 °C (252 K). What volume did the gas occupy when the nighttime temperature plunged to −43 °C (230 K)?

THERMODYNAMICS

CHAPTER 9

DOMINION SCIENCE PROBLEM

Renewable Energy

As of 2012, the world population had exceeded 7 billion people, and was growing at a rate just under 1.2 percent, adding 80 million people per year. Where will we get the energy to meet the needs of these additional souls? Many scientists believe that conventional energy sources like coal and petroleum are causing global warming when these fuels are burned in power plants. Most governments are trying to reduce reliance on these fuels. Nuclear power is far cleaner during operation than fossil fuel plants, but it generates radioactive wastes that must be processed or disposed of. And there is the ever-present fear of nuclear accidents that increases the costs of design and construction to make nuclear power plants safe. An ideal source of energy for producing electricity should be naturally replenished—it should be *renewable*. What source of renewable thermal energy is safe, nonpolluting, and economical?

9-1 Renewable energy sources for electricity include large wind farms such as this one off the coast of Sweden (left) and geothermal power plants such as the one in Iceland (right).

9A Objectives

After completing this section, you will be able to

- ✓ summarize the development of the theory of thermal energy.
- ✓ describe how Count Rumford discovered a problem with the caloric theory.
- ✓ explain temperature change in terms of the kinetic-molecular theory.
- ✓ compare and contrast thermal and internal energy.

9A Thermal Energy

9.1 *Introduction*

"[The sun's] going forth is from the end of the heaven, and his circuit unto the ends of it: and there is nothing hid from the heat thereof" (Ps. 19:6). The sun is one of the most glorious of all God's creations. Humans have always marveled at its beauty and power. In ancient times most people worshiped the sun, but the Bible teaches that the sun—glorious as it is—is no god. It exists not to be worshiped but to serve the purpose of its Maker. The Bible indicates that God's purpose for the sun is to benefit mankind. We were not made for the sun; the sun was made for us (Gen. 1:16–18; Deut. 4:19). As we use the sun's energy, we fulfill our God-given purpose to exercise good and wise dominion (Gen. 1:26, 28). Since God made the sun so that nothing could be hidden from its heat, we should learn how best to use this vast source of clean, renewable energy. But what is heat? How is it produced? How do we measure it? How do we use it? Is it possible that the sun could be a reliable, inexhaustible source of energy to solve our dominion science problem?

9-2 The sun is the source of the great majority of usable energy.

9.2 *Early Theories of Heat*

In the sixth century BC, the Greek philosopher Heraclitus claimed that there were three natural elements—earth, water, and fire. His writings greatly influenced later philosophers, such as Socrates and Plato. Because of the philosophical ideas of these men, people believed that heat and fire were the same for many centuries.

In the thirteenth century AD, many philosophers and scientists suggested that motion is the essence of heat, such as when you rub your hands together to produce warmth. This concept was reaffirmed around 1600 by the famous English philosopher and scientist Francis Bacon. In the early 1700s, Daniel Bernoulli suggested in his theories of fluid flow that matter was made of particles. His theories also stated that while gas pressure was the force of these particles colliding with matter, heat was their kinetic energy transferred to matter. It was difficult for scientists to accept this theory because most of them had not accepted a particle model of matter, which was its basis.

Later in the eighteenth century, Joseph Black suggested that heat was like an invisible fluid. Solids needed to be filled up with this heat-fluid until they melted. The great French chemist Lavoisier further developed this idea and called the heat-fluid "caloric."

9-3 Joseph Black (1728–99) was a Scottish physician and professor of chemistry. He is best known for suggesting the caloric theory of heat and discovering carbon dioxide.

The **caloric theory** of heat was quite successful in explaining observations and making predictions. Lavoisier considered caloric to be an actual fluid, so the matter in caloric could be neither created nor destroyed according to his law of conservation of matter. Caloric was thought to consist of self-repelling particles that adhered to matter. This theory explained why heat flowed from hot to cold. Hot objects had a lot of caloric while cold objects had little. Caloric naturally flowed by self-repulsion from areas of higher concentration to areas of lower concentration. Temperature was a measure of the density of the caloric fluid—the amount of caloric per unit volume. Nearly all the gas laws could be explained by the caloric theory. For example, when air molecules absorb caloric, they become warmer because of a greater concentration of caloric. They also grow in size as the caloric particles attach themselves to the air particles and occupy a greater volume, explaining Charles's law. Both physical and chemical processes could release or strip particles of their caloric, raising the temperature of their surroundings.

The most significant successes of the caloric theory involved the development of a scientific theory of steam engines in the early 1800s by Nicolas Sadi Carnot (see the facet on page 181). He explained the process by which these engines work, using only the caloric theory. In addition, the theory allowed significant progress in understanding how gases expand according to Newton's laws and in determining the speed of sound in air.

Nicolas Sadi Carnot (1796–1832) was a brilliant French engineer who first deduced the theoretical principles of the steam engine. He died in Paris at age 36 from cholera.

Carnot's work evolved into the science of modern **thermodynamics**, which is the study of thermal energy and heat and how they relate to other kinds of energy and work.

9.3 *The Kinetic-Molecular Theory of Heat*

The first serious blow to the caloric theory was dealt by Benjamin Thompson, who lived in the seventeenth and eighteenth centuries. Born in the Massachusetts colony, he was an interesting combination of spy, traitor, statesman, soldier, nobleman, and scientist. In 1797 he was the director of the Munich arsenal in Bavaria. He was known then as Count Rumford.

9-4 Benjamin Thompson (1753–1814), also known as Count Rumford, was an American-born British subject who led a colorful, if less than reputable, life. He is best remembered for his work that disproved the caloric theory.

During military training maneuvers, Rumford observed that cannons fired without cannonballs became much hotter than those that were fired normally. If the caloric theory were true, the release of caloric and the resulting temperature rise should have been unaffected by the presence of the cannonball when the gunpowder was fired. Later, in his role as director of the arsenal, one of his duties was to supervise the manufacturing of weapons. This allowed Count Rumford an opportunity to observe cannon-boring operations.

The cannon-boring procedure was relatively simple. A solid metal cannon casting was mounted on a rotating shaft turned by horse power. A cutter pressed against the end of the cannon drilled out the bore as the barrel rotated. As the boring progressed, great quantities of heat were released—enough, in fact, to boil the water that was poured into the barrel to cool the cutting tool. How was this heat produced?

9-5 Rumford observed that a dull boring bit could generate more heat than needed to melt the entire cannon barrel. His observation helped bring about the end of the caloric theory.

According to the caloric theory, the metal chips removed by the drill released their caloric. As more metal was removed, more caloric should have been set free. Rumford noticed, however, that the duller the cutter, the greater the amount of heat produced, even though *less* metal was cut away. This finding contradicted the caloric theory. Rumford devised an experiment in which he encased the cannon in water while it was being drilled. He used a dull bit and measured how much water could be boiled during the drilling process. He calculated that drilling released more caloric than should have been needed to melt the entire cannon! By repeating this experiment again and again with the same cannon, he showed that the caloric supply seemed inexhaustible. Clearly, the caloric theory was unable to explain these observations. Rumford concluded that motion was in some way responsible for heat, but he was unable to provide a clear cause-and-effect explanation.

Julius Robert von Mayer was the first to experiment with the idea of heat as energy (not as matter) in 1842. He used a horse-powered mechanism to stir a large pot of paper pulp, calculating the mechanical energy needed to heat the mixture. His findings demonstrated that mechanical energy could be converted to thermal energy. This conclusion was largely ignored by physicists because Mayer was a medical doctor and his experimental procedure was fairly crude. It was not until many years later that he was finally given credit for his discovery.

9-6 Julius Robert von Mayer (1814–78) was a German physician who worked extensively in physics. He was the first to discover that energy could be converted from one form to another.

James Prescott Joule is usually recognized for first showing a clear connection between mechanical energy and heat. He performed experiments using all sorts of mechanical devices that were dropped, shaken, and stirred to produce changes in temperatures of liquids and gases. He concluded that the equivalent of 4.18 N·m of mechanical work would raise the temperature of 1 g of water 1 °C. This finding clearly established that the cause of a temperature change in matter was not a material fluid but a form of energy. He later worked with William Thomson (Lord Kelvin) to calculate the speed of molecules in air based on their kinetic energies, helping to firmly establish the kinetic-molecular model of thermal energy (see Subsection 6.6 on page 114).

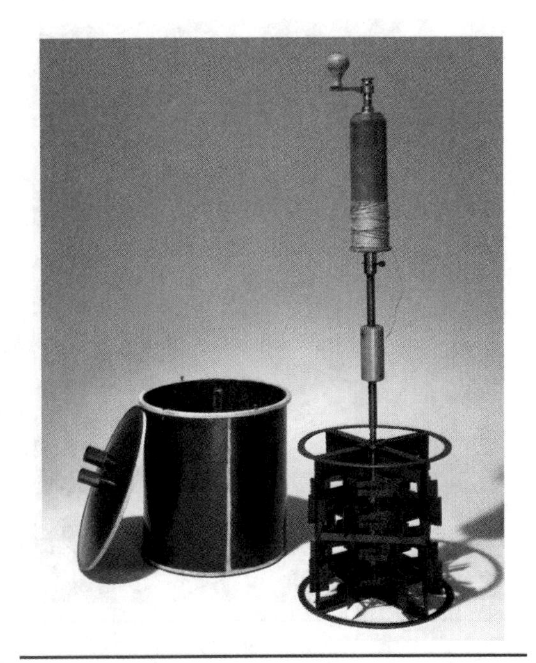

9-7 This is the actual apparatus that Joule used to demonstrate the equivalence between mechanical work and thermal energy.

FACETS OF SCIENCE

Nicolas Sadi Carnot

In 1712 a Baptist lay preacher, blacksmith, and plumber named Thomas Newcomen invented a steam engine. The earliest models of his engine could replace only two horses (2 horsepower) and were mainly used to pump water out of mines. By the time the Newcomen engines were replaced by more advanced designs, they could deliver 80 horsepower. In the late 1700s, an English mathematical instrument maker, James Watt, made several important improvements to the Newcomen engine, increasing its power fourfold. All new piston steam engine designs after his time relied on his basic improvements in some form.

One of James Watt's early steam engines

However, 100 years later, the best steam engines were still no more than about 3%–5% efficient. More than 95% of the fuel's energy was released as waste heat into the atmosphere.

The young French engineer Nicolas Sadi Carnot wondered whether there were some way to make steam engines more efficient. He also wondered whether there were some upper limit to the amount of work a steam engine could do.

At the age of 16, Carnot entered a respected engineering school in Paris where he was instructed by important scientists of his day. He graduated at the top of his class in 1814. He then took a two-year course in military engineering.

Carnot served twice in Napoleon's army as a military engineer. This time was very difficult for him. He felt discriminated against because his father, a former government employee, had been exiled to Germany by Napoleon's regime. Leaving the army, he studied many different industrial problems, especially those related to the theory of gases. In 1821 he began to study steam engines intensively.

At this point in the history of science, there were many misconceptions about the relationship between heat and the mechanics of Newton. Carnot dispelled these ideas in 1824 in a book called *Reflections on the Motive Power of Fire.* Machines called *heat engines* use the movement of thermal energy to do work. Carnot suggested that the ideal efficiency of a heat engine is related to the temperature difference between the engine's surroundings and the hot operating fluid, such as steam, inside the engine. He also noted that thermal energy can be converted to mechanical energy and vice versa but that there will always be some loss.

Carnot was also the first to recognize the practical applications of the laws of the conservation of energy and of entropy. (For a discussion of entropy, see the facet on page 199.) His careful, clear analysis was not initially appreciated by physicists. In 1832 he died from cholera in Paris at the age of 36. After his death, his unpublished scientific papers were discovered. Lord Kelvin and other scientists used his theories to understand the movement of thermal energy through a heat engine.

Carnot was the first to clearly define each step of the most efficient steam engine cycle. This cycle is now called the *Carnot cycle*. Carnot's contributions have greatly added to our understanding of thermodynamics, and his theories still govern the design and operation of gasoline, diesel, steam, and jet engines today.

Nicolas Sadi Carnot

9.4 *The Nature of Thermal Energy*

According to the kinetic-molecular model, atoms, molecules, ions, and their subatomic particles are in constant motion and thus have kinetic energy. In addition, they exert attractive or repulsive forces on each other that generate potential energies among the particles. The sum of all these energies is called the *internal energy* of matter. This energy cannot be measured.

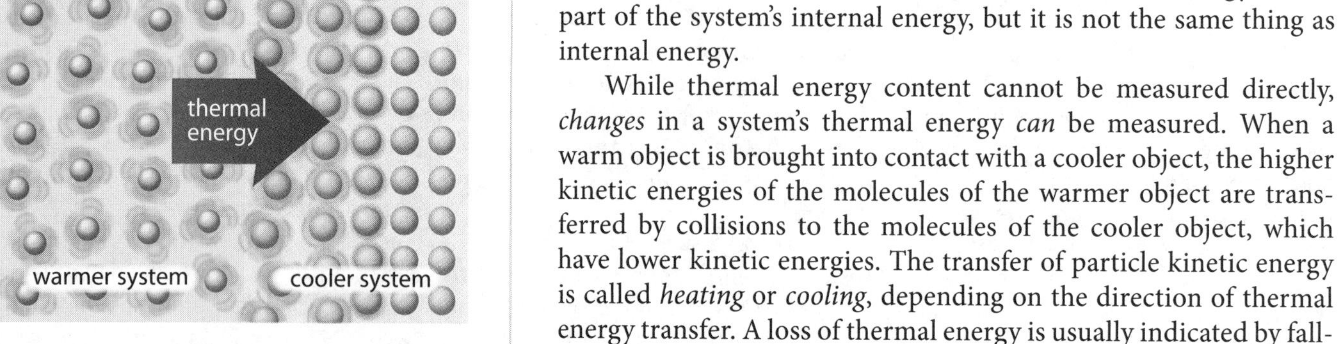

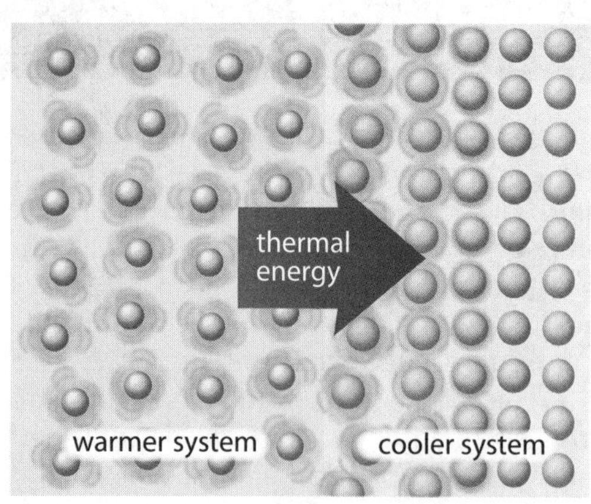

9-8 Thermal energy can be measured only when it is transferred from one system to another.

Internal energy can be affected by changes in a system's pressure and volume as well as by the exchanges of thermal energy with its surroundings. The *thermal energy* of that system is the average sum of the kinetic energies of its particles. Thermal energy is thus part of the system's internal energy, but it is not the same thing as internal energy.

While thermal energy content cannot be measured directly, *changes* in a system's thermal energy *can* be measured. When a warm object is brought into contact with a cooler object, the higher kinetic energies of the molecules of the warmer object are transferred by collisions to the molecules of the cooler object, which have lower kinetic energies. The transfer of particle kinetic energy is called *heating* or *cooling*, depending on the direction of thermal energy transfer. A loss of thermal energy is usually indicated by falling temperatures. A gain of thermal energy is usually accompanied by rising temperatures. You will learn more about the transfer of thermal energy in Section 9C.

9-9 This is a view of the Solar Energy Generating Systems facility in California. It is currently the largest solar thermal power plant in the world.

9-10 The Abengoa solar thermal plant near Seville, Spain, focuses sunlight on a steam-generating tower to boil water directly to high-pressure steam. The PS20 plant can produce up to 20 megawatts of electricity. This plant has a thermal energy storage system that permits brief power generation during nighttime operations.

9.5 *Using Solar Energy to Solve Problems*

The work of Joule, Mayer, Watt, Carnot, and many other scientists in thermodynamics is key to solving our dominion science problem. For a renewable energy source to be valuable, we must be able to convert its energy into a usable form. Can you think of a way to change the sun's energy safely and economically into electricity? How can the sun's energy be concentrated? Does the technology exist to do this?

Solar panels (properly called *photovoltaic* or *PV* panels) can change sunlight directly to electricity. The earliest PV panels were very expensive and inefficient. Though modern PV panels are more economical, they are still costly compared to other forms of energy. Even so, there are many PV electricity generation facilities being built around the world where sunlight is plentiful. These facilities produce safe, renewable energy from the sun.

Energy costs continue to rise, and these costs erode the wealth that a nation needs to build a financially sound society. Is there a more economical way of converting the free energy of the sun into electricity? One emerging technology is the *solar thermal (ST) power plant.* ST power facilities use large arrays of steerable mirrors that reflect sunlight to heat water or other kinds of fluids. Eventually, hot, high-pressure steam is produced, turning turbine generators to produce electricity. The largest ST power station in the world is the Solar Energy Generating Systems (SEGS) facility located in the Mojave Desert in California. Its installed generating capacity is 354 megawatts (MW) of electricity. At night, when solar energy is not available, it uses natural gas to run gas turbine generators. This permits the plant to generate electricity all day and night.

Dependence on daylight hours is a major weakness of solar energy sources. Some ST plants are experimenting with storing superheated water for generating steam during periods of darkness and cloudiness. Right now, some plants can store heated water for about an hour, but new approaches to this problem are being investigated. Solutions may include storing thermal energy in liquid sodium reservoirs or using the latent heat properties of phase changes in other materials. See Section 9C for a discussion of latent heat.

FACETS OF SCIENCE: JAMES PRESCOTT JOULE

James Prescott Joule

Often a scientist encounters data that leads him to challenge a widely accepted theory. But someone challenging accepted science usually meets with resistance. This is exactly what happened to James Prescott Joule.

Born on Christmas Eve 1818 to a wealthy English family, Joule took an early interest in science, especially electricity, which at the time was a new field of study. John Dalton, who proposed the first workable atomic model in modern chemistry, was one of his tutors.

Joule first worked to improve the efficiency of electric motors. He wanted to use batteries and electric motors to replace the steam engines that powered his family's brewery. However, he abandoned this project when he found that the cost of zinc for the batteries would be five times the cost of the equivalent coal.

Joule later became interested in the conversion of mechanical energy into heat. The prevailing view of the leading scientists of his day was that such a conversion was impossible. They accepted the *caloric theory* as the best model to explain heating and cooling. The caloric theory considered heat to be a mysterious invisible fluid stored in matter and released during burning and other chemical reactions. But Joule felt that this idea had problems.

In 1842, Joule began conducting a series of experiments involving various methods of stirring and agitating water. In the end, all his methods indicated that 772.55 **foot-pounds** of mechanical work would raise the temperature of 1 lb of water 1 °F. This finding contradicted the caloric theory by suggesting that immaterial energy could be converted into matter (caloric).

Since Joule was not a recognized scientist at the time, the prestigious Royal Society declined to publish his papers. Instead, he reported his findings in the *Philosophical Magazine*, one of the earliest scientific journals. He defended his work by showing critics that he was not supporting the idea that heat was destroyed when an object cooled. He said, "Believing that the power to destroy belongs to the Creator alone, I affirm . . . that any theory which, when carried out, demands the annihilation of force, is necessarily erroneous." Most scientists remained skeptical about his ideas until three scientists, George Gabriel Stokes, Michael Faraday, and William Thomson (Lord Kelvin) examined his work with new interest.

His careful and wide-ranging experimentation on the convertibility of mechanical and thermal energy was the key. It eventually enabled Lord Kelvin and other scientists to correctly apply the law of conservation of energy to thermal energy and thus replace the caloric theory with the more workable kinetic-molecular model.

Kelvin and Joule worked jointly on many experiments. They developed the absolute scale of temperature (the Kelvin scale). They also discovered the Joule-Thomson cooling effect, in which a gas cools as it expands. The modern air conditioner works on this principle. The effect also explains why a rising air mass cools.

Joule died in 1889. Inscribed on his gravestone is the number "772.55" and "I must work the works of him that sent me, while it is day: the night cometh when no man can work" (John 9:4). The SI unit for energy and work, the joule, is named in his honor.

A **foot-pound** (ft-lb) is the English unit of work (force × distance), just as the newton-meter is the SI unit of work. In Joule's day, scientists measured work in foot-pounds.

9A Section Review

1. What did scientists believe about caloric? State two examples of scientific phenomena that the caloric theory explained.
2. Who first successfully challenged the caloric theory? What observation could the theory *not* explain?
3. What was the significance of Joule's discovery of the relationship between heat and mechanical work?
4. How much energy (or work) is required to raise the temperature of a gram of water 1 °C?
5. What is the difference between the internal energy of an object and its thermal energy?
6. When *can* we measure thermal energy? What is actually measured?
7. DS ✪ Why is the sun a good energy resource? What challenges does a solar thermal energy power facility face in harnessing the sun's energy?
8. (True or False) There was no real difference between the caloric theory of heat and the kinetic-molecular theory.

9B Temperature

9B Objectives

After completing this section, you will be able to

- ✓ give examples of thermometric properties.
- ✓ compare and contrast the Fahrenheit, Celsius, and absolute temperature scales.
- ✓ explain how a fiducial point is chosen for a temperature scale.
- ✓ convert temperatures between all three temperature scales.
- ✓ describe how a thermometer works.

9.6 Thermometric Properties

You may feel "hot" on a steamy summer day or "cold" on a snowy, blustery day. These descriptions are *subjective*, which means that they depend on how you perceive hot and cold. However, they are scientifically useless. Scientists need an *objective* standard for measuring temperature, not one dependent on uncalibrated human perception.

The **temperature** of a substance is directly related to the average kinetic energy of its atoms and molecules. It is measured in a dimensional unit called the **degree** (°). The "change of hotness" per degree varies depending on the temperature scale (Fahrenheit, Celsius, or Kelvin). We measure temperature with an instrument called a *thermometer*. Thermometers work by exploiting a *thermometric property*, which is a physical property of matter that changes predictably with a given change of temperature. Most thermometers you're familiar with depend on the thermal expansion of liquids or metals or on the change of an electrical property to detect temperature.

This use of the term **degree** originated from the time when scientists called temperature the "degree of heat."

9-11 You may be cold, but saying so is not a scientifically meaningful statement.

Most thermometers work through direct contact with the substance whose temperature they are measuring. If the substance is warmer than the thermometer, thermal energy is transferred from the substance to the thermometric material. As the thermal energy of the thermometer's particles increases, it indicates a higher temperature. If the substance is colder than the thermometer, it removes thermal energy from the thermometric material and the thermometer indicates a colder temperature.

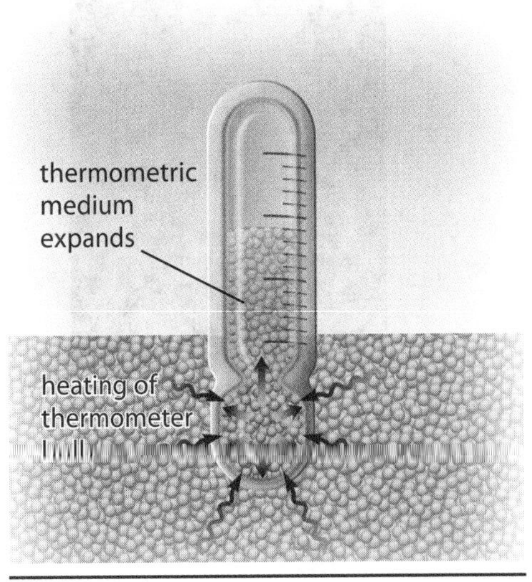

9-12 When thermal energy enters a thermometer, the indicated temperature rises. When it leaves the thermometer, the indicated temperature drops.

9.7 *Early Thermometers*

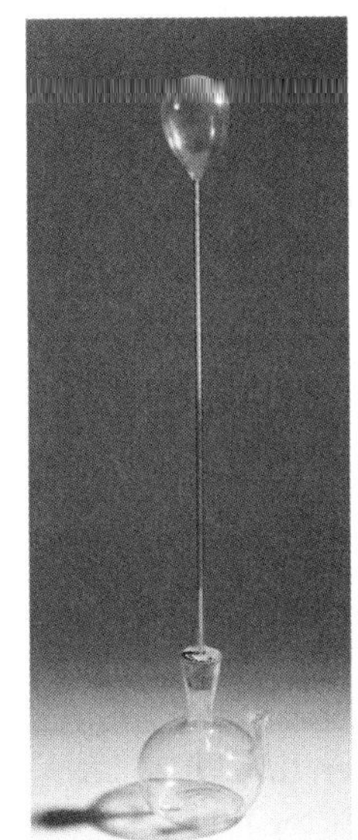

One of the first thermometers was built around 1600 by Galileo. The *thermoscope*, as Galileo called it, consisted of a long, slender glass tube with a sealed bulb at one end. It was inverted with its open end in a flask of water. The level of water in the tube rose and fell as the volume of air trapped inside fluctuated with its temperature. Later scientists revised Galileo's design by replacing the air, which was the thermometric substance, with a combination of alcohol and water in a sealed tube. Sealing the tube eliminated the effect of atmospheric pressure on the temperature reading.

9-13 Galileo's thermoscope. It was not always accurate because changes in air pressure affected the water level in the column.

9.8 *Temperature Scales*

A serious challenge for early scientists was the marking of their thermometers' scales. They needed to calibrate them so that the instruments reliably indicated the correct temperature. For scientists everywhere to obtain the same readings with their thermometers, they needed to agree on the anchor, or standard, points for a temperature scale. These needed to be fixed, precisely known, and easily reproducible temperature values. Such standards for a measuring scale are called *fiducial points.*

fiducial (fih DOO shul): (L. *fiducia*—trust, confidence)

The Fahrenheit Scale

In 1714 Gabriel Fahrenheit tackled the problem of developing a reliable thermometer. He preferred mercury to the alcohol-water mixture because it could be used for higher temperatures where alcohol would readily evaporate. He initially selected two temperatures as fiducial points for his thermometer: the freezing point of a mixture of salt, water, and ice and the temperature of the blood of a healthy man. After some further experimenting, he settled on a temperature scale with the fiducial points at the freezing and boiling temperatures of pure water. On the **Fahrenheit scale** there are 180 degrees between water's freezing point, 32 °F, and boiling point, 212 °F. The design of modern liquid thermometers is essentially the same as Fahrenheit's original.

9-14 Daniel Gabriel Fahrenheit (1686–1736) was a German instrument maker who invented the mercury thermometer and the temperature scale bearing his name.

9-15 Anders Celsius (1701–44) was a Swedish astronomer who carried out many different kinds of scientific investigations outside of astronomy. He is best remembered for the Celsius temperature scale.

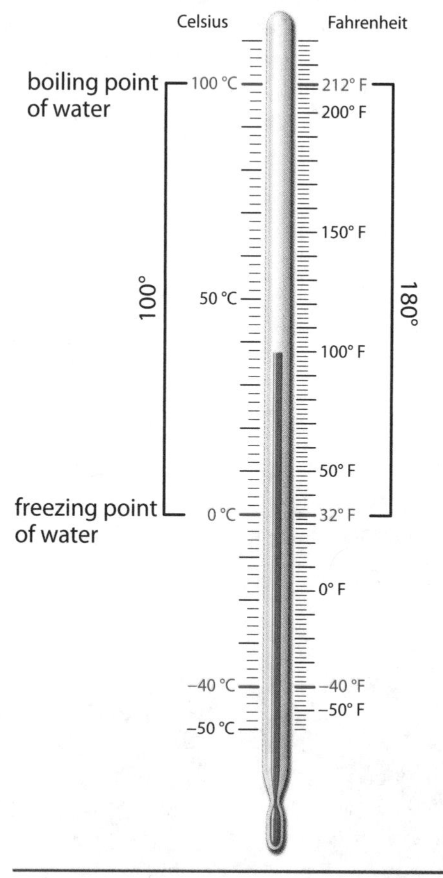

9-16 Comparison of the Fahrenheit and Celsius temperature scales. The fiducial points are fixed by the physical properties of water. The number of degrees between them was chosen by the scientists.

Some people think that remembering the conversion formulas is easier by using the fraction forms of the conversion factors. Note that

$$\frac{180°}{100°} = \frac{9}{5} \text{ and } \frac{100°}{180°} = \frac{5}{9}.$$

The Celsius Scale

In 1742 Anders Celsius devised a decimal temperature scale—the temperature range between the freezing and boiling points of water was 100 degrees. He also extended his scale above and below the fiducial temperatures. Interestingly, Celsius placed 0° at the boiling point of water and 100° at the freezing point, just the opposite of the standard arrangement today. In 1744 the scale was switched after Celsius's death so that values increased with temperature. Because of its convenience and decimal basis, the **Celsius scale** was adopted by the French commission that designed the metric system. Most countries and schools and almost all scientific laboratories currently use the Celsius scale.

You may have noticed that there are fewer Celsius degrees between the freezing and boiling points of water than Fahrenheit degrees (100 versus 180). Thus a Celsius degree is "larger" than a Fahrenheit degree. In addition, the scale values for the Fahrenheit and Celsius fiducial points differ. These differences make relating temperatures between the two scales inconvenient at best. Though most of the world uses the Celsius scale, the United States still depends on the Fahrenheit scale for most familiar purposes. Sometimes conversion between these scales is necessary.

Probably the easiest way to convert from one temperature scale to another is to use an Internet conversion service or an application downloaded to your computer, tablet, or smart phone. In many Web browsers, just typing a phrase like "convert 72 F degrees to C" in the Search field will yield the equivalent Celsius temperature.

For classroom work, you can use two formulas for this purpose. The most familiar one converts the Celsius temperature to Fahrenheit degrees using the conversion factor 1.8 °F/°C (from the ratio 180 °F/100 °C; see Figure 9-16) and then adding 32° to reference the temperature to the Fahrenheit value of water's freezing point,

$$t_F = 1.8t_C + 32°,$$

where

- t_F is the required Fahrenheit temperature and
- t_C is the given or known Celsius temperature.

The second and possibly less familiar method is probably easier to remember (for when you are converting temperatures while stranded on a desert island). It works because the temperature –40° is the same temperature on both the Fahrenheit and Celsius scales. It uses the same conversion factor for Celsius to Fahrenheit degrees (1.8 °F/°C).

$$t_F = (1.8)(t_C + 40°) - 40°$$

If you wish to convert from Fahrenheit to Celsius, just solve either formula for the Celsius (t_C) temperature using simple algebra. The new conversion factor is the reciprocal value of the original (that is, 1 °C/1.8 °F, or 5/9 °C/°F). The new formula is

$$t_C = (5/9)(t_F - 32°).$$

If you are using the second method, the only thing you will have to change is the conversion factor, which is

$$t_C = (5/9)(t_F + 40°) - 40°.$$

The Kelvin Scale

Scientists in the early 1800s discovered that gases had interesting properties at very low temperatures, far below the freezing point of water. Studies based on Charles's law indicated that if gases were cooled to a sufficiently low temperature, gas pressure would disappear. The Celsius temperature for this point was calculated to be about −273 °C. Scientists found that devising mathematical gas laws that worked at all temperatures was difficult because of the negative temperature values.

Lord Kelvin recognized that temperature is related to thermal energy content. He suggested that the temperature of matter in which no more thermal energy can be removed is the coldest temperature possible. In 1848 Kelvin proposed a temperature scale in which this temperature was assigned a value of zero, also called *absolute zero*. All higher temperatures were positive values. The degree in Kelvin's scale is equal to the Celsius degree in "size." On the Kelvin scale the freezing point of water is 273.15 K.

For the Fahrenheit and Celsius scales, two fiducial temperatures were used to "anchor" the scale. What points could be used for the Kelvin or absolute scale? Absolute zero is unattainable according to the laws of thermodynamics, so it could not be a measurable fiducial point.

After considering this problem, scientists decided to use only one fiducial point—the triple point of water. A substance's *triple point* is the temperature and pressure at which solid, liquid, and gaseous phases of the substance simultaneously exist in a stable condition. Water's triple point is 0.01 °C and 611.73 Pa. On the **Kelvin scale** the triple point of water is 273.16 K. Scientists then defined 1 K as 1/273.16 the triple point temperature of water.

Conversions between Kelvin and Celsius temperatures are easy to do. Since the Celsius degree and the kelvin are the same size, it is simply a matter of adding or subtracting 273.15° as appropriate. To convert Celsius to Kelvin, add 273.15°:

$$\boxed{T = t_{\mathrm{C}} + 273.15°}.$$

To convert Kelvin to Celsius, subtract 273.15°:

$$\boxed{t_{\mathrm{C}} = T - 273.15°}.$$

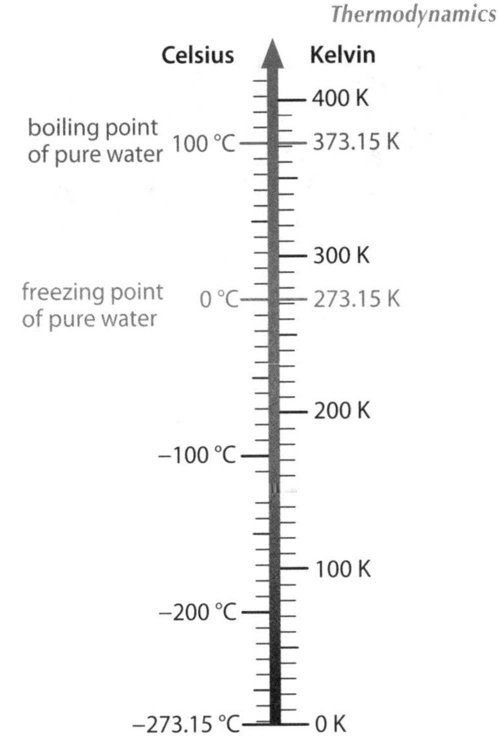

9-17 The Kelvin scale begins at absolute zero and has the same degree size as the Celsius scale.

Note that the degree symbol (°) is not used with the kelvin unit.

The symbol T is used to represent Kelvin, or absolute, temperatures in formulas.

You can remember the relationships between the Celsius and Kelvin scales by recalling that the Celsius temperature will always be a smaller number than the corresponding Kelvin temperature. For example, 0 °C is about 273 K and 100 °C is about 373 K.

9.9 Matter and Temperature

Temperature can have significant effects on important properties of matter. Engineers design products to take advantage of these temperature-dependent properties or try to avoid the problems they cause. This section covers a few of the important properties of matter that are affected by temperature.

Thermal Expansion

The center span of the Verrazano-Narrows Bridge in New York City soars 66 m (216 ft) above the water at its middle, but only on a cold winter day. The bridge and its suspension cables lengthen as they grow warmer. On a hot summer day, the concrete and steel span droops 3.7 m (12 ft)! Most materials have the property of *thermal expansion*. Bridge designers must leave gaps, called finger joints, between sections of the road surface so that the bridge can expand

9-18 The Verrazano-Narrows Bridge

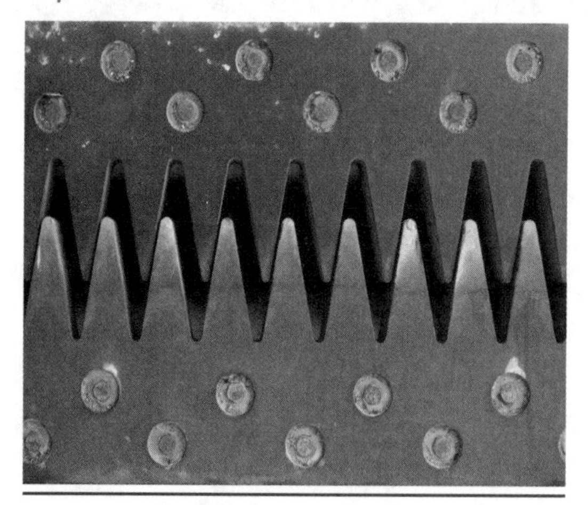

9-19 A road expansion joint

without buckling. Long bridges also have rollers at support points so that expansion and contraction will not damage the bridge foundations. Roads and sidewalks are built with compressible strips between the concrete slabs to allow the concrete to expand and contract with temperature changes.

Thermal expansion occurs at the particle level, where greater thermal vibrations at higher temperatures force particles farther apart. If an object is long and thin, like a wire, the expansion is most noticeable along its length. The increase in length is directly proportional to the number of particles lined up in a given direction. A long thin object expands much farther along its length because it has many more particles lined up in that direction than it has across its width: 1% of a length of 200 m is a greater value (2 m) than 1% of a diameter of 1 cm (0.01 cm). For this reason, high-voltage transmission wires suspended between electrical towers can sag a meter or more in summer, but the change in their diameters is hardly noticeable.

Every material has its own unique thermal expansion properties. For example, for the same temperature increase, brass expands almost seven times as much as porcelain. The solid and liquid phases of most substances expand at about the same rate for a given temperature change because their particles remain relatively close together. The gas phases, however, show a much greater volume change per degree change in temperature.

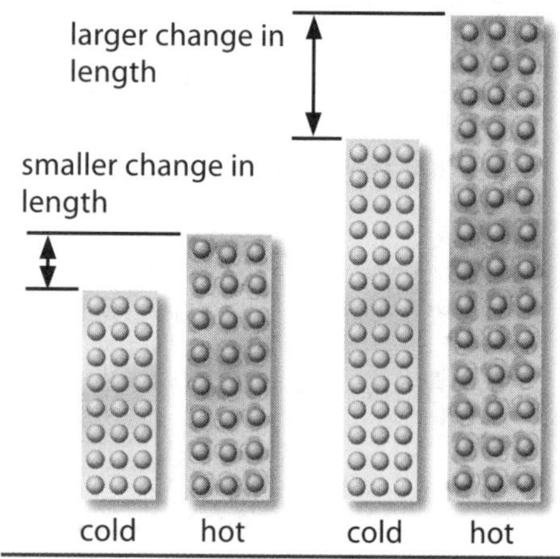

9-20 Thermal expansion is proportional to the number of particles lined up in a given direction.

Electrical Resistance

Certain materials that are excellent thermal conductors, such as metals, also conduct electricity well. One way to measure a material's ability to conduct electricity is to measure its *electrical resistance*. The greater its resistance, the poorer its conducting ability. See Chapter 10 for a more complete discussion of electrical resistance.

When the temperature of a substance increases, its electrons, which are involved in bonding atoms together, move more chaotically. They no longer flow easily in one direction, which is essential for carrying electrical current. The material becomes more resistant to the flow of electrical current as its temperature rises. Electrical transmission wires heated by the sun are a problem for electric companies because it takes more energy at the source to send electrical current through the higher electrical resistance.

9-21 Hot power lines increase electrical resistance, which increases costs to customers.

Viscosity

Another material property strongly affected by temperature is viscosity—a fluid's resistance to flow. Liquids like rubbing alcohol have low viscosities. Liquids like thick corn syrup and certain oil treatments are highly viscous. Viscosity generally decreases in warmer fluids. It decreases because the molecules vibrate faster and move farther apart so that they have less attraction for one another. Less viscous liquids are easier to pump, so engineers often try to warm liquids in piping distribution systems. On the other hand, lubricants depend on at least some viscosity to coat surfaces. Motor oil is used to lubricate car and truck engines, but if the oil gets too hot, it loses its viscosity and is less effective at reducing friction between moving metal parts.

9-22 Maple syrup is graded partly on its viscosity.

9B Section Review

1. What is the difference between the statements, "It is hot outside" and, "It is 103 °F outside"? What must one observe in order to make the second statement?
2. What property of a thermometer allows it to indicate temperature? Name two examples of this property.
3. What three properties must a fiducial point have? Name the fiducial point(s) for the Fahrenheit and Celsius scales.
4. What is one major advantage of the Celsius scale over the Fahrenheit scale? What major disadvantage do both scales have?
5. How does the Kelvin scale resolve the problem noted in Question 4?
6. State the temperature of the fiducial point(s) of the Kelvin scale. What defines the Kelvin scale's fiducial point(s)?
7. Name three properties of materials affected by temperature changes.
8. ✪ (True or False) The absolute temperature scale needs only a single fiducial point because the size of the degree was defined by the Celsius scale.
9. ✪ Convert the following temperatures to the indicated scale by any method.
   a. 100. °F to Celsius
   b. 22 °C to Fahrenheit
   c. 98.6 °F to Celsius
   d. −40. °C to Fahrenheit
10. ✪ Convert the following temperatures to the indicated scale by any method.
    a. 3652 °C to Kelvin (melting point of carbon)
    b. 1074 K to Celsius (melting point of table salt)
    c. −38.9 °C to Kelvin (melting point of mercury)
    d. 4.2 K to Celsius (boiling point of helium)

9C Objectives

After completing this section, you will be able to

- ✓ compare and contrast heat and thermal energy.
- ✓ describe the flow of thermal energy.
- ✓ describe the three methods of heat transfer and give an example of each.
- ✓ summarize the physical properties of insulators and conductors.
- ✓ describe how thermal energy affects matter's volume and its ability to allow electricity to pass through it.
- ✓ compare and contrast heat capacity and specific heat.
- ✓ explain the changes in thermal energy and temperature for an object that is heated from its solid state to its gaseous state.

9C Heat

9.10 *The Nature of Heat*

Remember the last time you opened a car door on a broiling summer afternoon and were met by a blast of hot air? After gingerly seating yourself on the hot seat, you quickly became aware of its ability to store energy! The longer you sat there, the more you felt the sun pouring additional energy into you and the car. All these sensations are evidence of *heat*—the flow of thermal energy from one place to another.

In everyday situations, the word *heat* is often loosely used for both thermal energy and temperature. You have probably heard it said that "a boiling pot has a lot of heat" (thermal energy) or "the oven is set to a high heat" (temperature). Neither of these statements is correct from a scientific point of view. The amount of thermal energy in an object is a property of the object—it *has* thermal energy. In science, and in this textbook, *heat* is used as a noun only to identify a quantity of thermal energy that moves between systems. Heat is not something that a system *has*.

9-23 A candle's flame has thermal energy and heats the person's hand.

The best way to use *heat* is as a verb, to describe processes in which thermal energy is added or gained. A stove *heats* a pot of water. The opposite process is *to cool*—to remove or lose thermal energy. Heating and cooling occur through one of three processes—conduction, convection, and radiation.

9.11 *Heat Transfer*

Conduction

When two objects of different temperatures touch, thermal energy moves from the hotter to the cooler object. This process is called **conduction**. How does conduction transfer thermal energy? When a warmer object contacts a cooler one, its faster-moving particles collide with the cooler object's particles, transferring kinetic energy. The object with more thermal energy will cool, and the second object will warm until both are at the same temperature. At this point the average kinetic energies of all their particles are the same—their temperatures are the same. These objects have attained *thermal equilibrium*.

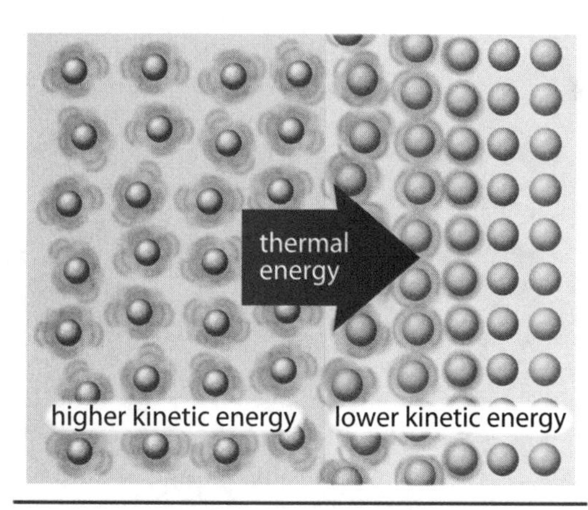

9-24 Thermal conduction occurs when thermal energy moves through matter as particle kinetic energy is transferred through collisions.

When one object is a continuous source of thermal energy, such as a stove burner, another object in contact with it continues to gain thermal energy until reaching thermal equilibrium with the first. Ideally, the temperature at which thermal equilibrium occurs would be equal to the source's higher temperature, but this does not normally happen. Thermal energy is continuously lost by the second object to its surroundings. Its surface remains cooler than the temperature of the hotter source, so thermal energy continues to flow.

Conduction is the chief process by which thermal energy moves through solids. Because the atoms of solids are tightly bound to each other, one strongly vibrating atom collides with adjacent atoms, causing them to vibrate more. Thermal energy thus moves quickly throughout the solid material.

All materials can conduct thermal energy, though some do so better than others. Diamond is the best natural *conductor* of thermal energy. Metals, because of the special way their atoms bond together, are generally excellent thermal conductors. Silver is the best metal for conducting thermal energy, but it is also relatively expensive. In most situations, less expensive conductors, such as copper and aluminum, are more practical. Substances that are good thermal conductors are usually good conductors of electricity also. Having an efficient thermal conductor is important when a large amount of thermal energy must be transferred. The heat exchangers and radiators in heating and cooling systems often have fins and coils made of thermally conductive metals.

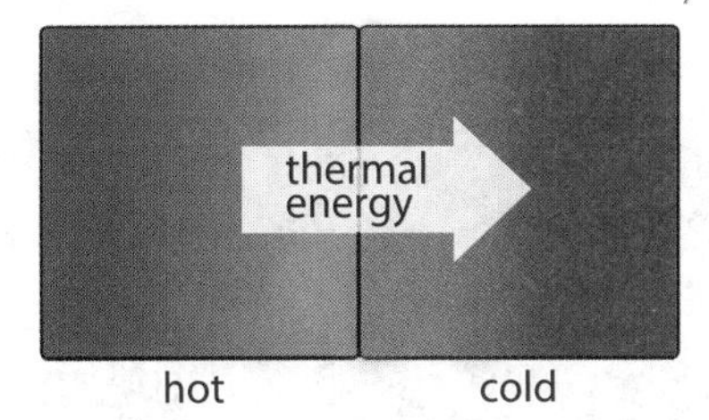

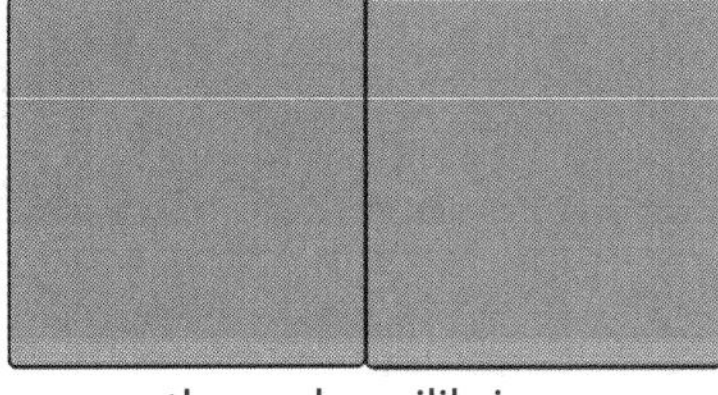

9-25 Ideally, two touching objects with different temperatures will eventually reach thermal equilibrium with each other.

9-26 In the real world, objects gain or lose thermal energy to arrive at thermal equilibrium with their surroundings.

Convection

That blast of hot air that greets you when you open a car door on a sunny summer day illustrates another method of thermal energy flow—**convection**. In convection, thermal energy is carried from one location to another by a fluid. Convection involves the movement of randomly vibrating atoms and molecules over distances that are large in comparison to the size of an atom. For this reason, convection does not occur in truly solid materials, but it is the most important way that thermal energy is transferred within fluids.

Natural convection occurs under the influence of gravity. Warm fluids rise and cold fluids sink due to differences in density. A heated fluid's particles move faster. The fluid occupies a greater volume of space as the more frequent collisions of its particles force them farther apart. Since the same number of particles occupies a larger volume, the fluid's density is lower and it becomes buoyant relative to the cooler surrounding fluid (see Chapter 8). The denser fluid flows in under the less dense fluid, forcing it upward. This process continues as long as there is a thermal energy source for the fluid. After the warmer fluid rises, it immediately begins to transfer thermal energy through conduction to the cooler fluid around it. This cools the rising fluid. Eventually, its density is greater than the fluid rising beneath it, and it flows to the side of the rising column and sinks down toward the heat source to begin the cycle over again. The fluid's cyclic path is called a *convection current.*

Convection currents can occur only in a gravitational field.

Hot water radiators and electric baseboard heaters heat rooms by setting up convection currents. The warm air in the vicinity of the radiator rises by convection and flows toward the ceiling in the middle of the room, cooling as it goes. It then sinks to the floor and

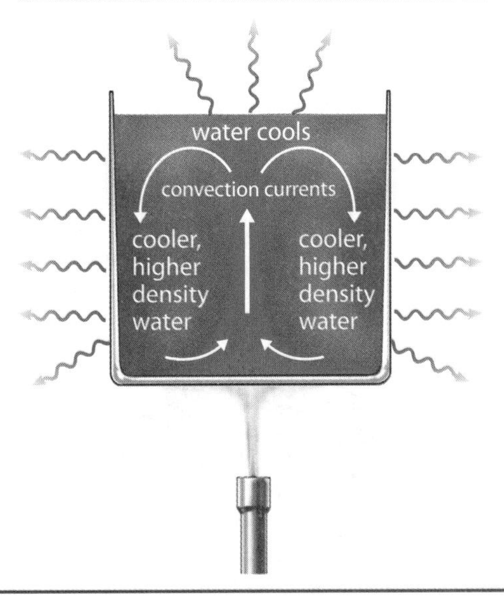

9-27 Formation of a convection current

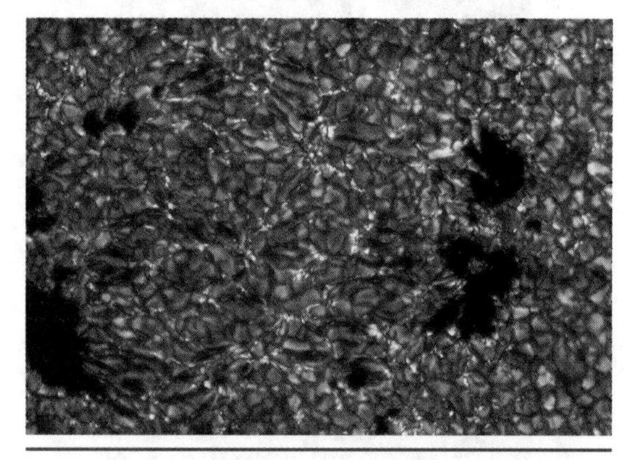

9-28 Plasma flows to the surface of the sun in convection currents, seen here as bumps, or *granules*, on the sun.

9-30 These special Alaskan oil pipeline supports conduct heat away from the ground to prevent melting the tundra permafrost. The fins help to radiate heat into the surrounding air.

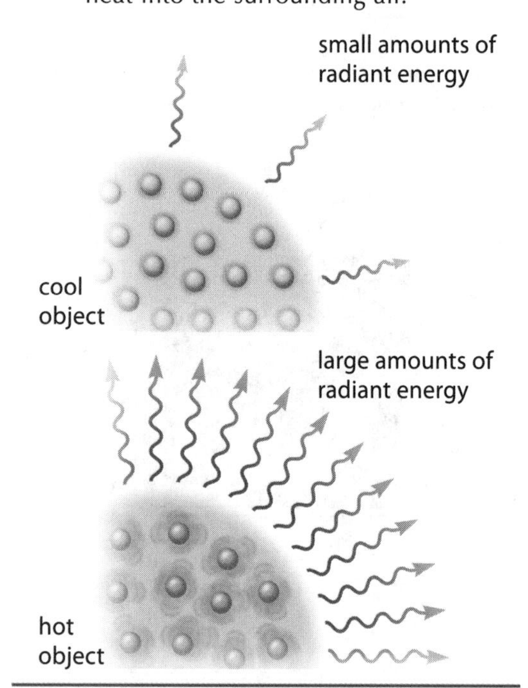

9-31 Radiant energy is emitted in proportion to the temperature of the object.

flows back toward the radiator to replace the less dense, warmer air that continues to rise. *Forced convection* is often used by engineers to increase the capacity and improve the efficiency of convection heating and cooling systems. Fans move air in ventilation systems and in convection ovens. Pumps move liquids through heating and cooling systems.

Atmospheric convection is responsible for many kinds of winds near the earth's surface and for large-scale atmospheric circulation. Local atmospheric convection currents include sea and land breezes as well as mountain and valley breezes (see Figure 9-29). Immense convection currents affect global weather by exchanging warm tropical air from near the Equator for the cooler air at higher latitudes and the frigid air at the poles. Global convection currents keep the earth's surface temperature within habitable limits. Without them, the equatorial regions would bake while the higher latitudes would be locked in ice.

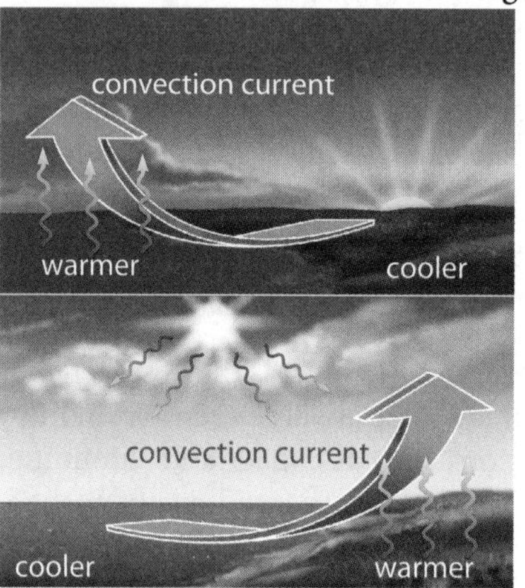

9-29 The differences in temperature of the land and sea surfaces drive the local breezes near the shore.

Radiation

How do the immense quantities of vital thermal energy move from the sun through 150 million kilometers of empty space to the earth? The density of the particles that *do* exist between the earth and the sun is too low to transfer thermal energy by conduction or convection.

Thermal energy moves most efficiently through a vacuum as *radiant energy*. It is converted to electromagnetic energy at its source and then converted back to thermal energy at its destination. As electrons move about the nucleus of a vibrating atom, the atom exchanges energy with its surroundings in the form of electromagnetic energy. All atoms emit this energy, particularly as infrared energy. The higher the temperature of the substance, the more electromagnetic energy is emitted. Electromagnetic energy moves in space through the process of **radiation** (see Chapters 6 and 14).

The hot plasma of the sun's surface contains great quantities of thermal energy. The highly agitated plasma ions emit intense radiant energy of all kinds into space. This electromagnetic energy races outward in all directions at the speed of light. A tiny fraction of this energy falls on the earth. Atoms in the atmosphere and on the earth's surface absorb the sun's radiant energy. They vibrate faster and their average kinetic energy increases, causing the earth's temperature to rise. Because thermal energy can be transferred by radiant energy, many people consider them to be the same thing; however, radiant energy is distinct from thermal energy in that it does not use matter to move between systems.

Radiant energy can transfer thermal energy between two objects that are not in contact. If they are not in thermal equilibrium, the infrared emissions of the warmer object will be greater than those of the cooler. The cooler object will absorb the excess infrared and grow warmer. This principle is the basis for medical heat lamps used to keep injured limbs warm, allowing healing to occur. Food displays in restaurants use infrared lamps to keep food hot.

You may have noticed that on a sunny day you feel warmer wearing a dark-colored shirt than you do wearing a light-colored one. The amount of radiant energy that is absorbed depends on certain properties of the material, one of which is color. For instance, dark-colored materials readily absorb radiant energy, while light-colored materials reflect most of it. Chapter 15 discusses how color affects absorbed and reflected light.

9-32 Thermal energy in one object can be transmitted by radiant energy to another that is not touching the first.

9.12 Insulation and Thermal Resistance

Some materials, rather than being good conductors, seem to hold thermal energy at bay. Materials that resist the flow of thermal energy are *thermal insulators.*

The atoms of good conductors are bonded closely together, and they have many loose electrons that can easily move among the atoms. When the atoms vibrate, motion is quickly transmitted throughout the material by moving electrons. Just as the properties of conductors arise from the arrangement and bonding of their atoms, so do those of insulators. The atoms of some insulators are bonded together in open, sponge-like structures, while other insulators have their atoms bonded close together with the electrons held tightly to the atoms. When thermal energy is transmitted to the surface of the material, few atoms or electrons are affected by the vibrations, and it takes a long time for thermal energy to move through the material. Insulators may be natural or man-made. The best artificial insulators are *aerogels.* Read the facet on the following page to learn about these fascinating materials.

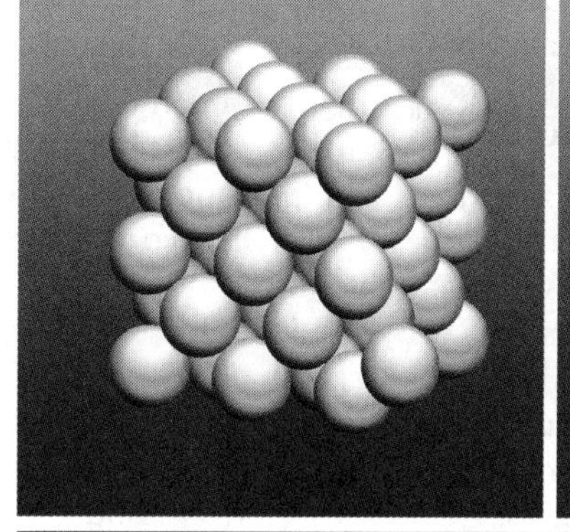

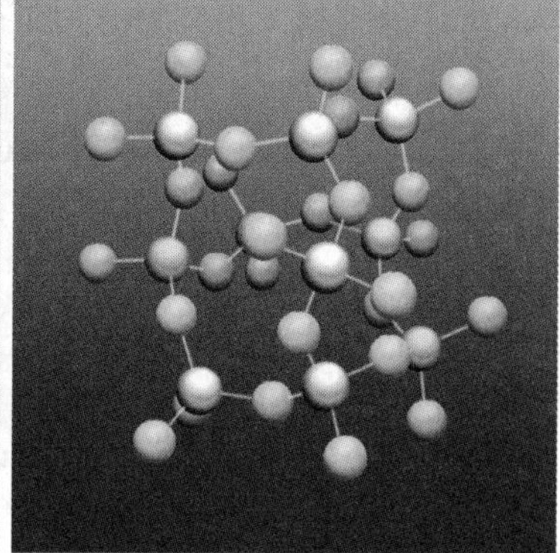

9-33 A good conductor (left). A good insulator (right).The arrangement of particles in matter has a lot to do with its thermal properties.

Gases can be good insulators because their particles are so far apart that they do not efficiently transfer thermal energy from particle to particle. Insulated windows consist of two panes of glass that confine a thin layer of dry gas between them. The gap must be less than a centimeter wide to keep a convection current from occurring, and the gas must be a relatively poor conductor. Gases that are confined so that they cannot produce convection currents prevent almost all thermal energy flow by conduction through the glass. The trapped air inside the cells of polystyrene foam or among the feathers in a down jacket achieves the same effect. The insulating ability of a material is increased by the air gaps it contains.

The best insulator of all is a vacuum because there are no particles present. The only way thermal energy can move across a vacuum is through radiation, which is not as efficient as conduction. Vacuum insulation is commonly used in insulated containers to keep food warm or to maintain extremely low temperatures in bottled liquefied gases.

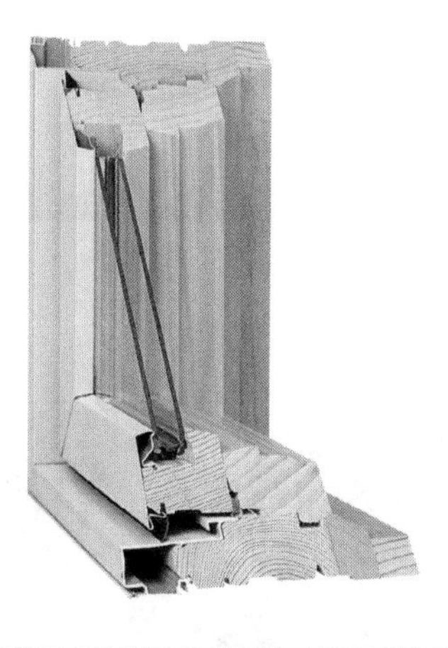

9-34 A double-pane insulated window has a thin gap of dry gas between the glass panes that minimizes conduction heat loss and prevents the formation of convection currents inside the window.

FACETS OF SCIENCE

Aerogels: Frozen Smoke

You may be thinking, "Frozen smoke? Are you kidding?" But that is what some people call aerogels. Others call them dried Jell-O, solidified air, and puffed sand.

An *aerogel* is the world's lightest solid substance, and it has an intriguing set of physical properties. Samuel Kistler discovered how to make this material about 1930, and several people have tried to make practical use of it since. We are just now learning how to make affordable, useful products exploiting its strange properties.

Aerogels can be made from many different raw materials, such as silica, iron oxide, cellulose, gelatin, and rubber. First, a mixture similar to gelatin is formed in a solvent. Within the gel the particles of the raw material interconnect to form a complex latticework. Liquid carbon dioxide is then used to replace the solvent. Finally, the mixture is transformed into an aerogel by raising the temperature and pressure until the carbon dioxide diffuses out, leaving a skeleton that is about 95% air. The density of an aerogel is less than 1/1000 that of glass, but it is exceptionally strong for its weight.

One type of aerogel makes the best insulation material known. You can safely touch one side of a thin slab of aerogel while the other side is being heated with a blowtorch. You will probably soon find new refrigerators and freezers insulated with it. Although it is four times as expensive as the polyurethane foam currently being used, it is twice as efficient and does not cause atmospheric pollution during production. It pays for itself in less than two years through conservation of electricity.

Probably the most exciting use of an aerogel was for NASA's Stardust Program. Launched in 1999, the *Stardust* spacecraft captured particles from the dust surrounding the head of comet Wild 2 and from interplanetary space. A mosaic of lightweight aerogel tiles, arranged in a disk about 40 cm in diameter, picked up thousands of comet dust grains on its front surface and many interstellar dust grains on its back surface. These high speed particles were safely slowed and trapped in place by the aerogel without damaging or altering them. The samples were returned to Earth in 2006.

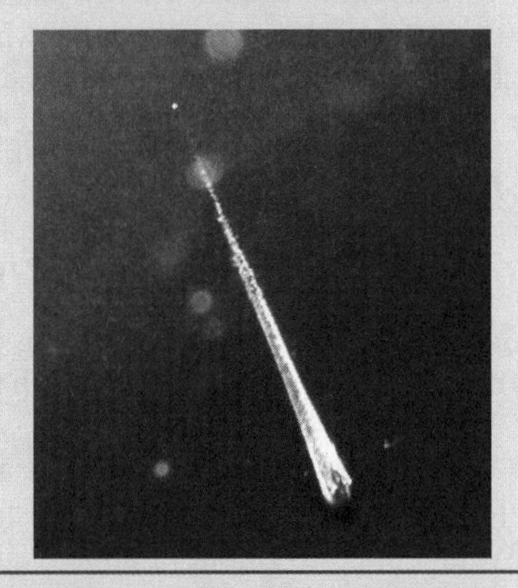

A piece of comet dust captured in aerogel by the *Stardust* probe

Aerogels have great promise for beneficial applications. They are exceptional fire retardants and could be built into homes, ships, and even clothing. They can insulate against sound and could have many applications in building and acoustic technologies. The military is interested in their great shock-absorbing and particle-trapping properties for stopping projectiles. It will be interesting to see how this amazing substance is used in the future.

Wild 2 (VILDT TOO)

9.13 Measuring Heat in Thermal Energy Changes

Heat Capacity

Good thermal conductors, such as metals, need to absorb only a small amount of thermal energy to noticeably raise their temperature. However, insulators, such as fiberglass, must absorb a lot of thermal energy before they show a similar temperature change. Every object has a particular relationship between the amount of thermal energy absorbed and the temperature change experienced. This property is its **heat capacity (*C*)**. Heat capacity is the thermal energy, in joules, that an object must gain or lose to cause a temperature change of 1 °C. The unit for heat capacity is J/°C (or J/K). This value depends on the amount

9-35 The heat capacity of this steam turbine must be determined experimentally.

and kinds of materials in an object. Heat capacities for objects that are made of more than one kind of material must be determined experimentally.

Because thermal energy is being transferred, the change in thermal energy that results in a change of temperature or state is called **heat (Q).** The heat capacity of an object is computed by dividing the heat gained or lost by the change in temperature:

$$C = \frac{Q}{\Delta t},$$

where

- C is heat capacity,
- Q is the thermal energy transferred to or from the object, in joules, and
- Δt is the associated change in temperature, in °C.

You can see from the formula that the larger the heat that causes a given change in temperature, the larger the heat capacity of the object.

If thermal energy is gained, the heat (Q) is positive. If energy is lost, the heat is negative.

Specific Heat Capacity

Good thermal conductors have low heat capacities, and insulators have large heat capacities. But mass as well as thermal conductivity determines an object's heat capacity. Comparing heat capacities of different materials is not very meaningful unless you compare similar masses of the materials.

A much more useful property for identifying materials is *specific heat capacity*, or just **specific heat (c_{sp})**. Specific heat is the heat capacity *per gram of material*—the amount of thermal energy that must be gained or lost to change the temperature of 1 g of the substance 1 °C. The SI unit for specific heat is J/g·°C. Specific heats of metals are much lower than those of other materials. The specific heat of the gaseous phase of a substance is much higher than for the same material in its liquid or solid state.

Sometimes the specific heat for bulk materials is given in the unit

$$\frac{\text{kJ}}{\text{kg}\cdot\text{K}}.$$

If an object has uniform composition, such as a solid brass cylinder or a liter of alcohol, you can determine its specific heat by dividing the heat (thermal energy exchanged with its surroundings) by the object's mass and its temperature change.

$$c_{sp} = \frac{Q}{m \times \Delta t}$$

An object's specific heat can be determined with a *calorimeter*, a device that measures transfers of thermal energy between objects contained in an insulated chamber.

calorimeter (KAL uh RIM ih ter)

One common activity of science is to measure physical properties, such as specific heat, for newly developed materials. This helps engineers know where the material could best be used. Once researchers know a material's specific heat, they can use it to compute temperature changes and thermal energy gain or loss for an object made of that material. Such calculations are important in designing systems with heat exchangers. Specific heat capacity is essential for optimizing the operation of solar thermal power facility collectors and steam plants, which are used to convert solar energy to electricity. For these purposes, the heat equation used is just the rearranged specific heat formula,

$$\boxed{Q = m \times c_{sp} \times \Delta t}.$$

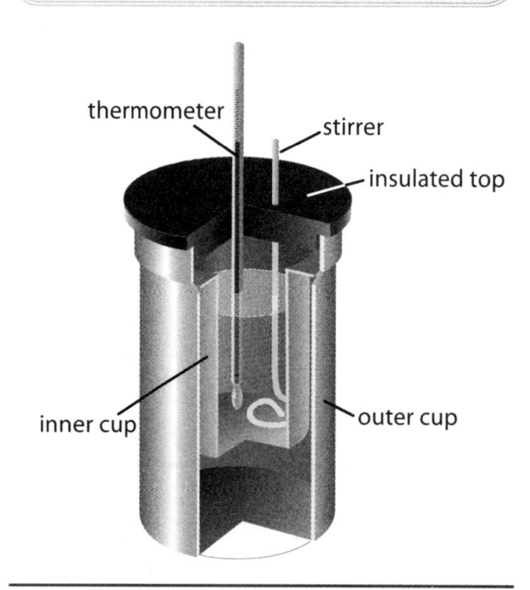

9-36 A calorimeter is used to determine the specific heat of an object by accurately measuring its temperature change for a known change in thermal energy content.

9-1 Specific Heats of Some Common Liquids	
Liquid	c_{sp}
ammonia	4.6
water	4.18
milk	3.93
seawater	3.93
alcohol, rubbing	2.47
propane	2.4
alcohol, ethyl	2.3
ethylene glycol	2.22
gasoline	2.22
acetone	2.13
paraffin	2.13
vegetable oil	1.67

$\Delta t = t_{final} - t_{init}$, where t_{final} is the last temperature measured in a time interval and t_{init} is the first temperature measured in the interval.

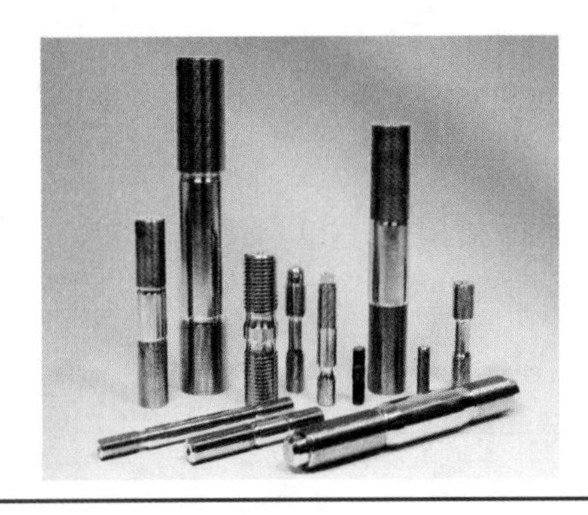

9-37 Heavy machine bolts

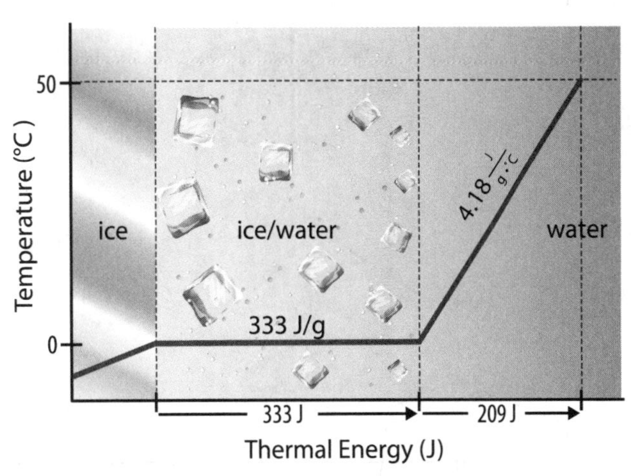

9-38 The temperature and thermal energy of a gram of ice change as it melts to water and is heated to 50 °C.

EXAMPLE PROBLEM 9-1

Using Specific Heat

An 850 g stainless steel bolt in an electricity-generating steam turbine heats from 27 °C to 109 °C as the turbine starts and warms up to operating temperature. How much thermal energy is gained by the bolt during this process? The specific heat for stainless steel is 0.50 J/g·°C.

Known: mass of bolt (m) = 850 g
$c_{sp} = 0.50$ J/g·°C
$\Delta t = t_{final} - t_{init} = 109\text{ °C} - 27\text{ °C} = 82\text{ °C}$

Unknown: heat gained by bolt (Q)

Required formula: $Q = m \times c_{sp} \times \Delta t$

Substitution: $Q = (850\ \text{g})(0.50 \frac{\text{J}}{\text{g·°C}})(82\ \text{°C})$

Solution: $Q \approx 34{,}800\ \text{J} \approx 35{,}000\ \text{J}$ (2 SDs allowed)

9-2 Specific Heats of Water	
Phase	**Specific Heat, c_{sp} (J/g·°C)**
solid (ice at 0 °C)	2.11
liquid (water)	4.18
gas (steam at 1 atm)	2.08

Of all familiar substances, water is unusual because it has a relatively high specific heat compared to other liquids and chemically similar compounds. Water's specific heat is approximately 4.18 J/g·°C. Compare water to the other familiar substances in Table 9-1. Water's large heat capacity allows perspiration to be an effective cooling mechanism for our bodies. It also allows large bodies of water to modify weather over adjacent continents. For example, the British Isles have a more moderate climate than other land masses at the same latitude because of the warming effect of the Gulf Stream in the Atlantic Ocean.

9.14 Heat and Phase Changes

When thermal energy is added to or taken from a material, the material's temperature usually changes because the average kinetic energy of its atoms and molecules changes. But sometimes the temperature doesn't change. For example, if you add many ice cubes to a glass of water at room temperature, the water quickly cools to near freezing. You slowly stir the water-ice mixture and monitor the temperature. The glass is uninsulated, so thermal energy from the room enters the water through the glass. You notice that the temperature of the mixture hovers near 0 °C. The only noticeable change is the melting of the ice cubes.

In ice, bonds between the molecules produce a potential energy that holds the molecules together. Work is required to pull the molecules apart, just as it takes work to pull apart attracting magnets. Thermal energy does this work as it flows from the room through the glass and water. The liquid water does not warm up while the ice is melting because the thermal energy is used to break the bonds between the molecules of ice.

It requires more than 333 J to melt 1 g of ice to 1 g of water at 0 °C. As 333 J/g is taken from the surroundings to melt ice, the same amount of energy per gram is transferred to the surroundings when water freezes. The amount of thermal energy exchanged per gram of material during melting or freezing is the material's **latent heat of fusion (L_f)**. Once all the ice is melted, the liquid water will continue to warm at a rate determined by its specific heat (about 4.18 J/g·°C).

The heat required to melt or freeze a mass of m with a latent heat of fusion of L_f is found by the formula

$$\boxed{Q = m \times L_f}.$$

Similarly, when a liquid begins to vaporize at its boiling point, the temperatures of the liquid and gas do not rise until the liquid is completely vaporized. In this phase change, thermal energy is used to boost individual molecules into the gaseous phase rather than to increase the average kinetic energy of the liquid phase. Every gram of liquid water requires 2256 J of thermal energy to vaporize at 100 °C. This is called the **latent heat of vaporization (L_v)**. When condensing, steam must lose the same amount of thermal energy to its surroundings.

The heat required to boil or condense a liquid with a mass of m with a latent heat of vaporization of L_v is found by the formula

$$\boxed{Q = m \times L_v}.$$

Every substance that can exist in more than one phase has its own latent heats of fusion and vaporization, which are characteristic for that material. Water has unusually large latent heats, especially for vaporization. This is why it is the preferred fluid for transferring thermal energy in powerful engines, such as steam turbines. Once water is turned into steam, it can give up a large amount of energy per gram for work as it condenses back into a liquid.

In summary, every substance has a distinctive specific heat—one each for its solid, liquid, and gaseous states, if applicable. In addition, each has a characteristic latent heat of fusion and of vaporization if it can exist in liquid or gaseous states.

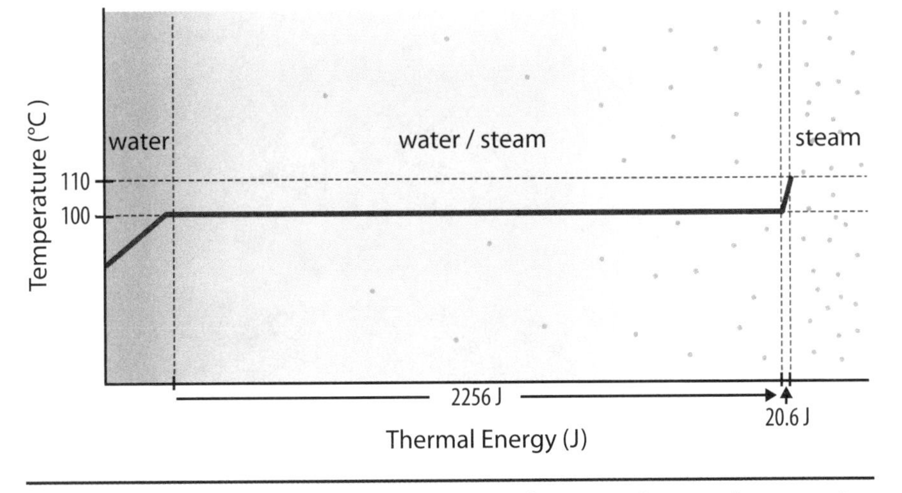

9-39 The temperature and thermal energy of a gram of water change as it boils to steam and is heated to 110 °C.

9C Section Review

1. Discuss the difference between thermal energy and heat.
2. Describe the three ways thermal energy can be transferred. Identify the most effective medium for each method of heat transfer.
3. DS ✪ On the basis of what you've learned about solar thermal power plants, name the most important process(es) of thermal energy transfer and storage these facilities may use.
4. State two ways that the arrangement of atoms and the bonding between atoms of a material can influence its insulating ability.
5. Compare and contrast heat capacity and specific heat capacity.
6. How does liquid water's specific heat affect the weather patterns of the world?
7. How much thermal energy must be lost by a gram of water at 0 °C to freeze to a gram of ice? What is this quantity of heat called?
8. How much thermal energy must a gram of water at 100 °C gain to vaporize to a gram of steam at 100 °C? What is this quantity of heat called?
9. (True or False) Thermal energy added to melting ice at 0 °C works to break the water molecules out of the ice crystal lattice.
10. ✪ How much thermal energy is needed to raise the temperature of 10.0 g of water 15.0 °C?
11. ✪ How much heat, in kilojoules, is required to melt an ice cube with a mass of 24.8 g at 0 °C?

FACETS OF SCIENCE / ENTROPY AND THE SECOND LAW OF THERMODYNAMICS

Lord Kelvin and other scientists in the early 1800s discovered an important aspect of thermal energy—the **second law of thermodynamics**. Physicists had to explain why thermal energy flowed from hot to cold and why higher pressure fluids flowed toward those with lower pressures. This law explains these observations. It applies to every process involving a transfer of energy or matter. Mathematical statements of the second law often include the quantity entropy. *Entropy (S)* is simply a measure of a system's disorder. Entropy could also be defined as a measure of how spread out energy is in a system.

We can express the second law this way: "The useful energy of a system tends to decrease, and the entropy of the system tends to increase." In this statement, "useful energy" is energy available to do work. The second law applies only to *natural processes* in *isolated systems*. Natural processes aren't influenced by any created intelligence. An isolated system is one where no energy is allowed to enter or leave the system boundaries.

surroundings

thermal isolation boundary

thermodynamic system

The second law of thermodynamics explains why thermal energy flows from hot to cold. Particles in matter at warmer temperatures are more disorganized (higher entropy) than in matter at cooler temperatures (lower entropy). Since entropy tends to increase in natural processes, energy flows in the direction that will tend to increase entropy, that is, from hot to cold. Eventually, the available thermal energy has spread out so that all parts of the system are at the same temperature—but with higher entropy.

Entropy doesn't *have* to increase in every natural process. When energy is allowed to enter from outside the process and is used specifically to counter the natural increase of disorder, entropy can decrease and available energy can increase. Photosynthesis is the principal example, but any living organism, as it grows, is temporarily decreasing the tendency of entropy to increase within the organism system. The cells of green plants use the sun's radiant energy to assemble sugar molecules containing high-energy bonds from carbon dioxide and water. Photosynthesis takes many small molecules and turns them into fewer larger molecules. Since fewer molecules exist at the end of the process (and energy is more concentrated), the entropy of the system is lower. This condition is temporary at best. When a plant dies or is eaten, the energy of the plant matter is released and the sugar molecules break down into carbon dioxide and water again. The energy in the sugar molecule's bonds is dispersed into the environment or used by the organism that ate the plant.

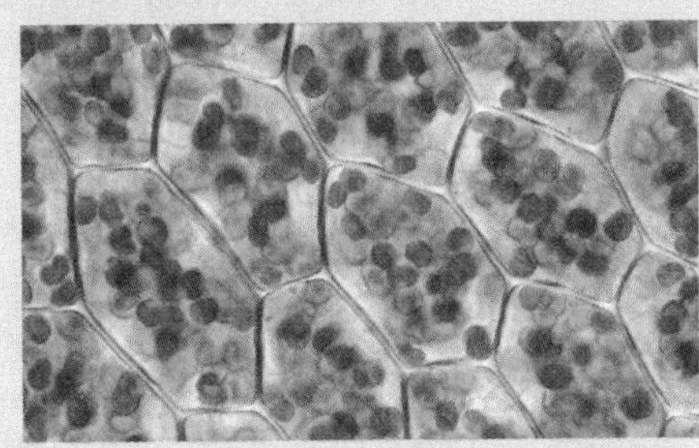

No violation of the second law is known to occur in any natural process. The second law of thermodynamics represents a serious hurdle to those who suggest that the order exhibited in the physical universe came into being by itself. Still, most astronomers believe that ancient dusty nebulas lost entropy as stars and planets formed millions of planetary systems in galaxies. Compared to the disorder of the original dust cloud, each of these processes would involve a significant reduction in entropy. A far greater reduction of entropy would have had to occur when the first living cell came into being.

Even a casual look around us affirms the second law of thermodynamics. Manufactured things wear out. The temporary increase in order when they are newly manufactured quickly degenerates, leaving broken and tattered objects that decompose and rust away. The mountains wear away and are washed into the sea. Living things age and die. Stars go supernova and are destroyed. Truly, the earth and the heavens "wax old as doth a garment" (Heb. 1:11).

It is apparent that without God's intervention eventually all available usable energy will be consumed. Some astronomers see evidence for the current formation of stars, but their interpretation of the data contradicts all other observable evidence. The evidence points to an inevitable "heat death" of the universe.

We can infer that there must have been a beginning that represented the greatest order and the least entropy. A big-bang type of explosion would have resulted in an immense *increase* in entropy followed by an immense *decrease* in entropy to produce the order we see today. God speaking the universe into existence day by day ("And God said…" [Gen. 1]) with ever-increasing order and decreasing entropy is far more consistent with today's universe.

Scientifically Speaking

Chapter Review

1. What important view about the structure of matter had to be fully accepted before scientists could agree on the nature of thermal energy and heat?
2. How did Count Rumford's cannon-boring experiment contradict the caloric theory of heat?
3. All matter contains internal energy. What part of this energy is identifiable as thermal energy? What kind of energy is the rest?
4. What measurable change in an object accompanies a change of thermal energy if no change of state occurs?
5. Why can we not use the terms *hot* and *cold* for scientific observations?
6. What factors could Gabriel Fahrenheit control when determining the size of the degrees on his temperature scale?
7. At what temperature are the Fahrenheit and Celsius scales equal? (*Hint*: See 9B Section Review Question 9.)
8. Would an electric company be more profitable, per unit of energy sold, in winter or in summer? Explain your answer.
9. Scientifically speaking, what is wrong with the statement, "You must be sick because your forehead has a lot of heat"?
10. A thunderhead develops over the eastern shore of Florida as warm air near the ground rapidly rises, lifting the cloud top almost into the stratosphere. What kind of heat transfer is this?
11. What condition or material is the best insulator? Give two applications of this kind of insulation.
12. Which should have a larger heat capacity, a 1 kg block of a conductor, such as iron, or a 1 kg block of an insulator, such as glass? Explain your answer.
13. Compare the amount of thermal energy lost by 50 g of water when it freezes to the amount of thermal energy gained when the same 50 g melts. What are these quantities of heat called?

True or False

14. Caloric was believed to be a material fluid that was invisible but could be sensed by temperature.
15. * A person can often produce valid evidence for a new scientific theory, even if he or she is not trained in that field of study. However, his or her findings will often be ignored by those who are proficient in that field.
16. ✪ A mechanical thermometer using human hair as its thermometric material would be an accurate instrument.
17. Initially, Anders Celsius's temperature scale increased with colder temperatures.
18. A boiling pot of soup on a stove is at the same temperature as the burner that is heating it.
19. Natural convection heating or cooling would not work inside the International Space Station.

20. Heat capacity (C) must be determined individually for each object.

21. Steam is useful for doing work because of its high specific heat capacity.

✪ 22. The coldest average temperature in the atmosphere, –90 °C, occurs around 80 km up in the mesosphere layer. What is this temperature on the Fahrenheit scale?

✪ 23. The temperature of the sun's photosphere, its outermost visible surface, ranges between 8700 °F and 10,800 °F. What are these temperatures on the Celsius scale?

✪ 24. Geologists believe that the center of the earth's inner core has a temperature as high as 5.0×10^3 °C. What is this temperature on the Kelvin scale?

✪ 25. A 250 g sample of Antarctic glacier ice was obtained from 800 m deep in an icecap using a deep-sampling core. Before it was sampled, the ice had a temperature of –48 °C. The geologists placed the sample in a storage freezer maintained at a constant –10. °C. After thermal equilibrium was attained, how much thermal energy did the ice sample gain? Assume that the specific heat of very cold ice is 1.85 J/g·°C.

✪ 26. A physical science student is performing a thermodynamics experiment in the school laboratory. She takes a 30.0 g ice cube at 0. °C and melts it to water in a beaker over a laboratory burner. She then heats the water to boiling at 100. °C. Finally she boils the water entirely to steam at 100. °C. How much thermal energy was gained by the water during this entire process?

✪ 27. If the laboratory burner used by the student in Question 26 released 540,000 J of thermal energy to melt the ice and boil the water, how efficient was this process?

DS ✪ 28. In a solar thermal power plant, a temperature difference exists between the working fluid in the solar collectors and the exhaust of the steam turbine generators. What do you think the relationship is between the plant's efficiency and this temperature difference?

DS ✪ 29. The efficiency of the collectors of some solar thermal plants like the SEGS facility is about 20%. If the efficiency is so low, what are the main advantages for using solar thermal power for electrical generation?

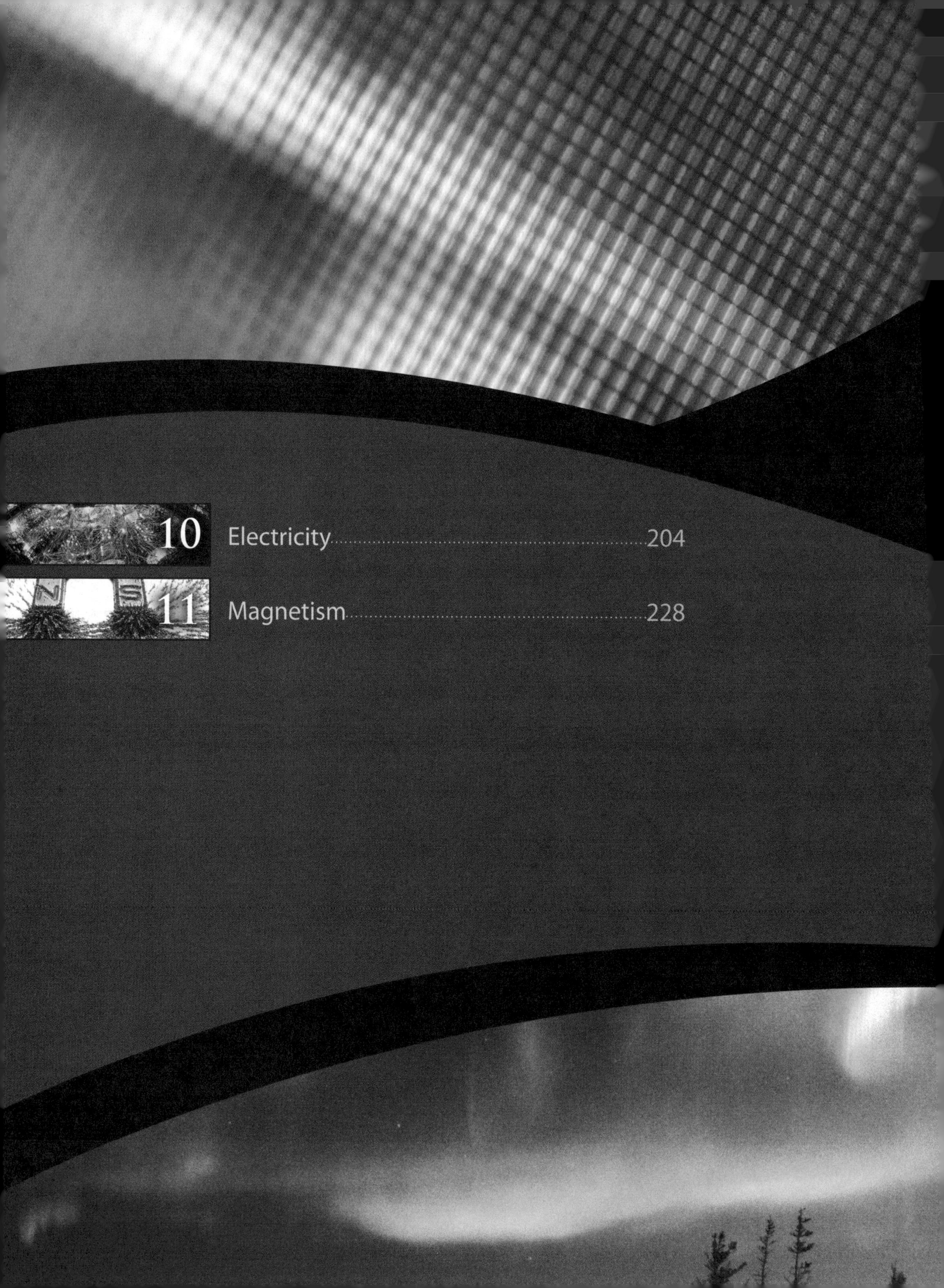

Modern technology is built on electromagnetism.

Without our current knowledge of electricity and magnetism, computers, cell phones, GPS, television, the electrical power grid, and a host of other technologies wouldn't be possible. Not only do technologies like these enrich people's lives, but many are considered essential to modern living.

You might be surprised to find that electricity and magnetism, which seem so different, are actually closely related. In fact, they're so interdependent that physicists call their combined action *electromagnetism*. Electromagnetism accounts for the very structure of the atom, chemical reactions, the transmission of light and radio waves, aurora displays (such as the one pictured below), and many other phenomena in the universe.

In Unit 3 you will learn the basics of electricity—electrical charge, electrical current, and sources of electricity—as well as magnetism. You will be able to relate these two forms of energy by studying how one can produce the other in generators and motors.

UNIT 3

ELECTRICITY

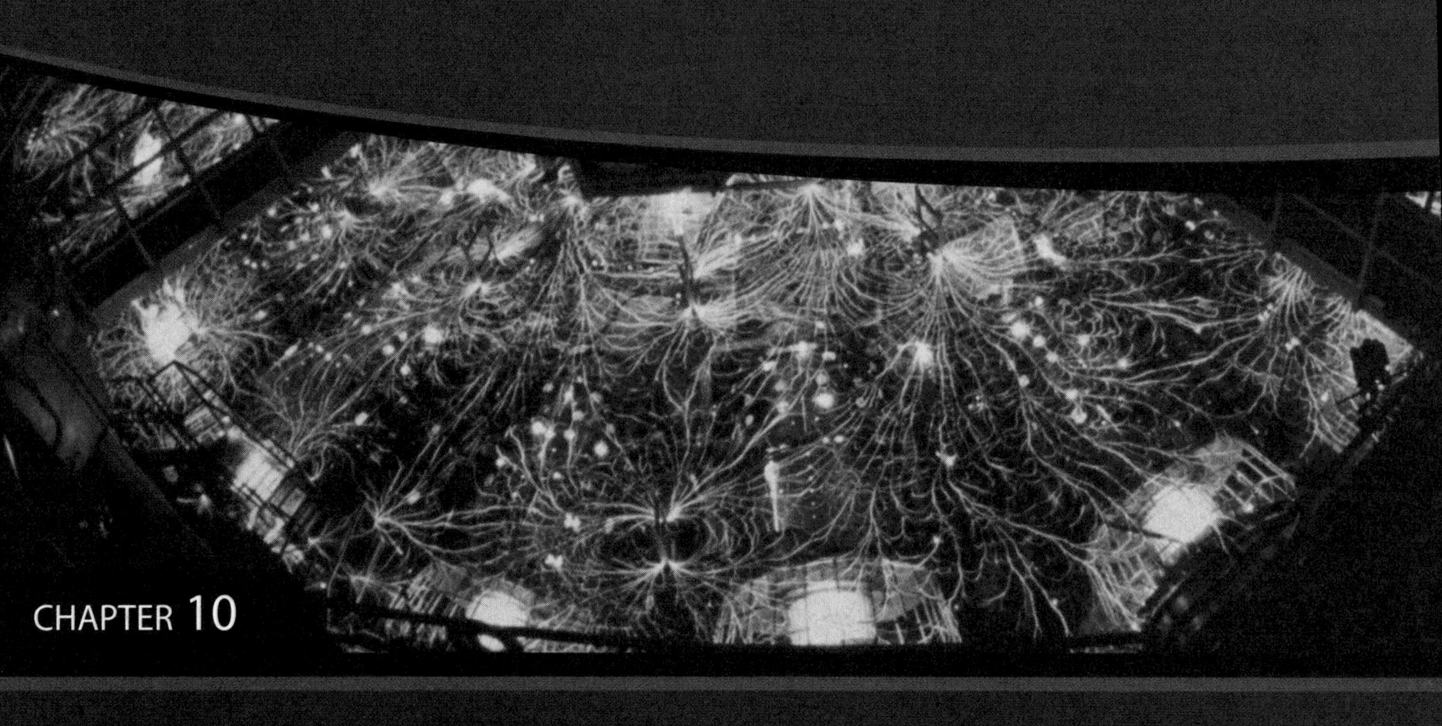

CHAPTER 10

10-1 Tall church steeples are particularly likely to be struck by lightning.

DOMINION SCIENCE PROBLEM

Avoiding God's Judgment?

During the Middle Ages, European towns and cities built many large churches. These churches were especially lofty, reflecting the importance of the Christian religion in the lives of Europeans. Because the churches were so tall, lightning strikes to steeples and bell towers became more frequent, causing great damage and death. Scores of bell ringers died, and hundreds of church towers were damaged, disrupting the religious life of the local populace. For example, peasants in Rosenberg, Austria, feared to attend church services because lightning so frequently struck the church. Many people believed lightning strikes were God's judgment on the disobedient or the workings of evil spirits. Those who held to pagan superstitions dealt with this problem by putting angels on church steeples and consecrating church bells, seemingly to no avail. How could people protect their churches and other buildings from lightning strikes?

10A Static Electricity and Electric Fields

10.1 Introduction

As stated repeatedly in this textbook, God created us to manage His world (Gen. 1:28). He has hidden many useful resources in His creation. Much of our dominion work is discovering and using these resources efficiently and safely. In doing so, we learn more about God's glory—as well as our own. As Proverbs 25:2 states, "It is the glory of God to conceal a thing: but the honour of kings is to search out a matter." God shows His greatness by hiding from us many of the marvels of His creation. We come to enjoy the glory of being human by finding what has been hidden and using it to benefit ourselves and others.

There is a form of energy that God has hidden throughout the universe for His glory and for our benefit. This energy, *electrical energy*, is all around us and is even essential to the proper functioning of our bodies, but only relatively recently in history have we been able to make good use of it. How can you tap into this valuable resource?

Believe it or not, you can begin by reaching for the comb in your pocket or purse. Try this experiment: Tear off a few small bits of paper and put them on the table. Run your comb through your hair and then hold it over the bits of paper. They will jump up and stick to the comb. Why do they do that? The answer is **static electricity**. The study of this type of electricity deals with stationary electrical charges and the forces that they exert.

10A Objectives

After completing this section, you will be able to

- ✓ explain how a static charge can exert forces.
- ✓ state the law of charges.
- ✓ describe the model for electric fields.
- ✓ summarize the process of electrical induction.

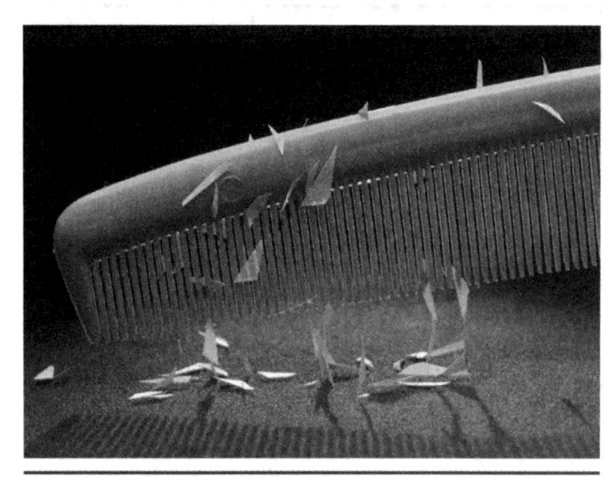

10-2 A comb with an electrical charge can pick up bits of paper.

10.2 The Electrostatic Force

People have observed static electricity throughout history. The most dramatic form has always been lightning discharge. But over 2000 years ago, the Greeks observed that a piece of amber they had stroked with wool or fur could "magically" attract certain lightweight objects.

Chapter 5 explains that the force between electrical charges, or the **electrostatic force**, is much stronger than the force of gravity. As the bits of paper in our experiment sit on the table, gravity pulls them down. The electrostatic force from the comb is able to overcome gravity and pick them up. How does the comb acquire this force in the first place? To answer this question, we must first examine the source of electrostatic attraction.

10-3 Lightning and its accompanying thunder can be one of the most awe-inspiring experiences in life.

10.3 Electrical Charge

When electrical charges were first investigated in the early 1600s, scientists soon discovered that electrically charged objects could exert forces of both attraction and repulsion. These two effects suggested that there must be two kinds of charge. This reasoning was based on experiments similar to the following, which you can re-create yourself.

Hang a rubber rod by a string. Rub one end vigorously with a wool cloth or a piece of fur. This will give the rod one kind of charge.

Hold another rubber rod in your hand and rub it with the same cloth so that the two rods will have the same charge. Bring this rod

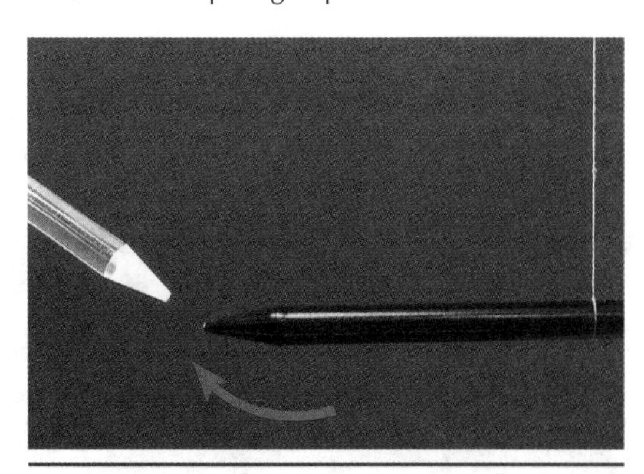

10-4 Demonstration of the electrostatic force. The hanging rod is made of hard rubber, and the other is glass.

10-1 Law of Charges Experiment		
Hanging object	Held object	Effect
rubber rod	rubber rod	repel
rubber rod	glass rod	attract
glass rod	glass rod	repel
glass rod	rubber rod	attract

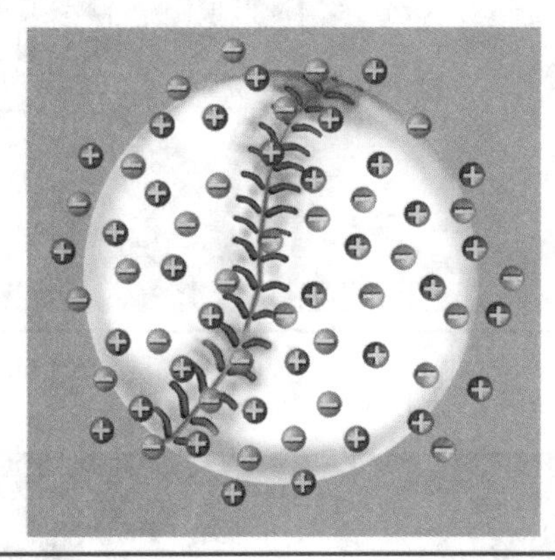

10-5 Most objects have a balance of charges. They are neutrally charged.

affinity (uh FIN ih tee): an inclination to attract or associate

coulomb (KOO lohm): named after Charles Coulomb (1736–1806), a prominent French mathematician and theoretical physicist

near the tip of the hanging one but do not let them touch. Your rod should repel, or push away, the hanging rod.

Now hold a glass rod in your hand and rub it with a silk cloth. This rod will also acquire a charge. Bring the glass rod near the hanging rubber rod. It should attract, or pull, the rubber rod. This effect is the opposite of what occurred with the like charges.

Next, hang a glass rod by the string. Rub it with silk to charge it. Repeat the experiment, first using a glass rod rubbed with silk, then a rubber rod rubbed with wool or fur. The results are listed in Table 10-1. The **law of charges** summarizes the table: *like charges repel; unlike charges attract.*

After extensive experimentation, physicists never found a charge that repels or attracts both rubber and glass rods. They concluded that there are only two kinds of electrical charges. We have Benjamin Franklin to thank for the names of these two kinds of charges—negative and positive.

We need to examine matter at the particle level to understand the source of electrical charge in objects. Chapter 2 describes the structure of atoms. They are made up of three important particles: protons, neutrons, and electrons. Protons, which each carry a single positive charge, are fixed in the atom's nucleus, so they cannot move freely through matter. Neutrons, which are neutral and do not contribute to electrical charge, are also in the nucleus. Electrons, which each carry a single negative charge, are found within the atom, around its nucleus. The outermost electrons that form chemical bonds are easily shared, gained, or lost. It is these mobile electrons that account for the movement and storage of electrical charge.

Normally, most objects have an equal number of electrons and protons. The opposite charges of these particles balance each other so that there is no detectable electrical charge on the object. Such objects are said to be *neutrally charged.* Sometimes, though, especially on low-humidity days, you receive a crackling shock when taking off your favorite wool sweater or when you reach for a metal doorknob. Where did that electricity come from?

In some materials that have a weak **affinity** for electrons, the outer, loosely held electrons of their atoms are easily removed through friction, such as simple rubbing. When you walk across a carpeted floor, your shoes pick up electrons and you become charged. When you touch a doorknob, the charge difference between you and the knob results in a shock.

When we rub two objects together, often one object loses electrons and the other gains them. The object that loses electrons has fewer electrons than protons and so has a net positive charge. The other object, which gained electrons, has more electrons than protons and so has a net negative charge. The net positive or negative charge is measured in coulombs. The ***coulomb (C)*** is the SI unit for electrical charge. One coulomb is the charge carried by approximately 6.24×10^{18} electrons or protons!

Chapter 9 discusses how thermal insulators have tightly held electrons

10-6 A shoe without a charge (left); a shoe after scuffing the carpet (right)

and how thermal conductors have loosely held, highly mobile electrons. In a similar way, *electrical insulators*, such as wool and glass, are much more effective at storing charges because electrons can't easily move toward or away from charged areas. *Electrical conductors*, such as metals, do not easily store charges unless they're totally surrounded by an insulator.

10.4 Electric Fields

The electrostatic force is a field force. That is, a charged object can exert a force on distant charges not in direct contact with it. An electric field can be described accurately only with mathematics, but physicists model these fields using *lines of force* radiating outward from a charged object. Lines of force are also directional. They align with the direction that the electrostatic force vector would move a tiny positive charge, called a *test charge*. If the source of an electric field is a strong positive charge, then the lines of force point away from the source because like charges repel. On the other hand, if the source of an electric field is a strong negative charge, then the lines of force point toward the source, because opposite charges attract. (Keep in mind that Figure 10-7 shows two-dimensional models of three-dimensional fields that surround the charges and that any neighboring charges are assumed to be too distant or too weak to noticeably affect the fields.)

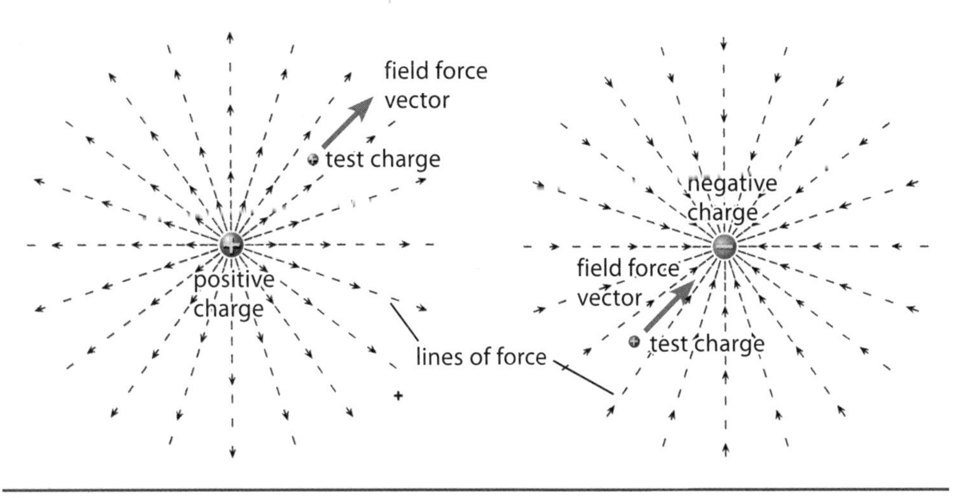

10-7 The electric field direction is defined by the direction that a small, positive test charge would move under the influence of the field. Note that like charges repel and opposite charges attract.

Lines of force in the field model are useful because they help us visualize both the direction and strength of an electric field. The *field strength* is represented by the distance between the lines of force. Near to the source charge the lines are very close together, indicating that the field force is strong. The lines spread apart farther away from the source, showing that the field weakens with distance from the source.

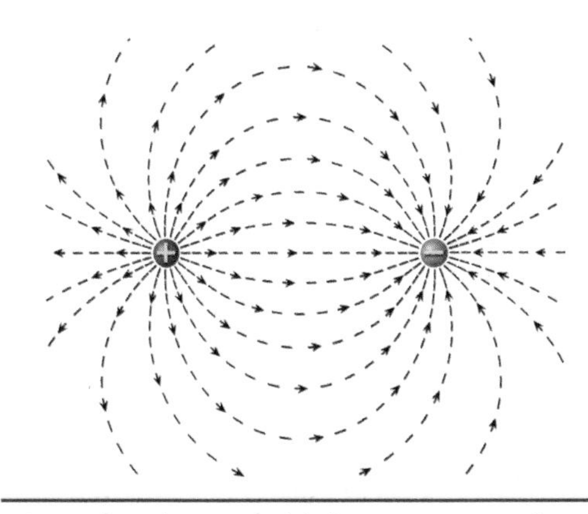

10-8 The electric fields between opposite charges bend *toward* one another in attraction.

Since each electrical charge exerts a field, electric fields interact when electrical charges are brought near one another. The electric fields between opposite charges bend toward one another in attraction. The electric fields between like charges bend away from one another in repulsion. Field lines cannot cross each other. If they could, the electric field vector at the location where the lines cross could point in more than one direction, which is impossible. The electric field vector can point in only one direction.

The field model is important for understanding many properties of matter, such as electrical and magnetic induction. We will discuss electrical induction in the next subsection and magnetic induction in Chapter 11.

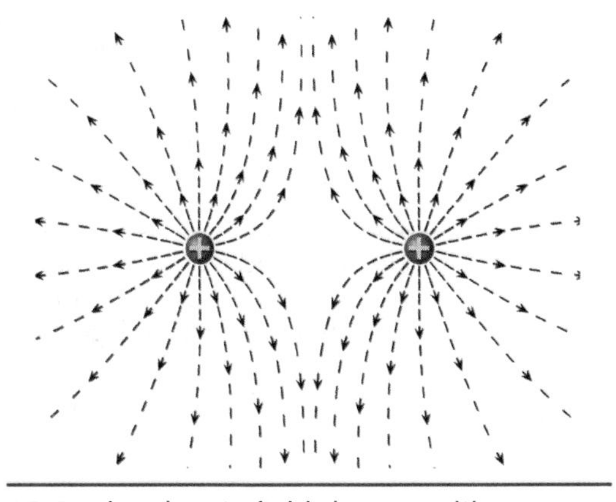

10-9 The electric fields between like charges bend *away* from one another in repulsion.

10.5 Electrical Induction

Recall the comb-and-paper experiment at the beginning of this chapter. Why did the comb attract the bits of paper? The comb became charged as you ran it through your hair. The bits of paper, on the other hand, were probably not charged. They were attracted to the comb, but not because of the attraction between two electric fields. There is another way that charged objects can exert the force of electrostatic attraction. **Electrical induction** is the creation of a charged region on a neutral object when exposed to a strong electrical field. As the comb approaches the bits of paper, it produces a temporary area of opposite charge on the part of the paper nearest the comb.

Remember that the paper contains equal numbers of protons and electrons since it is neutral. The charged comb, which was initially neutral, now has an excess of electrons, which have been removed from your hair. The electric field created by the net negative charge in the comb repels electrons in the bits of paper. Since paper is an insulator, it is extremely difficult to remove electrons from the paper fibers. Instead, the electrons in each molecule rearrange so that the molecule becomes an electrical *dipole*—a neutral molecule whose electrons have shifted to form positive and negative ends, or poles. All the molecules' positive poles point toward the negatively charged comb, and its negative ends point in the opposite direction. The result is that the interior dipoles neutralize each other as negative poles attract neighboring positive poles, but the paper's surface has a positive and a negative region (see Figure 10-10). This positive region on the bits of paper is attracted to the negatively charged comb as if the paper had a net positive charge.

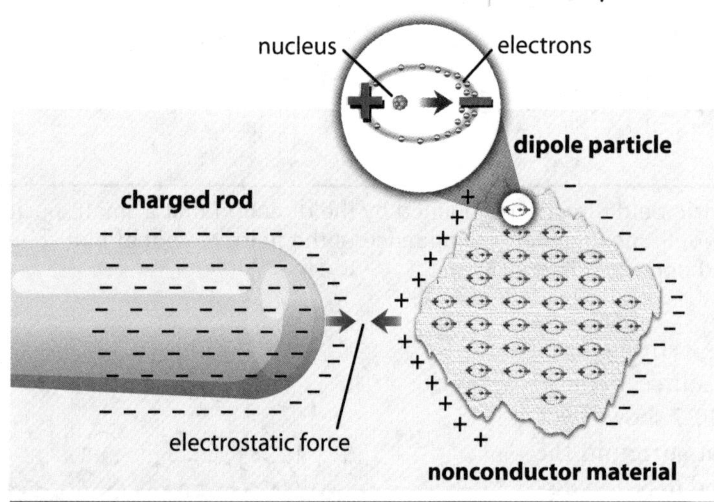

10-10 Electrical induction in an insulator, such as paper. The interior dipole molecules are shielded from the charge. The surface ones are not.

Water molecules are naturally dipoles. You can demonstrate this using a balloon and a water faucet. Charge the balloon by rubbing it with a cloth, preferably wool. What kind of charge should it have? Turn on the faucet so that a narrow, smooth stream of water flows out. Bring the balloon near this stream without touching the water. The charged balloon will attract the water molecules and cause the stream to bend (see Figure 10-12).

Water molecules play an important role in how objects lose their charge as well. In the comb experiment, the comb's charge will leak away, or *electrically discharge*, even if you don't touch it because of water molecules in the air. For this reason, all the demonstrations in this chapter work best in low-humidity weather.

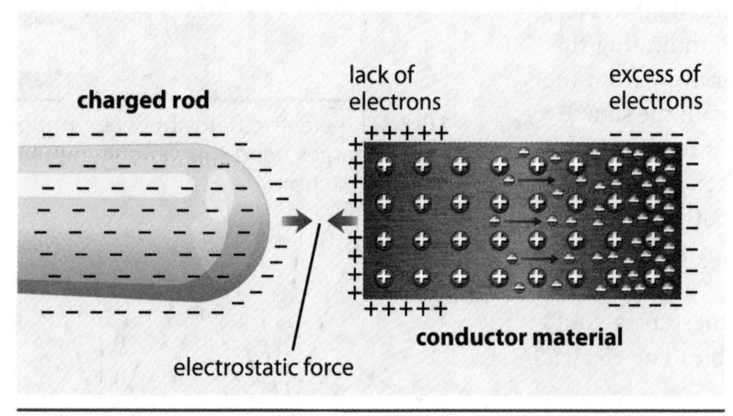

10-11 Electrical induction in a conductor. Because electrons can move in response to the charged object, the uncharged object itself becomes a dipole.

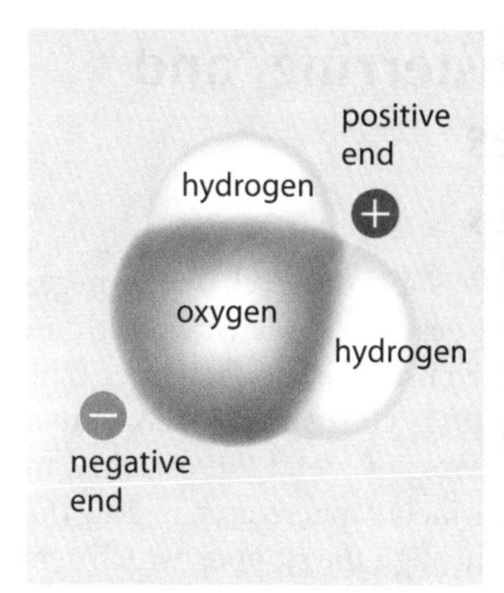

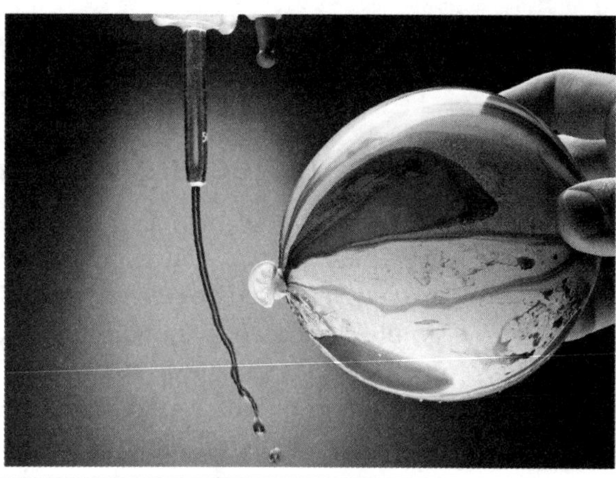

10-12 Because they are polar, the molecules of water are easily attracted by an electrical charge.

10A Section Review

1. What causes electrical charges to attract and repel? How does the strength of this force compare to the gravitational force?
2. How many kinds of electrical charges are there? How do we know this? What are they called?
3. State the law of charges. Copy the adjacent table onto your paper and fill in the effects column (repel or attract) for each pair of charges.

Object A	Object B	Effect
+	+	
+	−	
−	−	
−	+	

4. What makes most everyday objects electrically neutral?
5. What produces a negative charge on an object? What condition results in a positive charge?
6. In this section, you studied a nonmathematical model of electric fields. How does this model represent the presence, direction, and strength of electric fields?
7. As like charges move closer together, their lines of force bend farther away from the other charge and push together into a smaller space. What does this indicate about the size of the forces between them?
8. (True or False) Electrical induction can be used to identify an object's unknown charge.
9. A positively charged glass rod is brought close to one end of a neutral metal rod hanging from a string. Describe what happens. What causes this?

10B Objectives

After completing this section, you will be able to

- ✓ explain how an electroscope detects charge.
- ✓ compare and contrast the electrical properties of conductors, semiconductors, and insulators.
- ✓ describe how the Leyden jar and other capacitors store charge.
- ✓ give examples of how static electricity can be practically used.

10B Detecting, Transferring, and Storing Charges

10.6 Detecting Charges

How can you tell whether an object has a positive charge, a negative charge, or no charge at all? Charges can be detected with an instrument called an *electroscope*, which uses the law of charges. The simple electroscope shown in Figure 10-13 consists of a flask with a one-holed rubber stopper, two thin leaves of metal foil attached to the bottom end of a metal rod, and a metal sphere attached to the upper end of the rod. The rod passes through the stopper, which acts as both a support and an electrical insulator.

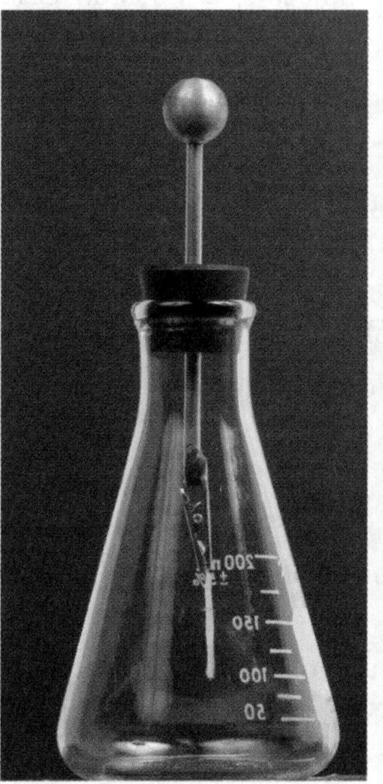

10-13 The positively charged rod attracts electrons from the electroscope and causes the foil leaves to spread apart.

The metal rod in the electroscope transfers charges between the sphere and the leaves. If we place a negative charge (extra electrons) on the sphere, the extra electrons instantly repel each other and spread out down the rod to the metal leaves, making both leaves negatively charged. The leaves, being very flimsy compared to the strength of the electrostatic force, are repelled and spread apart as the law of charges predicts, thus indicating that the electroscope is charged. The leaves will stay spread apart until the electroscope is *discharged.*

Touching a positively charged object to a neutral electroscope charges the electroscope by drawing electrons off the metal sphere. The remaining electrons in the metal parts redistribute evenly as they repel each other, leaving a net positive charge throughout. The foil leaves again spread apart as they repel each other, having the same positive charge.

These mechanical electroscopes can detect only relatively large charges because tiny electrical charges produce only tiny electrical forces of repulsion, which cannot move the metal leaves much. More sophisticated electroscopes use electronics to measure very small charges.

10-14 A needle electroscope uses a needle to indicate the charge on the instrument.

10.7 Transferring Charges

Some materials offer little resistance to electron motion. These materials are **electrical conductors** and are usually metals, such as copper, aluminum, silver, and gold. **Electrical insulators** do not allow electrons to move easily. Glass, wood, and rubber are all insulators. For example, electrical wires are made of conductors; the plastic coating on the outside of electrical wires is an insulator.

Most materials are either excellent conductors or excellent insulators. A few materials called **semiconductors** allow limited electron flow or conduct only under certain conditions. The elements silicon and germanium are examples of natural semiconductors. Scientists and engineers have developed thousands of semiconductor compounds that are used for microelectronic components in the modern computer and integrated circuit industries.

Static charges can move when conditions allow them. When objects with different charges touch, some of the electrons transfer from one object to the other. For example, when you walk across the carpet on a dry day and then touch a doorknob, the shock you feel is caused by electrons jumping from you to the doorknob.

As mentioned earlier, the loss of static charge as electrons move to another object is called electrical discharge. Static charges can discharge slowly, as when the charge on a comb leaks away, or they can discharge abruptly and dramatically. Lightning is an instantaneous electrical discharge produced when electrons in the earth or clouds are discharged through the insulating air. See the facet on the following page for a discussion of the formation of terrestrial lightning.

10-15 The shock you get from a doorknob is just one example of a static discharge.

10.8 Using Conductors to Solve Problems

Can you think of something we can use to protect buildings from lightning strikes? During the mid-1700s, Benjamin Franklin began experimenting with electricity. He showed that lightning is electricity, and he believed it could be controlled to some extent, like other electrical phenomena. He invented the *lightning rod*, a device we still use to protect buildings from lightning strikes. Lightning rods are made of conductors that are attached to the highest point of a building. The lightning's electrical current flows through a low-resistance wire to dissipate its charge harmlessly in the ground rather than passing through the building's structure. Lightning rods are often used in cell phone towers, high-tension power lines, and electrical substations.

Franklin's invention was widely held to be heretical! Many religious leaders in America and Europe who did not have a proper view of God and science said that lightning was God's tool for punishing disobedient people. To them, the invention of the lightning rod was an attempt to "play God." This resistance resulted in even more deaths. In 1767, shortly after the invention of the lightning rod, an Italian church, San Nazaro in Brescia, housing a large stock of gunpowder, was struck by lightning. The building exploded, killing 3000 people, and the city suffered incredible damage.

The Bible-believing Christian understands that God is indeed in control of every lightning strike. However, we do not always know why it strikes where it does. It's foolish to say that it is always for judgment, and it's foolish to claim that humans should not try to "tame"

FACETS OF SCIENCE

Lightning

Lightning strikes somewhere on the earth about 100 times every second. For something that happens so often, there are many unanswered questions about lightning.

With his famous kite experiment in 1752, Benjamin Franklin demonstrated that lightning is a form of electricity. During a thunderstorm, Franklin flew a kite from a long length of wet twine. Stationing himself under a shed, where it was dry, Franklin held a dry silk string attached to a key connected to the twine that held the kite. He observed that sparks flew from the key whenever he brought it near objects physically attached to the ground. He wondered if these sparks were identical to the electrical charges he could store in his Leyden jars. He tested his hypothesis by touching the key to a grounded Leyden jar. As Franklin suspected, charge flowed from the key to the jar just as it did from other sources of static electricity. His conclusion: lighting is static electricity.

But what causes lightning? Lightning occurs when charges separate in thunderclouds. Often, the base of the cloud acquires an overall negative charge, and the top of the cloud accumulates an overall positive charge. These opposite charges continue to accumulate in the cloud. At the same time, the negatively charged base of the cloud induces a positive charge on the earth. When the potential difference between the cloud and the earth approaches 3,000,000 volts/m, the insulating capacity of the air breaks down, and a lightning bolt discharges the excess electrons.

Many theories have tried to explain the formation of lightning but have failed to account for the data. In 1992 a new theory, called *Relativistic Runaway Electron Avalanche (RREA)*, was proposed. This model suggests that runaway free electrons bump into air molecules to produce more free electrons, as in an avalanche. This model appears to fit the data because the domino effect starts with only a few electrons, a condition that can easily develop in thunderclouds. Electrical potentials can become strong enough to move electrons long distances through the air, much like the sparks that you observe when you shuffle your feet across a carpet and reach for a doorknob. The RREA model predicts that runaway electrons will turn neutral air molecules into charged molecular ions and produce x-rays and gamma rays as a byproduct. If the RREA model is correct, then x-rays and gamma rays should be observed near lightning bolts.

How do you experiment with lightning? In Florida, where lightning strikes are very common, a laboratory called the *International Center for Lightning Research and Testing (ICLRT)* has been set up. This laboratory contains equipment to trigger lightning in thunderstorms by firing rockets into naturally occurring thunderclouds. The rockets trail thin wires connected to the ground. Sensitive, protected instruments on the ground measure any emissions from the thunderclouds overhead. *Scientific American*'s article, "A Bolt Out of the Blue" (May 2005), documents Joseph Dwyer's experimentation with lightning at ICLRT. He observed strong x-rays and gamma rays emitted from thunderclouds that could be sensed from the ground.

In the past decade scientists have gathered data that indicate lightning is often directly or indirectly initiated by cosmic rays. These high energy particles from space bombard the atmosphere. When a cosmic ray strikes an air molecule, an ion forms as one or more electrons are stripped away. These charges could be what starts the electron avalanche in the RREA model. Despite these findings, the process of lightning formation still leaves lots of unanswered questions.

Recent research also reveals that forked lightning, the lightning with which you are familiar, is not the only type. Rarer forms, such as *sheet lightning* and *ball lightning*, have been observed by scientists, but their causes remain a mystery. *Sprites* are great fountains of reddish lightning that have been seen only in the stratosphere. Sprites may occur at altitudes of 65–75 km (40–47 mi.) and are associated with large, severe thunderstorm systems. Photos of sprites from Earth's orbit are quite dramatic. A type of lightning called *blue jets* behaves in a similar fashion but starts lower in the atmosphere and seems to extend from the very tops of thunderstorms.

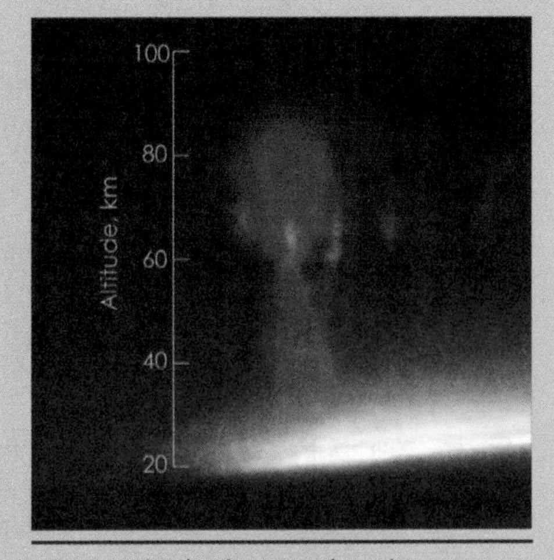

Sprites high above a thunderstorm

lightning to avoid God's judgment. Sometimes God uses seemingly bad things to show us His mercy (John 9:1–3)—in this case, His mercy through humans' use of science.

10.9 Storing Charges

The first scientists to study electricity could not produce enough electrical charge for more than a brief demonstration. After rubbing a rod with a cloth, the accumulated charge soon leaked away. They invented friction machines that could generate stronger sparks, but when the machines stopped, the charge still dissipated immediately. What they needed was a means of storing a large charge. The first solution to the problem was invented independently by Ewald von Kleist (late in 1745) and by Pieter van Musschenbroek (a few months later in 1746). The device was called the **Leyden jar** after Musschenbroek's hometown of Leiden, Netherlands.

Ewald Jürgen Georg von Kleist (1700–1748) was a Prussian inventor and cleric who invented the Leyden jar in 1745.

Pieter van Musschenbroek (1692–1761) was a Dutch physicist and professor who in 1746 independently developed a Leyden jar, inventing it shortly after Kleist but publishing his findings before him. His writings on Newtonian physics were influential in Europe.

The early Leyden jar consisted of a large glass jug coated inside and out with thin layers of lead, leaving the upper end of the jar uncoated. An insulated stopper plugged the jar. A metal rod was inserted through the stopper, which was attached to a metal chain resting on the inner lead coating at the bottom of the jar. The upper end of the rod supported a metal knob or ball (see Figure 10-16).

Kleist and Musschenbroek discovered that two layers of a conductive material separated by an insulator could store more charge than the conductor alone. The glass, a good insulator, separated the metal layers. When a source of static charge was applied to the metal rod, the charge was transferred through the chain to the inner metal coating. This charge induced an opposite charge on the outer coating, permitting even more charge to be deposited on the inside of the jar. In this way, very large charges could be stored. Kleist and Musschenbroek learned this the hard way when they both unexpectedly received severe shocks while experimenting! Later, scientists learned that if they connected the outer metal coating to a wire stuck into the earth, which became known as an *electrical ground*, far more charge could be stored in the Leyden jar. The ground acted as a source for new electrons or a place to receive repelled electrons, whichever tended to increase the induced charge on the outside of the jar.

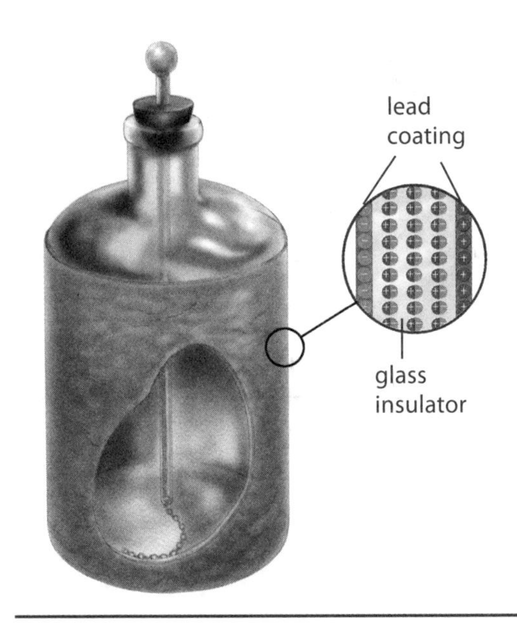

10-16 A Leyden jar was an early device that stored electrical charge.

Modern charge-storage devices are called **capacitors**. These compact devices store charge in most electronic circuits. The structure of a capacitor is similar to that of a Leyden jar. A capacitor is basically two layers of metal foil separated by an insulating material and sealed in an insulator. The amount of charge stored in a capacitor depends on the area of the metal foil and the kind of insulator.

The earth is essentially an infinitely large source or "sink" for electrons. Thus, it is the ideal point to attach an electrical ground.

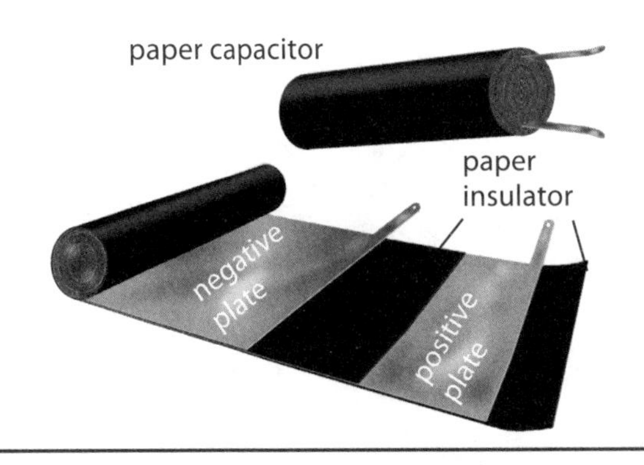

10-17 The structure of a capacitor. Capacitors perform many charge-storage functions in electronic circuits.

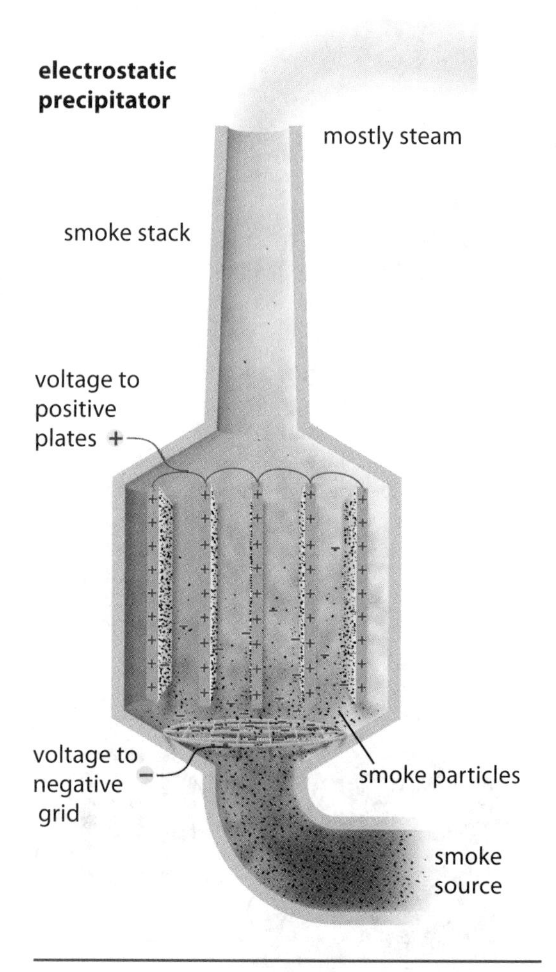

10-18 An electrostatic precipitator. The smoke particles are attracted to the charged plates. The device is washed frequently to remove the collected soot.

10.10 Applications for Static Electricity

Electrostatic principles help control pollution in some factories. As smoke travels from machinery to the smokestacks, it passes electrodes that charge the smoke particles. Just before the smoke enters the smokestacks, the charged particles are collected on oppositely charged metal plates. Much of the remaining vapor released through the smokestack is harmless steam. Frequently, the particles attracted to the electrodes contain valuable chemical substances that can be recycled. This type of antipollution device, called an *electrostatic precipitator*, substantially reduces certain kinds of air pollution.

Perhaps the most familiar application of static charges is the *laser printer*, a high-speed computer-printing technology used everywhere. The heart of the printer is a special drum coated with a semiconductor. The atoms in the drum's surface are given a charge that does not drain away because the surface normally doesn't conduct electricity; it acts as an insulator. If strong light strikes the drum's surface, however, the semiconductor changes to conduct electrons away from the surface. Wherever light falls on the surface, the charge disappears.

To print a page, a computer-controlled laser beam "paints" the document image onto the drum through a series of mirrors and lenses. The printer then charges the *toner*, a black, plastic-and-carbon powder used to create the image, with the same electrical charge as that on the drum. The toner sticks to the drum by static electrical induction wherever the laser beam has neutralized the charge. The drum then rolls over a sheet of paper, transferring the toner onto its surface. Hot rollers melt the toner onto the paper, completing the print.

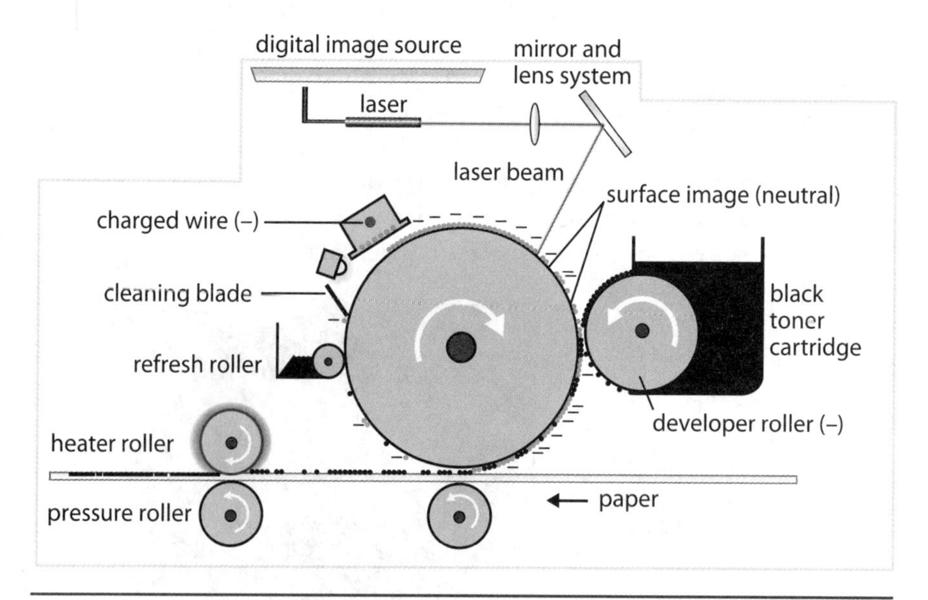

10-19 Electrostatic principles are essential to the operation of a laser printer.

10B Section Review

1. How can you tell that an electroscope has an electrical charge? Can you tell from an electroscope alone what kind of charge it has? Explain.
2. Describe electrical conductors and insulators and give one example of each. What do we call materials that conduct electricity only under special circumstances?
3. DS ✪ How does a lightning rod work?
4. DS ✪ The Bible teaches that God is in control of rain and lightning (Ps. 135:6–7). Is the use of a lightning rod an attempt to interfere with God's judgment on the disobedient? Why or why not?
5. What was the first device to successfully store electrical charges? How did it work?
6. ✪ Name two devices that use static electricity to complete a task. Give an example of static electricity (e.g., a biological one) not mentioned in the textbook.
7. Describe how static charge and induction are used to print a document in a standard laser printer.
8. (True or False) Lightning is a rapid electrostatic discharge.

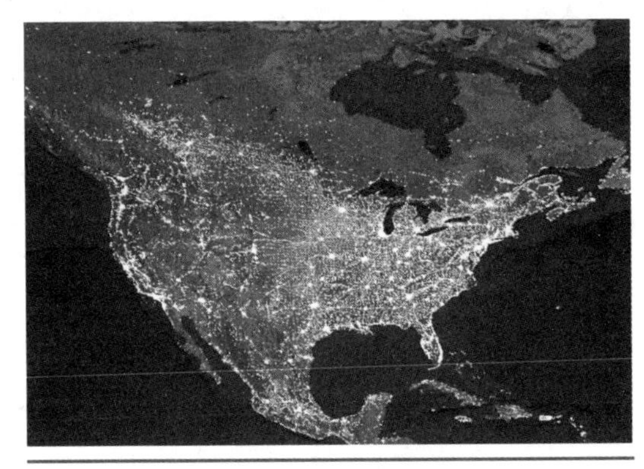

10-20 The importance of electricity to modern civilization is suggested by the illumination of city lights seen from space.

10C Electrical Current and Ohm's Law

10.11 Introduction

Electrical discharges are short-lived because an object has only a limited number of excess electrons to lose. To use electricity to perform useful work, we need a continuous flow of charge.

Electricity that involves continuously moving charges is called *electrical current.* Modern civilization could not exist without **current electricity**. Lighting, heating and air conditioning, computers, televisions and radios, cars, and hundreds of other necessities and conveniences require electricity to operate. In all these devices, current electricity provides energy.

10C Objectives

After completing this section, you will be able to

- ✓ summarize the two ways of defining current.
- ✓ list the types of components in a basic electrical circuit.
- ✓ give the SI units for charge, current, potential difference, and resistance.
- ✓ use Ohm's law and DC power formulas in calculations.

Electricity: Blessing or Curse?

The world we live in is very different from the world that our great-great-grandparents knew. Much of this difference is a result of our ability to use electricity. If we were transported back 120 years, we would have much difficulty adjusting to the living conditions.

We all enjoy the benefits of electricity and the technology it makes possible. This is proper according to the Creation Mandate. However, technology is often misused for sinful purposes. Electricity powers the medical devices that are used to save lives—and that are also used to perform abortions. Electricity has enabled us to overcome many hardships, but perhaps it has made us less capable of enduring "beneficial" hardships—those that increase our faith and dependence on God. What do you think? Is electricity a blessing or a curse?

10.12 Electrical Current

Current Models

Nearly all electrical current running through solids involves the flow of electrons. Electrical current flows most easily through conductors, such as metal wires. The most common type of electrical wire is made of copper, although some wires are made of aluminum or copper-coated aluminum. Electrons may also flow through ionized gases (plasmas), such as those in fluorescent lamps, or even through a vacuum. Older televisions and computer monitors contain a large vacuum tube screen called a *cathode-ray tube (CRT)* to display images "painted" by electrons.

Physicists use several models to describe electron flow through wires. One is the *electron diffusion theory.* In this model, electrons

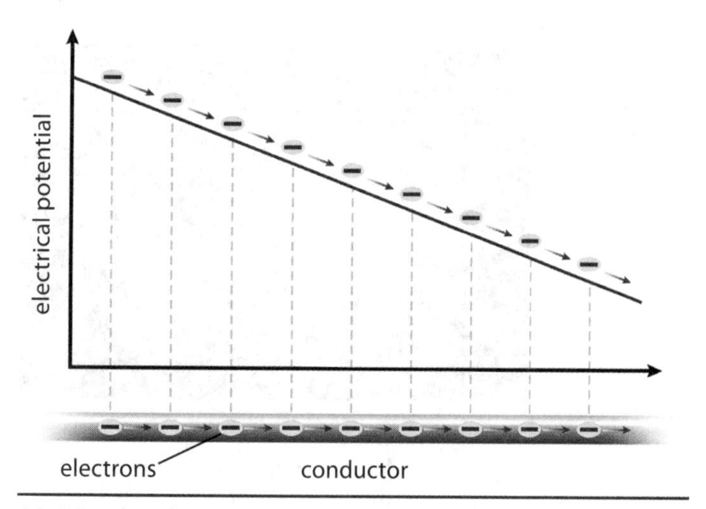

10-21 The electrical potential model accounts for how electrons flow through a conductor.

move from areas of high concentrations to areas of low concentrations just as gaseous perfume particles do when released into a room. Current is maintained in the conductor by a continuous supply of electrons at the source, which maintains this imbalance in electron concentrations.

Another model that is perhaps easier to understand is similar to a stream flowing down a hill. Water at the top of the hill has a higher gravitational potential energy than at the bottom, so gravity naturally pulls the water downhill. As it flows, it can convert its potential energy into work—by turning a waterwheel, for example. Similarly, electrons flow from a point of higher **electrical potential energy** in a conductor to a point of lower potential. A point that has a higher electrical potential is a source of charges. A point of low electrical potential is a "sink" for charges—a place that charges flow toward.

Conventional Current Flow

Electrical current running through wires and similar conductors is caused by the flow of electrons. However, that is not the way current was originally described. According to the model of electrical charges developed by Benjamin Franklin around 1750, electrical charge was a fluid. If a substance had a lack of fluid, it was negatively charged, and if it had an excess, then it was positively charged. In his model, current flowed from areas of positive charge to areas of negative charge.

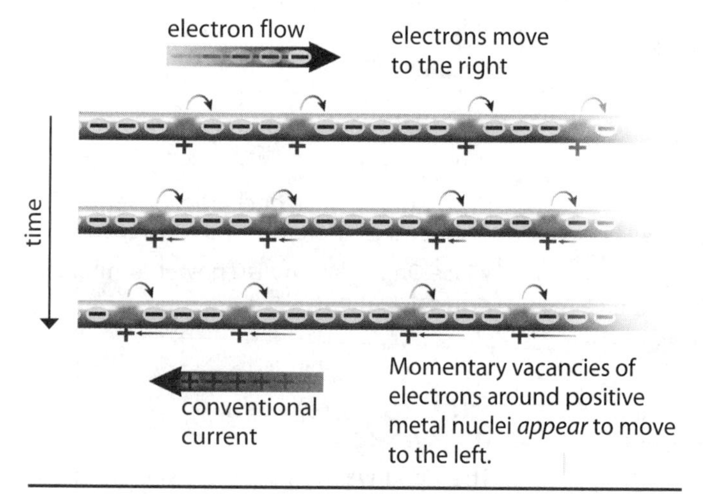

10-22 The conventional current of positive charges is in the direction opposite from electron flow.

When, more than a century later, electrons were discovered to carry current, it was far too late to rewrite the science textbooks and equations! Most scientists and engineers continue to describe **conventional current** as the flow of positive charges through a conductor, which is *opposite to the flow of electrons*. This definition is actually not a problem, although it may initially seem confusing.

Electrical current can also involve the flow of ions (charged atoms) in chemical current-carrying solutions and in some plasmas. In these currents, oppositely charged particles flow in opposite directions at the same time. Thus, defining conventional current as the flow of positive charges is satisfactory in all cases. You must keep this definition of current in mind as we continue our discussion of electricity.

Direct Current

Current that flows in one direction through a conductor is called **direct current (DC)**. You are familiar with many DC applications. Your cell phone, digital camera, and laptop computer are all powered by direct current, which is also the primary form of electrical energy for cars and most recreational boats and planes. Direct-current power sources allow many electrical devices to be portable. The remainder of this chapter discusses only direct current. The other important form of electrical current is *alternating current (AC)*, which we discuss in Chapter 11.

For current to flow, there must be a complete path from a source of electrons through a conductor and back to an electron sink (which may be the same place as the source). This path is called an **electrical circuit**. To make an electrical circuit useful, it contains an **electrical load**, such as a light bulb (see Figure 10-23), which converts electricity to some other form of energy. If a gap occurs in the circuit, the flow of electrons stops because they cannot flow across the gap. A *switch* allows you to open or close the circuit by opening or closing a gap in the current path.

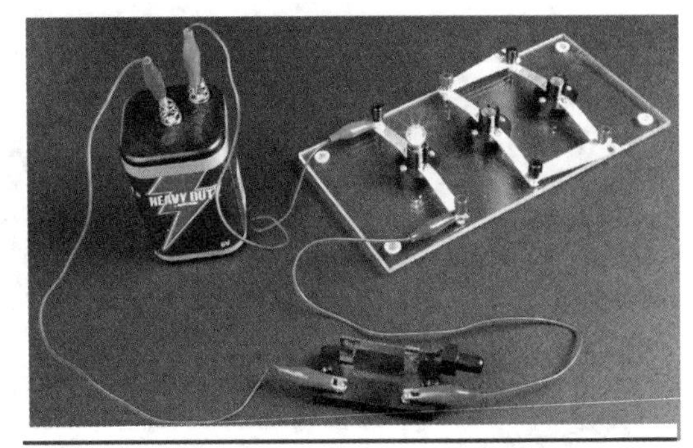

10-23 A simple electrical circuit is a continuous path from the current source, through the load, and back to the source. A switch can stop the flow of the current when desired.

10.13 Sources of Direct Current

Since electrons are tiny bits of matter, the law of the conservation of matter applies to them. They cannot be created or destroyed. A steady flow of electrical current requires a continuous supply of electrons, so they must come from somewhere. Electrons can be supplied by two methods—they can be chemically released from atoms to do work one time, or they can be reused over and over again.

The term *battery* is a military term for two or more cannons in a single location firing together to increase their effect. Scientists adopted the term for electrical batteries.

Electrochemical Cells and Batteries

Electricity from a battery is a current of electrons used one time. A **battery** contains one or more *electrochemical cells* that supply electrons released by a chemical reaction to a circuit. For example, a standard flashlight battery is a single electrochemical cell. It consists of a zinc cylinder filled with a moist chemical paste. The bottom of the cylinder forms the negative (–) end, or *pole*, of the cell. In the center of the cell is a carbon rod connected to the positive (+) pole.

A chemical reaction occurs between the paste and the zinc. This reaction releases electrons that flow through the metal can and become available at the negative pole. When you connect a battery to a complete circuit, electrons flow from the negative end through the conductor to the positive carbon rod. The rod carries the electrons into the chemical paste. A different chemical reaction at the surface of the carbon rod takes up the excess electrons to complete the reaction. Batteries that use solid or paste-like chemicals are *dry cells*. These cells are useful because they are portable, do not spill chemicals, and can operate in any position.

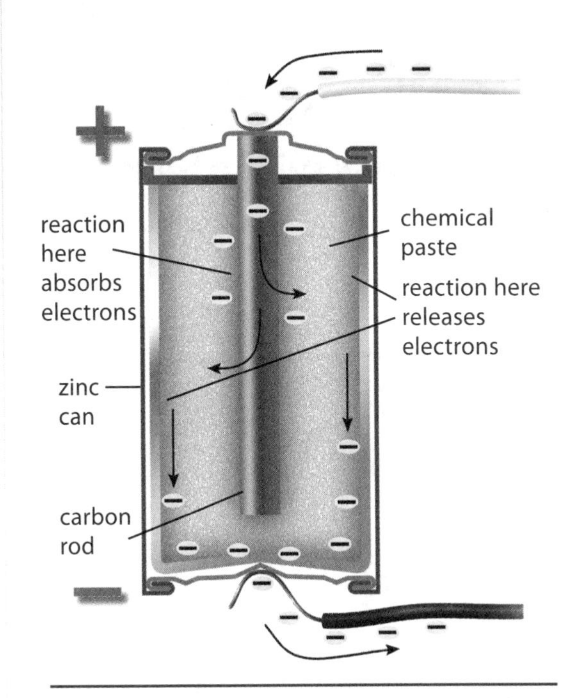

10-24 The structure of a common D-cell battery and its operation in a circuit

Modern batteries used in most digital devices, like cell phones and tablet computers, use a different kind of chemical reaction. These batteries are rechargeable. Called *lithium ion batteries*, they work on a reversible chemical reaction involving the movement of positive lithium ions within a paste contained inside the battery. As the battery discharges, the ions move toward the negative graphite pole; when charging, the ions move back into the lithium-metal-oxide paste. Lithium ion batteries last a lot longer than standard batteries and can deliver more current.

As long as the circuit is complete and the battery chemicals have not been used up in the battery, current will flow. When chemical reactions can no longer supply or absorb electrons, the battery is exhausted or "dead." Ideally, a battery shouldn't go dead unless used in an electrical device. But in reality, unused standard dry cells last only about one and a half years before they go dead. It's almost impossible to isolate the two chemical reactions for a long time inside the cell.

Other batteries use a combination of metals and acidic liquid solutions or gels to produce electricity. These are *wet cells* or *gel cell* batteries. Most wet cells must remain upright during use in order to avoid spilling the chemicals. The main advantage of wet cells is that they can produce higher currents. The most familiar wet cells are the

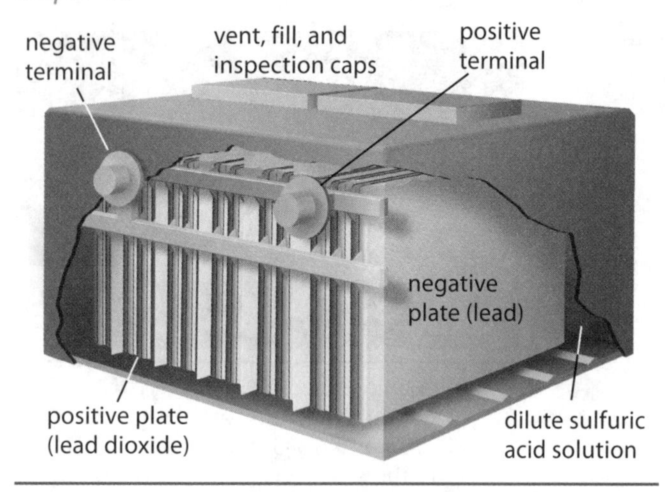

10-25 A common lead-acid battery consists of one or more wet cells. Such batteries can be recharged.

lead-acid batteries found in cars, trucks, and boats. A standard car battery consists of six wet cells connected together. Gel cells combine the liquid acid with other chemicals to produce a gelatin-like substance. These batteries *don't* have to be kept upright, so they're useful in motorcycles and in sailplanes, which operate at steep angles or even upside down! Like lithium ion batteries, most wet and gel cells can be recharged using an outside electrical source, which reverses the chemical reactions inside the cell.

Electrical Generators

Batteries are limited sources of electricity. Even in rechargeable batteries, the chemical reactions that produce electrons eventually stop because the chemicals are exhausted. Standard batteries then have to be replaced; rechargeable batteries must be charged from another electrical source. A more reliable source of continuous electrical current is the *electrical generator.* Generators are machines that use the special relationship between electrical and magnetic energy mentioned in Chapter 6 to move electrons through conductors. A generator basically consists of one or more coils of wire rotated in a magnetic field by some form of *prime mover,* such as a motor or a turbine powered by steam, water, or wind. You will learn more about how a generator works in Chapter 11.

As a generator produces current, you can think of it as lifting charges to a higher electrical potential, like a pump that refills a water tower. The charges then flow through the circuit, losing potential energy. Eventually, they return to the generator at its low-potential connection, which is like the low-pressure inlet of a pump. The advantage of a generator is that the charges can be continuously used over and over again.

10-26 Wind, water, motors, or engines can power a generator to produce electrical current. Shown above is a wind-powered DC generator.

Photovoltaic Cells

Electrons are strongly affected by electromagnetic energy. When a light ray is absorbed in some kinds of semiconductors, an electron gains enough kinetic energy to jump from its atom and do work. Engineers have developed devices called *photovoltaic (PV)* or *solar cells* that use this principle to power electrical devices located far from any standard electrical power source.

When connected to a circuit, PV cells can produce a continuous current as long as they are illuminated by the sun or other strong light source. The energetic free electrons flow through the circuit and then reoccupy the gaps left in the semiconductor when they return. As with a generator, the electrons are reused repeatedly.

10-27 A panel of PV cells provides electricity to power this emergency phone in a remote location. How can such technology preserve human life?

10.14 *Voltage, Current, and Resistance*

Potential Difference: The Volt

According to the conventional current model, the positive pole of a battery or generator is its point of highest electrical potential. Conventional current always flows from positive to a lower or negative electrical potential. The rate of current flow depends on the difference in electrical potential.

Determining the electrical potential at a particular point in a circuit is almost as difficult as trying to determine how much thermal energy is in an object. It is much easier and more important to measure the *difference* in electrical potential between two points. The potential difference between any two points in a field or circuit is the amount of work needed to move a charge between them. We measure potential difference with a unit called the *volt (V)*. A 1 V potential difference requires 1 J to move 1 C of charge between the points. The volt is defined in the equation

$$1\ \text{V} = \frac{1\ \text{J}}{1\ \text{C}}.$$

Batteries are rated by the potential difference between their positive and negative connections or poles. A standard D-cell battery is rated at 1.5 V. A lantern flashlight battery is usually 6 V. The battery that powers a smoke detector is 9 V. A car battery is 12 V. These potential differences, or *voltages*, determine the energy that the battery supplies to a circuit. Voltages are measured using a *voltmeter*.

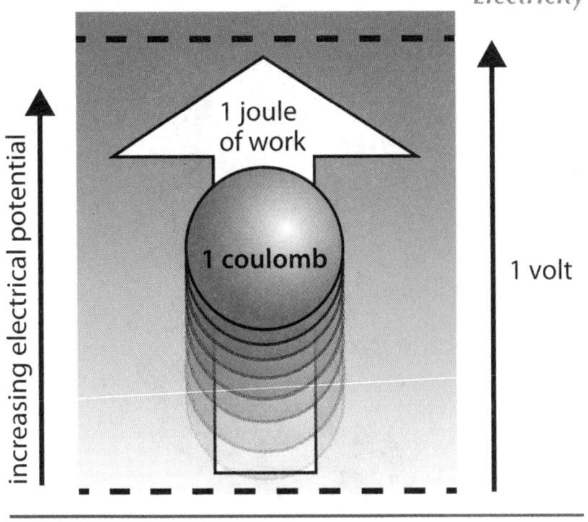

10-28 One joule is needed to move a coulomb of charges between two points that have a potential difference of 1 V. It does not matter how far apart the two points are.

Electrical Current: The Ampere

The more charges that flow through a circuit in a given amount of time, the greater the electrical current. Electrical current is measured in the unit called the *ampere (A)*, which is usually shortened to *amp*. Measuring electrical current in amperes is like measuring water flow by counting the number of liters of water that pass a point in a stream every second (liters per second). A current of 1 A is equal to 1 C of charge (6.24×10^{18} charges) moving past a point in the circuit every second, according to the equation

$$1\ \text{A} = \frac{1\ \text{C}}{1\ \text{s}}.$$

Electrical current is measured by instruments called *ammeters*.

ampere (AM peer): named after André Marie Ampère (1775–1836), another French physicist who made important early discoveries in electricity

How many coulombs of charge flow in 1 s past a point in a circuit with 2 A of current?

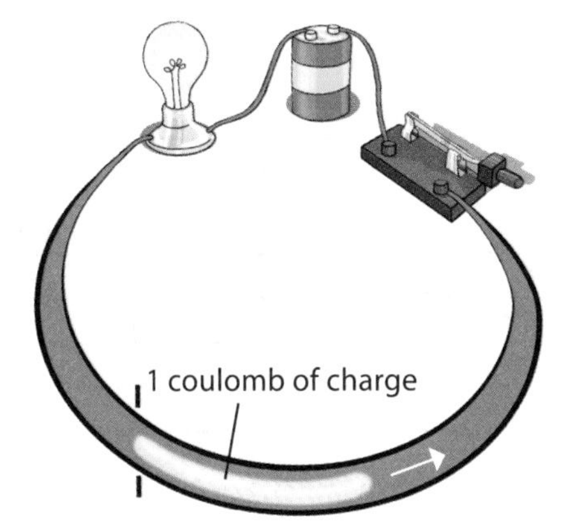

10-29 In this course, we define an amp as a coulomb of charges passing a point in a circuit in 1 s.

Electrical Resistance

How do moving charges do work in electrical devices? Consider water flowing down a stream using some of its kinetic energy to rotate a waterwheel. As the water exits the wheel, the water moves more slowly and has less kinetic and potential energy.

Similarly, every component of a circuit hinders the flow of charges to some extent. This property is called **electrical resistance**. As current flows through a conductor, collisions between the flowing electrons and the conductor atoms add kinetic energy to the atoms and sap energy from the electrons. These collisions reduce the flow of electrons and cause them to accumulate at the higher electrical potential in the conductor. The more resistance a conductor has, the higher the electrical potential is at the end supplied by the current source. Consequently, the more resistive the conductor is, the lower the voltage available downstream. A special device called a *resistor* can change the voltage within portions of a circuit.

In general, high electrical resistance is undesirable. Chapter 9 discusses how electrical resistance in power lines increases on hot days, raising the cost of electricity for consumers. However, in complicated electronic circuits, resistance can be useful. Electronic components in complex circuits require different voltages. Using different kinds of resistors allows the voltage from the electrical supply to be subdivided into various voltages within the circuit.

10-30 Electrical resistors come in numerous resistance values. Individual resistors are color-coded to identify the resistance they provide.

10-31 Georg Simon Ohm (1789–1854) was a German schoolteacher and college physics professor. He made many contributions to the theoretical understanding of electricity.

Because all electrical devices have some resistance, electrons always lose energy as they pass through a circuit. The law of energy conservation states that energy is never actually lost but is converted into other forms of energy. This energy conversion can be used for beneficial purposes. For example, if you are making a coffeecake, you use an electric mixer to convert electrical energy into the mechanical energy necessary to spin the beaters. An electric oven converts electrical energy into the thermal energy that bakes the cake.

Ohm's Law

In 1827 Georg Ohm published his discovery of the relationship between voltage, current, and resistance in electrical circuits. He observed two things in a simple circuit containing a DC voltage source and a resistance connected to it by wires.

- As the size of the voltage source (V) was changed and the resistance (R) remained constant, the current (I) measured was directly proportional to the voltage. In other words, if he doubled the voltage applied to a resistance, the current also doubled.
- As the resistance was varied and the voltage remained constant, the current was inversely proportional to the resistance. That is, if he doubled the value of the resistor, the current through the resistor was halved.

Combining these two observations into a single equation yields **Ohm's law.**

$$I = \frac{V}{R}$$

Direct current is directly proportional to voltage and inversely proportional to resistance.

This equation is commonly rearranged and written as

$$\boxed{V = I \times R}.$$

Resistance is measured in units called *ohms* (Ω). (The symbol Ω is the capital Greek letter *omega.*) The ohm is defined from Ohm's law. It takes 1 V of potential difference to draw 1 A of current through a resistor of 1 Ω. Rearranging the Ohm's law formula shows us that 1 ohm equals 1 volt per amp.

$$R = \frac{V}{I}$$

$$1\ \Omega = \frac{1\ \text{V}}{1\ \text{A}}$$

Electrical resistance is measured using an *ohmmeter.*

10-2 *Electrical Units*

Quantity	Unit	Symbol
charge	coulomb (C)	Q
voltage	volt (V)	V
current	ampere (A)	I
resistance	ohm (Ω)	R

EXAMPLE PROBLEM 10-1

Ohm's Law

A car's headlight draws a current of 4.2 A at the car battery's voltage of 12 V. What is the resistance of the headlight?

Known: headlight current (I) = 4.2 A
headlight voltage (V) = 12 V

Unknown: headlight resistance (R)

Required formula: $V = I \times R \Rightarrow R = \frac{V}{I}$

Substitution: $R = \frac{12\ \text{V}}{4.2\ \text{A}}$

Solution: $R \approx 2.\underline{8}5\ \text{V/A} \approx 2.9\ \Omega$ (2 SDs allowed)

10-32 A car's headlight is a DC load that obeys Ohm's law.

10.15 Direct Current and Power

Power measures how fast work is done (see Chapter 7). It can be calculated by dividing the total amount of work done, in joules, by the time it took to do the work, in seconds, to yield the power, in watts (W). In the same way, the faster a DC circuit uses energy or does work, the greater the power. For example, a 100 W light bulb uses twice as much energy per second as a 50 W light bulb. The extra energy produces more light from the 100 W light bulb. Industrial machines that use thousands of watts of power have power ratings expressed in kilowatts (kW).

We often refer to companies that sell electrical energy as "power companies"; however, the electric company usually charges you according to how much energy you use, not how fast you use it. The modern digital electric meter attached to your house displays the amount of electrical energy used since it was installed. Older models of electric meters use a spinning disk and a series of small dials with pointers. These dials serve the same purpose as the digital display but also provide a visual indication of how fast the energy is used. The faster the disk spins, the greater the power usage.

10-33 This steam turbine generator produces 1150 megawatts of electrical power.

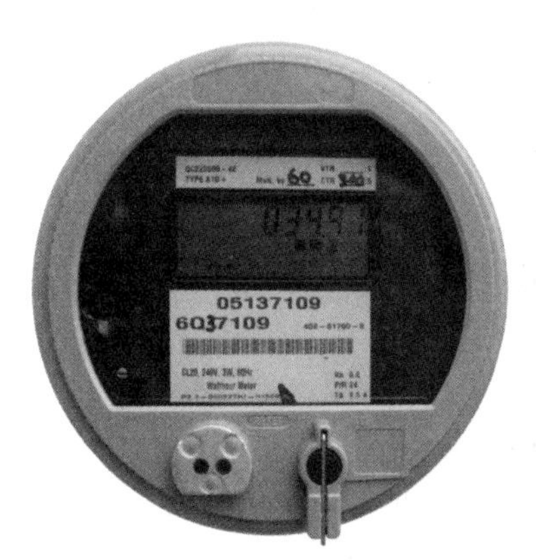

10-34 A digital electric meter (left); a mechanical meter (right). How many kilowatt-hours did your family use last month according to your electric bill?

A 100 W light bulb burning for ten hours uses as much energy as a 1000 W blow dryer running for one hour. The light bulb uses 100 J/s; the blow dryer uses 1000 J/s. The light bulb has to burn ten times longer than the blow dryer to use an equal amount of energy. For this reason, electric companies compute the energy consumed by multiplying the power used by the amount of time it was used:

$$\text{energy} = \text{power} \times \text{time interval}$$

$$E = P \times \Delta t.$$

Electrical energy is thus dispensed and billed in units of *kilowatt-hours (kWh).*

Electrical power in DC circuits is easily calculated from known quantities. Since a volt is related to work per charge and since power is the rate of doing work, we can combine these concepts to develop the DC power formula:

$$\text{power} = \text{voltage} \times \text{current}$$

$$P = V \times I.$$

DC Power Formulas

$P = V \times I$ $P = I^2 \times R$ $P = \frac{V^2}{R}$

10-35 A battery-powered lamp

Using Ohm's law to substitute for some of the variables in that equation, we can obtain other forms of the electrical power formula. As you can see, watts, volts, amps, ohms, and coulombs are all interrelated. Studying these relationships is the key to understanding electricity.

Example Problem 10-2

DC Electrical Power

A flashlight bulb draws 4.17 A when lighted. If it has a resistance of 1.44 Ω, what power does it draw from the 6.00 V battery?

Known: battery voltage (V) = 6.00 V
bulb current (I) = 4.17 A
bulb resistance (R) = 1.44 Ω

Unknown: electrical power (P)

Required formula: $P = V \times I$, $P = I^2 \times R$, or $P = \frac{V^2}{R}$

Substitution: $P = (6.00\text{ V})(4.17\text{ A}) = 25.\underline{0}2\text{ W}$;
$P = (4.17\text{ A})^2(1.44\ \Omega) \approx 25.\underline{0}4\text{ W}$;
or $P = \frac{(6.00\text{ V})^2}{1.44\ \Omega} = 25.\underline{0}0\text{ W}$

Solution: $P = 25.0\text{ W}$

10C Section Review

1. Contrast static electricity and current electricity.
2. Compare the two models of how electrons flow through a conductor.
3. How does conventional current compare to electron current? Justify the use of the conventional current model.
4. What is the name for current that flows in one direction? In what two ways can this kind of current be produced?
5. What are the minimum components needed for an electrical circuit to do useful work? What device allows us to control whether or not current flows in a circuit?
6. Name four man-made sources of electrons that could produce an electrical current. Which "recycle" electrons?
7. What is the unit for electrical potential difference? How is it defined in terms of other SI units?
8. What is the unit of electrical current? In this textbook, how is it defined in terms of other SI units?
9. What is the unit of electrical resistance? What formula defines this unit? How is it defined in terms of other SI units?
10. (True or False) The unit of electrical power is the kilowatt-hour (kWh).
11. ✪ A portable CD player uses one AA 1.5 V battery. When running, the player uses 57 mA (0.057 A) of current. How much electrical resistance does the CD player have?
12. ✪ How much power, in milliwatts, does the CD player in Question 11 use?

10D Electrical Circuits and Safety

10.16 *Series and Parallel Circuits*

A circuit may contain more than one electrical load, and those loads can be connected in different arrangements. Two basic arrangements have specific names that describe the current path through two or more loads.

In a **series circuit** the loads are connected one after the other in the circuit path. There is only one path for the current to follow. All the charges from the source flow through all the loads in sequence at the same rate. For example, if two or more light bulbs are connected in a series circuit (Figure 10-36), the charges that pass through the first bulb must also pass through the others. Therefore, the same amount of current flows through all the light bulbs. If any one bulb burns out, the circuit is broken and the current stops flowing through all the bulbs.

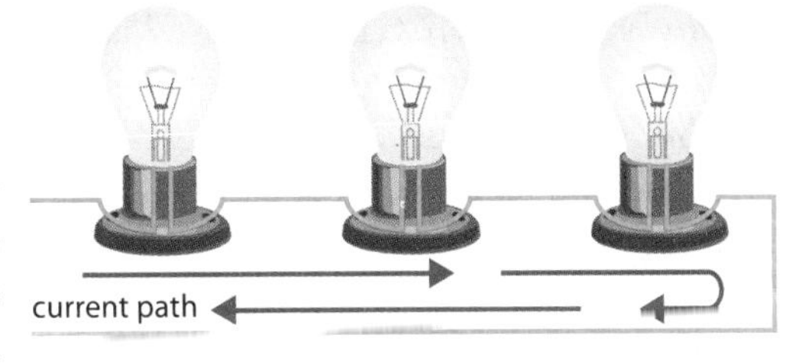

10-36 In a series circuit, all loads carry the same current because there is only a single current path.

In a **parallel circuit** the loads are connected in separate branches of the circuit. The current flows through more than one path, splitting up to flow through each load. The individual currents recombine on the other side of the parallel portion of the circuit. For example, if two or more bulbs are connected in a parallel circuit (Figure 10-37), each bulb represents a separate path for charges. The current from the supply conductor divides, and only a portion goes through each bulb.

The amount of current running through each bulb in a parallel circuit depends on its resistance according to Ohm's law. If the bulbs all have the same resistances, the currents are the same in each. If one of the light bulbs in the parallel circuit is turned off or burns out, current will continue to flow through the other branches and the bulbs will continue to burn. For this reason, building codes require that electrical fixtures and receptacles in houses be wired in parallel.

> **10D Objectives**
>
> After completing this section, you will be able to
>
> ✓ diagram an example of a series circuit and a parallel circuit.
>
> ✓ describe the dangers of a short circuit.
>
> ✓ compare and contrast circuit breakers and fuses.
>
> ✓ explain how circuit breakers cannot protect people from electrical shocks.
>
> ✓ summarize how ground-fault circuit interrupters work

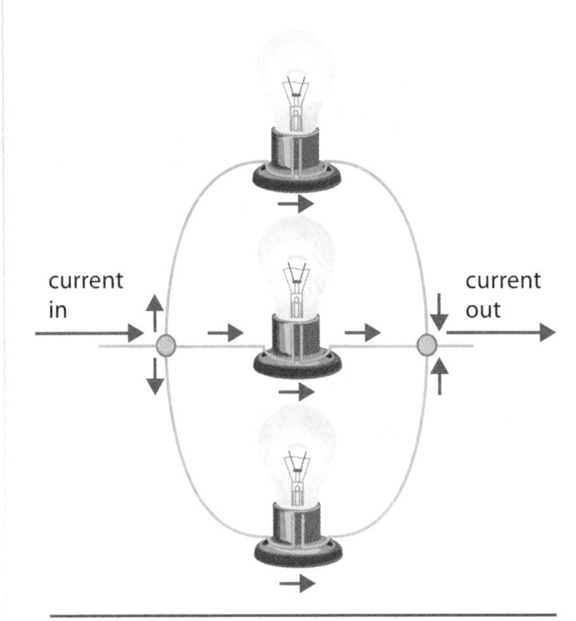

10-37 In parallel circuits, the current branches into parallel paths. The sum of the currents in the branches equals the current entering and leaving the branches.

10.17 *Short Circuits and Overcurrent Protection*

A current takes the path of least electrical resistance back to its source. This characteristic often causes electrical appliances to break down. The electrical supply cords to appliances have at least two wires in them. The wires are the conductors to and from the appliance. If the cord's insulation is damaged so that two wires touch, most of the current will bypass the appliance, taking a shortcut where the wires touch. This condition is a **short circuit**.

Short circuits are dangerous because the current bypasses the load, which normally keeps the current at safe levels in the wires. A wire can carry only a limited amount of current. In the previous section, we introduced a power equation that relates current and resistance: $P = I^2R$. Doubling the current running through the wire

10-38 Cuts in electrical supply cords can cause short circuits.

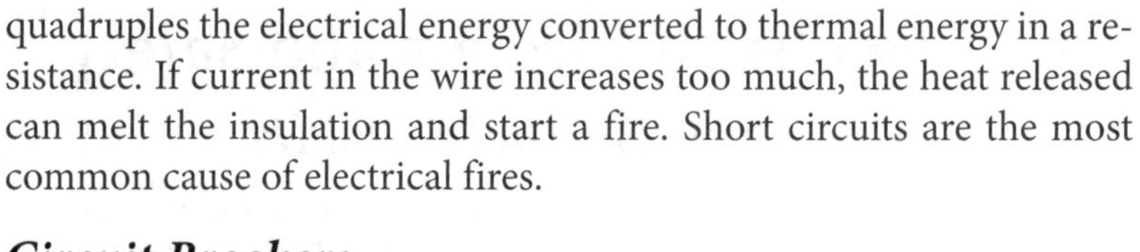

quadruples the electrical energy converted to thermal energy in a resistance. If current in the wire increases too much, the heat released can melt the insulation and start a fire. Short circuits are the most common cause of electrical fires.

10-39 Short circuits can cause wire overheating, which, in turn, can ignite a fire. This receptacle short-circuited when the furniture damaged it.

Circuit Breakers

To prevent currents in faulty circuits from starting fires, building codes require that every electrical circuit in a building have *overcurrent protection.* A typical house has between 10 and 40 circuits to protect. Each circuit has a *circuit breaker* connected in a series relation with the load. A circuit breaker is a switch that is manually closed but that automatically trips open if the current flowing through it exceeds a set value. Each device is rated for a certain number of amps (its *amperage*), clearly stamped on the breaker switch. Circuit breakers are installed in a central panel that must remain easily accessible at all times. Circuit breakers are reliable, easy to use, and reusable after the circuit fault has been corrected.

10-40 Circuit breakers are used to prevent overcurrent in wires due to short circuits. These breakers show, from top to bottom, the tripped, shut, and open positions (left).

A relatively new type of circuit breaker detects very high, short-duration currents caused by *arcing,* which usually accompanies short circuits. Electrical arcing occurs when electricity jumps tiny gaps between segments of a conductor. These faults are particularly hazardous because the hot sparks can ignite flammable materials that may not be touching the wires, and the wires may not draw sufficient current long enough to trip a standard circuit breaker. These protective devices are called *arc-fault circuit interrupters (AFCI).* They are "intelligent," often containing microprocessors that can detect an arcing short circuit. For homes built after 1999, all circuits supplying bedrooms are required to be protected by AFCI circuit breakers.

10-41 Arcing in the vicinity of flammable materials can lead to fires. An AFCI (right) quickly shuts off current in an arcing circuit.

Fuses

Older homes with original electrical wiring may have protective devices called *fuses* rather than circuit breakers. A fuse is a metal and glass container with a thin strip of metal inside that allows current to flow from the supply to the load. The thickness and material of the metal strip determine the fuse's amperage. It screws into a fuse panel near where the electricity enters the building in series with the circuit it is to protect.

If an electrical fault in a wire causes the current to exceed the rating of the fuse, the metal strip melts, creating a gap in the circuit. When the fuse "blows," the current stops flowing. Fuses are far less convenient than circuit breakers because the fuse has to be replaced after the problem is corrected. Also, it takes time for the current to melt the metal strip in the fuse. This delay may result in wiring damage and even allow a fire to start.

10.18 Electrical Shock and Ground Fault Protection

If you have ever experienced even a small electrical shock, you appreciate the importance of protecting people from such shocks. It may take tens or even hundreds of amps to start a house fire, but it takes only 0.2 A or less to paralyze your heart, which uses a complex set of tiny electrical signals to control its beat. You are exposed to hazardous voltages when insulation in a wire or an appliance fails. If you touch a faulty electrical appliance, your body could become part of a circuit!

Special devices called *ground-fault circuit interrupters (GFCI)* protect against shock in exposed locations. They monitor the currents between the wires supplying an appliance or outlet. As long as the currents remain balanced, the circuit is considered safe. When the GFCI senses a current imbalance, it instantly opens the circuit. A current imbalance occurs when a person, grounded to the earth by touching a metal fixture or standing on a wet surface, touches an active appliance with an electrical fault. He becomes a conductor of electricity and unbalances the current in the electrical circuit, activating the GFCI.

A GFCI device is usually a special outlet, but it can also be part of a portable extension cord. You may have seen GFCI outlets in your bathroom or kitchen, recognizable by the two small test and reset buttons. There are also GFCI circuit breakers that monitor entire building circuits, not just individual receptacles.

As the lightning rod has done, modern electrical safety devices continue to protect human lives from the consequences of living in a fallen world. God has given humans the knowledge to design these tools. Using that knowledge does not thwart God's judgment but instead shows love to one another and glorifies God.

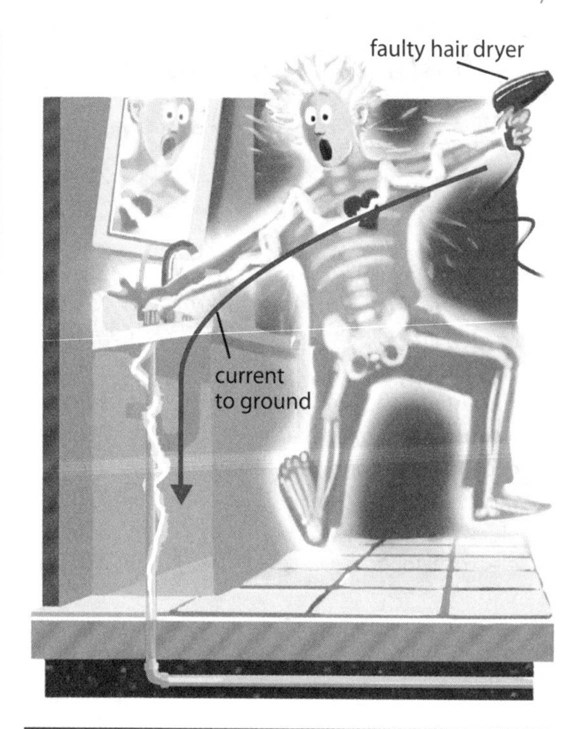

10-42 You can become part of an electrical circuit if the appliance you are using is faulty. The results can be lethal.

10-43 A GFCI receptacle. Once a month, press the test button to verify that the receptacle works. Then press the reset button.

10D Section Review

1. Describe a series circuit. Include in your description the current path and the magnitude of the current in each load.
2. Describe a parallel circuit. Include in your description the current path and how the magnitude of the current in each load is determined.
3. What happens to the current in a short circuit? What can potentially result from a short circuit?
4. What common electrical safety device is designed to prevent electrical fires? What newer device protects circuits specifically in bedrooms?
5. Give two reasons why circuit breakers are more effective and convenient than fuses.
6. Why can a person be shocked to death by less current than is required to start a fire?
7. What device protects users from electrocution? How is the invention of such a device an example of dominion science?

Scientifically Speaking

static electricity	205
electrostatic force	205
law of charges	206
electrical induction	208
electrical conductor	211
electrical insulator	211
semiconductor	211
Leyden jar	213
capacitor	213
current electricity	215
electrical potential energy	216
conventional current	216
direct current (DC)	216
electrical circuit	216
electrical load	216
battery	217
electrical resistance	219
Ohm's law	220
series circuit	223
parallel circuit	223
short circuit	223

Chapter Review

1. Why does rubbing a balloon on your sweater allow the balloon to stick to a wall or the ceiling?
2. What weather condition helps static charge build up in objects such as clothing? Why?
3. How can an insulator, like wool, lose electrons to acquire a positive charge?
4. Why do lines of force point away from positive charges?
5. Contrast the movement of charges in insulators and conductors during induction when a charged object is brought *near* them.
6. Describe how an electroscope becomes positively charged.
7. Why do static, or stationary, charges (electrons) move? Describe the duration of their motion.
8. How can you increase the amount of charge a Leyden jar stores?
9. When smoke passes through an electrostatic precipitator, the uncharged smoke particles are attracted to the charged plates in the device (see Figure 10-18). What electrical process causes the particles to act this way? How are electrostatic precipitators an example of dominion science?
10. List five electrical devices that don't need to be plugged into an electrical outlet to operate.
11. In the following figure, representing a section of wire viewed at four moments in time, which direction do electrons flow? Which direction does conventional current flow?

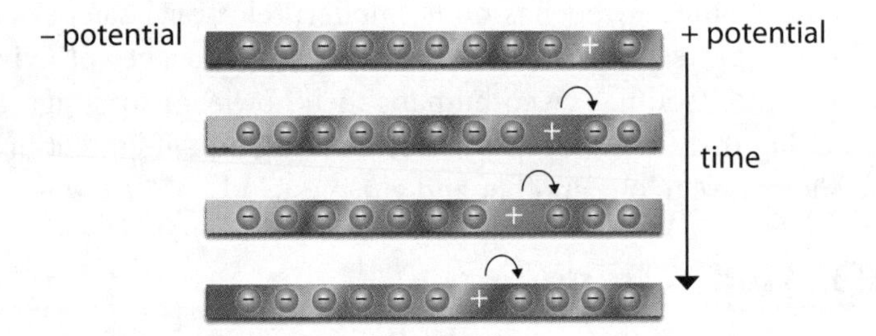

12. How does flow of electrons in an electrochemical cell differ from that in a PV cell?
13. List three kinds of energy to which electrical energy can be converted.
14. How much does electrical power absorbed in a wire ($P = I^2R$) change if the current triples? How does a wire eliminate the excess energy?

15. The labels "series" and "parallel" can apply to an entire circuit or a portion of a circuit. In the following diagrams, identify the type of circuit arrangement within the dotted lines.

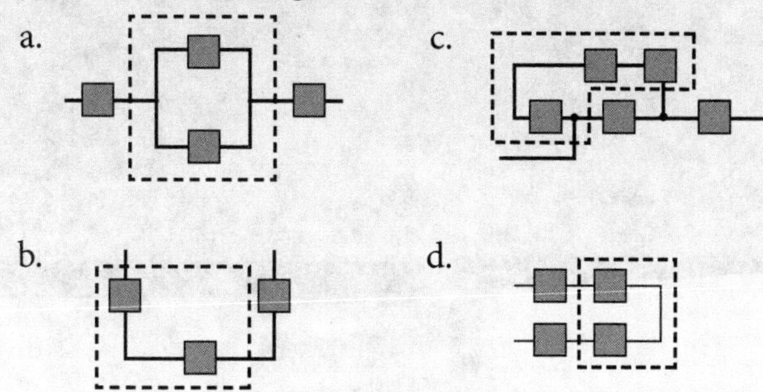

16. What three devices reduce the likelihood of an electrical fault starting a fire?
17. How do GFCI outlets protect from electrocutions?

True or False

18. Like charges attract; opposite charges repel.
19. In the electric field model, lines of force start at positive charges and end at negative charges.
20. You can be sure that two objects that electrostatically attract have opposite charges.
21. In a charged electroscope, one of the foil leaves has a positive charge and the other a negative charge.
22. Electrostatic charges build up when electrons are transferred from one object to another.
23. A lightning rod is made from a good insulating material.
24. * The earth can be considered an infinite source or sink of electrons.
25. A capacitor is a modern version of the Leyden jar.
26. Benjamin Franklin assigned the negative charge to the electron.
27. Modern wet cells can be recharged after they are "dead."
28. An electrical generator is like a water pump that pumps water to the top of a hill so that it can flow down again.
29. Opening a switch in a parallel branch of a circuit will turn off the entire circuit (see the adjacent figure).

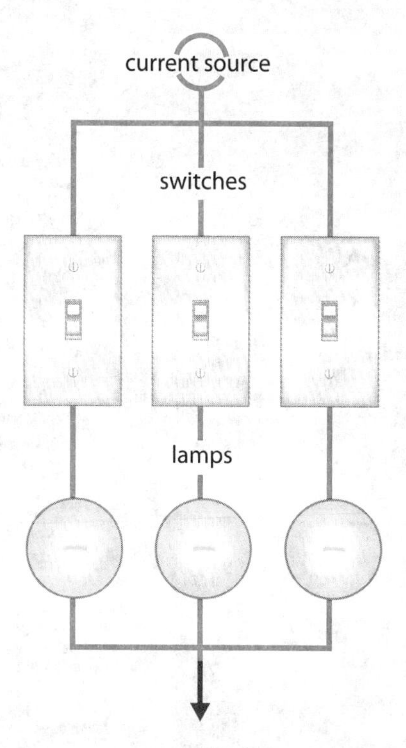

30. ✪ A typical flashlight bulb has a resistance of 60. Ω. If the bulb is used in a 6.0 V flashlight, how much current will it draw?
31. ✪ How much power does the light bulb in Question 30 use?

MAGNETISM

CHAPTER 11

11-1 Present-day methods for short- and medium-distance commutes are slow, cause city congestion, and are significant sources of air pollution.

DOMINION SCIENCE PROBLEM

Beating the Traffic—and Pollution

Air travel is ideal when people must travel quickly between locations hundreds or thousands of kilometers apart. But most times, people travel to destinations that are only a few kilometers away. Short-distance travel usually means traveling by car or bus. A large number of workers commute into and out of cities every day. Their travel contributes to traffic congestion, takes up parking needed for customers and visitors, and is the biggest source of urban air pollution. Another problem is that the hundreds of thousands of hours spent in traffic every year by a city's residents could be used more profitably.

One solution to these problems could include standard and high-speed rail services. These passenger trains offer more rapid point-to-point transportation than cars; however, even these can be relatively slow, and diesel-electric train engines are inefficient and polluting as well.

The perfect solution would be a means of transporting many people at once as fast as a passenger plane with the convenience of a bus or train. Oh yes—and it would be nice if commuters could travel quietly and smoothly too! Is there a way to provide our fellow humans with rapid short-distance mass transportation that avoids the common problems of existing methods?

11A Magnetism and Magnets

11.1 History of Magnetism

Magnets mystified the ancients. In Magnesia, Asia Minor, the early Greeks found lodestones, which baffled them with their strange ability to attract and hold iron particles. The Greeks were among the first to describe magnetism. The ancient Chinese made similar findings and were equally entranced.

To these people, magnets seemed magical. Some suggested that magnets were evil objects possessed by invisible spirits. Others claimed that magnets could heal sickness and cure headaches. In the Middle Ages the superstitious believed that owning a magnet endangered your health! Possessing a lodestone might brand you as a wizard or a witch and earn you a fiery reception from an angry crowd. Yet it was during this time that practical uses for magnets were first discovered.

As early as the fourth century BC, the Chinese recorded that pieces of lodestone seemed to align themselves in favored patterns. Fortunetellers made great use of these mysterious rocks. The Chinese eventually carved lodestones into spoon-like shapes so that their "handles" would point south when resting on a flat, smooth surface. Around AD 1100 they developed navigational compasses using magnetized needles. Until then, sailors had navigated using visual observations of the sun and stars or by measuring water depth. A sudden storm, an overcast sky, or deep water robbed sailors of these navigational aids and sometimes led them to their deaths.

11-2 The Chinese were the first to use a magnetized pointer as a navigational compass.

The magnetized-needle compass may have been imported to Europe along the trade routes from the Far East, or it could have been independently invented there. Inventive and learned Europeans and Arabians soon designed crude compasses that gave travelers more accurate direction. In these compasses, the needle was magnetized to point north rather than south as the Chinese compasses did.

In 1269 Petrus Peregrinus de Maricourt wrote in a letter to a friend that a spherical lodestone exhibited two separate spots where bits of iron collected. From this observation de Maricourt reasoned that the magnetic properties of the lodestone were concentrated at these two locations. He also noted that they seemed to be opposite in nature. Today these two areas of concentrated magnetic force are called **magnetic poles**. De Maricourt suggested that the earth's north and south magnetic poles resided in the heavens above the polar regions rather than on the surface of the earth.

Despite de Maricourt's findings, myths regarding magnetism abounded throughout the Middle Ages. In 1600 the English physician

11A Objectives

After completing this section, you will be able to

- ✓ summarize the early historical development of the study of magnetism.
- ✓ compare and contrast magnetism and electricity and describe their fields.
- ✓ compare and contrast the four different properties of magnetism.
- ✓ explain the domain model of magnetism.
- ✓ define a material's Curie temperature.
- ✓ identify the earth's magnetic North Pole.
- ✓ describe the difference between the geographic and magnetic North Poles.

The most ancient evidence for the use of magnetism seems to be from objects made by the Olmecs, one of the earliest peoples to occupy Central America, around 1000 BC.

Petrus Peregrinus de Maricourt was a thirteenth-century French naturalist who wrote the earliest existing paper on magnetism.

11-3 De Maricourt's magnetic lodestone had two poles.

11-4 William Gilbert was an English scientist and physician to Queen Elizabeth I. His book *De Magnete*, published in 1600, earned him the title Father of Magnetism.

William Gilbert put these ideas to rest. He conducted many experiments and compiled a book on the properties of magnets known or suspected at that time. He explained that a compass behaved as it did because the earth itself was a magnet. Some of the "facts" in Gilbert's book, such as the medicinal properties of magnets, were based more on his ideas that magnetism was the "soul" of a magnet than on observational evidence. Yet Gilbert brought the study of magnetism into the realm of science.

11.2 Magnetism and Magnetic Fields

Is magnetism produced by "magnetic charges" just as electricity is produced by electrical charges? This question directed the efforts of many early investigators. They supposed that these magnetic charges could be separated onto different objects, just as positive and negative charges can be isolated. Thus an object could have only one type of magnetic pole, called a *monopole*.

No one has ever observed such an object. Magnetic objects always have a north and a south magnetic pole—they are *magnetic dipoles*. If you cut a magnet in half, you will end up with two smaller magnets—each with a north and a south magnetic pole. Scientists have found that even when magnets are divided as many times as physically possible, each of the microscopic fragments has both a north and a south magnetic pole. Even the molecular structures that form magnets exhibit a magnetic dipole. Despite a continuing search, no one has demonstrated the existence of magnetic monopoles.

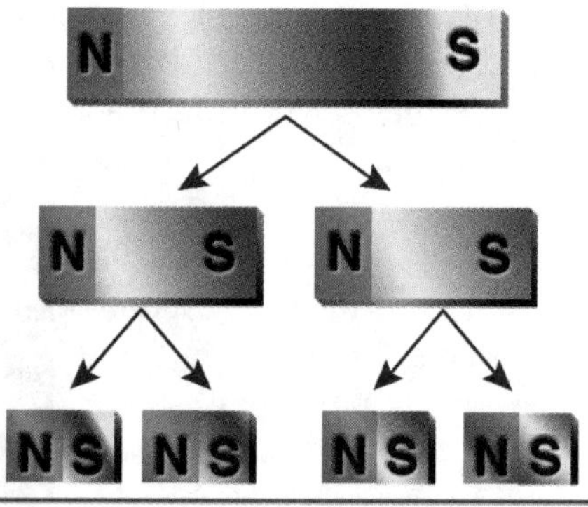

11-5 Magnets can be subdivided to the particle level. The pieces always have two magnetic poles.

Magnetism, like the electrostatic force, is a field force. A **magnetic field** can be modeled using magnetic lines of force, which are similar to an electrical field's lines of force. Field lines clump together at magnetic poles and spread apart farther from the poles. Unlike electric fields, which begin at positive charges and end in negative charges, magnetic lines of force are modeled as continuous loops both within a magnetized object as well as in the space around it (see Figure 11-6). The direction of the field line is determined by the direction a tiny compass would point if inserted into the field. Magnetic field lines point away from north magnetic poles and toward south magnetic poles. The distance between field lines indicates the strength of the field at that point. Thus, the field is strongest near the poles and weaker at points between the poles and at greater distances from them.

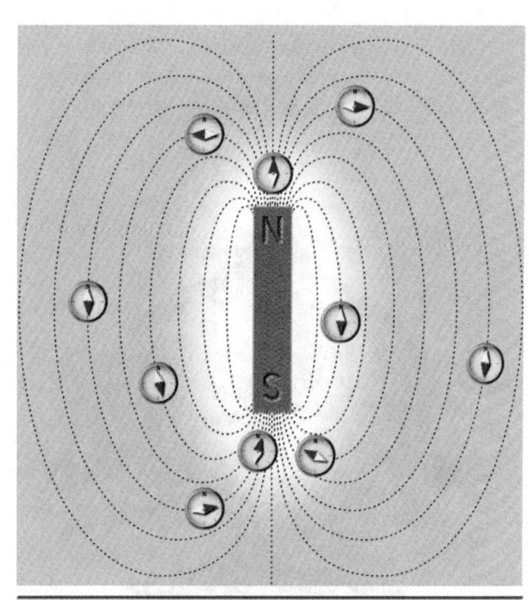

11-6 A magnetic field in a bar magnet. Just as in electric fields, a magnetic field extends around the bar magnet in three dimensions.

Magnetic fields interact with each other just like electrical fields do. When north and south magnetic poles are brought close to each other, the lines of force bend toward the approaching poles and the objects are attracted. The lines of force point from the north

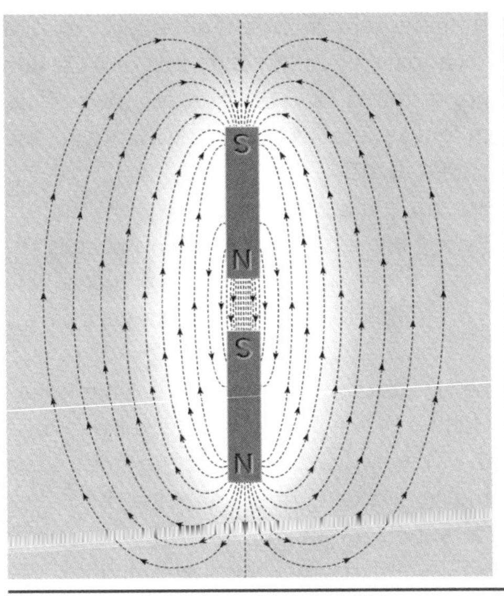

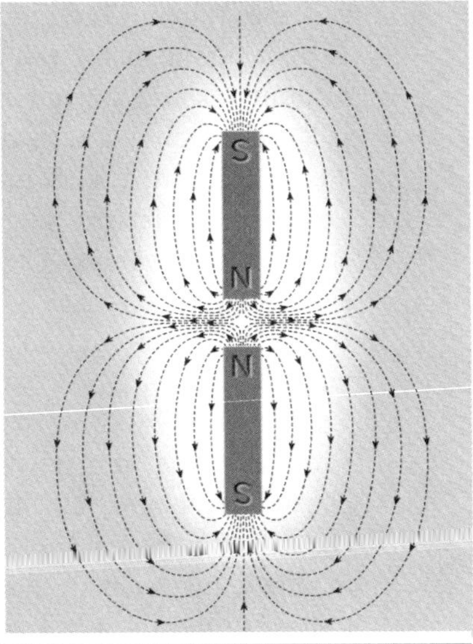

11-7 Magnetic fields attract and repel. The lines of force never cross one another.

pole of one object toward the south pole of the other. If like poles are brought near each other, the field lines bend away from the approaching poles and the objects repel. From these observations, we obtain the **law of magnetism**—*opposite magnetic poles attract; like magnetic poles repel.*

11.3 Magnetic Properties

Magnets that you are familiar with are made from a small group of metal elements and alloys. These exhibit the property of **ferromagnetism** and they are called *ferromagnetic materials.* Such materials include iron, nickel, cobalt, and a few other elements. Interestingly, liquid oxygen is also strongly ferromagnetic. Steel (an alloy of carbon and iron) and alnico (an alloy of aluminum, nickel, cobalt, and iron) are commonly used to form the strong permanent magnets used in motors and speakers. Alloys of certain rare metals, such as neodymium, make the strongest permanent magnets. Large versions of these magnets are hazardous to handle because they can severely pinch fingers and folds of skin. Smaller neodymium magnets are used to hold some jewelry, such as earrings, in place without piercing.

Identification of the various families of elements noted in this section can be found in the periodic table in Appendix G. Chapter 17 provides a detailed description of each.

neodymium (NEE oh DIME ee uhm)

According to the accepted model, magnetism originates at the particle level. Some atoms are tiny magnetic dipoles due to the arrangement of their electrons. Large groups of magnetic dipoles automatically align in the same direction to form microscopic regions called **domains**. In unmagnetized ferromagnetic materials the magnetic directions of the domains are oriented randomly, so their magnetic forces cancel each other throughout the material. When an external magnetic field is applied to ferromagnetic domains, they align with the lines

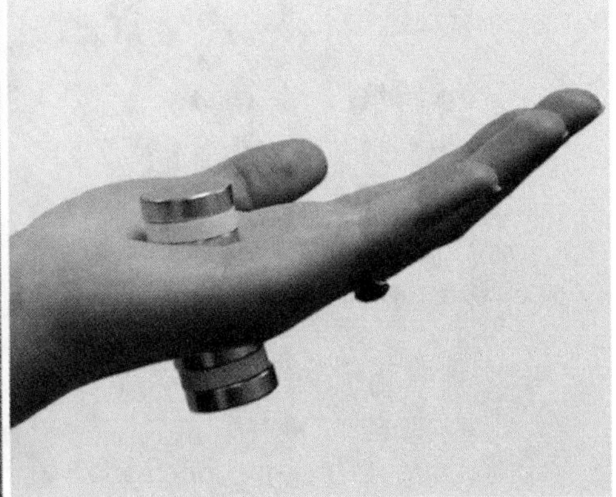

11-8 Neodymium magnets are made of extremely strong ferromagnetic neodymium alloys.

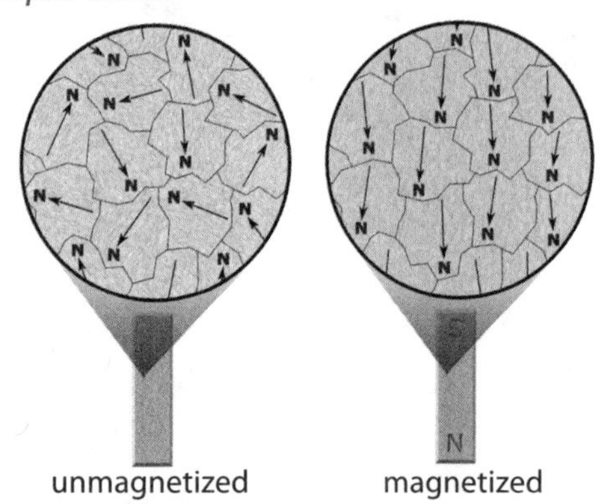

11-9 Ferromagnetic domains

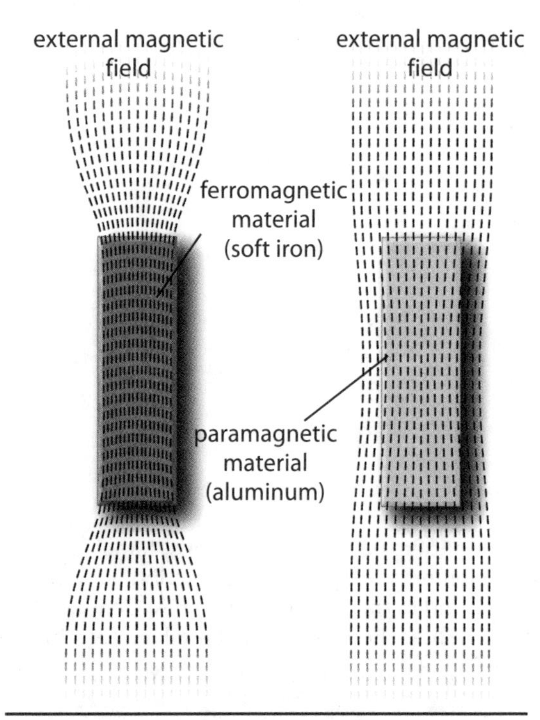

11-10 The effects of ferromagnetic (left) and paramagnetic (right) materials on an external magnetic field

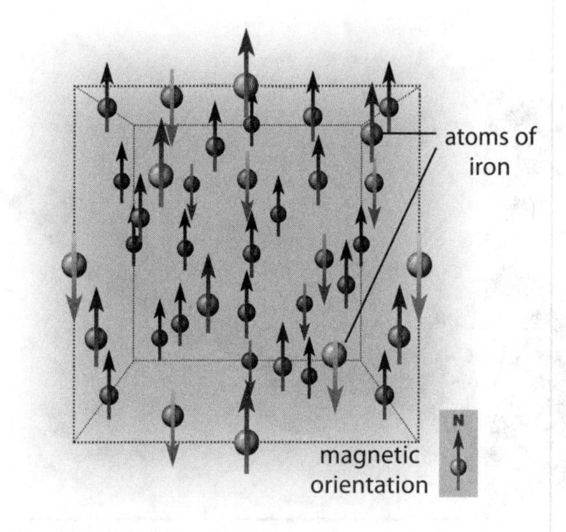

11-11 The ferrimagnetism of a natural magnet (lodestone) results from the arrangement of two kinds of iron atoms within the rock.

of force. When it is removed, the domains remain aligned to form a *permanent magnet*. The atoms of paramagnetic materials don't reside in domains and are randomly oriented. An external magnetic field will cause their atoms to align, but as soon as the field is removed, the atoms return to a random arrangement.

Paramagnetism refers to those materials that are only slightly attracted to magnets. These substances, such as aluminum, magnesium, sodium, and gaseous oxygen, are said to be *paramagnetic*. The strength of the attraction of paramagnetic materials often varies with temperature.

Figure 11-10 contrasts the effects of ferromagnetic and paramagnetic materials on magnetic fields. When a ferromagnetic iron bar is placed in a magnetic field, the lines of force are concentrated in the bar because the field is strengthened as it interacts with the ferromagnetic materials in the bar. When the magnetic field is removed, the bar usually retains some magnetism. This residual magnetism is why steel sewing needles or paper clips can be magnetized by being stroked with a magnet.

A paramagnetic material's effect on an external magnetic field is weaker. The field is not strengthened as much by its presence. Also, when the external magnetic field is removed, paramagnetic materials do not retain any residual magnetism. However, as long as the field is present, paramagnetic objects act like ferromagnetic materials.

Natural magnets, such as lodestones, display **ferrimagnetism**, which behaves somewhat differently from ferromagnetic materials. A lodestone is made up of compounds of oxygen and two different kinds of iron. One kind of iron atom has a magnetic field oriented in one direction. The other kind has a field oriented in the opposite direction. The two kinds of iron atoms are mixed together in the mineral. Since each kind is present in different amounts within a piece of lodestone and since each has a different magnetic direction, the two magnetic properties subtract. The resulting magnetism of the lodestone is the net difference of the two kinds of magnetism.

A fourth magnetic property is **diamagnetism**. Diamagnetic materials weaken an external magnetic field by generating an opposing field. In a field diagram the lines of force are dispersed slightly as they pass through the diamagnetic material. All forms of matter are naturally diamagnetic. However, many common substances also have natural atomic and molecular magnetic dipoles, which produce ferromagnetic and paramagnetic responses to an external magnetic field. Since the effect of diamagnetism is many thousands of times smaller than that of ferromagnetism or paramagnetism, diamagnetism is observed mainly in nonmagnetic materials.

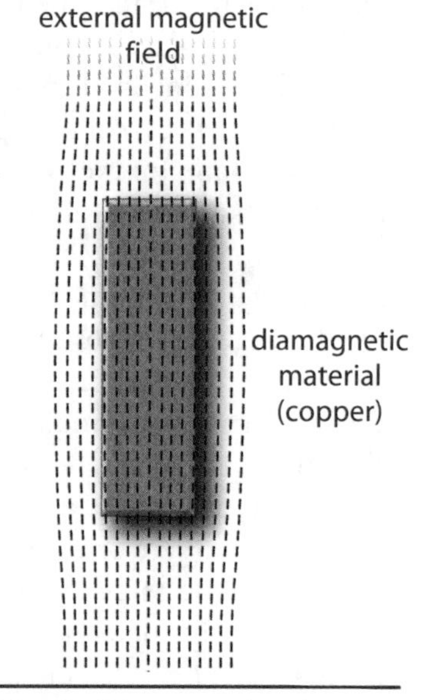

11-12 Diamagnetism in a bar of copper

Diamagnetism can produce some interesting effects. An external magnetic field causes diamagnetic atoms and molecules to form magnetic dipoles that align to oppose the external magnetic field. If the external field is strong enough, the diamagnetic effect can produce a measurable force that acts opposite the magnetic field force. Figure 11-13 shows a small (approximately 6 mm) sheet of artificial graphite that is highly diamagnetic. The strong magnetic field from the four neodymium magnets resting on a steel base act on the graphite to generate an opposing magnetic field in the graphite that lifts it above the magnets against gravity.

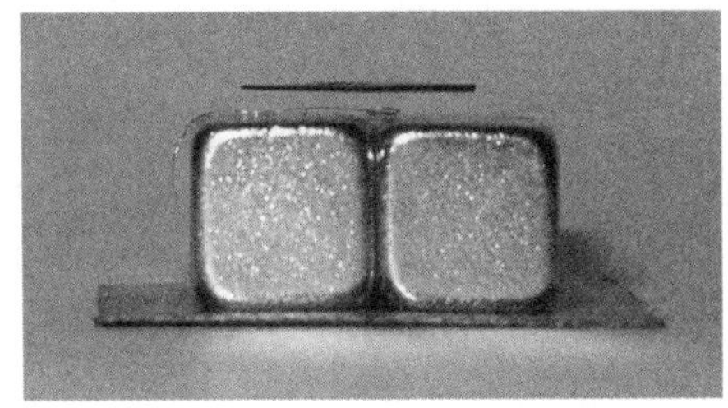

11-13 A piece of diamagnetic pyrolytic graphite suspended above four strong neodymium magnets

11.4 Magnetism and Temperature

Since magnetism originates at the particle level, temperature can significantly affect it. You know that particle motion increases with temperature. This motion tends to disrupt particle order. Ferromagnetic domains, such as those in iron, normally maintain a high degree of order due to alignment of the magnetic dipoles of the element's atoms. However, at 770 °C (1043 K or 1418 °F) iron abruptly loses its ferromagnetic properties. The domain boundaries dissolve, and the atoms assume completely random arrangements. This temperature is called iron's **Curie temperature**. Each ferromagnetic material has its own characteristic Curie temperature. A sample of iron above its Curie temperature will behave like a paramagnetic material when placed in a magnetic field.

The significance of temperature on magnets was discovered by the French scientist Pierre Curie (1859–1906), who conducted research in crystals, magnetism, and radioactivity.

11.5 The Earth's Magnetic Field

What makes a compass point north? William Gilbert showed that the earth itself is a gigantic magnet. Its magnetic field is aligned in a generally north-south direction. A compass contains a very light bar magnet mounted on a pivot point so that it can turn easily. This bar magnet aligns itself along the north-south direction of the earth's magnetic field as the needle's poles seek the opposite magnetic poles in the earth.

Did you know that the earth has two North Poles? The **geographic North Pole** is at the "top" of the earth at the axis of the earth's spin. This is sometimes referred to as "true north." All lines of longitude intersect at the geographic North and South Poles. The **magnetic North Pole** is where the earth's concentrated magnetic field lines are vertical at the earth's surface. It is nearly 470 km (292 mi) away from the geographic North Pole. Its position is not constant but is currently drifting nearly 40 km per year toward Siberia. Compasses are attracted to the magnetic North Pole, not the geographic North Pole. In some parts of the world, the difference is important. In Bangor, Maine, your compass would point 17° counterclockwise from the direction of true geographic north to magnetic north. If you lived in Barrow, Alaska, magnetic north would be 18° clockwise from true north.

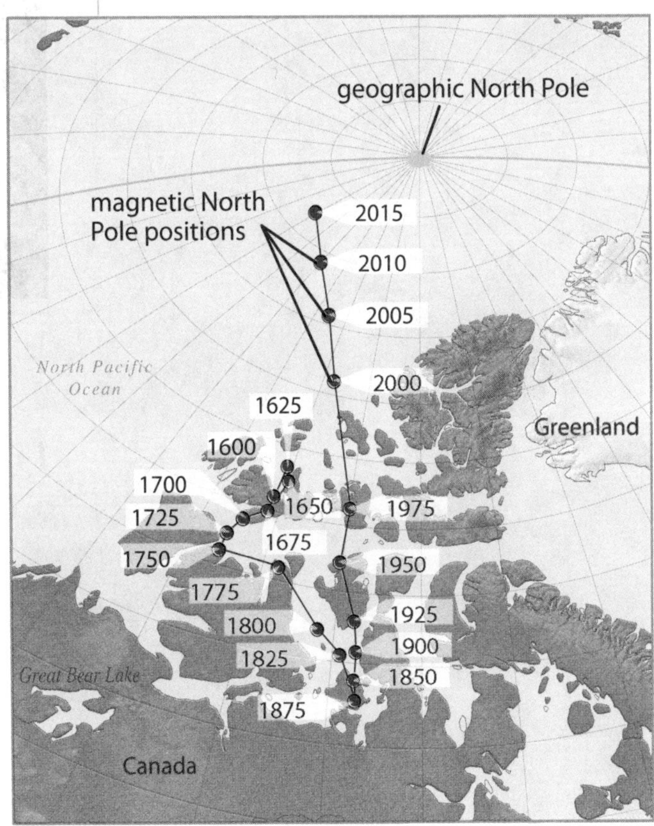

11-14 The geographic North Pole is fixed at the earth's rotational axis. The position of the magnetic North Pole is drifting slowly across the Arctic Ocean.

The difference in direction between the magnetic North Pole and geographic North Pole at a given location is called *magnetic declination*. It is measured in degrees. If the magnetic pole lies to the right of a line drawn from a location to the geographic North Pole, the declination is *east*; if to the left, the declination is *west*.

Is the magnetic North Pole of the earth really a north magnetic pole? Remember that it attracts the north poles of magnets. Are the north poles of magnets attracted to other north magnetic poles? No! The earth's magnetic North Pole is actually a south magnetic pole. This may require some mental adjustment to understand, but it has little impact on the use of magnetic compasses.

The earth's magnetic field originates deep inside the earth, probably due to the flow of liquid iron in the earth's outer core. This field extends far out into space. Near the earth the **geomagnetic field** deflects or traps the high-energy charged particles of "solar wind." Without the geomagnetic field, these particles would bombard the earth's atmosphere and surface, damaging or destroying life. Creationary scientists point to the existence of the geomagnetic field as just one of several characteristics of the earth favorable to supporting life that were designed by an omniscient Creator.

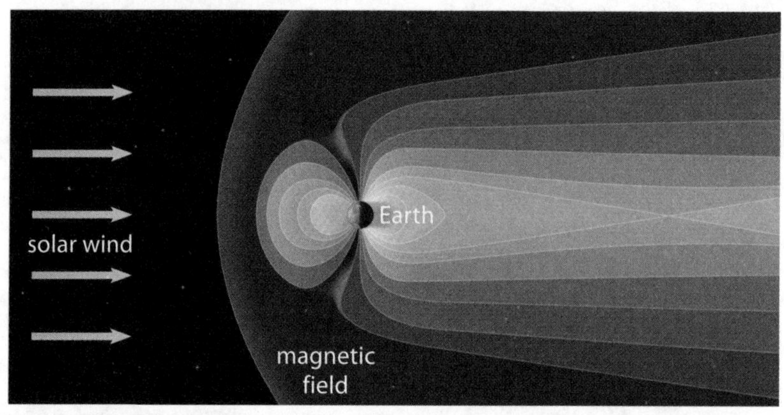

11-15 The earth's magnetic field protects living things by deflecting high-energy charged particles from the sun.

11A Section Review

1. Ancient discoveries of physical phenomena are poorly documented in history. Do you think this is because people were ignorant in those times or because we have sparse records of their knowledge? Taking a biblical view of human civilization, when do you think magnetism could have been discovered?
2. Describe a magnet and its field.
3. Permanent magnetism is a property of what kind of magnetic materials?
4. What feature of ferromagnetic materials makes them different from other materials? At what level of matter does one look to identify the source of ferromagnetism?
5. What could happen to a permanent bar magnet if you held it in a flame until it glowed?
6. What magnetic evidence do we have that the interior of the earth is not just solid rock acting as a bar magnet?
7. Why would the lack of a strong geomagnetic field be a problem for the earth? According to a biblical worldview, what accounts for the existence of the earth's magnetic field?
8. (True or False) All materials are diamagnetic to some extent, although some materials have other magnetic properties that mask their diamagnetism.

11B Electromagnetism

11.6 Electrical Current and Magnetism

In the late 1700s and early 1800s, scientists believed that electricity and magnetism were related, but they could not demonstrate how. On ships struck by lightning, sailors sometimes observed that their compasses were affected. Several scientists unsuccessfully attempted to duplicate this effect in the laboratory.

In 1820, while he was preparing for a classroom demonstration, the Danish physicist Hans Christian Oersted noticed that a compass on a nearby table deflected when he sent a current through a wire. He concluded that the current-carrying wire had produced a magnetic field.

Oersted immediately began experimenting with the wire and compass. From his observations he determined that a compass needle aligns to circular loops around a current-carrying wire. When he published his observations several months later, his report encouraged new experimentation in the scientific community.

11B Objectives

After completing this section, you will be able to

- ✓ orient the magnetic field around a current-carrying wire, using the right-hand rule.
- ✓ explain how to assemble an electromagnet.
- ✓ compare and contrast solenoids and electromagnets.
- ✓ explain how a magnetic field can create an electrical current.
- ✓ differentiate between alternating and direct current and the way they're generated.
- ✓ list and describe the types of turbines used to power generators.

Hans Christian Oersted (ER sted) (1777–1851) was a Danish scientist who studied electromagnetism. He also was the first to chemically produce aluminum.

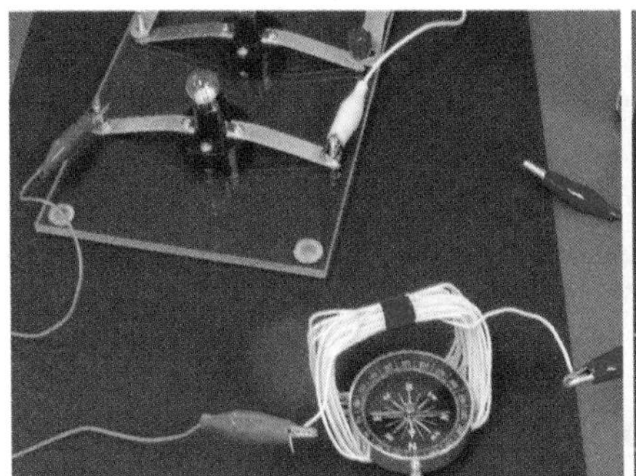

11-16 When current flows through the wires (right), the coil around the compass produces a magnetic field.

Figure 11-17 shows a current-carrying wire and the direction of the *conventional current*, which is opposite the flow of electrons. The direction of the magnetic field (the direction in which the magnetic north pole of a compass will point within the field) can be found by the **right-hand rule of magnetism**. If you grasp a wire so that your right thumb points in the direction of the flow of the *conventional* current, the magnetic lines of force caused by the current will point in the direction your fingers wrap around the wire.

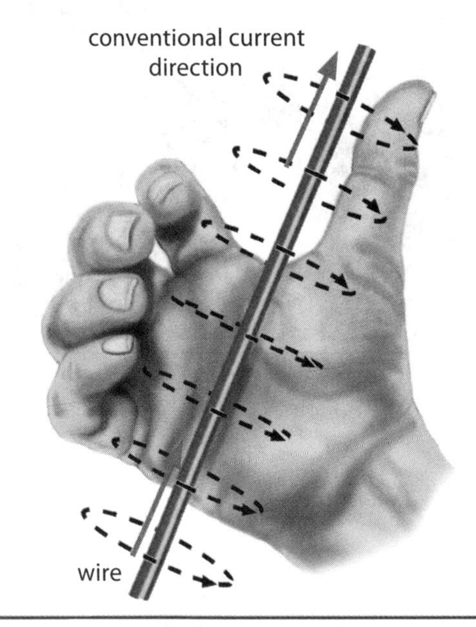

11-17 The right-hand rule of magnetism. Note that the current direction is the flow of conventional current.

11.7 Magnetism from Electricity

Only a week after Oersted's discovery was announced in France, the French physicist André Ampère announced a discovery of his own. He found that two parallel wires carrying current in the same direction were magnetically attracted. He further found that when he replaced one of the wires with a magnet, the remaining current-carrying wire was deflected away from or attracted to the magnet. Whether it was deflected or attracted depended on the direction of current flow in the wire and the orientation of the magnet. He also discovered that increasing the electrical current increased the strength of the surrounding magnetic field.

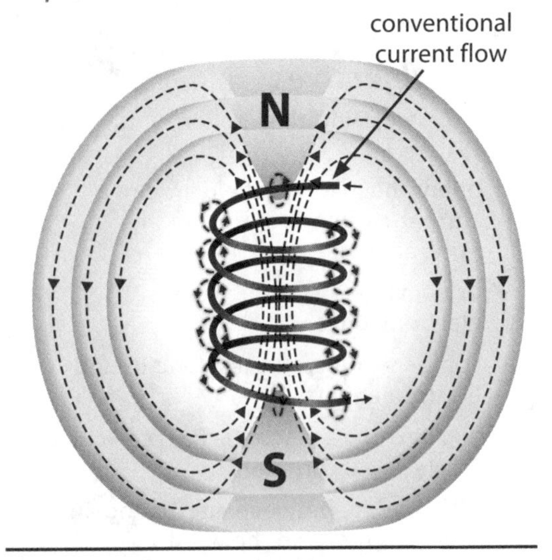

11-18 A current-carrying solenoid concentrates the magnetic field inside the coil.

solenoid (SO luh NOYD): solen- (Gk. *solen*—channel, pipe) + -oid (Gk. *oeides*—shape)

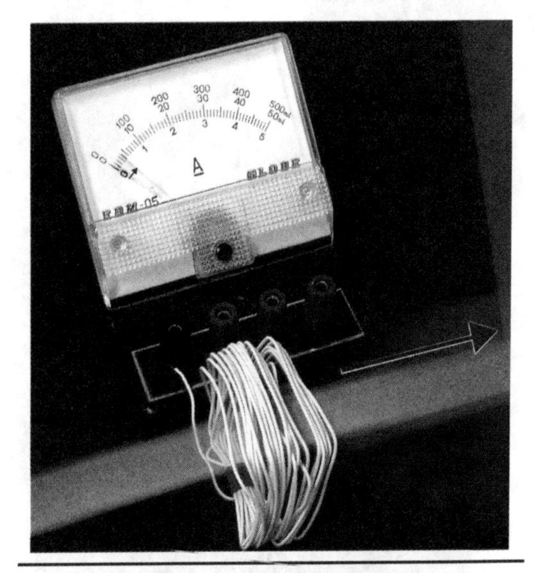

11-20 A changing magnetic field near a conductor produces electrical current. The direction of the field's motion through the conductor determines the direction of current.

Recall that the distance between the lines of force models the strength of a magnetic field.

The electrons are forced to move perpendicular to both the motion of the wire through the field and the direction of the magnetic field.

Later, scientists found that the magnetic field generated by a current-carrying wire is concentrated inside a loop of the wire. This arrangement produces distinct magnetic north and south poles, just as if a bar magnet were inserted through the loop. If the wire is looped multiple times into a coil, the magnetic field strength is multiplied by the number of loops. A coil of wire that acts like a magnet is called a ***solenoid*** (see Figure 11-18).

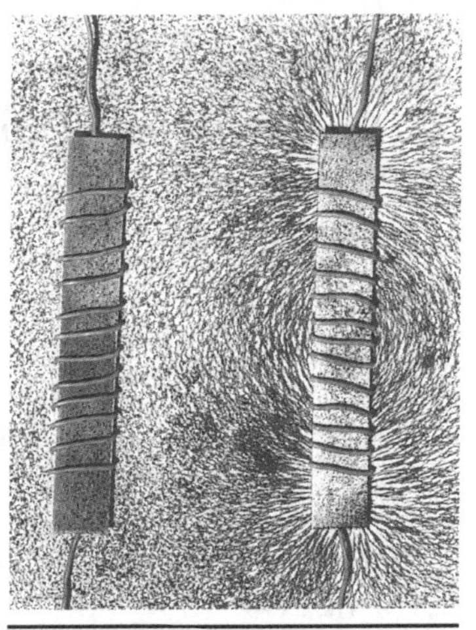

11-19 A simple electromagnet without current (left) and with current (right). The iron filings align with the field lines when the electromagnet is energized.

In 1823 an English experimenter named William Sturgeon observed that placing a ferromagnetic bar inside a solenoid greatly increased the magnetic field exerted by the solenoid coil. In fact, a solenoid with a pure iron core produced enough magnetic force to lift a piece of iron 20 times as heavy as the iron core! The solenoid's strong magnetic field aligns the ferromagnetic domains in the iron, thus further strengthening the external magnetic field. *Electromagnets* are critical parts of many appliances you use every day. Some of their important applications are discussed at the end of this chapter.

11.8 Electricity from Magnetism

Electrical current produces a magnetic field. Conversely, a magnetic field can produce an electrical current. If a bar magnet is thrust by hand through a coil of wire (Figure 11-20), a current is momentarily produced in the wire but only while the magnet is moving. When the magnet stops moving, the current stops. Sliding the wire coil along the magnet has the same effect as moving the magnet. Increasing the number of coils or the speed of motion increases the amount of current produced.

How does a magnetic field produce an electrical current? A magnetic field affects the motion of electrical charges in a conductor. When the concentration of magnetic field lines near a conductor increases or decreases, a force is exerted on the electrons in the conductor. The highly mobile electrons move through the conductor, creating a current if the conductor is part of an electrical circuit. The faster the magnetic field strength changes, the more electrons move. This effect is called **electromagnetic induction**. Even when a magnetic field surrounds a conductor, if the field does not change or the conductor does not move through the field, no current is produced.

11.9 Generators and Alternating Current

Electrical generators are important sources of current. They convert rotational mechanical energy into electrical energy. In a generator, some type of *prime mover*, such as a steam turbine or engine, rotates a shaft attached to one or more pairs of strong magnets. The outward-pointing poles alternate north and south in polarity. This assembly is called the *rotor.* The rotor spins inside a ring formed of tightly wound coils of wire, causing the magnetic field to sweep continuously through the coils. The ring of coils surrounding the rotor is called the *stator.* The ends of the wire supplying the stator connect to the electrical terminals on the generator. Study Figure 11-21 to understand the basic arrangement of a generator.

As the north and south magnetic poles alternately pass a coil, a current is induced in the wire if the generator is connected to a complete circuit. When a north pole passes by a coil, induction produces current in one direction; when a south pole passes, the current flows in the opposite direction to produce an **alternating current (AC)**. Voltage potential also alternates between positive and negative as the rotor passes each coil. This type of generator is the most important source of electricity used today.

Figure 11-22 shows a graph of current produced by a rotating AC generator. When the magnetic poles are aligned with the stator coils, the current is maximum. As the magnet poles rotate away from the coils, their field strength decreases and the current drops. When the rotor poles are between the pairs of coils, there are no field lines cutting through the coils and the current is zero, represented by the horizontal

stator (L. *sta-*, to stand); something that is a nonmoving part; stationary

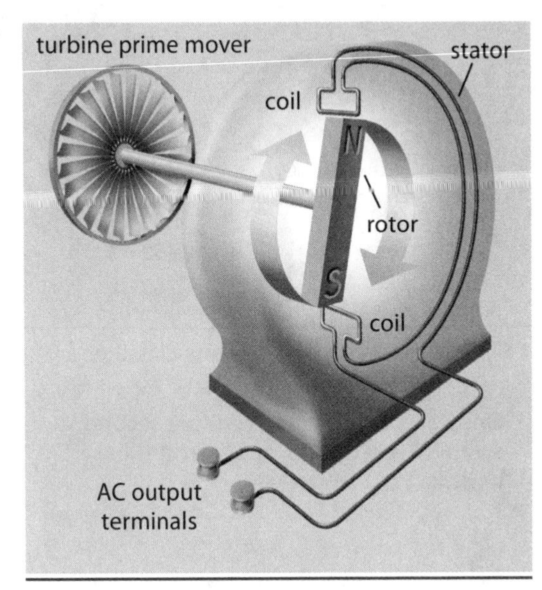

11-21 A simplified diagram of an AC generator showing only one pair of the many thousands of interconnected coils that form the opposite poles of a generator stator

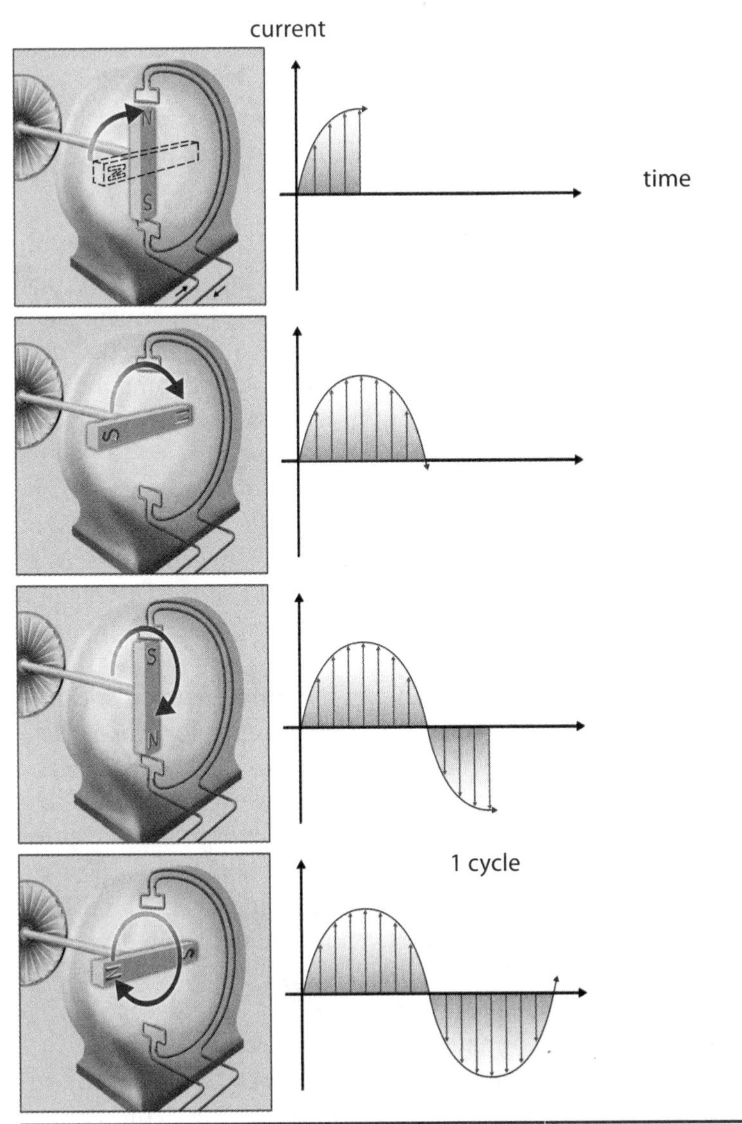

11-22 The generation of alternating current

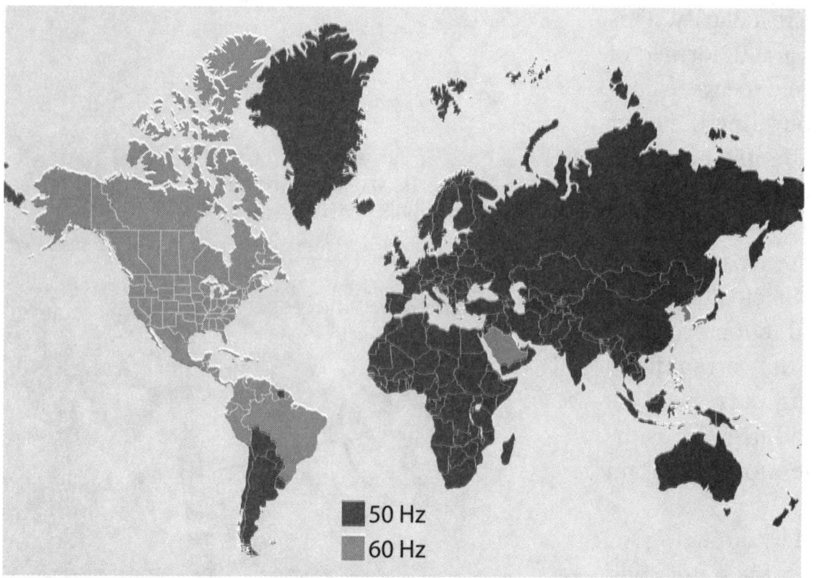

11-23 National power systems classified by frequency. See map legend.

axis. As the rotor continues turning, the magnetic fields increase in the coils and current starts rising, but in the opposite direction from the first half of the rotation. In the final panel the current again returns briefly to zero. This cycle repeats as the generator rotor spins within the stator.

The *AC frequency*, or the rate at which the current changes direction, is measured in cycles per second. The SI unit for frequency is the *hertz (Hz)*. In the United States and most of the Western Hemisphere, electric companies provide 60 Hz service to homes, businesses, and industries. So when you look at a light turned on in your house, it is really flickering 120 times a second, because current reverses twice each cycle. The flickering happens so fast that your eye cannot detect it. In most of the rest of the world, the frequency is 50 Hz. This difference is why you often need special electrical adapters when using portable AC appliances in foreign countries. Some home appliances, especially those with motors, are very sensitive to current fluctuations, so it's important that the electrical frequency be constant. Electrical utilities measure and control electrical frequency very carefully to avoid causing problems.

The concept of frequency applies not only to AC electricity but to other cyclical measurements as well. These are covered in later chapters.

One hertz equals one event per second. In symbols, 1 Hz = 1/s or s^{-1} (in the base SI unit, s).

Electricity is supplied to homes and businesses by electrical utilities—companies that specialize in generating and distributing electricity.

Generator Prime Movers

Most of the electricity we use each day is supplied by electrical generators. The mechanical energy necessary to turn generators usually comes from turbines. A *turbine* is a machine that converts the energy of a moving fluid into rotational motion. It is shaped like a propeller and can have just a few or many blades or vanes, depending on the fluid that propels it.

Steam turbines. Steam turbines are the most common prime mover for commercial electrical generators. They are turned by steam supplied from large boilers called steam generators. After the steam rotates the turbine, it flows out of the turbine and condenses back into water. It is then pumped back to the steam generator to be reheated.

There are three main sources of thermal energy that can be used to boil water to steam in a steam generator. Fossil fuels, especially coal, are the most common source of thermal energy. When they burn, chemical energy is converted to thermal energy. Nuclear energy from fissioning uranium is also converted to thermal energy and is far more efficient than fossil fuels. Heat from deep in the earth, or geothermal energy, boils water to steam within the earth. This steam is obtained from deep wells to turn the turbines. Geothermal energy is becoming more important worldwide. It seems inexhaustible. Its high costs are associated with finding and tapping the sources.

Water turbines. Hydroelectric plants use the gravitational potential energy of water stored behind dams or in natural waterfalls to provide energy to turn water turbines. Niagara Falls and

11-24 An electrical steam-turbine generator

11-25 Hydroelectric power uses water turbines to generate electricity.

Hoover Dam are two well-known hydroelectric plants in the United States. The Three Gorges Dam in China, the largest hydroelectric dam in the world, has more than 10 times the generation capacity of the Hoover Dam.

Wind turbines. Wind power is becoming a profitable method of generating electricity. Wind farms consist of hundreds of wind-turbine generators on towers. The generators are propelled by large fan blades that turn when the wind blows. The largest wind farms are capable of generating more than 1000 GW (gigawatts) of electricity, enough to supply 1,000,000 average households!

Gas turbines and diesel generators. The least economical method of producing electricity uses fossil fuels, such as oil or natural gas, in engines that are the prime movers of electrical generators. Electrical utilities may run large gas-turbine generators (basically jet engines) to supplement their normal generators during periods of peak electrical loads. Smaller utilities and industrial plants often drive supplemental generators using large diesel engines. At several dollars per gallon for diesel fuel, this is an expensive method of generating electricity.

11-26 Wind farms are becoming more common as a means of generating electricity from the energy in wind.

11-27 Engine-driven generators, such as this emergency diesel generator, are intended for short-term use. Using them is the most costly way to generate electricity.

11.10 Direct-Current Generators

Today nearly all the electricity we use comes from AC generators located at commercial power plants. However, there are many limited but important uses of direct current (DC; Chapter 10) that require a source of electricity larger than what can be supplied by the small batteries with which you may be familiar. These uses include emergency power for important or sensitive facilities, like nuclear power plants and nuclear submarines, and backup power for small installations where an AC generator just isn't practical.

Wind power is one growing area that uses DC generators almost exclusively. Wind turbine DC generators are necessary because in any wind farm with two or more turbines, each turbine turns at different speeds as wind speed varies across the facility. It would be impossible to get hundreds of AC wind turbines synchronized so that they operate at the same frequency.

Since all mechanical electrical generators use the same basic principle—the rotational motion of coils of wire and magnetic fields—you might expect all generators to produce alternating current. So how do we obtain direct current from a generator?

Most DC generators reverse the locations of the magnets and coils from where they are in an AC generator. The magnets are mounted in the stator, alternating north and south poles. The conductor coils are mounted on the rotor. Since the rotor spins, a means of drawing off the current induced in the rotor coils is provided by a split-ring *commutator* (see Figure 11-28, top). The end of each wire in a generator coil is attached to one segment of the commutator, which is insulated from the other segments. Electrical *brushes* connected to the circuit slip across the face of the commutator as it rotates underneath them.

As each coil passes a stationary magnet, a current is induced in the coil. Current flows through the coil in one direction and out the commutator to the circuit. Just as the coils begin passing under the

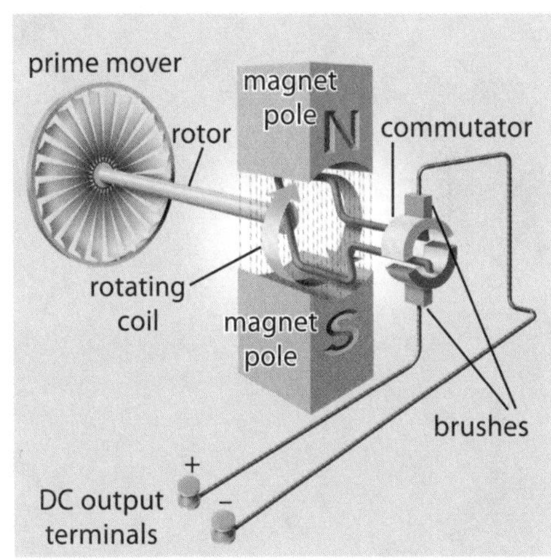

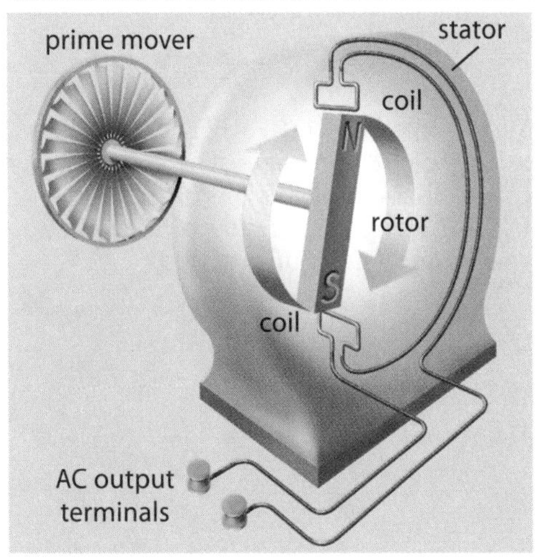

11-28 A simplified DC generator (top) compared to an AC generator (bottom)

opposite magnetic poles, the gap of the rotating commutator slips under the brushes, briefly cutting off current to the circuit. As the rotor continues to turn, the current in the coil reverses because the coil begins to pass under the opposite magnetic pole. *But because the connected commutator segment is now under the opposite brush, the current in the circuit flows in the same direction as before.* Thus, direct current flows though the circuit.

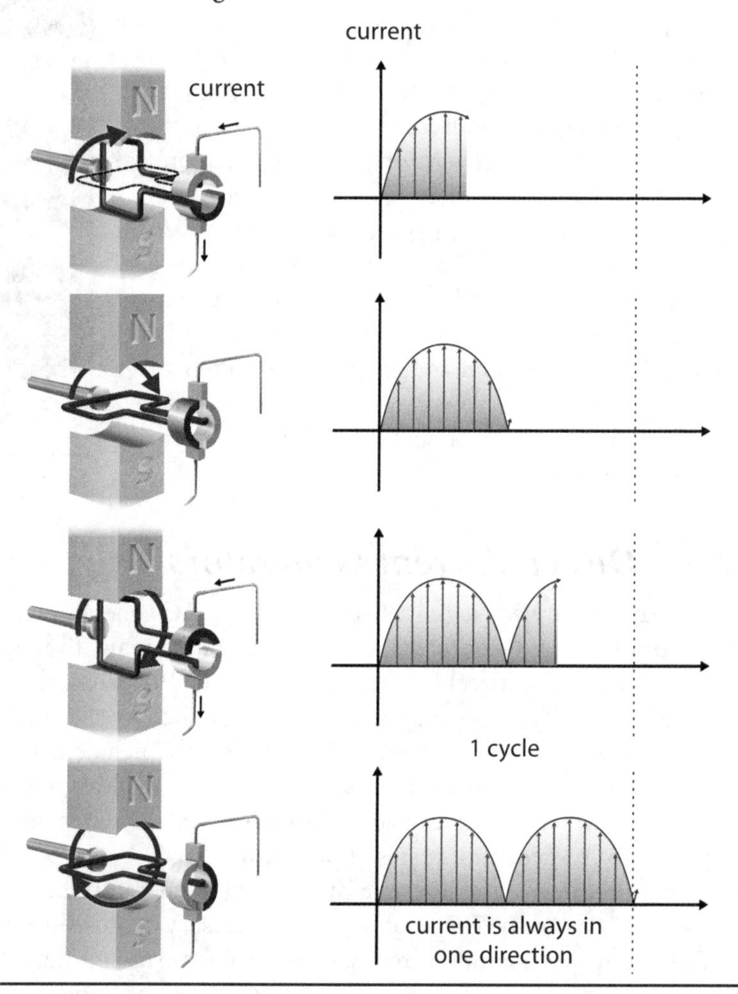

11-29 The generation of direct current

The first widely used generators produced direct current. The great experimenter Michael Faraday discovered the electromagnetic principles that led to the construction of the first DC generator in 1832. Thomas Edison tried to introduce DC generators as a power source for the electric lamps he invented in 1879. However, these generators were limited in their effectiveness because a high voltage was needed to supply current to customers at any distance from the generator. But a high voltage was dangerous and required thickly insulated wires. Late in the 1800s Nikola Tesla invented an AC generator that eliminated the restrictions of a DC distribution system. The first commercial AC generating plant and citywide electrical system went into operation in 1893 in Redlands, California.

Michael Faraday (1791–1867) was an English physicist and chemist who has been called the greatest experimental scientist. He was also a devout Christian and was active in his church.

11.11 Motors

Generators convert mechanical energy to electrical energy using rotating magnets. *Motors*, on the other hand, are machines that convert electrical energy into rotary mechanical motion. In 1831 Faraday invented a simple DC motor, but his pioneering device was unlike the motors we're familiar with today. That invention would take nearly 50 more years of work by many men on both sides of the Atlantic Ocean.

How a motor works differs depending on the kind of current supply. Both AC and DC motors are constructed like their corresponding generators. Discussion of motor theory is beyond the scope of this textbook; however, we can make these generalizations: the speed of AC motors depends on the frequency of the alternating current, and the speed of DC motors depends on the voltage of the direct current supplied to the motor. The structure of the motor, such as how many pairs of magnetic poles are installed, can also affect motor speed and power.

11-30 An AC motor used in industry

11-31 A large DC motor

11B Section Review

1. Describe the right-hand rule of magnetism in a conductor. Does a straight piece of current-carrying wire produce magnetic poles? Explain.
2. Which type of magnetic material is used as the core of an electromagnet? How does it make an electromagnet more efficient?
3. What must take place for electromagnetic induction to occur?
4. What two factors determine the direction of the current induced in the coil of an AC generator?
5. What is the standard AC supply frequency in the United States? How many times does a fluorescent bulb flicker per second?
6. Name three ways to produce steam to power a steam-turbine generator.
7. What are two major structural differences between AC and DC generators as described in this textbook? How do these differences affect the way the two machines work?
8. (True or False) The construction of AC generators and AC motors is similar, and the construction of DC generators and DC motors is similar.
9. What determines whether a machine acts as a generator or as a motor?

11C Objectives

After completing this section, you will be able to

- ✓ explain how transformers change voltage.
- ✓ give some familiar and unusual uses for electromagnets.
- ✓ discuss the properties and uses of superconducting magnets.

11C Using Electromagnetism

11.12 Introduction

Electromagnetism is vital to modern life. Hundreds of appliances use motors, including electric drills, laptop computers, and washing machines. Telephones, radios, computer monitors, and digital data-storage media all operate using magnetic and electrical fields. Let's examine some more uses for magnetism as it relates to electricity.

11.13 Transformers

A steady direct current flowing through a wire produces a constant magnetic field, but it can't produce any electricity in nearby wires by electromagnetic induction. An alternating current, on the other hand, is constantly changing. It produces a magnetic field that constantly changes strength as well. The changing magnetic field of an AC conductor can generate current in other nearby conductors through electromagnetic induction.

Suppose two coils of insulated wire are placed one on top of the other. Alternating current is supplied to one coil. The alternating magnetic field surrounding this coil induces an alternating current in the other coil of wire. The induction process is inefficient if the wire coils have no ferromagnetic cores. But if the insulated wires are wrapped around opposite sides of a rectangular iron frame, as in Figure 11-32, the strong ferromagnetic effect of the iron core greatly increases the efficiency of the induction between coils. When an alternating current is applied to one coil wrapped around the core, the alternating magnetic field set up in the frame core induces a current in the other coil of wire. This arrangement of coils on a common core is the basis for a **transformer**.

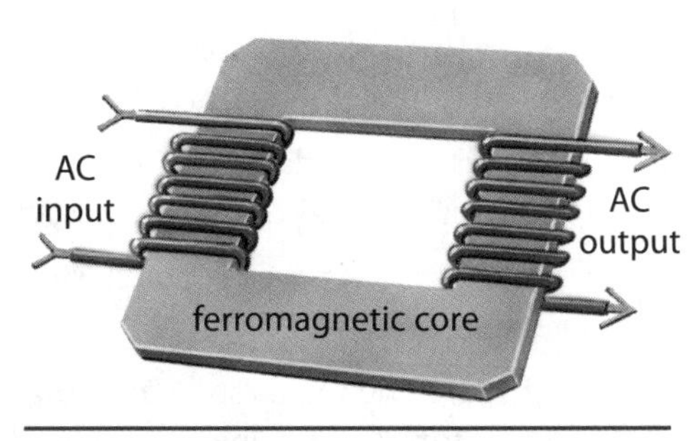

11-32 The structure of a transformer

What would happen if the wires were uninsulated? What would that electrical condition be called?

In Figure 11-33 (top), the left-hand, or *input*, coil of the transformer has more wraps than the right-hand, or *output*, coil. The alternating current in the input coil produces a magnetic field whose strength is related to the number of wraps of wire—the more wraps, the greater the field strength. The alternating field in the iron core induces a current in the output coil. Because there are fewer wraps in the output coil, a given change in magnetic field strength induces a smaller output current than the input current.

According to Ohm's law, current through a conductor is proportional to voltage for a given resistance. Thus, the smaller amount of current in the output coil is the result of a smaller induced voltage compared to the input voltage in the left-hand coil. This is the function of a transformer—to increase or decrease AC voltage as necessary. A transformer that decreases output voltage is called a *step-down transformer*. Note that even if there is no output current (an open

The strength of the induced field in the core of a transformer is directly proportional to the rate of change of the input voltage (and current) and the number of wraps of the input coil.

Similarly, the induced voltage at the output of a transformer is directly proportional to the rate of change of the magnetic field and the number of wraps of the output coil.

circuit), the induced output voltage in a step-down transformer is still smaller than the input voltage.

The bottom transformer in Figure 11-33 works just the opposite. The number of wraps in the input coil is smaller than the number in the output coil. The input alternating current induces a magnetic field in the transformer core. The alternating field induces a current and voltage in the output coil much larger than at the input coil because the output coil has many more wraps. A transformer that increases output voltage is called a *step-up transformer.*

A simple mathematical relationship allows us to compute the output voltage of a transformer if we know the input voltage and the number of wraps in the input and output coils:

$$\frac{V_{out}}{V_{in}} = \frac{N_{out}}{N_{in}},$$

where

- V_{out} is the voltage of the output circuit,
- V_{in} is the voltage of the input circuit,
- N_{out} is the number of wraps in the output coil, and
- N_{in} is the number of wraps in the input coil.

If you know any three of these values, you can easily calculate the other.

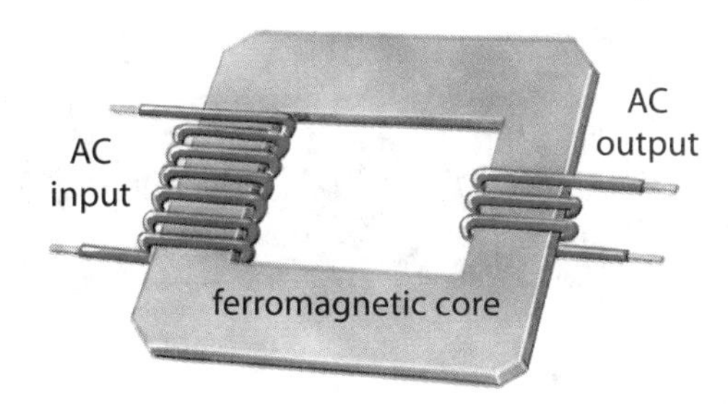

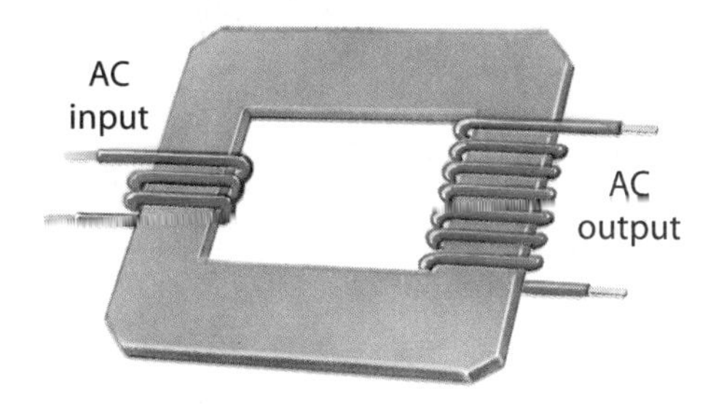

11-33 A step-down transformer (top); a step-up transformer (bottom)

EXAMPLE PROBLEM 11-1

Electrical Transformers

In most homes built after 1999, every bedroom is required to have a smoke detector wired directly to the house's electrical circuits with a battery backup. Some of these detectors operate on an AC voltage well below the 120 V standard in the United States. A small step-down transformer wired into the house circuit is used as the power source. Assume the input coil of the transformer contains 480 wraps and the output coil has 100 wraps. What is the output voltage supplied to the smoke detector?

Known: input voltage (V_{in}) = 120 V
output coil wraps (N_{out}) = 100
input coil wraps (N_{in}) = 480

Unknown: output voltage (V_{out})

Required formula: $\frac{V_{out}}{V_{in}} = \frac{N_{out}}{N_{in}}$

Substitution: $\frac{V_{out}}{120\text{ V}} = \frac{100}{480}$

Solution: $V_{out} = 120\text{ V} \times \frac{100}{480} = \frac{12{,}000\text{ V}}{480}$

$V_{out} = 25\text{ V}$

Multiply both sides by 120 V.

The fact that the output voltage is lower than the input voltage confirms that the transformer is a step-down type.

11-34 A small transformer used to supply residential smoke detectors

11-35 An electrical utility substation

Power companies use step-up and step-down transformers to conserve energy. The wires that carry electricity have low resistance, but they may be many miles long, resulting in large total resistances. Energy is wasted in running current through these long wires. Recall that thermal power lost in a conductor is related to the square of the current and the resistance ($P = I^2R$). If the current can be reduced in transmission wires, less power is lost in the process. Power in DC circuits is computed by $P = VI$. Power in AC circuits can be calculated in a slightly different but related way. If the voltage in a wire can be increased significantly, then the current is reduced proportionately. Lower current greatly lowers the power lost to electrical resistance in the wires. This is where step-up transformers come in. Electric companies use extremely high voltages of 110,000 V or higher to transmit electricity over long distances. Step-down transformers are used at the destination's electrical substations before the electricity arrives at your residence.

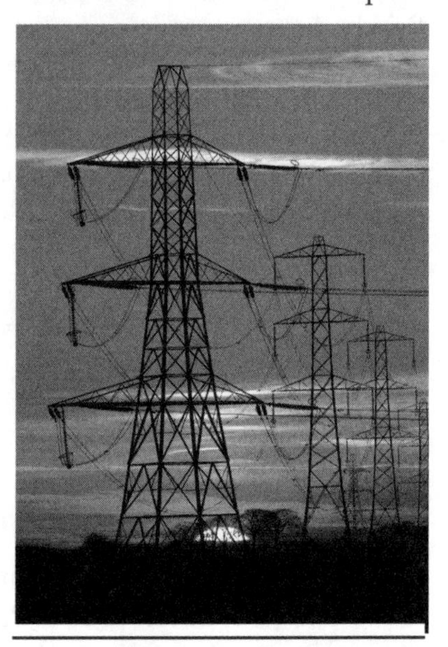

11-36 High-voltage transmission towers carry electrical energy over great distances. The wires are kept far apart by the towers so that no arcing occurs.

11.14 Applications of Electromagnets

Since Sturgeon developed the first practical electromagnet in 1823, electromagnets have been used for many applications in which an easily controlled, variable force is needed.

Electromagnets have proved especially useful for lifting heavy sections of iron and steel. If you visit a junkyard, you might see a crane using a large flat disk to lift pieces of cars and trucks. This disk is part of the ferromagnetic core of an electromagnet. When direct current is allowed to flow through the electromagnet, the disk produces a strong magnetic force that can pick up heavy pieces of iron and other ferromagnetic metals. When the metal has been moved to the proper position, it can be released by turning off the current in the electromagnet.

11-37 An electromagnet can lift a large amount of metal when the current is flowing.

Electromagnets are also useful for sorting commonly discarded metals, particularly aluminum and iron. If a mixture of iron and aluminum materials is carried on a moving belt through a magnetic field, the iron pieces will be strongly deflected to one side, while the aluminum materials will pass on unaffected. Aluminum refineries use such electromagnetic sorters to keep iron products from contaminating aluminum reclaimed from recycled metal.

Particle accelerators, which are large machines used in nuclear research, use electromagnets to focus beams of charged particles and to bend their paths as necessary to keep them from colliding with the walls of the accelerator. The electromagnets herd the particles back into the path, where they can be accelerated.

FACETS OF SCIENCE

Magnetic Digital Media

You probably use a computer every day without understanding just how it stores and retrieves your valuable information. Yet storage is one of the most important parts of a computer system. Early computers stored data in a variety of ways, but the most successful ones relied on magnetism. The majority of computers built between 1955 and 1975 used a type of main memory called *magnetic core memory*, which stored data in tiny ceramic rings (see below). Semiconductor memories replaced magnetic core in the mid-1970s, but other forms of magnetic storage are still in use today. The *hard drive* is one such example.

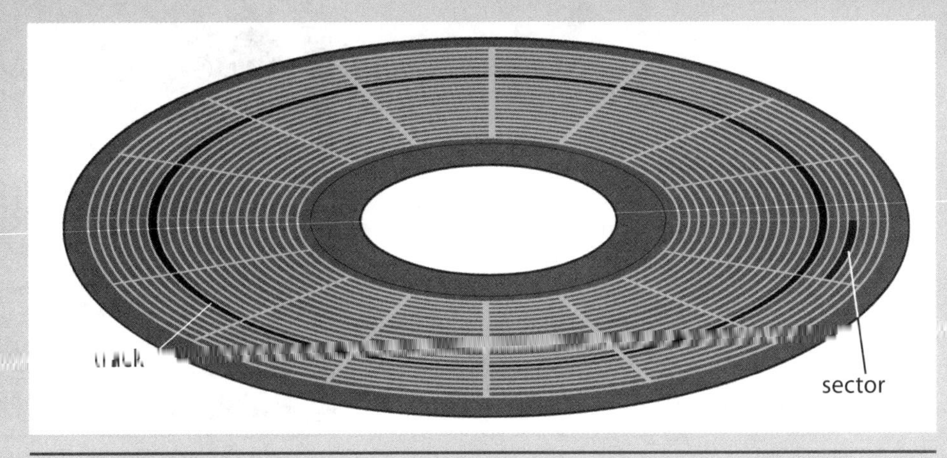

Arrangement of data on a hard disk

An early computer memory core

Components of a modern hard disk drive

A personal computer hard drive is enclosed in a flat metal box that can easily fit in your hand. It contains the motors, mechanical parts, and electronics to control the drive and one or more disks mounted to a single shaft. The disk itself is a thin, highly polished *platter* of glass. A magnetic medium of extremely small particles coats both surfaces. These particles can be easily magnetized. The magnetic medium is an excellent material for rapidly reading and writing data and can retain its contents for several decades.

All digital information is encoded using the *binary system*, a pattern of 1s and 0s called *bits*. To represent a bit on the magnetic surface, a tiny area is magnetized one way for a 1 and the opposite way for a 0. A group of eight bits forms a *byte*. The computer represents information as groups of bytes. For example, a digital photograph consists of individual pixels (colored dots) that require one to three, or even more, bytes each.

Modern hard drives can hold up to several terabytes (TB) of data (1 TB is approximately 1 trillion bytes) spread across several platters. Each platter is provided with a pair of tiny electromagnetic devices called *read/write heads*, one for each surface of the disk. These heads are mounted on arms that pivot so that the head can move across the disk surface. All the arms are assembled in a single unit moved by an electromagnetic controller. Each read/write head skims over the spinning surface of the disk 10–20 nm away (much smaller than the diameter of a bacterium). The magnetic pickup in the head magnetizes or reads the digital bits at an incredibly fast pace. Disks rotate at up to 15,000 revolutions per minute. Data can be recorded or read at hundreds of megabytes (millions of bytes) per second.

Information is stored on disks in thousands of concentric tracks on each surface of the platters. To organize the data so that it can be retrieved quickly, the disk surfaces are subdivided into pie-shaped sectors. Within these tracks and sectors, the data is organized into files. The computer's operating system maintains a map that keeps track of the location of all the digital data.

Hard drives continue to be the most popular form of mass storage, especially for giant *servers* that hold information for thousands of people. But other types of storage are gaining popularity, particularly on personal computers.

Solid-state disks (SSDs) use semiconductor memories to mimic the behavior of traditional hard drives. They have the advantage that they have no moving parts to malfunction. Unlike hard drives, they are also *not* susceptible to magnetic fields. SSDs are becoming popular in notebook computers and other portable electronic devices. And researchers are developing other forms of storage that may one day make the hard drive look as antiquated as magnetic core memory does today.

11-38 These superconducting magnets are cooled to extremely low temperatures to provide the focusing energy needed in this particle accelerator. A worker is circled to indicate the size of this facility.

Some particle accelerators contain special electromagnets called *superconducting magnets*. Superconductors are certain materials that lose all electrical resistance at extremely low temperatures. An electrical current in a superconductor loop flows almost indefinitely with no additional input of electrical energy. Superconducting magnets are wrapped in specially coated wires and have cores of niobium and titanium or niobium and tin. They are cooled by liquid helium to maintain superconductivity. These materials can sustain extremely high magnetic fields at temperatures below 20 K (–253 °C).

More commonly, you use an electromagnet every time you ring an electric doorbell. A doorbell circuit is basically a switch (the doorbell button) connected in a series with an electromagnet. When you push the button, the switch closes and electricity flows through the wire coil of the electromagnet. The resulting magnetic field around the electromagnet pulls a movable metal core, causing it to strike a bell. In two-tone doorbells the core compresses a spring. When the button is released, the spring shoves the core in the other direction and it strikes the other bell.

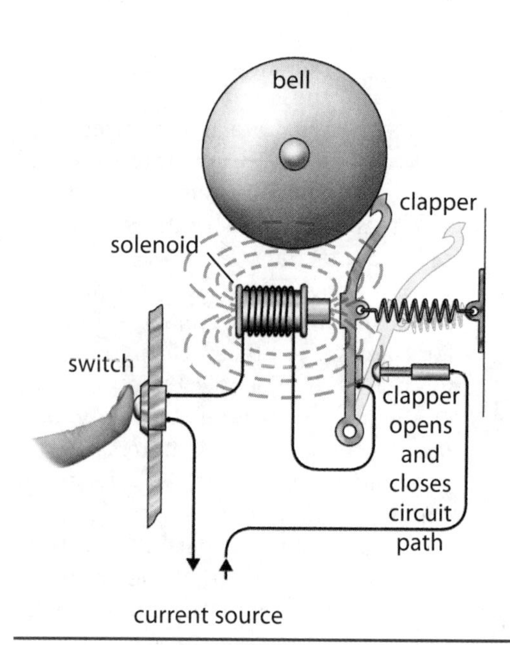

11-39 A schematic of a buzzer circuit. Note that the clapper acts as a switch in the electromagnet's circuit.

In buzzers the movable core or clapper forms a switch in the electromagnet's circuit. When the button is depressed, the energized magnet pulls the clapper. When the clapper moves, it opens the circuit (see Figure 11-39), but its inertia allows it to strike the bell. A spring pulls the clapper back to its starting position, shutting the switch and activating the electromagnet again, starting the cycle over. As long as the button is depressed, the device continues to buzz.

Electromagnets are used in special electrical switches called *relays* to turn electrical circuits on and off. When electrical current passes through a relay, the magnetic force produced by the electromagnet pulls a switch closed, turning the circuit on. Electromagnets are also important parts of remote door locks, television and computer monitors, and the acoustic pickups for electric musical instruments.

11.15 Using Electromagnets to Solve Problems

To solve the commuter problem in large cities, a number of countries are experimenting with a high-speed train system called *maglev*, a shortened name for *magnetic levitation*. These trains differ from the old rail-and-wheel systems because they have no engine and hover 1–10 cm over the guide rails using electromagnets. Since the train is not in direct contact with the rails, friction is greatly reduced, allowing the train to approach speeds of 500 km/h (310 mi/h)!

There are several maglev designs being developed around the world, primarily in Germany, Japan, and China. China currently has the only operational maglev system providing service between cities. Maglev uses electromagnets installed in a guideway, a concrete chute or rail on which the maglev train runs (see Figure 11-41). Depending on the design, either electromagnetic repulsion or aerodynamic lift at high speed lifts the maglev car off the ground. Other electromagnets switch on and off to pull and repel the vehicle along the guideway. These strong magnetic fields (some systems use superconducting magnets) interact with large magnets on the undercarriage of the train.

The main hindrance to the widespread construction of maglev trains is expense. The technology is still maturing. Other forms of transportation are still more economical overall, but time and more experimentation on full-sized systems are needed to determine if maglev will become the best commuting system. Imagine yourself as one of the engineers who develops the breakthrough to make this exciting technology practical and cost-effective!

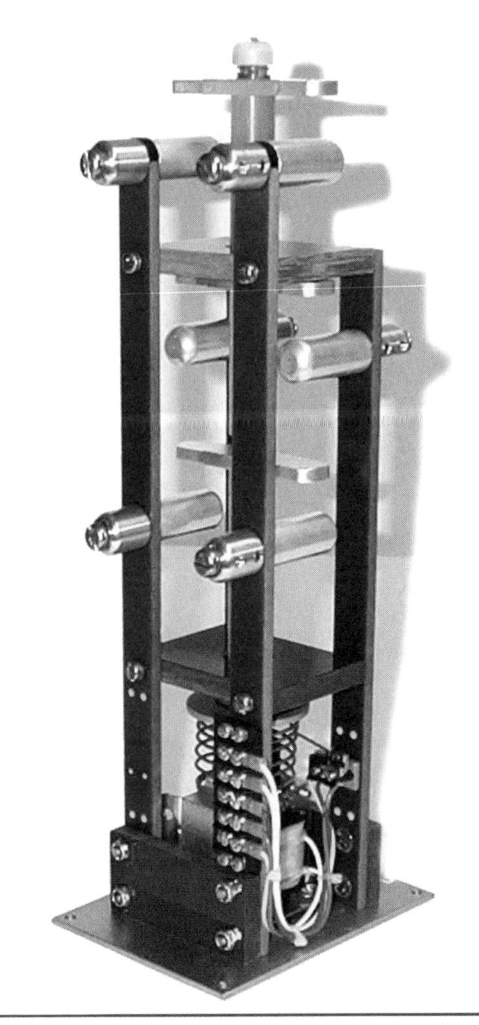

11-40 A large electromagnet (bottom) opens and closes this high voltage electrical switch, called a *relay*.

11-41 An operational maglev passenger train in Shanghai, China

11C Section Review

1. Describe the basic construction of a transformer. What transfers electrical energy between the two coils?
2. In which coil of a transformer does the current determine the strength of the magnetic field in the core?
3. When are superconducting electromagnets required? Discuss several applications mentioned in the text. What environmental condition is necessary for these kinds of magnets to work?
4. List at least four everyday applications of electromagnets.
5. (True or False) The strength of the magnetic field created by an electromagnet depends on the type of material in the core, the number of wraps of wire, and the amount of current in the wire.
6. DS ✪ On September 22, 2006, 23 people died and 10 were injured when an experimental maglev train in Germany moving at 200 km/h collided with a repair car accidentally operating in the guideway. How should a Christian respond to such tragedies?
7. ✪ An electrical substation for your neighborhood includes a transformer that supplies your street. Its input coil has 2450 wraps and its output coil has 490 wraps. If the input voltage is 115 kV, what is the voltage of the electrical supply to your street?

FACETS OF SCIENCE

Samuel F. B. Morse—American Inventor

"What hath God wrought!" These were the first words ever sent by intercity telegraph. Can you name the source of the quotation—Numbers 23:23—or tell how it was selected? This fascinating story begins with the inventor Samuel F. B. Morse. In his mind was born the first means of communication that could transmit messages instantly from city to city.

Samuel Morse was born in 1791 in Charlestown, Massachusetts. As the son of the town pastor, he was reared in a Christian home, but he did not accept Christ as his Savior as a child.

Though he enjoyed science, he was more interested in studying art. After he graduated from Yale, Morse traveled to England to study with Benjamin West, the most respected artist of that time. The training he received during his four years abroad catapulted him into the ranks of America's foremost artists. While in England he met the English politician and abolitionist William Wilberforce and his group of evangelical Christians known as the Clapham Sect.

On Morse's second trip to Europe, the idea of the telegraph began to take shape in his mind. During the return voyage, Morse drew sketches of his first crude electric telegraph. It was based on the idea that opening and closing an electrical circuit could control an electromagnet located at a distance. The brief movements of the electromagnet core punched impressions on a moving paper tape. Providentially, Morse received an appointment as professor of sculpture and painting at the University of the City of New York. This position provided him with space to conduct experiments and with people to help him. After many months of labor, he first demonstrated his invention to his colleagues at the university. Working with another inventor, Alfred Vail, Morse developed a code consisting of dots and dashes to represent numbers. At first, codes of a series of numbers were used to stand for words, but soon after, these men realized that codes for individual letters that could be memorized would be much simpler and faster. We know this system today as *Morse code*.

One of the earliest public demonstrations of Morse's telegraph took place in Morristown, New Jersey. It elicited great public delight. Morse's demonstration in 1838 at the Franklin Institute in Philadelphia brought him the acclaim of the scientific community.

The next demonstration took place in Washington for, among others, President Martin Van Buren and several members of his cabinet. Sadly, the country was suffering a severe depression, and no governmental funds were available to build a trial intercity telegraph.

Morse, however, kept persevering. Late in 1842 he set up a demonstration line between the rooms of the House Committee on Commerce and the Senate Committee on Naval Affairs. He remained faithfully on duty, sending messages for the legislators and answering questions. By the end of the year, the Committee on Commerce had submitted a favorable report on Morse's telegraph and recommended that $30,000 be appropriated to set up a trial intercity telegraph system.

The committee's resolution passed fairly easily in the House. But the bill reached the Senate on the last day of the session, and when Morse retired for the day, he had no hope that the bill would pass.

The next morning Morse was visited by Annie Ellsworth, the daughter of the Commissioner of Patents, who brought good news—his bill had passed around midnight and had been quickly signed by President Tyler. In appreciation for her message, Morse made Miss Ellsworth a promise—that after the lines were finally completed from Washington to Baltimore, she could choose the text for the first message.

On 24 May 1844, the famous four words from the book of Numbers were transmitted from Baltimore to Washington and back again. Morse was delighted that these words of Scripture had been chosen for the first intercity telegraph demonstration. In a letter to his brother he wrote, "It is in my thoughts day and night, 'What hath God wrought!' It is His work, and He alone could have carried me thus far through all my trials and enabled me to triumph over the obstacles, physical and moral, which opposed me."

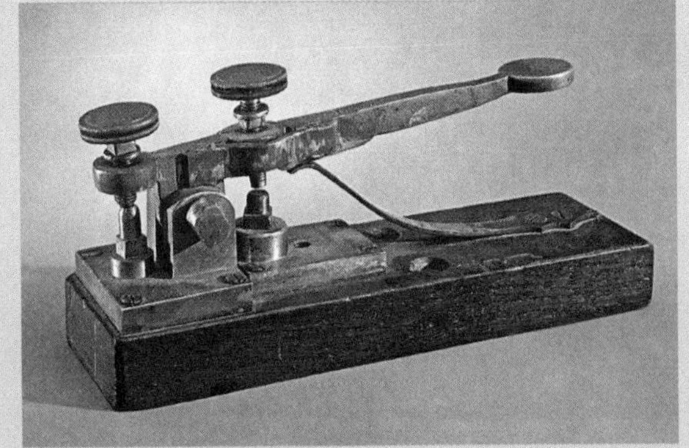

A Morse code key similar to the one originally made by Alfred Vail for Samuel Morse

Scientifically Speaking

magnet	229
magnetic pole	229
magnetic field	230
law of magnetism	231
ferromagnetism	231
domain (magnetic)	231
paramagnetism	232
ferrimagnetism	232
diamagnetism	232
Curie temperature	233
geographic North Pole	233
magnetic North Pole	233
geomagnetic field	234
right-hand rule of magnetism	235
electromagnetic induction	236
alternating current (AC)	237
transformer	242

Chapter Review

1. Who was one of the first Europeans to accurately describe magnetism? What was his most significant observation?
2. In what ways are electrical and magnetic fields similar? How do they differ?
3. What is the difference between a magnetic pole and a magnetic dipole?
4. What property of ferromagnetic materials sets them apart from other kinds of materials? How does this property affect an external magnetic field?
5. Suppose you pick up a steel bolt with a permanent bar magnet and then heat the magnet until it glows. What do you expect will happen? Explain your answer.
6. In New York City the magnetic declination is about 14° west. Does the magnetic North Pole lie in a clockwise or counterclockwise direction from the geographic North Pole as observed from New York City?
7. Examine the diagram of a coil of wire. The arrow indicates the direction of conventional current in the wire. Using the right-hand rule of magnetism in conductors, in which direction is the north pole of the magnetic field created by the coil?

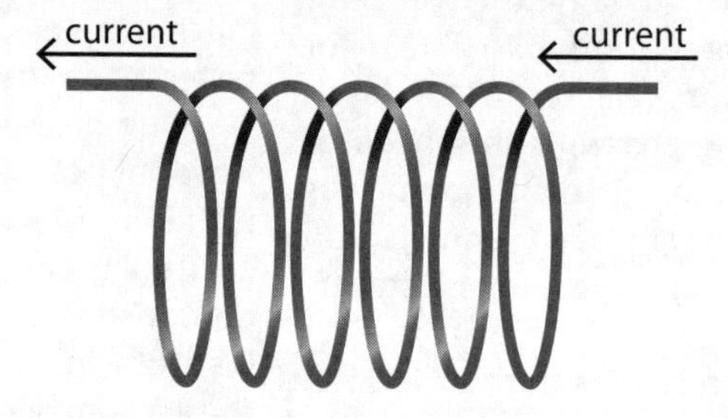

8. A solenoid containing a ferromagnetic core is the basis for what device?
9. What must occur between a magnetic field and an electrical conductor before magnetically induced current can flow?
10. What causes the current to alternate directions in the coils of an AC generator?
11. Which of the prime movers for generators can draw on seemingly inexhaustible sources of energy?
12. What is the purpose of the commutator in a DC generator?
13. What determines the speed of an AC motor? a DC motor?
14. What electrical quantity is a standard transformer designed to increase or decrease?
15. Why would electromagnetic sorting be effective in a scrap metal recycling plant? What metals couldn't be sorted this way?
16. ✪ Is the current that directly energizes the electromagnet in a doorbell alternating or direct? How can you tell?

True or False

17. The earliest archaeological evidence for magnets was found in the ruins of ancient Greece.
18. Ferrimagnetic materials contain two kinds of magnetic particles or substances that work against each other to produce a net magnetic field.
19. Except for certain locations on the earth's surface, the direction to magnetic north rarely coincides with the direction to true geographic north.
20. The amount of magnetically induced current in a coiled conductor is related to the strength of the field, the rate of change of the field's strength, and the number of loops in the coil.
21. Gas turbines (jet engines) are the principal generator prime movers for many electrical utilities.
22. Electricity flows in only one direction *inside* a DC generator.
23. The coiled wires in a transformer are uninsulated.
24. Step-up transformers are required in the substations of cities distant from electrical plants to provide the voltage customers need.

✪ 25. A power dam supplies customers more than 500 mi away. It must use step-up transformers to transmit electricity this distance. The water-turbine generators produce electricity at 440 V. The transformer input coil contains 4800 wraps and the output coil contains 230,000 wraps. What is the output voltage of the step-up transformer in kV?

✪ 26. Research and present a report on the behavior of magnetotactic bacteria.

✪ 27. Research and present a report on the evidence for the use of magnetic materials by the Olmec civilization in Central America.

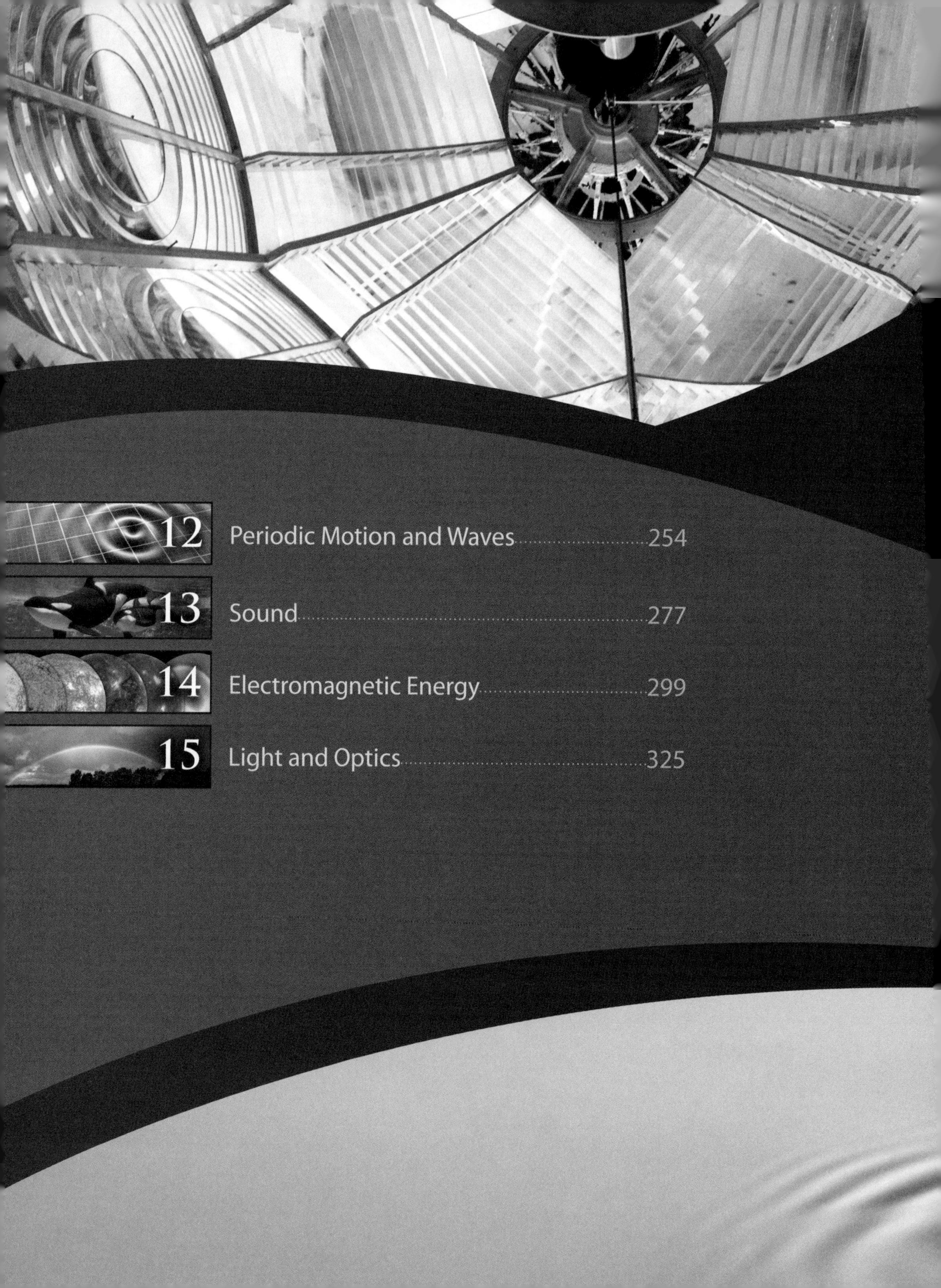

PERIODIC PHENOMENA

Energy is essential for life.

If energy were not conserved in certain processes, every time it was used it would be lost forever. The universe would quickly cool into a desolate place. No light, no warmth, no life.

But energy *is* conserved. Potential energy is converted to kinetic energy and back again under the influence of forces that conserve energy. Electric and magnetic fields oscillate repetitively as they interact to support each other. The ordinance of periodic motion through vibrations and oscillations is another example of how God sustains His creation.

In Unit 4 you will learn how matter vibrates and oscillates. You will discover that mechanical energy moves through matter as waves—mainly as sound and other kinds of mechanical waves. Electromagnetic energy also moves in a wavelike manner but without the need for matter. The properties of electromagnetic energy are "illuminated" as you examine the characteristics of visible light.

UNIT 4

PERIODIC MOTION AND WAVES

CHAPTER 12

DOMINION SCIENCE PROBLEM

The Waves of Time

Have you ever observed your watch running "fast" or "slow"? Clocks keep time by counting the rhythmic motions of an object such as a pendulum, a spring-powered balance wheel, or a vibrating quartz crystal. But a pendulum's motion varies slightly with temperature or humidity. Clock mechanisms wear out. Electronic circuits change with voltage and age. Even the earth's daily and annual motions, from which the units of time were originally defined, are not constant. Because many technologies, such as the Internet, cell phones, and GPS, rely on accurate timekeeping, a universal time standard is vital. So how do we obtain accurate time around the world?

12-1 Historically, scientists have found that a lack of precision in time measurements was one of their greatest limitations.

12A Periodic Motion

12.1 Introduction

Earlier chapters explain that matter in motion has kinetic energy and describe the way that that energy is transferred. When energy is transferred between systems, work is done by one system on another. Work may be accomplished in two ways. A contact force acts directly, like a baseball bat's hitting a ball. A field force acts without touch, like a magnet's attracting a paper clip. Work can also be done on individual particles through collisions or in response to field forces.

Linear, one-time motion is not the only kind of motion that does work. Work may also be done by *repetitive motion*, motion that occurs again and again. Vibrations and oscillations are two kinds of motion that can be repetitive or random. *Vibrations* are very small and rapid repetitive motions, and *oscillations* can be larger or slower movements. There is no distinct boundary between the two. These kinds of motions are observable at every size scale imaginable. Atoms and molecules randomly vibrate in response to thermal energy. Mosquito wings beat in rapid oscillations. You've seen the oscillations of a playground swing. Astronomers have observed ripples in stellar dust light-years in width. This chapter is particularly concerned with motion that repeats at a constant rate, called **periodic motion**.

Some kinds of periodic motion require matter while other kinds do not. In matter, periodic motion occurs when its energy alternates repetitively between mechanical potential energy and kinetic energy. One example of this periodic energy exchange is a clock pendulum. The properties of periodic motion apply to many different areas of physics, especially sound and light, which are the topics of Chapters 13–15. Thus, it is important to understand how systems experience periodic motion and how that motion can transfer energy from one place to another.

12A Objectives

After completing this section, you will be able to

- ✓ differentiate between kinds of repetitive motion.
- ✓ define periodic motion.
- ✓ explain why knowledge of periodic motion is as important to the musician as it is to the physicist.
- ✓ explain the relationship between rest position and restoring force.
- ✓ compare period and frequency.
- ✓ describe conditions that affect the period of an oscillating system.

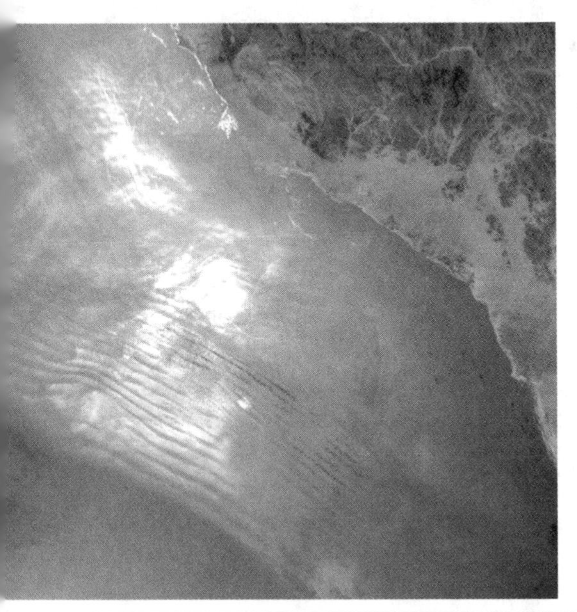

12-2 Periodic motion can be found on nearly any scale of matter. Ocean density waves viewed from space (left); a clock pendulum (middle); and ripples in stellar dust (right) are just three examples.

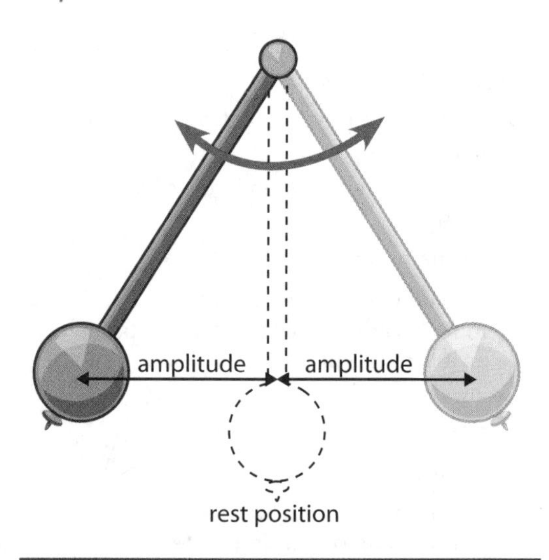

12-3 A pendulum is an example of a periodically oscillating system.

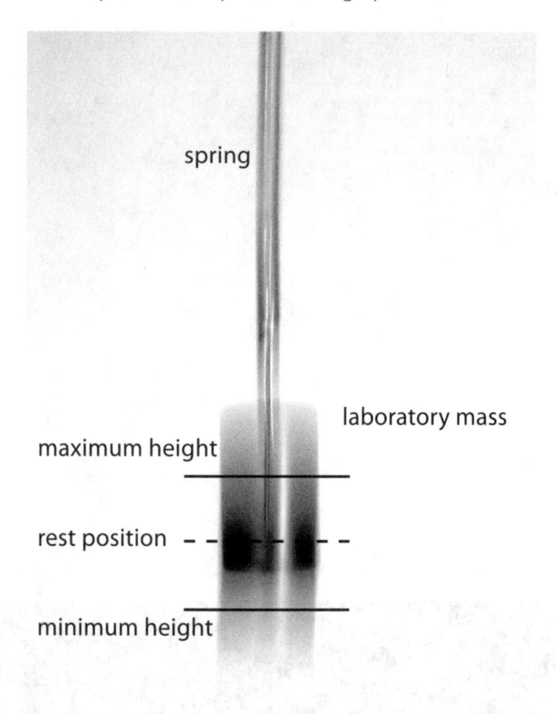

12-4 A laboratory mass oscillates on a spring that acts against gravity.

12.2 Describing Periodic Motion

Position

What does periodic motion look like? An oscillating system such as a clock pendulum swings back and forth repeatedly. The pendulum is at its *rest position* when it hangs straight down. If you pull the pendulum to the side and then release it, it swings through its rest position until the pendulum's bob (its weight) is at an equal height on the opposite side. The pendulum stops for an instant, then swings the other way. This motion continues repeatedly and indefinitely if no outside forces stop it.

Similarly, if you pull a laboratory mass hanging from a spring down from its rest position and release it, the weight and spring bounce up. The elastic potential energy (EPE) in the spring is converted to kinetic energy. The mass then rises to its maximum height above its rest position and stops momentarily. The kinetic energy of its motion is converted into gravitational potential energy (GPE). The mass then drops down to a point below the rest position. Gravitational potential energy is converted back into kinetic energy and EPE. These oscillations repeat around the rest position of the mass and spring.

You can see faster vibrations if you hold a plastic ruler over the edge of your desk and thrum it so that it vibrates. It vibrates so quickly that you can't follow its motion with your eye, but you can see that it bends above and below its rest position.

The maximum distance that a system moves from the rest position during each oscillation is called the **amplitude** of the oscillation. The system swings to its amplitude distance at opposite sides of the rest position.

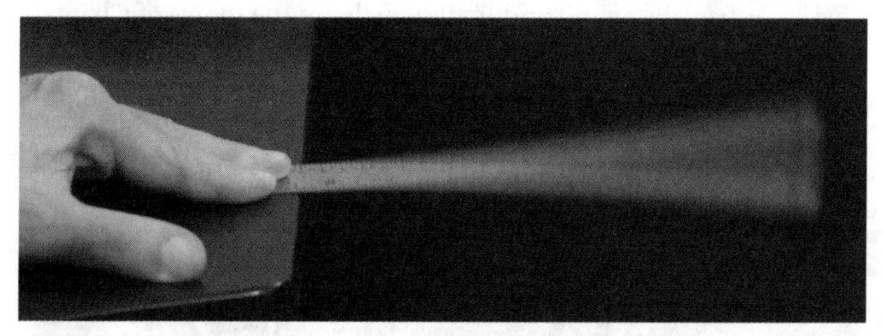

12-5 Rapidly vibrating systems move faster than the eye can follow.

Period

The most important property of periodic motion is its fixed rhythm. Each repeated motion, or *cycle*, takes the same amount of time. The word cycle comes from the Greek word for circle. If you repeatedly trace a circle with your finger, it travels the same path over and over. Cyclic motions include the spinning of a wheel, the orbital motion of a planet, and the swing of a pendulum. In fact, all periodic motions can be modeled by similar equations based on circular motion. Though constant circular motion is a general type of periodic motion, this chapter focuses specifically on oscillations and vibrations.

The time it takes to complete one cycle of a system is its **period** (*T*). Physicists measure the period of vibrations and oscillations in seconds (s), the base SI unit for time. For example, if the back-and-forth motion of a playground swing takes two seconds, we express the swing's period as $T = 2$ s.

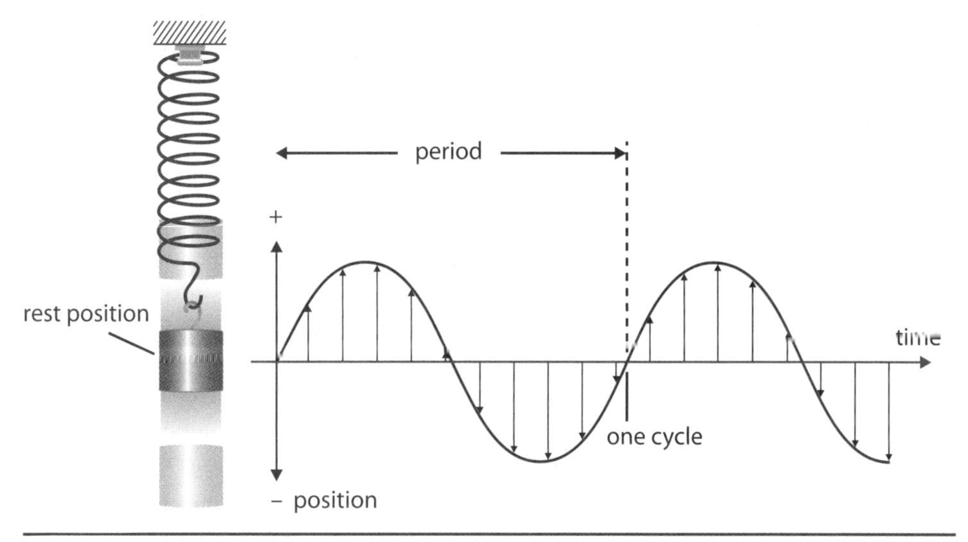

12-6 Position-time graph of a system showing periodic motion. The period is the time needed to complete one cycle of motion.

Frequency

The rate at which cycles repeat is called the **frequency (*f*)** of periodic motion. Frequency is measured in cycles per second or hertz (Hz). For example, alternating current (AC) in the United States changes direction twice each cycle, 60 times per second, so its frequency is 60 Hz. Frequency can also be expressed in units of "s^{-1}." This is the formal mathematical way of writing "$\frac{\text{cycles}}{\text{s}}$," but the word *cycles* is omitted since it is understood when measuring frequency. Therefore,

$$\frac{\text{cycles}}{\text{s}} = \frac{1}{\text{s}} = \text{s}^{-1} = \text{Hz}.$$

Frequency is the inverse of period, and vice versa. If the period of a water wave passing by a pier is 2 s, then its frequency can be figured as follows:

$$f = \frac{1}{T} = \frac{1}{2\text{ s}}$$

$$f = \frac{1}{2} \times \frac{1}{\text{s}} = 0.5\text{ s}^{-1} = 0.5\text{ Hz}$$

Similarly, the wings of a ruby-throated hummingbird (*Archilochus colubris*) beat on the average 52 times per second (52 Hz). The period of each wing beat can be figured as follows:

$$T = \frac{1}{f} = \frac{1}{52\text{ Hz}}$$

$$T = \frac{1}{52\text{ s}^{-1}} = \frac{1}{52} \cdot \frac{1}{1/\text{s}} = \frac{1}{52} \cdot \frac{\text{s}}{1}$$

$$T \approx 0.019\text{ s}$$

Understanding this relationship between frequency and period will be important when you study wave motion later in this chapter.

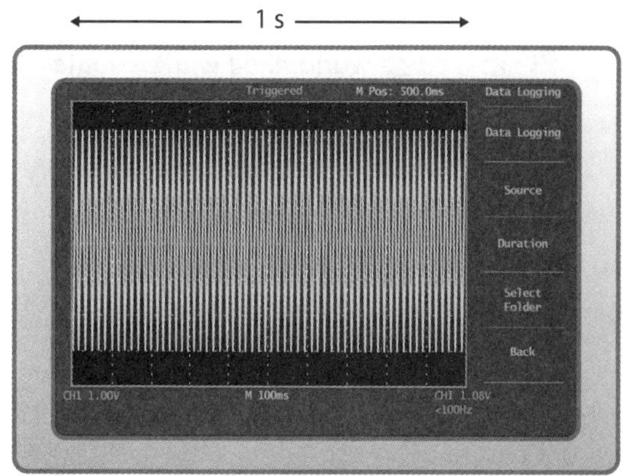

12-7 A 60 Hz signal visible on an oscilloscope (uh SIL uh SKOPE) screen—an instrument designed to display periodic electronic signals

When you see the unit s^{-1}, think "cycles per second."

12-8 A ruby-throated hummingbird

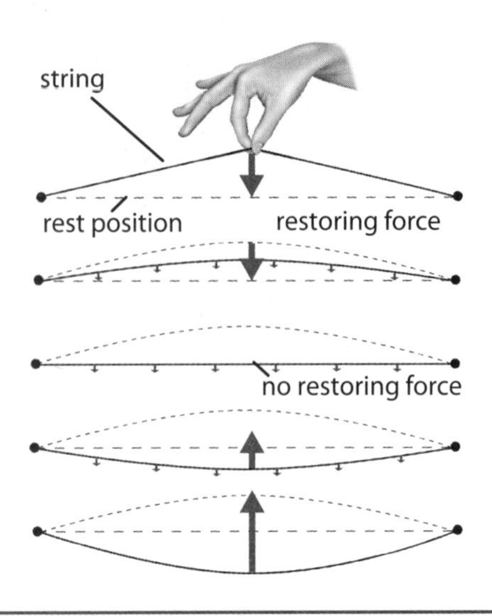

12-9 The restoring force is generated by the tension in the string when the string is displaced from its rest position.

Periodic Motion and the Arts

God has made a world filled with periodic motion. We tend to think of such motion as something that is important to science. But as shown in this chapter, it is just as important to music. What does this reveal about our God?

The period of an oscillating system is proportional to the square root of the system's mass.

$$T = k\sqrt{m}$$

12-10 The speed of vibration of musical string relates to its mass (its thickness) and to its tension (which determines the restoring force).

12.3 The Cause of Periodic Motion

Let's see what happens when you make a tight string vibrate. When the string is stationary, all forces on it are balanced. If you pull the string away from its rest position, you feel a pull as the string stretches. This pull is called the *restoring force* because it tends to restore the string to its rest position. The farther you pull the string away, the stronger the restoring force grows. When you release the string, the restoring force accelerates the string back toward its rest position. During the vibration cycle, the restoring force decreases to zero as the string approaches its rest position. However, the string's speed and thus its momentum are large, so the string keeps moving past its rest position. The restoring force begins to grow in the opposite direction. It eventually stops the string and accelerates it back toward its rest position. The whole process repeats itself again and again, producing vibration in the string.

When an object moves with true periodic motion, it experiences a restoring force whenever it is not at its rest position. The farther the object is from its rest position, the greater the force. The restoring force creates potential energy in the object. This potential energy may be elastic, as in the taut string or a spring; gravitational, as in a pendulum; electrical, as between charges in an electric field; or magnetic, as between magnets and iron objects. The potential energy is maximized when the object is farthest from its rest position. As the object moves toward its rest position, the restoring force converts the potential energy into kinetic energy. As the object, due to its inertia, moves away from the rest position, the restoring force converts kinetic energy back to potential energy with each oscillation. Thus, periodic motion is yet another example of the law of the conservation of energy.

12.4 Factors Affecting Periodic Motion

Mass

A vibrating system's mass affects its period. The more massive a system is, the more slowly it will vibrate and the longer its period will be when other factors are held constant. Consider a laboratory mass attached to a spring that is connected to a wall. Assume that the mass rests on an ideal, frictionless surface. There are no net forces acting on the mass other than the spring. When you pull the mass and release it, the elastic spring force accelerates the mass through its rest position. On the other side the spring compresses, slowing the mass until it stops. But the spring's elastic compression force accelerates the mass in the opposite direction. This motion repeats indefinitely in the absence of friction.

What happens if you replace the laboratory mass with one that has more mass? The greater mass has more inertia that opposes the force of the spring. For the same force exerted by the spring, it accelerates more slowly (Newton's second law). Though the kinetic and potential energies may be the same, the maximum speed of the mass is much lower, so it takes longer to complete a single cycle. The period will be longer, and the frequency will be lower. Strings on a musical instrument vary in thickness. The thicker, more massive strings vibrate more slowly than the thin strings to produce a sound with a lower frequency.

Damping

Oscillations don't continue forever because all real systems are subject to friction. Energy is applied to a system to begin an oscillation. During each cycle of the system's motion, some energy is removed by air friction, internal resistance to motion, or friction with other objects. This energy is converted to thermal energy, lowering the energy available to move the system. The system's amplitude decreases, though the period remains relatively unchanged. Eventually, oscillations become so small that the system stops moving. The effect of friction on periodic motion is called **damping**.

Damping is desirable in most oscillating systems. Without natural or artificial damping, vibrations in objects would just continue without stopping. The ground would never stop shaking after an earthquake. A plucked string would continue vibrating. Most musical instruments would sound odd if their notes were not rapidly damped. A car's shock absorber is designed to maximize damping so that, after going over a bump, the car doesn't keep bouncing.

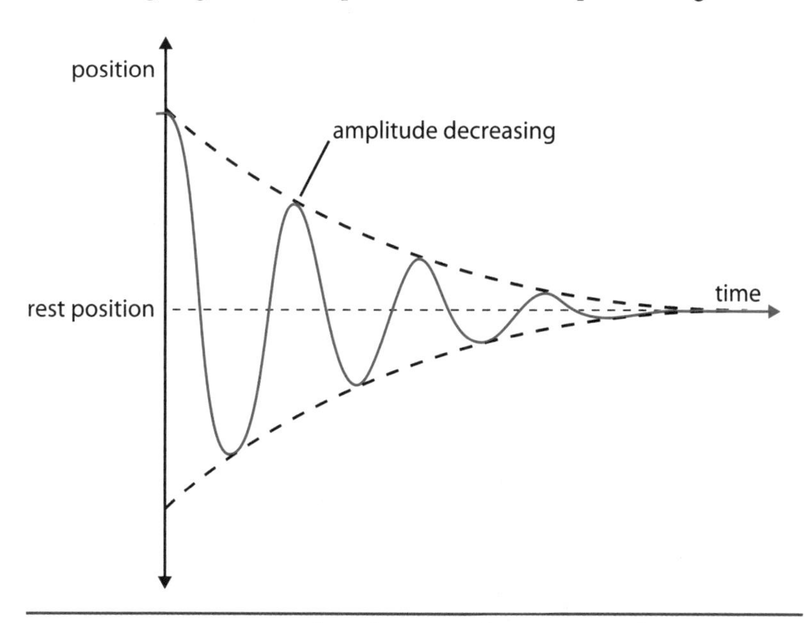

12-11 Damping causes the vibrations of a system to diminish and finally cease after an interval of time.

Damping in an oscillating system is unwanted when a constant amplitude is important, such as in mechanical clocks or playground swings. To overcome damping, some source of energy periodically provides a boost to sustain the amplitude. To continue swinging, you have to "pull" on the chains or have someone push you. Mechanical clocks use suspended weights, springs, or electromagnets powered by batteries to boost each cycle.

12-12 The potential energy of the clock spring is converted to kinetic energy to overcome frictional damping in the pendulum mechanism.

Resonance

The opposite of damping is **resonance**. All objects that can oscillate have a *natural frequency*. Resonance occurs when an outside source adds energy to an oscillating object at its natural frequency. The amount of energy exceeds that which is needed to overcome natural damping in the system. This makes the oscillation amplitude increase. Eventually, either the structure of the object limits the amplitude, or the object flies apart.

An object that is experiencing resonance is *resonating*.

12-13 This photo shows how a loud sound applied at the natural frequency of this goblet eventually caused it to vibrate to destruction.

12B Objectives

After completing this section, you will be able to

✓ describe the parts of a pendulum.

✓ summarize the history of the pendulum.

✓ explain how a pendulum moves.

✓ calculate a pendulum's period given its arm length.

✓ calculate the value of g given a pendulum's period and arm length.

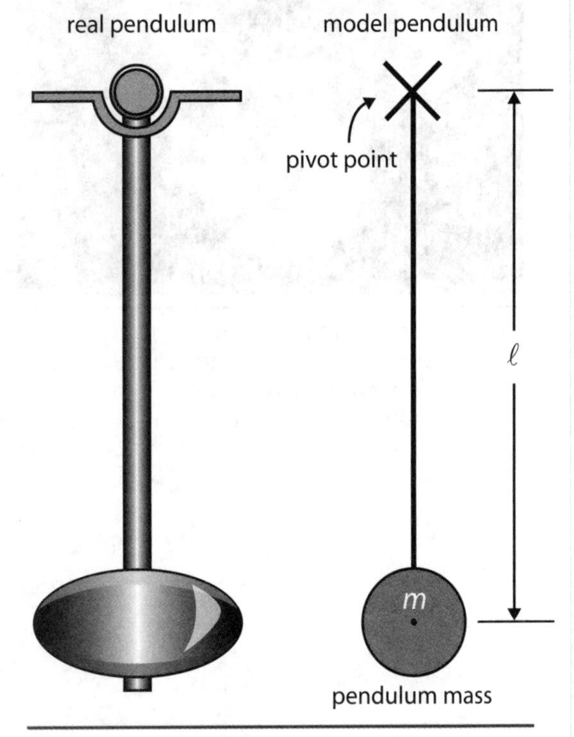

12-14 The components of a pendulum

You may be familiar with some examples of resonance. A person pushing a child in a playground swing exerts a force on the downward side of the swing cycle, allowing the swing to go higher on the upward side. With each push the swing goes higher. In another example, the annoying buzzing you may hear in your car, especially at high speeds, is generated by resonance in car parts. An imbalance in a wheel or the road causes the whole car, including objects in the car, to vibrate. The frequency of the vibrations depends on the speed of the wheels. Only specific speeds generate resonant frequencies. On the other hand, horns and other musical instruments are designed to resonate to produce rich musical tones.

12A Section Review

1. Give an example of repetitive motion that is not periodic.
2. What do periodic oscillations look like? What characteristic makes them especially different from other kinds of motion?
3. What is the mathematical relationship between an oscillating system's period and its frequency? What are the standard units for period? for frequency?
4. What force produces periodic oscillations? How does this force occur?
5. How does the mass of an oscillating system generally affect its frequency?
6. What happens to the amplitude of real-world periodic oscillations over time? Why does this occur?
7. What happens to an oscillating system's amplitude when it begins to resonate? What limits a resonating system's amplitude?
8. (True or False) Periodic motion includes regular back-and-forth oscillations and vibrations as well as circular motion.

12B Pendulums

12.5 *What Is a Pendulum?*

When you think of a pendulum, you probably think of a tall, intricately carved grandfather clock containing a heavy brass disk swinging slowly behind a glass pane. In physics, a **pendulum** is simply a mass attached to an arm at a pivot point that swings under the influence of gravity. A pendulum may be a rigid arm-and-weight arrangement as in most clock pendulums, but it may also be a laboratory mass hooked to the end of a string or a birdcage hanging from a chain attached to the ceiling. A pendulum's appearance may look very different from the grandfather clock pattern, but all pendulums work the same way.

In an ideal pendulum the mass (m) is assumed to be concentrated in a single point at the end of the pendulum arm. The length (ℓ) of the pendulum arm is the distance between the *center of mass* and the pivot point. The ideal pendulum arm is assumed to have negligible mass. In real pendulums, the mass is not a point and the pendulum arm does have mass. These differences from the ideal case affect the modeling of pendulum motion slightly.

12.6 *History of the Pendulum*

The origin of the pendulum is clouded in uncertainty. Pendulum-like objects were probably known from the earliest days of the human race. Excavations in African caves have uncovered extremely ancient illustrations of pendulums used for **dowsing**.

Pendulums were likely used long before scientists could mathematically describe their motion. The first documented practical use of a pendulum was in AD 132, when a Chinese philosopher named Zhang Heng invented a **seismoscope** believed to contain a heavy pendulum. Many written sources claim that around AD 1000 an Egyptian mathematician named Ibn Yunus was the first to scientifically describe a pendulum. However, this claim is disputed by others.

Some records suggest that the pendulum gained the interest of several European philosophers in the 1400s, but it wasn't until the sixteenth century when Galileo established that the motion of a pendulum is regular. As a teenager, he noticed the swaying of chandeliers in a cathedral and timed their periods using his pulse. Toward the end of his life, he sketched a design for a mechanical clock regulated by a pendulum.

In 1656 the great scientist Christiaan Huygens had a clock built based on Galileo's design and used it in his astronomical observations. His clock was accurate to within 15 s per day and was the first to include a minute hand. After the seventeenth century, the pendulum clock was highly refined, and exceptionally accurate timepieces were made.

The motion of a pendulum has been understood differently through the ages depending on the scientific paradigm used to explain it. (See the Science and Truth facet in Chapter 1 for a discussion of scientific paradigms.) In Aristotle's time a pendulum was simply a rock tied to a string. Philosophers assumed that the rock was trying to move toward the center of the earth according to its "nature." The string inhibited that motion, and only after a great deal of effort involving ever-decreasing arcs did the rock finally arrive at the lowest point in space allowed by the string.

Galileo saw the periodic motion of a pendulum as the result of an "impetus" that set an object in motion. The impetus was used up as the oscillating object reached its maximum amplitude, and then the impetus was restored as the object returned to its resting position. This understanding was completely different from the Aristotelian view.

After Isaac Newton developed his laws of motion, scientists had new tools to describe mechanical motion. The pendulum was seen to respond to gravitational force and to gain momentum in each half of its swing. With more precise timepieces, the formulas for describing pendulum motion were developed. The new paradigm of mechanics allowed scientists to view pendulum motion in a new way.

As you can see, the way humans changed their understanding of pendulums parallels changes in many areas of scientific knowledge through history. These changes weren't just the accumulation of new observations but involved revolutionary changes in the way the same observations were interpreted.

Dowsing is a mystical practice that is even today believed to reveal hidden information, such as the location of water in the ground or the sex of an unborn child.

A **seismoscope** is a device that provides indication that an earthquake has occurred, but it does not create a record of the quake as does a *seismograph*.

12-15 Zhang Heng's seismoscope

12-16 The first pendulum clock, designed by Christiaan Huygens and based on a design by Galileo

FACETS OF SCIENCE

THE FOUCAULT PENDULUM

From the time that Copernicus's heliocentric theory began to gain acceptance in the sixteenth century, critics continued to argue that there was no direct evidence that the earth actually turned on its own axis. They believed that if the earth did rotate, people and loose objects would be flung from the surface of the earth like mud from a quickly spinning wheel. Birds would be swept away when they took flight as the earth rotated under them. Even after Newton's mechanics showed that the earth's motions were too subtle to cause these kinds of problems, there still was no direct evidence that the earth actually rotated.

The required proof came from the *Foucault pendulum*. This special kind of pendulum was named for the French physicist Jean Foucault, who made the first one in 1851. Foucault hung a 67 m (220 ft) wire inside the dome of the famous Panthéon building in Paris. At the end of the wire, he attached a 28 kg (62 lb) iron ball. Soon after he started the ball swinging back and forth, observers could see that the direction of its swing relative to the floor was changing. Scientists already understood the principle of inertia, that a moving object keeps moving in the same direction unless an outside force acts on it. Foucault had hung his wire by a single-point suspension so that the only forces acting on the pendulum were gravity and the force exerted through the wire by the attachment point. Thus, only the rotation of the earth could explain the direction and pattern in which the pendulum's swing changed relative to the floor of the Panthéon.

This motion is most obvious when a Foucault pendulum is set up directly over the geographic North or South Pole. As the pendulum swings back and forth in the same direction, the earth turns under it at the rate of 15° per hour. The pendulum's inertia maintains a constant direction of swing since there are no external forces acting on it to change its direction. After six hours the earth has turned 90° to the east, and the pendulum is swinging perpendicular to the line of its original path. At 12 hours, the earth has rotated 180°, and the pendulum swing is again in line. After 18 hours and a rotation of 270°, it's swinging perpendicular to the line again. And after 24 hours and 360°, its swing aligns with its original path. In 2001 scientists actually performed this experiment at the South Pole.

If a Foucault pendulum were set up at the Equator, the pendulum would not change direction in its swing relative to the ground at all. The inertia of the pendulum's motion relative to the earth's rotation is always in the same direction, so no forces act on the pendulum to change its swing.

As the location of a Foucault pendulum is moved farther away from the Equator, either north or south, the change in direction begins to be noticed over time. The swing changes because as the earth turns, the attachment point of the pendulum, the inertia of the pendulum mass, and the earth's motion under the pendulum don't lie in the same plane. This arrangement creates a force on the pendulum's mass that shifts its swing.

Interestingly, the rate at which the swing changes direction relative to the ground varies with latitude. At the Equator it doesn't change at all, but at 10° latitude it takes 137 hours to complete one cycle. At 45° from the Equator, it takes about 34 hours; at 60° latitude, 28 hours; and at the poles, only 24 hours. In addition, the direction of the shift in the swing differs depending on which hemisphere the pendulum is located. In the Northern Hemisphere the swing shifts clockwise, and in the Southern Hemisphere it shifts counterclockwise.

Once and for all, the Foucault pendulum silenced arguments against the rotation of the earth. Not until man first visited the moon and could look back at the rotating earth was there any more obvious or direct evidence. Planetariums and museums often include a Foucault pendulum in their public displays.

12.7 Modeling Pendulum Motion

The period of a pendulum's swing is an essential part of a model of the motion of a pendulum. A pendulum's motion is the result of two forces continuously acting on the mass at the end of the arm—the tension force in the arm and the weight of the pendulum's mass. The tension in the pendulum arm varies with the angle of the pendulum. It is at a minimum at either side of each swing and at its maximum at the bottom of the swing. The tension force in the pendulum arm is due in part to the weight of the mass and in part to the Newton's third law reaction force that the mass exerts on the arm as it swings around the pivot point. These forces combine to act as a restoring force on the pendulum's mass. They are not proportional to the pendulum's distance from its rest position, as in systems that show true periodic motion. A pendulum approximates true periodic motion only when its swing covers a small amplitude.

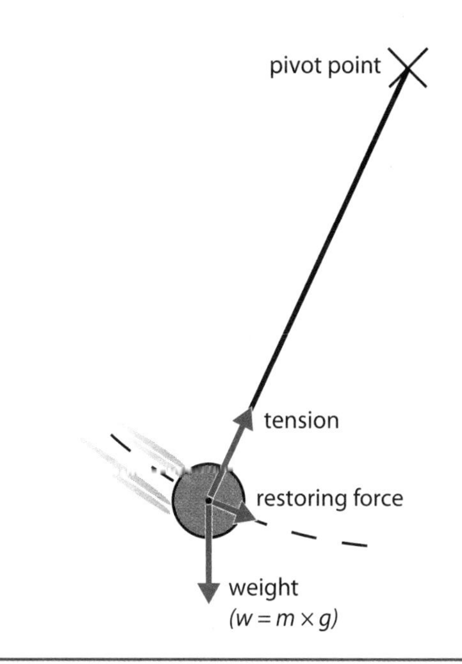

12-17 The two forces acting on the pendulum's mass are the tension in the pendulum arm and the force of gravity.

The weight of the pendulum's mass generates GPE as the mass moves from its rest position. Recall that GPE = $m \times g \times h$. It is the GPE of the pendulum that ultimately drives the device. When the pendulum is moving, some of this GPE has been converted to kinetic energy, which also depends on the mass of the pendulum. Note that all the forces acting on a pendulum are related to the weight of the pendulum's mass. Thus, it would seem that the mass of the pendulum is an important variable in its motion. However, the weight of an object's mass depends on the magnitude of gravity (g). Since gravitational acceleration is independent of mass, the period of a pendulum actually depends on gravitational acceleration rather than on its mass.

The only two factors that affect a pendulum's period are the length of the pendulum arm (ℓ) and the acceleration due to gravity. For any given place on the earth's surface, gravitational acceleration is essentially constant, so the length of the pendulum arm is the only part of a pendulum that determines its period. The longer the pendulum arm, the longer the period of its swing. The formula for the period of a pendulum is

$$T = 2\pi\sqrt{\frac{\ell}{g}},$$

where

- T is the period of the pendulum swing, in seconds,
- ℓ is the length of the pendulum arm, in meters,
- g is the acceleration due to gravity at the earth's surface, which is assumed to be 9.8 m/s^2, and
- 2π is a constant factor approximately equal to 6.28.

With such a simple formula, scientists and engineers were able to build clocks and other time-keeping devices to produce any period they needed. In addition, by rearranging the formula, they were able to precisely determine gravitational acceleration anywhere in the world. Before the launching of gravity-measuring satellites, data for global maps of surface gravity was generated using pendulums to compute gravitational acceleration. These scientific pendulums had carefully measured arm lengths, and their periods were precisely timed at each location.

By rearranging the equation for the period of a pendulum, we obtain the formula for determining g.

$$g = 4\pi^2\frac{\ell}{T^2}$$

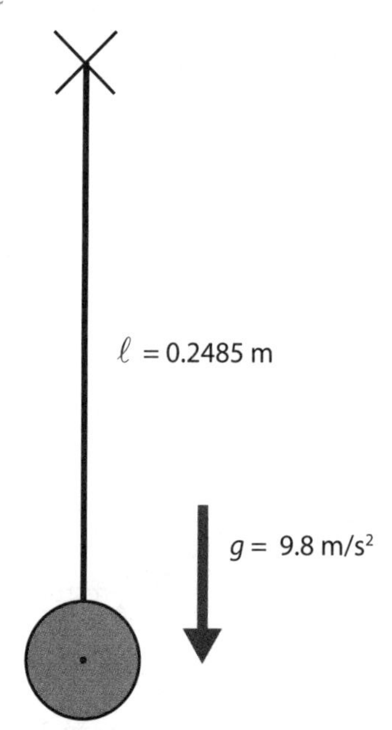

12-18 The pendulum in Example Problem 12-1

EXAMPLE PROBLEM 12-1

Period of a Pendulum

What is the period, in seconds, of a pendulum that has an arm 0.2485 m long? For what purpose would such a pendulum be useful?

Known: pendulum arm length (ℓ) = 0.2485 m
gravitational acceleration (g) = 9.8 m/s²

Unknown: pendulum period (T)

Required formula: $T = 2\pi\sqrt{\frac{\ell}{g}}$

Substitution: $T = 2\pi\sqrt{\frac{0.2485\ \cancel{m}}{9.8\ \cancel{m}/s^2}}$

Solution: $T \approx (2\pi)(0.15\underline{9}2\ s) \approx 1.00\ s$

Answer: This pendulum has a period of 1 s. It could be used for regulating a clock.

12B Section Review

1. Describe the parts of an ideal pendulum.
2. What was the first recorded use of the pendulum for a non-occult purpose? Who was the first European scientist who studied the pendulum?
3. What were three historical views of the pendulum? Why is the current paradigm the most useful?
4. What factors affect the period of a pendulum? What property of a pendulum does not affect its period?
5. How does a pendulum's period vary with the length of its arm? with its mass? with gravitational acceleration?
6. (True or False) A pendulum works in the International Space Station just as it would on Earth.
7. The gravitational acceleration on the surface of Mars is 3.7 m/s². What period would a 0.64 m long pendulum have on Mars?

12C Objectives

After completing this section, you will be able to

- ✓ define what a wave is.
- ✓ label the parts of a wave on a diagram.
- ✓ calculate wavelength or frequency using the wave speed equation.
- ✓ classify the different types of waves by particle motion.
- ✓ describe how waves reflect.
- ✓ explain constructive and destructive wave interference.
- ✓ compare and contrast wave refraction and diffraction.
- ✓ describe the Doppler effect.

12C Waves

12.8 Defining Waves

The sound waves of cheers throb in your ears at a baseball game. Water waves wash gently ashore or cause great devastation as tsunamis. The ground shakes as waves from earthquakes pass through it. Waves surround us, moving energy from here to there. We hear because of sound waves and see because of light waves. **Waves** are periodic changes that transmit energy from one place to another. Waves in matter—*mechanical waves*—periodically displace particles around a relatively stationary position. Though the individual particles periodically oscillate as the wave passes, there is little or no overall motion of matter. *Electromagnetic waves* (see Chapter 14) transmit energy without matter.

12-19 Sound and water waves are mechanical waves.

12.9 Describing Waves

Waveform

Figure 12-20 shows a cross-section of a typical waveform. The highest points of a wave are *crests*, and the lowest points are *troughs*. Midway between these two points is the wave's rest position. The distance of a crest or trough from the rest position is the amplitude of the wave, similar to the amplitude of any other form of periodic motion. One cycle of a wave consists of one complete crest and trough. The length of a wave cycle, or its **wavelength**, is the distance between corresponding parts of two sequential waveforms, such as from one crest to the next or from one trough to the next. Wavelength is represented by the Greek letter *lambda* (λ).

Wave Frequency

The frequency of a wave is the number of wave cycles that pass a given point in 1 s. Similar to other periodic motion, wave frequency is measured in hertz. Multiples of hertz using SI prefixes are commonly used for waves with high frequencies. One kilohertz (kHz) equals one thousand cycles per second, 1 megahertz (MHz) equals one million cycles per second, and 1 gigahertz (GHz) equals one billion cycles per second.

All mechanical waves require a *medium*, the matter through which waves travel from place to place. However, the particles of the medium do not move very far from a given spot compared to the motion of the wave. How can this be? Figure 12-21 illustrates one form of particle movement found in water waves. As the wave surface oscillates up and down, the particles within the wave move in a circular path. As the wave passes, they return to near their starting point. Their nearly circular motion clearly demonstrates the cyclic nature of this periodic motion.

This idea is easier to visualize on a vibrating string. When you pluck a long string, the vibrations clearly travel down the string, but do the string particles (the medium) travel with the wave? No, they only move up and down or side to side. Other than the up-and-down motion of the wave, the particles of the string remain in place. The motion of a wave is different from the motion of the individual particles that are disturbed. The wave carries energy by transferring oscillations to adjacent particles, not by carrying the particles along with the wave.

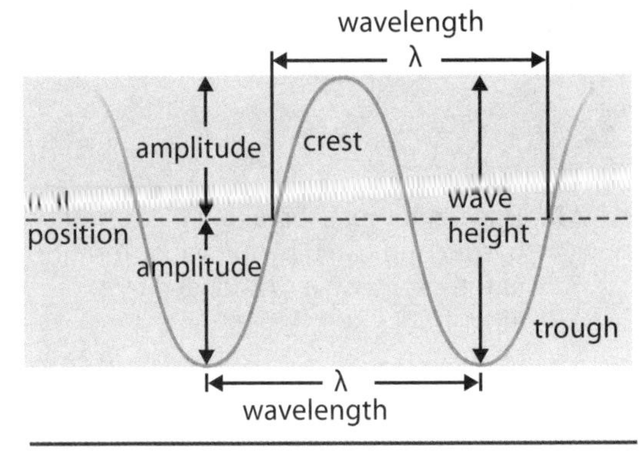

12-20 The parts of a waveform

trough (TRAWF)

lambda (λ) (LAM duh)

Don't confuse wave amplitude with wave height, which is the distance from trough to crest. Amplitude is one-half wave height. Physicists relate the energy of a wave to its amplitude. Mariners are concerned about water's wave height because of its effect on ship motion.

medium (L. *medius*—middle, between)

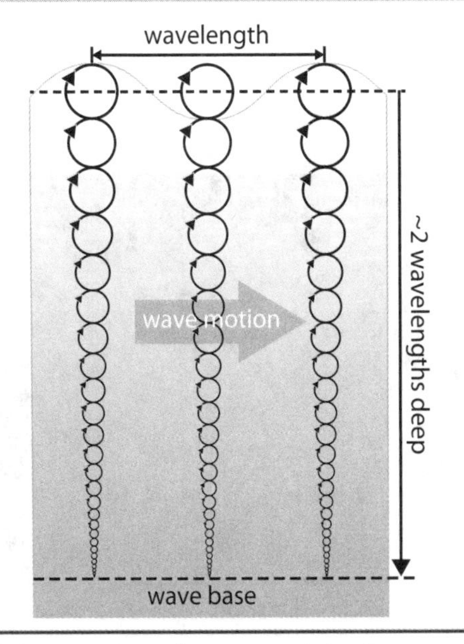

12-21 Motion of water particles in a passing water wave

12-22 This fighter jet is breaking the sound barrier, indicated by the vapor forming at the shock waves produced by the aircraft.

Wave Speed

How fast do waves travel? Wave speed depends on the wave type and its medium. For example, sound waves travel at about 300 m/s in air, 1533 m/s in seawater, and 5130 m/s in iron. The speed of sound is also subject to factors such as temperature, relative humidity, air pressure, and density, depending on the medium.

Surface waves in water occur where air and water meet. They are different from sound waves in the way the water particles move as the wave passes through. The speed of surface waves depends mainly on the depth of the water compared to their wavelength. Waves in deep water can move faster than waves in shallow water. Typical wind-driven waves have speeds of a few meters per second. Earthquake-generated tsunamis can travel at over 200 m/s (more than 700 km/h)!

The fastest waves in the universe are electromagnetic waves. These waves zip through space at an incredible 300,000,000 m/s—fast enough to get from the earth to the moon in 1.3 s! Electromagnetic waves slow down when they pass through matter.

The speed of a wave is mathematically related to wavelength and frequency. The following equation expresses this relationship for most waves when all other properties in the medium are constant:

$$\boxed{v = \lambda \times f},$$

where

- v is the wave speed, in meters per second,
- λ is the wavelength, in meters, and
- f is the wave frequency, in cycles per second (Hz).

12-23 Ocean waves formed in a ship's wake

Recall that $1\ \text{Hz} = \text{s}^{-1} = \frac{1}{\text{s}}$.

EXAMPLE PROBLEM 12-2

Wave Speed

What is the speed of a deep-water ocean wave that has a frequency of 0.31 Hz and a wavelength of 16 m?

Known: wave frequency (f) = $0.31 \cdot 1/\text{s}$
wavelength (λ) = 16 m

Unknown: wave speed (v)

Required formula: $v = \lambda \times f$

Substitution: $v = (16\ \text{m})(0.31 \cdot 1/\text{s})$

Solution: $v \approx 4.\underline{9}6\ \text{m/s}$
$v \approx 5.0\ \text{m/s}$ (2 SDs allowed)

A single wave cycle or a very short burst of waves is called a *pulse*. If you stretch a rope out and then pluck it once, you can see the pulse move rapidly down the rope. Sound wave pulses occur when animals produce clicks or chirps. Sharp volcanic explosions are pulses of sound and seismic energy. Humans use pulsed sounds for alerting others (a whistle or a yell, for instance) or in sonar to track fish or submarines. Similarly, radar pulses are used to determine the range and direction of moving objects in the air and on the ground.

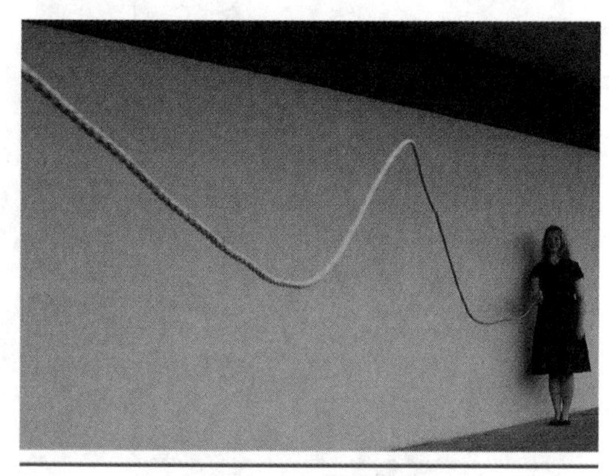

12-24 A pulse in a rope

Wave Energy

Mechanical waves exist because they transfer energy through matter. So how much energy does a mechanical wave contain? This question is not easily answered, just as we could not state how much thermal energy an object contains. However, the *rate* at which a wave transfers energy, or its power, is proportional to its amplitude squared, assuming that the wave's period is constant. Thus, if the amplitude doubles, the power quadruples. This exponential relationship is one reason why large ocean waves can damage beaches much more quickly than small lapping waves.

The energy-transfer rate of a mechanical wave is proportional to the square of its amplitude.

The energy in electromagnetic waves can be measured because this kind of nonmechanical wave comes in separate packets of energy. The energy of each packet is directly proportional to the wave's frequency. So, as electromagnetic frequency increases, its energy increases. See Chapter 14 for more information on the energy in electromagnetic waves.

12.10 Classifying Waves

Waves are classified according to the direction the particles in the medium move as the wave energy passes through the medium. If the vibration is at right angles to the direction of travel, the wave is a **transverse wave**. The vibrations of a string are an example of a transverse wave. Note in Figure 12-25 how the points marked on the string do not move along the string, just perpendicular to it. The wave crest is formed as particles spread out in one direction perpendicular to wave direction. The trough is formed as particles spread out in the other direction.

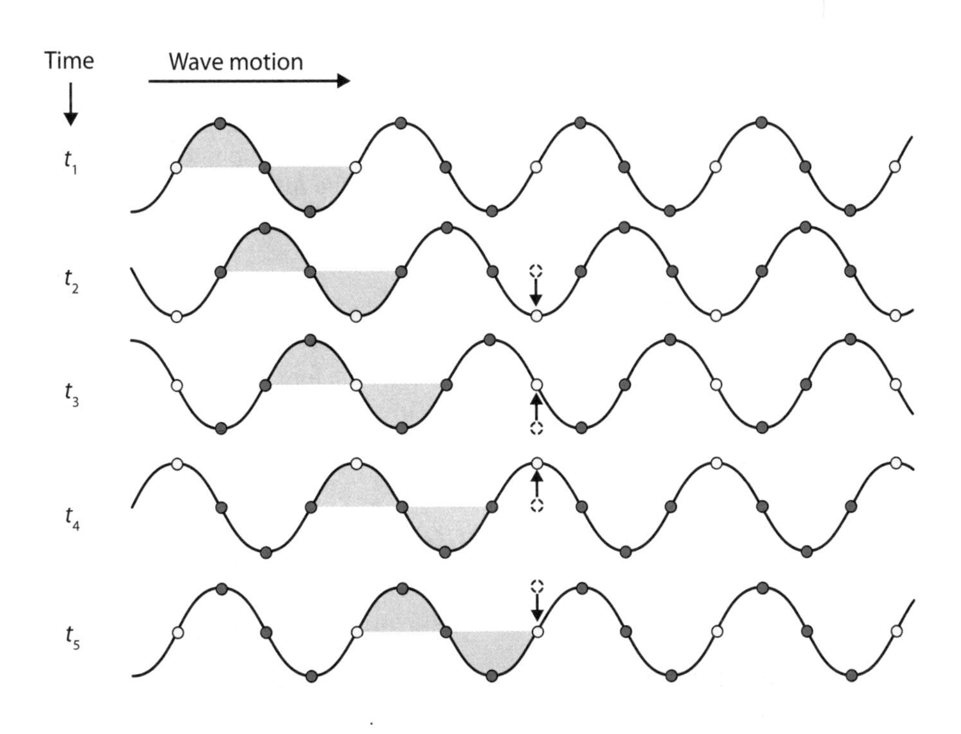

12-25 In a transverse wave the particles of matter oscillate perpendicularly to the direction of wave motion.

longitudinal (LON ji TOOD un ul)

rarefaction (RAIR uh FAK shun): rare- (L. *rarus*—rare) + faction (L. *facere*—to make)

Light and other forms of electromagnetic energy are transverse waves, although these waves don't involve particles of matter. The electric and magnetic fields change in a periodic way that appears like a transverse wave pattern.

If the vibration of a wave is parallel to the direction of wave movement, it is a **longitudinal wave**. Sound waves are the best examples of longitudinal waves. These waves form as particles are shoved together in *compressions* or spread apart in ***rarefactions***. The compressions correspond to wave crests, and the rarefactions are like wave troughs. If you graph the density of the particles in the longitudinal wave's medium with time, you can observe the familiar wavelike shape of the graph (see Figure 12-26).

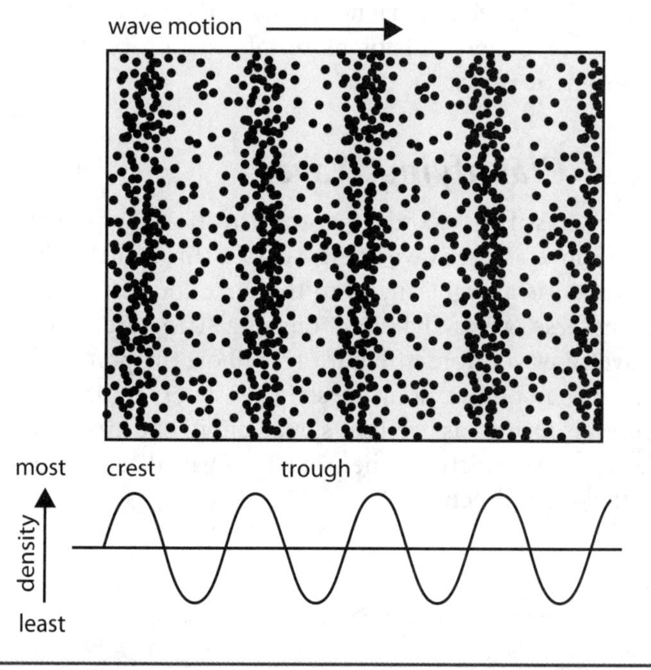

12-26 Longitudinal waves in a medium. Note that the density of the particles of the medium determines "crests" and "troughs."

Water waves and certain kinds of earthquake waves are not simple transverse or longitudinal waves. These waves occur at the boundary between two different media, for example, between air and water. The individual particles in these waves move in circles or ellipses and have both transverse and longitudinal motion.

12.11 Wave Phenomena

Nearly all waves have similar behaviors. Let's consider the ways that waves interact with matter and with each other.

Reflection

If you stretch a rope taut and then pluck it once, you can see a pulse move rapidly down the rope. If the far end of the rope is fixed, the pulse will bounce back. The wave is reflected along the string in the opposite direction. **Reflection** occurs when a wave bounces off a surface that doesn't completely absorb its energy. The reflected wave may change depending on how it reflects. If it reflects off a hard surface, the outgoing wave is flipped, though its amplitude is unchanged. If the reflecting surface absorbs some of the wave's energy, then it is flipped and its amplitude is smaller. In every case, energy is conserved.

Reflection is used in many ways. Light waves generated by a flashlight bulb reflect from the concave mirror behind the bulb to produce a beam of light. The dish-shaped antennas used for radar and television waves, the concave mirrors of large optical telescopes, and the acoustic shells placed behind orchestras all concentrate waves in a more useful direction.

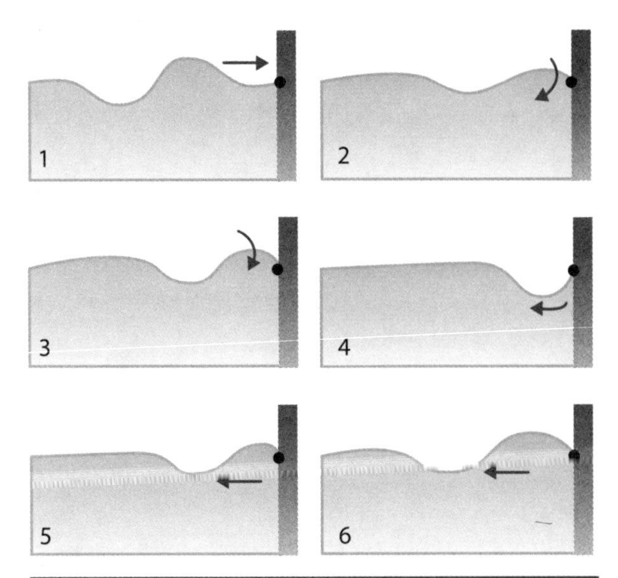

12-27 Wave reflection. When totally reflected, a wave is inverted.

Refraction

When a wave moves from one medium to another, its speed changes. This phenomenon is called **refraction**. Under most circumstances, a wave bends from its original path when it enters the new medium. For example, light waves bend as they pass from air into glass or from glass into air. Bending does *not* occur if a wave enters the new medium perpendicular to the boundary between the media.

Waves with shorter wavelengths are refracted more than waves with longer wavelengths. For example, when visible light passes from air into a glass prism, it is refracted. The different wavelengths of light are refracted at slightly different angles. Blue light has a shorter wavelength than red light, and it is refracted at a greater angle. This is why prisms disperse light.

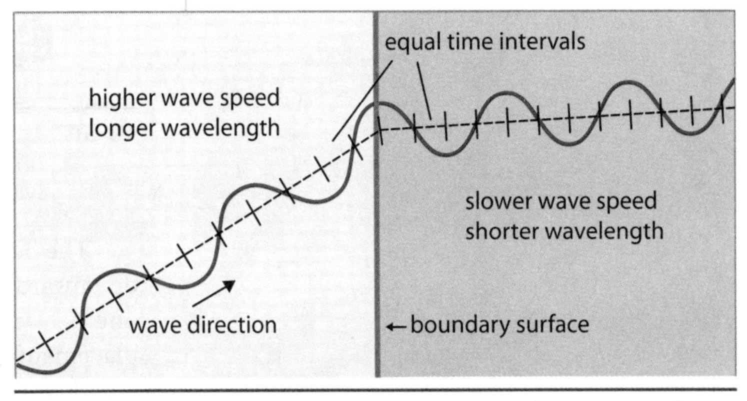

12-28 Refraction occurs when a wave passes between media where its speeds are different.

Sound waves are also refracted. It's far from quiet beneath the ocean's surface. Besides the biological noises of shrimp snapping, fish grunting, and whales singing, man-made noises from ships and other marine activities can be heard. Water temperature and pressure differences alter the acoustic medium, and these differences affect the way sound waves spread out through the ocean basin. Thus, sound waves bend down and up in deep water, sometimes repeatedly, and can travel hundreds of kilometers. Sound refraction in the ocean is especially important to submarines and to those who hunt them.

When the speeds of a wave in two media are greatly different, some of the wave's energy may be *reflected* as well as refracted at the boundary between them. This is especially true for shorter wavelengths, which are more easily refracted. It's also especially true as the wave's angle decreases as it approaches the boundary. For example, a swimmer at the bottom of a swimming pool can hear the voice of someone shouting directly above him at the surface much more clearly than someone shouting at the far end of the pool. This is because the sound waves from the person overhead are closer to perpendicular as they move from air into the water.

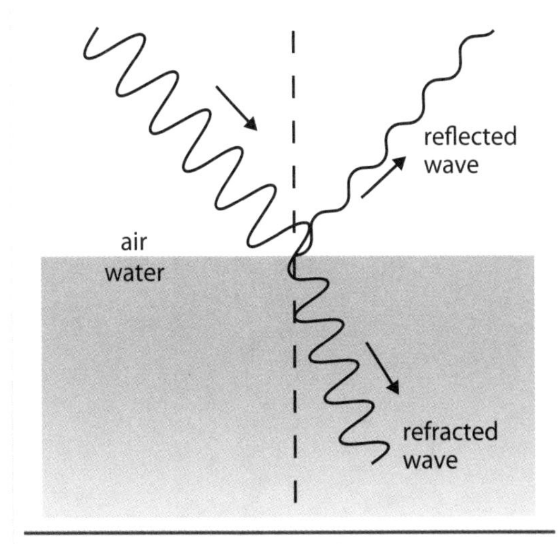

12-29 Under certain conditions, waves usually both refract and reflect at the boundary between two media.

Wavelength changes from one medium to another because wave speed changes. Recall that $v = \lambda \times f$ (see Figures 12-28 and 12-29). The frequency of a wave is determined by its source, and it remains unchanged as long as the wave exists. However, denser media slow waves down, so the wavelengths must shorten in proportion to wave speed.

Wave Interference

Picture two boys holding a long, slack rope between them. At the same instant, each boy flicks his end of the rope upward, sending a pulse toward the other. What happens when the two pulses meet at the center? Since both pulses are upward displacements of the rope, they are added together for a brief instant as they pass through each other. At that moment there is a single pulse whose amplitude is the sum of the amplitudes of both pulses. They do not stop but travel on as before. The ability of two pulses to encounter each other, survive, and move on as before comes as a surprise to many people. This phenomenon, though, fits the predictions of the law of the conservation of energy.

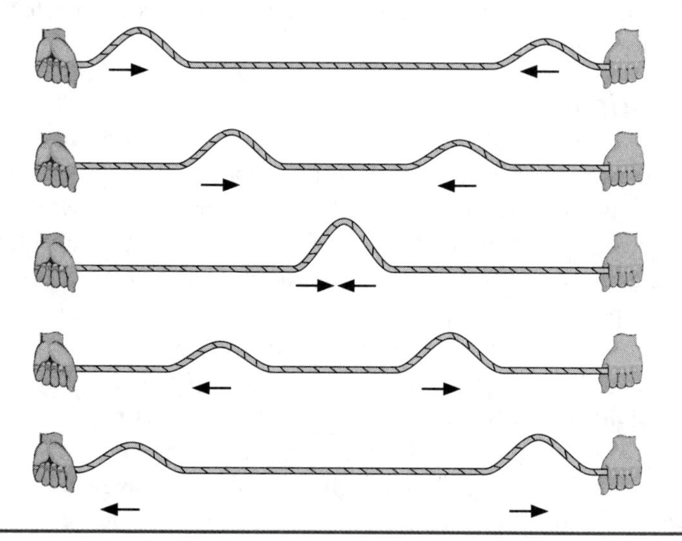

12-30 Approaching wave pulses pass through each other unimpeded. Note that their amplitudes add when the waves are superimposed.

The boys repeat the experiment, but one of them flicks the rope downward as the other flicks it upward. When the two pulses meet, one has an upward displacement and the other a downward displacement. What do you think will happen this time? If the pulses are of equal amplitude, there will be a brief moment during which they cancel each other. The rope will appear to be approximately

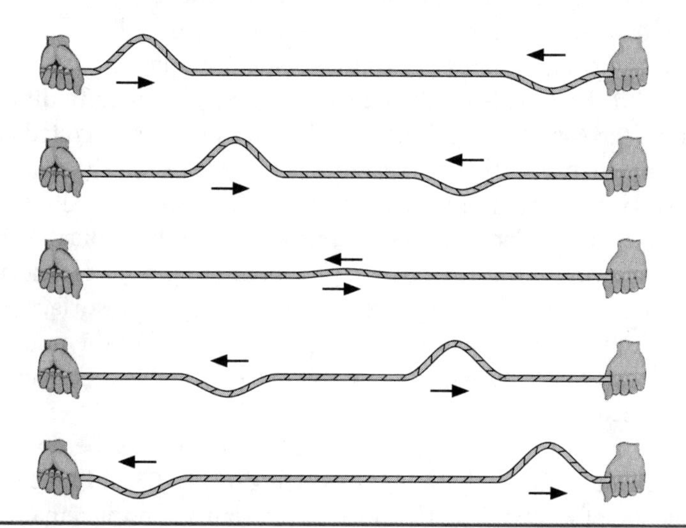

12-31 Wave pulses with opposite orientations momentarily subtract from each other when superimposed. The energy in the waves is not cancelled, just the displacement of the medium.

straight, though both the waves and their energy are still present because the rope is moving and not still. Then the pulses will pass by each other and move on unchanged.

Many waves can combine in this way. When waves from different sources cross paths, their interaction is called **interference**. When the crests align with crests and the troughs with troughs, the waves are *in phase*. Their wave heights are added together, producing high crests and deep troughs. This is called *constructive interference*. If the crests align with troughs and the troughs with crests, *destructive interference* occurs. The waves are *out of phase*, and their wave heights are subtracted from each other, producing smaller crests and troughs or even briefly canceling the waves completely. When waves with different wavelengths and amplitudes interact, both kinds of interference occur at any given moment.

Constructive interference can be dangerous to ships at sea. Ship captains have seen small waves come together in phase to form waves 15–30 m high. Without warning, these waves can combine to capsize a small vessel.

Destructive interference can also cause problems. Some auditoriums have "dead spots," where it is difficult to hear even if the sound is amplified. Dead spots are caused by the destructive interference of the sound waves from two different loudspeakers or from direct sound waves combining with waves reflecting off hard surfaces. What do you think happens when constructive interference occurs in an auditorium?

When two traveling waves with exactly the same wavelength and frequency move through each other in opposite directions, a *standing wave* is formed. A standing wave *appears* to be stationary in the medium because the positions of the crests and troughs do not move. Instead, they rise and fall in place as the two sets of waves interfere with each other. Points in a standing wave that experience no vertical displacement are called *nodes*. The spacing of the nodes is half the wavelength of the waves.

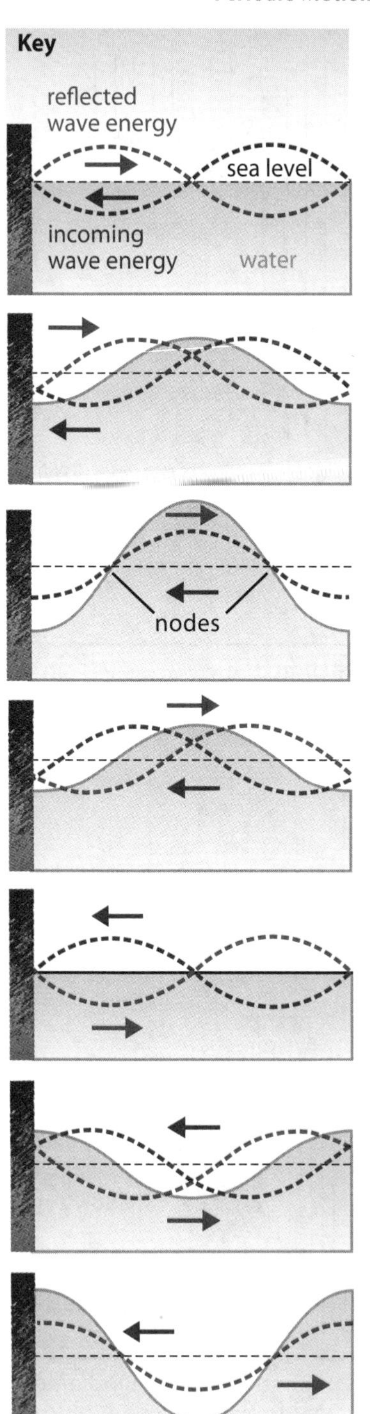

12-32 A standing wave doesn't seem to move through the medium. Standing waves are often seen when water waves reflect off a flat surface.

Beats

One form of wave interference is especially important to acoustics and structural engineering. When two or more continuous waves interfere, they are replaced by a set of waves with exceptionally large amplitudes. These waves have a lower frequency than the two sets of interfering waves and are called *beats*. Beats are especially noticeable when two audible tones interfere, sometimes producing an annoying, slowly pulsating sound. You may have heard such sounds during an airplane flight or boat cruise when the engine speed settings were not identical. The sound of the beat frequency can be very loud.

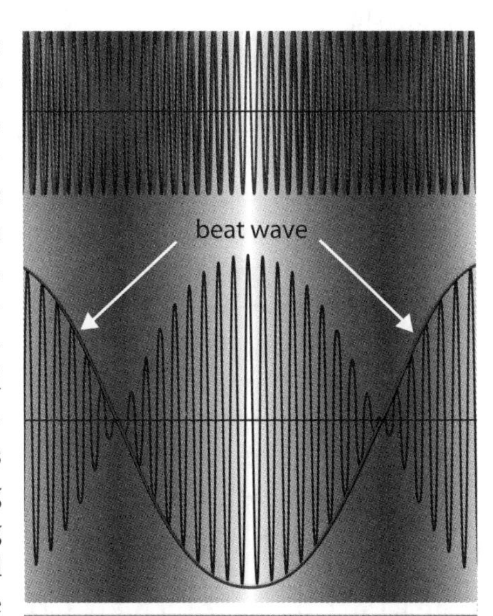

12-33 Beats form when two sets of waves with different frequencies interfere. The beat frequency is lower than the frequency of either set of interfering waves.

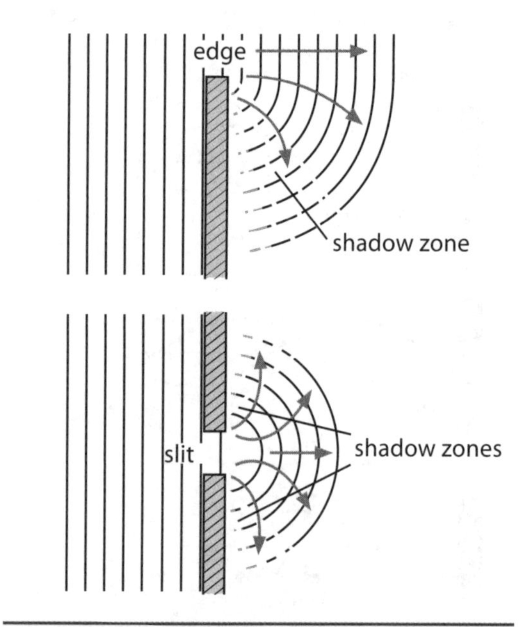

12-34 Diffraction of a wave around an edge and through a slit

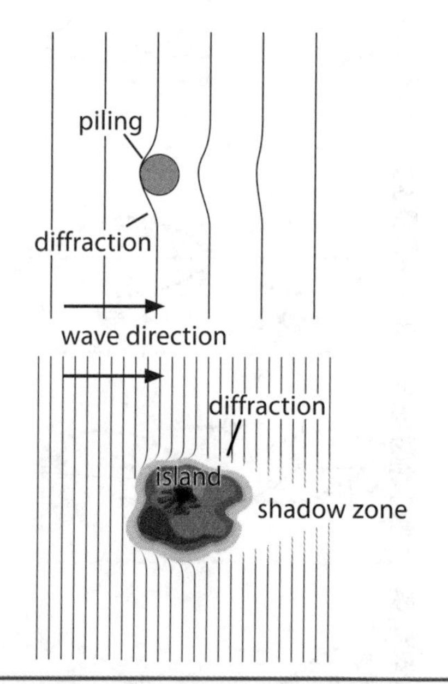

12-35 Diffraction of waves around an object is related to the size of the object in comparison to wavelength.

Diffraction

Did you ever wonder why you can hear the voice of someone who is around a corner in a hallway? You might guess that the sound bends around the corner, and you would be correct. However, there is no change in the medium, so the sound waves don't bend because of refraction. What *is* happening?

When a wave moves past a corner in its medium, it naturally folds or flows around the corner. This phenomenon is called **diffraction**, and every kind of wave experiences it. To understand diffraction, let's refer to a model of waves first proposed by Christiaan Huygens in 1690. His model is still useful today for explaining certain wave properties. Huygens recognized that waves are three-dimensional and that most start at a source that can be modeled as a point. The waves spread out in a spherical pattern from the source. He suggested that every point on the spherical surface of a wave acts as a source of a new set of waves. Each of those new waves then spreads out in turn. When a wave surface moves past an edge of an opening or a corner, the spherical waves generated at the edge move outward, giving the appearance of the wave bending around the corner (see Figure 12-34). Since the diffraction directs more of the wave's energy straight ahead, less energy wraps around the edge and thus a *shadow zone* forms behind the barrier.

The effect of diffraction appears significant when the wavelength is large compared to the size of an obstacle or opening. For example, water waves that are several meters from crest to crest flow around a dock piling half a meter in diameter and merge back together. However, the same water waves will bend only partially around a small island with a diameter many hundreds of times larger than the water wavelengths. A "shadow zone," an area lacking waves, will form behind the island. Sound waves are shorter than water waves, but they will still bend around most average-sized objects, explaining why you can hear that voice from around a corner. Light waves also experience diffraction, but because the wavelengths of visible light are exceedingly short compared to most obstacles, the effect is hardly noticeable. Light beams always form shadows when objects block them.

Doppler Effect

The frequency of a wave affects how we perceive it. For example, the frequency of a sound wave determines its *pitch*—how high or low the sound is. The higher the frequency, the higher the tone. Our eyes and brains interpret the different frequencies of visible light as color. Higher frequencies produce shades of green and blue, and lower frequencies produce oranges and reds. The source of a wave determines its frequency. However, an observer may sense a different frequency from that emitted by the source if he or the source is moving relative to the other.

In general, as the source of waves or the observer move toward each other, the observer receives them at a higher rate than the source is emitting them, which means that the observed frequency is higher. Similarly, if the source or the observer is moving away from the other, the observer receives the waves at a lower rate, so the observed frequency is lower.

This phenomenon was first predicted in 1842 by the Austrian mathematician Christian Doppler. His astronomical observations led him to predict that scientists would eventually be able to tell the speed of a star by its color change. If the star was moving toward the observer, the waves would appear to be closer together and thus bluer in color. If the star was moving away from the observer, the waves would appear to be farther apart and the star's color would be redder. This prediction was remarkable because no instrument existed then to measure the predicted color shifts.

Christian Andreas Doppler (1803–53) was an Austrian physicist and mathematician who is best known for discovering the wave phenomenon that bears his name.

Doppler confirmed his theories with experiments using sound waves. He placed musicians on a train. They played single notes on their instruments as the car passed other highly trained musicians on the ground next to the track. These individuals wrote down the musical notes they heard as the train car passed. Doppler was able to demonstrate that relative motion between sound source and observer did indeed change the frequency perceived by observers. His findings were not immediately accepted by the scientific community, but eventually the **Doppler effect** was recognized as a property of all waves. You will learn more about the Doppler effect in later chapters.

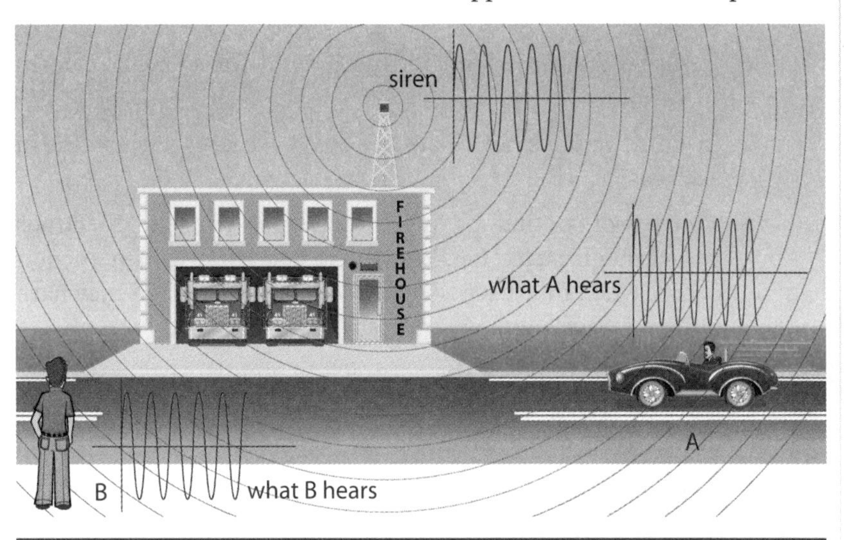

12-36 In the Doppler effect an observer receives a different frequency than that emitted by the source of the wave if he or the source is moving relative to the other. Here, A hears a higher frequency than the siren; B, who is stationary, hears the siren's frequency.

12.12 Using Waves to Solve Problems

Is there a way to use waves to accurately keep time? Can you think of the most accurate way to tell time? In 1955 Louis Essen developed the first *atomic clock*, which was more accurate than any pendulum or quartz clock had ever been. His clock produced an error of less than 1/300 s per year (or, an error of only one second in three hundred years).

Many atomic clocks use a gaseous sample of the element *cesium*. When the gas absorbs radio waves at a particular frequency, all the cesium atoms vibrate together, emitting their own waves. The periodic signal, which is very steady, is used to drive the electronic circuits that indicate the time. The atomic clocks located at the US Naval Observatory in Washington DC are used to report standard time for the United States. Those clocks are so accurate that they will neither gain nor lose one second in more than 60 million years. The current definitions for the SI units of time, frequency, and length depend on the waves generated by the cesium atom.

12C Section Review

1. What do we call waves in matter that transmit energy through particle motion? What nonmaterial waves consist of pure energy?
2. Sketch a single cycle of a waveform and label the parts and dimensions of the wave.
3. What do we call matter through which mechanical waves travel? How far does an individual particle of matter move in comparison to the distance through which a wave travels?
4. What two factors determine wave speed?
5. If the frequency of a wave in a given medium is twice that of another wave in the same medium, how do their wavelengths compare?
6. Describe the three kinds of mechanical waves mentioned in this section.
7. In what two ways can a wave bend? What causes the bending in each case?
8. Two sets of continuous waves traveling in opposite directions pass through each other. What wave interference phenomenon occurs when they have the same wavelengths? when they have different wavelengths?

DS ✪ 9. One way to improve the accuracy of atomic clocks is to use lasers to cool the cesium atoms to almost absolute zero. How does cooling affect the motion of the cesium atoms?

DS ✪ 10. What term could we use to refer to the 9,192,631,770 Hz at which cesium atoms resonate? (See Subsection 12.4.)

11. (True or False) Mechanical waves can be classified as transverse, longitudinal, or combinations of the two.

✪ 12. The speed of sound in room-temperature air is 346.5 m/s. A musical instrument produces the note A above middle C, which corresponds to a frequency of 440. Hz. What is the tone's wavelength?

Chapter Review

1. Compare the way a ringing bell moves and does work on the air particles in its surroundings with the way a hammer does work when striking a nail.
2. How does the amplitude of an oscillating object relate to its rest position?
3. Express the unit hertz (Hz) in terms of one of the seven base SI units.
4. Explain briefly how a vibrating string illustrates the law of the conservation of energy.
5. Explain why natural damping is often desirable.
6. What kind of important scientific information have pendulums been used to collect?
7. How is energy transported through matter using periodic motion? How is this kind of particle motion different from particle motion due to thermal energy?
8. How does wave height compare to wave amplitude?
9. How is the power (or rate of energy transfer) of a mechanical wave related to its amplitude? What determines the energy of an electromagnetic wave?
10. When transverse waves move through matter, you can see a familiar wave pattern with crests and troughs. What parts of longitudinal waves correspond to crests and troughs?
11. What properties of a wave change when it is refracted?
12. What are the stationary points in a standing wave called? How many are there in each wavelength?
13. Light diffracts around an object far less than sound does. What is formed behind the object as a result?
14. (Multiple Choice) The tone of an ambulance siren will sound higher
    a. when the ambulance drives toward you.
    b. when you drive toward the ambulance.
    c. the instant when the ambulance passes by you.
    d. when the ambulance drives away from you.
    e. choices a and b

True or False

15. The frequency of an oscillating system is the reciprocal of its period.
16. A mechanical oscillating system converts kinetic energy into potential energy and back again.
17. Damping in a vibrating object is related to the mass of the object.
18. In its simplest form, a pendulum consists of a mass attached to a string or rod that swings from a pivot point.
19. The explanation of how a pendulum works has been the same throughout recorded history.

Scientifically Speaking

20. The period of a pendulum is related to the length of the pendulum arm, the acceleration due to gravity, and the pendulum's mass.
21. Mechanical waves require a medium.
22. For a mechanical wave to travel far, the medium must be able to move along with the wave.
23. As the wavelength of a mechanical wave doubles, so does the wave's power.
24. Water waves and certain types of earthquake waves are combinations of transverse and longitudinal waves.
25. The adjacent image illustrates wave diffraction.
26. Wave interference occurs when two different waves pass through each other.
27. The Doppler effect was first observed in sound waves.

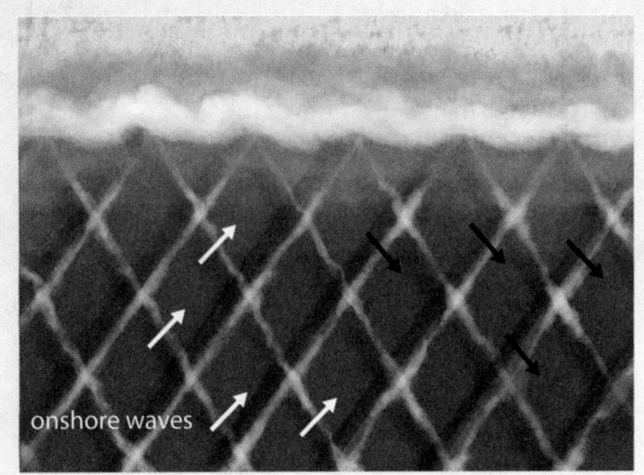

✪ 28. The pendulum of the Great Clock of Westminster has an arm 4.4 m long. What is the period of this pendulum's swing?

✪ 29. Astronaut Fizzix goes through a space warp and gets lost. He makes an emergency landing on an unknown planet. Feeling sort of heavy, he decides to measure the planet's gravitational acceleration to see if he can safely attempt a launch back into space. He pulls out his government-issued 1.0 m pendulum and finds that its period is 1.28 s. What is the planet's gravity?

✪ 30. The speed of light in a vacuum is 3.00×10^8 m/s. If the frequency of a certain color of red light is 4.55×10^{14} Hz, what is its wavelength in meters and nanometers?

✪ 31. Astronaut Fizzix has resigned himself to living on his unknown planet. He decides he needs an accurate pendulum clock. How long a pendulum arm does he need to regulate the clock so that it measures 1.00 s for each cycle of the pendulum? Refer to Question 29 for needed data.

SOUND

DOMINION SCIENCE PROBLEM

Bracing for the Big One

California is a land of earthquakes, laced with a network of active geologic faults. Earthquake waves violently shake buildings and bridges. Most deaths during earthquakes occur when structures collapse. How can engineers prepare California, with its huge population and architectural marvels, for the "big one"?

13-1 Low-frequency earthquake waves can cause severe damage in cities and populated areas.

13A Objectives

After completing this section, you will be able to

- ✓ define sound waves.
- ✓ describe how sound waves transmit energy.
- ✓ list and describe how certain properties affect how fast sound travels.
- ✓ summarize the different properties of sound and how we detect them with our ears.
- ✓ explain how important sound phenomena occur.

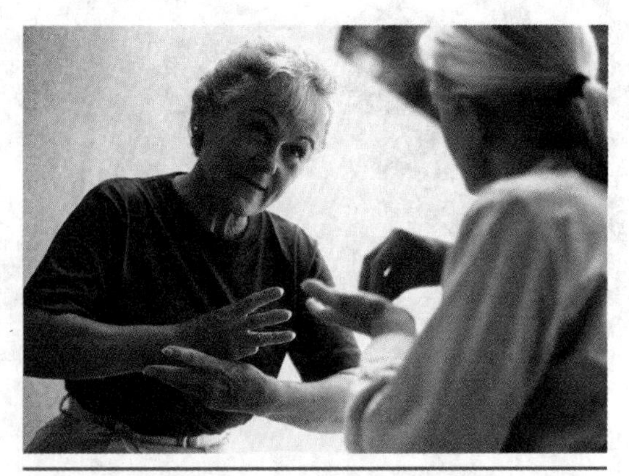

13-2 Sign language is one means of communication used by people who cannot hear.

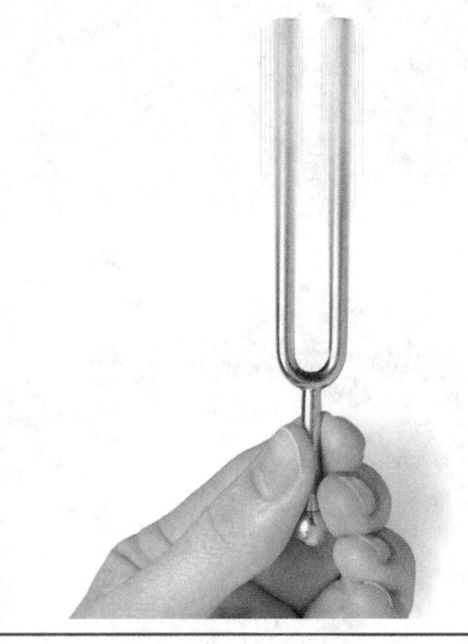

13-3 A tuning fork's tines vibrate, setting up sound waves in the air.

13A The Science of Sound

13.1 Introduction

The human ear is a very important organ. Our ears provide us with information about the world around us. Even before birth, babies can hear the sounds of their environment. We learn to talk by listening to others. The delicate, sensitive hearing organ is easily damaged by injury, disease, or improper development.

Perhaps you know someone who is deaf or hard of hearing. People who lose their hearing later in life often feel very isolated from those around them. They may use a variety of methods to communicate with each other and with the hearing world. Many use sign language, read lips, or even learn to speak without hearing their own voice. Some hearing people are surprised when deaf people react to a loud noise. How did they "hear" it? Could it be that sound can often be detected *without* hearing?

Have you ever felt a sound? Maybe it was a low-flying jet over your home that caused the pictures on the wall to rattle, or perhaps you have stroked a cat and felt its purr as it welcomed your attention. In this chapter we examine the fascinating properties of sound—how it is produced, transmitted, and interpreted.

13.2 Sound Waves

As a starting point, **sound** can be defined as the form of energy detectable by our ears. This definition is not entirely adequate because sound energy, or *acoustic energy*, does not have to be heard to be sound.

Sound waves are carried through a medium. They can pass through gases, such as the atmosphere; liquids, such as oceans; and solids, such as metals and the interior of the earth. Sound waves are more difficult to picture than water waves because they are longitudinal waves. In other words, the atoms and molecules of the medium oscillate back and forth, parallel to the direction of the wave motion. Sound waves are transmitted as periodic changes of particle density. A sound wave "crest" is a point of high particle density, or a **compression**. As the crest passes, the natural repulsion of particles forces them apart. The "trough" of a sound wave is the point of lowest particle density, or a **rarefaction**. As the trough passes, normal random particle motion evens out their distribution, and then the next compression pushes them together again. Alternating compressions and rarefactions distinguish acoustic energy from other kinds of wave energy.

How do you create a sound wave? To make a sound, something must vibrate in a medium. A tuning fork is one of the simplest ways to make sound. When a tuning fork is tapped, the tines or prongs of the fork vibrate, making a sound with a single frequency or pitch. If you look closely, you may be able to see the tines vibrate. These vibrations push on the air, first in one direction, then in the other, sending out waves of compressions and rarefactions. To see the periodic motion of the tines more clearly, immerse their tips into a tray full of water. The fork will cause ripples in the water, illustrating those it causes in the air.

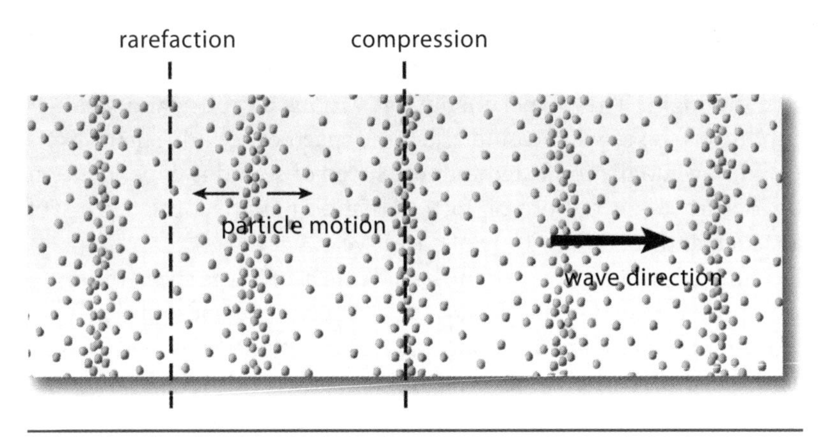

13-4 A longitudinal sound wave. Particle motion is parallel to the direction of motion of the wave.

13-5 A tuning fork's vibrations are clearly seen in a ripple tank.

While we often use the word *sound* to refer to vibrations of matter that we can detect with our ears, we can also detect sound through our sense of touch. You have been able to "feel" as well as hear a cell phone when it vibrates. On a much larger scale, earthquakes are vibrations in the earth's crust. During an earthquake, sometimes one can hear deep rumbling sounds. But as you well know, the destructiveness of earthquakes does not come from the sound they produce—it comes from the waves that shake the ground. You will learn more about these kinds of vibrations later in this chapter.

13.3 The Sound Medium

Robert Boyle conducted an experiment on sound in 1660. He suspended a watch with an alarm from a thread in the glass receiver of his famous vacuum pump (see Figure 13-6). He then pumped the air out of the vessel to form a vacuum. He and his assistants waited for the sounding of the watch's alarm. Time passed, and they did not hear it.

Boyle had suspected that air was necessary to transmit sound. To confirm his hypothesis, Boyle opened a valve and allowed air to reenter the chamber encasing the watch. As the air rushed in, the alarm grew louder and louder. Boyle had successfully demonstrated that sound needs a medium. Further experimentation showed that sound moves quite well through liquids and solids also. In fact, they generally are better acoustic media than air.

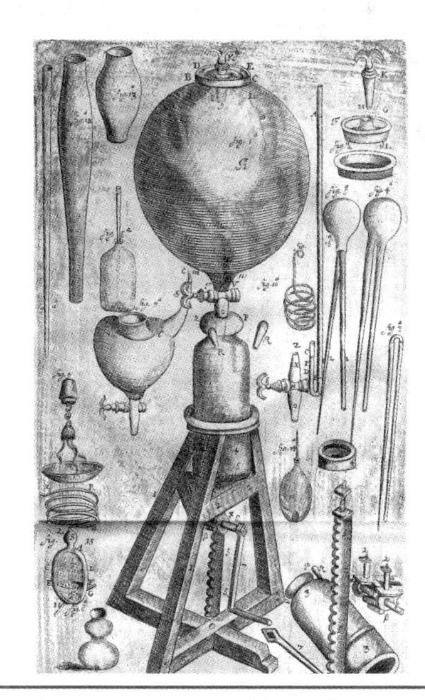

13-6 The apparatus used by Robert Boyle to demonstrate that sound required a medium

13.4 The Speed of Sound

How fast does sound travel? From 1708 to 1709, William Derham first accurately measured the speed of sound through air. He provided assistants with synchronized watches and pistols and stationed them at various distances, some several kilometers away. When they fired their guns at specified times, Derham recorded the time the sound was heard. Knowing their distances and the length of time between the scheduled firing and the sound of its boom, Derham calculated the speed of sound. He averaged the results of several trials and came up with a value that was surprisingly accurate.

William Derham (1657–1735) was an English clergyman who held a lifelong interest in the natural sciences. He wrote many scientific papers and was a member of the Royal Society.

13-1 Speed of Sound in Selected Materials

Material/Condition	Speed (m/s)
dry air, 0 °C	331
dry air, 22 °C (room temperature)	345
dry air, –60 °C (mesosphere)	293
seawater, 35 ppt salinity, 4 °C, 0 m	1467
seawater, 35 ppt salinity, 4 °C, 9900 m	1638
steel	5900

13-7 Many experiments were done to determine the speed of sound in air.

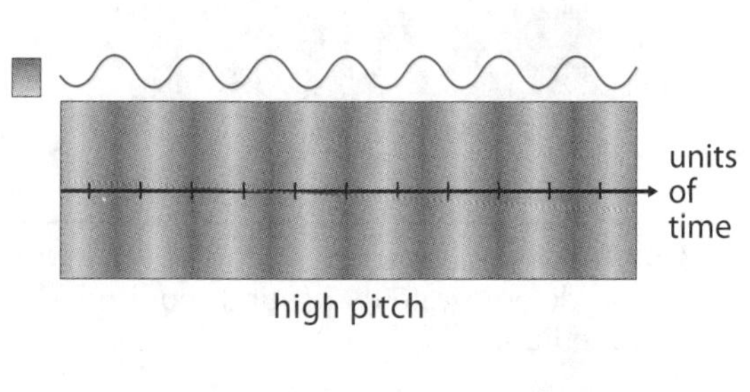

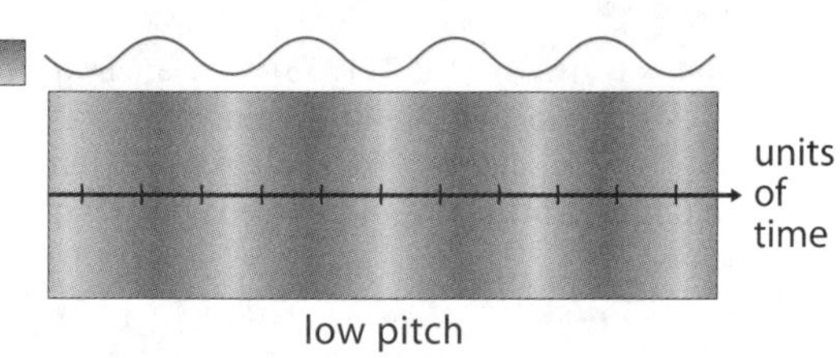

13-8 Comparison of high-frequency (top) and low-frequency (bottom) sounds. Difference in shading indicates amplitude.

Scientists immediately tested Derham's value for the speed of sound. They fired guns and cannons of various sizes during the day and the night. They experimented at various altitudes and weather conditions to see what would affect the speed of sound. Their observations eventually indicated that the speed of sound is dependent on the density of air but not on factors that do not affect the density of air, such as direction or time of day or year.

Since Derham's time, scientists have measured the speed of sound with precise instruments. Today, the accepted speed of sound in air is 331.4 m/s under standard atmospheric conditions (0 °C and 1 atm). The speed of sound in air typically increases with temperature. At room temperature, the speed of sound is 345 m/s, within 1% of the speed first determined by Derham!

The speed at which sound travels through various media depends on the "stiffness" and temperature of the medium. In general, the stiffer the medium, the faster sound travels. Thus sound generally travels fastest in solids, slower in liquids, and slowest in most gases. On the other hand, as density increases in a given material, the speed of sound decreases. Thus, the speed of sound in air is higher on a hot summer's day than on a cold winter's day.

13.5 Characteristics of Sound Waves

Chapter 12 discusses some basic properties of all waves—wavelength, frequency, amplitude, and interference. God created us with sensory organs that can receive different types of wavelike information from our environment and transmit this information to our brains. For each type of wave, the basic properties listed above are interpreted differently. Our eyes sense electromagnetic waves, and our brains interpret the frequency, amplitude, and interference of these waves as color, brightness, and pattern. Our ears sense sound waves. Pressure sensors in our skin can detect very-low-frequency sounds as well. How do our ears and brains interpret the properties of sound waves?

Musical instruments and other objects produce distinct sounds. The shape of the instrument, its materials, and the room or space in which the sound is produced all can affect the sound that reaches the ear. Our ears can detect several measurable properties of sound waves:

- Pitch (high or low tones?)
- Intensity (soft or loud?)
- Quality (sounds like a violin or a steam whistle?)

All these factors affect every sound we hear.

Pitch and Frequency

Sound waves form a continuous **acoustic spectrum** of energy. This spectrum consists of waves with differing frequencies, or vibrations per second, including frequencies too low to hear, audible sounds, and frequencies too high to hear. Frequencies below human hearing are called *infrasonic* sounds, and those above are called *ultrasonic* sounds.

Pitch is how high or how low an audible tone sounds to the human ear. The term refers to how we perceive a sound's frequency. The higher a sound's frequency is, the higher its pitch. Recall from

Chapter 12 that frequency is measured in hertz (Hz). The bottom note on an 88-key piano has a frequency of 27.5 Hz. The highest note has a frequency of about 4186 Hz. Young children can hear pitches ranging from about 20 Hz to 20,000 Hz. As a person grows older, he typically loses the ability to hear higher frequencies.

Do different pitches travel at different speeds through a uniform medium? What would it sound like if the lower notes from the choir traveled faster than the higher notes? You would hear them at different times, and the choir wouldn't sound unified. Since this doesn't happen, we can conclude that sounds of all pitches travel at the same speed. Experiments have confirmed this hypothesis.

Loudness and Intensity

Loudness is the response of your ear to the intensity of a sound wave. **Intensity** is a measure of the power contained in an acoustic wave, that is, the rate at which it transmits energy. The loudness of a sound depends on the sound's energy or power. (Recall that the power of a mechanical wave is proportional to the square of its amplitude.) Loudness also involves interpretation by the human brain. A sound of a certain intensity that seems painfully loud to one person may be just uncomfortably loud to another. Each person perceives sound intensity differently.

The intensity of a sound wave is compared to the energy of a standard audible sound. The softest sound detectable by the average young person's ear is called the *threshold of hearing (TOH)*. It corresponds to a specific *sound-pressure level*, which is the maximum pressure exerted by a passing sound wave. This value is about twenty-millionths of a newton per square meter (20 μPa). All other sound intensities are usually measured as a ratio compared to that sound amplitude. Because the range of audible sounds is so large, this ratio is expressed on a *logarithmic scale* (as powers of 10). A logarithmic scale describes a wide range of values using a small range of numbers. Intensity is measured in units called *decibels (dB)*, which were named for Alexander Graham Bell.

The TOH has a loudness of about 0 dB near 1000 Hz. A whisper is about 20 dB and has 100 times as much energy as the TOH. Normal conversation is about 60 dB, about a million times the energy of the TOH. At 120 dB, called the *threshold of pain*, sound becomes painfully loud and can permanently damage hearing. A 120 dB noise has a trillion times as much power as a sound at the threshold of hearing!

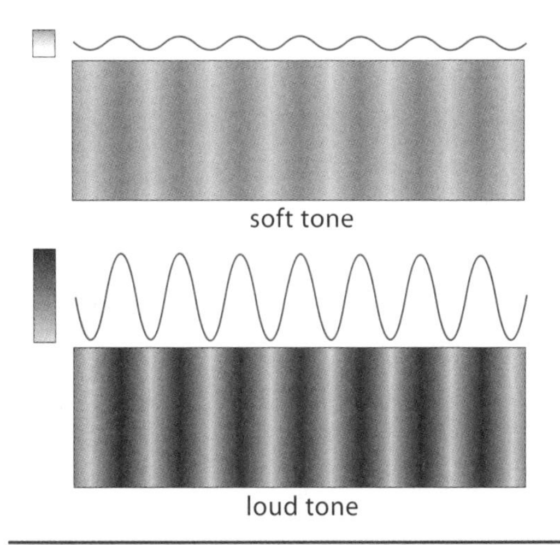

13-9 Comparison of a soft (top) and loud (bottom) sound of the same pitch

The bel (B) was a unit created by Bell Telephone Laboratories for the standard amount of signal energy loss measured over 1 mile of telephone wire. The unit was too large for most other power scale purposes, so the decibel (dB) was selected instead.

The human ear is most sensitive to sounds of 1–5 kHz (1000–5000 Hz).

At a frequency of 1000 Hz, a tenfold increase in sound energy generally equals a doubling of loudness.

Quality and Harmonics

Tuning forks produce sound with mainly one frequency. When a musical instrument plays the same note, the sound is different from that of the tuning fork because a musical instrument plays mixtures of notes, or acoustical tones. When a trumpeter, for example, plays a middle C, the instrument produces a mixture of frequencies that are multiples of middle C's frequency. This mixture gives the trumpet its unique sound. The particular sound of an instrument is its **quality**. Acoustic quality is what makes any sound distinctive. We make judgments on whether a sound is pleasant based on its quality.

clarinet

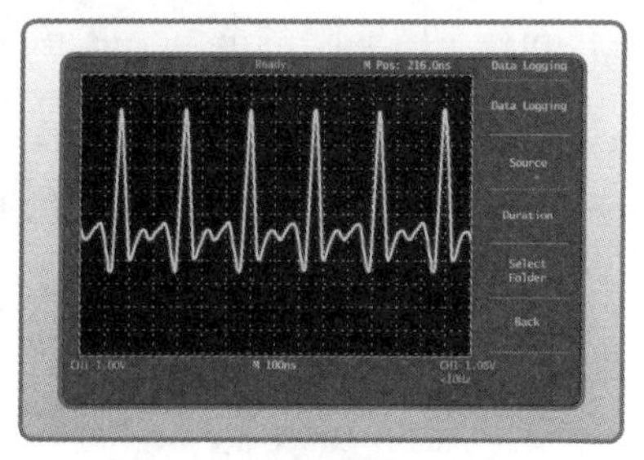

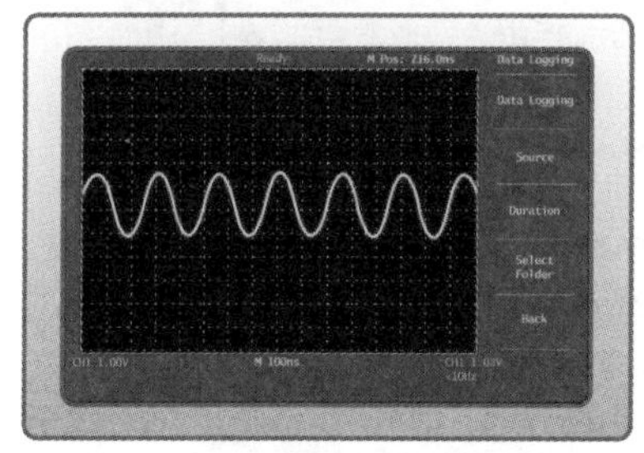

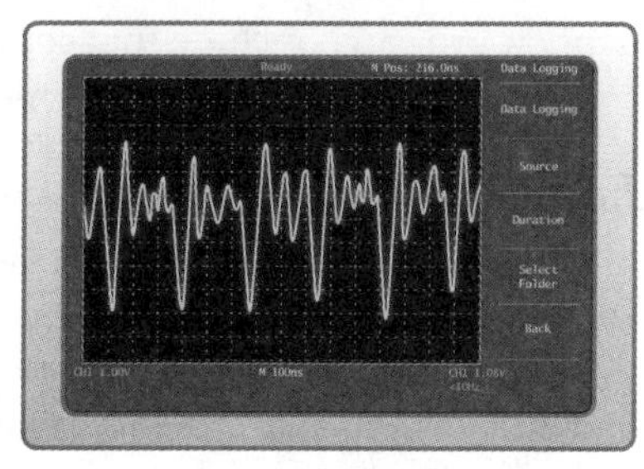

13-10 Waveforms showing the combination of overtones for selected instruments

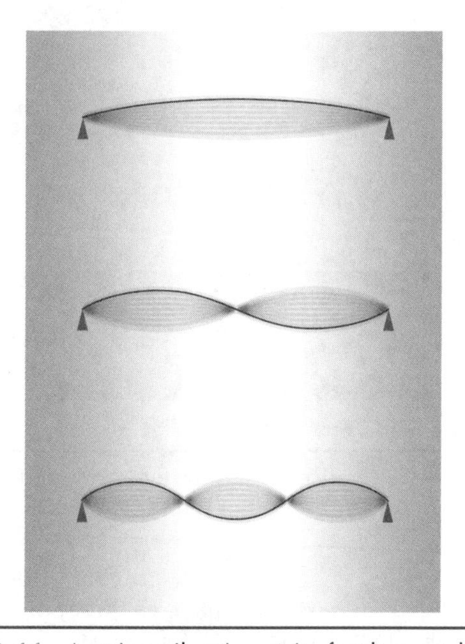

13-11 A string vibrating at its fundamental frequency (top), its first overtone (middle), and its second overtone (bottom)

Vibrating strings illustrate how instruments produce mixtures of notes. If you pluck a guitar string, for example, it vibrates as one big wave from end to end (see Figure 13-11). The string produces this same tone each time it is plucked. This sound is called the fundamental tone, or simply, the *fundamental*. It is a *standing half-wave* anchored at either end by two *nodes* (see Chapter 12 for definitions of these terms). If the string is tight enough, it also vibrates in other standing-wave patterns at the same time, with more than one crest along the string. These shorter, faster vibrations produce higher pitches called *overtones* or *harmonics*. Harmonic frequencies are whole-number multiples of the fundamental frequency. For example, if the fundamental frequency of 440 Hz is 1 × 440 Hz, then the first overtone of the 440 Hz tone is 2 × 440 Hz, or 880 Hz. When you pluck a guitar string, it typically produces several of these harmonics simultaneously. The fundamental tone is the loudest and determines the note people hear, but the quieter overtones give the guitar its unique quality. The construction and materials of an instrument determine its dominating harmonics, making each instrument sound distinct.

13-12 The Whispering Gallery in the US Capitol

13.6 Sound Phenomena

The National Statuary Hall of the United States Capitol is sometimes called the Whispering Gallery. The building's architect unintentionally designed this elliptical room so that its hard plaster walls reflect sound from one end of the room to the other. Although no one in the center of the room would notice, a person at one end of the hall can easily hear the hushed conversation between persons at the other end. Why does this happen?

Acoustics

A room or a building's acoustics is its effect on sound. Any type of enclosure will reflect sound. How well it does that is another matter. The science of **acoustics** is the study of controlling the

13-13 The Walt Disney Concert Hall shows how acoustics is an important factor in performance hall interior design.

sounds that reach our ears. Using this science, architects can build auditoriums with good acoustics so that the entire audience can hear adequately. They also try to minimize *reverberations*—multiple echoes of a sound formed when it reflects off distant surfaces. Reverberations may distract or irritate listeners because the same sounds (the original and the echo) arrive at slightly different times. Reverberations can make speech unintelligible and music jarring if the delays are more than a few milliseconds.

In general, materials with hard surfaces are used where good sound reflection is desired. Chapter 12 discusses how reflection of a wave off a hard surface returns nearly all the wave's energy. Sometimes hard surfaces are curved to reflect sound to specific areas to balance the overall sound. Often the sides of a stage are angled to reflect sound outward to the audience. However, if sound is reflected back too strongly from different directions, reverberations may occur. They can be prevented by covering walls and ceilings with sound-absorbing materials. Soft, porous materials make good sound absorbers. Numerous small holes or openings greatly improve the sound-muffling qualities of almost any material. Acoustic tiles are designed with such holes to trap sound energy.

It's important to remember that sound is a form of energy. The greater the intensity, the farther sound waves travel. However, sound cannot travel forever because, like all energy, it changes to other forms when it interacts with matter. Though good acoustics are important, nothing compares to sheer volume for ensuring a performance is heard. Even the most acoustically advanced auditorium has some type of *amplification* system to increase the amplitude of sound waves (see Section 13C). The excellent acoustics of some ancient amphitheatres, however, are nearly legendary. For example, the Arena in Verona, Italy, is one of the largest Roman amphitheaters remaining in the world today. Its acoustics are so good that 15,000 people can listen to an outdoor opera performance without artificial amplification.

13-14 Jesus used the excellent natural acoustics of the shoreline of Lake Gennesaret to preach to the multitudes from Peter's boat (Matt. 13:2; Luke 5:3).

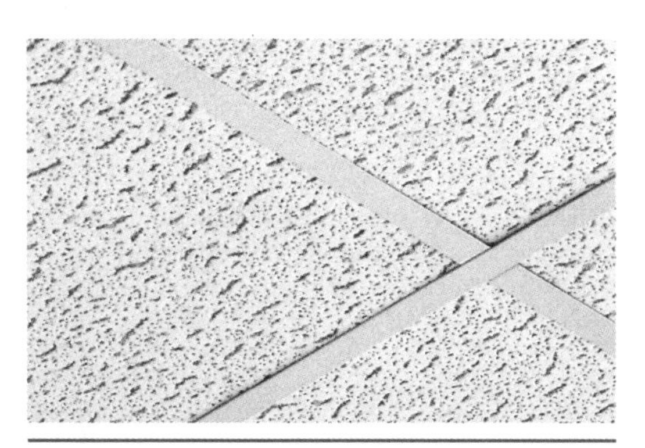

13-15 Sound-absorbing acoustic tiles help reduce undesirable sounds.

13-16 The tone obtained from blowing across a bottle mouth is due to a resonating air column in the bottle.

Resonance

Most objects have a frequency at which they naturally vibrate. If a sound matching an object's *natural frequency* is produced nearby, the object itself begins to vibrate. You may have noticed a car that rattles only at certain speeds or a radio that buzzes only when playing certain sounds. These are examples of resonance.

Acoustic resonance affects our daily lives. *Resonance* in your sinuses amplifies the sound of your voice. The air column in your ear canal amplifies incoming sounds so that you can clearly hear. The ornate wooden bodies of stringed instruments resonate to give these instruments their distinctive qualities. The air rushing at the proper angle over the opening of a bottle creates vibrations that resonate in the bottle. You may have heard the wind moaning as it blows through a building. These are resonant tones created as wind blows by gaps in windows or across chimney openings. The deep-toned blast of an ocean liner's horn depends on resonation in the horn. Organ pipes also resonate to produce loud, distinct tones with a small amount of air.

The resonance of buildings is a major cause of building collapse in earthquake zones. As the ground shakes, the vibrations pass up into the building. If the frequency of the earthquake waves is similar to the building's natural frequency, the amplitude of the building's motion increases until the structure falls apart.

Beats

When two waves with different frequencies interfere, they produce a new wave. The resulting waves, called *beats*, contain periodic crests and troughs that are much larger than either of the contributing waves. (See Figure 12-33 on page 271.) In sound waves, beats are heard when two sources are at slightly different frequencies. For example, when two musical instruments are not quite tuned together, you can hear a warbling or buzzing noise. To avoid this problem in a concert, the concertmaster plays a pure tone on his instrument, and the other musicians play the same note, listen for beats, and adjust their instruments to bring them into tune with the concertmaster's note.

Digital devices used by musicians to tune instruments also employ the acoustic beat concept. The monitor is placed on or near the instrument, and a note is played. The monitor displays a series of lights or a digital scale indicating the presence of a beat frequency created by the interference between its internal tone and the instrument's note. The musical instrument is adjusted until the two sounds have exactly the same pitch, which is indicated by the beat display's disappearing or the illuminating of green lights on the device.

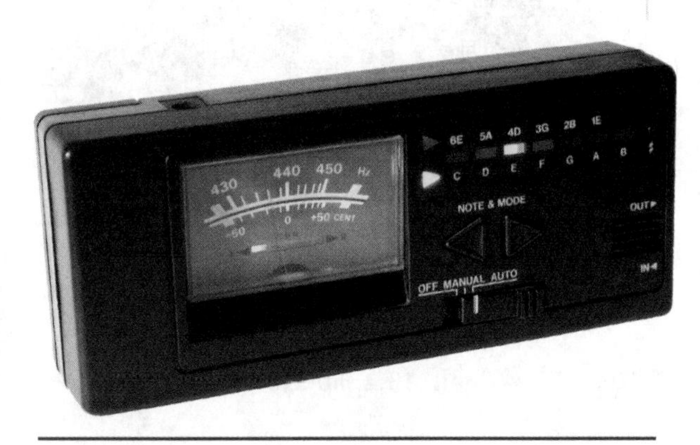

13-17 A digital musical-instrument tuner showing a 436 Hz tone

> The Doppler effect can occur when the sound's source, the listener, or both move relative to each other.

The Doppler Effect

In acoustics, the Doppler effect occurs when either the sound's source or the listener moves relative to the other. The source's (or listener's) motion changes the rate at which the sound waves arrive at the listener, although the sound itself is being produced at a constant frequency. If the source is moving toward the listener, the sound waves are generated in the air closer together in the direction of motion, and the hearer receives more sound waves per second. The heard sound has a higher pitch than the source's. In the opposite case, sound waves are produced farther apart in the direction opposite that of the source's motion. If the source is moving away, the waves are farther apart as they arrive. Their pitch, as heard by the listener, is lower than the source's frequency.

You have probably heard the Doppler effect in action. For example, the sirens of emergency vehicles change pitch as they whiz by you on the road. You can hear the same effect at auto raceways, at air shows, and near major highways. In each case, the droning of vehicle tires or the hum of engines changes pitch as cars and airplanes move relative to the listener.

The Doppler effect on received sound depends on the speed of the sound's medium as well as the relative motion of the source and the hearer. If the wind is blowing toward you from the direction of a stationary source of sound, the sound's frequency will be slightly higher than if the air is still. For example, a 20 km/h (12.4 mi/h) wind will increase the pitch of a 1500 Hz tone heard by a stationary listener about 24 Hz! This is why distant sounds like train horns or airplane engines seem to vary in pitch on gusty days.

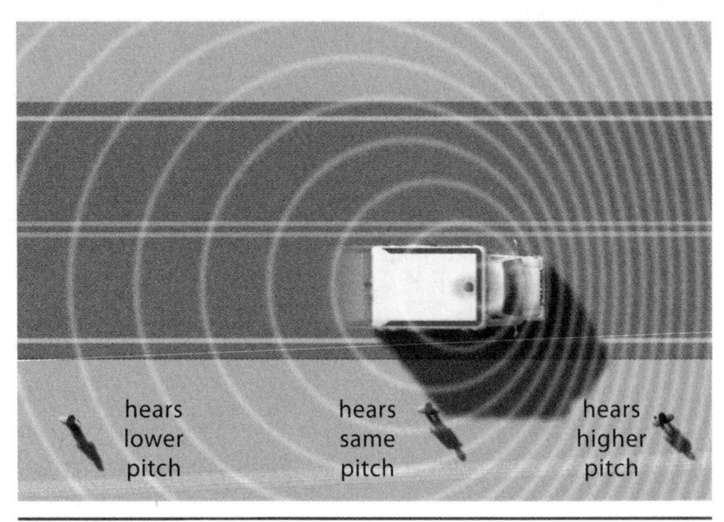

13-18 The Doppler effect alters the way we hear sound from a moving source.

13A Section Review

1. What is sound? How do sound waves differ from water waves?
2. Summarize the experiment that proved that sound requires a medium.
3. Why did scientists test Derham's value for the speed of sound in air? Is this a normal aspect of science?
4. What factors affect the speed of sound in a medium? In what kind of medium is sound generally fastest?
5. What audible property of sound is related to its frequency? What range of frequencies is detected by a person with good hearing?
6. Distinguish between sound intensity and loudness.
7. What makes various sources sound different, even if they are playing the same pitch? Why does this happen?
8. What are reverberations? Why do they occur? When are they undesirable?
9. What factors affect the frequency of a sound heard by a listener compared to the sound's frequency at its source?
10. (True or False) Two convertibles are driving in a line in the same direction at 105 km/h (65 mi/h) on a straight highway. The driver of the following car honks his horn, producing a pure 1500 Hz tone. Since both cars are traveling at the same speed, the driver of the lead car hears a 1500 Hz tone from the horn.

13B Objectives

After completing this section, you will be able to

- ✓ describe the various parts of the human body related to the voice.
- ✓ explain how human anatomy contributes to vocal production.
- ✓ define vocal range and explain why it differs among people.
- ✓ describe the structure and function of the parts of the human ear.
- ✓ discuss design features of the ear that work around acoustic limitations.

13B The Human Voice and Hearing

13.7 Anatomy of the Human Voice

Many creatures can make sounds through calls or voices, but God supplied humans with a unique combination of physical features that permit us to do more than just make a few distinctive sounds. We can communicate complex thoughts using words, whether through speech or singing. Unlike the rest of God's creation, we are able to vocally praise Him. How does this wondrous ability work?

Larynx

Vocal sounds originate in the larynx. The *larynx* is a box-like structure located at the top of the trachea, the "wind pipe" that leads to the lungs. You probably know the larynx better as the "voice box" or "Adam's apple." It consists of nearly a dozen pieces of cartilage held together by ligaments to form a flexible container to support the two vocal cords.

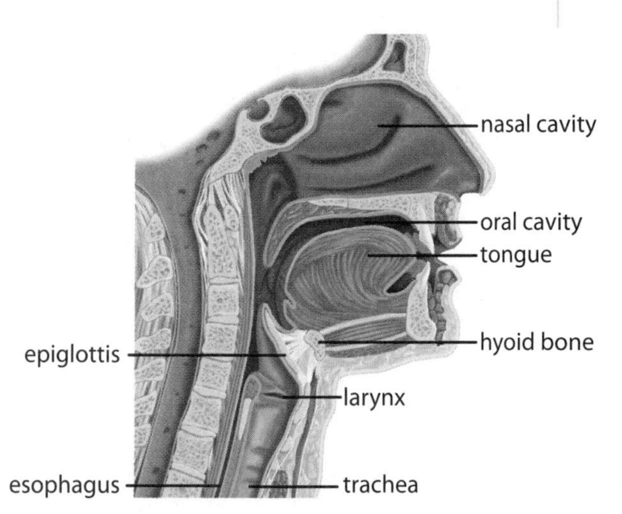

13-19 The internal structures involved in producing vocal sound

Vocal Cords

The *vocal cords* are two flat pieces of mucous membrane tissue stretching across the upper portion of the larynx. The opening between them is called the *glottis*. The muscles surrounding the larynx control the tension in the vocal cords. When you breathe normally, the vocal cords are relaxed and the glottis is open. When you hold your breath, the glottis is closed and the vocal cords are tight. When you speak or sing, a small gap exists, allowing the vocal cords to vibrate with the passage of air. The tension in the cords determines their resonant frequency and thus the pitch of the sound they produce.

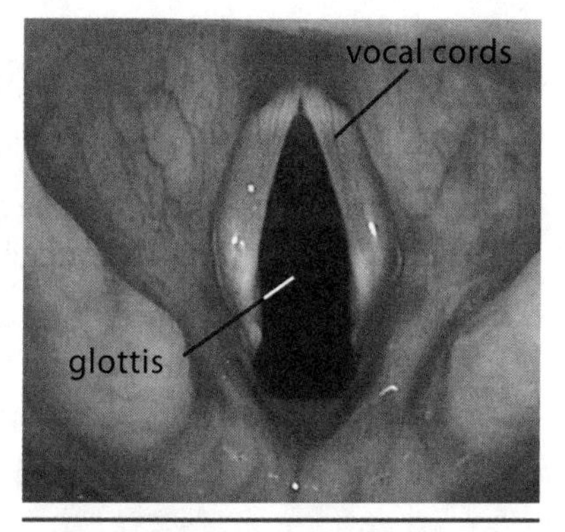

13-20 The human vocal cords

The most important and complex musical instrument is the human voice. To be musical, a voice requires continuous interplay between itself, the ear, and the brain. This exchange allows the singer to monitor and adjust the tone of each note.

Throat and Sinus Passages

Just above the larynx, your throat divides. Air travels by one passage (the trachea) to and from your lungs. The other passage is for food on its way to your stomach. This is the *esophagus*. A flap of cartilage called the *epiglottis* covers the trachea when you swallow to prevent food and drink from entering the lungs.

The back of your throat also opens upward into the nasal and sinus passages behind your nose and surrounding your eyes. Sinuses are spaces and passages in your skull lined with mucous membrane tissue. Besides reducing the mass of bone in your skull, the sinuses also assist your sense of smell, and warm and moisturize incoming air. As a side effect, sound produced by the vocal cords resonates within the sinuses. We are used to hearing people speaking with clear sinuses. When they are blocked due to a cold, a person's voice sounds flat and "nasal." Together, the throat and sinuses form a resonating chamber for the voice.

Diaphragm

The loudness, or power, of your voice is determined by the volume of air that moves by your vocal cords per unit of time. The rate of flow can be controlled by one of two means. You can vary the tension in your vocal cords (which not only determine the pitch of the sound you make but also how large the glottis becomes), or you can use your diaphragm. The diaphragm is a large, arched muscle that forms the floor of the chest cavity, which contains the lungs and heart. The diaphragm works in conjunction with the muscles that raise and lower the ribs to allow breathing. When the diaphragm contracts, it increases the volume of the chest cavity, allowing air to enter the lungs as they expand. When it relaxes, the lungs contract.

Voice teachers and choir directors know that trying to control the flow of air with the vocal cords alone quickly fatigues the vocal cords and their associated muscles. You can obtain much better breath control while singing by using the diaphragm and chest breathing muscles.

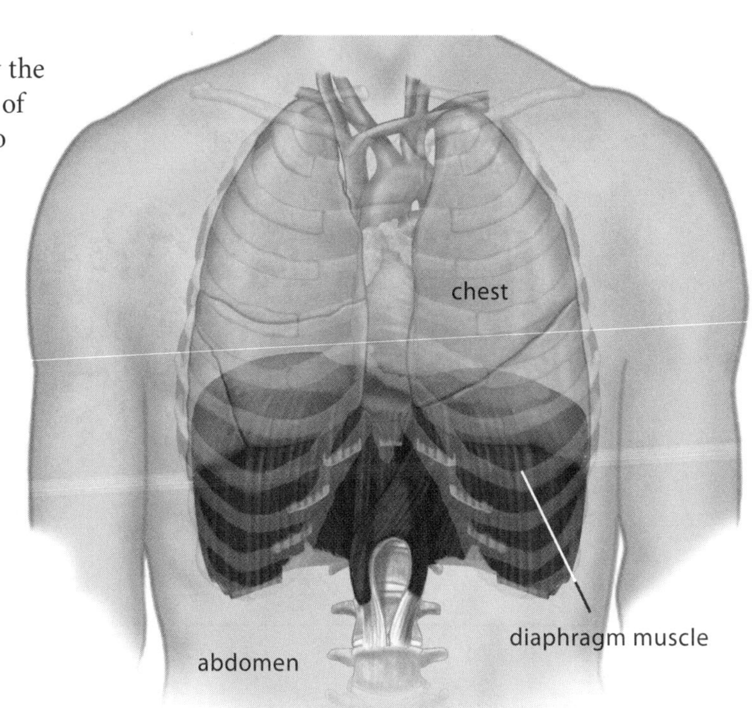

13-21 The diaphragm is the major muscle that controls breathing.

Tongue, Teeth, and Lips

The mouth is where the voice finally bursts forth. The interior of the mouth forms yet another resonance chamber whose shape is completely flexible. The tongue, teeth, and lips are the chief anatomical parts that contribute to the formation of individual sounds. They work together with the cheeks and other mouth muscles not only to produce the exact sound desired but also to give it meaning.

Vocal Ranges

The human voice has amazing flexibility. The useful range of a person's singing voice typically extends between 80–1100 Hz, or over one and a half to two octaves (eight-note sequences). That range varies from person to person, depending on gender and the structure of the larynx. This variance is why choral music is subdivided into parts—soprano, alto, tenor, bass, and so on. Operatic singers recognize four women's parts and four men's parts. Some gifted singers have much wider-than-average vocal ranges. One contemporary singer with the largest recorded range in the *Guinness Book of World Records* has the ability to sing over a ten-octave range! In fact, he holds the record for the lowest note produced by a human voice.

There was a long-standing story that some opera singers could break a wine glass by singing a high pitch equal to the resonant frequency of the glass using only their unamplified voice. However, there was no documented evidence that this had ever been done—until 2005. During a television program to test this claim, a voice coach broke a wine glass on his twentieth try—the first time such an event had been authenticated.

13.8 Anatomy of the Human Ear

The structure of the ear is extraordinarily complex and is a model of efficient design. It provides us with a means of balance and our sense of hearing. Let's examine how the ear receives and transmits sound.

Outer Ear

Sound enters through the external ear and the external auditory canal. The *outer ear* helps us determine the direction of a sound's

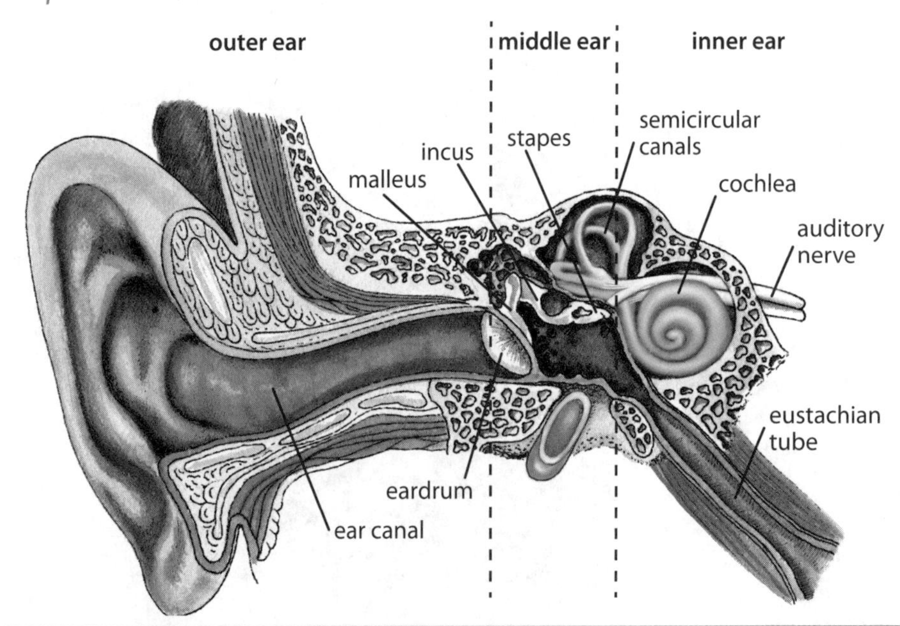

13-22 Cross section of the human ear

source by blocking sounds coming from behind. The *external auditory canal* serves as a resonating chamber, amplifying frequencies from the middle range of the audible spectrum.

The inner end of the auditory canal is formed by the *tympanic membrane*, or eardrum, which separates the outer ear from the middle ear. This thin, flexible membrane converts acoustic energy to mechanical motion by vibrating in response to extremely tiny sound vibrations of air in the auditory canal. A normal eardrum responds to pressure changes that move it as little as a billionth of a centimeter. Ordinary sounds move the eardrum about a millionth of a centimeter.

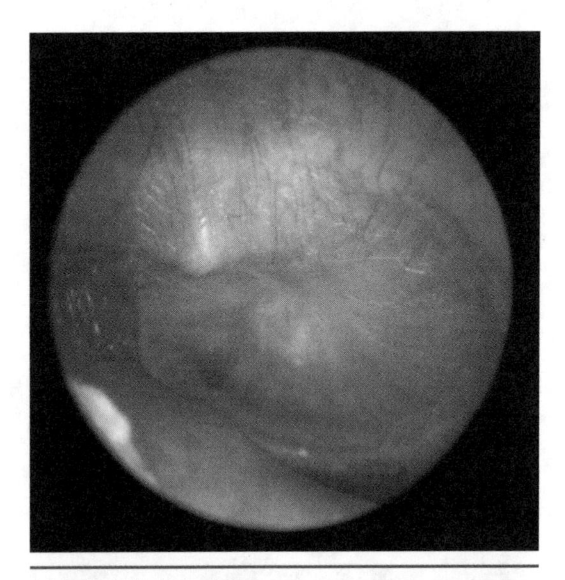

13-23 The eardrum, or tympanic membrane

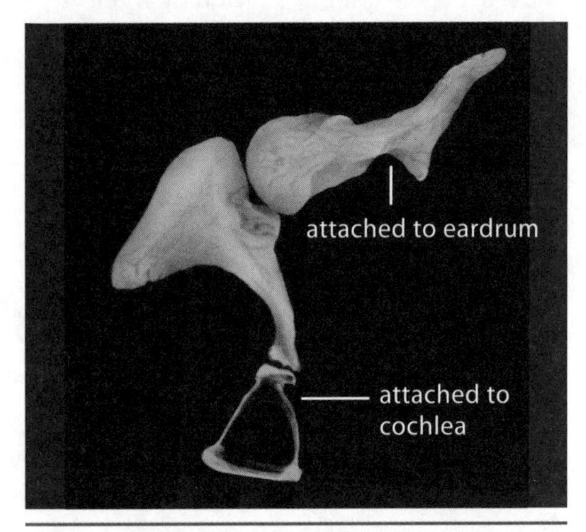

13-24 The bones of the middle ear—the hammer (*malleus*), anvil (*incus*), and stirrup (*stapes*)

Middle Ear

Before sound impulses in the ear can be converted into nerve impulses and sent to the brain, they must enter the liquid that contacts the nerve endings in the inner ear. Transmitting sound from air to a liquid presents a challenging problem. Imagine sitting in a boat and trying to talk to someone swimming underwater. The speeds of sound in the two media are very different. Waves tend to reflect off such boundaries (see Subsection 13.13). Since most of the sound energy would simply bounce back from the water's surface, you may not be heard underwater at all. This problem is solved in the ear by the mechanism of the *middle ear.*

The middle ear consists of the eardrum and three bones called the *hammer*, the *anvil*, and the *stirrup* (named for their shapes). Suspended by ligaments, these bones function as tiny levers. The hammer transmits the mechanical vibrations from the eardrum to the anvil. The anvil in turn passes the vibrations along to the stirrup. The flat portion of the stirrup covers the *oval window*, the entrance to the fluid-filled inner ear, or *cochlea.*

The fulcrums of the three bones are positioned in such a way that the force of their vibrations increases about one and a half times from the eardrum to the oval window. A much greater gain in force is created by pressure amplification (see Chapter 8). Because the area of the eardrum is approximately 22 times the area of the oval window, the pressure of a vibration is increased 22 times as it is transmitted through the oval window. Together, the lever action and pressure amplification in the middle ear increase the force of vibrations about 37 times!

Several structures in the middle ear protect it from injury. Two tiny muscles, controlled by the brain, restrict the vibration of the eardrum and the stirrup as a safeguard against uncomfortably loud sounds. In addition, the *eustachian tube* connects the middle-ear cavity with the throat and helps to equalize the air pressure on both sides of the eardrum. If your ears "pop" as you ascend or descend rapidly in an elevator or airplane, you know that your eustachian tubes are doing their job. You yawn or swallow, opening your eustachian tubes;

and the eardrum, which has been pushed aside by unequal pressure, snaps back into place, causing the popping sound.

Inner Ear

The *inner ear* is the most intricate and least understood part of the ear. It contains two fluid-filled organs: the *semicircular canals* (the organs of balance) and the cochlea, the snail-shaped tube in which sound is converted into nerve impulses.

Running almost the entire length of the cochlea is a delicate partition that divides it into two liquid-filled parts. At the oval window the stirrup transfers sound vibrations to the liquid of the cochlea. The vibrations travel through the liquid to the *organ of Corti*, which runs the length of the partition and contains thousands of sensory hair cells and nerve endings. Differences in structure allow only certain sensory cells in the organ of Corti to be moved by a certain frequency of vibration. The sensory cells relay thousands of separate signals to the *auditory nerve*. These signals indicate the pitch, loudness, and quality of a sound. The auditory nerve conducts these signals to the brain, which interprets them as sound.

The human ear is an amazing organ. Man has never created an artificial model that can duplicate its complex anatomy and physiology. The ear gives clear evidence of a Designer.

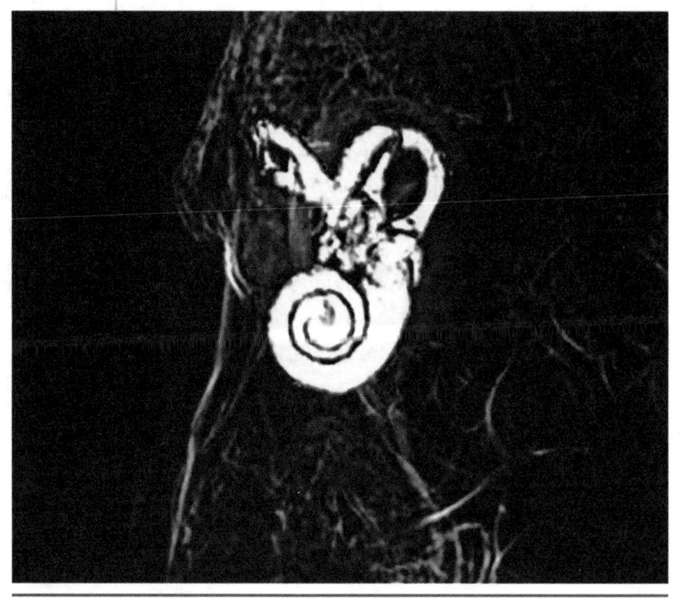

13-25 A micro CT scan of the semicircular canals, which are the sensors for balance, and the cochlea

13B Section Review

1. Name six parts of the human anatomy that affect the voice and explain their significance.
2. Explain why it's better to control your breath while singing by using your diaphragm and breathing muscles rather than your vocal cords.
3. ✪ What anatomical part do you think determines the vocal range of a singer? What properties of this part most influence vocal range?
4. What are the three main functions of the external ear?
5. What design feature of the middle ear increases the amplitude of the sound before it is processed by the inner ear?
6. ✪ What is the state of matter of the substance inside the cochlea? This substance delivers the acoustic signal to the nerve endings that transmit sound to the brain. Considering what you learned about sound-transmission media in Section 13A, explain why this reflects good design.
7. (True or False) The ability of certain opera singers to break delicate wine glasses by pitching their voice at the correct resonant frequency is by and large a myth.

13C Objectives

After completing this section, you will be able to

✓ list several types of sound technologies and explain how they work.
✓ describe the properties of infrasonic sound.
✓ describe the properties of ultrasonic sound.
✓ explain how animals and people use infrasonic and ultrasonic waves.
✓ describe some methods developed to minimize earthquake damage to buildings.

A *synthesizer* is an instrument that can mimic a specific sound by producing the appropriate frequencies and quality of the sound and then combining them together.

13-26 A megaphone focuses a sound in one direction.

13C Applications of Sound

13.9 Acoustic Synthesizers

After World War II, electronic audio systems were developed that could produce musical notes. These sounds emitted from electronic speakers. Electrical and acoustic engineers worked together to refine these systems so that they could imitate almost any musical instrument. With the invention of high-speed digital technology, *acoustic synthesizers* can produce nearly the full range of tones of most instruments as well as realistic imitations of other sounds.

Electronic piano keyboards are very popular and are much less expensive than an equivalent stringed piano. Electronic keyboards are commonly used to supplement traditional pianos and organs with specialized sounds that mimic various classes of musical instruments, even the human voice.

13.10 Acoustic Amplification

Acoustic **amplification** is the process of making a sound louder. *Mechanical amplification* focuses sound energy in one specific direction. This is how a simple megaphone works. Sound waves enter the small opening of a funnel-shaped tube, reflect off the sides of the cone, and exit the large opening. This reflection increases the amplitude of the sound waves leaving the megaphone. Interestingly enough, if you point the large end of a megaphone at the source of sound and listen at the smaller end, the same thing happens. This is why people sometimes cup their ears to hear better. Mechanical amplification occurs in caves and in rooms with hard surfaces, where sound reflection occurs easily.

Electronic amplification uses electronic technology to make sound louder. A sound receiver, such as a microphone, converts sounds into a small electronic signal. This signal is sent to an amplifier, which magnifies the electrical signal. The output signal is identical to the original, but its amplitude is much larger. The amplified signal is then sent to a speaker. A speaker contains a cone of stiff material that vibrates. The vibrating speaker cone produces large amplitude vibrations that we hear as sound. If the system is well designed and

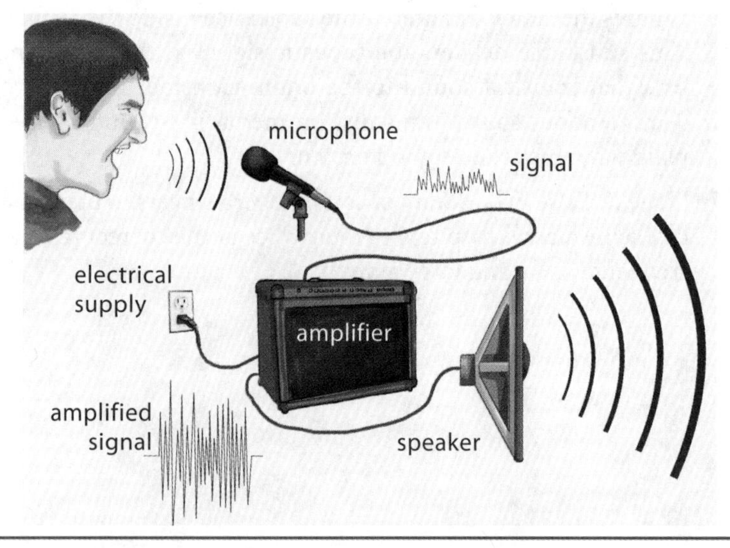

13-27 A simplified schematic of an electronic amplification system

uses a variety of speaker sizes, the output sound is identical to the input but much louder. In the performing arts, sound amplification has become essential as the size of performance halls has increased.

13.11 Echolocation and Sonar

Echolocation

Sound travels relatively fast through any medium, and its waves easily reflect off many materials. Because sound behaves this way, it can be used to measure the distance to an object. The time interval between an outgoing pulse of sound and its returning echo can be used to figure the distance to the sound-reflecting surface. The direction from which the strongest echo returns gives the direction to the object. Using sound this way is called **echolocation**. God designed many of His creatures to use echolocation to "see" long before humans came up with the idea.

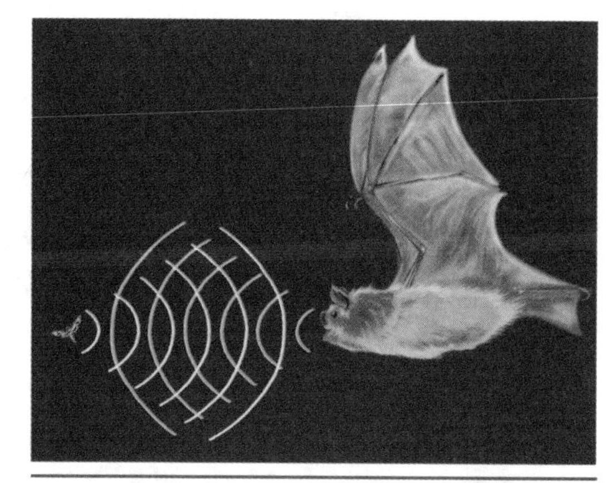

13-28 Some bats use echolocation to hunt and to avoid obstacles.

Some types of bats use echolocation to find food at night and to orient themselves in pitch-black caves. They emit high-frequency chirps beyond the range of human hearing. These ultrasonic waves reflect off insects, cave walls, and other objects. The bats' brains can interpret the data provided by the sound to determine distance, speed, and even the size and type of insect! Since the 1700s scientists had suspected that bats used their ears to navigate and hunt. It wasn't until 1938 that an experiment confirmed that bats use ultrasonic echolocation. Biologists are continually amazed by bats' ability to "see" using sound.

Cetaceans, a group of mammals which includes porpoises, dolphins, and whales, use sound to navigate, communicate, and hunt in the murky ocean depths. In the early 1950s researchers noticed that captured porpoises could easily find their way out of nets through small openings. From these observations they suspected that cetaceans used echolocation. Research of underwater recordings was published in 1957 that confirmed these observations. Cetaceans emit and amplify acoustic signals in their heads to send out powerful sounds into the water. For hunting, they generate high-frequency chirps that can indicate the size and kind of prey. They can also sense whether an approaching animal is a danger. Recent evidence indicates that porpoises use powerful sound pulses to stun and disorient their prey before capture. The use of sound by these animals shows the marvels of God's design, equipping each animal with the tools it needs to succeed in its unique environment.

13-29 Porpoises and other cetaceans use echolocation for hunting as well as communication.

Sonar

Scientists became interested in sound in water as early as 1822, when Daniel Colloden used a submerged bell to calculate the underwater speed of sound. This experiment spurred scientific interest in how sound travels in water. After the RMS *Titanic* collided with an iceberg and sank in 1912, researchers considered ways to avoid future collisions with ice. They suggested using devices in the water to listen to the sounds icebergs make. It was the need for wartime technology, however, that spurred the development of **sonar**. Sonar was frequently used during the two world wars to detect enemy submarines.

The word *sonar,* adopted by the Americans in World War II, is an acronym for *so*und *na*vigation and *r*anging.

13-30 Daniel Colloden determining the speed of sound in water (1822)

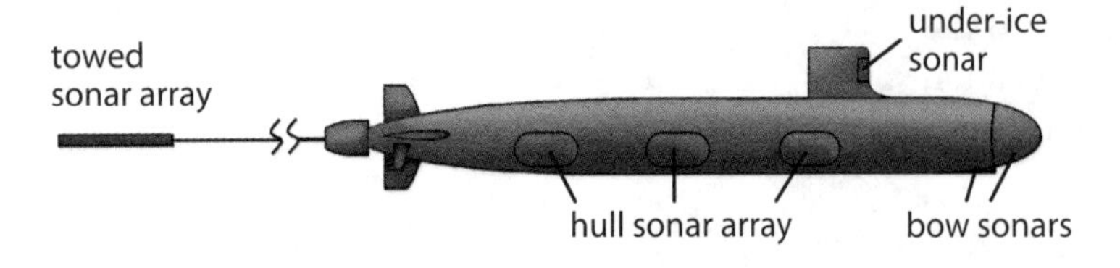

13-31 Passive sonar sensors on a modern military submarine. Using active sonar may give away a submarine's position. Active sonars, like the under-ice sonar, are used only when necessary.

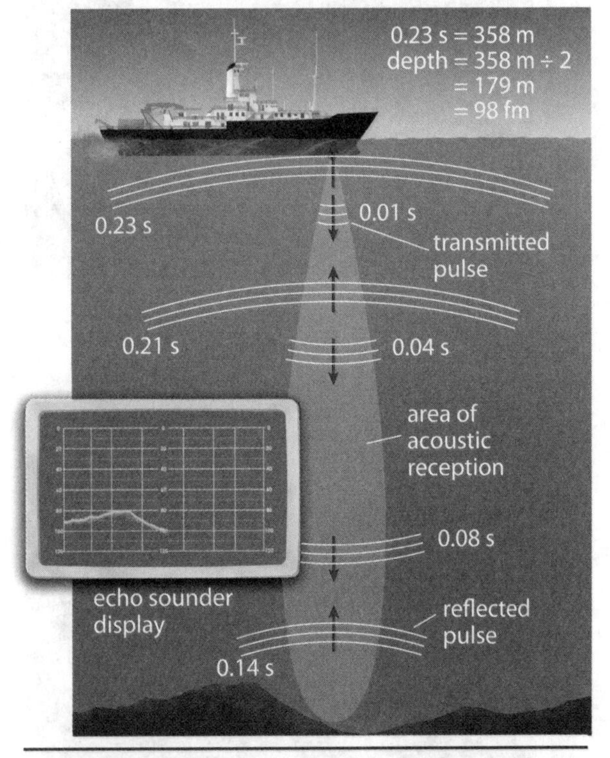

13-32 Active sonar is used to map the ocean bottom and to locate objects on and in the bottom sediment.

This name paralleled another new word—*radar*, an acronym for *ra*dio *d*etection *a*nd *r*anging. The earliest sonars were only receivers of underwater sounds. They were *passive* instruments that used underwater microphones called *hydrophones*. The sensors of *active* sonars, developed during and after World War I, used *transducers* to produce and listen for short pulses of sound—the "pings" often heard in submarine movies.

Military sonars primarily detect enemy submarines. Modern submarines are extremely quiet and require a passive array of many ultrasensitive hydrophones to detect them. Active sonar can betray the location of a ship hunting enemy submarines. Since submarines assigned to hunt other submarines depend on their stealth to be successful, they almost never use active sonar.

Modern sonar technology also has nonmilitary uses, such as mapping the bottoms of oceans and lakes and even revealing geologic formations hidden beneath the bottom sediment. The transducer unit in a ship emits a pulse of sound into the water from the bottom of the hull. The ping travels downward at about 1560 m/s. Sound waves reflect if they hit a boundary between two very different media. Thus, a sandy ocean bottom will strongly reflect the sound upward toward the ship. The transducer receives the pulse, and the sonar receiver calculates the time interval from transmission to reception. The sonar screen displays an image of the surface that reflected the pulse.

Sonar has been an important tool for finding lost wrecks, such as the *Titanic*. It is also used to keep floating deep-sea drilling rigs in position while drilling. By placing active sonar beacons on the ocean floor and hydrophones underneath the rig, an onboard computer controls thrusters that keep the massive vessel directly over the drill hole.

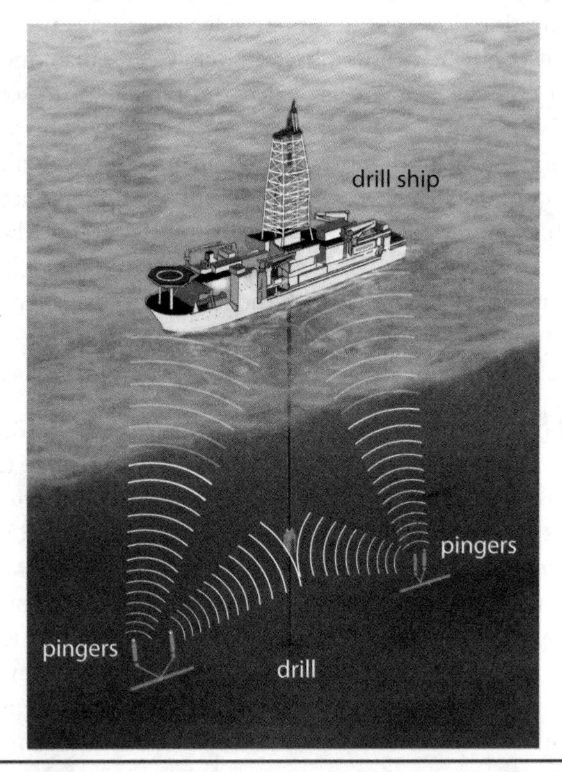

13-33 Drill ships use active sonar pingers to keep themselves positioned directly over the drill hole in deep water.

13.12 Infrasonic Waves

Infrasonic waves are sound waves of low to very low frequency. These waves have many natural and artificial sources. They are too low for humans to hear, but they can often be felt. For example, distant heavy machinery can produce infrasonic vibrations that make people feel nervous or restless. The invention of technology enabling humans to study infrasound is fairly recent.

Infrasound carries large amounts of energy over long distances through a variety of media. The large amplitudes possible with these kinds of waves make them potentially very damaging. Earthquakes are complex combinations of waves of varying frequencies, both audible and inaudible. However, it is the lower frequency earthquake waves that damage buildings and other structures. Volcanoes and avalanches can also produce damaging infrasonic earth waves. Infrasonic seismic instruments can help detect early signals of some of these dangerous phenomena so that people in areas affected by these waves can be warned.

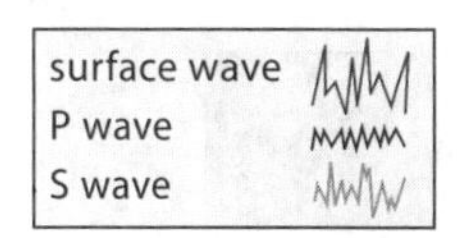

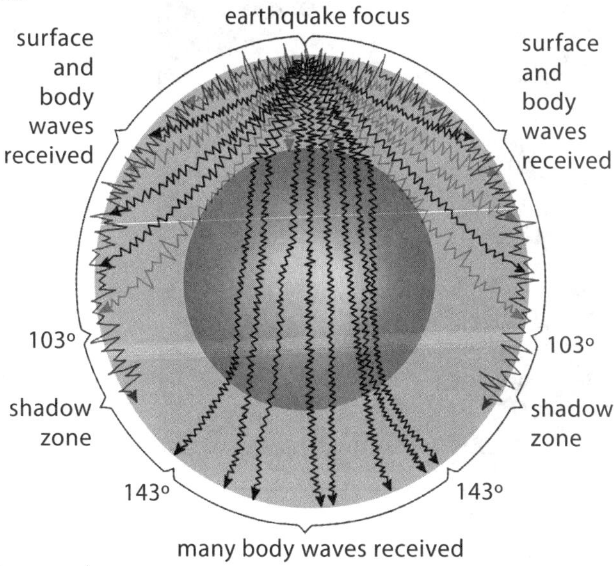

13-34 Earthquakes are disturbances in the earth consisting mainly of infrasonic waves.

Recent studies reveal that infrasound is important to some animal communication. Elephants, for instance, can reportedly communicate with each other up to 10–15 km away at frequencies of 12–35 Hz. Rather than using their ears, they seem to receive these waves through ground vibrations via their feet. Giraffes have been observed to vocalize at frequencies of 14–40 Hz. Scientists believe that they use these sounds to inform others of their distress or location. Infrasound can easily pass through jungle foliage and other obstructions. Cats use infrasonic sound not only to communicate but also to help other cats recover from injury. Research indicates that increase of bone density, tissue repair, and pain relief occur more rapidly when exposed to infrasonic purring near 20–50 Hz. So next time you're sick, don't shove your purring cat off the bed!

Infrasound is also a potential battlefield weapon. Large amplitude infrasonic waves cause disorientation, nausea, and even unconsciousness in humans. Military engineers are studying the feasibility of developing non-lethal weapons that can stun or incapacitate enemy soldiers in combat. As unpleasant as war is, it is an all-too-true reality of our world under the Curse. An effective weapon that would help minimize death in warfare is a God-honoring application of dominion science.

13-35 These tigers (*Panthera tigris*) can use infrasonic waves for communication and to comfort injured members of their group.

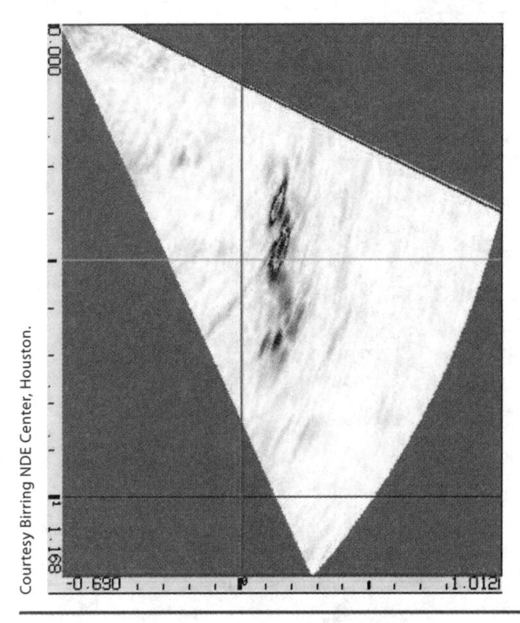

Courtesy Birring NDE Center, Houston.

13-36 Ultrasonic waves can penetrate short distances into metal to detect welding flaws that could weaken the structure.

13.13 Ultrasonic Waves

Ultrasonic waves are sound waves that have frequencies above the range audible to humans—typically greater than 20,000 Hz. Waves with shorter wavelengths and higher frequencies are refracted more than waves with longer wavelengths. Because ultrasonic waves have very short wavelengths, they are strongly affected by the media through which they move. Whenever a wave refracts at a boundary between two very different media, some of the wave's energy is also reflected. Thus, ultrasonic waves easily reflect at boundaries between different materials. This reflectivity makes ultrasonic waves useful in sonar-like technologies. However, higher-frequency waves lose energy to the medium much more quickly than lower-frequency waves. Ultrasonic sonar is not practical for long-distance work in oceans and lakes because ultrasonic waves are quickly absorbed in just a few meters of water. Ultrasonic waves are ideal for short-range sensors.

Ultrasonic waves are useful, not only because they reflect well, but also because they do not strongly diffract around objects. Diffraction depends on wavelength. The shorter the wavelength, the less diffraction occurs for a given-sized opening or object. Longer sound waves tend to wrap around small objects and are not affected by their presence. But if the wavelength is much shorter than the dimensions of the object, the waves cannot wrap around it. Instead, they tend to form shadow zones behind it and reflect off it. Both of these factors make ultrasound a good way to image surfaces and objects inside other objects.

The sensitivity of ultrasonic waves to differences of size and reflectivity makes them ideal for imaging the interior of the human body. Ultrasound is also used in remotely piloted submersibles to do short-range searches for uniquely shaped objects on the ocean floor. Because of these properties of ultrasound, the echolocation used by bats and cetaceans is very effective in detecting food and in navigating.

seismic isolators
flexible mounts
seismic damping
shock absorbers
concrete-filled steel tube columns
concrete
steel tube columns
angled trusses

13-37 Examples of various devices used to reduce damage caused by earthquakes

13.14 Using Knowledge of Geologic Sounds to Solve Problems

Seismology is the study of earthquakes and the waves that cause them. Engineers and seismologists work together to understand how earthquakes damage buildings and other structures. What can be done to make existing buildings less prone to earthquake damage? How can new buildings and bridges be designed to be earthquake resistant?

Engineers have learned that the effects of earthquake vibrations can be minimized by damping building oscillations. Pre-existing buildings can be *retrofitted* to make them more earthquake resistant. Retrofitting can include implementing new technology or strengthening through added supports. Skyscrapers can have trusses, or angled beams, fixed to the outside to dissipate earthquake energy. Sometimes a large tank of water called a *slosh tank* can be put in an upper story of a building to increase the building's inertia and absorb earthquake vibrations. Bridges and overpasses can have concrete-anchored cable supports added to strengthen them. Even houses can be retrofitted.

New buildings can be built on top of large structures called *base isolators*. These structures are shaped like hockey pucks and consist of layers of steel sheets with layers of rubber between them. A lead core penetrates downward through the center of the layers and anchors the building to the bedrock below. Base isolators absorb vibrations caused by earthquake waves to minimize the structure's movement. When the Golden Gate Bridge was renovated to make it more earthquake resistant, base isolators were put under its footings.

Sound is a natural phenomenon of energy moving through matter in the form of waves. God put in His creatures the ability to sense and use this form of energy in many different ways. While we cannot hear all the possible frequencies of sound, God has given us the capacity to devise instruments that can detect them. In this way we can learn about their properties and sources. There is still much to learn! Acoustics remains a wonderful area of study for the Christian physicist or biologist to serve his or her fellow humans and return glory to God.

13C Section Review

1. What musical instruments have been developed that are purely electronic?
2. State the two primary types of sound amplification. Give an example of each.
3. Describe the basic principle of echolocation. What two pieces of information must be processed to determine distance and direction to the target?
4. What animals use echolocation for hunting and avoiding unseen obstacles?
5. What are the two different ways sonar operates? What is the main purpose for each method?
6. What is the entire range of possible sound waves called? How is this range divided?
7. List three sources of infrasonic sounds. What are some effects of infrasound on humans?
8. What factor limits the usefulness of ultrasonic waves?

*9. What applications benefit from ultrasonic waves?

DS ✪ 10. Buildings that are 10 to 12 stories high are especially prone to earthquake damage because their movements resonate with the frequency of earthquake waves. What is this frequency of the buildings called?

11. (True or False) The speed of sound in water was determined more than a century after the speed of sound in air.

FACETS OF SCIENCE

Ultrasound in Medicine

Your relative who is expecting a baby excitedly pulls our her smartphone and shows a picture posted on a social media site, producing "oohs" and "aahs" from onlookers. It is a recent ultrasound image of her unborn baby. Such technology is an amazing application of ultrasound, those sound waves with frequencies too high for humans to hear.

Physicians use ultrasound as a diagnostic tool during pregnancy because it has no known harmful effects with limited use. An ultrasound technician takes a small hand-held device called a *transducer* and places it over the mother's abdomen. The transducer emits ultrasonic waves and receives their echoes as they reflect off the tissues of the unborn child. The machine's computer constructs an image of the fetus.

The latest in medical ultrasound technology, *4D ultrasound*, strings images of the unborn child together to make a movie of the fetus in the womb. It should be noted that some doctors are concerned that ultrasound is being used unnecessarily to obtain keepsake photos of babies before they're born. Some studies suggest that the higher-energy ultrasonic waves needed for 3D and 4D images may be harmful to the baby's development.

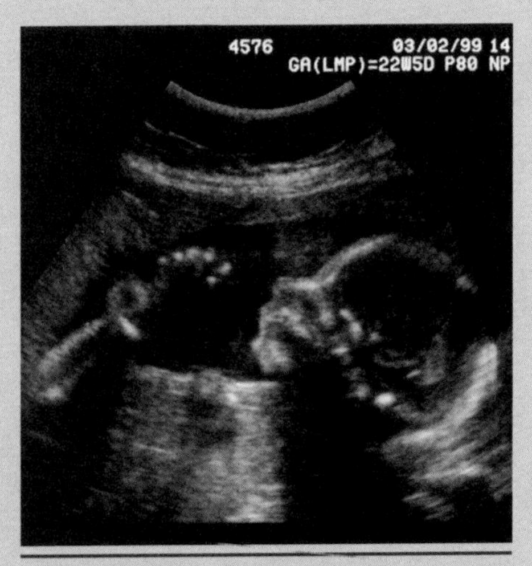

Older, 2D ultrasound images were difficult to interpret.

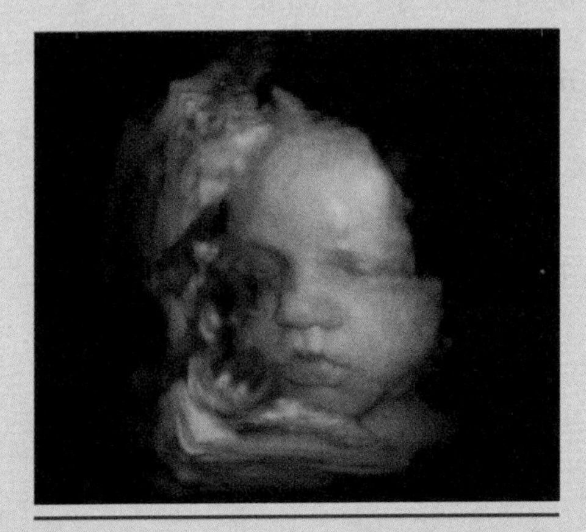

A 4D ultrasound still-frame image of a 33-week-old human fetus

Ultrasonic waves have other uses in medicine as well. Doctors use ultrasound as a diagnostic tool for many illnesses, such as cancer. The technique is similar to ultrasounds performed on pregnant women. The images generated by these ultrasonic waves, called *sonograms*, can give valuable information about the condition of organs, tendons, and muscles. They can also detect the presence of abnormal tissue, indicating a possible tumor.

A special ultrasound image of the heart, known as an *echocardiogram*, gives an image of the heart from the inside out, showing the source of heart murmurs, tissue damage, and coronary artery blockage. Since heart disease is still the number one cause of death in the United States, engineers are working to make ultrasound technology more detailed and accurate, providing physicians an important window to the internal workings of the human body.

Doctors also use focused or high-intensity ultrasound to rid the body of unwanted tissue or objects, such as cancerous and benign tumors, cataracts, or kidney stones. High-intensity ultrasonic waves are focused on the target area to destroy the harmful object.

High-intensity ultrasound also works with antibiotics to kill bacterial infections and can be used to heat areas of the body for physical therapy. Ultrasonic waves can even be used to clean teeth!

Both ultrasonic imaging and treatment illustrate how we can use inaudible sound to help our fellow man.

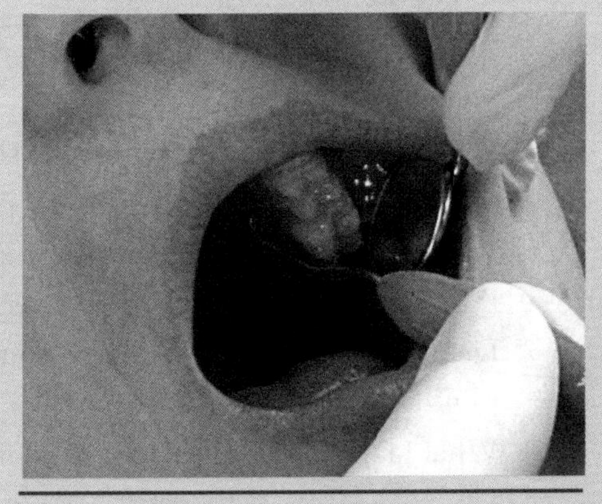

An ultrasonic dental tool used to clean teeth

Chapter Review

1. What is sound?
2. What did Robert Boyle's experiment prove about the nature of sound?
3. What determines the speed of sound through a material? Does sound travel faster through hot steel or cold steel?
4. What property of sound is measured in decibels? Why are these logarithmic units needed?
5. What property of a sound wave is related to its pitch?
6. What is the principal frequency produced by a vibrating object? What are whole-number multiples of this frequency called?
7. How do acoustic tiles reduce reverberations?
8. What physical process occurs when a venetian blind buzzes as wind blows through an open window?
9. Will the pitch that you hear of a blaring radio in an approaching car be higher or lower than what the driver hears? What phenomenon does this illustrate?
10. Which part of the human anatomy is acting when you hold your breath? Which part is most effective at controlling your breathing while singing?
11. What factors determine the usable range of a person's singing voice?
12. What physics problem is overcome by the design of the middle ear? Why is this necessary? How does the middle ear accomplish this feat?
13. What principle of fluid mechanics is utilized when you "pop" your ears while ascending or descending on a mountain road?
14. If an acoustic synthesizer is to sound like a violin playing a certain note, what two properties of the violin must it imitate?
15. List the three pieces of equipment necessary to electronically amplify a sound. Describe the function of each.
16. Why are ultrasonic waves (instead of infrasonic waves) used for detecting welding flaws?

True or False

17. Losing one's hearing does not mean a total loss of communication with other people.
18. Sound travels through air by the up-and-down motion of air molecules in a way similar to that of a water wave.
19. A sound's speed does not depend on the intensity of its source.
20. In the normal audible range, a sound's speed does not depend on its frequency.
21. A 100 dB sound has ten times the energy of a 10 dB sound.
22. The second harmonic of a 300 Hz tone is 900 Hz.
23. Reverberation and resonance are two different phenomena.

Scientifically Speaking

24. The acoustic Doppler effect depends only on the relative speed and direction of the motion between the sound's source and the hearer.
25. There is no evidence that humans can produce ultrasonic pitches with their voices.
26. The ear and all its parts are involved solely in the sense of hearing.
27. Most nonmilitary applications of underwater sound use active sonar.
28. Infrasonic waves can have both damaging and beneficial effects on humans.
29. Ultrasonic waves in water are a good way to detect enemy submarines over long distances.

✪ 30. Suggest two reasons why the human voice is more versatile than nearly any other musical instrument.

✪ 31. Research and write a short paper on the physics and historical development of your favorite musical instrument.

✪ 32. Research the original methods used by scientists to determine the speed of sound in air and in water. Design an experiment that duplicates their efforts using modern materials.

✪ 33. Suggest a way that echolocation could be used to help people who are blind.

ELECTROMAGNETIC ENERGY

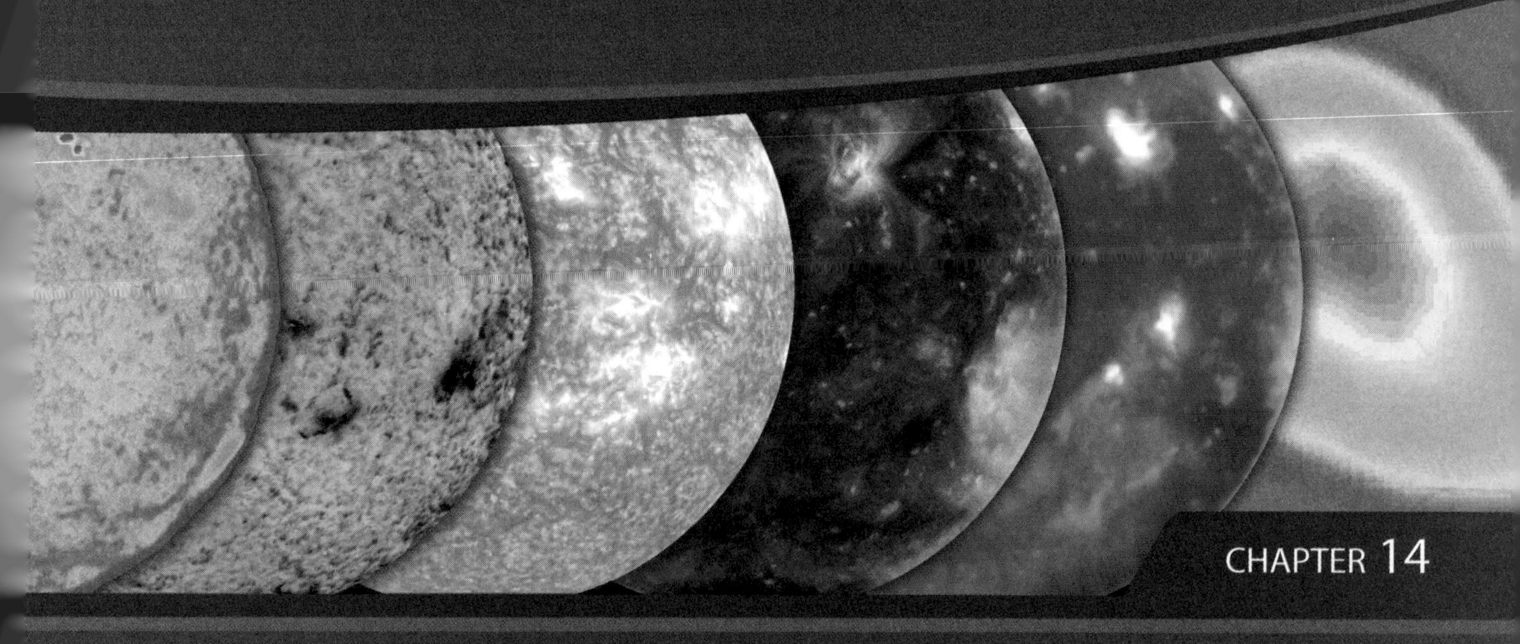

CHAPTER 14

DOMINION SCIENCE PROBLEM

Killing Cancer

One of the consequences of the Fall was the appearance of fatal diseases, such as cancer. Cancer was not part of God's original "very good" creation. Fossilized cancerous tumors have been discovered in many kinds of vertebrate fossils, including dinosaurs, so cancer has been with us since before the Flood. Cancer causes many to suffer.

This disease causes cells to stop their designed function and begin multiplying rapidly. The process often produces toxic substances. Left untreated, cancerous cells eventually destroy and crowd out healthy cells, causing organs and bodily functions to fail.

Today, cancer is one of the most common diseases. Anyone can develop it. According to the American Cancer Society, it is the second-greatest killer in the United States after heart disease. Cancer is responsible for one out of every four deaths. It is often difficult to detect and treat because it grows inside our bodies and often progresses quite far before producing symptoms that people notice. How can we detect cancer at its earliest stages and eliminate it?

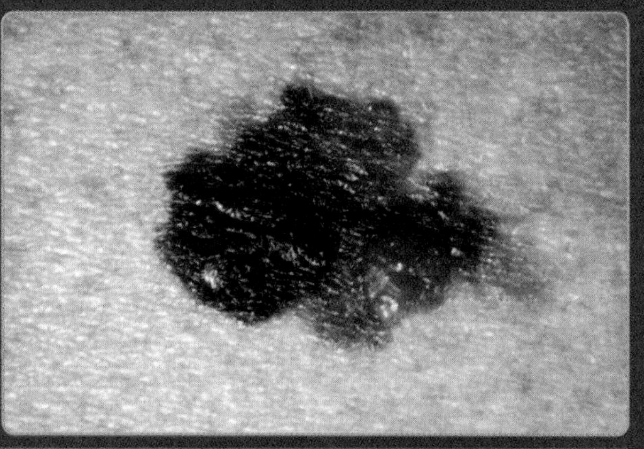

14-1 Cancer, as well as other diseases, is a consequence of God's curse on human sin.

14A Objectives

After completing this section, you will be able to

✓ discuss how EM waves differ from mechanical waves.

✓ describe the structure and motion of an EM wave.

✓ state the speed of light (*c*) to three significant digits.

✓ describe how EM frequency and wavelength are related.

✓ define a photon and discuss its properties.

✓ relate photon energy and frequency.

✓ give examples of how EM waves exhibit typical wave-like behavior.

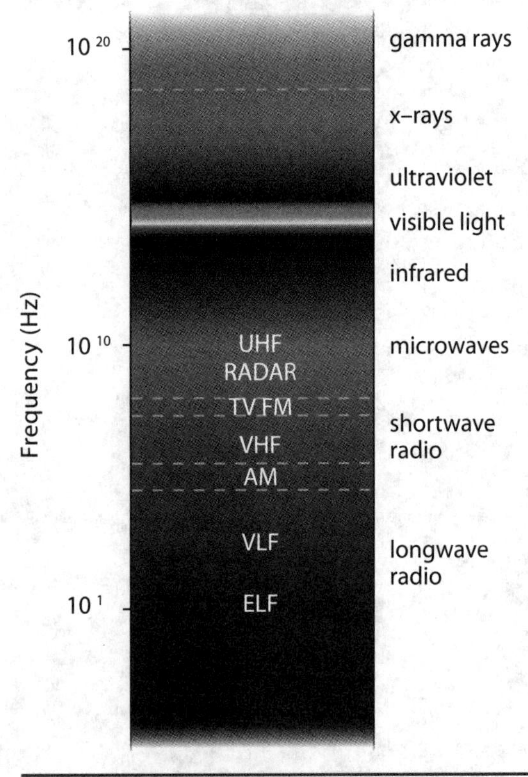

14-2 The EM spectrum is a continuum ranging from extremely-low-frequency radio waves to extremely-high-frequency gamma rays.

14A Electromagnetic Waves

14.1 Introduction

Imagine that you're floating peacefully in a canoe on a lake. A powerboat speeds by. The large waves following the boat tip your canoe, and over you go. This is obviously an energy transfer! Water waves are rhythmic disturbances that transfer energy from one place to another through water. When you hear the roar of a jet flying overhead, you sense sound waves. Sound waves, like water waves, involve the motion of matter. Sound waves, water waves, and seismic (earthquake) waves are examples of mechanical waves.

There is another type of wave. These waves perform the same function as sound—they transfer energy. They also have amplitude and frequency as sound does, but there are some dramatic differences. These waves cannot be heard and can travel through a matterless vacuum. Some have wavelengths or energies that can penetrate solid concrete walls! Although you cannot hear these waves, you may be able to see and feel some of them. Others you can use to cook food. Some are even powerful enough to kill living organisms!

Think of the energy we receive from the sun. How does it get here? By light waves, you might say—but there is no medium to carry the waves to us. Light waves are a small part of the family of *electromagnetic waves*. The entire frequency continuum of these waves forms the **electromagnetic (EM) spectrum**. This type of energy transfer requires no matter at all. In fact, matter hinders the transmission of EM energy. Besides visible light waves, the EM spectrum also includes radio waves, microwaves, x-rays, and infrared and ultraviolet light. Just as there are sounds that we cannot hear, we cannot sense most of the EM spectrum without the aid of instruments.

In the 1600s Sir Isaac Newton discovered that sunlight could be dispersed into a rainbow of colors using glass prisms. He was completely unaware of the relationship of the electrostatic force to the magnetic force, much less its relationship to light. In 1820 André Ampère suggested a theory of electromagnetism, which mathematically compared electrical and magnetic forces. However, in 1864 it was the great Christian physicist James Clerk Maxwell who published a set of comparatively simple equations describing the relationships of electric fields and currents to magnetic fields. He also showed that both electric and magnetic fields interact in a wave-like fashion and suggested that light moved as a wave disturbance in electric and magnetic fields. Maxwell coined the term *electromagnetic energy* to account for both of these phenomena.

14.2 Properties of Electromagnetic Waves

The Waveform

Maxwell's equations describe light as a wave phenomenon—in some ways similar to sound and water waves. However, light doesn't move by vibrations of matter. Its waves occur as oscillations in magnetic and electric fields oriented perpendicular to each other (see Figure 14-3 on page 302). Just as a changing magnetic field in an electrical generator produces electrical potential, so a change in the magnetic field of an EM wave induces a change in its electric field. Similarly, the changing electric field produces a changing magnetic field. The

FACETS OF SCIENCE

JAMES CLERK MAXWELL

Undoubtedly, James Clerk (KLARK) Maxwell was one of the greatest geniuses in the history of science. His mathematical equations predicted the existence of radio waves many years before their discovery. They also served to unite the separate sciences of electricity, magnetism, and optics in a single comprehensive theoretical model. Albert Einstein called this achievement "the most profound and fruitful that physics has experienced since the time of Newton." At the same time, Maxwell was a humble and devout man of God.

Maxwell was born in Edinburgh, Scotland, in 1831. He was homeschooled by his mother, a dedicated Christian, until her death in 1839. She often urged him to "look up through nature to nature's God." By the time he was eight years old, he had memorized all of Psalm 119. After his mother's death, Maxwell continued his education at Edinburgh Academy, the University of Edinburgh, and Trinity College, Cambridge. Because of poor health, he got off to a slow start at the Academy; but he caught up and passed his classmates, graduating first in math and English. At the university his instructors expected and received great things from him.

His brilliant career included teaching at three different institutions of higher learning. Two years after accepting his first teaching position at a college in Aberdeen, Scotland, he married Katherine Mary Dewar (DOO ur), who was a lifelong source of strength and encouragement. In 1860 he became professor of physics and astronomy at King's College in London, and in 1871 he took over the newly formed chair of experimental physics at Cambridge University. In connection with this position, he was selected to supervise the planning and construction of the famous Cavendish Laboratory.

In addition to discovering the close relationship between electricity and magnetism, embodied in his famous equations, one of the best-known of Maxwell's contributions to astronomy was his mathematical analysis of the rings of Saturn. He found that if they were composed of tiny particles of solid matter in a stable orbit, they could never coalesce into satellites as some had predicted. These findings were confirmed by the Cassini space probe in 2005.

Maxwell's spiritual life set an example for all around him. Those who knew him best described him as humble, devoted, and fearless in promoting the truth. His beliefs were biblical, and his scientific research consistently strengthened his conservative stand. He strongly disliked the preaching of morality without the gospel. Through constant study he developed a thorough knowledge of the Scriptures, and he regularly shared that knowledge with the sick and shut-ins in the community. He also served as an elder in a church that had been founded through his spiritual leadership and financial assistance.

Maxwell viewed science as a God-ordained means of subduing the earth. He wrote in his notes, "Almighty God, Who hast created man in Thine own image, and made him a living soul that he might seek after Thee, and have dominion over Thy creatures, teach us to study the works of Thy hands, that we may subdue the earth to our use, and strengthen the reason for Thy service; so to receive Thy blessed Word, that we may believe on Him Whom Thou hast sent, to give us the knowledge of salvation and the remission of our sins. All of which we ask in the name of the same Jesus Christ, our Lord."*

Several biographers have credited Maxwell with disproving Pierre Simon de Laplace's hypothesis that the sun and solar system condensed from a large disk-shaped nebula. Some evolutionary writers question whether Maxwell ever openly claimed that he had disproved Laplace's hypothesis and even whether he allowed his Christianity to influence his science. While it may be difficult now to prove that he directly refuted the nebular hypothesis, to say that his faith did not influence his science contradicts his own written testimony.

Maxwell died of cancer at age 48. His friends Lewis Campbell and William Garnett collected his letters, notes, and papers and wrote his biography, *The Life of James Clerk Maxwell*. They quote G. W. H. Taylor, Vicar of Trinity Church, Carlisle: "Maxwell has indeed left us a very bright memory and example. We, his contemporaries at college, have seen in him high powers of mind and great capacity and original views, conjoined with deep humility before his God, reverent submission to His Will, and hearty belief in the love and atonement of that Divine Saviour Who was his portion and comforter in trouble and sickness, and his exceeding great reward."†

*Lewis Campbell and William Garnett, *The Life of James Clerk Maxwell* (London: Macmillan, 1882), 323.

†Ibid., 174.

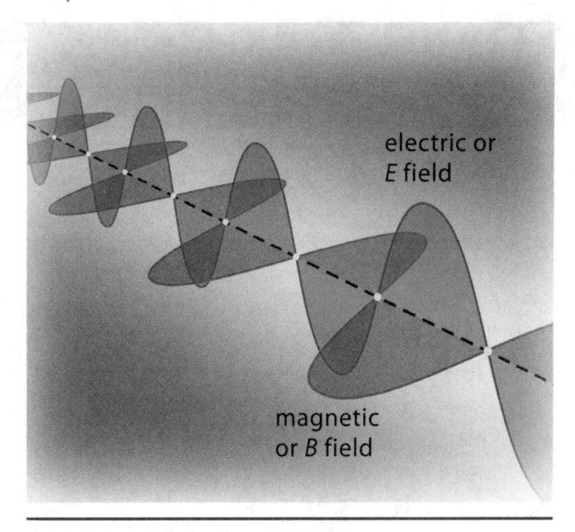

14-3 A model of an EM wave

two fields are completely dependent on each other. Each field exists because of the other and nothing else. This relationship explains why an EM wave moves without a material medium.

In Figure 14-3 notice that the electric field (E) of the EM wave is oriented vertically, and the magnetic field (B) is oriented horizontally. In reality, EM waves may travel in any orientation. The wave may even twist as it moves through some materials. However, the electric and magnetic fields always remain perpendicular to each other. The orientation of these fields is especially important when studying certain optical phenomena.

Speed

All forms of EM energy, including visible light, move at the same speed. However, this speed depends on the matter through which the EM energy travels. The **speed of light (*c*)** in a vacuum is exactly 299,792,458 m/s. For most educational purposes, c is rounded to 3.00×10^8 m/s.

When any form of EM energy moves through matter, its speed is slower than c because the EM waves interact with the matter. The speed of EM waves in air is 99.97% of c. However, in water, their speed is 75% of c. In diamond, it is 41%, and in the semiconductor compound gallium phosphide, it's a comparatively tortoise-like 29% of c. Because the speed of light varies with the matter through which it travels, the constant c is always defined as the speed of light in a vacuum.

Frequency and Wavelength

Electromagnetic waves have all the measurable properties of mechanical waves, including frequency and wavelength. In Chapter 12 you learned that the speed of a wave through a given material is constant. Wave speed equals the product of frequency and wavelength. Therefore, the relationship of these quantities for EM waves can be expressed as

$$\boxed{c = f \times \lambda},$$

where

- c is the speed of EM waves in a vacuum, in meters per second,
- f is the frequency of the wave, in hertz or cycles per second, and
- λ is the wavelength, in meters.

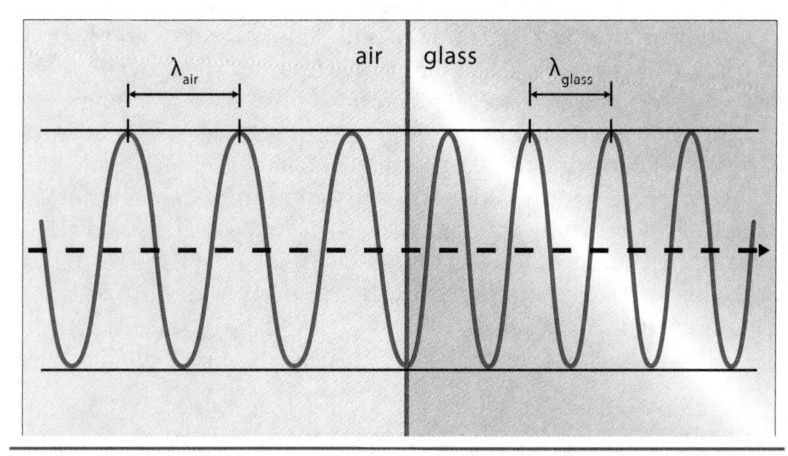

14-4 When an EM wave enters a denser material, its speed slows. Since frequency remains the same at all times, its wavelength shortens.

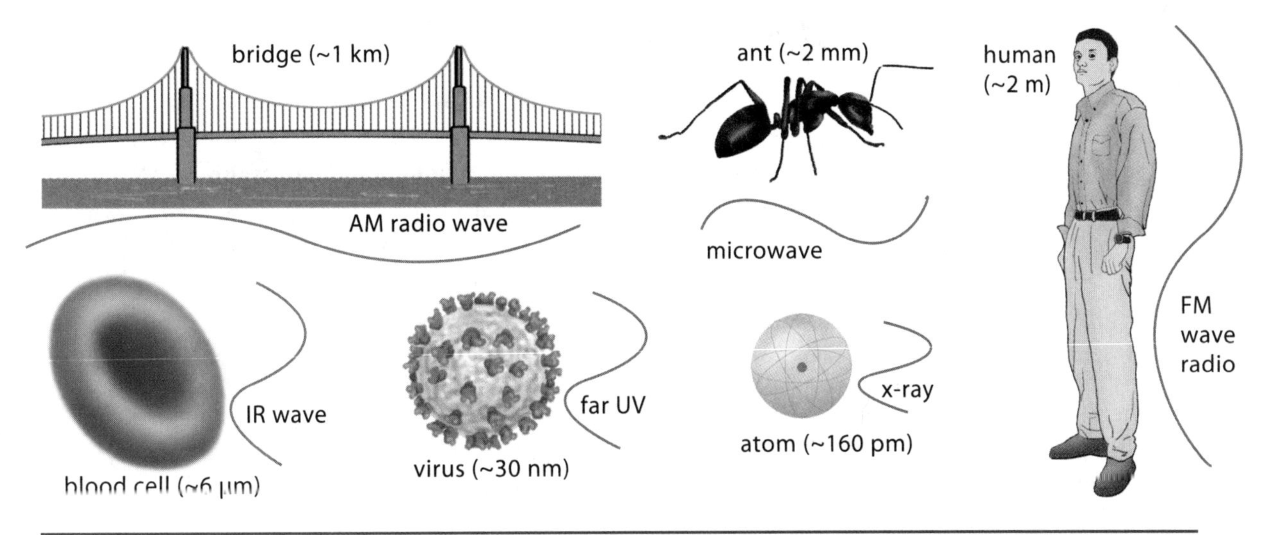

14-5 The wavelengths of EM energy vary from extremely long to extremely short.

Since c is constant in a vacuum, frequency and wavelength are inversely proportional to each other. Higher-frequency waves have shorter wavelengths, and vice versa. In the EM spectrum, wavelengths range from many kilometers for low-frequency radio waves to wavelengths shorter than the diameter of an atom's nucleus in extremely-high-frequency gamma rays.

Example Problem 14-1

Frequency and Wavelength

What is the wavelength of a 300. Hz radio wave? a 3.00×10^{18} Hz gamma ray? Choose appropriate units for each answer.

Radio Wave

Known: wave frequency (f) = 300. Hz = 300./s
speed of light (c) = 3.00×10^8 m/s

Unknown: radio wavelength (λ)

Required formula: $c = f \times \lambda$

Substitution: 3.00×10^8 m/s = (300./s)λ

Solution:

$$\frac{3.00 \times 10^8 \text{ m/s}}{300./\text{s}} = \lambda$$

$\lambda = 1.00 \times 10^6$ m
$\lambda = 1.00 \times 10^3$ km
Kilometers are appropriate for this kind of radio wavelength.

Gamma Ray

Known: wave frequency (f) = 3.00×10^{18} Hz
speed of light (c) = 3.00×10^8 m/s

Unknown: radio wavelength (λ)

Required formula: $c = f \times \lambda$

Substitution: 3.00×10^8 m/s = (3.00×10^{18}/s)λ

Solution:

$$\frac{3.00 \times 10^8 \text{ m/s}}{3.00 \times 10^{18}/\text{s}} = \lambda$$

$\lambda = 1.00 \times 10^{-10}$ m = $100. \times 10^{-12}$ m
$\lambda = 100.$ pm

Recall that 1 Hz = 1/s = 1 s^{-1}.

Physicists have estimated that the longest EM wavelengths could be equal to the size of a galaxy or even larger! A wave this size would be very difficult to detect.

The frequency of the gamma ray using the proper SI prefix would be 3.00 EHz, or *exahertz* (10^{18} Hz).

Picometers (PEE kuh MEE turz) (10^{-12} m) are appropriate for gamma ray wavelengths.

The Photon

Sir Isaac Newton believed that light consisted of particles called *corpuscles*. However, one of his contemporaries, the Dutch scientist Christiaan Huygens, suggested that light acted more like waves. For more than a century and a half, scientists disagreed on the best model of light. One important experiment in 1801 clearly demonstrated the wave nature of light through *interference* and *diffraction*. However, other experiments in the late 1800s showed that light acted like tiny particles.

How could light act as both a wave and a particle? Albert Einstein finally answered this question in 1905. Based on the work of other contemporary physicists, he suggested that light travels in a wave bundle called a **photon**. All EM energy is transported as photons. In some circumstances, photons act like waves. For example, photons can refract, diffract, and interfere just like other waves. But they can move in straight lines, affect individual atoms, and have momentum just like tiny particles. Thus, photons are difficult to model in ways that are familiar to us.

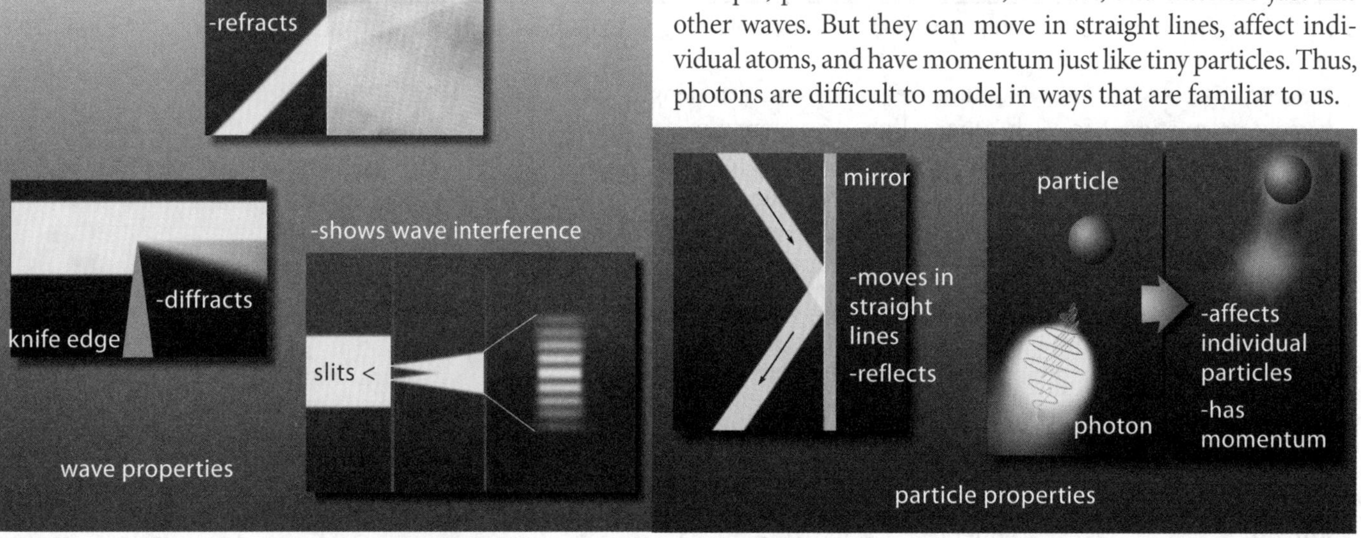

14-6 Electromagnetic energy has properties of both waves and particles.

Intensity and Energy

The power and intensity of a mechanical wave is proportional to the square of its amplitude. This relationship applies to the intensity of EM waves as well. Since the EM energy of a photon consists of electrical and magnetic waves, their amplitudes technically determine the individual photon's intensity. However, the amplitude of an EM wave cannot be measured directly because it is not matter. It must be calculated from other properties of a photon.

Einstein suggested that photons were immaterial particles because another physicist, Max Planck, had recognized that EM energy comes in distinct packets. These packets are called *quanta* and are related to the EM energy's frequency. Einstein saw no difference between Planck's energy quanta and immaterial photons. The energy of a photon is calculated by the formula

$$\boxed{E = h \times f},$$

where

- E is the photon's energy,
- h is an extremely small proportionality constant called Planck's constant (its value is not important for our purposes), and
- f is the frequency of the EM waves in the photon.

Max Planck (1858–1947) was a German physicist who altered the course of physics with his discoveries about the nature of EM radiation. He won the Nobel Prize in Physics in 1918 for his development of the quantum theory.

quanta (KWON tah): plural of *quantum*

Photons with higher frequencies contain more energy. Thus, gamma ray photons are far more energetic than radio wave photons.

14.3 Electromagnetic Wave Phenomena

Electromagnetic waves exhibit all the phenomena of mechanical waves plus some unusual ones. Electromagnetic waves bend, or refract, because their speed changes in different materials. For example, radio waves refract as they move through the atmosphere because air density changes with altitude and temperature.

Electromagnetic waves can constructively and destructively interfere just like sound waves. Low-frequency radio waves diffract around the surface of the earth, permitting extremely long-range radio communications. X-ray diffraction and wave interference are used to determine the crystalline structure of solids.

The Doppler effect is also observed in EM waves. As an EM wave source and an observer move toward or away from each other, the observed frequency increases or decreases respectively, just as for sound waves. However, unlike in mechanical waves, the relative speed of the material that EM waves travel through has no effect on EM Doppler shift because EM waves do not depend on a medium to be transmitted.

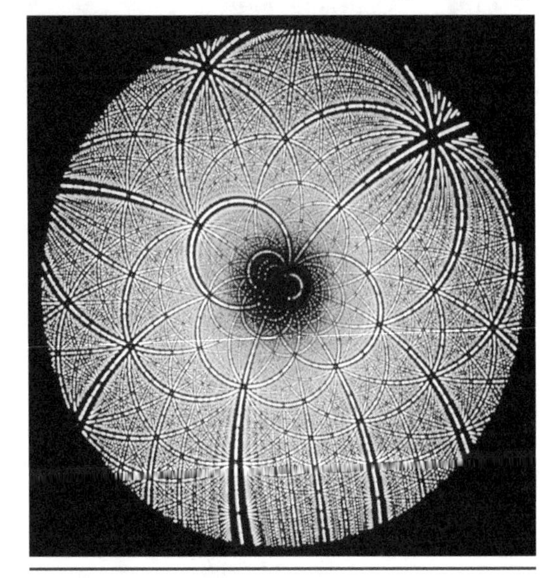

14-7 Atom spacing in crystals is measured using x-ray diffraction.

After Einstein developed his *general theory of relativity*, he predicted that gravity could bend EM waves. This was an unexpected prediction since EM waves have no mass for gravity to act on. However, when EM waves travel through an intensely strong gravitational field, such as around a neutron star or a dense galaxy, their paths bend. Einstein's hypothesis was proven correct when astronomers observed this effect, called *gravitational lensing*. Visible light from distant galaxies that are hidden behind nearer, massive objects can be seen as distorted images, focused by the gravity of the nearer object.

14-8 In gravitational lensing, light from a distant, unseen galaxy bends around the extremely massive galaxy cluster marked by an arrow. The images (*I*) of the distant galaxy form arcs around the nearer object.

14A Section Review

1. What is the EM spectrum? Why is it a continuum?
2. What scientific breakthrough revealed the nature of EM energy and waves?
3. Why can an EM wave be transmitted without matter?
4. When we use the speed of light as the constant c, what condition are we assuming? What is its value under this condition?
5. Look at Figure 14-2 on page 300. In general terms, compare the frequency and wavelength of visible light to the same properties of a television signal.
6. How is EM energy like a wave? like a particle? What model accounts for both of these characteristics?
7. The energy of a photon is proportional to which property? Which type of EM photons are the most energetic? Why?
8. What unexpected phenomenon has been observed in EM waves?
9. (True or False) The speed of light is always the constant value c.
10. ✪ What is the wavelength of a 30. GHz radio wave in a vacuum?

14B Objectives

After completing this section, you will be able to

- ✓ list the six major bands of the EM spectrum.
- ✓ identify significant subregions of the EM spectrum.
- ✓ describe the sources of each kind of EM energy.
- ✓ explain what determines the boundary between adjacent regions of the EM spectrum and whether any overlap exists.
- ✓ give at least two applications for each portion of the EM spectrum.

14B Electromagnetic Spectrum

14.4 Introduction

Light waves, television waves, and microwaves are members of the EM spectrum. Since these families of EM waves tend to travel as rays directly away from their sources, they are classified as *radiant energy*. There is amazing variety in the wavelengths and the frequencies of these waves, the ways that they are generated, and how they affect matter.

God has designed tremendous diversity into His creation. He could have made just one kind of star, one kind of rock, one kind of plant, one kind of animal, and human beings who all look exactly the same and have the same abilities. If there were just one kind of EM wave, for example, we would not be able to both see and be warmed by the sun, because light and infrared are two different kinds of EM waves.

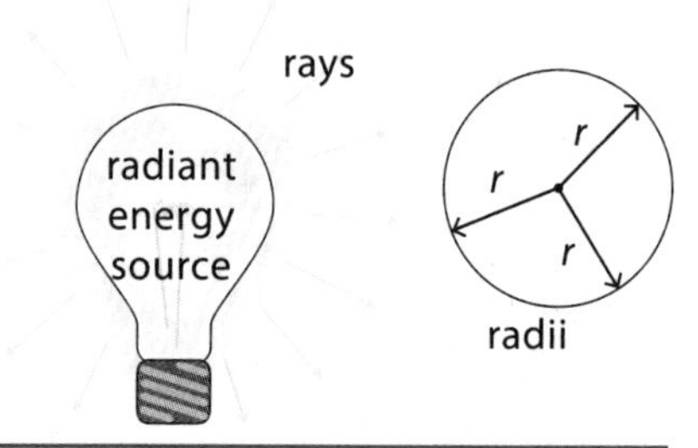

14-9 The terms *ray*, *radiant*, and *radius* all relate to a line emitting from a point.

14.5 Radio Waves

Radio waves in the EM spectrum include the wave frequencies that carry both radio and television signals. If you could see this energy, you would be amazed at how many waves are passing through the room right now! The idea that they travel right through your body might disturb you, but they are completely harmless. Even the relatively strong waves from nearby radio and television stations aren't strong enough to cause dangerous side effects.

14-10 Life is tough having to deal with all these radio waves!

The radio portion of the EM spectrum extends from far below 300 Hz to 300 GHz (3×10^{11} Hz). Radio waves are subdivided into sections based on their usage and frequency. Figure 14-11 shows the names and the frequency boundaries of the various radio bands. The names of the lower radio-frequency bands were determined during the early twentieth century, when radio technology was still developing. This is why the medium-frequency (MF) band is not in the middle of the radio portion of the EM spectrum. Frequencies below the MF band were designated the low-frequency (LF) band, and those above were designated the high-frequency (HF) and very-high-frequency (VHF) bands.

The frequency boundaries of each band begin with a 3 so that the wavelengths for the band can be easily computed by dividing the speed of light—3×10^8 m/s—by the frequency. For example, the MF band extends from 300 kHz to 3 MHz with wavelengths from 1000 m to 100 m, respectively.

The acceleration of electrons by a magnetic or electric field produces radio waves. These electrons may be parts of a solid conductor, such as a radio antenna, or they may be free electrons moving through a vacuum, such as outer space. The radio-frequency emissions of radar track distant objects. We use radio waves for data, voice, and visual communications. In astronomy, special dish-shaped telescopes detect radio waves from space. At the end of the chapter, we discuss in greater depth these and other applications of radio-frequency energy.

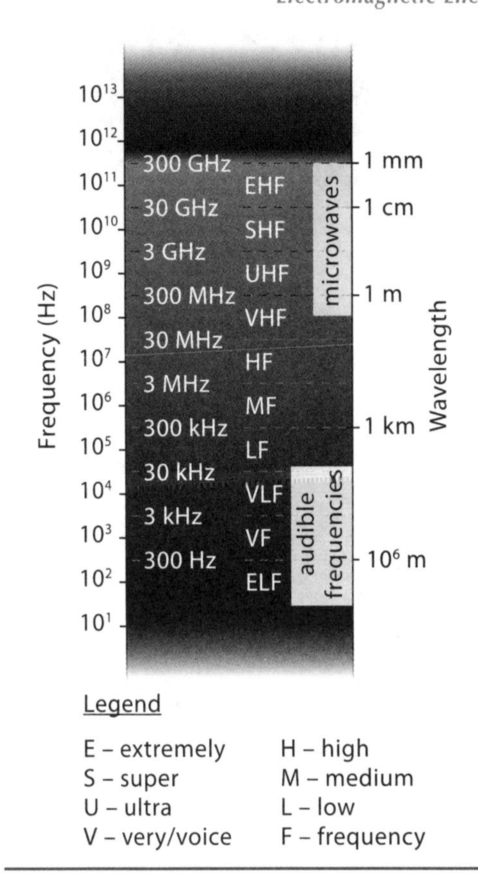

14-11 The radio-frequency band overlaps with audible frequencies on the low end of the spectrum and microwaves on the upper end.

14-12 This galaxy radiates in the radio-frequency portion of the EM spectrum. No antennas here!

14.6 Microwaves

The radio frequencies used for VHF communications merge into the next region of the EM spectrum—**microwaves**. Officially, microwaves span the upper end of the radio-frequency band, but they have some interesting properties that also set them apart from the lower frequencies. The frequencies of microwaves range from 100 MHz to 300 GHz with wavelengths decreasing through the band from approximately 100 cm to 0.1 cm.

Microwaves are produced in the same way as radio waves. They differ from the rest of the radio-frequency spectrum mainly because they can travel only in a *line of sight*, or in a straight line. The lower-frequency radio waves can follow the curvature of the earth to some extent, so they can communicate with other ground stations over the horizon. Microwaves can communicate only if the receiver is situated above the horizon of the transmitter.

Not only can microwaves aid in communications, but they can also cook your food. When operating, a microwave oven sets up within its shielded enclosure a pattern of microwaves. The oscillating electric fields affect water, fat, and protein molecules in the food. The changing energy fields cause the food molecules to rapidly oscillate first one way, then the other. As the molecules vibrate, they collide

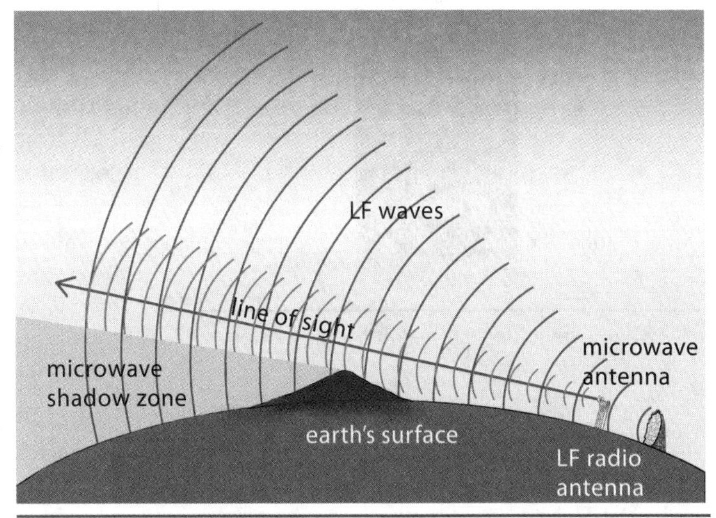

14-13 Most radio waves can diffract around obstacles. Microwaves can radiate only in a straight line (line of sight).

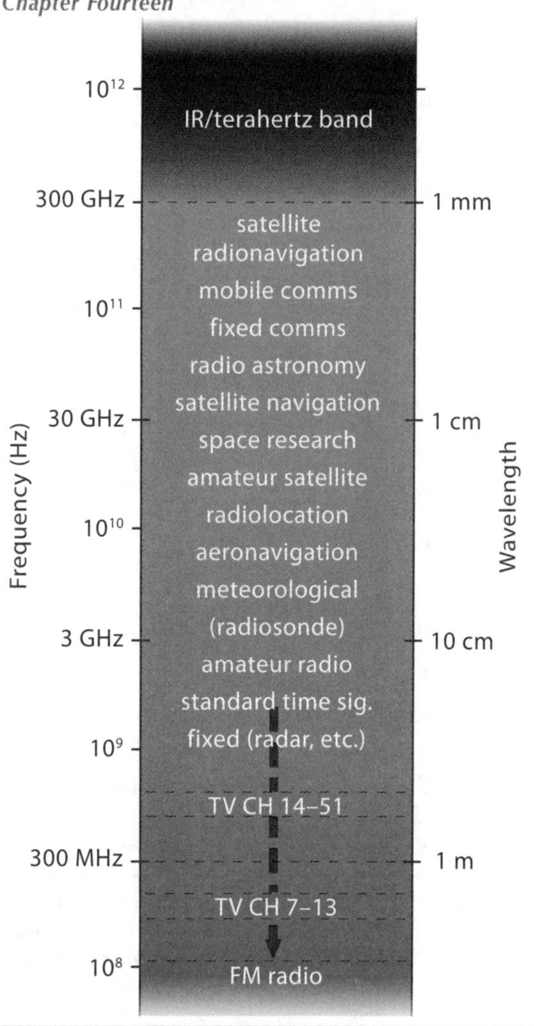

14-14 Usage of the microwave-frequency band

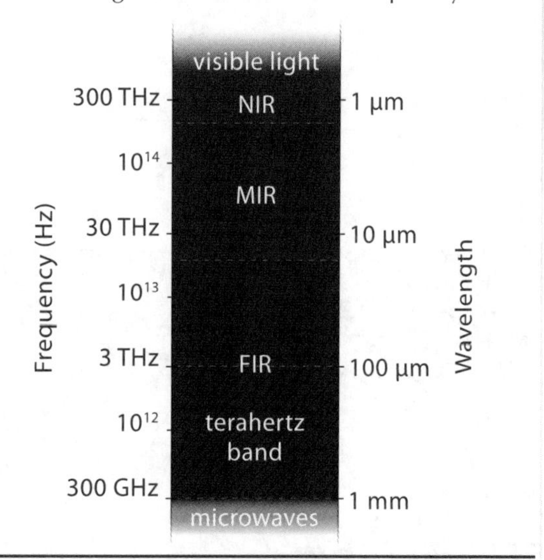

14-16 Subdivisions of the infrared-frequency band

Fredrick William Herschel (1738–1822) was an English astronomer who discovered Uranus and built over 400 telescopes in his life, including a 40-ft refracting telescope.

incandescent (IN kun DES unt)

with other molecules, rapidly converting EM energy into thermal energy. Since the waves penetrate the food, the inside of the food is heated almost as quickly as the surface. Cooking time is greatly reduced because it's not necessary to wait for the heat to move from the surface to the interior of the food as in a standard oven.

The microwave band is used for the rapid transmission of large amounts of information. This band includes the ultrahigh-frequency (UHF) band where the higher frequencies of standard TV channels and cell phones transmit. Microwaves are also used to transmit telephone signals across continental networks as well as by satellite. In fact, the information in nearly all satellite data links and orbital communications is carried by microwaves. Radars for tracking civilian aircraft and for military targeting also operate in the upper portions of the microwave band.

14-15 Microwave horn antennas are used to relay telephone circuits from point to point, avoiding the need for installing miles of cable.

14.7 *Infrared Waves*

Moving up the EM spectrum to still higher frequencies and energies, we come to **infrared (IR)** waves. The prefix *infra-* means "below." The frequency band for this type of energy is just below the red end of the visible-light band. Officially, the IR band extends from 300 GHz to about 384 THz (terahertz, or 10^{12} Hz), which is the lowest EM frequency visible to the human eye. Infrared wavelengths span the range of 1 mm to 780 nm (10^{-9} m). The IR spectrum is subdivided into three bands: far infrared (FIR), just above the microwave band; medium infrared (MIR); and near infrared (NIR), which is just below the visible-light spectrum. Infrared is produced mostly by the thermal vibration of atoms and molecules. All atoms above absolute zero vibrate to some extent, so infrared is produced by all matter.

Sir William Herschel discovered IR radiation and reported his results in 1800. He refracted rays from the sun through a prism to separate the different colors. Herschel placed a thermometer in the region of each color and recorded its temperature. Then he placed the thermometer in the unlighted region next to the red end of the spectrum. The temperature he recorded in this IR region was greater than in any other part of the spectrum. He called the energy he discovered "calorific rays" or "rays of heat."

An **incandescent** object is an object heated to glowing. It is a good source of both infrared and visible light. You have probably noticed that incandescent lights give off considerably more heat than fluorescent lights. Because fluorescent lights convert less electrical energy to thermal energy, they're more economical to operate. Another example of an incandescent object is the burner of an electric stove. If the burner gets hot enough, it emits a red-orange glow and you can feel the heat at a distance. What happens when the burner is turned off? The color disappears, but you can feel the excess IR energy that the burner emits until it has cooled to room temperature. Human skin contains sensitive IR receptors that are plentiful on the face and the back of the hand.

14-17 The stove burner in the left photo is off; the burner in the middle photo is radiating only IR waves; in the right photo, the burner is radiating both IR and visible-light frequencies.

The *terahertz band*, a distinct subregion of IR energy, extends from 300 GHz to 3 THz. This portion of the EM spectrum was essentially ignored until recently because detectors of this radiation were difficult to construct. Since the 1980s, scientists have developed detectors and instruments that can receive and emit radiant energy in this frequency band. Astronomers found the study of astronomical terahertz radiation to be very useful for filling in details of lower-energy processes occurring in space. In the past decade newer applications have been discovered for medical imaging and materials analysis. Scientists from all areas of research are excited about the possibilities of terahertz science, which is still in its infancy.

Special-purpose IR receptors that provide directional information have been discovered in snakes, in insects, such as beetles and butterflies, and even in mammals, such as some species of bats. Man-made devices for detecting and using IR energy include thermometers, thermostats, IR remote-control devices, and wireless computer mice and keyboards. Miniaturized IR imaging systems are used by firefighters and in military applications like weapon sensors and night vision goggles. Infrared sensors and cameras are also important in purely scientific pursuits, like imaging earth resources from satellites and aircraft and observing deep space from orbital telescopes.

14-18 The pits below the nostrils of this rattlesnake *(Crotalus adamanteus)* are IR receptors for hunting warm-blooded prey at night.

14.8 Visible Light

Visible light includes all the EM energy frequencies humans can see. Because a person's vision depends on the unique sensitivity of his retina, not everyone sees the same range of frequencies. In general, though, the visible-light frequency band extends from about 384 THz (deep red light) to about 789 THz (deep violet light). The wavelengths in this band are very short, ranging from about 780 nm for deep red light to about 380 nm at the deep violet end of the visible spectrum.

Visible light is emitted by the electrons of atoms when they lose energy and drop from higher energies to lower energies *within the atom*, rather than by the physical movement of electrons through a conductor or by vibrations. Energy is conserved as their excess energy is given off as light photons. Some NIR light is produced the same way, so the sources of visible light overlap the upper limits of the IR band slightly.

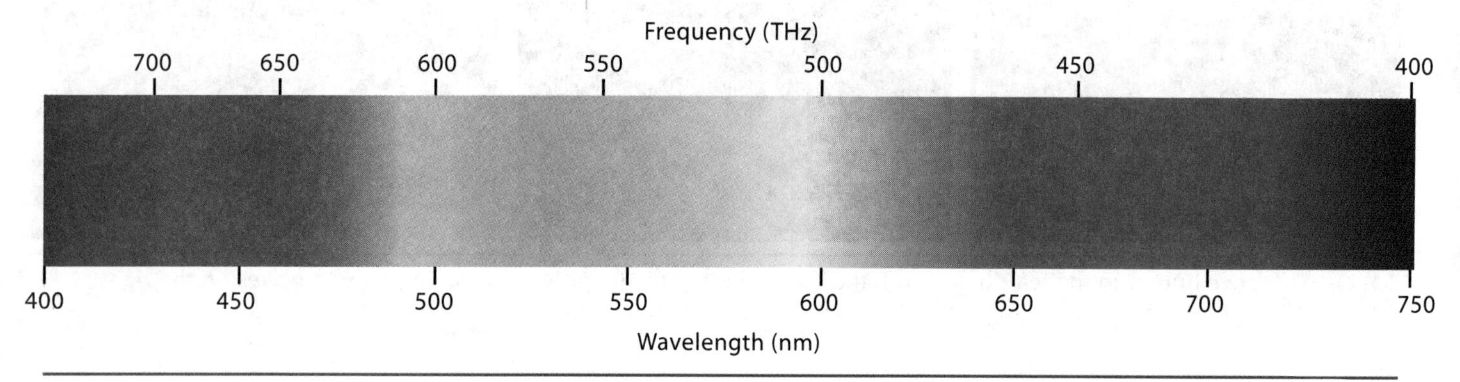

14-19 The visible-light spectrum

14.9 Ultraviolet Rays

The prefix *ultra-* means "beyond." The frequencies of **ultraviolet (UV)** light are just beyond those of deep violet visible light. Its frequencies are too high for the human eye to detect. Because it cannot be seen, it is sometimes called "black light." It's also referred to as *UV rays* or *UV radiation*.

The UV band spans the frequencies from about 789 THz to 30 PHz (petahertz, or 10^{15} Hz). Physicists have subdivided the UV band into three parts. The near-ultraviolet (NUV) band includes lower energy frequencies just above those visible to the human eye. The far-ultraviolet (FUV) band contains medium-energy UV waves with higher frequencies. The extreme-ultraviolet (EUV) band contains very energetic UV waves with frequencies just below those of x-rays. Wavelengths of UV waves extend from 380 nm for the NUV band to 10 nm for the EUV band. Ultraviolet waves are produced just like visible light waves, but the energy changes of the electrons in the atoms as they drop to their lowest energy states are extremely large, emitting very energetic UV photons.

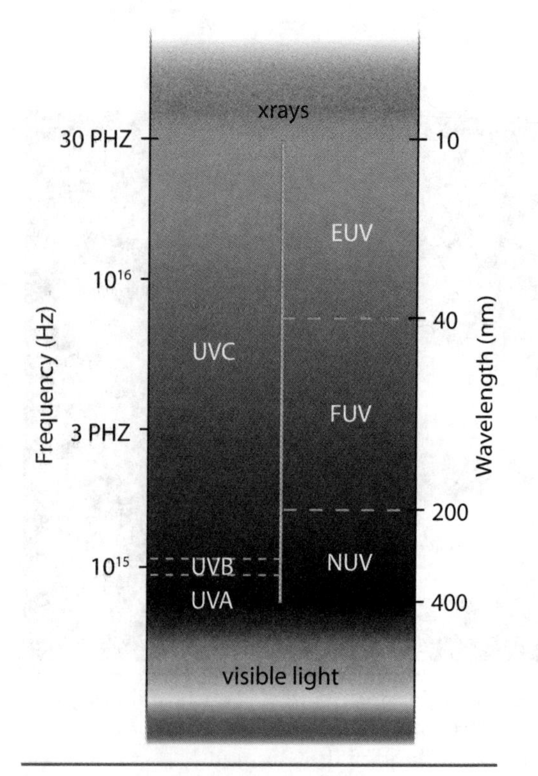

14-20 The UV spectrum and its subdivisions

Ultraviolet radiation was discovered by J. W. Ritter. Soon after Ritter's discovery, W. H. Wollaston found that when sunlight passed through a prism, the energy adjacent to the violet end of the visible spectrum could darken certain silver compounds. The compounds decomposed to form dark-colored metallic silver. The researchers reasoned that the invisible UV light was the energy that caused this reaction.

Johann Wilhelm Ritter (1776–1810) was a German chemist who helped discover the UV portion of the EM spectrum. He also developed the first dry cell battery. He did many experiments with electricity, even using his own body, which may have contributed to his early death.

William Hyde Wollaston (WOOL uh stun) (1766–1828) was an English chemist who discovered two elements—palladium and rhodium. He also studied the spectrum of the sun's light.

Ultraviolet light can have both beneficial and hazardous medical effects. Medically, it can be divided into three frequency bands—UVA, UVB, and UVC—according to its ability to do biological damage. All three forms are produced by the sun. The most energetic and dangerous, UVC, is filtered out by the atmosphere. Of the three, UVA has the longest wavelengths and can penetrate deep into the skin. Prolonged exposure can cause premature skin aging and even skin cancer. It is present in sunlight at all times of the day. Sunblock creams and glass stop most UVA. The last form, UVB, is almost completely filtered out by the ozone layer, but it is present at the sunniest part of the day. It can cause severe sunburn and can easily damage skin DNA. Abnormal DNA can lead to mutations that result in various skin cancers (see Figure 14-1 on page 299). Glass filters out UVB, so windows provide protection against this form of ultraviolet. Wearing UV filtering sunglasses to protect the eyes and wearing sunblock and clothing to protect the skin against the sun's UV rays are important for good skin health.

However, natural and artificial UVB rays can promote the formation of vitamin D in the skin. This vitamin is essential for immunity from certain cancers, for strong bones, and for correcting several ailments of newborn infants. Ultraviolet light is useful for some forms of medical treatments, for killing bacteria, and for setting certain oral adhesives. It is also used for classifying some materials and minerals by the property of fluorescence. It can sterilize food and disinfect drinking water. More uses for UV light are continually being discovered.

14-21 If you spend a lot of time in the sun, you should wear clothing that covers your limbs, and a broad-brimmed hat and sunglasses, to protect against UVA and UVB. Sunblock can also filter out solar UV light.

14.10 X-rays

W. C. Roentgen, the first winner of the Nobel Prize, discovered **x-rays** in 1895 while studying cathode rays (streams of electrons) in an early cathode ray tube (CRT). X-rays have frequencies ranging from 30 PHz to 30 EHz. Their wavelengths are extremely short, ranging from 10 nm to 10 pm, much smaller than the diameter of an atom. X-rays are generated when high-speed electrons strike an atom and slow down rapidly. The lost kinetic energy is converted into high-energy x-ray photons.

Wilhelm Conrad Roentgen (RENT gun) (1845–1923) was a German physicist known primarily for his discovery of x-rays. He is often called the Father of Medical Radiology, which is the science of using the EM spectrum in medicine.

The best-known property of x-rays is their ability to penetrate matter to produce images. This has made them useful in medical work for viewing internal structures of the body and in airport

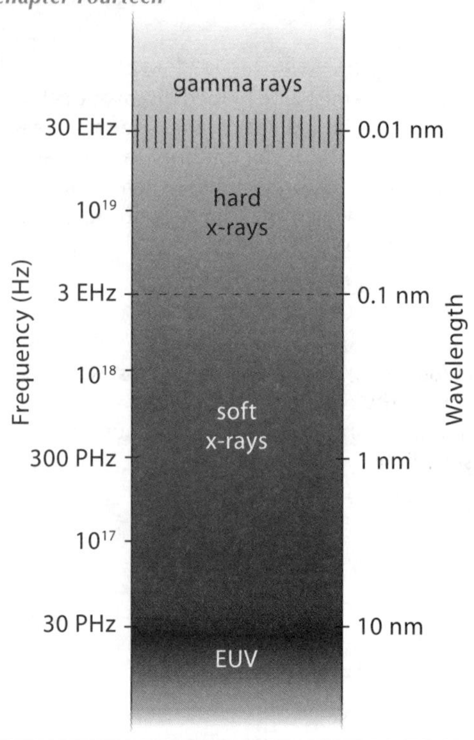

14-22 The x-ray portion of the EM spectrum

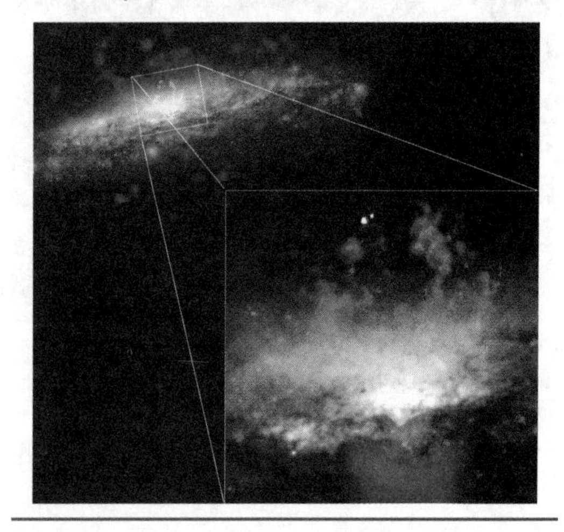

14-24 An x-ray photograph of galaxy NGC 3079 taken from the satellite Chandra X-ray Observatory

femtometer (FEM tuh MEE tur)

security for inspection of people as well as luggage. Similarly, x-rays are used in industry to inspect solid objects for internal defects. They have also been used to determine the pedigree of important historically significant paintings.

Intense exposure to x-rays can cause burns, cancer, and even death. Prolonged exposure to even low-level x-rays can cause cancers, though brief, infrequent exposures are believed to be safe. X-rays can destroy body cells and cause mutations in genetic material. However, because each x-ray photon carries a large "punch," they are useful for treating some forms of cancer. Inaccessible tumors or cancer that is too widespread for surgery can be killed by producing chemical changes in the diseased cells. Cancer cells are generally more sensitive to x-rays than healthy cells are. These rays can also change the DNA in cancer cells so that they cannot grow and multiply.

X-ray astronomy (observing the x-rays emitted by astronomical objects instead of their visible light) is an important area of investigation. Since the earth's atmosphere filters out x-rays, special telescopes orbiting the earth gather information on their emissions from the sun and other astronomical bodies. X-rays are produced by gases at very high temperatures, such as those that exist in the vicinity of neutron stars, black holes, and interstellar nebulas. Thousands of x-ray sources in deep space have now been cataloged.

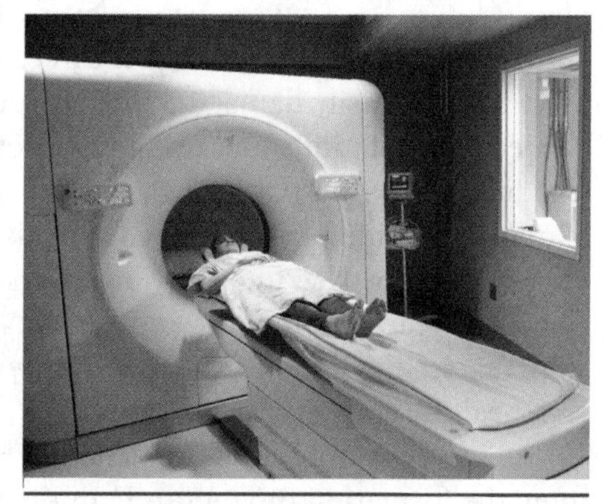

14-23 A modern CAT scanning machine uses x-rays to create detailed images of the inside of the body.

14.11 Gamma Rays

At the highest frequencies of the EM spectrum lie **gamma rays**. Gamma rays differ from x-rays in that they are produced by changes in atomic nuclei rather than from electron collisions. Because they come from a different source than do x-rays, their frequencies overlap those of high-energy x-rays but extend even higher, from approximately 30 EHz and up. Their wavelengths are the shortest known, ranging from 10 pm for the lowest-frequency gamma rays to as short as 1 fm (femtometer, or 10^{-15} m). Scientists are generally more interested in the energies of these waves than their wavelengths, which are unimaginably small. Some gamma rays from celestial objects have been observed to have an estimated frequency of 10^{30} Hz!

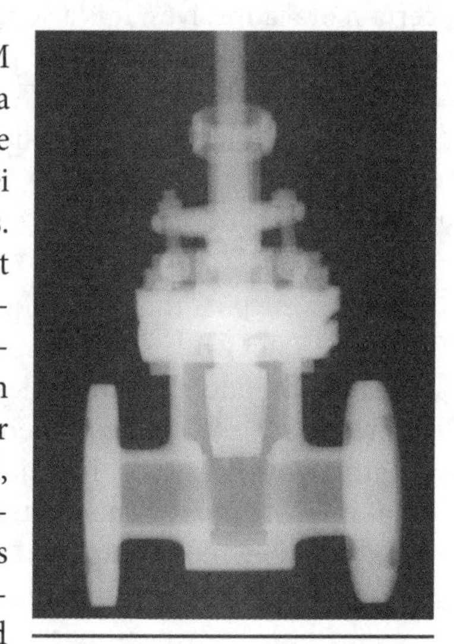

14-25 This inspection photo showing the interior of a metal valve was taken using a gamma ray illumination source.

Scientists and engineers use gamma rays for many of the same applications as x-rays. In medicine, gamma radiation is used to treat cancer. Exposure to gamma rays can destroy cells almost instantly and provides a very useful procedure when combating tumors, especially in the brain.

Gamma rays represent the most destructive form of EM radiation because they have the highest energy per photon. They can travel right through humans, steel, and concrete walls, and can cause great damage to matter, including destroying chemical bonds and even changing atoms from one element to another. Most radioactive elements release gamma rays as their atoms undergo nuclear changes (see Chapters 2 and 16). Gamma rays are not normally a significant hazard because their occurrence on Earth is relatively scarce.

Gamma rays can penetrate most kinds of matter. Standard glass mirrors and lenses don't work with gamma rays, so gamma-ray-detecting instruments, especially telescopes, must depend on other kinds of processes to trap and focus gamma rays. These rays can be stopped effectively only by very dense materials made of large atoms. For this reason lead is the preferred shielding material for the great amounts of gamma radiation produced by nuclear reactors. Workers in nuclear facilities wear special monitoring devices and take special precautions at all times to minimize exposure to gamma radiation. Radioactive materials in a nuclear reactor produce gamma rays even when the reactor is not operating.

14-26 This radiation monitor measures the amount of gamma radiation that a nuclear worker receives from working in a nuclear plant.

14.12 Using Electromagnetic Energy to Solve Problems

As you can see, several parts of the EM spectrum are used in the battle against cancer, especially x-rays and gamma rays. Different diagnostic x-ray tests, including standard film x-rays similar to those that Roentgen developed more than a century ago, can detect cancerous tumors before they produce symptoms. Newer x-ray technologies include *computed axial tomography* scans, better known as *CAT* or *CT* scans. These can produce digital three-dimensional images of the body's interior or thin image "slices" of an organ to detect minute abnormalities.

Radiotherapy is the use of **ionizing radiation**, such as x-rays and gamma rays, to kill cancer cells or to shrink tumors before surgery. Radiotherapy is used for half of all cancer patients, sometimes combined with chemotherapy or other cancer treatments. A cancer patient's dosage and type of radiation depends on the area and type of cancer. Radiotherapy can also be used to reduce pain when a cure is not possible or to prepare the body for a bone marrow transplant.

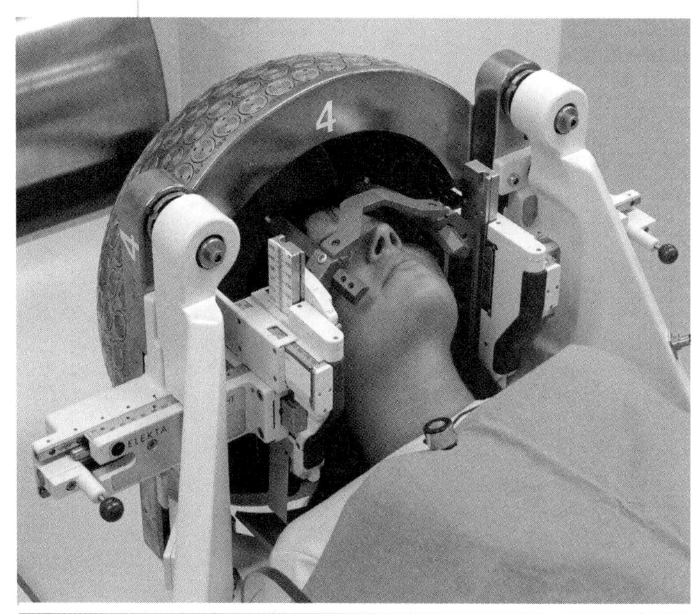

14-27 A gamma knife uses high-energy gamma rays to destroy deadly cancer cells in a very small region of the brain.

Ionizing radiation is any form of EM radiation or nuclear particles that have enough energy to ionize atoms and molecules.

14B Section Review

1. Why is EM energy also called radiant energy?
2. How do the frequencies of radio waves compare to all other bands of EM energy? How do radio waves form?
3. How are microwaves similar to radio waves? How are they different?
4. What defines the upper limit for the IR band? How is IR radiation formed?
5. What is the range of frequencies for visible light in terahertz? How are the photons of visible light formed?
6. How is UV light produced? What are the main hazards from frequent exposure to UV light, such as from the sun and in tanning salons?
7. How are x-rays produced? Why do you sometimes wear a lead-lined apron when you receive a dental x-ray?
8. How are gamma rays formed? Can certain frequencies of EM waves include both gamma rays and x-rays? Explain.
9. DS ✪ How is EM radiation used to treat cancer? What is this type of treatment called?
10. DS ✪ Though radiation treatment can kill cancer, it can hurt healthy tissue too. Suggest some ways that doctors can avoid damaging healthy tissue during radiation treatment.
11. ✪ (True or False) The terahertz band is a region within the FIR spectrum that seems to have great potential for medical imaging.

14C Objectives

After completing this section, you will be able to

✓ explain how radio waves are transmitted and received.
✓ give examples of radio technology.
✓ summarize how the GPS works.
✓ analyze the benefits and drawbacks of radio astronomy.
✓ list some discoveries that have taken place as a result of radio astronomy.

14C Radio-Frequency Technology

14.13 Introduction

Humans were created to use and sense visible and IR light. Using the intelligence God gave us, we have discovered the unobservable parts of the EM spectrum. Every portion of the spectrum is useful in some way, but the radio-frequency portion has become an especially fertile source of technological applications.

14.14 Radio

Man-made radio waves are generated by accelerating electrons back and forth many times each second in a conductor called an *antenna*. The antenna may be a wire or a metal pole, tower, or more elaborate geometric shape. Electrons in the antenna are accelerated over a space of centimeters or meters by an electronic device called a *transmitter*. The number of electron oscillations per second determines the frequency of the emitted wave. Although a portion of the electrons' energy is used in moving through the antenna, most of it is given off as radiant energy. The generation of radio waves is an amazingly efficient use of energy—a few watts of power can produce waves that travel thousands of miles.

All modern radio communications take place in the radio-frequency band. *AM radio* broadcasts occur in frequencies within the MF band. *FM radio* occurs in the VHF band, along with the lowest TV channels. The HF band, also known as shortwave radio, supports long-range radio communications. It is used by amateur radio operators and mobile organizations that communicate frequently by radio, such as police departments and cab companies. Users are often part of emergency communications networks that assist in disaster and emergency aid coordination. Many Christian missionary radio broadcasts also use shortwave radio frequencies to reach large areas of the world with the gospel.

The lower end of the radio-frequency spectrum is used for very-long-range communication and navigation. The very-low-frequency (VLF) and voice-frequency (VF) bands are reserved mainly for military communications with submarines and between military command centers. For several decades VLF signals transmitted by a long-range navigation system provided latitude and longitude positions to ships and aircraft around the world, day and night and in all weather.

AM stands for "amplitude modulation." In this technique, the amplitude of a constant-frequency carrier wave changes to transmit voice and information.

FM stands for "frequency modulation." In this technique, the frequency of a constant-amplitude carrier wave changes to transmit voice and music.

14-28 This missionary radio antenna broadcasts over 500,000 km² of the Caribbean Sea and Atlantic Ocean. The antenna is the pole perched on top of the metal tower.

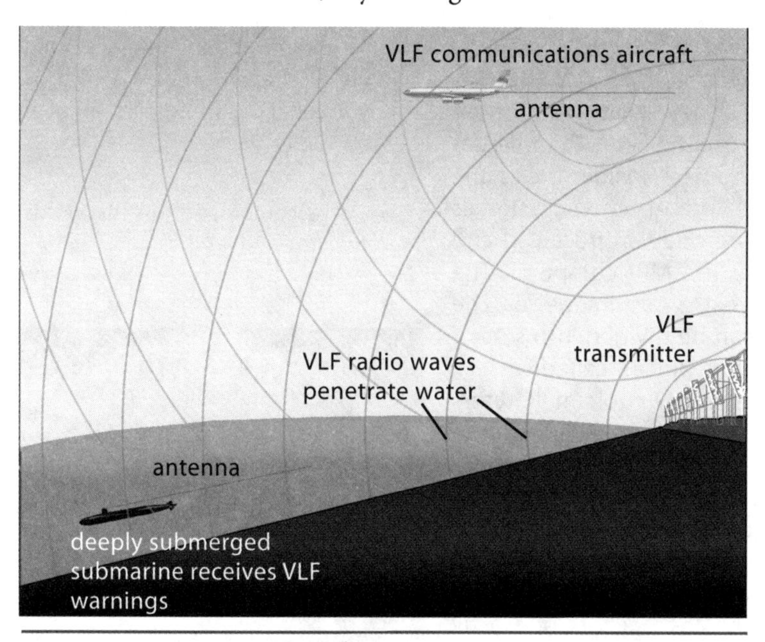

14-29 Very-low- and extremely-low-frequency radio waves are used to communicate with submarines that are several hundred meters below the water's surface.

14.15 Radar

The microwave portion of the radio-frequency spectrum is used primarily for communications and data links, but it is also used in **radar** (from *ra*dio *d*etection *a*nd *r*anging). Radar waves can be reflected by relatively small dense objects, such as aircraft, ships, and even automobiles. Large objects, such as land masses and storm clouds, can also reflect these waves.

Radar transmitters use specially shaped antennas to beam strongly focused waves at an object. If the compass direction is important, the beam is shaped into a tall, thin vertical lobe and swept horizontally in a circle. If the altitude of the object is important, the beam is shaped into a wide horizontal lobe that is vertically thin and swept up and down. If both direction and altitude are important, the beam may be shaped like that of a flashlight and moved in a tight, circular path around the target. The reflected radar pulses are received by the antenna, and the receiver computes the time interval between the transmission and the reception of each pulse. As in sonar (see Chapter 13), the distance to the object is found by multiplying half the time interval by the speed of the wave—the speed of light in the case

FACETS OF SCIENCE: Redeeming Resonance

Resonance is a phenomenon of vibrations. It is the sustained vibration of an object caused when another source vibrates at that object's natural frequency. Except in the case of musical instruments, resonance just seems to be a plain nuisance. To the average person the annoying buzz of a damaged radio speaker, the sustained rattle in a car, and the drone of a plastic part on your refrigerator are all examples of irritating resonance. What good could resonance be?

Physicists turned to resonance to solve a problem of medical imaging—peering into the human body without making incisions, exposing the body to hazardous radiation, or otherwise damaging tissue.

What if we could get all the billions of atoms in a certain location within the body to orient themselves in the same direction so that their minute magnetic poles were in phase and were reinforcing each other? We can do that by putting the body inside a strong magnetic field, because each atom and most molecules act somewhat like tiny magnets. As they twist to align to the magnetic field, they generate some weak high-frequency radio signals. By repeatedly causing the molecules to rotate within an oscillating magnetic field, the resonating molecules become continuous sources of radio-frequency energy.

If we send a radio signal with a slightly lower frequency through the tissue, it constructively or destructively interferes with the signal produced by the oscillating molecules, creating an EM beat signal (see Chapters 12–13). This signal is strong enough and has a low enough frequency that we can receive, amplify, and interpret it with the help of a computer. These principles are combined in a process called *magnetic resonance imaging (MRI)*.

Suppose that a person has a tumor on his liver and his doctor wants to know whether it is cancerous. His body is placed inside a ring of computer-controlled magnets that produce an oscillating magnetic field. The MRI detector is mounted inside the same ring and the radio-frequency signals emitted by the body are received and processed by the MRI computer. The processor creates an image of the tumor that can be divided into sets of tiny volumetric pieces called *voxels*, which are just a few cubic millimeters each. When analyzed, the MRI signals generated by fat, bone, muscle, or cancerous tissue each resonate at a different frequency.

Using the different beat frequencies of the resonating tissues, the MRI computer can generate a picture of exactly what the inside of each voxel looks like. It puts all the voxels together in various arrangements to provide a series of cross-sectional "slices" or a three-dimensional image of the entire tumor. The tumor image can be rotated so that surgeons can examine its size and understand how it affects the surrounding tissue. The MRI allows this kind of sophisticated inspection of the interior of the body without the use of chemicals, surgery, or exposure to harmful radiation.

Although the theory of magnetic resonance has been known since the 1930s, the medical applications of MRI were developed in the 1970s by the Christian physician and researcher Raymond Damadian. He and his team of graduate students built the first operable MRI scanner in 1977.

Even though the mind of man has been affected by Adam's sin, people made in the image of God can still do amazingly creative things. Damadian's work is a "radiant" example of dominion science. Perhaps you, too, will be able to invent such an important and useful device.

voxel: a combination of the words *volumetric* and *pixel*

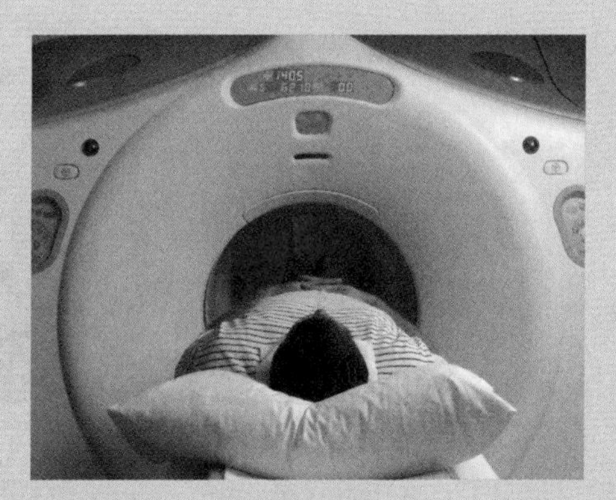

A modern MRI system (top). Raymond Damadian (bottom, left), was the principal inventor of the MRI scanner.

of radar. For example, when an air traffic controller wants to know the location of an airplane, the radar computes the direction (e.g., northeast) and the distance (e.g., 15 km) and displays its location as a bright dot on a screen.

Radar uses various frequencies in the radio and microwave spectrum for specific purposes. Most radars use microwaves to track small objects. The smaller the intended target, the higher frequency and shorter wavelength the radar signal must be. Police cruiser radars and hand-held radar "guns" use very-high-frequency microwaves that are just a few centimeters in wavelength. Long-range radars, on the other hand, may use lower radio-frequency signals that can follow the curvature of the earth somewhat.

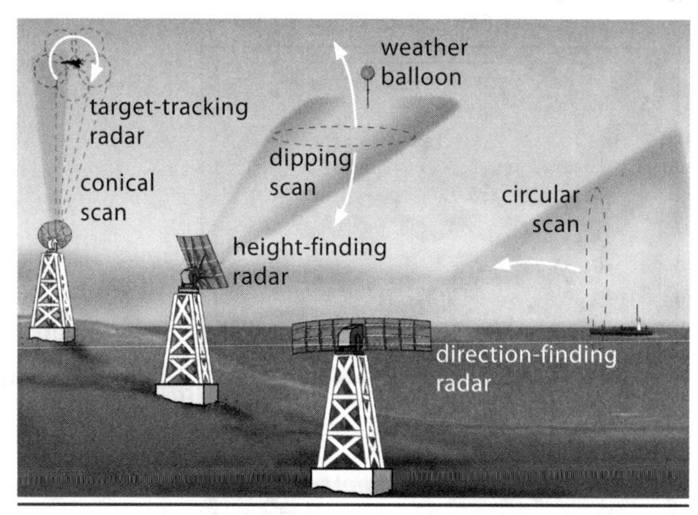

14-30 Radar beams can take several shapes and scan in several ways to obtain required information.

14.16 Television

Television as we know it developed slowly at first, beginning in the late 1800s. The idea of transmitting images over long distances occurred shortly after development of the telegraph. Several inventors used spinning disks containing holes to break up a projected image into small pieces. The light fell on a light-sensitive detector that created electrical impulses that could travel over wires. As interesting as these efforts were, none proved practical. In 1927 a young inventor named Philo Farnsworth designed and built the first practical television (TV) system.

Using current technology, any kind of moving artificial image depends on two properties of the human visual system. First, your eye and brain have a limited ability to separate a pattern of dots into separate objects. If the dots are small enough and close enough together, they appear as a continuous smooth image. Engineers call each of these image-forming dots a picture element, or *pixel.* Millions of pixels can make up the images in magazines, textbooks, billboard advertisements, tablet screens, and TV images.

The human visual system also has a limited ability to sense rapid movement or events. The average human eye can see separate images if they occur at more than 50 milliseconds (1/20 second) apart. If images occur faster than that, they begin to blur together. So if the position of something in a series of images gradually changes, then viewing these images rapidly creates the appearance of motion. The more images per second, and the smaller the change of the subject in each image, the smoother the motion appears. "Flip books" and older film projectors in movie theaters apply this principle.

Television technology uses both of these human visual system properties to create an illusion of moving images. A digital TV transmission begins when a lens system focuses an image on a digital image-forming surface called a *charge coupled device (CCD).* The CCD codes an entire image as pixels many times per second. This data is formatted by connected digital electronics into a radio frequency signal that can be transmitted great distances.

14-31 Philo T. Farnsworth (1906–71) was an American engineer and inventor who developed the first fully functional television system (shown).

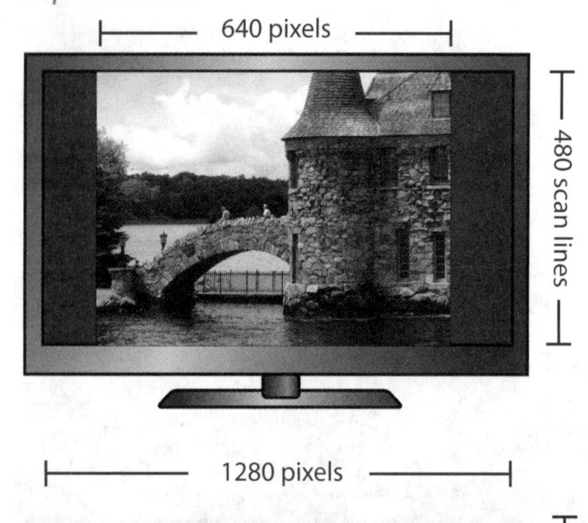

14-32 Two common resolutions for digital TV screens

The signal formatting determines the kind of TV image that will be formed at the receiver—*digital TV (DTV)* or *high definition TV (HDTV)*. TV receivers are designed to present images in various formats determined by several properties.

The TV's *resolution* (the number of pixels and scanlines, such as 1280 by 720 respectively) is probably the most important to the viewer because these numbers determine the clarity and size of the image. The TV's *aspect ratio*, or the ratio of its width to its height (e.g. 4:3 or 16:9), allows for viewing standard TV narrow-screen images, or widescreen content, like theater movies.

Some TVs build their images *progressively*, line by line, from top to bottom of the screen (designated by a "p"); others create images by scanning alternating lines of pixels twice per image frame, called *interlacing* (designated by an "i"). To create the illusion of motion, TVs refresh images many times a second—called the *frame rate*—at 24, 30, or 60 frames per second (fps). The higher the frame rate, the smoother the TV action. A complete description for a TV image might be "720p in a 4:3 format." The various combinations of image format properties have their advantages and disadvantages.

14-33 Two common TV display formats: 4:3, or standard (top), or 16:9, widescreen (bottom)

Radio waves originally carried TV broadcasts through the air. Signals originated at TV towers connected to powerful TV station radio transmitters. Small antennas located on the roofs of homes and connected directly to a TV set by a long flat antenna cable received these radio waves. Today, even with digital TV, many homes still receive TV signals this way. However, a large percentage of households subscribe to *cable TV* (CATV) services, where the radio frequency signals are delivered to a house via a specially constructed *coaxial cable*. A smaller share of the commercial TV market receives programming beamed from *geostationary satellites*. A small dish-shaped antenna at your house receives these radio signals. You can also view TV programs delivered by the Internet, using signals carried by a different portion of the radio frequency spectrum.

Modern digital TV broadcasting creates seemingly endless possibilities for controlling displayed images, recording and editing them, and replaying program segments as the viewer desires. TV broadcasting and personal computer technology are merging to such an extent that they may be hard to tell apart in just a few years.

14-34 Progressive scan (top); interlaced scan (bottom). Scan lines are included for illustrative purposes only.

14.17 Other Applications

Radio-frequency technology has opened up new ways of communicating that weren't even imagined a generation ago. The invention of the cellular telephone, for example, has completely revolutionized personal communications.

The modern *cell phone* is a far different device from early mobile phones. Besides the essential mobile telephone service, which allows calling from even remote areas (by satellite phone), the most capable cell phone is a message center, alarm clock, Internet browser, calculator, e-mail client, voice recorder, camera, music and video player, personal computer, and GPS receiver. Its range of functions seems to be ever expanding. From a dominion science perspective, the ability to communicate when in danger, injured, or lost is immensely helpful for preserving human life and health. This entire technology depends on radio-frequency transmissions.

Highway systems are using several versions of **radio-frequency identification (RFID)** technology to monitor truck weights and collect tolls on the Interstate highway system. In the case of toll systems, such as FasTrak and E-ZPass, a small radio transmitter-receiver is installed in the vehicle. The owner is assigned a code that is transmitted when the vehicle passes through specially equipped tollgates. The owner's name, the date, and the time are electronically recorded and his toll is automatically deducted from a prepaid account.

Radio-frequency identification is also used for inventory control, which is especially important for large retail stores. Paper-thin tags contain an antenna and a tiny programmable microprocessor chip, but no battery. These tags can be used to identify products just like barcodes. The energy from an external radio signal induces enough current in the tag to activate the chip and generate a response. The advantage of an RFID system over barcodes is that radio scanning can be done more rapidly and from a greater distance. Some people believe that the RFID inventory control system will be extended to automating store checkouts, eliminating the need for cashiers. This technology has been incorporated into ID cards, debit cards, passports, and shipping container identification. It is anticipated that eventually personal identification and medical information can be stored on a tiny RFID slip implanted under the skin. Some consider such use an invasion of privacy. Such technology, though, could be a proper exercise of dominion science by giving us the ability to identify lost children and pets and immediately access a person's medical history in the event of an accident.

14-35 A modern cell phone tower. Each tower may carry dozens of antennas and thousands of signals.

Chip Implant: Mark of the Beast?

Many Christians find RFID technology disturbing. To them, it sounds similar to the biblical prophecies concerning the mark of the Beast recorded in Revelation 13. Read Revelation 13:16–18. Does this passage indicate that RFID technology should be avoided today?

14-36 Examples of RFID tags and technology

14.18 Radio Navigation

Any method of navigating a ship, plane, or other vehicle that uses radio-frequency signals is a form of **radio navigation**. There are two major categories of radio navigation: terrestrial and *global navigation satellite systems (GNSS)*. We will discuss only two terrestrial systems here.

Omega

The Omega system was the first terrestrial global radio navigation system placed in service. Set up in the early 1970s to guide nuclear bombers for the US Air Force, it used land-based VLF radio transmitters. The long-range signals provided position with an accuracy of about 4 NM (7 km), which was sufficient for aiming atomic bombs. Eventually, civilian ships and aircraft were allowed to use the system. The US shut down the Omega system in 1997 because other available navigation systems were far more accurate and reliable.

Recall that NM stands for *nautical mile*. A NM is exactly 1.852 km or about 1.15 statute miles.

LORAN-C

Another terrestrial navigation system still in use in European waters, the Indian Ocean, and the Far East is the LORAN-C system, which operates in the LF region of the radio spectrum. The *Chayka* navigation system in the Russian Arctic is similar. The stations in this system broadcast between the frequencies of 90 and 110 kHz. Each station transmits a signal with a distinct time delay from a reference time synchronized across the world. This delay identifies each specific station. At the receiver, the time delays from different stations are compared and it computes a position shown on its display. LORAN-C positions are accurate to within 0.25 NM (460 m), but weather and the location of stations can decrease accuracy significantly at times. US and Canadian LORAN-C operations ceased in 2010, though we can still use signals in North America from the operational stations overseas.

LORAN is the acronym for LOng RAnge Navigation.

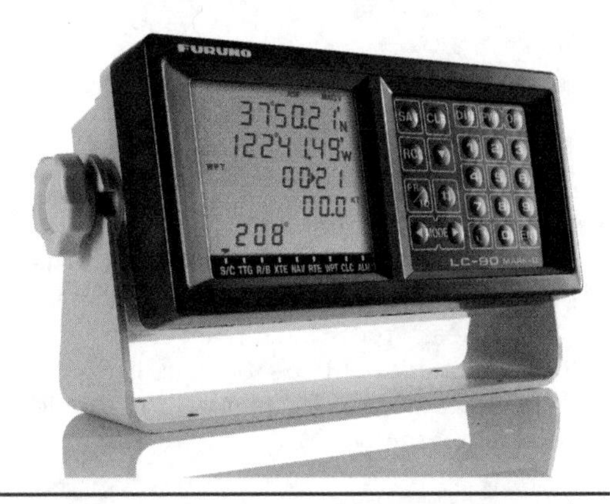

14-37 A LORAN-C navigation receiver

Use of LORAN-C declined in the 1990s with the increasing dependence on the GPS. However, upgrades to LORAN-C and its integration with the GPS seem to ensure that it will continue to operate for the foreseeable future.

The Russians operate a positioning system similar to the GPS called the GLONASS (*Gl*obal *N*avigation *S*atellite System).

GPS

The *Global Positioning System (GPS)* is the most sophisticated and accurate global navigation system yet devised. It consists of a "constellation" of 24 operational satellites. The orbits of the satellites are placed around the globe so that the signals from a minimum of four satellites can be received at any location on the earth at a given moment. The first few GPS satellites were operational at the start of the Gulf War in 1991, when their positioning information was effectively used for combat purposes. Since that time, the entire complement of satellites has become operational. The system is maintained entirely by the US government, though it may be used by anyone around the world.

14-38 A modern surveyor's GPS receiver system can obtain positions that are accurate to within a few millimeters.

A GPS satellite is relatively uncomplicated compared to other satellites. It consists of a simple computer, radios, and an atomic clock. Its actual position in orbit is verified twice daily by US Air Force ground stations. The onboard computer accurately estimates the satellite's position between updates. The radios transmit coded signals that identify the satellite and provide its position to GPS receivers.

A GPS position is determined by **trilateration**. In this process three known distances are used to obtain a position. The distances from three GPS satellites intersect at a unique point on the earth's surface. If a fourth satellite is used in the calculations, an accurate altitude can also be obtained. The accuracies of these positions are within 13 m or better under normal system conditions. Surveyors can obtain accuracies within a few millimeters by using special equipment and techniques. Cars can be equipped with GPS devices that can provide map positions and directions. In an accident some instruments can automatically connect the car's occupants by radio to an advisory center and transmit the vehicle's GPS location. Many lives have been saved by the rapid and accurate emergency responses initiated by these services.

trilateration: tri- (L. *tri*—three) + -later- (L. *lateralis*—side) + -ation

14-39 A GPS map display provides drivers with an accurate location in unfamiliar cities.

14.19 Radio Astronomy

Radio astronomy is a relatively new branch of astronomy involving the observation of radio emissions from celestial objects, including the planets, the sun, stars, our galaxy, and distant galaxies and nebulas. While traditional astronomy involves observation of the heavens using the visible-light spectrum, such data reveals only a part of the astronomical picture. Higher-frequency visible light emissions are easily absorbed by interstellar dust and gases. Long-wave radio signals, on the other hand, can easily penetrate these obstacles. The earth's atmosphere is completely **transparent** to radio waves from space. The main disadvantage of this kind of astronomy is that very large dish-shaped telescopes are required to receive the long waves, and often many telescopes must be used together. Also, the many man-made radio-frequency sources on earth sometimes interfere with the astronomical signals.

The term **transparent**, in addition to meaning "clear," can also be used, as here, to describe something that allows any form of EM energy, not just visible light, to pass through it.

Radio astronomy had its official beginnings in 1931. A physicist from Bell Laboratories, **Karl Jansky**, was investigating potential sources of radio interference for trans-Atlantic radio communications when he discovered radio waves coming from the Milky Way. The first radio telescope intended for astronomy was built by a young engineer, **Gröte Reber**, for his personal study in 1937 just before World War II. After the war, radio astronomy took advantage of the much-advanced knowledge of radio technology and became established as a distinct branch of astronomy. Many countries built increasingly larger radio telescopes. The largest individual steerable dish antenna is the 100 m diameter Robert C. Byrd Green Bank Telescope in West Virginia. Multiple radio telescopes working together to observe the same object can produce sharper images of the object at

Karl Guthe Jansky (1905–50) was an American physicist and radio engineer who first detected radio waves from space.

Gröte Reber (1911–2002) was an American electrical engineer who pioneered the science of radio astronomy. He produced the first complete sky map of radio sources with his home-built radio telescope dish.

14-40 The 100 m Green Bank radio telescope is the world's largest steerable antenna.

14-41 The EVLA is currently the world's largest radio telescope system.

radio frequencies. The largest system of linked radio telescopes is the Expanded Very Large Array (EVLA). This system of 27 antennas, each 25 m in diameter, is arranged in a Y-shaped pattern in the desert 50 mi west of Socorro, New Mexico. Even larger systems are being built or are planned in Europe and in the Southern Hemisphere.

Astronomers have discovered amazing things in the distant regions of space that are completely outside the visible spectrum. One such discovery was *pulsars*. Believed to be the remnants of supernova stars, pulsars are neutron stars spinning at hundreds of revolutions per second. Their motion accelerates electrons in their vicinity, creating radio signals shaped like rapidly rotating flashlight beams. **Quasars** were discovered using radio telescopes. In visible light, they look somewhat like stars, but they radiate more energy in the radio and visible spectrums than entire galaxies or even hundreds of galaxies! Radio astronomy has revealed many new and fascinating aspects about our universe that continue to puzzle astronomers.

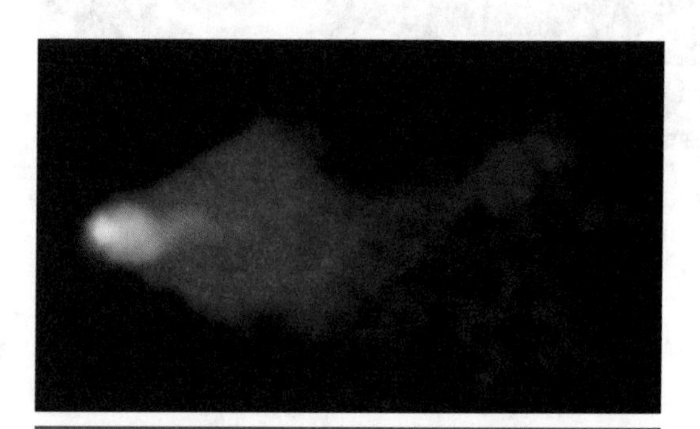

14-42 A radio telescope image of a pulsar

Quasar stands for *quasi-stellar*, meaning that quasars are sort of like stars but differ in important ways. They are also called *QSOs*, which stands for *quasi-stellar objects*.

14C Section Review

1. What portions of the EM spectrum can humans detect without instruments?
2. What device is essential to form man-made radio waves? Which part of the radio spectrum is used almost exclusively by the military?
3. How is radar similar to active sonar?
4. What two characteristics of the human visual system does television use to create the illusion of moving images?
5. How is the invention of the cell phone an example of dominion science?
6. What are the advantages and disadvantages of using RFID chips for personal identification?
7. Name the two major operational radio-frequency systems currently used for navigation over a large portion of the earth's surface. Which one is the most accurate?
8. Why are radio telescopes generally so large? Why is it unnecessary to place radio telescopes in orbit?
9. (True or False) Miniature RFID technology suffers from the need to provide a tiny battery in the tags.

Scientifically Speaking

Chapter Review

1. Describe the major difference between the way mechanical waves and EM waves move.
2. Is it possible to split the electric field wave from the magnetic field wave in an EM wave? Explain.
3. Why can we assume that the speed of EM waves in the atmosphere is essentially the value of *c*?
4. If a photon of light enters a piece of glass and its speed decreases to half the value of *c*, what happens to its frequency? to its wavelength?
5. Explain the statement, "Photons are both particles and waves."
6. Which photons have more energy, 30 GHz microwaves or 30 PHz x-rays?
7. What happens to the received frequency of an EM signal when the source is moving away from the observer? What is this phenomenon called?
8. ✪ What kind of EM emission would you expect if an electron speeding through space near the earth were rapidly accelerated as it entered the earth's magnetic field?
9. Why would an owner of a cell phone have problems receiving a signal deep in the Appalachian Mountains even though a microwave cellular tower was located nearby on the opposite side of a mountain ridge?
10. Why can IR radiation be considered a form of heat?
11. Which color of visible light has the highest energy photons, deep violet or deep red? Explain.
12. What kind of radiant energy do sunblock creams and lotions absorb or reflect?
13. DS ✪ Why are x-rays so effective at killing cancer cells?
14. Which natural substances on Earth are likely sources of gamma radiation? Why are these sources, in their natural setting, not particularly dangerous to humans?
15. How could you learn to use radio communications in a way that supports the model of dominion science presented in this text?
16. What shape of radar beam is used to track and control the movements of ships in a busy harbor? (*Hint:* See Figure 14-30.)
17. How does a digital TV camera produce the many still images of a scene per second needed to produce the illusion of motion?
18. ✪ How do you think cell phone technology has affected our culture? Overall, has it been more positive or negative?
19. What pattern do you notice between operating frequency band and positional accuracy for the three navigational systems described in Subsection 14.18? (Note that the GPS satellites transmit in the microwave portion of the radio spectrum.)

20. The sharpness (resolution) of radio telescope images is relatively poor at long radio wavelengths. What two things have astronomers done to improve radio images?

True or False

21. Electromagnetic energy can be mathematically described by relatively simple equations that demonstrate the wave nature of this energy.
22. A photon is a discrete packet of EM energy.
23. Radio waves must be formed in a metal conductor called an *antenna*.
24. Far infrared has higher frequencies than near infrared.
25. Ultraviolet rays are formed in the same way as visible light is formed.
26. X-rays are the only form of EM energy that can penetrate matter.
27. Some gamma rays have so much energy that they can change an atom into a different element.

DS ✪ 28. Radiant energy can be used for disease diagnosis as well as treatment.

✪ 29. Information transmitted by radio waves is formatted in basically the same way for all radio frequencies.

30. Radio-frequency EM energy is emitted inside a digital flat screen to produce TV images.

DS ✪ 31. Do some research and make a list of all the ways EM energy is used in medical diagnosis and disease therapy.

DS ✪ 32. Not all human tissues are equally susceptible to radiation damage. Do an Internet search to identify 10 types of human tissues most affected by gamma radiation. List them in order from the most susceptible to the least.

✪ 33. Is the problem with defining the photon a lack of understanding concerning the phenomenon, or is it that our language is inadequate to describe it?

LIGHT AND OPTICS

CHAPTER 15

DOMINION SCIENCE PROBLEM

Cutting Down on Surgery

Until late in the twentieth century, medical diagnostic options were limited when a person became critically ill with no clear indication of the cause. Chapters 13 and 14 discuss various medical imaging techniques available today that use sound and EM energy to examine the body's interior apart from any kind of surgery. Sometimes the results of these tests are inconclusive, and doctors have to look directly inside the body to understand the problem. Normally, this would require major exploratory surgery with large incisions, a long hospital stay and recovery, and an increased risk of complications. And these measures are taken only when there is already a problem. Is there a way to look directly inside the body to diagnose, treat, and prevent disease without major surgery?

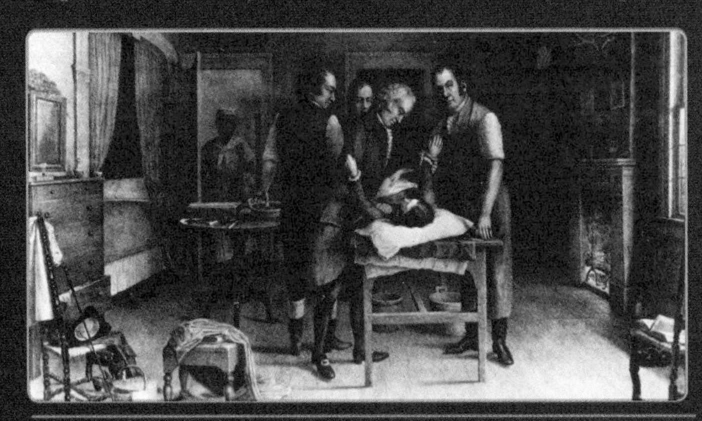

15-1 In the not-so-distant past, surgeons often resorted to exploratory surgery to understand the nature of an undiagnosed ailment. The injury from surgery was frequently worse than the ailment!

15A Objectives

After completing this section, you will be able to

- ✓ describe the general properties of the visible spectrum and the distribution of colors within it.
- ✓ state the speed of light to three SDs and discuss how the unit of astronomical distance is derived from this value.
- ✓ state how the intensity of light is measured and the unit with which it is measured.
- ✓ describe how illumination from a light source varies.
- ✓ list five kinds of light sources and give at least one example of each.

15A Visible Light and Its Sources

15.1 *Introduction*

God gave humans and nearly all animals the ability to see. Though more of the human brain is devoted to processing touch than any other sense, we depend far more on sight to interact with the surrounding world. The eye is sensitive to a relatively small portion of the EM spectrum, yet through its architecture and the connecting nerves and brain cells (the visual system), we can see three-dimensional images in millions of colors! Our study of visible light and optical phenomena begins with an examination of the nature of light.

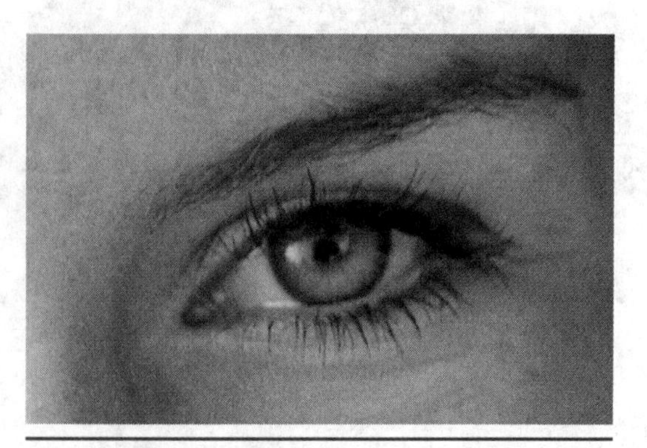

15-2 The human eye is just one of the great evidences of God's design.

15.2 *The Visible Spectrum*

Recall from Chapter 14 that the visible portion of the EM spectrum lies in the middle of the terahertz band, about 384 THz for deep red light to about 789 THz for deep violet light. The wavelengths of visible light range from about 780 nm for deep red light to about 380 nm for deep violet light. We say "about" because the limits of the visible spectrum are different for each person. One's eyes and brain determine how this EM energy is perceived.

We see light as color. Even sunlight, which appears to have no color, is really a mixture of nearly all possible colors in the visible spectrum. Sunlight's secret is revealed in a rainbow. The colors of a rainbow seem to blend continuously into each other with no obvious breaks between them. People commonly divide this color spectrum up into familiar *index colors*—red, orange, yellow, green, blue, indigo, and violet. However, specific wavelengths are not assigned to each of these colors because everyone perceives color differently. Colors can be uniquely named only by their frequency or wavelength. For example, EM waves with 700 nm and 690 nm wavelengths are slightly different colors in the red portion of the spectrum.

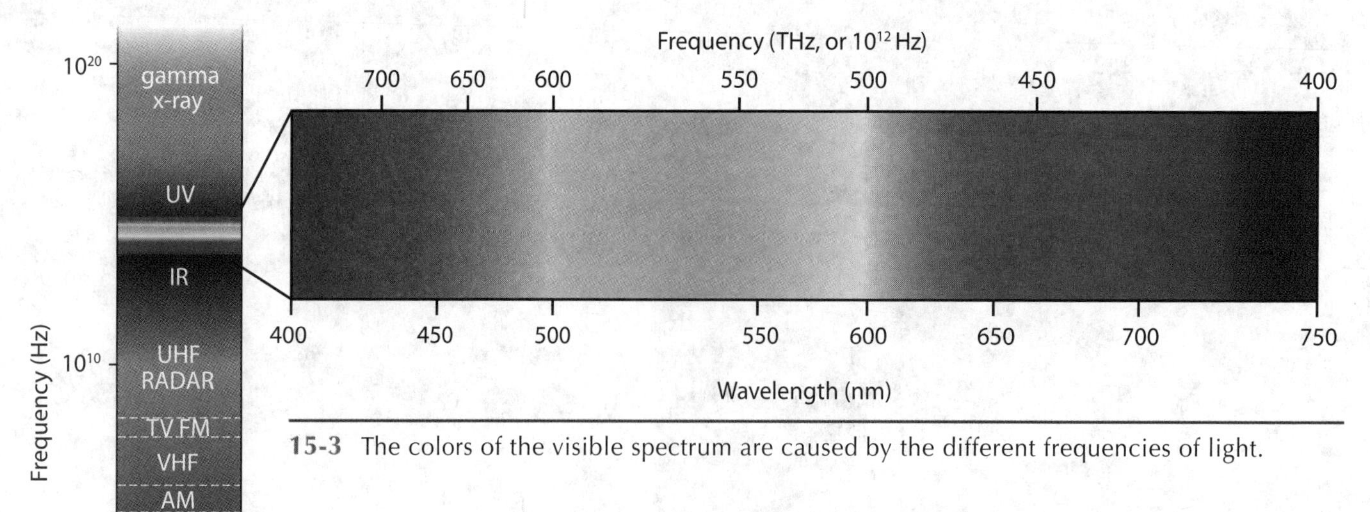

15-3 The colors of the visible spectrum are caused by the different frequencies of light.

Luminous objects produce visible light. Ideally, a luminous object produces all possible wavelengths of visible light to form a *continuous spectrum*. The sun comes very close to emitting a continuous spectrum, but there are numerous very fine gaps throughout its spectrum that are detectible only with sensitive instruments. To the unaided eye, our sun produces a continuous spectrum, though it doesn't radiate all colors at the same intensity.

Some luminous objects, such as lasers or **light-emitting diodes**, produce single colors or wavelengths of light. These luminous objects are ***monochromatic***. All light photons emitted by a monochromatic source have the same frequency and wavelength. Other luminous sources produce multiple distinct colors. If light from these sources is passed through a prism, you see only narrow bands at specific wavelengths, not a blended continuum. These sources produce line spectra. A *line spectrum* contains distinctly visible lines of color aligned along a scale in a spectrograph. A *spectrograph* is a photograph, printout, or other display of a spectrum produced by a ***spectroscope***. Many fluorescent lights emit five main visible frequencies of light, two in the blue region, one in the green, a weaker one in the yellow, and one in the red region. Our eyes sense this combination of frequencies as blue-tinged white fluorescent light.

A **light-emitting diode** (LED) is a semiconductor device that emits light when current passes through it in one direction but not the other.

monochromatic (MAHN uh kro MAT ik): mono- (Gk. *mono*—one, single) + -chroma- (Gk. *chroma*—color) + -tic

A device that breaks up light into its component colors is called a **spectroscope**. Both continuous spectra and line spectra can be [illegible] with a spectroscope.

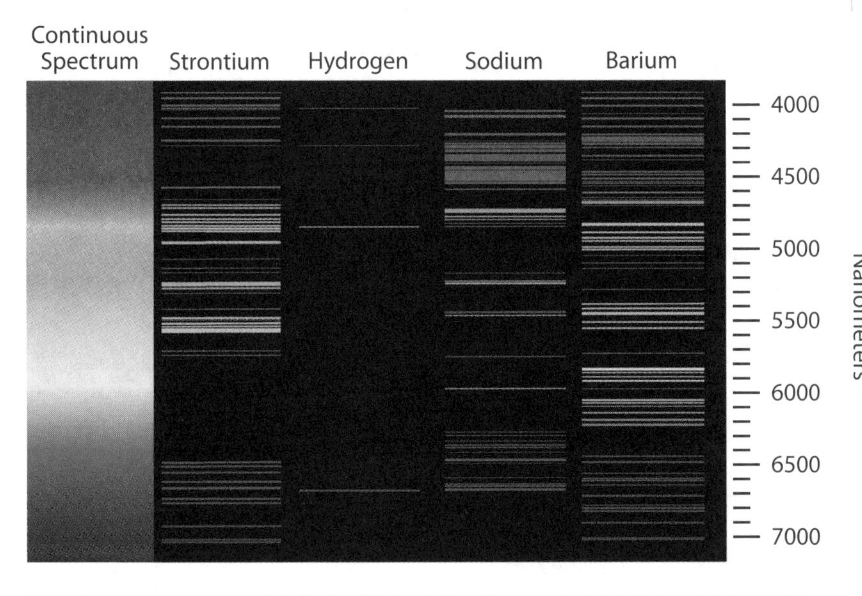

15-4 A continuous spectrum and several line spectra as seen in a spectroscope

Light also comes from *illuminated* objects. These objects are not light sources; they merely reflect light. While the sun is a luminous object, the moon is an illuminated object. It reflects the light of the sun. Illuminated objects have many different colors, and the same object can appear to have different colors under different colors of illumination. See Section 15B for more about colors.

15.3 The Speed of Light

Philosophers and scientists have considered the nature and speed of light for thousands of years. Some ancient Greeks believed that the speed of light (*c*) was infinite, while others thought it was very fast but finite. Though these two views were held, no one attempted an experiment to prove either theory until the time of Galileo in the seventeenth century. He suggested an experiment in which two people on hills separated by a known distance uncover lanterns to measure the amount of time light takes to be seen by each observer. By this method, he realized that the speed of light must be very high.

15-5 The Danish astronomer Ole Rømer made the first reasonably accurate determination of the speed of light from observations of Jupiter's moons.

15-1 **Determinations of the Speed of Light**

Year	Experimenter	Method	c (× 10^8 m/s)	Uncertainty (± m/s)
1600	Galileo	lanterns	"fast"	
1676	Rømer	moons of Jupiter	2.14	unknown
1729	Bradley	aberration of light	3.08	unknown
1849	Fizeau	spinning slotted wheel	3.14	unknown
1879	Michelson	rotating mirror	2.9991	75,000.0
1926	Michelson	rotating mirror	2.997 98	22,000.0
1950	Essen	microwave cavity	2.997 925	1000.0
1958	Froome	interferometer	2.997 925	100.0
1972	Evenson	laser method	2.997 924 57	1.1
1974	Blaney	laser method	2.997 924 59	0.6
1976	Woods	laser method	2.997 924 59	0.2
1983		definition	2.997 924 58	0.0

In the centuries following this experiment, the speed of light was determined by different methods more than 100 times. Table 15-1 lists the data for some of the more significant determinations of *c*. Note that the current accepted speed of light is defined as an exact value rather than a measured one.

The speed of light depends on the material it is traveling through. Light travels fastest in a vacuum. In optically dense materials, light tends to move more slowly. The way a material affects the speed of light also depends on the kinds of atoms and the bonds between them.

At 300 million meters per second, light can travel about 18 million kilometers per minute. Since the sun is 150 million kilometers away, its light takes approximately 8.3 min to get to Earth. Astronomers say that the sun is about 8.3 *light-minutes* away. A light-minute is the distance light travels in a minute. It allows us to use relatively small numbers to measure large distances within the solar system. Sunlight takes about 4.2 h to reach the outermost planet, Neptune. We could, therefore, say that Neptune is more than 4 *light-hours* from the sun. In discussing distance to stars, astronomers use a larger unit, the light-year. The *light-year (ly)* is defined as the distance light travels in a year, approximately 9.5 trillion kilometers. The three stars nearest to the sun, all of the Alpha Centauri system, are 4.2–4.4 ly away.

15-6 The great Orion Nebula (the hazy patch just below the "belt" in the constellation Orion) is about 1600 ly away.

15.4 The Intensity of Light

A light source's power, the rate at which it radiates energy, is called its brightness or *intensity*. To tell how bright a light is, we must compare it to a standard of known brightness. In the 1800s physicists used a specific type of candle as a standard of intensity. However, candles cannot be made to burn with a perfectly steady intensity, nor can they be made to a uniform quality.

When electrical illumination sources were developed in the late 1800s, greater uniformity became possible, although it was still not good enough for advanced scientific research. In the 1930s physicists agreed to adopt a theoretical light intensity standard based on the emission of light from solidifying platinum (at 2045 K). In 1948 the unit of light intensity was called the *candela (cd)*, which is an SI unit. Because of the difficulty in establishing the conditions in a laboratory to use the molten platinum as an intensity standard, physicists adopted a new standard in 1979. It is based on the same light intensity as the solidifying platinum standard but uses other more easily measurable dimensions. The full definition of the candela is provided in Appendix B.

An **incandescent** object is so hot that it glows, emitting visible light.

The intensity of a light source can be measured and compared against the SI standard to determine the number of candelas it produces. Ordinary **incandescent** light bulbs produce about 0.8 cd of light per watt of electrical power they use. Fluorescent tubes are considerably more efficient—they produce about 4 cd per watt.

When you read a book indoors by the light of a lamp, the amount of light, or *illumination*, you receive on the printed page depends on two factors—the intensity of the light source and the distance of the light source from the page. The illumination of an object is directly proportional to the intensity of the light source. If two lamps are the same distance from the page, a lamp that produces 100 cd will illuminate the page twice as much as a 50 cd lamp.

On the other hand, illumination is *inversely* proportional to the *square* of the distance from the light source. Moving twice as far from the lamp reduces the illumination to one-fourth the original amount. Moving three times as far from the lamp reduces the illumination to a mere one-ninth. This type of relationship is called an **inverse square law**, a situation encountered many times in physics. Distance from the light source, therefore, is a critical factor in illumination.

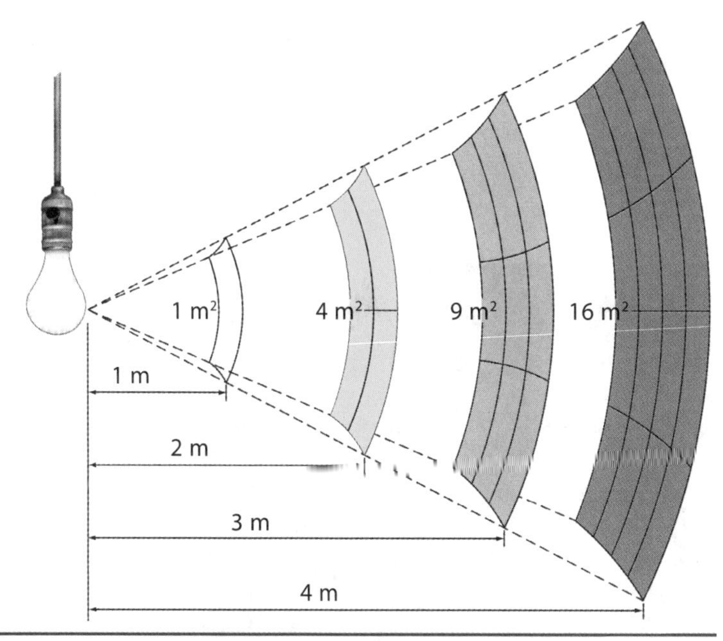

15-7 Illumination from a source decreases with the inverse square of the distance from the source.

Another example of the inverse square law: The force of gravity between objects varies according to the inverse square of the distance between them.

photometer (foe TAHM ih tur)

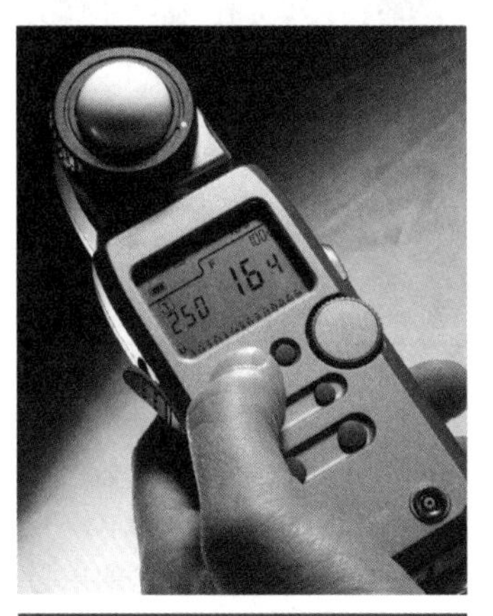

15-8 A photographer's light meter is one kind of photometer.

Photographers use *light meters*, or **photometers**, to measure the exact amount of illumination falling on an object to be photographed. Inside a photometer is a photoelectric cell that changes light energy into an electrical current. The brighter the illumination, the more current is produced by the photoelectric cell. This current is used to produce a measurement of the illumination that is then displayed by the meter. Using the reading from the photometer, the photographer adjusts the lens opening and shutter speed of his camera so that he will take the kind of picture he wants given the available light. He may also adjust any artificial light sources. Most modern cameras have built-in photometers that adjust camera settings automatically.

Humans have their own built-in photometers. The human eye can detect an amazing range of light intensities. Although it can function in bright sunlight, it can also adjust to operate in nearly total darkness—only one-billionth the intensity of full sunlight!

15.5 Types of Light

There are many natural and artificial types of visible light. Probably the most familiar is **incandescence**, the light produced by materials heated until they glow. Incandescence results when atoms and their electrons rapidly vibrate with thermal energy. Some thermal energy is converted to EM energy, which radiates from the object. In relatively cool objects this energy radiates in the IR part of the EM spectrum. As an object grows hotter, light in the red portion of the visible spectrum becomes apparent. You see this when you start a

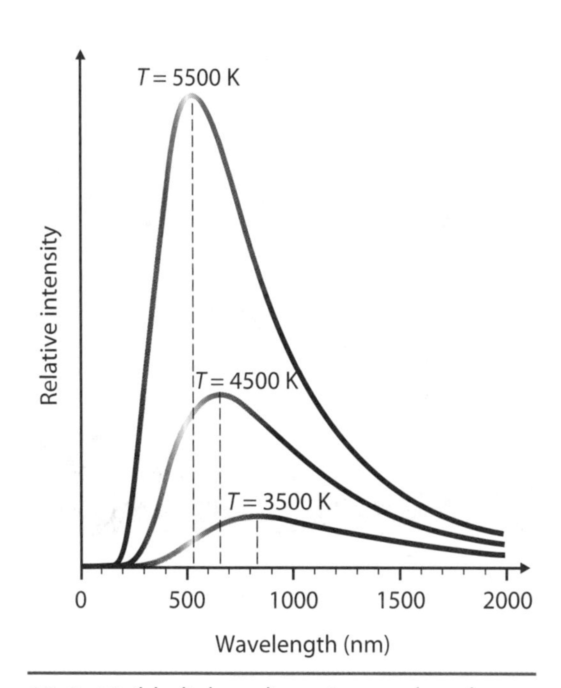

15-9 Visible light radiates in wavelengths that depend on its kelvin temperature. At a given temperature one wavelength of light has the greatest intensity compared to all the others.

fire or turn an electric burner on high. As the temperature increases, the most intense light emitted shifts wavelengths from yellow to blue and then into the violet portion of the spectrum. Overall, very hot objects appear as an intense blue-white glow as all the colors blend. The wire filaments in an incandescent light bulb heat up because of electrical resistance, reaching a temperature at which they glow brilliantly. The sun is also an example of incandescence.

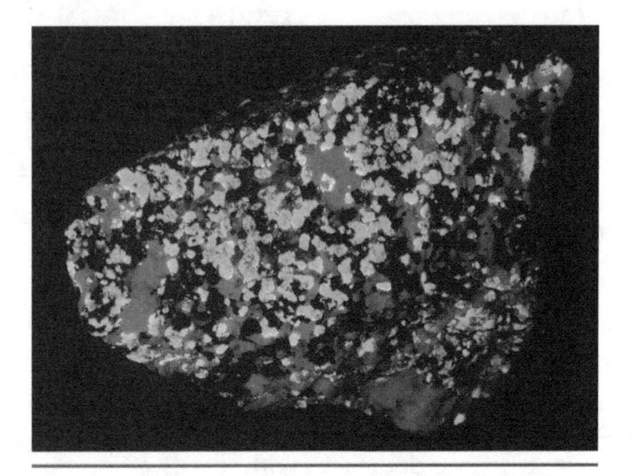

15-10 Some minerals can be identified by the way they fluoresce under UV light.

Fluorescence is another familiar type of light. It occurs when a substance radiates visible light as it absorbs a higher-energy form of EM energy (like UV rays). When the higher-energy source is removed, the fluorescence stops. Fluorescent lamps consist of a tube coated with several *phosphors*, compounds that fluoresce. Inside the tube is low-density mercury gas, which becomes a plasma when electricity passes through it. This plasma emits UV rays and deep blue visible light. The high-energy UV rays are absorbed by the phosphors, which fluoresce in visible light in the green, yellow, and red wavelengths. This combination of red, green, yellow, and deep blue is perceived as white light. Some geologic minerals can fluoresce when illuminated by near-ultraviolet lamps, and this property is used to identify them.

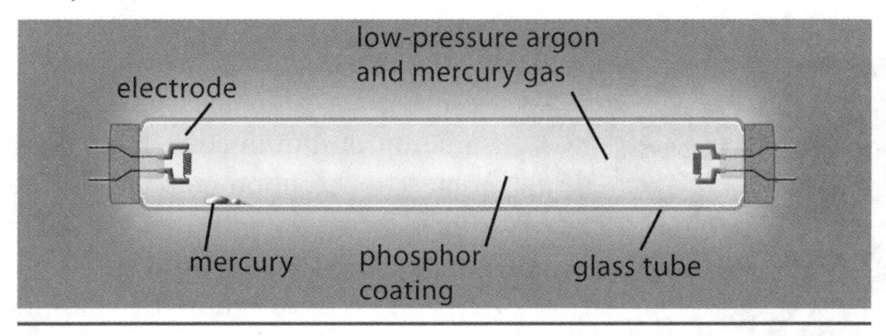

15-11 This light bulb uses fluorescence to provide illumination.

Fluorescent materials glow only when a source of photons is supplied; however, substances that show or display **phosphorescence** continue to emit visible light after the energy source is removed. Glow-in-the-dark stickers and mechanical watch dials use phosphorescent materials.

15-12 This LED emits coherent invisible IR light as it reads a compact disc.

Coherent light sources produce monochromatic light. Monochromatic light consists of photons that all have the same wavelength, that is, the same color. The photons also radiate from the source at the same time so that their waves are all "in step." Coherent light is extremely bright and has many uses in medicine, industry, communications, and the military. Lasers and certain light-emitting diodes (LEDs) produce coherent light. See the facet on the facing page for more information on lasers.

15-13 Light sticks produce cold light through low-energy chemical reactions.

Some sources of light provide illumination through chemical reactions at temperatures far below those needed for incandescence. This **cold light** can be produced artificially or naturally. Man-made cold light sources include light sticks, which use *chemiluminescence*. These familiar lights are found in road emergency kits as well as party favors. Many living things produce cold light, known as **bioluminescence**. Have you ever caught fireflies at night and watched their little "lanterns" flash on and off? Fireflies (beetles in the family Lampyridae) and many other organisms, such as ocean plankton and jellyfish, produce cold light.

Bioluminescence is any form of light produced by animals or plants through biological processes.

FACETS OF SCIENCE: LASERS

What do LASIK surgery, flat-screen TVs, grocery scanners, surveying equipment, and Blu-ray (BD) players all have in common? They use lasers. The word *laser* stands for *l*ight *a*mplification by *s*timulated *e*mission of *r*adiation.

What makes the light from a laser so special? An ordinary light bulb emits photons of many different wavelengths. Even waves with the same wavelength are out of step, or out of phase, with each other. A laser is different because it produces a coherent, concentrated, monochromatic beam of light. Every photon has the same wavelength and is synchronized with every other wave. Crests align with crests and troughs with troughs. These properties make lasers very useful.

The first visible-light laser

Laser beams travel in straight lines, scattering very little compared to other kinds of light. This trait makes them useful for leveling surfaces or aligning things. Lasers are also used to measure shifts in large buildings or movement in the earth's crust across geologic faults.

Lasers can also be used to measure distances. If a scientist knows the time it takes for a laser pulse to cover a distance and return, he can use the speed of light to calculate the distance. For instance, Apollo 11 astronauts placed a laser reflector on the moon in 1969. NASA scientists can direct laser pulses toward this reflector and time them to very accurately measure the distance between the earth and the moon. This technology allows scientists to accurately track the earth's and the moon's movements relative to each other. Surveyors also use lasers to measure distances on the earth.

Lasers can rapidly store and retrieve data. Pulsing laser beams are used to carry large amounts of information, such as audio (sound), video (television images), and computer data across long distances.

Certain types of laser beams generate a tremendous amount of heat wherever they strike. These lasers are used in industry to weld sheet metal and to fuse sections of glass or plastic. Applying lasers, ophthalmologists have revolutionized eye surgery using a procedure called *LASIK*. A laser is used to reshape the cornea of the eye by vaporizing tiny bits of tissue. When successful, this procedure gives patients nearly 20/20 vision.

Every DVD and BD player contains at least one laser. Many computer printers now rely on lasers. Low-power lasers are being developed for everything from taillights for vehicles to large flat-panel screens for TVs and computers.

For several decades the US military has been working on developing battlefield weapons based on high-energy lasers. One program that was canceled in 2010 would have allowed shooting down ballistic missiles from an aircraft. Other systems still in development include defensive laser weapons to destroy incoming battlefield rockets and artillery shells. Another system will intercept anti-ship cruise missiles at sea. The Advanced Tactical Laser (ATL) is intended to destroy ground targets from an aircraft. A prototype was successfully tested in 2009.

This is one design for a tactical sea defense laser installation onboard a ship.

Lasers have become a vital tool in many technologies. Scientists continue to explore the applications of this powerful form of light. Be alert for reports of even more spectacular uses for laser light.

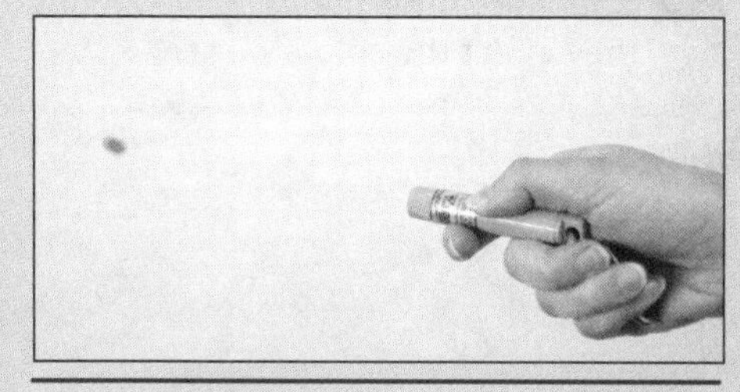

A coherent LED used in a laser pointer

15A Section Review

1. Why is it impossible to define distinct limits of the visible light spectrum?
2. How do physicists identify a particular color to avoid confusing it with other similar colors?
3. What two terms describe kinds of light sources? Which kind produces its own light?
4. Refer to Table 15-1 on page 328. Why does the speed of light's accepted value have an uncertainty of zero?
5. What is light intensity? How does light intensity relate to the energy of a light source? What is the SI unit for light intensity?
6. How does the illumination on a surface vary with distance from a light source? Which would provide a greater increase of a lamp's illumination of a book: doubling the wattage of the light bulb or moving the book to half the original distance from the bulb?
7. (Matching) For objects a–d, match the kind of light, using the abbreviations from the choices given. Choices may be used once, more than once, or not at all.

a. LED	(I) incandescent
b. hot stove burner	(P) phosphorescent
c. glowing jellyfish	(M) monochromatic
d. mineral glowing green while under UV light	(F) fluorescent
	(CL) cold light

8. (True or False) A light-year is the time it takes light to travel a certain distance.

15B Objectives

After completing this section, you will be able to

- ✓ explain how people perceive color.
- ✓ compare and contrast how additive and subtractive colors mix.
- ✓ state the relationship between the additive and subtractive primary colors.
- ✓ state the properties of color and how they affect the colors we see.

15B The Nature of Color

15.6 Color Perception

Is a red ball red because that is how you perceive the wavelengths of light reflecting off the ball, or is the ball itself red? In other words, do the properties of something depend on how we observe them, or are they a basic characteristic of the object apart from our observation?

Philosophers have struggled with questions such as this for thousands of years. Bible-believing Christians understand that God created the universe, and everything He created is real. The ball reflects certain wavelengths of light, and those wavelengths exist whether there is someone looking at the ball or not. God created the human visual system to interpret certain wavelengths of light as a certain color, and humans have agreed to call that color *red*. So the color that you observe an object to be is a combination of what it is and your perception of it.

15-14 Is the ball red because you perceive it as red or because it *is* red?

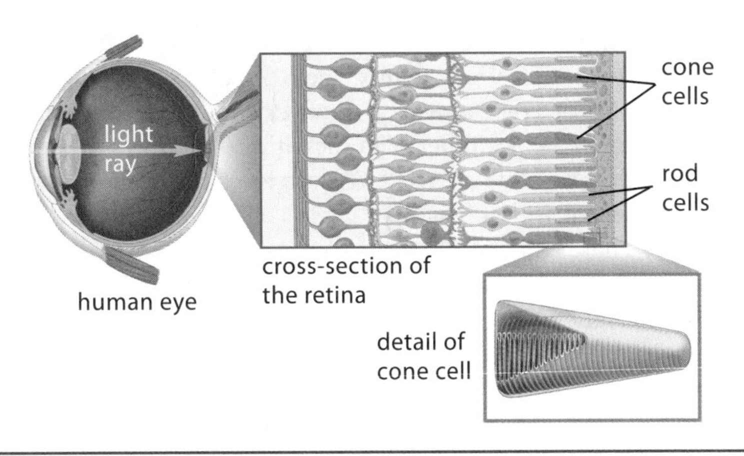

15-15 A cone cell of the eye's retina. The chemicals present in the cell determine the wavelengths of EM radiation that are detected.

Color perception is a complicated process involving the shape, structure, and composition of our eyes (see the facet on page 335), which produce nerve impulses that travel to the brain. The brain receives millions of nerve impulses every second and assembles them into full-color, three-dimensional patterns that are images of the world around us. The intricate structure and marvelous function of the human visual system are extraordinary—unquestionable evidence for a supernatural Designer.

15.7 *Color Mixing*

There are many colors that aren't in the rainbow. For example, what wavelength produces brown light? None—no wavelength of visible light corresponds to brown or to a myriad of other colors. These colors are mixtures of wavelengths of light. When our eyes and brains receive a mixture of wavelengths, they translate this information into a color that is a combination of those wavelengths. The most familiar example is white. There is no frequency of visible light that produces the color white. White light is a mixture of the entire rainbow of visible wavelengths.

After much study, scientists have learned that the human eye and brain can distinguish three colors of light, called *primary hues*, that can be mixed to produce most other hues. The word *hue* identifies the basic color of light, such as red, orange, yellow, green, and so on. The primary hues of light, or **additive primary colors**, are red, blue, and green. Light is *additive* because the brain adds the nerve stimuli of different wavelengths together so that the wavelengths appear to be only one color. If you mix these three colors together at proper intensities, the mind perceives the mixture as white light. Millions of colors can be sensed by mixing appropriate combinations of the three additive primary colors. For example, a mixture of green and blue light reflecting off a white background appears cyan. No wavelength of the visible spectrum corresponds to cyan; it is a color perceived only by the brain.

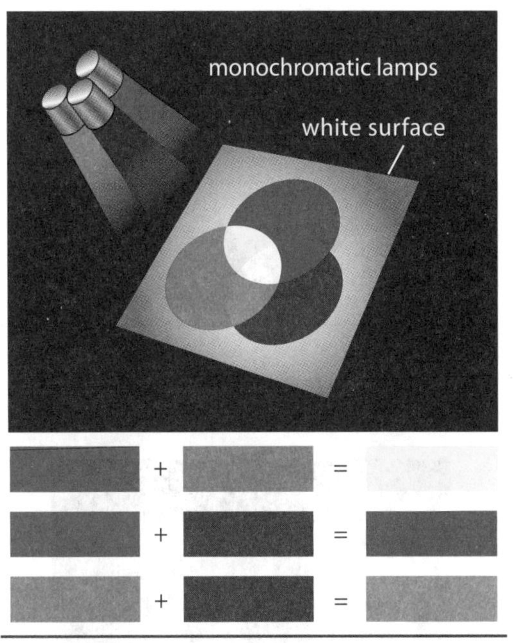

15-16 Red, green, and blue are the primary additive colors. Red and green form yellow, red and blue form magenta, and green and blue form cyan. When all three primaries are mixed, they're perceived as white.

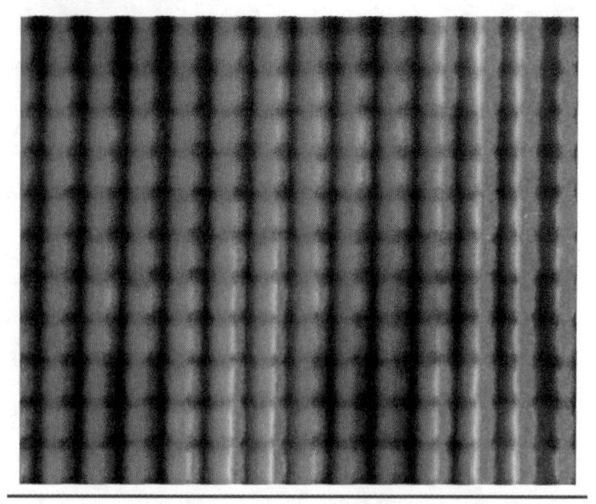

15-17 A magnified view of a color TV screen

The detail that can be displayed on a color screen depends on how small the dots are. Since three color dots are required to produce a picture element (pixel), the smaller the dots, the smaller the pixel. Better screen resolution is available with a larger number of pixels per centimeter or per inch.

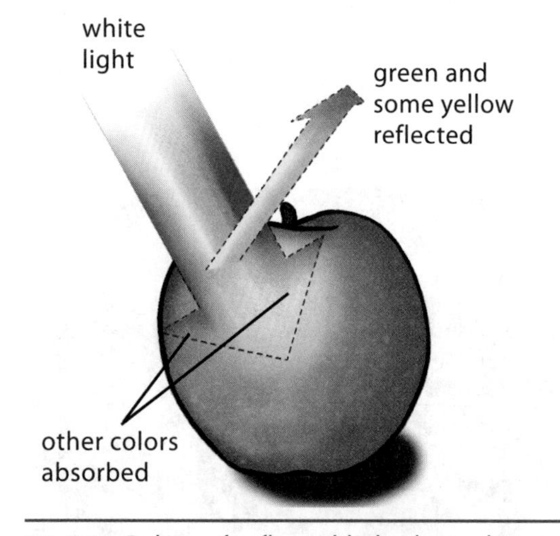

15-18 Colors of reflected light depend on the colors that are absorbed in the material.

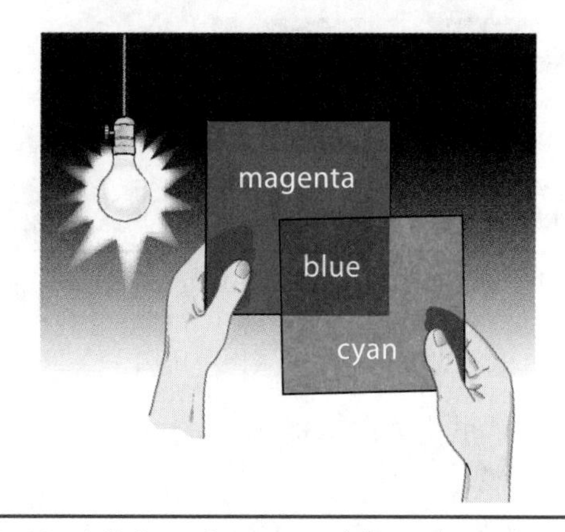

15-19 Mixing pigments of the subtractive primary hues

TVs, computer monitors, and cell phone screens take advantage of the eye and brain's ability to mix colors. A color screen (see Figure 15-17) consists of a repeating pattern of very fine red, green, and blue dots. To display a color other than these three, mixtures of different *intensities* of these dots are used. A close look at a cyan object on a television screen with a magnifying glass will reveal only green and blue dots. There's no cyan there at all!

Colors in *pigments* can be mixed, such as the solids found in paints and light filters. Pigments produce colors by absorbing certain wavelengths of color and reflecting or transmitting others. When a substance reflects all wavelengths of white light, it appears white. When it absorbs all wavelengths, it appears black. Primary pigments can produce a large range of colors when mixed in appropriate proportions. Different colors are produced by mixing primary colors that absorb or subtract certain wavelengths of light. For illuminated objects, we observe only the reflected wavelengths of light. Historically, artists considered red, yellow, and blue the primary color pigments. As color theory developed, especially in the publishing industry, printers adopted the **subtractive primary colors**—cyan, magenta, and yellow—as the primary pigments because their mixtures produced more brilliant colors. Notice that each of the subtractive primary colors is a mixture of two additive primary colors (see Figure 15-16 on the previous page). Subtractive primary colors are the *complementary colors* to additive primary colors.

So what color will we get if we mix magenta and cyan pigments? Magenta light can be produced by a mixture of red and blue light. Magenta pigment reflects these two colors and absorbs green light. Similarly, cyan pigment reflects green and blue light and absorbs red light. When we mix magenta and cyan pigments, the mixture absorbs green and red light. The only light reflected (or transmitted) by both magenta and cyan pigments is blue, so a mixture of magenta and cyan produces blue (see Figure 15-19).

When we mix all three subtractive primary colors, theoretically all the colors in light are absorbed and we should see black. This doesn't happen with real colors because no subtractive primary pigment absorbs a given color completely. Mixing all three subtractive primary colors often produces a muddy brown or dark gray.

15.8 Properties of Colors

Hue is just one property of color. Millions of colors can be produced from the base hues by varying a second property—their intensity, or brightness. Our eyes have different sensitivities to different wavelengths of light. For a given intensity, yellow-green light will appear brighter than any other color in the rainbow. Our eyes are far less sensitive to red and blue light, so those colors must have more intensity to appear as bright as yellow-green light. The intensity of a color is called its *saturation*. The saturation of red can vary from an extremely faint pink to brilliant fire-engine red.

A third property of color is how dark or light the color appears. Any hue can vary from nearly black to nearly white. This property is the *value* of the color. Value is especially important when mixing pigments. Colors are darkened by adding black pigment and lightened with white pigment.

FACETS OF SCIENCE

THE EYE

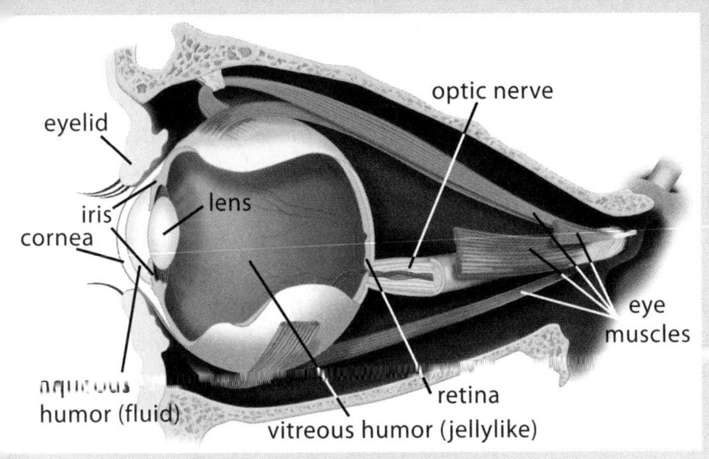

Digital cameras are the products of impressive technology. They can focus automatically and produce movies. Yet our eyes are even more impressive. They automatically aim, focus, and adjust and are far more sensitive than anything artificial. The medium on which they record develops instantly and self-renews. The normal eye's focusing ability is so good that it can clearly distinguish an object only 0.05 mm across—the diameter of a fine human hair. Our two eyes work together to give us stereoscopic vision, depth perception by the use of two different vantage points. On a typical day, our eyes make some 100,000 separate motions. Functioning for 16 or more hours with seldom a complaint, our eyes perform their own maintenance during the few hours we sleep.

As light enters the eye, it passes through the cornea, the aqueous humor, the lens, and the vitreous humor. Each of these parts refracts the light to focus a clear image on the retina. It is at the *retina* that light energy is changed to electrical impulses that are carried by means of the optic nerve to the brain. Only recently have we begun to understand in any depth how the brain interprets these impulses.

Within the retina are two kinds of cells that transform light into neural signals. They are the rods and the cones, both of which are named for their shapes. The *rod cells* respond to rapid changes of light and dark. They are the most important cells for tracking motion and for adapting to the dark. There are nearly 20 times as many rods as cones. The *cone cells* are mainly responsible for color vision. They contain one of three different kinds of chemical pigments that are sensitive to red, green, or blue light.

The orientation of the rods and cones is an interesting structural detail. They are *behind* the supporting tissues of the retina and point *away* from the incoming light. Scientists have wondered about this, thinking that the interfering support tissues should reduce the eye's sensitivity; however, recent studies show that this arrangement actually assists in preparing the optical image for interpretation by the brain.

Each eye has six external muscles, which serve to aim it at whatever is being viewed. These muscles are synchronized with the six muscles of the other eye by nerve impulses from the brain. Tiny muscles inside each eyeball change the thickness of the lens and the size of the pupil.

The eye is designed to protect and maintain itself. The cornea not only is the main optical part that focuses light into the eye, it is also a transparent protective shield. The eye must carefully regulate its own internal pressure. Tear glands, eyelid muscles, and nerves that control blinking and eyeball movement are all independent factors that support human vision. The eyelid, also controlled by two sets of special muscles, provides a "windshield wiper" that keeps the cornea moist and dust free.

Evolutionary scientists begin with Charles Darwin's theory of natural selection and point to the eye as an example of parallel evolution in many different species. They believe that the earliest light-sensitive tissues in organisms could have evolved as long as 600 million years ago. They point to the vast variety of eye types and shapes and the similarity of eyes in different animals as evidence that animals had to evolve eyes to survive. They presume that a creature that developed any kind of vision, however poor, would be selected for survival over competitors that had less vision.

Creationary scientists begin with presuppositions based on biblical Creation and interpret the scientific data quite differently. They see the variation of types of eyes and forms of vision as an economy of design and function in the original creation. The eye's design is so complex that the probability of its evolving on its own can't be computed. In sighted creatures, every form of eye is functional and appropriate for the animal's needs. There is no evidence that animals' eyes are evolving into more complex or capable forms.

Because of the vastly different underlying assumptions on the origin of all things, evolutionary and creationary scientists will never agree on how such a wonderful organ was formed. Neither side can present compelling evidence that will convince the other because their biases cannot allow such agreement. However, a Christian can be assured that the God of truth has given us the true account of the origin of the eye in His Word in Genesis 1. We can also say with the psalmist, "I will praise thee; for I am fearfully and wonderfully made" (Ps. 139:14).

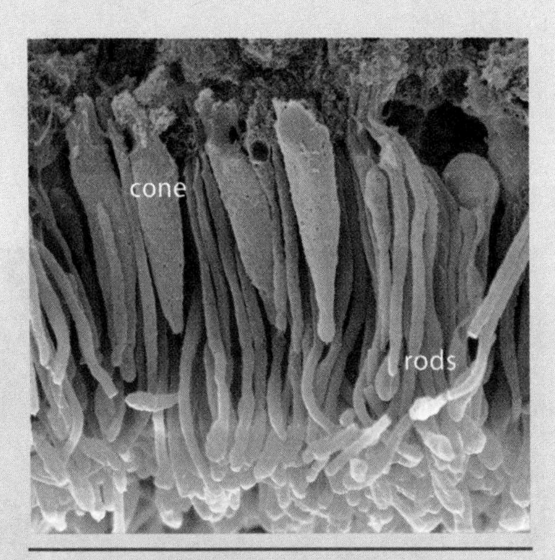

Rod and cone cells in the retina of the human eye

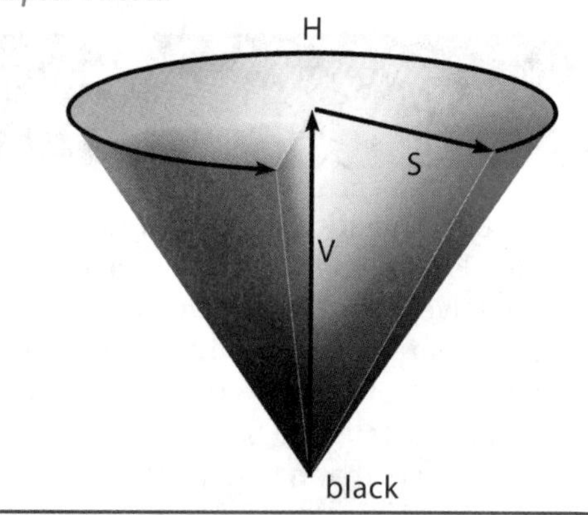

15-20 The HSV color space

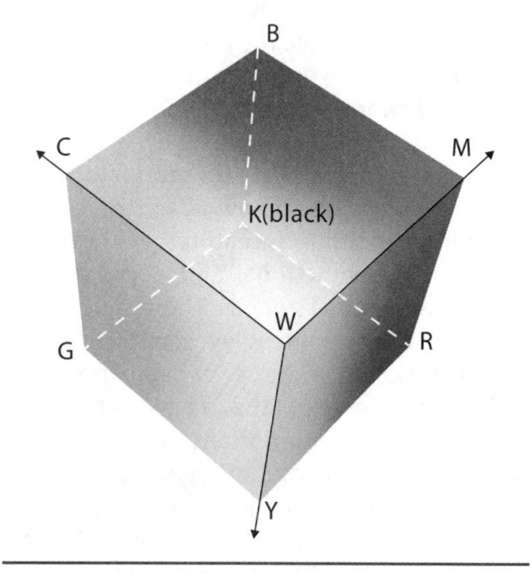

15-21 The CMYK color space

Color perception can be modeled using three properties—hue, saturation, and value (HSV). Some authorities produce a three-dimensional map of color using these quantities as variables. The resulting three-dimensional HSV *color space* identifies the colors that can be produced by the HSV color system. A **color system** is identified by the primary colors mixed together to produce the colors in the system.

Many color systems exist because of the different kinds of media in which color is used (e.g., newspapers, books and magazines, billboards, cloth, plastics, television, digital displays, photography, paints, and food coloring). No one color system can produce all the colors that can be perceived by the human visual system. Certain colors can be more easily seen than others, depending on the medium using the colors. For example, this textbook was printed with the *c*yan-*m*agenta-*y*ellow-blac*k* (CMYK) color system to produce images with true-to-life colors.

15B Section Review

1. Explain why the color of something depends both on the nature of the object itself and the way that we perceive it.
2. What are the additive primary colors for human vision? What group of colors is produced by mixing pairs of the additive primary colors?
3. Explain the difference between additive color mixing and subtractive color mixing.
4. Name and briefly describe the three properties of color.
5. What is a "color space" when referring to a color system?
6. Since a mixture of all subtractive primary colors cannot produce black, how do computer printers produce black in color images and in text?
7. (True or False) Artists usually use red, blue, and yellow as primary colors for painting.
8. What color would be produced by mixing yellow and cyan pigments viewed in white light?

15C Objectives

After completing this section, you will be able to

✓ define ray optics.

✓ state the law of reflection and define the optical terms included in it.

✓ compare and contrast virtual and real images.

✓ differentiate between plane and curved mirror reflections.

✓ compare the images produced by plane, concave, and convex mirrors.

15C Reflection and Mirrors

15.9 ***Light Rays***

Have you ever seen rays of light streaming through holes in thick clouds? Have you seen how a flashlight or the headlight of a car shines into a foggy night? These observations suggest that light normally travels in straight lines.

A *beam* of light is made of lots of photons moving in the same direction. Physicists use this definition to help describe the behavior of light when it interacts with matter. The path a light photon takes is represented by a ray. A **ray** is a line with an arrowhead showing direction. The study of the movement of light photons is called *ray optics.* Ray optics greatly simplifies the analysis of how images form and how optical instruments are designed.

15-22 A lighthouse beam illustrates how light rays move in straight lines.

15.10 Types of Reflection

We see objects when either they emit their own light or they reflect light from another source. Most things we see are nonluminous (do not produce their own light). We see them by reflected light. How they reflect light often helps us identify what they are made of.

There are two kinds of **reflection**. Both depend on the kind of reflective surface present. The most common type of reflection is called *diffuse reflection*. Rough or uneven surfaces, especially at the microscopic level, reflect light rays in all directions. Fabric, brick, concrete, paper, and most natural materials reflect light diffusely. It's not possible to see an image of the light source reflected in the surface of such materials.

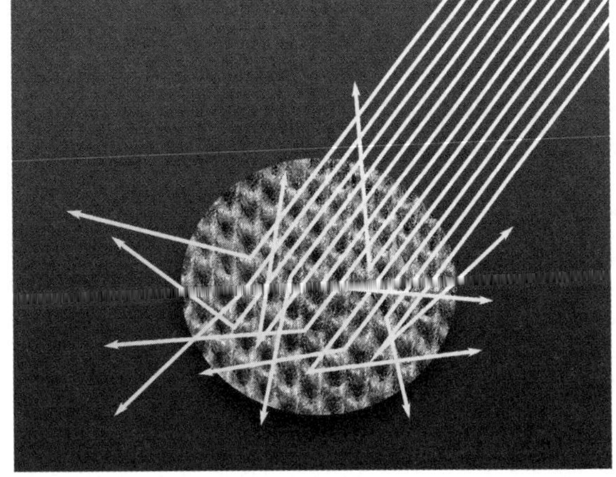

15-23 A diffuse reflection results from rough surfaces and from smooth objects with a matte finish.

Smooth, highly polished surfaces, especially metal and glass, produce *specular reflections*. When a surface is so smooth and uniform that it reflects all the light rays from a source in generally the same direction, they form an image of the light source. A perfectly specular reflection is mirror-like, reflecting all rays and wavelengths uniformly. Materials like glass and water reflect only a portion of the light rays from the surface and present a *glassy* specular reflection. The other rays are reflected from the interior of the material, pass through it, or are absorbed.

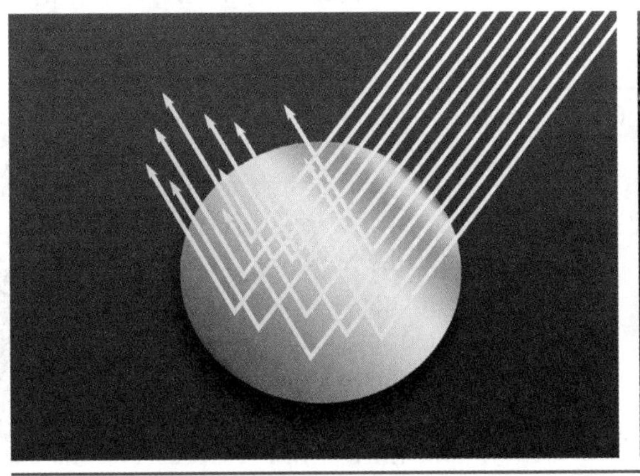

15-24 A specular reflection can produce an image of the light source.

incident (L. *incidere*—to fall into or upon)

The light ray approaching a reflective surface is called the **incident ray**. The outgoing ray is the **reflected ray**. When a light ray reflects off a surface, the angle of the incident ray, called the *angle of incidence,* equals the angle of the reflected ray, called the *angle of reflection*. This principle is called the **law of reflection**. The law of reflection holds true for every light ray in both diffuse and specular reflection. Physicists measure the angles of incidence and reflection from an imaginary line that is perpendicular to the surface at the point of incidence. This line is called the *normal,* which means *perpendicular to*. (You will learn of many physical phenomena that are measured from a normal in this and other courses later in your education.) These angles are measured from the normal because it can always be defined, even on a curved or uneven surface. It would be nearly impossible to measure these angles from the surface itself.

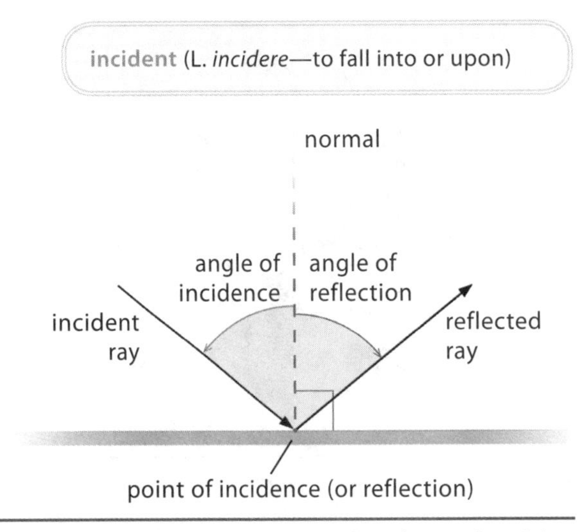

15-25 The law of reflection states that the angle of the incident ray equals the angle of the reflected ray measured to the normal at the point of incidence.

15.11 Plane Mirror Reflection

When you get ready for the day each morning, you look in your bathroom mirror. The standard bathroom mirror is a flat, **plane mirror**. You see what appears to be an upright image of yourself an equal distance away on the opposite side of the mirror. But that image exists only as a construction made by your eyes and brain from the reflected light rays. Physicists call this a **virtual image** because it is seen in a position where no real object exists—behind the plane of the mirror.

Every point on a luminous or illuminated object emits light rays in all directions. Light rays that fall on a plane mirror are reflected according to the law of reflection. Only the reflected rays from a mirror enter our eyes, not the incident rays. Since our visual system assumes that light rays travel in straight lines from their source, we interpret reflected rays as coming from behind the mirror in straight lines through the mirror to our eyes. As our eyes gather all the light rays reflected from the mirror, our minds construct an image just as if we were looking at the object itself.

15-26 Reflected rays from a plane mirror form a virtual image.

There is one catch—although the image appears to be reversed, it isn't. It is a "mirror image" of the real object. A mirror image shows the right side of an object where we would expect to see the left side if we were looking directly at it, and vice versa. When you raise your right hand, your image raises its "left" hand (if it were a real person). Actually, the hand on the right side of the image is raised. The illusion is created by subconscious expectations when looking at the human figure. In all other ways, though, plane mirror images preserve the size, shape, and distances of objects reflected in a mirror.

15-27 A mirror image is not reversed.

15.12 Curved Mirror Reflection

Real Images

Specular reflections off flat mirrors can produce, at best, only virtual images. This is because light rays diverge, or spread apart, from their sources and continue to diverge after they are reflected. The opposite of a virtual image is a real image. A **real image** is formed when the rays of light from a source are made to converge or come together by a focusing optical device, such as a special mirror or lens. Where the rays intersect, they form an image, which is a representation of a point on the source's surface. This image exists even if there is no observer to see it or instrument, like a camera, to record it. If the optical device arranges all the image points from the source in their correct relationship, an image of the entire source can be reconstructed. When the image points fall

15-28 Light rays reflecting from concave mirrors can produce real images. A real image is inverted in relation to the object that the mirror is reflecting.

on a flat surface, such as a movie screen, the image can be viewed. Thus, real images can be projected; virtual images cannot.

Concave Mirrors

Converging *reflected* light rays from an object can be produced only by a special curved mirror called a concave mirror. A **concave mirror** is slightly dished in on the reflective side. The deepest point of the dished surface is the center of the mirror. An imaginary line normal to the mirror's surface at this point is called the *principal optical axis.* The optical axis is helpful to understand how concave mirrors form images.

Let's examine what happens when light rays from a single point fall on the mirror. The light from a distant star aligned with the optical axis is a good example because the star is so far away that its rays are essentially parallel. When its light rays fall on the mirror's surface, they reflect according to the law of reflection and converge on a point in front of the mirror called the *principal focus* or **focal point** (see Figure 15-29). The focal point lies on the optical axis at a distance from the mirror's center called the **focal length** of the mirror. If you were to place a small piece of white paper at this location, you could see an image of the star where all the reflected light rays intersect (Figure 15-30). Because the image of the star can be projected and exists apart from our minds, it is a real image.

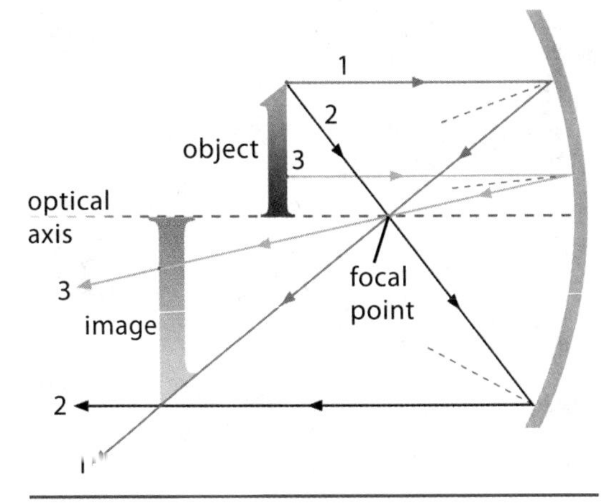

15-29 The real-image reflection of an object in a concave mirror is inverted both top to bottom and left to right.

The more sharply curved the concave mirror is, the closer the focal point is to the mirror and the shorter the focal length.

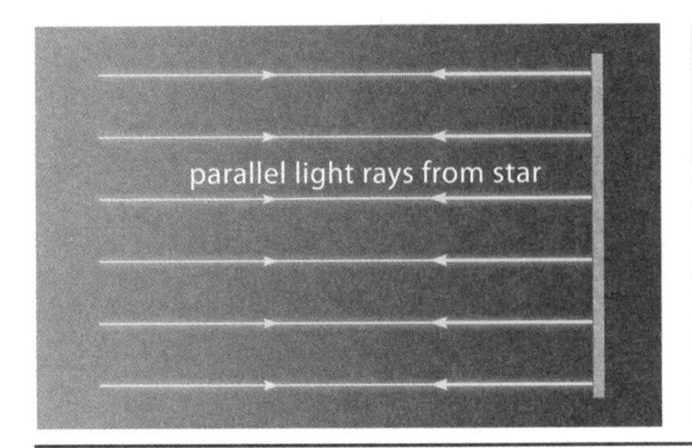

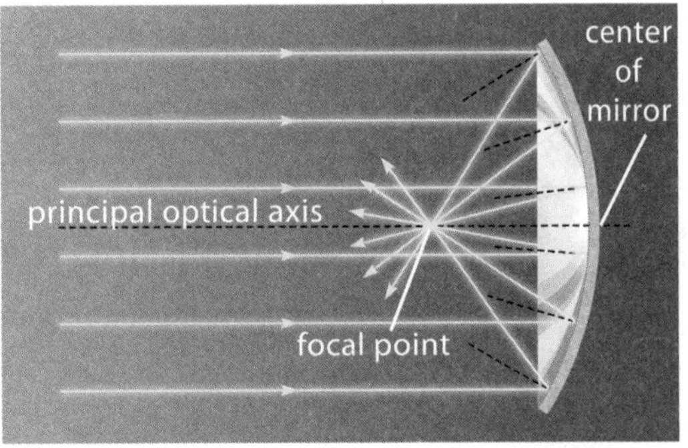

15-30 Parallel light rays from a star reflecting off a plane mirror (left) and a concave mirror (right). The short dashed lines are the normals for each ray.

For objects much closer than a star, their light rays incident on a concave mirror are not parallel but diverge slightly from every point on the object. Multiple rays from each point strike the mirror at different locations. The curved surface of the mirror reflects the incident rays at angles that may cause them to converge, depending on how far the object is from the mirror. If these rays do intersect, a real image of the original point on the object is produced.

A real image always appears on the same side of the mirror as the object. A virtual image appears to be on the opposite side of the mirror from the object. The real image produced by a concave mirror is reversed from the actual source object. The image appears upside down and reversed left to right, unlike a virtual "mirror image," which is upright with right and left sides reflected as right and left (Figure 15-27). A concave mirror *can* produce an upright virtual image when one observes a reflected object from a position closer than the mirror's focal point. Magnifying makeup mirrors use this principle.

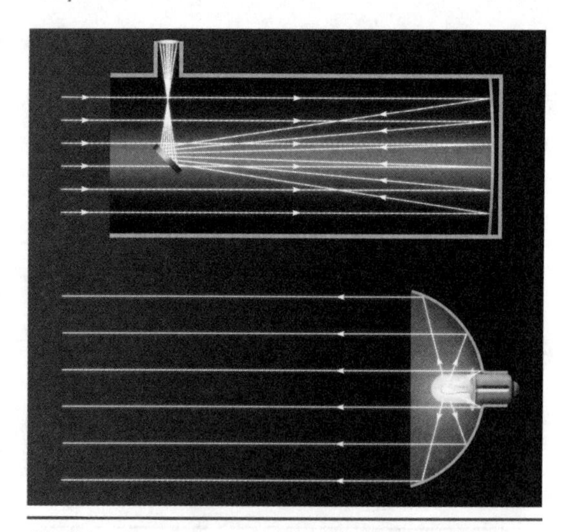

15-31 An astronomical telescope collects light from distant objects (top). A flashlight reverses the process by projecting rays from a light source at the focal point (bottom).

15-32 Convex mirrors are useful for observing large areas and intersections.

Concave mirrors used in *reflecting telescopes* collect parallel rays of light from a star and direct them to the focal point. Near this point there is usually a smaller mirror that reflects the image to the side or rear of the telescope. There the image may be photographed or viewed by an observer. In very large telescopes the image may be directly viewed by an astronomer who sits in a special enclosure right at the focal point.

Concave mirrors may also reflect light in the opposite direction. If a light bulb is placed at the focal point of a concave mirror, the incident rays of light hit the mirror and reflect outward in a tight beam of light. This arrangement is used to produce concentrated beams of light in flashlights, searchlights, and automobile headlights.

Convex Mirrors

The surface of a mirror may also bulge outward. This kind of mirror is a **convex mirror**. It cannot produce a real image because all light rays reflecting off a convex surface diverge even more than from a plane mirror. Every image seen in a convex mirror is a virtual image. Because of the exaggerated divergence of the reflected light rays, convex images appear unnaturally smaller than the objects they reflect and the reflected image includes more background area. For this reason, convex mirrors are used to provide wide-angle views. They are useful for security monitoring, in the passenger-side sideview mirrors of cars, and at blind corners of hallways and street intersections.

15C Section Review

1. When light photons are considered to be particles, what kind of path do they usually travel? Why is this useful?
2. What *cannot* be produced by a diffuse reflection but is possible with a specular reflection?
3. Describe the law of reflection. To which form(s) of reflection does the law apply?
4. How does a plane mirror form an image?
5. What kind(s) of image(s) can a plane mirror produce? Describe the image properties.
6. What defines the location of the focal point in a concave mirror? What do we call the distance of this point from the mirror?
7. What are the properties of a real image?
8. Name some applications for convex mirrors.
9. (True or False) Light reflects off our skin as a diffuse reflection.

15D Refraction and Lenses

15.13 Light Refraction

When light waves pass from one transparent material into another, they refract if their speed changes. The speed of light in a material is related to the material's *optical density*. One measure of optical density is the material's index of refraction. The **index of refraction (*n*)** is the ratio of the speed of light in a vacuum to the speed of light in the material.

$$n_{\text{mat.}} = \frac{\text{speed of light in vacuum}}{\text{speed of light in material}}$$

$$\boxed{n_{\text{mat.}} = \frac{c}{v_{\text{mat.}}}}$$

The value of n is always greater than 1 for all forms of matter because the speed of light is highest in a vacuum. The value of n thus increases with the optical density of the material. For example, the index of refraction for diamond, one of the most optically dense natural materials, is 2.42.

Light rays that cross the boundary between two materials at an angle are bent as they pass into the second material. This bending is the most obvious result of refraction. Why do light rays bend when they're not perpendicular to the boundary? Figure 15-33 illustrates a ray of light and its wave properties. It moves through air ($n = 1.0003$) and approaches a piece of glass ($n = 1.5$) at an angle. The ray bends as each photon's speed slows and its wavelength shortens in the optically denser material. When the wave passes through the glass back into air, the ray bends in the opposite direction, stretching the wavelength as it speeds up.

Notice also in Figure 15-33 that normal lines indicate the angles at which the ray enters and leaves the glass. During refraction a ray will bend *toward* the normal if the second material is optically denser and *away* from the normal if the second material is less optically dense. Refraction can be mathematically described by comparing a ray's angle of incidence to its *angle of refraction*. Both angles

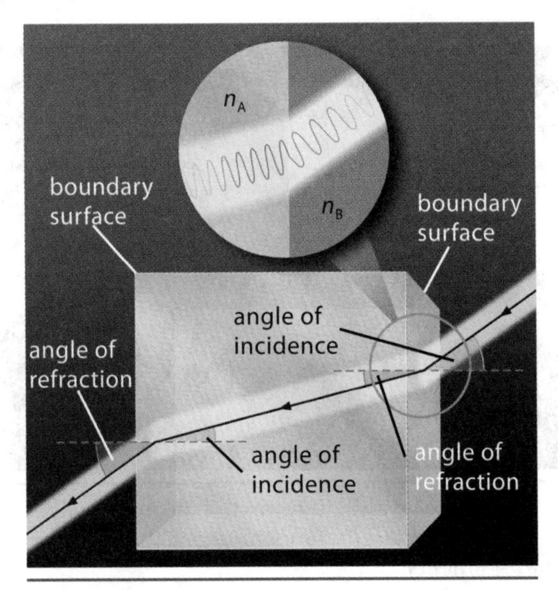

15-33 Light-ray refraction. Note that n_A is greater than n_B.

15D Objectives

After completing this section, you will be able to

- ✓ calculate a material's index of refraction.
- ✓ explain the law of refraction using correct terms.
- ✓ describe the conditions necessary for total internal reflection.
- ✓ explain what causes a rainbow.
- ✓ explain how lenses magnify and refract an image in different ways.
- ✓ describe how lenses correct deficient eyesight.
- ✓ state several current applications of refraction optics.

A light wave's frequency is constant. Recall that the product of frequency and wavelength is wave speed:

$$c = \lambda \times f$$

Therefore, the wavelength in the denser material must decrease in proportion to the decrease in wave speed.

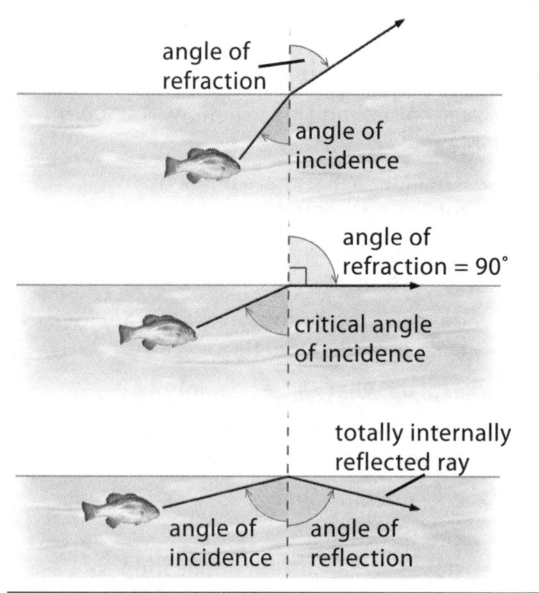

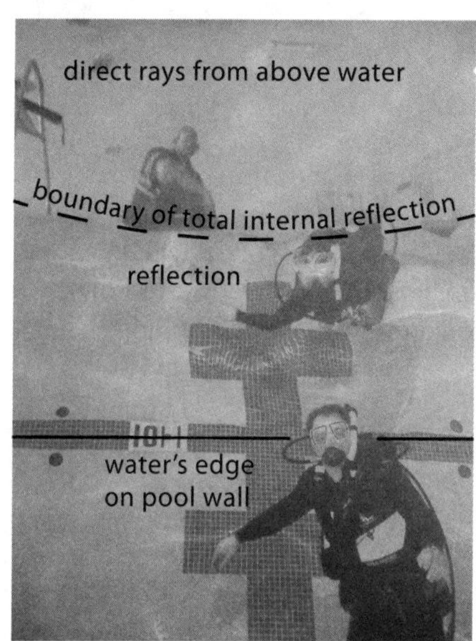

15-34 Illustration of total internal reflection in a body of water (left). The photograph on the right shows the circular boundary of total internal reflection as viewed from within a pool.

are measured from a line normal to the boundary between the two materials at the point where the ray crosses the boundary.

The greater the angle of incidence at which a light ray enters a less dense material, the closer its angle of refraction will be to 90°. This angle of refraction makes the refracted ray travel parallel to the boundary. If the angle of incidence is increased further, the light ray bounces off the boundary back into the original material (see Figure 15-34). This phenomenon is called **total internal reflection**. The angle of incidence where this phenomenon begins is called the *critical angle of incidence*. You may have observed total internal reflection when you dived to the bottom of a swimming pool and looked up toward the sky. You saw a disk of sky above you, but surrounding that, all you saw were the totally reflected rays from the inside of the pool.

The refraction of light as it enters water causes many illusions. Have you ever tried to net a fish in an aquarium? When you lowered the net into the water, did you find that you had missed the fish? Refraction of light fools you by making the fish seem to be in a different position than it really is.

Another interesting effect of refraction is the bending of the sun's rays, which is especially noticeable near sunrise or sunset. Layers of warmer or cooler air can distort the sun's shape, flattening it or breaking up its disk into flattened segments. The density differences between two layers of air cause the light rays from the sky above the horizon to bend gradually upward. This refraction of light often fools people in the desert into thinking that they see pools of water ahead. Similarly, as you drive along a road in a car on a hot

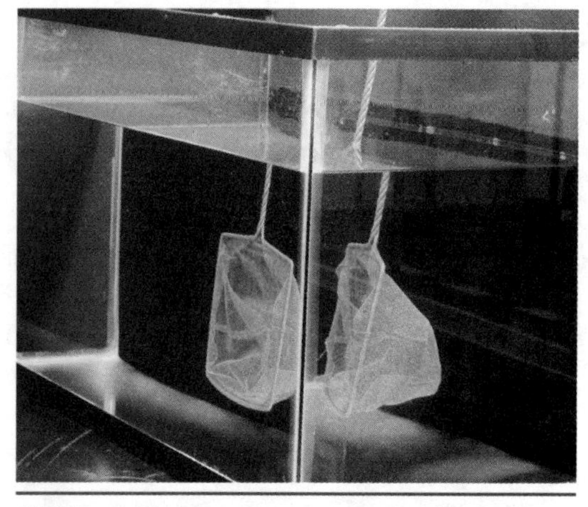

15-35 Refraction in an aquarium makes catching fish with a net challenging!

15-36 In extreme cases, sunlight refracted through layers of the atmosphere at different temperatures can break the sun up into segments.

day, the highway may appear wet in the distance. These optical effects are *mirages*. All mirages are due to the refraction of light.

15.14 Using Reflection to Solve Problems

Alexander Graham Bell, inventor of the telephone, also invented the lesser-known *photophone* in 1880. With this gadget, he used a beam of light to transmit sound waves. This invention paved the way for observation and communication using light. In the early 1900s scientists began using long, extremely thin rods of glass, or *optical fibers*, to transmit light for television. In the 1930s lenses in rigid, rodlike tubes were used to peer into the stomach and intestines, but these instruments were not easy to use or comfortable for the patient. In the 1950s a vast improvement was made with the invention of endoscopes, viewing instruments containing flexible bundles of optical fibers. By the 1970s endoscopes were commonly used for surgery. Today, **fiber optics** is an important and growing field of technology with diverse applications.

endoscope: endo- (Gk. *endon*—inside) + -scope (Gk. *skopein*—to view or look)

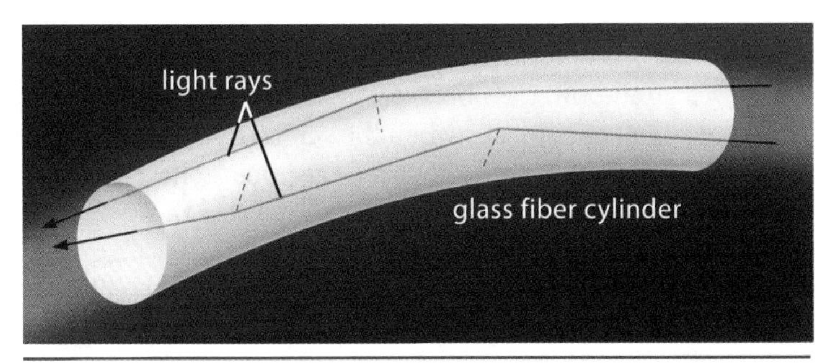

15-37 Total internal reflection is responsible for transmitting light rays through optical fibers.

Optical fibers work on the principle of total internal reflection. When light strikes the side of a glass fiber, its angle of incidence is so large that it reflects back into the fiber rather than escaping. To transmit light, these glass fibers must be optically pure. They are coated with layers of tough, flexible substances to protect them from damage.

Fiber optics has many uses in *endoscopy*—the medical procedure in which the body's interior is viewed directly by use of an endoscope inserted through a natural opening or small incision. Endoscopies can be done in the intestines, stomach, lungs, larynx, joints, and even a pregnant woman's uterus. Fiber optics has revolutionized surgery. Optical fibers deliver lasers to operation sites or provide detailed, on-site images in real time using remote-control surgical instruments. Large incisions are often no longer needed for diagnosis or treatment. Inspections that used to involve major surgery, weeks of recovery, and high risk of infection can often be done on an outpatient basis with little risk and full recovery after just a few days.

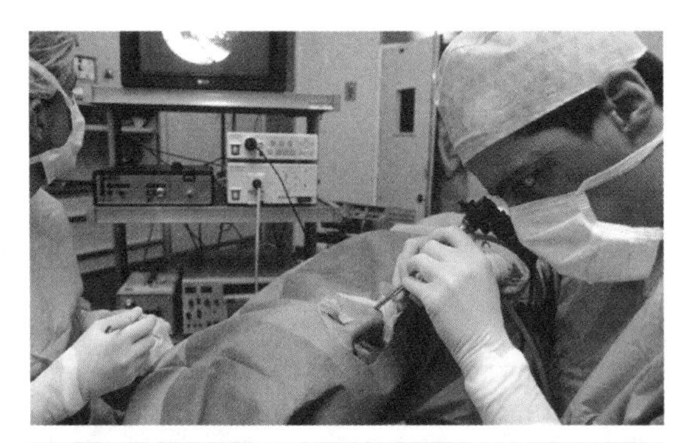

15-38 An endoscope in use during a surgical procedure

15.15 Light Dispersion

When a light ray passes through the boundary between two materials, its angle of refraction depends on its frequency as well as its angle of incidence. For a given combination of materials and the same angle of incidence, a high-frequency light ray will bend more than a low-frequency ray. For example, a green beam of light entering glass at an angle will deflect more than a red beam of light under the same conditions.

Isaac Newton was the first to study this phenomenon. He discovered that when a beam of sunlight passed through a triangular wedge of glass, it *dispersed*, or split, into a band of colors similar to those in a rainbow. The colors toward the red end of the spectrum bent the least, and the colors toward the blue end bent the most. Today, this phenomenon is called light **dispersion**.

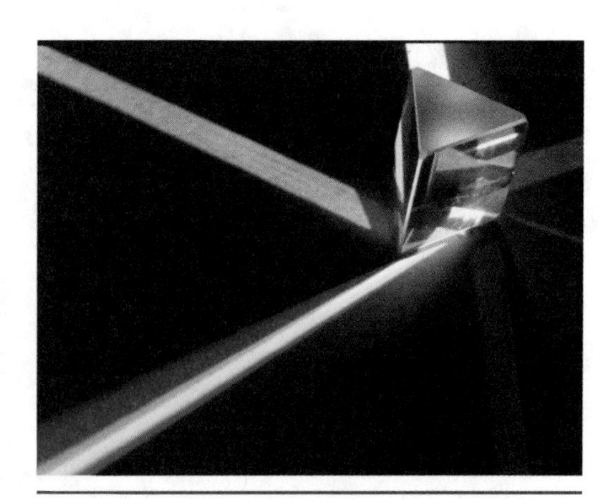

15-39 A prism separates the colors in white light according to their frequencies.

Whenever a beam of light enters a piece of glass, no matter what its shape, some dispersion takes place. When the glass has nonparallel surfaces, as in a prism, the dispersion increases. You can observe light dispersion when light shines through beveled glass in a window or the pieces of a crystal chandelier.

Light dispersion produces the colors seen in weather-related rainbows. To see a rainbow, the observer must be positioned between the sun and the falling rain, ideally when the sun is relatively low in the sky. Sunlight entering the raindrops first refracts, then reflects off the back of the raindrop, and finally refracts again as it leaves the drop. The colors in the sunlight are dispersed as in a prism. The dispersion pattern forms a circle of colors. However, under normal circumstances, the ground cuts off the lower part of the circle, leaving only the familiar upper arch of the rainbow. From some locations, such as high on a mountain, you may be able to see more of the circle. People in airplanes can often see the entire circle of a rainbow.

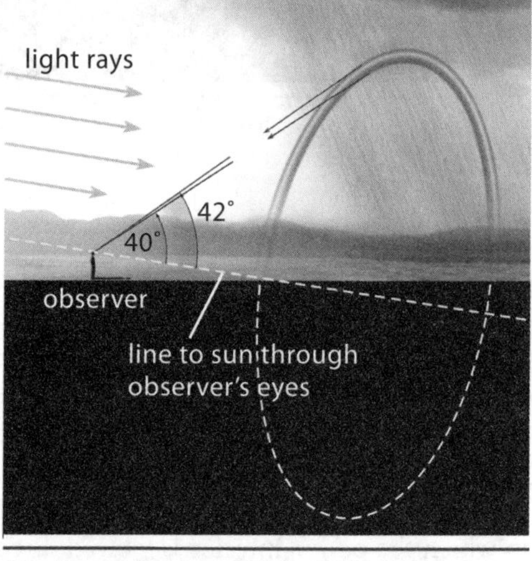

15-40 The relationship between an observer and a rainbow

What Is a Rainbow?

The rainbow is first mentioned in the Bible in Genesis 9:13–15. God declared that it was to be a sign of His promise that He would never again destroy the world by water.

Is a rainbow just light dispersed through raindrops, or is it something more than this?

15.16 Lenses

The most common and useful application of optical refraction is a lens. A **lens** is a disk of optically clear material that refracts light to produce a real or virtual image. An object may be magnified so that it appears larger than it actually is, or it may be reduced, depending on the purpose of the lens.

Lenses can be classified as converging or diverging. A *converging lens* collects the incoming rays of light and focuses them at a point, just like a concave mirror. A *diverging lens* spreads the rays of light apart similar to a convex mirror.

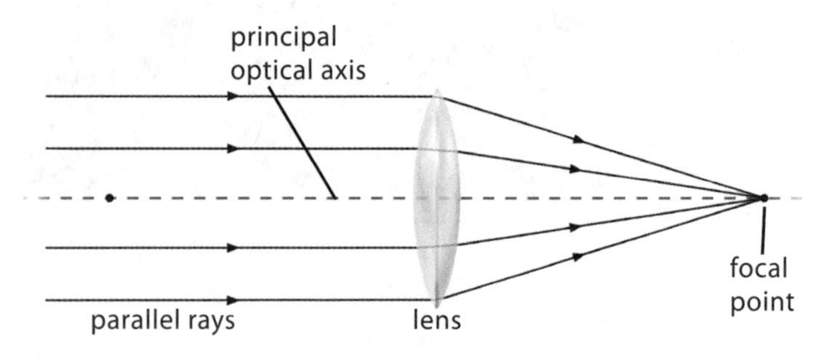

15-41 Optical terms for a lens

Many of the optical terms you learned for mirrors apply to lenses. The imaginary line through the center of a lens is the principal optical axis. Parallel light rays pass through the lens and converge at the focal point, located on the principal optical axis on the *opposite side* of the lens from the object. A lens has two principal focal points because light can pass through the lens in either direction. The distance of a focal point from the center of the lens is the focal length, just as for concave mirrors. The focal length shortens as the curvature of the lens increases. A converging lens has at least one convex surface. A diverging lens has at least one concave surface. Figure 15-42 shows the possible shapes for simple lenses.

Recall from the description of concave mirrors that the focal point is where the image forms.

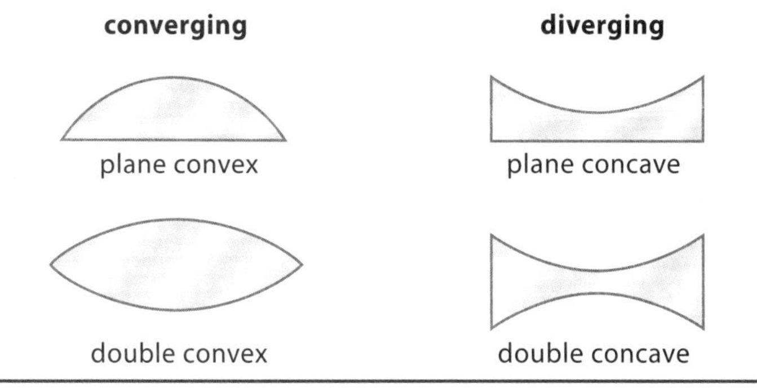

15-42 Simple lens designs

The kind of image produced by a lens depends on the distances of the object and observer from the lens. The lens is always placed between the object and the observer. If the object is closer to a converging lens than its focal point, the observer will see a virtual upright image that is larger than the object. This is why a magnifying glass works only if placed close to the object. If the object is much more distant than the focal point, the image seen by the observer will usually be smaller than the object as well as upside down and reversed. Such images are real images and can be projected onto a screen.

If the eye of an observer is closer to a lens than the distance at which the image is focused, the image he sees is always a virtual image.

Diverging lenses always produce upright virtual images that appear smaller than the object. These lenses are used in optical instruments, such as cameras and projectors, to adjust the shape of a light beam before it passes through a converging lens.

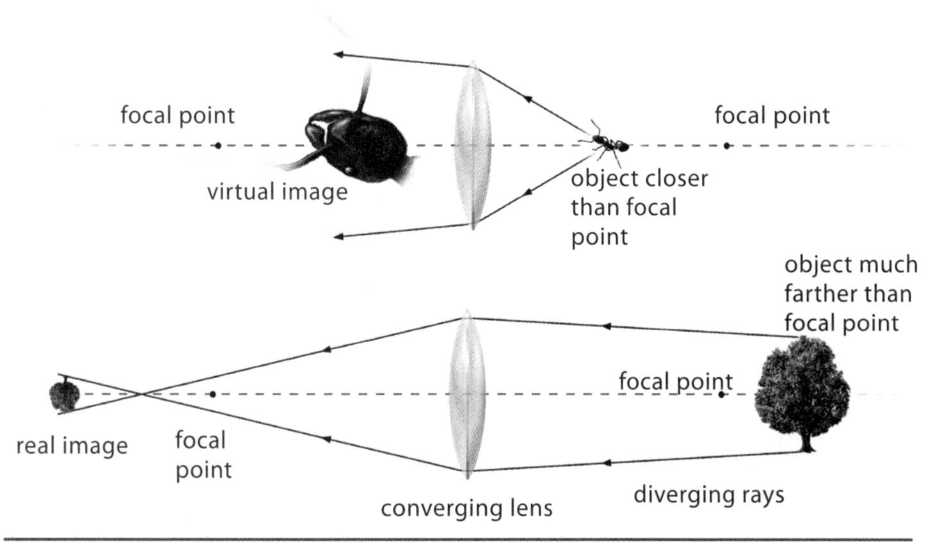

15-43 Converging lens used as a magnifier (top) and as a projection lens (bottom)

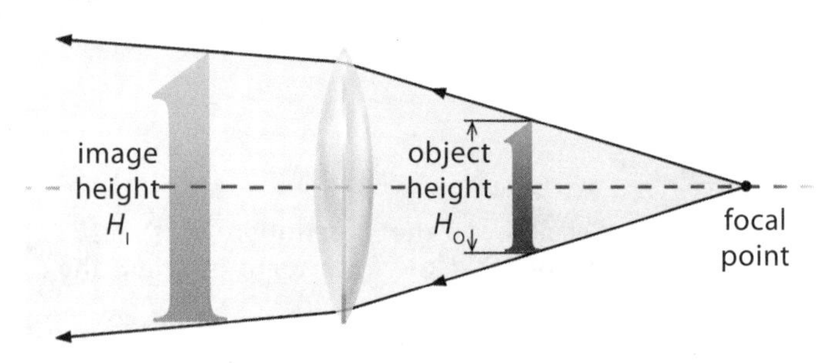

15-44 Magnification is found by dividing the height of the focused image by the height of the object.

Magnification

Most lenses produce some magnification of an object. The shape of the lens and its focal length determine how much magnification occurs. Magnification can be calculated by the ratio of the image size to the object size:

$$\text{magnification } (M) = \frac{\text{image size } (H_I)}{\text{object size } (H_O)}$$

When the image size is larger than the object, the magnification is greater than 1. When the image size is smaller than the object, the magnification is less than 1. For instruments that have multiple lenses, such as a telescope, the total magnification is determined by multiplying the magnifications of the separate lenses together. For instance, if the main lens of a telescope has a magnification of 20 and the eyepiece magnifies 10 times, the total magnification of the telescope is

$$20 \times 10 = 200 \text{ times.}$$

The magnification of a lens is sometimes called its *magnification power* and given the symbol ×.

Eyesight Correction

It's quite possible that you are benefiting from lenses as you read this textbook. Nearly two-thirds of Americans wear glasses or contact lenses because their natural lenses can't adjust enough to focus an image on a retina that is too close or too far away. This may be due to a problem with the lens, but it is more likely due to the eyeball being too long or too short compared to a "normal" eye. Their eyes are like a camera or a projector that needs focusing. When an eyeball is longer than the distance to the focal point of the natural lens, it focuses the image in front of the retina, causing a person to see a blurry image. This condition is called *nearsightedness.* If the eyeball is shorter than the distance to the focal point of the lens, the image focuses behind the retina. This condition is called *farsightedness.* Corrective lenses compensate by focusing light rays exactly on the retina.

The lenses in eyeglasses are called *meniscus lenses.* Both sides of the lenses are curved in the same direction, similar to the curved surface of water in a graduated cylinder, called a *meniscus.* The word comes from the Greek, meaning "crescent." Farsightedness is corrected by a converging lens, a meniscus lens with a curve that is flatter on the inner surface than on the outer surface. The edges are thinner compared to the middle of the lens. Nearsightedness is corrected by a diverging lens, a meniscus lens in which the inner concave curvature is greater than the outer convex surface of the lens, making it thick along the edges and thin in the middle. Contact lenses are shaped similarly. If you wear glasses, what shape of lens do you have?

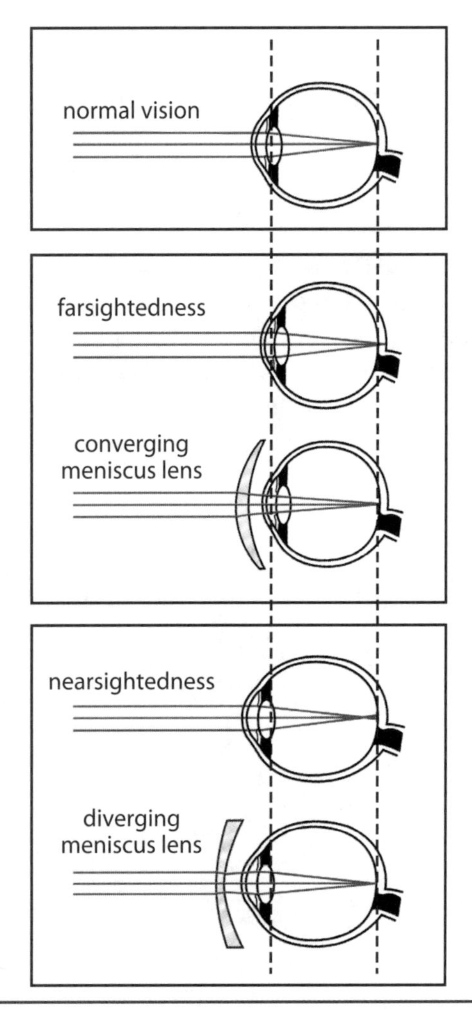

15-45 With normal vision, the image is focused on the retina. In farsightedness the image is focused behind the retina, so only distant objects focus properly. Converging lenses correct farsightedness. In nearsightedness the image is focused in front of the retina, so only close objects are in focus. Diverging lenses correct nearsightedness.

Other Lens Applications

There are many lens applications other than eye correction that are important to both scientific investigation and to everyday life. When the first microscopes were invented in the 1590s, people became aware of the previously unseen world of the extremely small. A microscope is basically a tube with lenses at both ends. The lower lens, the *objective*, magnifies the object. The objective is designed to focus on very close objects, so it has a very short focal length. The upper lens, the *eyepiece*, enlarges the image of the objective. The microscope's total magnification is the product of the powers of the objective and the eyepiece. Better-quality microscopes have multiple objectives of different powers mounted on a rotating nosepiece, or *turret*. Often eyepieces with different powers can be swapped to provide a larger range of optical magnifications. Microscopes have been invaluable in the battle against disease-causing microorganisms and in aiding our understanding of life at the cellular level.

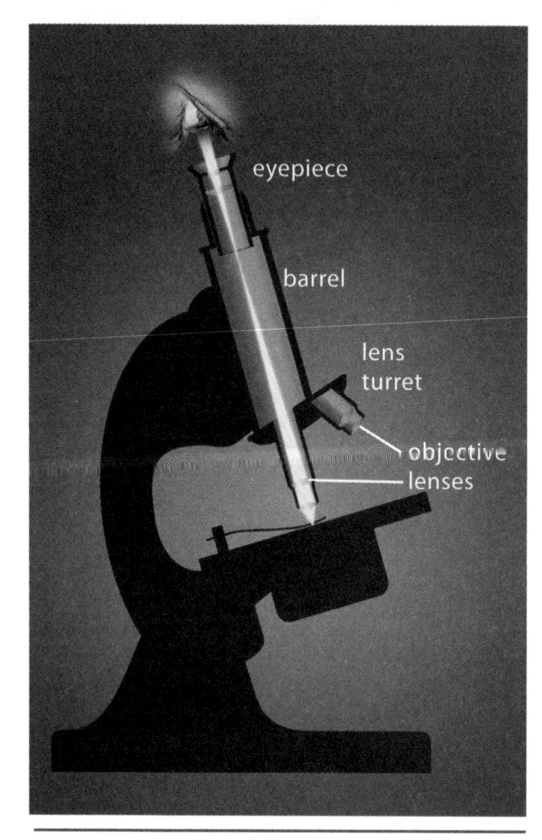

15-46 The optical path of a microscope

People began to understand the immensity of the universe after Galileo directed his telescope toward the heavens in 1609. Telescopic discoveries confirmed the heliocentric model of the universe and drastically changed man's philosophical view of himself and his planet. A telescope makes distant things appear near by increasing their apparent size. As with a microscope, the *refracting telescope* consists of a tube with an objective and an eyepiece containing one or more lenses. The objective in a telescope is designed to focus images that are very distant, so it has a relatively long focal length. The diameter of the objective determines how much light can be collected from the object. In astronomical telescopes the image is inverted, but in terrestrial telescopes the inverted image is turned right-side up by a separate lens or prism. *Binoculars* are a pair of compact terrestrial telescopes that allow both eyes to be used so that viewing is more comfortable.

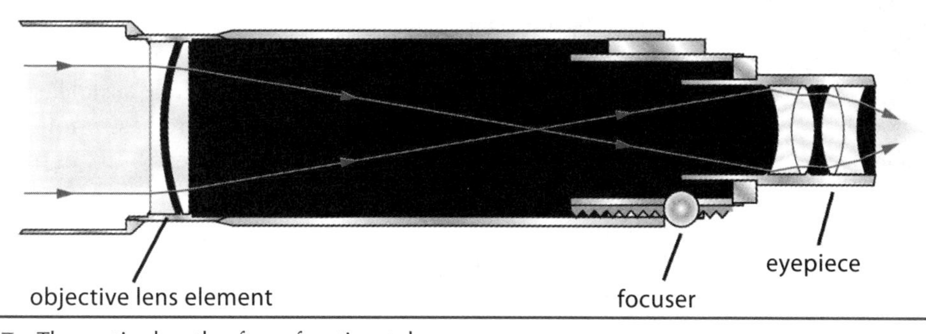

15-47 The optical path of a refracting telescope

Cameras are optical devices that record images. A camera basically is a light-tight box with a lens in an opening. The lens projects an image onto a light-sensitive surface such as photographic film or, more commonly, on a *charge-coupled device (CCD)*. (See Chapter 14.) The image is captured when a *shutter* behind the lens rapidly opens and shuts, exposing the light-sensitive surface. Some CCD cameras use an electronic shutter that "reads" the CCD at a given instant to take the image and transfers it into digital memory. Modern camera lens systems have been designed with computers to reduce their weight and to make them more compact. Many cameras now use lenses made of high-quality plastics that reduce the camera's weight even further. Miniature digital cameras can now be found in cell phones, tablets, and personal computers.

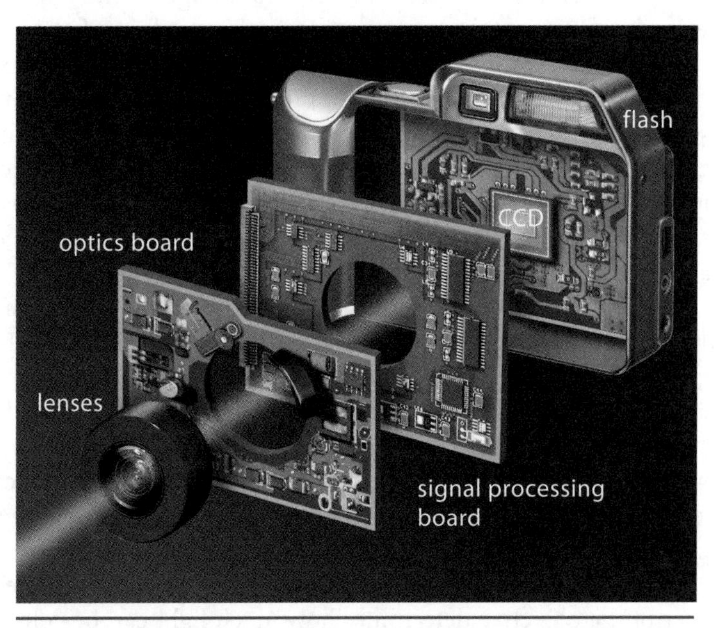

15-48 A digital camera showing the lens system and light-sensitive charge-coupled device (CCD) surface that records the digital image

15.17 The Future of Optics

Many new optics-based technologies are being developed. New sensors that extend the range of vision into the other bands of the EM spectrum use optical principles. Examples include aerial laser mapping from aircraft and orbital infrared and x-ray astronomical observatories. Researchers are investigating the feasibility of inserting artificial retinas in the human eye to provide a simple form of vision to the blind. Some of Jesus's miracles during His earthly ministry involved the restoration of sight. As a Christian and a scientist, you could imitate the Savior in using scientific knowledge to do the same.

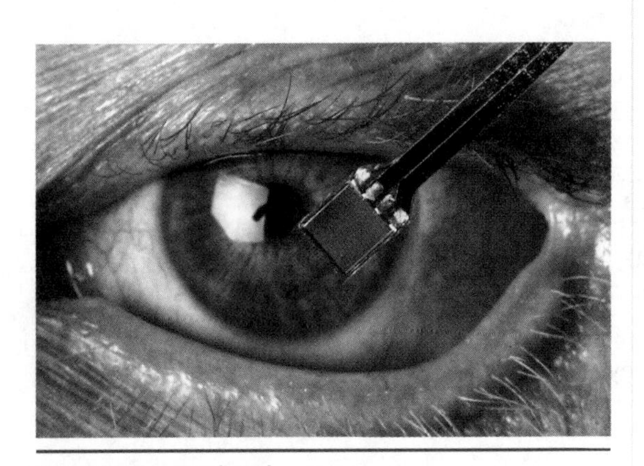

15-49 An artificial retina

15D Section Review

1. How is a material's index of refraction related to its optical density?
2. A light ray in optically dense glass approaches the boundary between the glass and the air at an angle. Will the ray's angle of refraction be larger or smaller than its angle of incidence? What happens if the ray's angle of incidence exceeds the critical angle?
3. How many times is a beam of light refracted when it passes through a triangular prism? What happens to the colors in the beam each time it is refracted?
4. DS ✪ Why would using an endoscope sometimes be preferable to using noninvasive methods, such as ultrasound or x-ray?
5. Explain why more of a rainbow can be seen when the sun is low as opposed to when it is high in the sky.
6. The real images produced by a concave mirror are always on the same side of the mirror as the source object. Where is the real image formed by a lens in relation to its object?
7. When a lens is used as a magnifying glass, is the image seen by the lens a real or virtual image? How do you know?
8. What do all optical instruments and devices have in common?
9. (True or False) During an early morning rain shower, Jim could see a rainbow to the north as the sun's rays in the east broke through the clouds, illuminating the landscape.
10. ✪ The speed of light in a certain kind of newly developed clear plastic is 1.98×10^8 m/s. What is the plastic's index of refraction?

Scientifically Speaking

inverse square law	329
incandescence	329
fluorescence	330
phosphorescence	330
coherent light	330
cold light	330
additive primary color	333
subtractive primary color	334
color system	336
ray	336
reflection	337
incident ray	337
reflected ray	337
law of reflection	337
plane mirror	338
virtual image	338
real image	338
concave mirror	339
focal point	339
focal length	339
convex mirror	340
index of refraction (*n*)	341
total internal reflection	342
fiber optics	343
dispersion	344
lens	344

Chapter Review

1. Is it correct to say that the sense of vision is in the eyes? Explain your answer.
2. Why can we *not* accurately define the seven colors in the visual spectrum as red, orange, yellow, green, blue, indigo, and violet?
3. Why is an astronomical object that an astronomer claims is more than 6000 ly away an apparent problem for the Bible-believing Christian? What unprovable assumptions make this seem like a problem?
4. Which kind of light source must be heated before it glows?
5. How does a TV screen produce all the colors of the rainbow?
6. Evolutionists claim that the human eye evolved to be most sensitive to yellow-green light because the sun shines most intensely in the yellow-green portion of its spectrum. How could this relationship be explained from a biblical creation viewpoint?
7. How can you model the motion of a photon? What do you call a large number of photons moving together in the same direction?
8. Why do physicists measure the angles of incidence and reflection from the normal to a surface rather than from the surface itself?
9. Describe the properties of an image seen in a plane mirror.
10. When a concave mirror produces a reflected image of a distant object, what kind of image (real or virtual) exists inside the focal point of the mirror? beyond the focal point?
11. How would a light ray approaching an optically denser material refract if its angle of incidence was 0°? What happens to the light *wave* in this case?
12. DS ✪ What optical principle is the basis for fiber optics? Do you think an endoscope with more and finer fibers would be more flexible than one with fewer, thicker fibers? Explain.
13. As light passes through a prism, how does the amount of refraction relate to the light's frequency?
14. How must the thickness of a lens vary from edge to edge for it to be a converging lens? a diverging lens?
15. Compare the purposes of the objective in a microscope and a telescope. What is adjusted to meet these purposes?

True or False

16. More of the human brain is devoted to the sense of sight than to any other sense.
17. A rainbow consists of seven distinct colors—red, orange, yellow, green, blue, indigo, and violet.
18. A light-minute is a unit of distance useful within the solar system.
19. A 50 W fluorescent lamp will be brighter than a 50 W incandescent bulb.
20. Every possible color has a specific wavelength in the visible spectrum.
21. Every color we can see can be assigned a hue, saturation, and value in the HSV color space.
22. A reflection is specular if you can see a clear image of the light source in the surface.
23. Plane mirrors have little use since the images they produce are always virtual images and not real ones.
24. A sharp, distinct boundary or surface between two materials is required for refraction to occur.
25. DS ✪ Endoscopes can eliminate the need for any surgery when it becomes necessary to look directly inside the human body.
26. Meteorological rainbows are actually the visible portions of a complete circular pattern of color.
27. When an image produced by a lens is *smaller* than the associated object, the magnification of the lens is negative.
28. ✪ Inventing artificial retinas would be working against God's will for those He has allowed to be blind.

29. ✪ An insect's image projected by a lens on a screen is 13.0 cm long and the insect is 2.0 cm long. What is the magnification of the lens?
30. ✪ Refer to Appendix B to list the three measurable properties stated in the definition of a candela. What is the dimension of each of these units?
31. ✪ What is the speed of light in a transparent plastic having an index of refraction of 1.65?

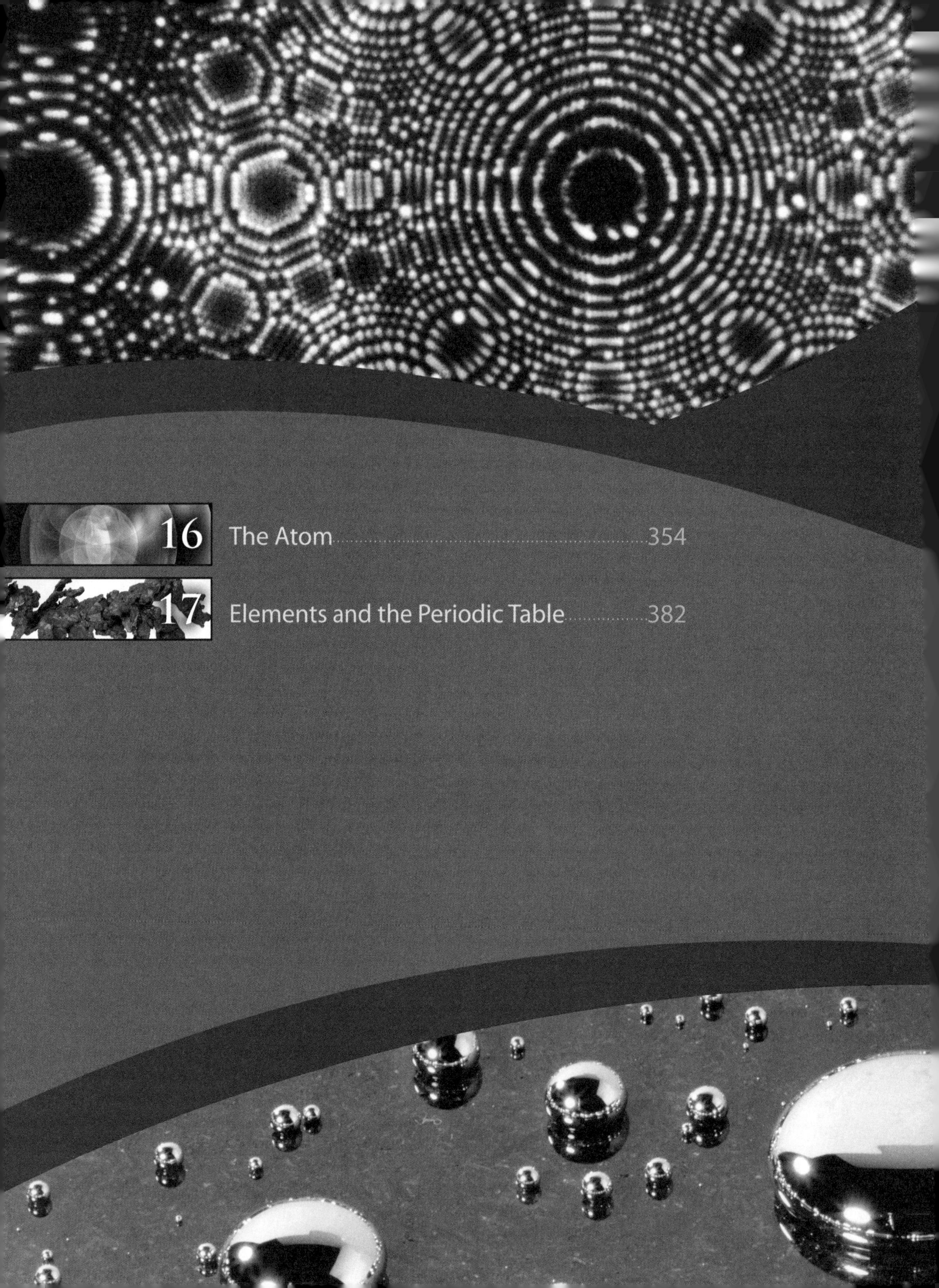

The Structure of Matter

What is the universe made of?

Where do we look to answer this question? The Bible has a lot to say about *how* the universe came into existence. Hebrews 11:3 says that "the worlds were framed by the word of God." From this passage and others in Genesis 1, we understand that God spoke all things into existence, a process which no longer occurs. He formed Adam from the dust of the ground and breathed life into him. However, the Bible says little about *what* the universe is made of, except in very general terms.

Remember that the Bible, not science, is the ultimate source of truth. But science can give us workable models to understand the substance of the physical creation. In Unit 5 you will examine the early models of the atom—the smallest distinct particle. Studying the current model, which is no more permanent than the earlier ones, will allow you to understand what is currently believed to be the arrangement of the atom's major components and how they work together. You must understand the structure of the atom to understand the patterns of chemical and physical properties possessed by the elements in the periodic table.

Unit 5

THE ATOM

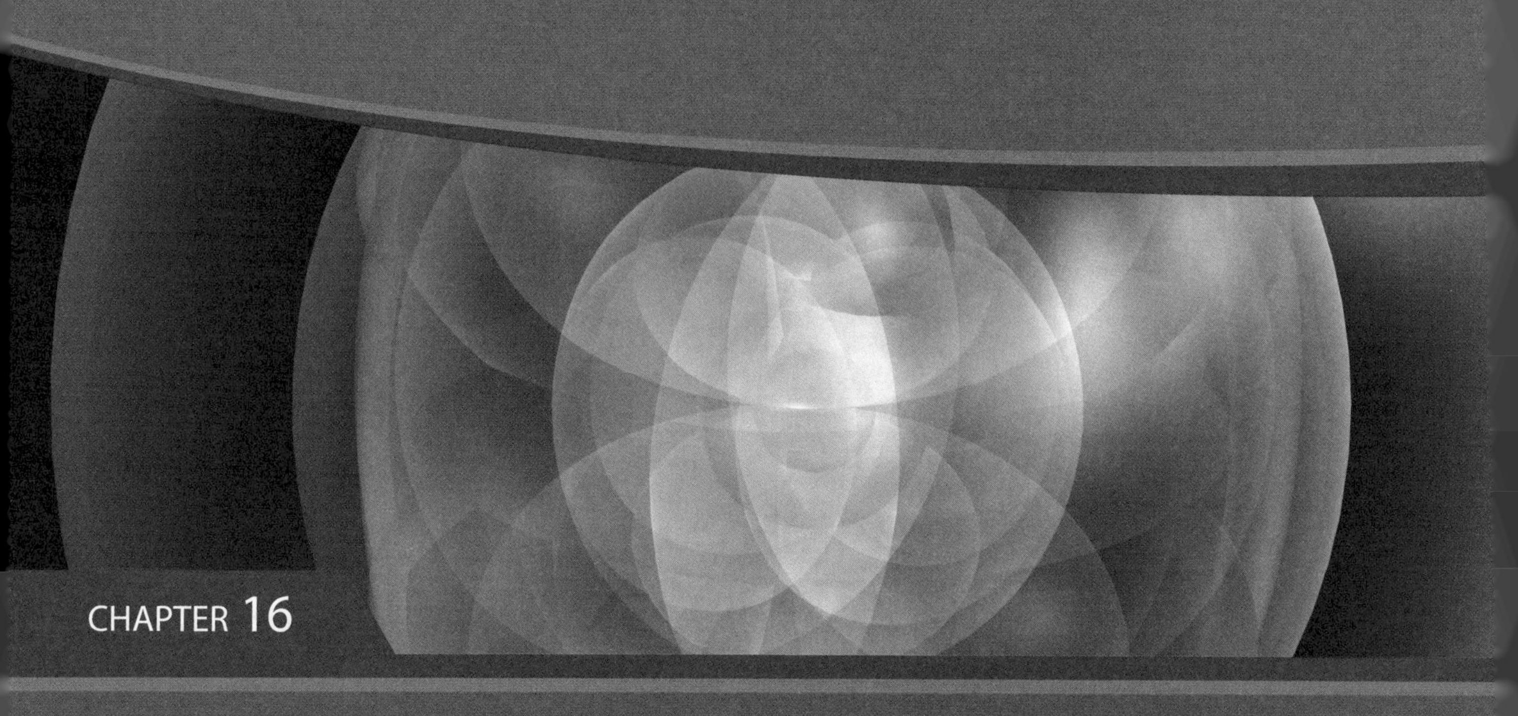

CHAPTER 16

DOMINION SCIENCE PROBLEM

House Fire Detection

Between 2007 and 2009, house fires in the United States killed 11,035 people and injured 39,410. About half of all fatalities from house fires in the United States happen while people are sleeping. Domestic fires can spread quickly. Within minutes, fires fill houses with poisonous gases and block possible exit routes. You may think that smoke from a fire would wake the people in the house, but instead the poisonous gases send them into a deeper sleep. How can we use technology to wake people so that they can evacuate a burning house before it is too late?

16-1 House fires kill hundreds of people every year in the United States.

16A The Atomic Model

16.1 *Introduction*

Scientists use models to understand the natural world and create theories about how it works (Chapter 1). For this reason, scientific models have to be *workable*—they must provide useful answers to the questions scientists ask.

To understand how atoms act in the presence of other atoms—the major concern of chemistry—we have to understand their structure. Section 16A teaches the significant models of the atom that scientists have developed throughout history. It's important to realize that each model was the best available in its time. When a different model followed, its acceptance caused a revolution in the way scientists thought about matter. Even with our current understanding, a new discovery could overturn the accepted model of the atom in favor of a more workable one.

16A Objectives

After completing this section, you will be able to

- ✓ compare and contrast the modern atomic models to the atom as understood by ancient Greek philosophers.
- ✓ summarize the current understanding of the atom.
- ✓ describe the major discoveries leading to the modern atomic model.
- ✓ show how each atomic model was more workable than its predecessor.
- ✓ associate the key scientist with each advance in the atomic model.

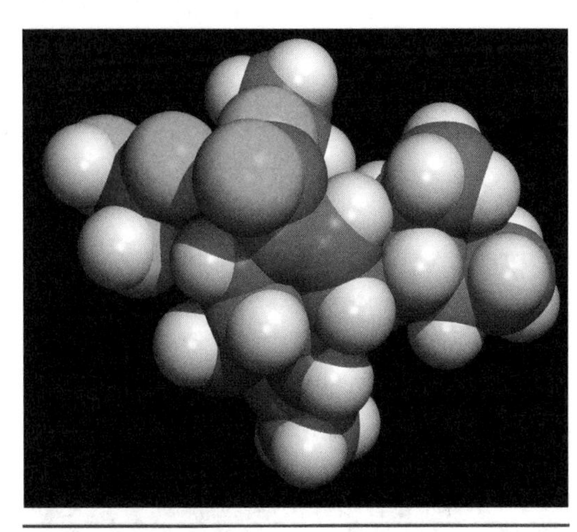

16-2 Different models of atoms are useful for their particular purposes. Here, atoms are represented by solid spheres.

16.2 *Ancient Atomic Models*

People must have thought about the structure of matter very early in history. The earliest record of anyone's considering that matter was made of tiny particles is in ancient Hindu writings more than 2600 years old. During the fifth century BC, the Greek atomistic concept was originated by Leucippus and further developed by his student Democritus. Democritus coined the word *atomos*, or *atom*, meaning "indivisible." He stated that everything is made of atoms, even thoughts and the soul. He taught that atoms were the only things that exist, that they had always existed, and that they would always exist.

Consistent with the Greek philosophers' bias that reasoning alone was the path to knowledge, Democritus did not experiment to test his ideas. His concepts were completely naturalistic and denied any possibility of supernatural influence or design. His ideas were so radical that even his contemporaries rejected his theory. The teachings of another Greek philosopher, Aristotle, who said that matter was continuous, influenced scientists for nearly 2000 years.

16-3 One way to illustrate atoms of the atomistic Greek philosophers. Democritus believed that atoms of solids had hooks holding them together, that atoms of liquids were slippery, and that atoms of salts and spices were sharp.

Evaluating Atomic Models in the Light of God's Word

Today we realize that many of Democritus's ideas don't provide workable explanations for observations. We know from later observation, for example, that some things that exist, like light, are not made of material particles like atoms. More importantly, we know from Scripture that atoms have not always existed. God created them at the beginning of time (John 1:3).

16.3 The Atom Rediscovered

The fourteenth and fifteenth centuries in Europe saw a renewed interest in studying the world. Scholars studying early Greek and Roman documents preserved by Arab cultures rediscovered the Greek thoughts on the atom. Initially the atom was considered just an interesting idea that ran counter to the accepted view of matter handed down from Aristotle. As time passed, more scholars saw evidence in the emerging science of chemistry that the atomic model of matter might have merit. By the late 1700s scientists such as Boyle and Lavoisier had laid the foundation for a new atomic theory based on experimentation and observations that Aristotle's model of matter could not explain.

16.4 John Dalton: The Indivisible Atom

In the 1790s a French chemist named Joseph Louis Proust (1754–1826) discovered an important principle of chemistry called the **law of definite proportions**. It established that the masses of chemical substances combine in definite, repeatable ratios when forming compounds. For example, when 18 g of water is decomposed, it always yields 2 g of hydrogen and 16 g of oxygen. If 9 g of water is decomposed, the results are 1 g of hydrogen and 8 g of oxygen. The ratio of the masses of hydrogen to oxygen is always 1:8. This ratio holds true for pure water from any source.

16-4 John Dalton (1766–1844) was an English public schoolteacher and private tutor of mathematics and chemistry. He was also colorblind and wrote several papers on his disability.

An English schoolteacher named John Dalton realized that this predictable behavior of compounds exists because particles of the elements, or atoms, combine in fixed ratios. This theory was strengthened from his own experimentation with the chemical reactions of gases. Dalton suggested the following:

- An atom cannot be created or destroyed in chemical reactions.
- An atom cannot be subdivided into smaller particles.
- Elements are made of only atoms.
- The atoms of an element are all alike.
- The atoms of one element are different from the atoms of all other elements, especially their masses.

Dalton believed that the elements have different masses because of the presence of a "heat envelope" that varied in thickness with each element. His **core-envelope model** was based on the caloric theory, which was challenged a short time later by Count Rumford. However, most of John Dalton's atomic model has proved to be very durable over the past two centuries.

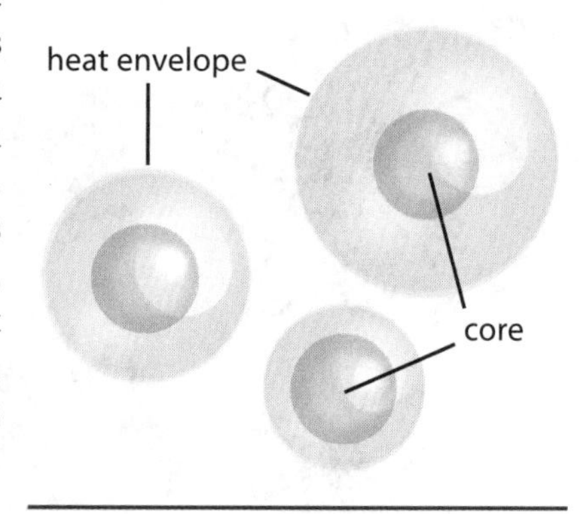

16-5 John Dalton imagined atoms as solid spheres of different sizes and weights.

16.5 J. J. Thomson: The Divisible Atom

A significant discovery in 1897 by Englishman J. J. Thomson revealed that atoms were not quite indivisible after all. Physicists in the second half of the nineteenth century discovered that if a heated metal was placed inside a vacuum tube and connected to an electrical source, it would give off streams of something that was negatively charged. Did these streams consist of waves or tiny particles? Thomson proved that the metals were giving off negatively charged particles, which he called *corpuscles*. Another physicist suggested calling them *electrons*, which is what we call them today. Thomson dared to suggest that atoms were not indivisible but consisted of charged particles.

16-6 Sir Joseph John Thomson (1856–1940) was an English physicist who discovered the electron and isotopes of elements and invented the mass spectrometer. He won the Nobel Prize in Physics in 1906 for his discovery of the electron.

Thomson reasoned that if atoms contained negatively charged electrons, they must also contain positive charges to hold the negative charges together. He pictured the atom as a positively charged mass with negatively charged electrons embedded in it like raisins in a plum pudding (a traditional British dessert). Lord Kelvin called Thomson's theory the **plum pudding model** for that reason. The positive charge in the atom's material ("pudding") balanced the negative charge of the electrons ("raisins"), which were held in place by the attractive forces between these opposite charges. While Dalton's theory was still accurate regarding the interaction between atoms, Thomson's theory better showed how atoms themselves are constructed. For his day, Thomson's model was quite remarkable.

Electrons don't seem to vary in mass or charge but are the same in every element. Early experiments indicated that the masses of all the electrons in an atom form only a tiny fraction of the total mass of the atom, so scientists still questioned what composed the rest of the matter in an atom.

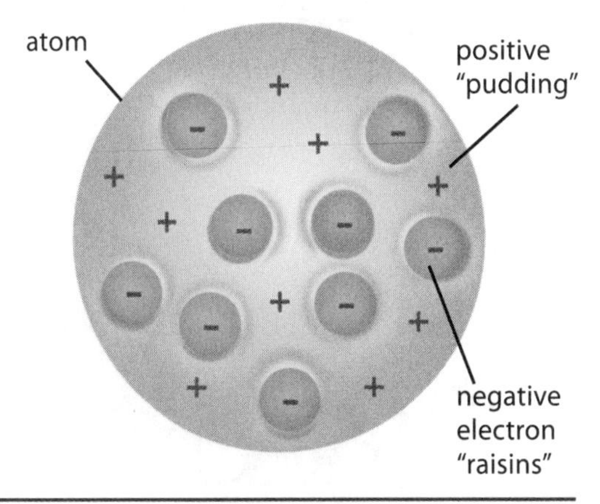

16-7 The atomic model developed by Thomson suggested that negatively charged electrons were embedded in a positive substance.

16-8 Ernest Rutherford (1871–1937) was an English physicist born in New Zealand. He is best known for his discovery of the atomic nucleus and is known as the Father of Nuclear Physics. Several prominent nuclear physicists started their careers as his students. He won the Nobel Prize in Chemistry in 1908.

16.6 Ernest Rutherford: The Nuclear Atom

In 1908 Ernest Rutherford had one of his students and an assistant design an experiment to investigate the structure of atoms. The experimental apparatus consisted of a source of high-energy *alpha particles*—essentially helium atoms stripped of electrons—directed in a narrow beam at a thin strip of gold foil.

Figure 16-9 shows a cutaway of the apparatus Rutherford's research team used in this experiment. The heavy, positively charged alpha particles were beamed from a shielded source at a stationary target of gold foil inside a vacuum chamber. A microscope with a phosphorescent screen swiveled around the target to detect the presence of alpha particles, which caused tiny glowing dots where they hit the screen. The research team found that most of the particles went straight through the foil as if nothing were in the way. A few, however, were deflected at large angles from the streaming path. The most astonishing thing was that some bounced almost straight back toward the source. Rutherford was astounded and compared this to shooting a cannonball at a piece of paper and having it bounce back to you!

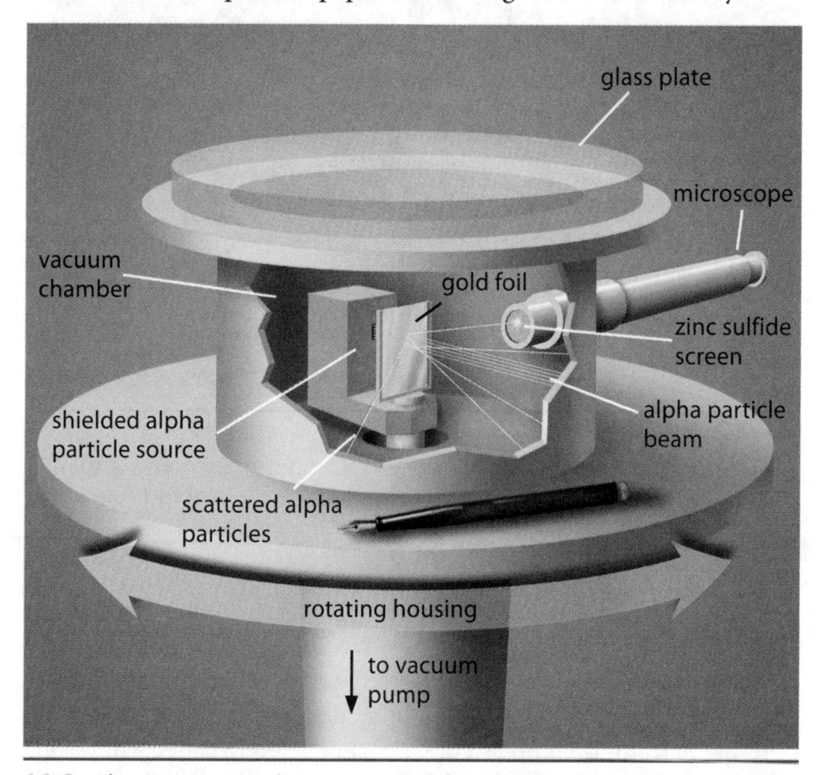

16-9 The experimental apparatus used to discover the atomic nucleus

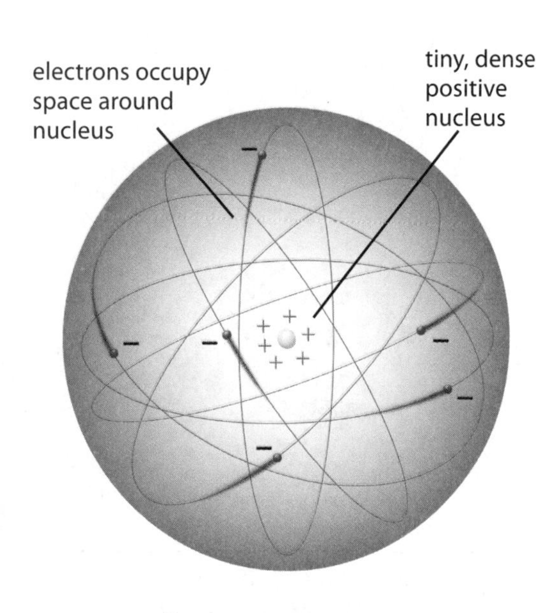

16-10 Rutherford's model of the atom had a tiny, positively charged, massive nucleus surrounded by electrons.

He concluded, first, that there must be lots of empty space in the atom. If this were not true, the relatively large alpha particles wouldn't have been able to pass through the foil without colliding with something and being deflected. Second, there must also be a very dense, tiny, positively charged region in each atom that deflected some of the alpha particle "cannonballs."

By 1919 Rutherford had devised an atomic model to account for these observations. He calculated that the positive charge and practically all the atom's mass were concentrated in a tiny region of space called the *nucleus*. The electrons were thought to whirl around the nucleus at high speeds that kept them from being drawn into the positive nucleus. Further research by Rutherford's staff revealed positively charged particles in the nucleus. Rutherford called them *protons* at the suggestion of Eugene Goldstein, who unknowingly had first detected them in 1886.

At this point Rutherford's **nuclear model** was still incomplete since it had been determined that the mass of the nucleus was greater than the total mass of its protons. The nucleus had to contain something besides protons. It was not until 1932 that James Chadwick discovered *neutrons*—particles that have approximately the same mass as protons but no electrical charge. They reside in the nucleus along with the protons and account for the additional mass found there.

Rutherford's modified atomic model provided a workable structure of the atom and its nucleus, but it did not explain the motion or location of electrons. How far away from the nucleus were they? How were they arranged? These questions began to be answered by a contemporary of Rutherford.

16-11 James Chadwick (1891–1974) was an English physicist who discovered the neutron and conducted basic research leading to the controlled fissioning of uranium. He won the Nobel Prize in Physics in 1935.

16-12 If the nucleus were the size of a period, this is how large an average atom would be in comparison. The nucleus is 1/100,000 the diameter of the atom!

16.7 Niels Bohr: Electron Energy Levels

Any respectable fireworks display must have brilliant colors. How are the colors emitted? In fireworks, color-producing compounds are added to an explosive mixture. The energy from the explosion heats the atoms of certain elements in these compounds. As these atoms cool, they release energy in the form of visible light. Each element emits light of a characteristic wavelength or color. Sodium, for example, burns yellow; strontium, red; and barium, green. Changing the compounds containing these elements alters the color of the fireworks.

Niels **Bohr** knew that certain elements emit specific colors of light when heated to incandescence. When he passed these *emissions* through a prism, he was able to identify elements by the specific wavelengths of light in their *line spectra*. Bohr suspected that these emissions were caused by the electrons in atoms.

He suggested that when atoms were heated, their electrons gained enough energy to "jump" farther from the nucleus, entering an excited state. They then returned quickly to their original, lower energy positions, releasing the excess energy as light waves. The frequency of the light, and thus its energy, depended on the electron's excited

16-13 Danish physicist Niels Henrik David Bohr (1885–1962) made many contributions to the understanding of atomic structure. He won the Nobel Prize in Physics in 1922 and also helped develop the first atomic bombs during World War II.

Bohr (BOR)

16-14 Line spectra of several incandescent elements showing the unique combinations of light emitted

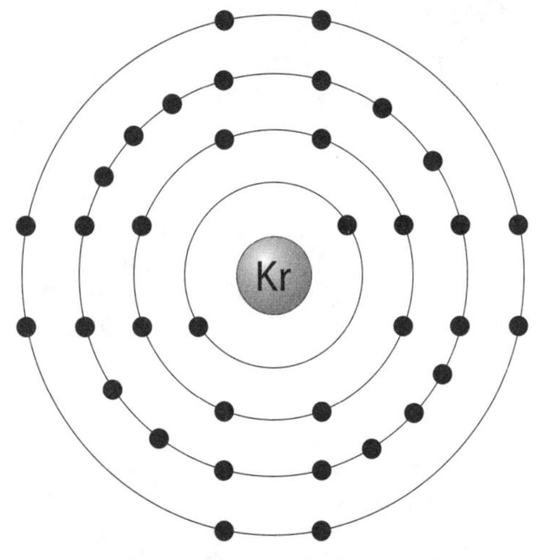

16-15 The atomic model based on Bohr's theory places electrons in orbits at specific distances from the nucleus.

When a group of measurements is **quantized**, they are related to each other as multiples of a basic quantity.

distance from the nucleus and the distance it dropped as it returned to a lower energy state.

Different elements give off different colors when heated. This means that the changes of energy involved when electrons jump between levels vary from element to element. These differences suggest that each element has a unique electron structure. That the same colors of light are always obtained for the same element indicates that electrons in an atom must be associated with specific, "normal" energies, not random unpredictable ones.

Since light frequency and energy are related (see Chapter 14), Bohr suspected that the emitted photon's energies were tied to the changes of the electron's potential energy in relation to the atom's nucleus. He used the single hydrogen electron's potential energy to calculate its distance from the nucleus. He then developed a model for the hydrogen atom that accounted for his observations of hydrogen's line spectra. In his model, electrons occupied specific orbits, or *energy levels*. They could jump to higher energy levels if they absorbed energy, releasing exactly the same amount of energy when they returned to their original state. Electrons could not inhabit areas of the atom between these energy levels. Thus, the differences of energy between electron orbits were fixed—they were **quantized**. Bohr's model looked like a miniature solar system with the nucleus in the center, like a sun, and electrons orbiting it at fixed distances, like planets orbiting a sun. It is often called the **planetary model**.

16.8 The Quantum Atom

A flurry of discoveries during the early twentieth century revealed that the electron acts like both a wave and a particle. Physicists accounted for this behavior by refining Bohr's planetary model. Instead of well-defined orbits or pathways, the electron energy levels within the atom are regions where the electrons could *probably* be found. Because the electron inside an atom acts as a wave, its location cannot be precisely fixed at any one moment.

This **quantum model** of the atom has been continually refined over the past century. It now accounts for many properties of atoms, including how they behave in the presence of other atoms and how and why they form bonds. Within the electrons' principal energy levels, pairs of electrons occupy subregions of space, called *orbitals*. Physicists have devised three-dimensional, geometrically shaped "clouds" to represent orbitals. Their shapes vary from simple spheres to complex combinations of lobes and rings. In the quantum model, physicists describe the distribution of electrons in the atom as a cloud because it has no clearly defined boundaries.

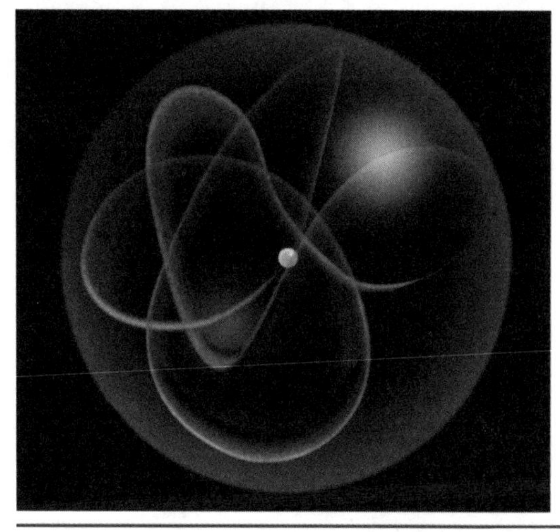

16-16 An electron behaves like an oscillating wave of matter inside the quantum atom in a way that is almost impossible to conceive.

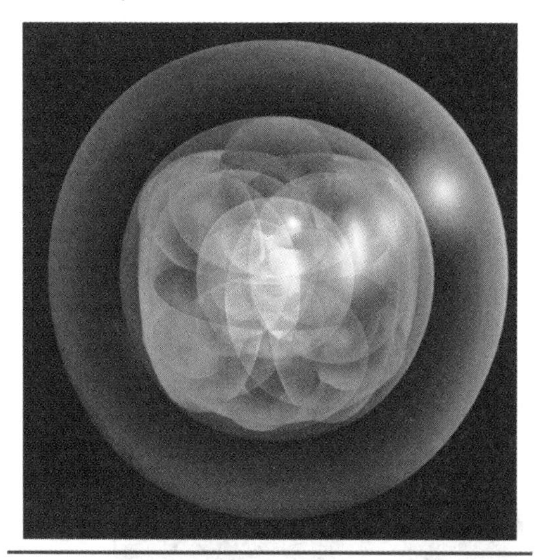

16-17 An illustration of the orbitals within an iron atom

16A Section Review

1. What is the most important aspect of a scientific model? Why has the model of the atom changed?
2. What part of the Greek model of the atom has been set aside by current scientific knowledge? What part of the Greek model can be refuted by Scripture?
3. How did the ancient Greeks differ from European scientists in their pursuit of knowledge about the world, especially the atom?
4. Discuss one aspect of John Dalton's atomic model that was set aside by a later discovery.
5. What was the most significant question about the atom after Thomson's discovery of the electron?
6. What made Rutherford conclude that the nucleus was an extremely small part of the atom?
7. What phenomenon did Niels Bohr observe that convinced him that every atom has a unique electron structure?
8. How do orbitals in the quantum model of the atom differ from orbits in the planetary model?
9. (True or False) The Bohr model of the atom is wrong, and the quantum model is correct.

16B The Orderly Atom

16.9 Subatomic Particles

The development of the modern atomic model revealed that atoms are made of three principal particles—protons, neutrons, and electrons. Each affects the structure of the atom.

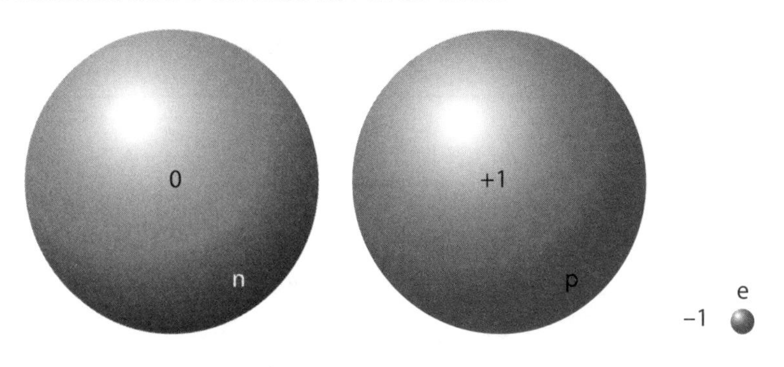

16-18 Models of the neutron, proton, and electron, assuming that their sizes are proportional to their masses. The numbers represent the charge on each particle.

16B Objectives

After completing this section, you will be able to

- ✓ describe the three major subatomic particles and their locations in an atom.
- ✓ define atomic number and explain its significance.
- ✓ discuss the significance of isotopes and how a mass number identifies them.
- ✓ use isotopic notation to identify isotopes of an element and to determine the number of neutrons in an atom's nucleus.
- ✓ define atomic mass and explain its significance.
- ✓ compare and contrast the mass number and the atomic mass of an element.

The **fundamental electrical charge** (*e*) is equal to 1.602×10^{-19} coulomb (C).

The protons and neutrons in a nucleus are called **nucleons**, the largest particles that occupy the nucleus.

Protons

A proton (p) is a subatomic particle located in the atom's nucleus. It has a mass of 1.6726×10^{-27} kg, which is about 1836 times the mass of an electron. It carries a single positive **fundamental electrical charge** (+1 *e*). Isolated protons are stable, so it is probable that none have broken down into simpler particles on their own since Creation. Protons are very important in chemistry and physics because they ultimately determine most of an atom's major properties.

Neutrons

Neutrons (n) were discovered last because they carry no electrical charge, and so they were difficult to detect. A neutron has slightly more mass than a proton, about 1838 times that of an electron. Neutrons are also located in the nucleus. Neutrons and protons in the nucleus are called **nucleons**. Neutrons outside the nucleus ("free neutrons") break down after an average of 15 minutes into an electron and a proton. However, within the nucleus, neutrons are bound to protons by the strong nuclear force, which stabilizes them. They help hold the nucleus together despite extreme electrostatic repulsion between the protons.

Electrons

Electrons (e^-) are the smallest of the main subatomic particles, having a mass of 9.11×10^{-31} kg. They occupy a spherical region surrounding the nucleus. Their arrangement within this region is defined by their quantum energies. Each electron carries a single negative fundamental electrical charge (–1 *e*). Electrons are responsible for all the chemical properties of an atom, including how bonds form with adjacent atoms. Many physical and electromagnetic properties of an atom are also determined by the number and arrangements of electrons around the nucleus.

16-19 These properties of matter—chemical (left), physical (middle), and magnetic (right)—and many others are directly related to the arrangement of electrons around the nuclei of atoms.

FACETS OF SCIENCE — ELEMENTARY PARTICLES

After Ernest Rutherford discovered the nucleus of the atom in 1911, a new branch of science called *nuclear physics* immediately developed. Scientists began probing the nature of the nucleus. Rutherford himself discovered one of its inhabitants, the proton. He suspected that there was another kind of particle to account for the remaining mass in the atom, but he could not detect it.

Other scientists investigated the changes in the nucleus, particularly alpha and beta decays. One physicist, Wolfgang Pauli, discovered in 1931 that during beta decay the energies of the beta particle (electrons ejected from the nucleus) and the remaining nucleus did not add up to the original mass and energy of the nucleus before decay. This was especially troubling because it seemed to be a violation of the law of the conservation of matter and energy.

Pauli suggested that a neutral, high-energy particle with almost no mass must be emitted during beta decay. He called this hypothetical particle a *neutrino*, or "little neutral one." In 1932 James Chadwick discovered the neutron, the other principal nucleon predicted by Rutherford, but there was still no experimental evidence for the neutrino.

The science of *quantum mechanics*, which had produced the quantum model of the atom, predicted that other subatomic particles should exist. In 1932 a positive electron (*positron*) was discovered in cosmic rays. This led to the theory that each type of subatomic particle would have an opposite particle called an *antiparticle*. These antiparticles would have the same mass as the "normal" particles, but their other properties, such as charge, spin, and magnetic properties, would be opposite. For example, the theory predicted the existence of antiprotons, which were discovered in 1955.

In the 1930s scientists began discovering many different subatomic particles that result when cosmic rays collide with the nuclei of atmospheric atoms. Just before World War II, particle accelerators, also known as atom smashers, revealed these particles as well as numerous other small particles in the atom. All these particles were classified as *elementary particles* because they were believed to be the building blocks of protons, neutrons, and electrons.

The onset of World War II halted most theoretical nuclear research except the development of the atom bomb. When research resumed, many new elementary particles were discovered. A pattern began to develop, and physicists were able to classify these particles into three broad categories: *leptons*, *quarks*, and *bosons*. These classes define what physicists call the *standard model* of the universe. Elementary particles compose all matter in the universe and also transmit most of the fundamental forces.

Leptons and quarks are the basic building blocks of all matter. They form the principal subatomic particles. The electron is a lepton, as is the neutrino, which was finally detected in 1956. Quarks combine to form heavy particles called *baryons*. Protons and neutrons (and their antiparticles) are baryons. Six kinds of quarks and their antiquarks have been identified. They have fanciful names, such as up, down, charm, strange, top (or truth), and bottom (or beauty). A proton is made of two up-quarks and a down-quark. A neutron is made of one up-quark and two down-quarks.

The third group of particles, the bosons, transmit the fundamental forces (see Chapter 5) between leptons and quarks. These particles include the *photon*, which transmits EM forces; the *gluon*, which transmits the strong nuclear force; the *Z* and *W particles*, which transmit the weak nuclear force; and the *graviton*, which some believe transmits the gravitational force. No direct evidence for the graviton has been observed.

All told, there are more than 300 elementary particles and antiparticles. Physicists are still searching for many more particles predicted by quantum mechanics. They expect to detect several of the "missing" elementary particles with the Large Hadron Collider. This collider began operation in 2008 at the CERN high-energy physics facility, near Geneva, Switzerland. It generates particle collisions with immense energies that break subatomic particles into their smallest parts.

While much of the motivation for high-energy physics research is to determine the conditions shortly after the big bang, we can be sure that this research will produce many more questions than answers. In any case, we will learn more about the mysteries of God's creation as physicists delve further into the nature of elementary particles.

Wolfgang Pauli (1900–1958) was an Austrian physicist who later became a naturalized US citizen. He received the Nobel Prize in Physics in 1945 for his discovery of a fundamental rule of atomic electron structure called the *Pauli exclusion principle*.

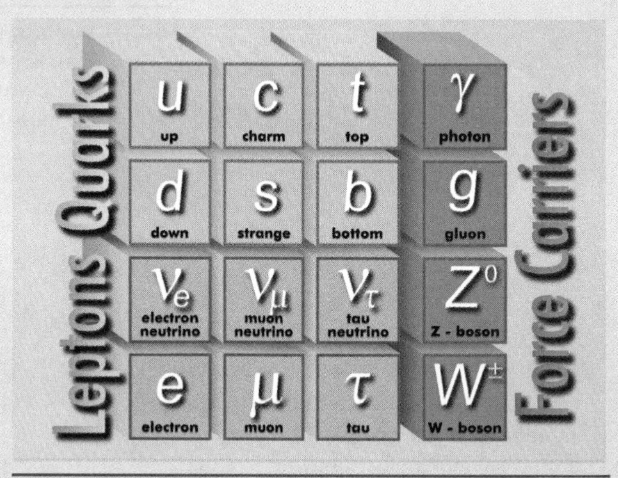

The standard model consists of these 16 basic particles from which all other subatomic particles are made or by which they interact with each other.

16.10 Atomic Number

Since each element is composed of its own kind of atom, something must make an element's atoms unique. But what is that something? Each type of atom has a distinctive number of protons in its nucleus. For example, every gold atom has 79 protons. If a proton could be removed from a gold atom, the remaining atom would no longer be a gold atom. Removing one proton would form an atom of platinum, because all atoms with 78 protons are platinum atoms. All the atoms of a given element must have the same number of protons.

The number of protons in the nucleus determines the kind of atom; this number is known as the **atomic number (*Z*)**. The atomic number also tells us the number of electrons surrounding the nucleus in a neutral atom. Since protons carry a single positive fundamental electrical charge and electrons each carry a single negative charge, an atom is neutral only if it has an equal number of protons and electrons. For example, the atomic number of carbon is 6, so a neutral carbon atom has six protons and electrons each. You can think of the atomic number as an element's "ID number," since the atomic number is unique to each element. The atomic numbers for all the elements are listed in the periodic table (Appendix G).

Atoms with atomic numbers from 1 to 118 have been identified in nature or as artificial products of nuclear reactions.

16-20 Helium ($Z = 2$) is a colorless, odorless, tasteless, lightweight gas used to fill balloons (left). Lithium ($Z = 3$) is a soft, silvery metal that reacts violently with water (right). A lithium atom has only one more proton than an atom of helium, but what a difference that one proton makes!

16.11 Mass Number and Isotopes

Except for the lightest form of hydrogen, every kind of atom has both protons and neutrons in its nucleus. Interestingly, every element can have atoms with different numbers of neutrons. Atoms of an element that have different numbers of neutrons are called **isotopes** of that element. To tell isotopes apart, physicists assign them a special number called the **mass number (*A*)**, which is the sum of the protons and neutrons in the nucleus.

isotope (EYE suh TOPE): iso- (Gk. *isos*—equal) + -tope (Gk. *topos*—place); thus, different atoms having the same place (on the periodic table)

The mass number of an atom allows us to compare it to other isotopes without determining its actual mass—the amount of matter. The mass number assumes that protons and neutrons have the same mass and that electrons have insignificant mass. The mass number is unitless since we are just counting particles.

Since many elements can have the same mass number, the isotope of a specific element is identified by its name or symbol followed by its mass number. For example, the simplest form of hydrogen has only one proton in its nucleus, so this isotope is called hydrogen-1. Another hydrogen isotope has one proton and one neutron, making its mass number 2; it is called hydrogen-2. A hydrogen isotope with two neutrons is called hydrogen-3.

The three isotopes of hydrogen have special names:

Hydrogen-1 is *protium*.

Hydrogen-2 is *deuterium*.

Hydrogen-3 is *tritium*.

Very few other elements have special names for their isotopes.

What is the mass number of a helium atom that has two protons and two neutrons?

Sometimes we need to know the number of neutrons in the nuclei of an isotope. This is easily calculated by finding the difference between the mass and atomic numbers. For example, the most common form of sulfur has a mass number of 32. Since sulfur's atomic number is 16, we know every isotope of sulfur has 16 protons in its nucleus. The number of neutrons in a nucleus of sulfur-32 is found by the following operation:

$$\text{number of neutrons} = A - Z$$

$$16 \text{ neutrons} = 32 \text{ (protons and neutrons)} - 16 \text{ protons}$$

The mass number is always a whole number.

An isotope can also be indicated by *isotopic notation*, which shows both the atomic and mass numbers of the element with its chemical symbol. Starting with the element's symbol, which can be found in the periodic table (Appendix G), the isotope's mass number is placed to the upper left and the element's atomic number is placed to the lower left. The general form is

$$^{\text{mass number}}_{\text{atomic number}}\text{X},$$

where X is the chemical symbol of the element. Thus, the isotope of sulfur mentioned above can be written

$$^{32}_{16}\text{S}.$$

16-21 The mass number allows a quick comparison of the relative masses of isotopes.

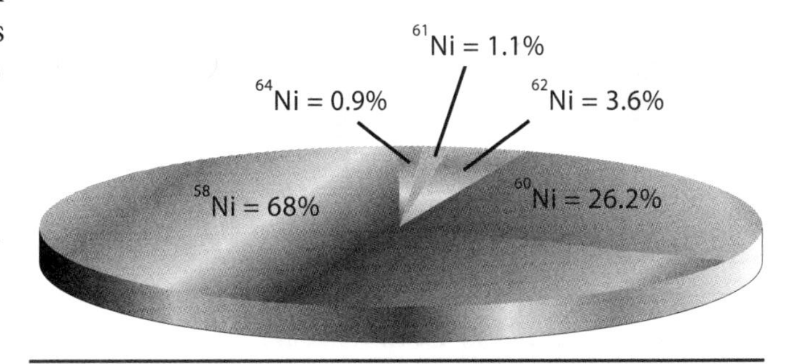

16-22 This pie chart shows the percentage occurrence of the five nickel isotopes in nature.

Example Problem 16-1

Isotopic Notation

Carbon-14 is a relatively rare and slightly radioactive form of carbon that is used to date ancient artifacts. Using the periodic table, write its isotopic notation. How many protons, neutrons, and electrons are in one neutral atom of carbon-14?

Answer: The symbol for carbon is C, and its atomic number is 6, making its isotopic notation

$$^{14}_{6}\text{C}.$$

One neutral atom of carbon-14 has six protons, eight neutrons, and six electrons.

When comparing isotopes of *different* elements, scientists sometimes refer to them as *nuclides*.

The decay of the carbon-14 isotope is used to "date" ancient organic matter, such as bones and wood.

16.12 Atomic Mass

It is often necessary to know how much matter an atom contains, especially when trying to predict the quantities of products from a chemical reaction. We need to know the actual masses, not just the mass numbers, of atoms in elements and compounds.

Determining the **atomic mass** of an atom is not a simple addition of the known masses of its protons, neutrons, and electrons. Nucleons are held together by *nuclear binding energy* (see Subsection 6.12), which is produced by the conversion of matter in the nucleus into energy. So an atom's actual mass is less than the sum of the masses of its parts. The most accurate way to obtain the atomic mass of a given isotope is through experimentation. However, elements in their natural state don't exist as a single isotope. For example, the element chlorine naturally occurs as a combination of about 76% chlorine-35 and about 24% chlorine-37. The atoms of each of these isotopes have different masses. So how do we determine an element's atomic mass?

The minute amount of matter in a single atom adds to the difficulty of the problem. For example, the mass of a chlorine-35 atom is 5.80×10^{-26} kg, and the mass of a chlorine-37 atom is 6.14×10^{-26} kg. To simplify representing an atom's mass and to facilitate working with the masses of isotopes, scientists defined a new unit of mass, called the atomic mass unit. The *atomic mass unit (u)* is the mass equal to *one-twelfth the mass of a carbon atom with six protons and six neutrons (carbon-12).* Thus, 1 u is *approximately* the mass of a proton or neutron in a nucleus. The atomic mass unit eliminates the difficulty of working with extremely small numbers of kilograms or grams.

The atomic mass unit that is accepted by both chemists and physicists is the *unified atomic mass unit*, which is where the unit symbol "u" comes from.

The atomic mass unit is a *relative mass.* Masses of other atoms are compared to one-twelfth the mass of a carbon-12 atom. It is approximately equal to 1.660×10^{-27} kg.

The atomic mass of an element is determined by finding the average mass of its atoms in nature, taking into account the percentage occurrence of each of its isotopes, and expressing that mass in atomic mass units. The atomic mass of natural chlorine, for instance, is 35.453 u. The known or estimated atomic masses for all the elements are displayed under the element symbols in the periodic table (see Figure 16-24).

Unlike mass numbers, which are whole numbers, atomic masses are always expressed as decimals because they represent averages of measured quantities. Scientists continually update these numbers. The percentage occurrence of isotopes is not uniform around the world, so the average atomic masses vary slightly as the values are refined.

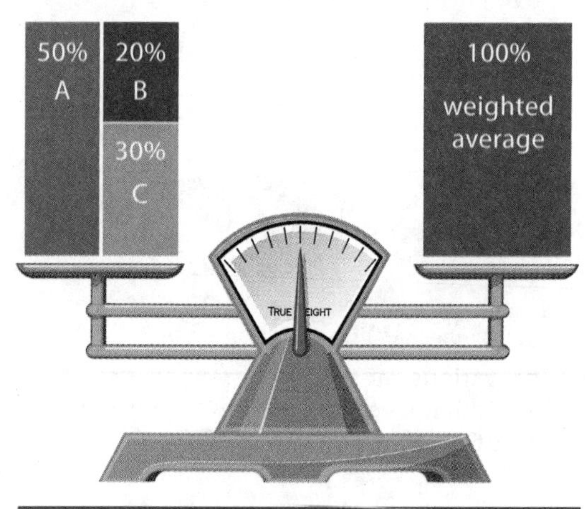

16-23 Chemists work with the average atomic mass for a given element.

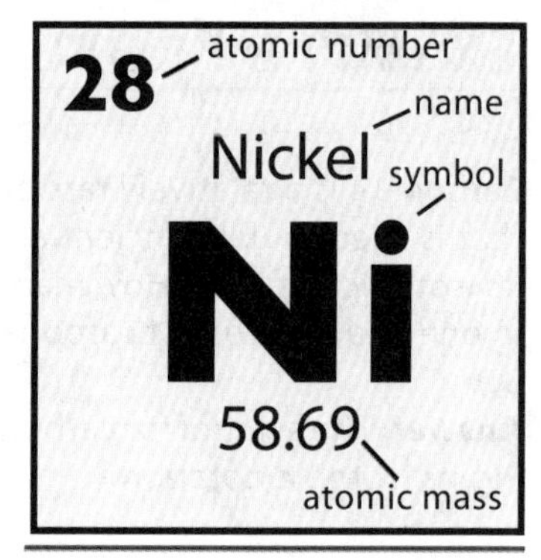

16-24 The element nickel as represented on the periodic table of the elements

16.13 Electron Arrangement in Atoms

An atom of hydrogen has only one electron around its nucleus. That one electron is relatively simple to locate. But an atom of uranium has 92 electrons in its electron cloud. As the quantum model of the atom was developed, mathematical equations revealed that electrons are arranged in orderly patterns. Understanding electron arrangement in atoms will be important for your study of the chemistry of matter in later chapters.

Energy Levels

Niels Bohr's model of the atom suggests that electrons in an atom have different energies. An atom's electrons with lower energies exist closer to its nucleus; those with more energy are farther from the nucleus. This idea is similar to gravitational potential energy. An apple high in a tree has more potential energy than one on the lowest branch.

Using conceptual models, physicists realized that viewing the atom as a miniature solar system of electrons orbiting the nucleus was too simplistic. Atoms modeled as spheres were more workable. Physicists also discovered that they could better model electrons within an atom as waves rather than as particles. It's impossible to locate an electron precisely anywhere inside an atom. According to the quantum-mechanical model of the atom, the best they can do is to calculate the probability of where they would *most likely* find an electron with a certain amount of energy. Within the spherical envelope of an atom, physicists arrange these three-dimensional probability regions as shells or layers at various average distances from the nucleus (see Figure 16-17 on page 361). Their location depends on the energies of the electrons they contain. Physicists and chemists call these layers electron **energy levels**.

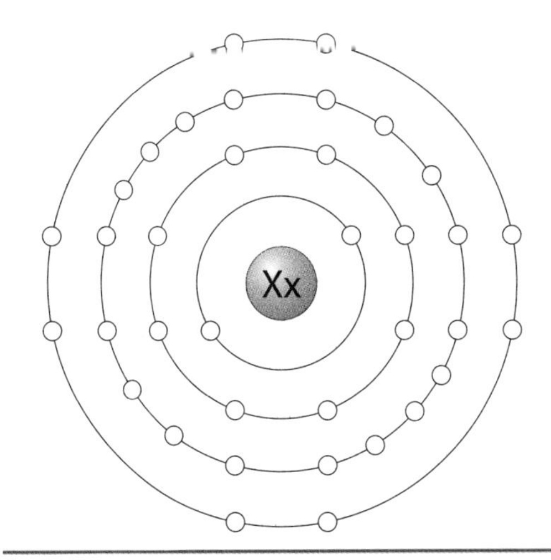

16-25 This is the complete Bohr model electron structure for elements with one to four energy levels. If spots are filled in, they represent occupied electron positions. Empty spots are available but normally unfilled positions in the energy level for a given element.

Electron Structures

Bohr's discovery that electrons occupied energy levels began a series of rapid discoveries about atomic electron structure. Most importantly, physicists realized that the lowest energy levels could contain only a few electrons, and higher energy levels could contain many more. Furthermore, the location of an electron with any given energy depended on the overall energy of the atom. The electrons of atoms with more energy spent their time in higher energy levels than those of less energetic atoms. To simplify our discussions of atoms, their electron structures, and their properties, we will assume that atoms are at the lowest possible energy condition—their *ground state.* No known element has more than seven energy levels when at its ground state.

The various models of electron arrangements that we use in this textbook always show the atoms in their ground state.

According to the quantum mechanical model of the atom, each energy level can hold a maximum number of electrons. Table 16-1 illustrates how this number increases with each of the seven energy levels. In general, as atomic number increases, the electrons fill the energy levels from the innermost to the outermost but there are important exceptions to this rule that you will study in a chemistry course further on in your education. We will use a Bohr model similar to Figure 16-25 to help illustrate how the electrons are arranged in an atom. To keep things simple, we will consider only low-numbered elements having no more than four energy levels.

16-1 Maximum Electrons per Energy Level

Level	Electrons
1	2
2	8
3	18
4	32
5*	50
6*	72
7*	98

*No known elements have enough electrons to fill these energy levels completely.

The actual number of electrons in any given level depends on the atomic number of the atom. Recall that the number of protons in the nucleus determines the number of electrons in the neutral atom. For smaller atoms, as atomic number increases from element to element, the additional electrons appear in the innermost unfilled energy levels. This pattern breaks down for medium-sized and larger atoms. You will learn the details of electron structure in more advanced science courses.

The most important energy level from a chemistry point of view is the highest or outermost energy level that contains electrons. These electrons are the **valence electrons**. The number and arrangement of valence electrons determine the important chemical properties of an element. The valence energy level may be any of the main energy levels, depending on the total number of electrons that the atom has. The number of valence electrons may be between 1 and 8. Figure 16-26 shows four different elements and their electron structures. The valence electrons are colored gold.

valence (VAY lenss): (L. *valentia*—capacity); relates to the bonds an element can make with other elements

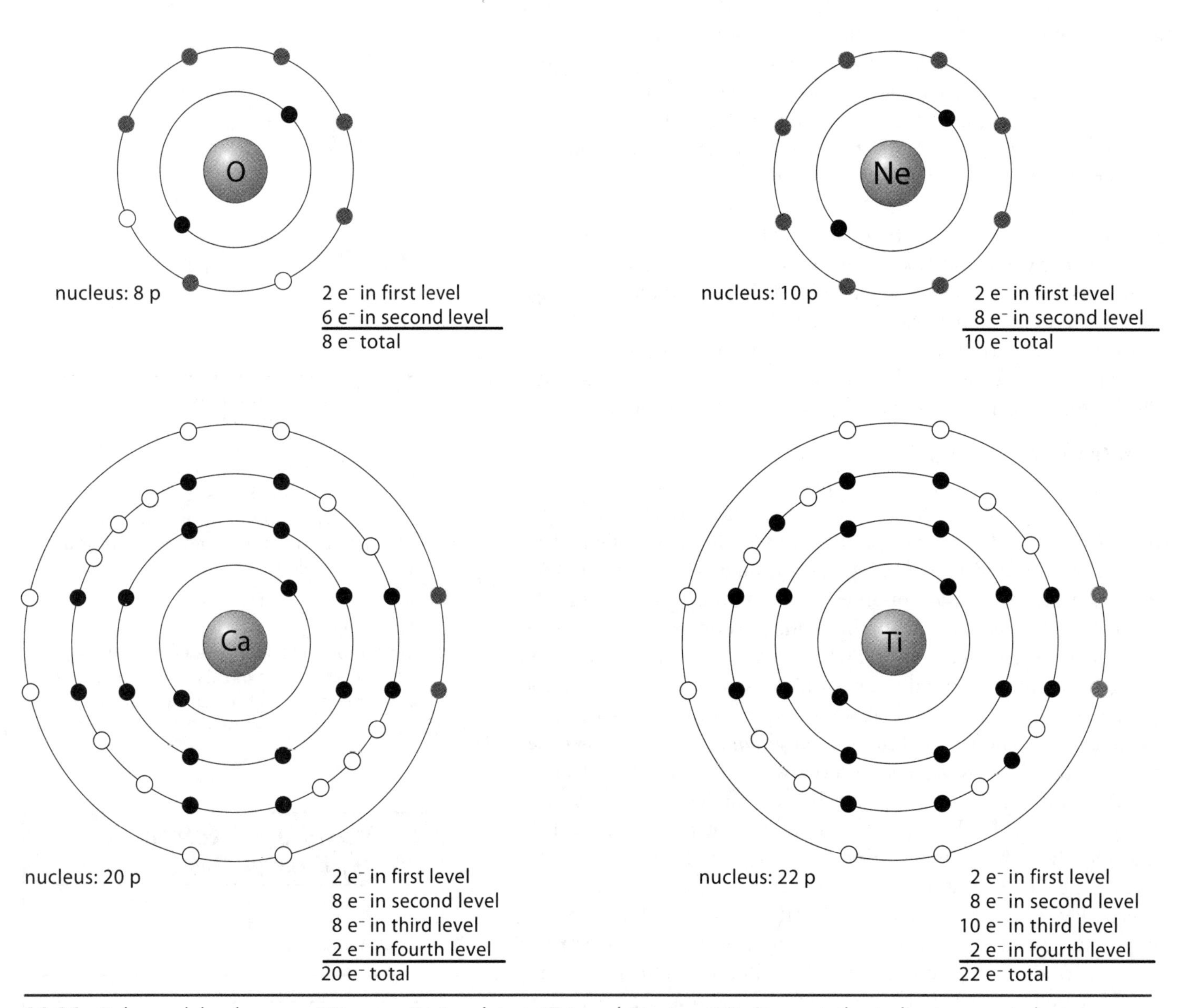

16-26 Bohr models of oxygen (O), neon (Ne), calcium (Ca), and titanium (Ti). Open circles indicate vacant electron positions in energy levels. Gold dots represent valence electrons. Notice that the outermost energy level can't hold more than eight electrons until another energy level is added to hold additional electrons.

The chemical stability of an element depends on the number of valence electrons it has. With the exception of hydrogen and helium, the most stable chemical configuration exists when an atom has eight valence electrons. This is called the *octet rule*. You will see how this rule affects chemical reactivity and stability in later chapters.

Hydrogen and helium have only one electron energy level that can hold a maximum of only *two* electrons, so these elements can't meet the octet rule.

16B Section Review

1. Make a table of the three principal subatomic particles that make up the atom, comparing their relative mass, charge, and location within the atom.
2. What is the most important use of an element's atomic number? What else can we know from a neutral atom's atomic number?
3. How do atoms of the same element differ? What are these different types of atoms called? What abbreviated form is used to represent them?
4. Explain the difference between an atom's mass number and its atomic mass. Why are both needed in chemistry?
5. What is an energy level, and what particle occupies it? How are energy levels arranged in an atom?
6. What do we call the electrons occupying the outermost energy level? Why are they significant?
7. (True or False) The atomic mass unit (u) is equal to the average mass of a proton and a neutron.
8. ✪ Draw the following table on your paper and fill in the blanks using Appendix G, assuming that all isotopes are neutral. The first row has been completed for you.

Isotopic notation	Protons (p)	Neutrons (n)	Electrons (e)
$^{207}_{82}\text{Pb}$	82	125	82
$^{\_}_{\_}\_$	79	118	
$^{79}_{35}\text{Br}$			
$^{127}_{\_}\text{I}$			
$^{\_}_{\_}\text{Fe}$		30	

16C Objectives

After completing this section, you will be able to

- ✓ define radioactivity and identify the important kinds of nuclear radiation.
- ✓ describe the various processes of nuclear decay and how each affects the atom.
- ✓ read nuclear decay equations that illustrate each kind of decay process.
- ✓ discuss one application of nuclear decay that helps save lives.
- ✓ compare and contrast nuclear fission and fusion, including the conditions under which each occurs.
- ✓ discuss the difficulties of developing electrical generation using fusion energy.

Most elements in their natural state are not radioactive or hardly so. However, starting with polonium ($Z = 84$) through uranium ($Z = 92$), the most common isotopes of these naturally occurring elements are unstable and thus radioactive.

16C The Nuclear Atom

16.14 Nuclear Radiation

The nuclei of most atoms around us are relatively stable because they have roughly equal numbers of protons and neutrons. For atoms with larger atomic numbers, more neutrons are needed to overcome the strong repulsion between the protons. But if there are too many (or too few) neutrons, the nucleus becomes less stable. Under certain situations these nuclei emit high-energy particles or EM rays as they change into a more stable form. The study of changes that occur in atomic nuclei is called **nuclear chemistry**. Nuclear chemistry began with the accidental discovery of *radioactivity*, which is the emission of **nuclear radiation**, the rays and particles emitted by unstable nuclei.

In 1896 Henri Becquerel found that photographic film could be exposed by natural uranium ores, even though the film was wrapped in paper to protect it from light. It was as if light had penetrated the opaque covering. This discovery was completely unexpected. It followed the discovery of x-rays by only two months. Many questions immediately surfaced about radiation. What is it? What causes it? How can it be used? The idea that nuclear radiation might be hazardous wasn't even considered until after World War I. A detailed analysis of radiation was needed. Many researchers immediately dropped what they were doing to concentrate on this new and mysterious form of energy.

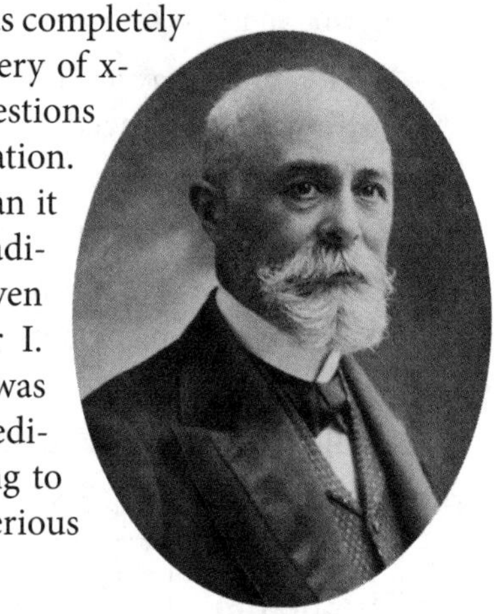

16-27 Antoine Henri Becquerel (1852–1908) was a French physicist who was one of the original discoverers of nuclear radiation and radioactivity. He won the Nobel Prize in Physics in 1903.

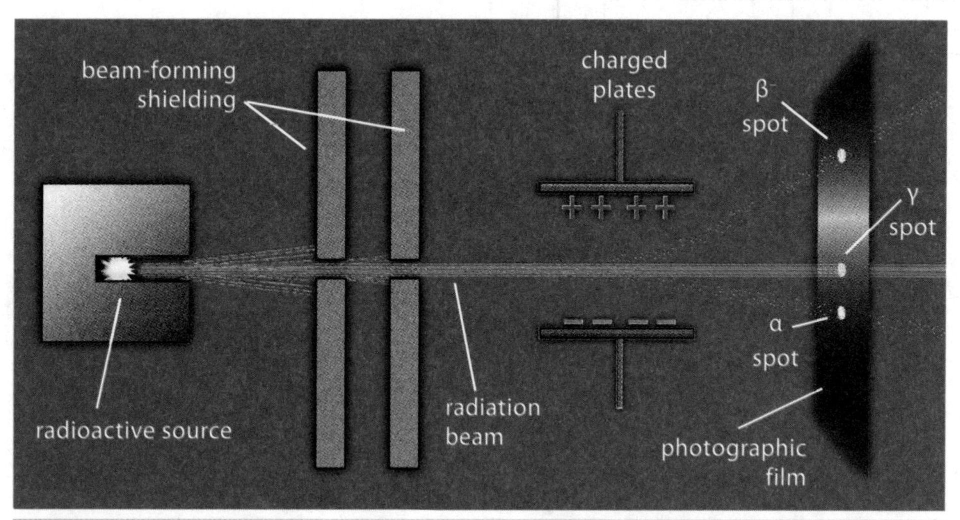

16-28 An experiment where three kinds of nuclear radiation can be identified

In early experiments researchers used a sample of a uranium compound as their source of radiation. It was housed in a lead container that had a small opening to screen out all radiation except for a narrow beam in one direction. The beam passed into an apparatus between positively and negatively charged plates. The particles and rays struck a photographic plate, whose pattern of exposure recorded any effects on the radiation beam. The researchers found that an electric field between the plates separated the initial beam into three separate beams that formed spots where they exposed the film.

One beam was deflected only slightly in the direction of the negatively charged plate. The slight deflection indicated that the beam was composed of fairly massive particles having a large amount of inertia. The direction of the deflection indicated that the particles were positively charged. These particles, called **alpha (α) particles**, were later found to be composed of two protons and two neutrons—the same composition as helium nuclei.

Particles in the second beam were greatly deflected. Because the beam easily bent toward the positive plate, these particles must have had little mass and a negative charge. These **beta (β) particles** were later found to be *free nuclear electrons* produced by the decay of neutrons into protons and electrons.

The third beam was unaffected by the electric field, indicating that it had no electrical charge. Eventually, this beam was found to consist of EM waves rather than particles. These waves are called *gamma (γ) rays*. They are similar to x-rays but usually carry much more energy.

Even as the identity of nuclear radiation was being determined, physicists were only beginning to unravel where these particles and rays were coming from.

What is the mass number of an alpha particle?

Alpha particles, beta particles, and gamma rays are represented by the Greek letters alpha (α), beta (β), and gamma (γ).

16.15 Nuclear Decay

Most people think of the nucleus as a motionless mass of nucleons. This view is not entirely accurate. The nucleons in small stable nuclei are likely to have low excess energy and little motion. Such a nucleus forms a stable, compact ball, but larger nuclei with many nucleons tend to wobble and oscillate like a drop of liquid in free fall, especially if they gain excess energy from outside the atom.

Gamma Decay

One way for a nucleus to shed excess energy is to emit a gamma ray through **gamma decay**. There is no change in the atom except for reducing the amount of energy in its nucleus. The following expression, written like an equation, shows the gamma decay of uranium-235.

$$^{235}_{92}\text{U}^* \longrightarrow {}^{235}_{92}\text{U} + \gamma \text{ (gamma ray)}$$

The arrow, which is like an equal sign in an equation, indicates the direction in which the decay occurs.

Gamma rays leap from their nuclei at the speed of light and are very penetrating. The typical gamma ray can pass through several meters of concrete, and it takes more than 20 cm of lead to stop nearly all gamma rays! A nucleus can potentially experience numerous gamma decays without permanently changing the atom. Gamma decays often accompany other kinds of nuclear decays. They can be very damaging, especially to organic molecules found in living things. Molecules that absorb gamma rays can be destroyed as the energy disrupts bonds between atoms. Nuclear changes can even occur in the atoms of the molecule.

Effects of Nuclear Radiation

This subsection discusses various kinds of nuclear radiation and their effects. All forms of nuclear radiation can disrupt chemical bonds between atoms. This is why they are also called *ionizing radiation*—exposure to them can result in tissue damage by forming ions of atoms that were part of normally-functioning biological molecules.

Some people use the word "radiation" for other forms of radiant energy, like radio waves and UV light. But these forms of energy don't normally cause ionizations, and are therefore classified as *non-ionizing radiation*. So it's perfectly safe to "nuke" last night's leftovers in a microwave oven—there is no ionizing radiation hazard there!

In nuclear decay equations, an asterisk indicates that the nucleus has excess energy.

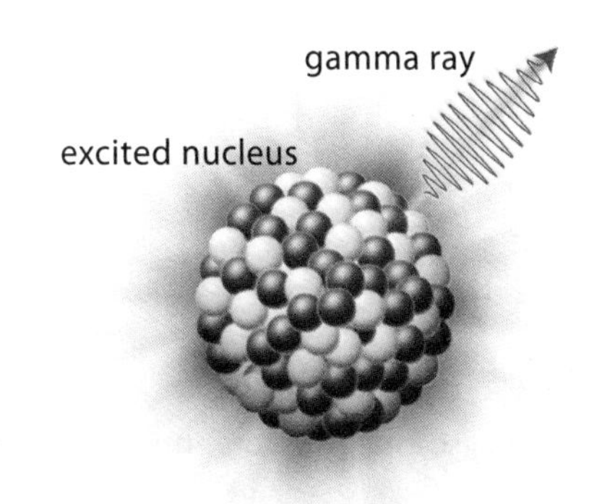

16-29 A gamma decay of a uranium-235 nucleus

Alpha Decay

A more significant nuclear change occurs when a nucleus ejects an alpha particle. During **alpha decay** the atom's nucleus loses two protons and two neutrons, reducing its atomic number by 2 and its mass number by 4. Because its atomic number has changed, the remaining atom is now an isotope of a *different* element.

Isotopic notation is useful for writing nuclear changes in a brief form. For example, one isotope that undergoes alpha decay is uranium-238 ($^{238}_{92}U$). Since the alpha (α) particle is the same as a helium (He) nucleus, we may represent an alpha as $^{4}_{2}He$ in a decay equation. The complete nuclear change in the following equation shows that no mass or energy is lost.

$$^{238}_{92}U \longrightarrow \, ^{234}_{90}Th + \, ^{4}_{2}He + \gamma$$

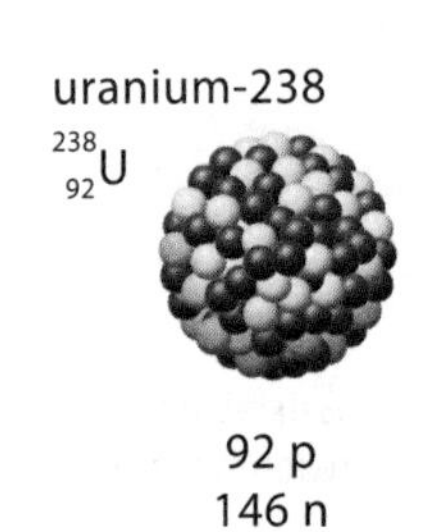
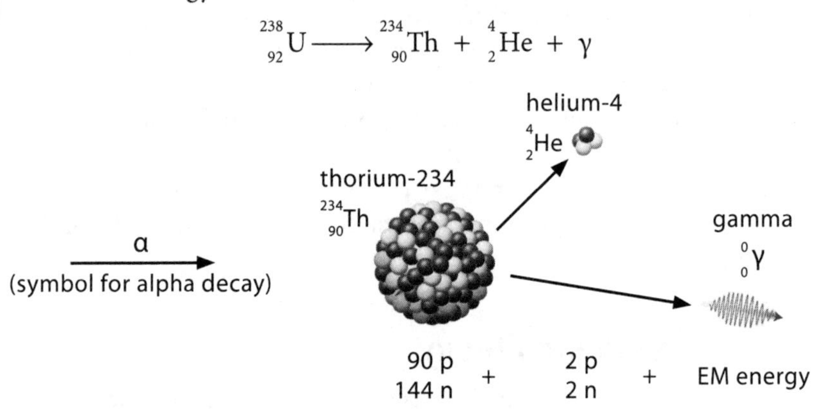

Notice that the sum of the mass numbers and of the atomic numbers on the right of the arrow equals the uranium's mass and atomic numbers on the left. This relationship must exist according to the law of mass and energy conservation.

Once the free alpha particle hits something, it gains electrons and becomes a normal helium atom. The large particle slows down quickly and thus has low penetrating power. Alpha particles can be stopped by a sheet of paper. However, because of its large mass and electrical charge, it tends to interact with other atoms very easily, damaging chemical bonds as it rips electrons away from other atoms and molecules.

Beta Decay

Many radioactive nuclei are unstable because the ratio of protons to neutrons in the nucleus is not ideal. For example, the first product of uranium-238 alpha decay is thorium-234 ($^{234}_{90}Th$). The ratio of 144 neutrons to 90 protons in thorium-234 is too large to be stable, so almost immediately one of its neutrons changes to a proton and ejects a high-energy electron, or beta particle. The atom produced by this action, known as **beta decay** of thorium-234, is protactinium-234 ($^{234}_{91}Pa$). Beta decays do not change the number of nucleons present, so the mass number remains the same. However, the change of a neutron into a proton increases the atomic number by 1, changing the atom into an isotope of another element. The beta decay of thorium is written in the following equation.

Radioactive isotopes are sometimes called *radioisotopes*.

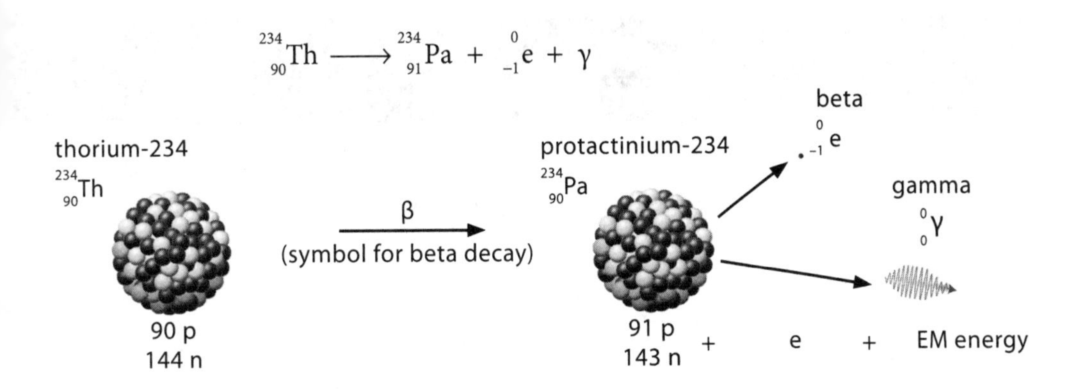

Notice in the equation how an electron is represented in isotopic notation. It has no nucleons, so its mass number is 0. Its "atomic number" (normally, an atom's total nuclear charge) is −1, the charge of an electron. As before, the sum of the mass numbers on one side of the arrow equals the mass number on the other. The same goes for the atomic numbers, illustrating once again that nuclear decays conserve mass, energy, and charge.

The beta particle, a free electron, is subject to capture by any atom it encounters, making this form of radiation also easily stopped. Beta particles can be stopped by wood or thin sheets of aluminum. They travel farther than alpha particles because they are much smaller and have a smaller electrical charge. Although beta radiation is far less damaging than alpha or gamma radiation, it can disrupt the sensitive chemical bonds in living cells, so some biological hazard is associated with it.

Is Nuclear Decay "Very Good"?

When God completed His work of creation, He declared that it was all "very good" (Gen. 1:31). How can nuclear decay be a part of a "very good" creation?

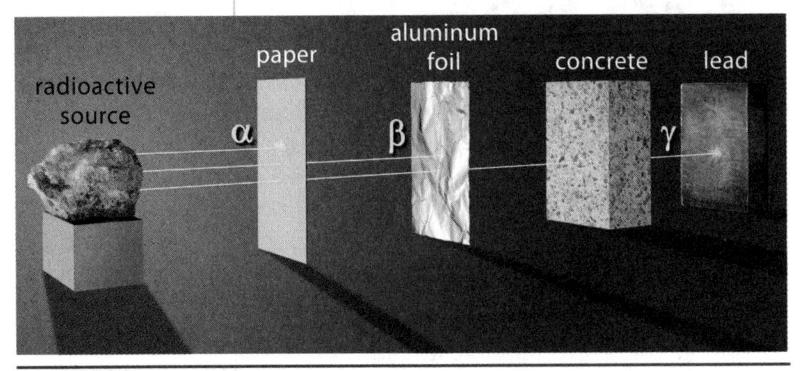

16-30 The relative penetration abilities of alpha, beta, and gamma radiation

16.16 Using Nuclear Decay to Solve Problems

How can the atom's nuclear properties help warn people about a fire in their home? You probably already know the answer—smoke detectors. They are cheap and easy to install and save many lives. How do smoke detectors relate to nuclear decay?

One common type, an ionizing smoke detector, contains 0.0002 g of the radioactive isotope **americium**-241 (Am-241). This isotope emits alpha particles when it decays. The alpha particles collide with air molecules and strip away their electrons, forming ions. The ions carry a small current between electrodes connected to the detector's battery (see Figure 16-31). When smoke enters the detector chamber, it disrupts the current flow by neutralizing the ionized air particles. When the detector senses a drop in the electrical current, it triggers the alarm.

Though almost 90% of American homes are equipped with smoke detectors, and though they are required for all new homes, almost a third don't function because of dead batteries or defects. How can you ensure that your smoke detector could save your family?

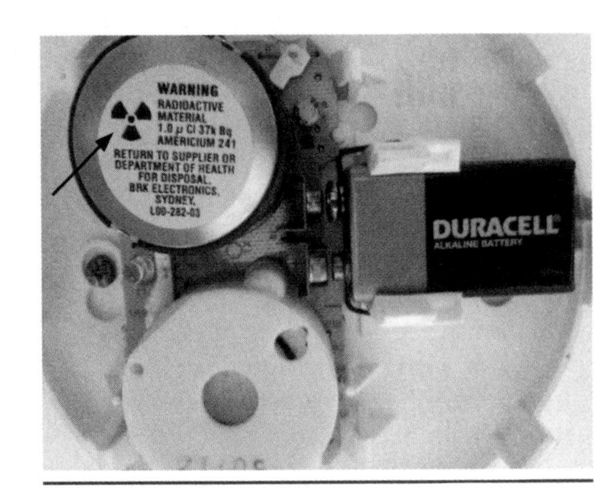

16-31 The location of the americium isotope in a smoke detector

americium (AM uh REE see um)

FACETS OF SCIENCE

A Question of Time: Radioactive Dating Methods

Mount Ngauruhoe in New Zealand, where lava deposits formed within the past century were dated as old as 3.5 million years using the potassium-argon decay method

Numbers enjoy a special status in our culture. They have an almost magical way of authenticating the statements they adorn. A published report containing statistics and measured data will have better success and command greater respect from its readers than one without. The problem is that some people use questionable statistics to bolster their ideas and pass them off as fact to unsuspecting people.

Have you ever read a headline such as "50-Million-Year-Old Bat Fossil Confirms Theory"? How do researchers arrive at such a large figure, and how certain are they of its accuracy? If you study the biblical account of Creation, evaluate the biblical genealogical record as a literal history, and match certain historical events recorded both in Scripture and elsewhere, the analysis seems to indicate an earth that is about 7000 years old. These factors also imply that living things have been here for the same amount of time. Why such a discrepancy?

Even devoted evolutionists admit that the theory of evolution is doomed if it does not have one very important ally on its side—lots of time. They take the approach that given enough time (roughly 4.5 billion years) anything can happen, including evolution. Evolutionary scientists use dates of billions of years as if they were common knowledge—and anyone who challenges these dates is "unscientific" or "uneducated." (One scientist goes so far as to say that such a person is "wicked.") So evolutionary scientists have asserted that the earth is billions of years old and have devised *radiometric dating techniques* to "prove" the ages of fossils and rocks.

Deep-Time Dating Methods

Estimates concerning the ages of rocks and fossils run into the millions and billions of years based on the so-called deep-time dating methods—uranium-lead, potassium-argon, and rubidium-strontium. Each method refers to a known sequence of radioactive decays that starts with an unstable isotope of the first element and eventually ends with a stable isotope of the second. Uranium decays into the element lead in a series of alpha and beta decays. Potassium decays into argon in a single step, and rubidium decays directly into strontium. It takes a certain amount of time for this decay process to occur at a known rate. When a scientist measures the amount of argon in a rock sample, he can calculate the approximate age of rock, assuming that when the rock was formed it contained only potassium. This is like measuring the amount of gas in the tank of a car after a trip, knowing how fast the car uses gas, and then estimating how much time elapsed since the beginning of the trip.

These deep-time dating methods are not deceptive in themselves, but the scientist who uses them must make some basic *assumptions* that he cannot verify. For example, the scientist using the potassium-argon dating method has no way of knowing how much argon was originally present in the sample. Assuming that no argon was present when the rock was formed is foolish, since it has been proved that rocks of known age and formation (volcanic rocks) have been found with an abundance of argon.* The problem with these dating methods is that a key ingredient of scientific methodology—*documented observation*—is missing.

Another assumption made is that the radioactive decay rate is constant. Scientists use the principle of *half-life* to describe how long it takes for a sample of an isotope to decay. A 1.0 g sample of iodine-131 will have only 0.5 g of the isotope after eight days. So the half-life of iodine-131 is eight days. Many radioactive isotopes have half-lives that have never been directly observed but have been *extrapolated* (graphing a projected trend). Uranium-238 is a good example of this—it is said to have a half-life of 4.51 billion years, the estimated age of the earth! Obviously, scientists had to extrapolate that figure assuming that uranium has decayed at a constant rate since the origin of the universe.

Recent evidence uncovered by a team of Christian researchers suggests that sometime in the earth's past there was a brief, intense period in which the half-lives of all radioactive isotopes were very short. This period could have produced most of the decay products measured in the earth's crust today. It also would mean that half-lives cannot be assumed to be constant for the purpose of determining the age of the earth and its rocks.

As you can see, these dating methods rely on some questionable assumptions. The main problem, however, with these deep-time dating methods is in the interpretation of the results. Evolutionary scientists say that these dating methods prove that the earth's rock layers are old and, therefore, biological evolution has had the time to occur. But all that these methods ever really show is the amount of end product present within a rock, and that result can be interpreted many different ways. In fact, there are many scientists claiming to believe the Bible who say that some dating methods are valid. They have various theories to reconcile old-earth estimates and the Creation week described in Genesis 1. However, those who compromise a straightforward reading of Genesis to accommodate deep-time dating methods undercut the very authority that establishes the major doctrines of the Christian faith.

As you learned in Chapter 1, interpretation of scientific data depends directly on the scientist's worldview. If his worldview forces him to discount any evidence (especially biblical) other than direct observation or what can be reasoned from observations, then an evolutionary interpretation is reasonable. If he accepts the Bible as the Word of God and its historical statements about the origin of the world, he has to reject the old- earth assumptions that lead to an interpretation of billions of years. There cannot be any middle ground, which is where compromise lies.

Carbon-14 Dating

Carbon-14 dating is a dating method that is often used to determine the ages of certain fossils, manuscripts, and other artifacts made of organic materials. Cosmic rays from the sun strike atmospheric molecules that eject high-energy neutrons. One of these collides with the nucleus of an atmospheric nitrogen-14 atom, which then ejects a proton to produce a heavy isotope of carbon—carbon-14. It decays at a constant rate and has a half-life of 5730 years (almost as long as the earth has existed as inferred from the Bible).

All living things, animals and the plants they eat, always contain a minute amount of carbon-14 because carbon-14 is fixed in plant matter through photosynthesis. When an organism dies, it no longer takes in carbon-14. The isotope gradually beta decays to nitrogen-14. By measuring the amount of carbon-14 left, scientists can determine the approximate time of the organism's death *if an initial amount of carbon-14 in the material is assumed.*

Using human artifacts that can be accurately dated by other means, scientists have determined that the amount of carbon-14 in the atmosphere has varied considerably during the past 5000 years. Other evidence using tree-ring data of very old trees confirms this finding. If one considers that the Genesis Flood about 5500 years ago was probably accompanied by worldwide volcanic activity releasing great quantities of carbon-12 into the atmosphere, it is not surprising that the proportion of carbon-14 in organic matter, which does not come from volcanic activity, varied considerably during that period. Once again, the interpretation of radiometric dating data ultimately rests on presuppositions derived from one's worldview. Carbon-14 dates must be evaluated in view of what the Bible says about history.

*Andrew A. Snelling, "The Cause of Anomalous Potassium-Argon 'Ages' for Recent Andesite Flows at Mt. Ngauruhoe, New Zealand, and the Implications for Potassium-Argon 'Dating'" (paper presented at the Fourth International Conference on Creationism, Pittsburgh, PA, August 1998).

16.17 *Nuclear Bombardment Reactions*

Some nuclear changes occur when a nucleus is struck by a high-energy particle or another nucleus. Physicists call such processes *nuclear bombardment reactions*. While these reactions are rare in nature on Earth, scientists use them to release huge amounts of energy and particles. If well controlled, they can even be used to make new and useful elements not occurring in nature.

How is this type of reaction different from nuclear decay? Bombardment reactions release millions of times more energy and many more particles than alpha or beta decay. Also, alpha and beta decays usually occur spontaneously, that is, without human action. For most radioactive isotopes naturally existing on Earth, these kinds of decays occur infrequently. For example, it would take around 4.5 billion years for half of a sample of uranium-238 to alpha decay to thorium-234. Bombardment reactions, on the other hand, are almost always artificially induced in special nuclear reactors and particle accelerators. These devices carefully control the amount and rate of particle release. Bombarding nuclei is a very exacting and complex process, but the results are staggering.

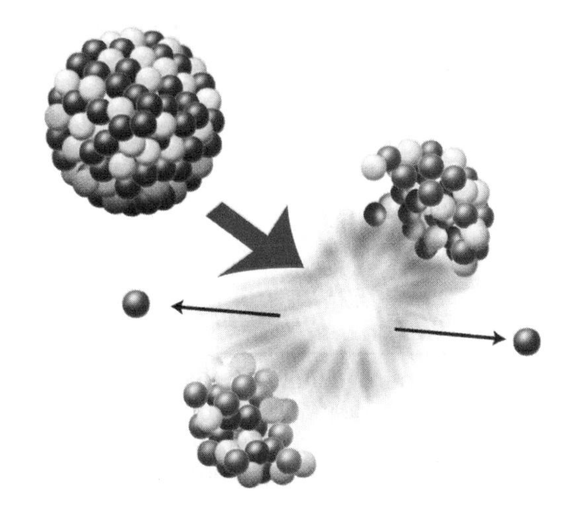

16-32 Nuclear fission is the splitting of a large nucleus into smaller ones with the release of several neutrons and a large amount of energy.

Nuclear Fission

In the 1930s physicists found that they could make a massive nucleus unstable by striking it with an energetic free neutron. The unstable nucleus soon broke into two smaller nuclei and several more free neutrons. This kind of nuclear change is called **nuclear fission**. Fission occurs when a large, unstable nucleus splits into smaller, more stable nuclei, releasing energy. This can be forced to occur artificially, such as in a nuclear reactor, and it can also occur spontaneously, although such occurrences are infrequent. If neutrons released during fission are absorbed by other nuclei, these can also fission, releasing more free neutrons. As this process of fissions causing fissions continues, a nuclear *chain reaction* occurs. In a controlled nuclear chain reaction, an immense amount of energy can be released to do useful work.

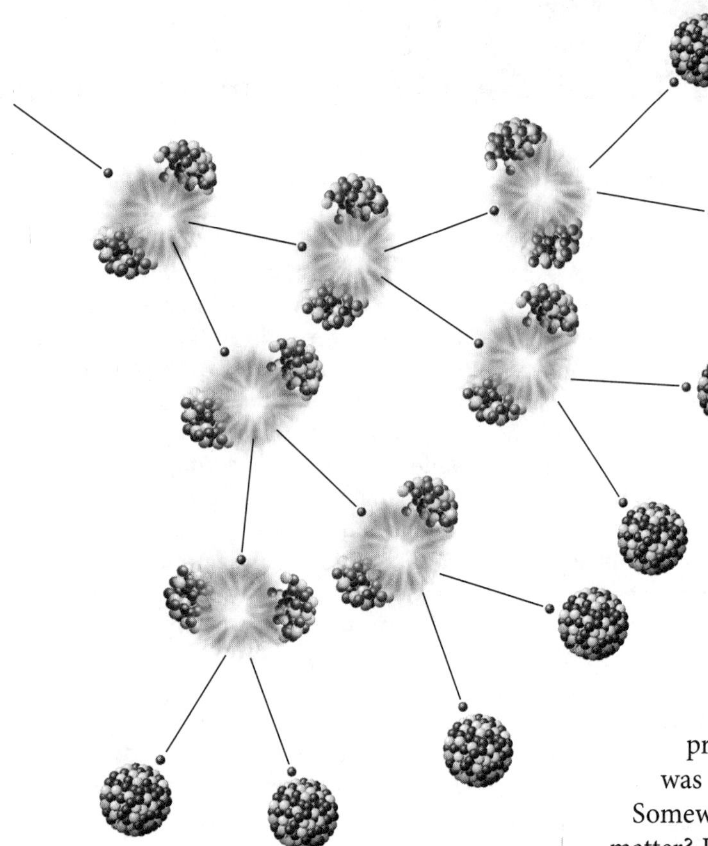

16-33 A nuclear chain reaction

Where does all this energy come from? When physicists measured the masses of the new nuclei and neutrons produced by fission, they found that the sum of the masses was less than the mass of the original nucleus before fission! Somewhere, some mass had been lost. What had happened to this matter? Earlier, while developing his special theory of relativity, Albert Einstein demonstrated that matter was equivalent to energy in his famous equation $E = mc^2$. According to this equation, the amount of energy released during fission corresponded exactly to the mass lost. Even tiny bits of matter can produce extremely large quantities of energy. If you could convert 1 g of matter, about the mass of a paper clip, into energy, you could launch a 24,000 kg (26.4 ton) rocket to the orbit of the International Space Station!

The power of fission was demonstrated to the world in the two bombs that ended World War II. Fission bombs are also known as atomic bombs, A-bombs, or nuclear bombs. Only eight nations have tested fission weapons, although several others are suspected of having them or are attempting to acquire them. Most technologically advanced countries have also learned to harness the same kind of energy to generate electricity. Today, about 20% of the electricity generated in the United States is produced by nuclear power plants. The United States is committed to increasing nuclear power because of growing concern that carbon dioxide emissions from fossil fuel plants are contributing to global warming.

16-34 The second atomic bomb, dropped by an American bomber on the industrial city Nagasaki, Japan, brought an immediate end to World War II.

Nuclear Fusion

The opposite nuclear bombardment process is smashing small nuclei together to form larger ones. **Nuclear fusion** produces even more energy than fission. The sun and stars get their energy from fusion reactions by combining hydrogen nuclei to form helium. There is astronomical evidence that larger stars that have nearly exhausted their hydrogen and helium fuel can form much heavier elements through fusion, such as carbon and iron.

As with fission energy, application of fusion energy first began with military hardware. Fusion bombs, better known as hydrogen bombs,

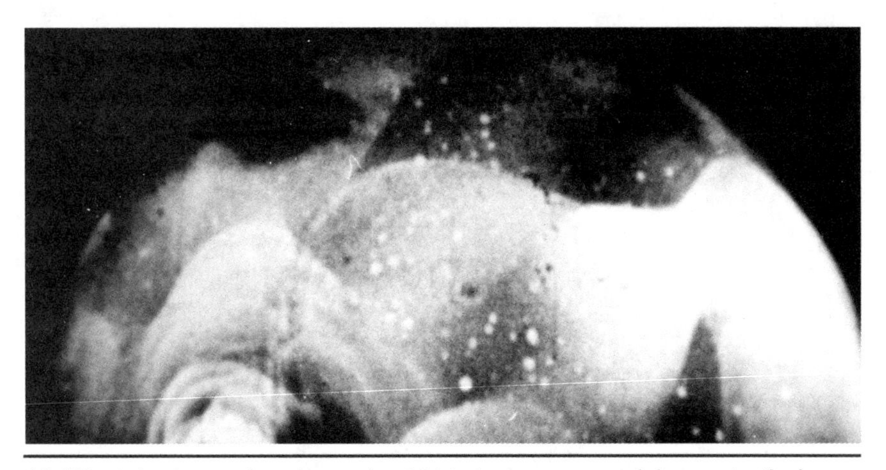

16-35 A hydrogen bomb can be 1000 times as powerful as a similarly sized fission bomb.

H-bombs, or thermonuclear bombs, were quickly developed after World War II, first by the United States, then by the Soviet Union, in the early 1950s. A fusion bomb can be 1000 times as powerful as a similarly sized fission bomb. A mere six nations have tested fusion weapons. The former Soviet Union conducted the largest fusion bomb test in history. It yielded a blast equivalent to 50 million tons of dynamite.

Nuclear fusion has rightly been called the energy of the future. It has many compelling selling points. First, it requires no hard-to-find fuels; the oceans contain all the hydrogen atoms we would ever need in the form of water (H_2O). Second, while the burning of fossil fuels produces harmful pollutants and fission power generation produces long-lived, highly radioactive wastes, fusion would probably produce only low-level radioactive wastes that could be easily managed. Third, the fusion process produces an extremely large amount of energy from what is "invested" without depleting nonrenewable natural resources.

Several major obstacles must be overcome, however, before fusion can be controlled and used to produce electricity. In fact, controlling nuclear fusion is one of the greatest technological challenges in the history of mankind. For this reaction to occur, hydrogen nuclei must be under conditions similar to those in stars, where the temperatures and pressures are incredibly high.

One fusion reaction that has received a fair amount of study because it requires the lowest temperature—fusion of two hydrogen isotopes, tritium and deuterium—still requires temperatures around 50–100 million kelvins! As you might imagine, matter could be only in a plasma state under these conditions. One method being investigated to produce such high-temperature plasma involves aiming a bank of lasers at the material to be fused, and heating it rapidly to very high temperatures by bursts of laser light.

Another significant challenge is maintaining and containing such an extremely hot plasma. Certainly, contact with any physical container by the plasma would cool it to an extent that it would no longer be a plasma. However, since plasma is composed of charged particles, it can theoretically be contained with electric and magnetic fields in a "magnetic bottle." Because putting this theory into practice is difficult, however, sustained fusion isn't commercially feasible yet. Recently, several partner nations began construction of the prototype

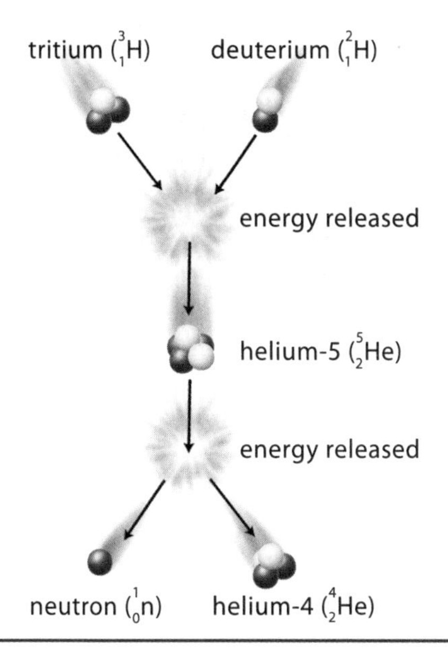

16-36 In deuterium-tritium fusion, the most common form of fusion, nuclei of hydrogen-2 and hydrogen-3 fuse within the extreme heat of a hydrogen plasma. The reaction forms a helium nucleus and releases a neutron and an immense amount of energy.

ITER originally stood for *International Thermonuclear Experimental Reactor*. Since *iter* means "the way" in Latin, the ITER organization has adopted just the acronym to symbolize the way to the future peaceful use of nuclear energy.

fusion energy demonstration plant in Cadarache in southern France. The ITER fusion plant is expected to begin plasma generation in 2020 and its first fusion energy operations in 2027.

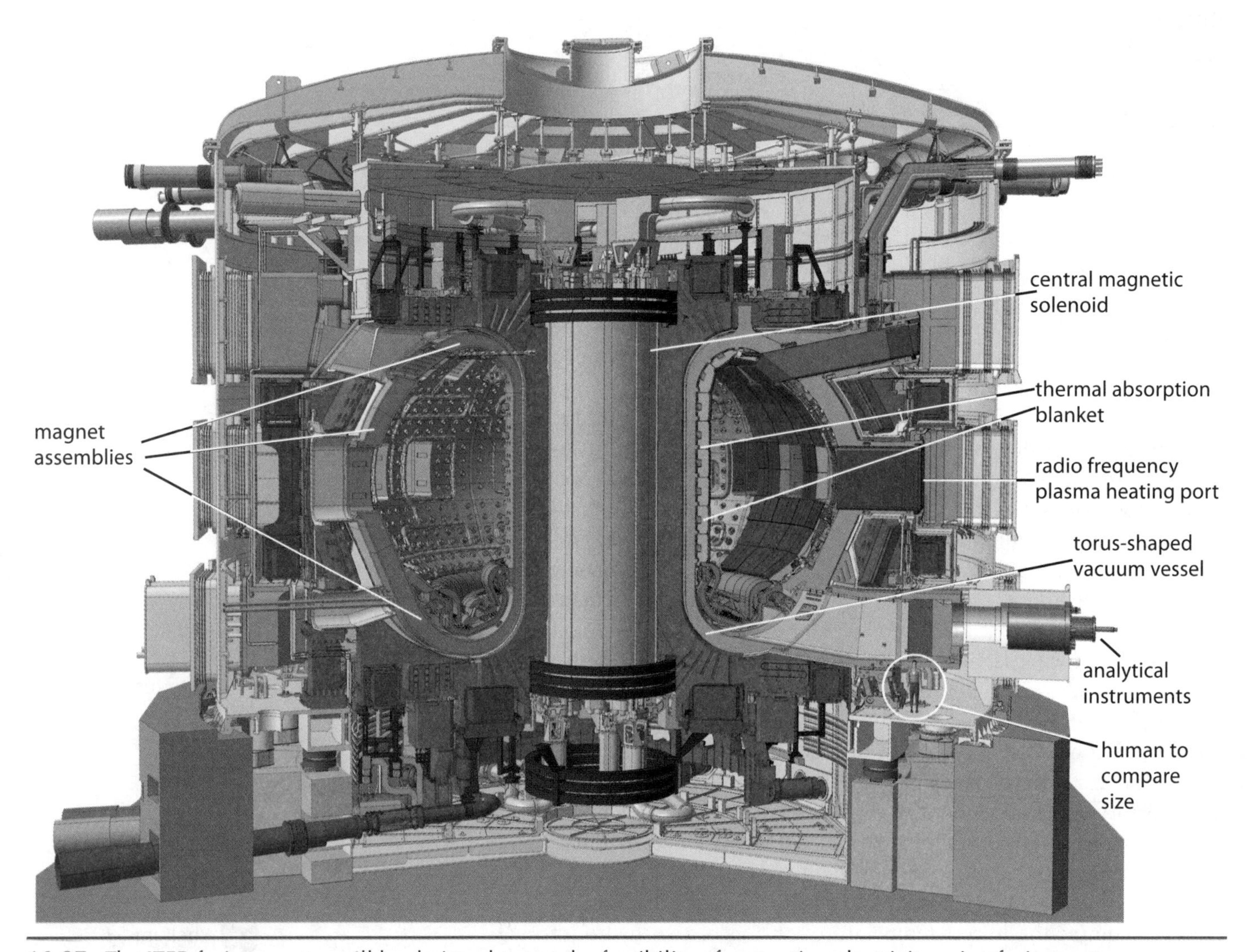

16-37 The ITER fusion reactor will be designed to test the feasibility of generating electricity using fusion energy. Compare the size of the enclosure with the size of a man (circled).

16C Section Review

1. What makes an element radioactive? Which elements found in nature have mainly radioactive isotopes?
2. Name the three main kinds of nuclear radiation and give the characteristics of each.
3. What kinds of changes to a nucleus occur for each kind of nuclear decay?
4. How can nuclear radiation affect living organisms?
5. DS ✪ What kind of nuclear decay occurs within most smoke detectors? How does this show that not all kinds of degeneration are necessarily bad?
6. In what important way do fission and fusion differ from nuclear decays? How common are natural fission and fusion on Earth?
7. What is the source of energy released in fission and fusion reactions?
8. What are the two most difficult aspects of producing fusion energy for electrical generation?
9. (True or False) Beta decay is produced by the change of a neutron in a nucleus.
10. ✪ Write the nuclear decay equations for the following reactions. Refer to the periodic table in Appendix G for atomic numbers and the names of the resulting isotopes.
    a. gamma decay of iodine-128 ($Z = 53$)
    b. beta decay of plutonium-246 ($Z = 94$)
    c. alpha decay of radon-210 ($Z = 86$)
    d. beta decay of cesium-138 ($Z = 55$)

Scientifically Speaking

law of definite proportions	356
core-envelope model	356
plum pudding model	357
nuclear model	359
planetary model	360
quantum model	360
nucleon	362
atomic number (*Z*)	364
isotope	364
mass number (*A*)	364
atomic mass	366
energy level	367
valence electron	368
nuclear chemistry	370
nuclear radiation	370
alpha (α) particle	371
beta (β) particle	371
gamma decay	371
alpha decay	372
beta decay	372
nuclear fission	376
nuclear fusion	376

Chapter Review

1. If an atomic model cannot explain a scientific observation, what should happen to it?
2. Why did early scientists in Europe first disregard the idea that matter was made of atoms?
3. What two factors restored interest in the atomic nature of matter in the 1700s?
4. What inspired John Dalton to develop his atomic theory? What was his reasoning for his conclusion?
5. What aspects of Dalton's atomic model are still useful?
6. Why do you think that J. J. Thomson suggested the plum pudding style model to account for the electrons in atoms?
7. ✪ Look at Figure 16-9 on page 358. What could account for flashes in the alpha particle detector at small angles away from the main beam of particles in Rutherford's experiment?
8. What element did Bohr's original atomic model apply to? In Bohr's model, what does an energy level determine for the electrons it contains?
9. What aspect of electrons did the quantum model of the atom account for that Bohr's planetary model could not? What was the key difference between the two models?
10. What seems to be the main purpose of neutrons in nuclei?
11. Compare the electrical charge of the proton to that of the electron. What is the *magnitude* of these charges called?
12. What property of an atom determines what element it is? What else does this property determine in a neutral atom?
13. Why are atoms of the same element not always identical? What do we call such atoms? What two ways can we use to identify them?
14. Compare and contrast mass number and atomic mass.
15. What do we call the lowest-energy electron arrangement of an atom? In the largest atoms, what is the maximum number of energy levels that electrons may occupy in this condition?
16. ✪ How many valence electrons does a neutral atom of hydrogen ($Z = 1$) have?
17. What makes an element radioactive? What are the products of radioactivity?
18. List the penetrating ability of each kind of nuclear radiation from greatest to least. Explain the properties of each that account for this characteristic.
19. DS ✪ How is alpha decay used in the operation of a typical household smoke detector? What problem could occur if a particle of an alpha-emitting material such as americium-241 were breathed into a person's lungs and lodged there?
20. Compare and contrast nuclear fission and fusion.

True or False

21. Scientific models don't need to be accurate in every detail; they just need to be workable for their intended purpose.
22. When the caloric theory of heat was overturned, Dalton's model of the atom was essentially destroyed.
23. Electrons in different elements are the same.
24. According to Bohr's model, electrons can exist only in specific orbits at certain distances from the nucleus.
25. The neutron is the heaviest main subatomic particle.
26. The mass number of an atom is the sum of the atomic number and the atomic mass.
27. Carbon-5 is a rare isotope of carbon.
28. Alpha particles are helium atoms.
29. Nuclear decays can change one element into another.
30. All kinds of nuclear radiation pose some risk of biological damage with direct exposure.

DS ✪ 31. Nuclear decays are dangerous and result in degeneration, and thus they are not useful in day-to-day life.

32. Nuclear fission normally occurs spontaneously, or it can be caused by the collision of a high-speed particle with a nucleus.
33. Commercial power generation using fusion energy is expected to become a reality within the next few years.

✪ 34. Write the following elements in isotopic notation.
   a. lithium-7
   b. lead-207
   c. iron-56
   d. carbon-14

✪ 35. Write the nuclear decay equations for the following reactions:
   a. beta decay of iodine-128 ($Z = 53$)
   b. alpha decay of gold-182 ($Z = 79$)
   c. gamma decay of thorium-215 ($Z = 90$)

DS ✪ 36. How should smoke detectors be installed and maintained to make sure that your family is protected?

✪ 37. Research and write a report on the problems with the assumptions used in radiometric dating of rocks and fossils.

✪ 38. Research and present a report on the differences between reactor plants using boiling water and pressurized water. Include in your report the advantages and disadvantages of each design.

ELEMENTS AND THE PERIODIC TABLE

iRocks.com phot

DOMINION SCIENCE PROBLEM

Fueling Pollution

No one likes traffic. People grumble about the late hour, the gridlock, the fumes, and the noise as they gaze out of closed car windows at smoggy urban skies. Did you know that driving a car is probably the most polluting thing the average American does? Car emissions contribute up to 60% of urban pollution in some areas of the world, and pollution causes health problems, like cancer and asthma, and environmental problems, like acid rain. But these fumes don't come just from the exhaust pipe. When a car sits or refuels, it releases toxic fumes as fuel evaporates. How can we minimize car emissions and thus urban pollution?

17-1 Vehicle emissions are one of the largest sources of air pollution in the United States.

17A A Brief History of the Elements

17.1 *Introduction*

Chapter 2 defines an element as any pure substance that is made of only one kind of atom. Almost every chapter in this textbook deals with elements, but this chapter provides a focused discussion, answering questions such as how elements are identified in nature, how chemists represent them in writing, and how elements relate to each other.

17A Objectives

After completing this section, you will be able to

- ✓ summarize the historical development of the idea of a chemical element.
- ✓ describe the role of alchemy in the discovery of chemical elements.
- ✓ explain how the historical problem of element symbols was solved.
- ✓ identify the naturally monatomic and diatomic elements.

17.2 *The Earliest Known Elements*

Since the earliest days of recorded history, certain substances were considered pure, distinct materials. The ancients knew of at least nine substances that we call elements today, seven of which are mentioned in the Bible (see the facet on the next page). All of them are solids that occur naturally in either pure form or in combination with other elements in a rocky substance called *ore*. Most metals are separated from their ores and worked by being melted.

Ancient Greeks were the first to define elements as the basic building blocks of matter. Many believed that all matter was composed of four "elements"—air, fire, earth, and water. They reasoned that nearly all substances have some properties of these "elements." This model was accepted for thousands of years.

17-2 The four classical elements

Human curiosity about nature motivated a study called **alchemy**, which arose many centuries before the birth of Jesus. Alchemy included aspects of chemistry, philosophy, religion, metallurgy, medicine, and art. It also acquired mystical and magical characteristics that often resulted in disreputable and even harmful activities. *Alchemists* are often viewed as dishonest, shifty, greedy, or deluded because the main objective for many of them was to find ways to turn valueless materials like lead into gold. However, many were serious

17-3 Alchemists often sought to convert base materials like lead into gold. They believed that this was possible according to Aristotle's ideas.

Solid elements, like gold and silver, that naturally occur in their pure form are called *native minerals*.

alchemy (AL kuh mee): (Gk. *chumeia*—to pour, as in pouring pharmaceutical juices and infusions). The *al-* prefix was added by Arabians during the Middle Ages.

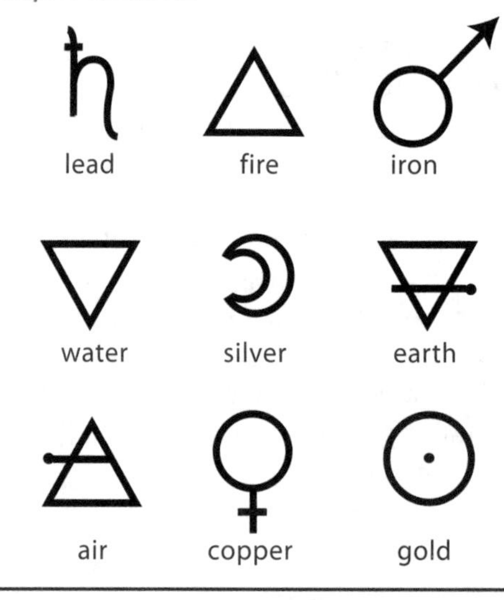

17-4 Some alchemy symbols for common substances and elements

about better understanding nature so that they could help their fellow humans. Alchemists discovered and purified many substances that were later understood to be chemical elements. Since alchemists believed in only the four classical elements, they did not recognize these as pure elements.

Alchemists developed a complex system of symbols to describe their work. The symbols for some common chemicals known to them are shown in Figure 17-4. These ancient symbols represented chemicals and often had mystical meanings. Different schools of alchemists developed their own symbols, often to hide their knowledge from others. This variation stifled the progress of studying elements and compounds because it made sharing information very difficult.

17.3 Studying Elements in the Seventeenth and Eighteenth Centuries

Alchemy was transformed into a real science in the seventeenth and eighteenth centuries by men like Isaac Newton, Robert Boyle, and Antoine Lavoisier. They believed that true scientific knowledge was

FACETS OF SCIENCE: ELEMENTS IN THE BIBLE

Seven of the 92 natural elements are mentioned in the Old Testament. Several are used to illustrate spiritual truths, and a few give evidence of the historical accuracy of the Old Testament.

Copper

The words *bronze* and *brass* appear many times in the Bible. They are often used figuratively to represent strength or hardness (Job 40:18) and judgment (Lev. 26:19).

Today, the term *brass* is used to indicate an alloy of copper and zinc, but Bible scholars believe that the brass mentioned in the Authorized Version of the Bible is simply copper or sometimes bronze, an alloy of copper and tin. Deuteronomy 8:7–9 makes an important reference to copper, stating that God's chosen people could mine it from the hills of the Promised Land. For a long time, no one could find any copper deposits in Israel, and many liberal scholars tried to use these verses to deny the inspiration of the Bible. However, in 2006 an international team of archaeologists excavated an ancient fort and nearby copper mines in western modern-day Jordan (30.6800°N, 35.4365°E). The place is called Khirbat en-Nahas (Arab.—*ruins of copper*). It consists of 100 buildings including living quarters and smelters on a 24-acre site. The researchers concluded that the radiocarbon age of the industrial-scale copper mining and smelting operation suggests that activities at this site spanned 1200–500 BC. This indicates that it existed during the time frame of the Israelite kings David and Solomon. While these mines probably were located within the boundaries of ancient Edom, Israel's bitter enemy, the existence of the nearby fort suggests that King Solomon may have built the fort to protect the extended frontier of Israel during his reign as well as this source of copper.

The Four Horses of St. Mark's Basilica (Venice, Italy), a Roman statue made of copper

Gold

Job 23:10 says, "When he hath tried me, I shall come forth as gold." Gold is a remarkable element. Because it is so beautiful and so easily worked, and because it does not easily tarnish or corrode, gold has been a symbol of value and riches since the beginning of history. Scripture often uses gold to compare with the believer's faith, which should be refined and is more precious than gold (1 Pet. 1:7).

Iron

The statue in King Nebuchadnezzar's dream had "legs of iron, his feet part of iron and part of clay" (Dan. 2:33). The iron is generally thought to represent power and dictatorship. (See also Job 40:18.) Joshua 17:16–18 mentions the Canaanites' chariots of iron, which is evidence for the high level of technology achieved by some ancient societies.

Lead

Jeremiah 6:29 illustrates the character of lead when it says, "The lead is consumed of the fire." Most people would agree that gained by observing, experimenting, and reasoning. Newton actually published more works on alchemy than on physics or optics.

Robert Boyle, best known for Boyle's law, published *The Sceptical Chymist* in 1661. In this important book he criticized scientists for poor experimental methods and for not sharing results. He was interested in identifying pure substances, especially the elements, and in distinguishing between compounds and mixtures. Boyle concluded that many substances could not be composed of the classical elements. His writings indicate that he favored defining an element as any substance that could not be decomposed into simpler substances.

Antoine Lavoisier also helped advance chemistry. He dealt a death blow to the classical idea of four elements when he broke down water into hydrogen and oxygen. He showed that oxygen was part of air. He also was the first to define an element as a simple chemical substance that could not be broken down into simpler substances by chemical analysis. He emphasized experimenting as part of chemistry. Lavoisier contributed much to science before he was executed in 1794 during the French Revolution.

17-5 Antoine Lavoisier and Marie-Anne, his devoted wife and chief laboratory assistant, were responsible for placing chemistry on the same scientific footing as physics in the eighteenth century.

Lead was used in ancient plumbing systems because it could be easily shaped into pipes.

lead is not an attractive metal. Because it corrodes easily, its silvery surface quickly becomes dull gray when exposed to air. Lead also lacks strength and so has only limited uses in building. Scripture often uses lead to symbolize the corruptible and temporary things of this world.

Silver

Silver is mentioned often in the Bible. It has been used as currency since Old Testament times. Abraham paid four hundred shekels (about 5 kg) of silver for the cave of Machpelah and the land around it (Gen. 23:16–18). This metal was also used to make trumpets, platters, and bowls. In Solomon's time it was used to make tables and candlesticks for the temple.

Proverbs 25:4–5 uses silver to illustrate a principle of government: "Take away the dross from the silver, and there shall come forth a vessel for the finer. Take away the wicked from before the king, and his throne shall be established in righteousness." Dross is the layer of impurities that rises to the top of molten metal. The metal is pure when the dross is removed; in the same way, a government will be strong and righteous if wicked men are removed from it. It is profitable for national leaders to surround themselves with those who seek the righteousness of Christ.

A silver Roman coin

Sulfur

Sulfur (translated "brimstone" in the Authorized Version) is a symbol of God's judgment (Gen. 19:24). He used a supernatural form of it to destroy the cities of Sodom and Gomorrah, and He will use it for judgment in the future. Brimstone is actually molten sulfur, which has a relatively low melting point of 115 °C. Many sulfur deposits can be found around the Dead Sea. Occasionally, these deposits catch fire, filling gullies with flaming liquid that produces choking smoke—an appropriate symbol for judgment. Brimstone is mentioned no fewer than six times in the book of Revelation.

Tin

Tin is mentioned several times in Scripture (Num. 31:22; Isa. 1:25; Ezek. 22:18, 20; 27:12). During Bible times the Phoenicians probably mined it in the British Isles. Some suggest that the name *Britain* comes from the Phoenician *barat-anac*, "land of tin." As mentioned previously, tin was often alloyed with copper to make bronze.

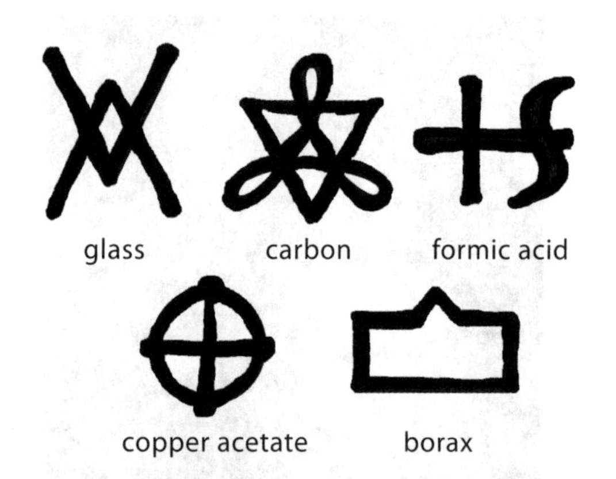

17-6 Some chemical symbols used by European chemists in the 1800s

Jöns Jakob Berzelius (1779–1848) was a prominent Swedish chemist and educator. He firmly established Dalton's atomic theory and introduced the modern system of element notation.

$$\mathrm{CuO} + \mathrm{S}\overset{3}{\mathrm{O}}$$

= copper sulfate ($CuSO_4$)

$$\dot{\mathrm{N}}\mathrm{H}^{6}$$

= ammonia (NH_3)

$$\dot{\mathrm{N}}$$

= nitrogen (N_2)

$$\overset{2}{\mathrm{C}}\overset{2}{\mathrm{H}}\overset{4}{\mathrm{O}}$$

= oxalic acid ($C_2H_2O_4$)

$$\mathrm{H\!\!\!-}$$

= water (H_2O)

17-7 Examples of Berzelius's notation. He often used superscripts rather than the subscripts used today to indicate the number of atoms in a chemical formula.

chlorine: chlor- (Gk. *chlōros*—green) + -ine (having the property of)

Recently discovered, very heavy, artificial elements are so short-lived that they are given temporary placeholder names until their discovery can be verified. These names come from the Latin or Greek words for their atomic numbers. For example, the name for element 113 is *ununtrium* and its symbol is *Uut*.

The International Union of Pure and Applied Chemistry (IUPAC), which is responsible for naming elements, has made good progress recently assigning permanent names to these heavy elements. Only elements 113, 115, 117, and 118 retain their temporary names.

By the end of the eighteenth century, 32 of the modern elements had been identified. However, their symbols were difficult to memorize and use. This problem would not be corrected until the next century.

17.4 Berzelius's Element Notation

Though John Dalton proposed his atomic theory in 1803 (see Chapter 16), it was not immediately accepted by many scientists for a number of reasons. One problem was that his system of symbols was just as difficult to use as the systems used by other chemists at the time. The great chemist Berzelius developed a much simpler and more logical symbolic system based on the first two letters of an element's Latin name.

Berzelius's system was quickly accepted by most European scientists. His system is essentially the same as the one used today. Berzelius also discovered the elements thorium, selenium, silicon, and cerium. When he died in 1848, there were 57 known elements.

17.5 Modern Elements and Their Symbols

Today 118 elements have been discovered or at least detected. The heaviest elements are all radioactive and have extremely short half-lives (see the Radioactive Dating Methods facet on page 374), so they're difficult to analyze before they decay to lighter elements. Scientists represent elements by their *chemical symbols*—Berzelius's notation consisting of one or two letters derived from the element's name.

The symbols of common elements are simply the first letter of their names. Oxygen's symbol is O, nitrogen's is N, and hydrogen's is H. Some elements' symbols are the first letter plus one other letter in the name. For example, silicon's symbol is Si, magnesium's is Mg, and platinum's is Pt. The first letter of a symbol is always capitalized, and the second is always lowercase. Many symbols come from an element's Latin name. Ag represents silver (*argentum*); Na, sodium (*natrium*); Fe, iron (*ferrum*); and Pb, lead (*plumbum*). Elements may also be named for famous scientists (einsteinium, Es), places (europium, Eu), heavenly bodies (helium, He), as well as colors (chlorine, Cl) or other properties of the element. Pages 392–93 show the periodic table of the elements, which lists all the elements, their names, and their symbols. You should begin to study the periodic table and to learn the symbols for the elements listed in Table 17-1.

17-1 Familiar Elements and Their Symbols

aluminum	Al	copper	Cu	mercury	Hg	radium	Ra
argon	Ar	fluorine	F	neon	Ne	radon	Rn
arsenic	As	gold	Au	nickel	Ni	silicon	Si
barium	Ba	helium	He	nitrogen	N	silver	Ag
bromine	Br	hydrogen	H	oxygen	O	sodium	Na
calcium	Ca	iodine	I	phosphorus	P	sulfur	S
carbon	C	iron	Fe	platinum	Pt	tin	Sn
chlorine	Cl	lead	Pb	plutonium	Pu	tungsten	W
chromium	Cr	magnesium	Mg	potassium	K	uranium	U

Of the 92 naturally occurring elements, all but 13 are solids in their pure forms at room temperature (25 °C). Mercury and bromine are liquids, and the remaining 11 elements are gases—hydrogen, helium, nitrogen, oxygen, fluorine, neon, chlorine, argon, krypton, xenon, and radon.

Most elements occur in nature combined in compounds, as masses of identical atoms, or in molecules of two or more atoms. Some elements, called *monatomic* elements, occur as single atoms. The only elements that exist naturally in this form are helium, neon, argon, krypton, xenon, and radon. These elements are often called the *noble* or *inert gases*. This special group is discussed in Section 17C.

Some elements naturally occur as molecules of two atoms. These *diatomic* elements include hydrogen, nitrogen, oxygen, fluorine, chlorine, bromine, and iodine. They always exist as diatomic molecules under standard conditions. You will need to memorize the seven diatomic elements for writing chemical equations in Chapter 19.

Some elements naturally exist as molecules of more than two atoms. For example, oxygen can occur as three-atom molecules called *ozone* (O_3) or four-atom molecules called *tetraoxygen* (O_4). Sulfur can occur as molecules of eight atoms in a ring (S_8) as well as rings of seven (S_7) or twelve atoms (S_{12}). Carbon usually occurs as an element in massive combinations of numerous atoms bonded together. Chemists have been able to coax carbon atoms to bond together into large net-like spherical or cylindrical molecules called *fullerenes*. They are named after Buckminster Fuller, who popularized the geodesic dome. The first of these specialized molecules was aptly called the *buckyball* (C_{60}).

You can see the incredible complexity and variation possible in matter God has created! There are many millions of compounds and other substances, both natural and artificial, that are composed of different combinations of relatively few (118) elements. As you learn more, you will see further evidences for the purposeful design of God's universe.

The metals gallium, rubidium, and cesium liquefy at temperatures just slightly above room temperature.

The Diatomic Elements

hydrogen	H_2	chlorine	Cl_2
nitrogen	N_2	bromine	Br_2
oxygen	O_2	iodine	I_2
fluorine	F_2		

The molecules of elements that contain more than two atoms are called *polyatomic molecules*.

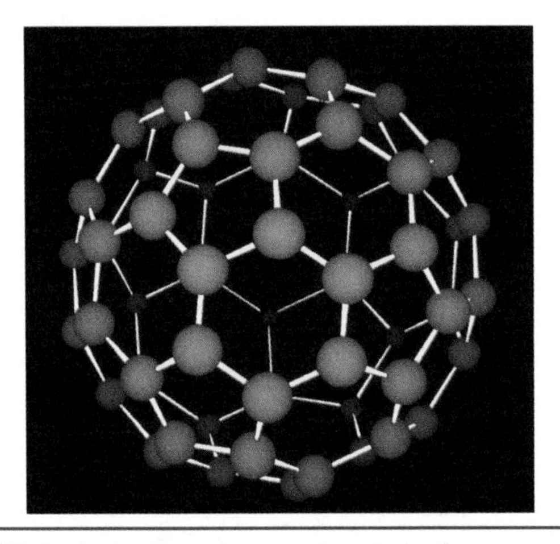

17-8 A single carbon molecule in the shape of a geodesic sphere, called a *buckyball*

17A Section Review

1. The Bible mentions several substances known today as chemical elements. Were these recognized as elements at that time? Explain.
2. What is alchemy? What did alchemists try to accomplish?
3. What did Robert Boyle contribute to the development of chemistry?
4. How did Antoine Lavoisier develop the modern definition of a chemical element?
5. What widespread problem in chemistry did Berzelius solve? How does his solution apply today?
6. List four sources from which elements' names are taken.
7. Name the elements that are liquid at or slightly above room temperature.
8. Which elements are naturally monatomic?
9. Describe the composition and arrangement of a buckyball.
10. (True or False) When alchemists believed that they could rearrange the ratios of the elements (air, earth, water, and fire) in lead to change it into gold, they were working within the scientific paradigm of their time.

17B Objectives

After completing this section, you will be able to

- ✓ summarize the early attempts to organize the elements.
- ✓ describe the organization and features of Mendeleev's periodic table.
- ✓ identify the problems with Mendeleev's periodic table and how they were solved.
- ✓ state the modern periodic law of the elements.

17B The Periodic Table

17.6 Organizing the Elements

How would you like to memorize all the physical and chemical properties of every element? If you were a student taking chemistry in the early nineteenth century, you would have had to do just that. Just like today, chemists and chemistry students needed to know what the properties were so that they could predict how the elements would behave.

Your list of properties would seem endless. Some elements would be gases, some would be liquids, and many would be solids. There would be metals, elements that were definitely *not* metals, and some with characteristics of both. You would have to remember which elements reacted explosively with water, which combined easily with other elements, and which ones failed to react at all. And your task would keep growing, because new elements were being discovered all the time. By 1860 scientists had identified more than 60 elements, but no one had been able to arrange them into a useful order.

Nineteenth-century scientists knew that they had to find a way to organize the elements. Sorting them into groups with similar properties would make them much easier to study. Eventually, after attempts by many scientists, the Russian chemist Dmitri Mendeleev finally succeeded. He arranged the elements into what would become the periodic table used today. The **periodic table of the elements** is one of the most useful tools of science. Knowing how it was developed will help you to see how it enables scientists (and students) to remember the chemical properties of elements.

17-9 Early chemists had many bad days before the invention of the periodic table!

17.7 The Discovery of Periodicity

In 1829 **Johann Döbereiner**, a chemist, found several small groups of elements with similar properties. When placed in order of increasing mass, the middle element often displayed properties that were midway between the other two. Since almost every group had three elements, Döbereiner called them *triads*. One contained the elements chlorine, bromine, and iodine. These are colored gases and all have similar chemical properties. As more elements were discovered, the groupings gained more members and became larger than triads. Although an interesting concept, it was not generally accepted by

Johann Wolfgang Döbereiner (1780–1849) was a German chemist who investigated the properties of chemically similar elements.

the scientific world. Döbereiner's idea was important because as new elements became better understood, chemists started to recognize patterns in properties. These discoveries gave rise to the idea of *periodicity*—the repetition of certain properties at regular intervals when elements were placed in order of atomic mass.

The first scientist to test the theory of periodicity was **A. E. Beguyer de Chancourtois**. In 1863 he graphed the known elements in order of increasing atomic mass on paper. He noticed that if his graph was rolled into a cylinder so that 16 mass units marked off its circumference, every eighth element often had similar physical and chemical properties to the elements above and below it (see Figure 17-10). Though de Chancourtois was not able to explain why this worked, his finding lent more support to the idea of periodicity.

A. E. Beguyer de Chancourtois (buh GEE ay · duh · SHAN coor TWAH) (1820–86) was a French geologist who held many governmental positions related to geology and mines.

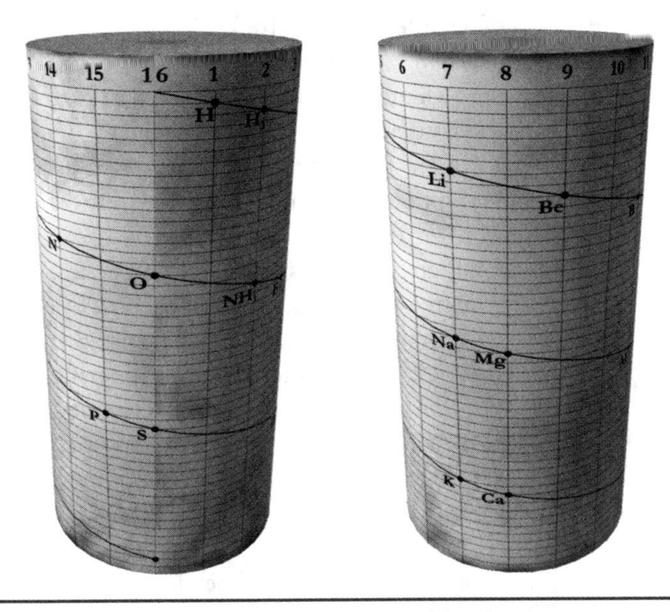

17-10 Two views of de Chancourtois's table after it was wrapped in a cylinder. His table showed that properties repeated with increasing atomic mass.

In 1866 a chemist named **John Newlands** proposed a similar organization of the elements. He arranged the elements in a table in order of their increasing atomic masses. When displayed in seven columns he noted that in most cases every eighth element had similar properties. He called these repeating patterns *octaves*—a term he took from his musical training. A musical octave is an interval that consists of seven notes and repeats as a harmonic (see Chapter 13) on the eighth note in a musical scale. In Newlands's octaves, the eighth element had properties similar to the first, the ninth to the second, and so forth. He placed 49 of the elements in seven rows of seven and called his system the *law of octaves*. This arrangement worked better than Döbereiner's had, but most scientists were still dissatisfied. In fact, Newlands's attempt to relate a physical property of matter to a musical concept earned him ridicule among scientists.

John Alexander Reina Newlands (1837–98) was an English chemist who illustrated the eight-element periodicity. He also predicted the discovery of the element germanium.

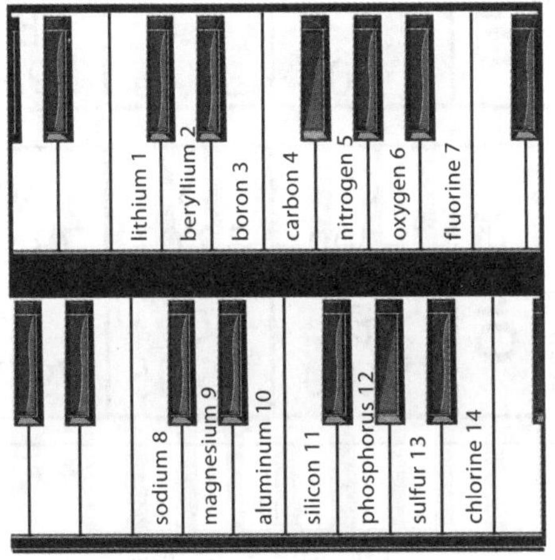

17-11 John Newlands organized the elements into octaves similar to the scales on a piano.

17-12 Dmitri Ivanovich Mendeleev (1834–1907) was the preeminent Russian chemist of the nineteenth century.

17.8 Mendeleev and His Periodic Law

Three years after John Newlands proposed his law of octaves, Dmitri Mendeleev proposed a similar system. He also organized elements by increasing atomic masses. However, Mendeleev did not believe that the properties of the elements always repeated every eight elements. He wrote all the elements and their properties on cards and filed them according to their atomic masses. Then he arranged the cards until he had grouped similar elements in rows. The rows were then arranged by increasing atomic mass. Mendeleev first placed hydrogen and lithium together. He then set hydrogen off to the side in its own column because it didn't really fit into a pattern. The second column held seven elements. The remaining six columns followed a similar pattern. Figure 17-13 is an early reproduction of Mendeleev's original.

Mendeleev's table had an important unique feature—blank spaces for undiscovered elements. As he grouped similar elements, Mendeleev came across elements that did not fit. He realized that if he left a blank space and skipped to the next row each time he found a "misfit," the element lined up correctly with other elements sharing its properties. He predicted that the gaps would be filled by elements yet to be discovered. He even listed their possible chemical and physical properties!

At first his predictions shocked a skeptical scientific world; however, researchers soon began to discover the elements he had predicted. His *periodic table* clearly shows the periodicity of the elements. Mendeleev called the principle that the properties of elements vary in a periodic or recurring pattern with their atomic masses the *periodic law*.

	Typische Elemente						
			K = 39	Rb = 85	Cs = 133	—	—
			Ca = 40	Sr = 87	Ba = 137	—	—
			—	?Yt = 88?	?Di = 138?	Er = 178?	—
			Ti = 48?	Zr = 90	Ce = 140?	?La = 180?	Th = 231
			V = 51	Nb = 94	—	Ta = 182	—
			Cr = 52	Mo = 96	—	W = 184	U = 240
			Mn = 55	—	—	—	—
			Fe = 56	Ru = 104	—	Os = 195?	—
			Co = 59	Rh = 104	—	Ir = 197	—
			Ni = 59	Pd = 106	—	Pt = 198?	—
H = 1	Li = 7	Na = 23	Cu = 63	Ag = 108	—	Au = 199?	—
	Be = 9,4	Mg = 24	Zn = 65	Cd = 112	—	Hg = 200	—
	B = 11	Al = 27,3	—	In = 113	—	Tl = 204	—
	C = 12	Si = 28	—	Sn = 118	—	Pb = 207	—
	N = 14	P = 31	As = 75	Sb = 122	—	Bi = 208	—
	O = 16	S = 32	Se = 78	Te = 125?	—	—	—
	F = 19	Cl = 35,5	Br = 80	J = 127	—	—	—

17-13 Mendeleev's original periodic table. Horizontal rows contained elements with similar chemical properties.

17.9 Revising the Periodic Table

Mass Order				
	O 16.00	S 32.07	Se 78.96	I 126.90
	F 19.00	Cl 35.45	Br 79.90	Te 127.60

Property Order				
	O 16.00	S 32.07	Se 78.96	Te 127.60
	F 19.00	Cl 35.45	Br 79.90	I 126.90

17-14 Iodine and tellurium seemed to fall in the wrong rows when arranged by atomic mass. Switching them put iodine in a row with similar elements.

Mendeleev's table had problems, however. Arranging the elements in order of increasing atomic mass did not always produce a table with similar elements next to each other. For example, iodine chemically resembled chlorine and bromine, and was therefore grouped with those elements. In this order, iodine comes after tellurium according to its chemical properties. However, iodine's atomic mass is less than tellurium's.

That the masses of some elements were not well known was another problem, causing their positions on the table to be only approximate. For example, the atomic masses of nickel and cobalt were so close that Mendeleev was unsure which to list first. Their properties were nearly identical as well.

Henry Moseley, one of Ernest Rutherford's assistants, suggested a solution. He discovered how to use x-rays to count the protons in an atom's nucleus to find an element's atomic number. He arranged the elements in order

FACETS OF SCIENCE

ADDING TO THE PERIODIC TABLE

In 1939 the periodic table showed only 92 elements. Uranium was the most complex element that could be found on the face of the earth, and it held the highest position on the table. No matter how carefully scientists analyzed mineral ores, they couldn't find any atoms that had more than 92 protons. Had they discovered the limit, or were higher elements possible? They thought that new elements might still be added and that the study of nuclear reactions would be the key.

Scientists knew that elements 84 through 92 were radioactive. It seemed that the "heavier" the atom, the more unstable its nucleus. They also knew that neutrons could break up to form a proton and an electron, or beta particle, in a process called *beta decay*. When this nuclear decay happened, the proton stayed in the nucleus and the electron was emitted. An extra proton in the nucleus would give the atom a higher atomic number and make it a higher element. What would happen if a neutron in a uranium atom split into a proton and an electron? The new proton would raise the atomic number to 93. It might be possible to make an entirely new element this way!

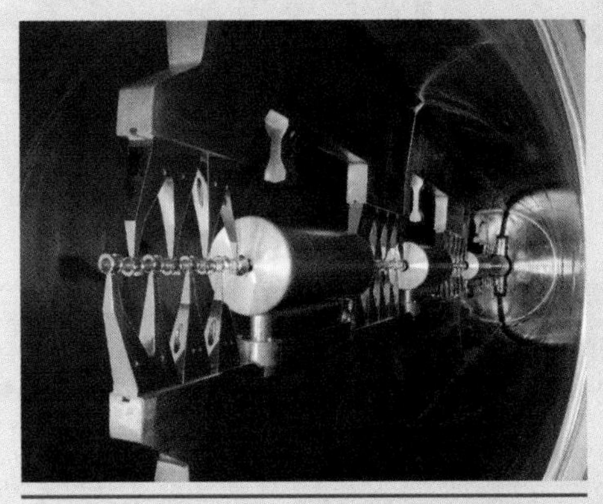

New elements are discovered using high-energy particle accelerators, such as this unit operated by the Heavy-Ion Research Center (GSI) near Darmstadt, Germany.

The idea of forming new elements fascinated scientists. But to make uranium emit a beta particle, they had to find a way to upset its nucleus. They could do that by adding extra neutrons to it. In 1940 scientists bombarded a uranium sample with neutrons and then painstakingly analyzed the resulting atoms. They found a new element! It had an atomic number of 93 and was named neptunium. Only a tiny number of neptunium atoms had been made, but they were a start. The next year, using the same technique, scientists were able to produce an even higher element, plutonium ($Z = 94$).

After the first two transuranic ("beyond uranium") elements had been made, nuclear "projectiles" bigger than neutrons had to be used. In 1944 alpha particles (helium ions) were fired at uranium, resulting in americium. When helium ions were aimed at samples of the new transuranic elements, even higher elements were made—curium (1944), berkelium (1949), and californium (1949).

Several of the transuranic elements were first produced in hydrogen-bomb tests.

Several years later, groups of scientists from the University of California at Berkeley and Argonne National Laboratories published sketchy reports about two new elements. At the end of the reports, appended notes said that not all the data about the newer elements could be printed. Why had some information been held back? What were the new elements like? How had they been made? These questions remained unanswered for a full two years.

In 1954 the reason for all the secrecy became obvious. The new elements had been formed inside the thermonuclear fireball of an experimental hydrogen bomb that was exploded in 1952. Until the secrecy that shrouded the test was lifted, only the barest details could be given to the public. It turned out that when tremendous streams of neutrons hit uranium during the explosion, seven to eight protons were added to the nucleus. The two elements were named einsteinium and fermium to honor two great physicists, Albert Einstein and Enrico Fermi.

The quest for new elements continues. Higher-numbered heavy elements are expensive because the small amounts that can be produced decay quickly.

The latest element to be manufactured has an atomic number of 118. The temporary IUPAC name for this element is *ununoctium*. Just three atoms of element 118 were reportedly formed from the fusion of californium-249 and calcium-48. However, they decayed to lighter elements in less than a millisecond. Some physicists believe that superheavy atoms with mass numbers of around 300 may be very stable. It remains to be seen whether their production will be possible.

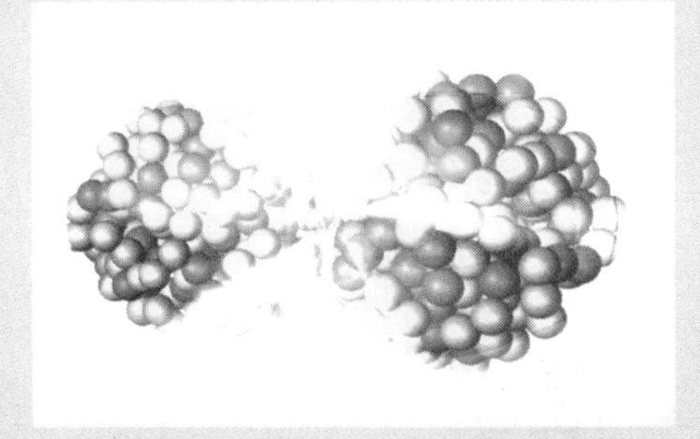

Element 118 was detected indirectly by observing the formation of its predicted decay products.

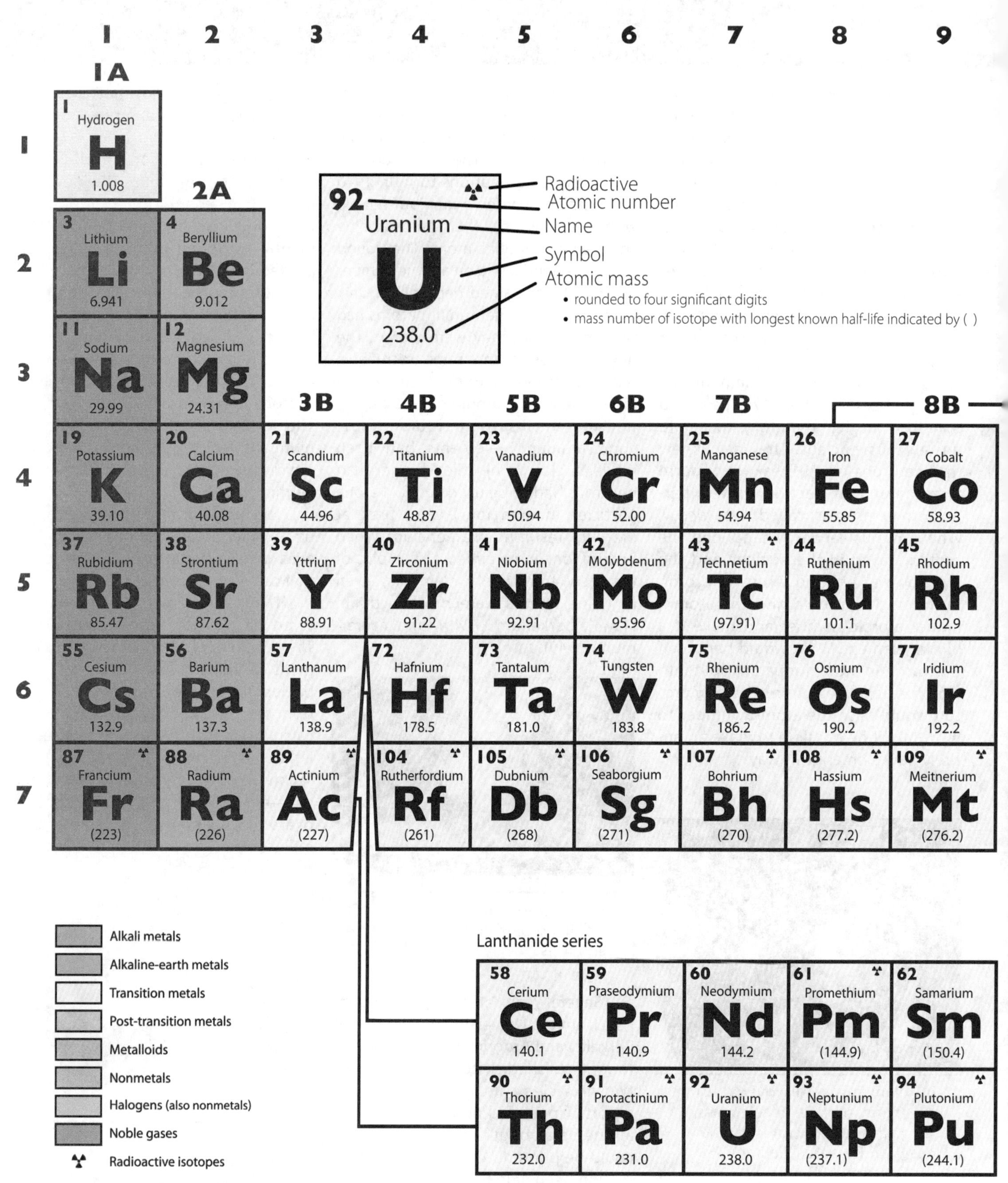
1 2 3 4 5 6 7 8 9
1A
2A
3B 4B 5B 6B 7B 8B
1 2 3 4 5 6 7
92 Uranium U 238.0
Radioactive
Atomic number
Name
Symbol
Atomic mass
• rounded to four significant digits
• mass number of isotope with longest known half-life indicated by ( )
1 Hydrogen H 1.008
3 Lithium Li 6.941
4 Beryllium Be 9.012
11 Sodium Na 29.99
12 Magnesium Mg 24.31
19 Potassium K 39.10
20 Calcium Ca 40.08
21 Scandium Sc 44.96
22 Titanium Ti 48.87
23 Vanadium V 50.94
24 Chromium Cr 52.00
25 Manganese Mn 54.94
26 Iron Fe 55.85
27 Cobalt Co 58.93
37 Rubidium Rb 85.47
38 Strontium Sr 87.62
39 Yttrium Y 88.91
40 Zirconium Zr 91.22
41 Niobium Nb 92.91
42 Molybdenum Mo 95.96
43 Technetium Tc (97.91)
44 Ruthenium Ru 101.1
45 Rhodium Rh 102.9
55 Cesium Cs 132.9
56 Barium Ba 137.3
57 Lanthanum La 138.9
72 Hafnium Hf 178.5
73 Tantalum Ta 181.0
74 Tungsten W 183.8
75 Rhenium Re 186.2
76 Osmium Os 190.2
77 Iridium Ir 192.2
87 Francium Fr (223)
88 Radium Ra (226)
89 Actinium Ac (227)
104 Rutherfordium Rf (261)
105 Dubnium Db (268)
106 Seaborgium Sg (271)
107 Bohrium Bh (270)
108 Hassium Hs (277.2)
109 Meitnerium Mt (276.2)
Lanthanide series
58 Cerium Ce 140.1
59 Praseodymium Pr 140.9
60 Neodymium Nd 144.2
61 Promethium Pm (144.9)
62 Samarium Sm (150.4)
90 Thorium Th 232.0
91 Protactinium Pa 231.0
92 Uranium U 238.0
93 Neptunium Np (237.1)
94 Plutonium Pu (244.1)
Actinide series
Alkali metals
Alkaline-earth metals
Transition metals
Post-transition metals
Metalloids
Nonmetals
Halogens (also nonmetals)
Noble gases
Radioactive isotopes

10	11	12	13	14	15	16	17	18
								8A
								2 Helium **He** 4.003
			3A	4A	5A	6A	7A	
			5 Boron **B** 10.81	6 Carbon **C** 12.01	7 Nitrogen **N** 14.01	8 Oxygen **O** 16.00	9 Fluorine **F** 19.00	10 Neon **Ne** 20.18
	1B	2B	13 Aluminum **Al** 26.98	14 Silicon **Si** 28.09	15 Phosphorus **P** 30.97	16 Sulfur **S** 32.07	17 Chlorine **Cl** 35.45	18 Argon **Ar** 39.95
28 Nickel **Ni** 58.69	29 Copper **Cu** 63.55	30 Zinc **Zn** 65.38	31 Gallium **Ga** 69.72	32 Germanium **Ge** 72.63	33 Arsenic **As** 74.92	34 Selenium **Se** 78.96	35 Bromine **Br** 79.90	36 Krypton **Kr** 83.80
46 Palladium **Pd** 106.4	47 Silver **Ag** 107.9	48 Cadmium **Cd** 112.4	49 Indium **In** 114.8	50 Tin **Sn** 118.7	51 Antimony **Sb** 121.8	52 Tellurium **Te** 127.6	53 Iodine **I** 126.9	54 Xenon **Xe** 131.3
78 Platinum **Pt** 195.1	79 Gold **Au** 197.0	80 Mercury **Hg** 200.6	81 Thallium **Tl** 204.4	82 Lead **Pb** 207.2	83 Bismuth **Bi** 209.0	84 Polonium **Po** (209)	85 Astatine **At** (210.0)	86 Radon **Rn** (222.0)
110 Darmstadtium **Ds** (281.2)	111 Roentgenium **Rg** (280.2)	112 Copernicium **Cn** (285.2)	113 Ununtrium **Uut** (284.2)	114 Flerovium **Fl** (289.2)	115 Ununpentium **Uup** (288.2)	116 Livermorium **Lv** (293)	117 Ununseptium **Uus** (294)	118 Ununoctium **Uuo** (294)

The names given to elements 113, 115, 117 and 118 represent the Latin and Greek names for their arabic numbers.

63 Europium **Eu** 152.0	64 Gadolinium **Gd** 157.3	65 Terbium **Tb** 158.9	66 Dysprosium **Dy** 162.5	67 Holmium **Ho** 164.9	68 Erbium **Er** 167.3	69 Thulium **Tm** 168.9	70 Ytterbium **Yb** 173.0	71 Lutetium **Lu** 175.0
95 Americium **Am** (243.1)	96 Curium **Cm** (247.1)	97 Berkelium **Bk** (247)	98 Californium **Cf** (251.1)	99 Einsteinium **Es** (252.1)	100 Fermium **Fm** (257.1)	101 Mendelevium **Md** (258.1)	102 Nobelium **No** (259.1)	103 Lawrencium **Lr** (262)

17-15 Henry Gwyn Jeffreys Moseley (1887–1915), a young, brilliant English physicist, discovered a method to determine the atomic numbers of the elements. This led to the establishment of the modern periodic table. He died while fighting in World War I.

of increasing atomic number, making the problems in Mendeleev's table disappear. His work in 1912 led to a revision of the **periodic law of the elements**. In its modern form it states that *the properties of the elements vary with their atomic numbers in a periodic way.*

Though a very useful tool, the periodic table is constantly changing. For instance, Mendeleev's table (Figure 17-13) was rotated to appear as the periodic table shown on pages 392–93. Atomic numbers increase from left to right in rows. Columns represent elements having similar properties. This format improves the table's readability and usefulness. The periodic table also changes as new elements are discovered. Every year, the atomic masses of the elements are updated with the latest measurements, so you may find slightly different values or even different numbers of elements in other tables.

Why do chemical properties vary with atomic number? Remember that the number of electrons in a neutral atom equals its atomic number. Thus, a neutral atom with nineteen protons ($Z = 19$) must have nineteen electrons. The electrons in an atom's highest energy level—its valence electrons—affect how it bonds with other atoms. In Section 17C you will begin to see how valence electrons determine the periodic properties of elements.

17B Section Review

1. Why was it especially difficult to study chemistry in the early 1800s?
2. Although Döbereiner's and Newlands's organizations of the elements were not very useful, what did they show about elements?
3. State Mendeleev's periodic law.
4. What feature of Mendeleev's periodic table prompted new discoveries?
5. What problems did Mendeleev's periodic table have? How were these solved?
6. State the modern periodic law of the elements. How has the periodic table changed since Mendeleev proposed it?
7. Why does an element's atomic number affect its properties?
8. (True or False) An element's atomic mass usually equals its atomic number.

17C Objectives

After completing this section, you will be able to

- ✓ describe the properties of metals and nonmetals.
- ✓ compare metalloids with metals and nonmetals.
- ✓ define a family or group of elements.
- ✓ name the major element families and state how many valence electrons a neutral atom of each element has.
- ✓ define a chemical period and state the significance of the period number.

17C Classes of Elements

17.10 Introduction

Elements can be categorized into three broad groupings: metals, those that are not metals, and those that are somewhere in between. Elements in each broad grouping share a general set of properties and occupy a certain location in the periodic table.

17.11 Metals

Almost three-fourths of the elements are **metals**. We use metals in automobiles, bridges, cooking utensils, and home appliances. Our modern way of life depends on metals and their properties. Because

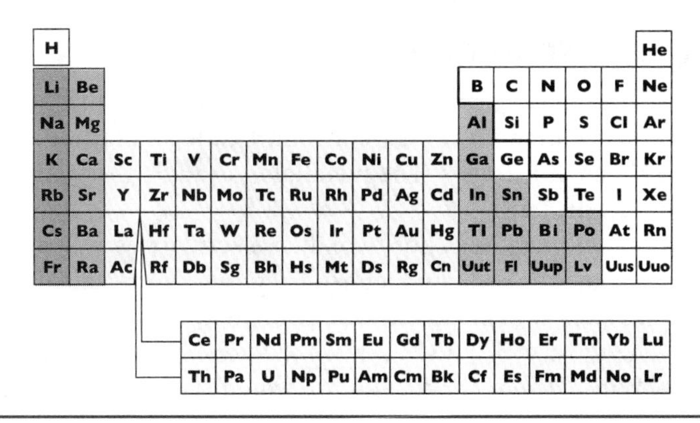

17-16 Most elements are metals.

most of the metals we use are hard, strong, and dense, many people think that *all* metals have these qualities. Yet many pure metals are so soft that they can be cut with a knife! Some are very weak. Some are lightweight. In fact, many metals in their pure state do not have physical properties anything like the metals we use every day.

Metals have common properties that give clues to their electron structure. They belong to families with relatively few valence electrons. These valence electrons are not held very strongly and are easily removed. It is because of these "loose" electrons that metals have a metallic luster, are malleable and ductile, and conduct electricity and heat.

Metals reside in the two-thirds of the periodic table left of the "stairstep" dividing line. They include the two rows below the main part of the table. If these rows were placed in the table in atomic-number order, it would be quite wide (see Figure 17-17). The table is usually displayed as in this textbook to save space.

Common Characteristics of Metals

1. Clean surfaces of most metals have a silvery or metallic *luster*.
2. All metals except mercury (Hg) are solids at room temperature.
3. Most metals are *malleable*—they can be rolled or hammered into a shape.
4. Most metals are *ductile*—they can be drawn into wire.
5. Most metals are good conductors of electricity and heat.
6. Metals tend to be reactive, easily forming bonds.

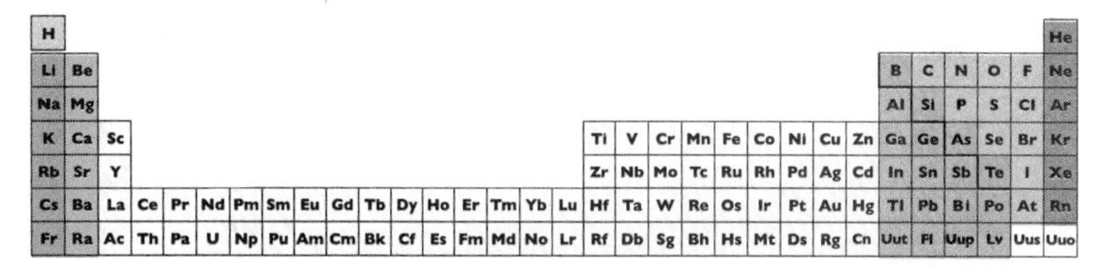

17-17 This is the way the periodic table would look if the two separate rows of metal elements were placed in the table in numerical order.

17.12 Nonmetals

Hydrogen ($Z = 1$) and the elements above and to the right of the stairstep on the periodic table are called **nonmetals**. These elements hold their electrons tightly, so their properties are very different from metals. Compare their properties listed in the characteristics box with those of the metals. The properties of nonmetals vary much more than those of metals because there is a greater variety in electron structures among these elements. Most metals have one or two valence electrons, but nonmetals can have anywhere from one to eight valence electrons.

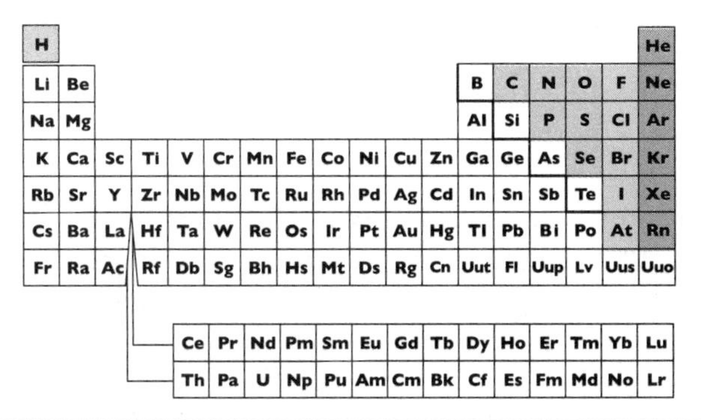

17-18 Nonmetals on the periodic table

Common Characteristics of Nonmetals

1. Nonmetals exist as solids, liquids, and gases at room temperature; most are gases.
2. Solid nonmetals exist as brittle crystals that shatter easily.
3. Nonmetals are poor conductors of electricity and heat.
4. Nonmetals have a variety of colors.
5. Six nonmetals called the *noble* or *inert gases* do not readily react with other elements.
6. Nonmetals other than the noble gases generally form bonds by sharing or taking electrons from other atoms.

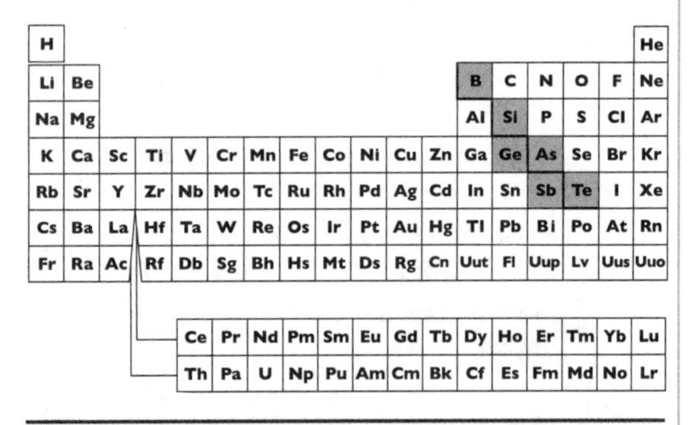

17-19 Metalloids on the periodic table

17.13 Metalloids

The third type of element has properties somewhere in between a metal and a nonmetal. These elements are called **metalloids** or *semimetals* ("almost metals"). The metalloids straddle the stairstep line on the periodic table.

Although metalloids don't conduct electricity as well as metals, they are poor-to-good conductors. They can be manufactured into *semiconductors*, which can conduct electricity under certain conditions. The most widely used semiconductors are made primarily of silicon. Without semiconductors, we wouldn't have computers, portable music players, calculators, cell phones, or a host of other items using digital technology.

17.14 Element Families

In general, neutral elements in the same *column* of the periodic table have the same number of valence electrons. For example, oxygen ($Z = 8$) has eight electrons. It has two electrons in the first energy level and six in the valence energy level. Sulfur ($Z = 16$), located just below oxygen in the periodic table, has 16 electrons. Its electron configuration is two electrons in the first level, eight in the second, and six in the valence energy level. Both oxygen and sulfur have six valence electrons. Both are located in the column labeled 6A (or 16). Just as brothers and sisters in a family have similarities, elements with the same number of valence electrons often share similar physical and chemical properties. Because of this, each column on the periodic table is called a **family** or *group*.

While the periodic table is good at predicting most properties of atoms, there are exceptions. The principles in this section are generally true.

At the head of each group on the periodic table is a number-letter combination: 1A, 2A, and so on. This is the way periodic tables published in the United States identify the element groups. The elements in the B groups have very similar properties distinct from all the other elements.

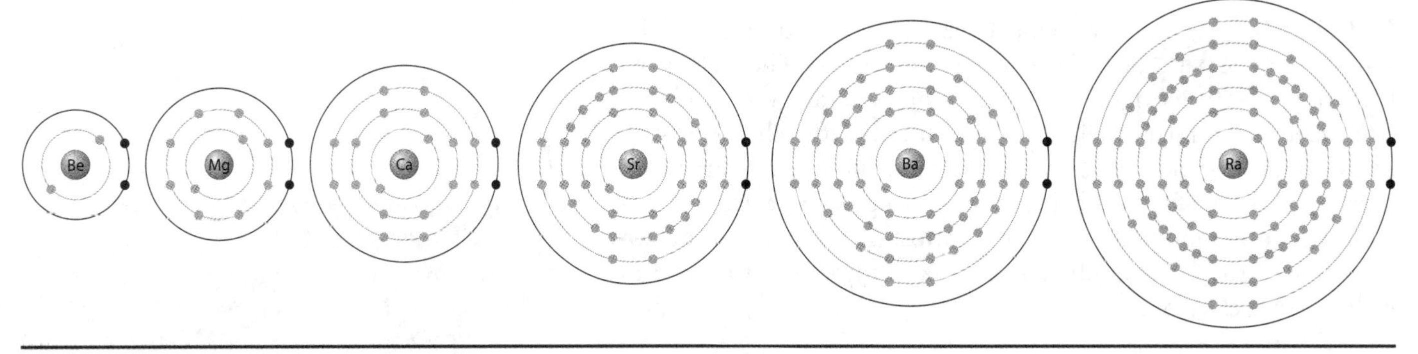

17-20 All elements in a family (2A in this case) have the same number of valence electrons. The Bohr models show this pattern. (The horizontal arrangement of this figure is to save space on the page. The family is arranged vertically in the periodic table.)

Above the number-letter group identifiers is a row of numbers from 1 to 18. Each of these numbers also corresponds to a chemical group. This method is used in countries other than the United States and is preferred by the International Union of Pure and Applied Chemistry (IUPAC). This textbook refers to element groups by the number-letter designation but includes the IUPAC group number in parentheses. You should be familiar with both systems.

The Alkali Metals (Group 1A)

Group 1A (or Group 1) contains six elements: lithium (Li), sodium (Na), potassium (K), rubidium (Rb), cesium (Cs), and francium (Fr). This family is called the **alkali metals**. Each of them has one electron in its outermost energy level. Notice that even though hydrogen, a nonmetal, appears in the first column, it is set off by itself.

alkali (AL kuh LIE): (Ar. *al-qaliy*—the ash of certain plants obtained by roasting at a high temperature)

The single valence electron of the alkali metals is removed easily, making these metals very reactive. If you drop a small piece of sodium into water, it reacts violently, producing heat and flammable hydrogen gas. No alkali metal has ever been discovered as a pure native mineral in nature. In fact, alkali metals are so reactive that they will combine with oxygen in the air if they are not stored in a container of oil (such as kerosene).

Two alkali metals, sodium and potassium, are an essential part of our diet. They play a part in the movement of bodily fluids, the transmission of nerve impulses, and muscle control. They're also common metals in the minerals of the earth's crust and form a large part of the dissolved solids in the ocean.

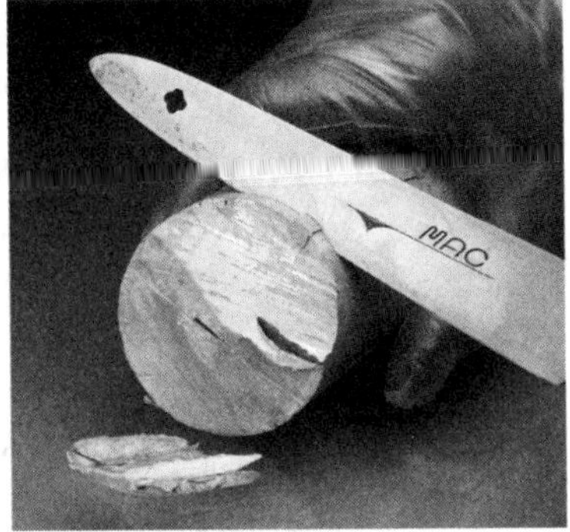

17-22 Sodium is so soft that it can be cut with a knife.

H																	He
Li	Be											B	C	N	O	F	Ne
Na	Mg											Al	Si	P	S	Cl	Ar
K	Ca	Sc	Ti	V	Cr	Mn	Fe	Co	Ni	Cu	Zn	Ga	Ge	As	Se	Br	Kr
Rb	Sr	Y	Zr	Nb	Mo	Tc	Ru	Rh	Pd	Ag	Cd	In	Sn	Sb	Te	I	Xe
Cs	Ba	La	Hf	Ta	W	Re	Os	Ir	Pt	Au	Hg	Tl	Pb	Bi	Po	At	Rn
Fr	Ra	Ac	Rf	Db	Sg	Bh	Hs	Mt	Ds	Rg	Cn	Uut	Fl	Uup	Lv	Uus	Uuo

Ce	Pr	Nd	Pm	Sm	Eu	Gd	Tb	Dy	Ho	[illegible]	[illegible]	[illegible]	[illegible]
Th	Pa	U	Np	Pu	Am	Cm	Bk	Cf	Es	Fm	Md	No	Lr

17-21 The alkali metals

The Alkaline-Earth Metals (Group 2A)

Beryllium (Be), magnesium (Mg), calcium (Ca), strontium (Sr), barium (Ba), and radium (Ra) make up Group 2A (2), the **alkaline-earth metals**. They are "close cousins" of the alkali metals, but they each have two valence electrons. This extra electron makes them slightly less reactive than the alkali metals but not stable enough to remain free (uncombined) in nature. Instead, they are found in many common minerals.

Alkaline-earth metals have many uses. Beryllium is used to harden *alloys* (metal mixtures). A copper alloy that contains 2% beryllium is six times as strong as pure copper. Because magnesium is so light, its alloys are used to construct airplanes. Calcium compounds strengthen substances like teeth, bones, and modern construction materials such as concrete, mortar, and drywall board. Strontium compounds give fireworks and flares their brilliant crimson color. Radium compounds—all of which are radioactive—were formerly

H																	He
Li	Be											B	C	N	O	F	Ne
Na	Mg											Al	Si	P	S	Cl	Ar
K	Ca	Sc	Ti	V	Cr	Mn	Fe	Co	Ni	Cu	Zn	Ga	Ge	As	Se	Br	Kr
Rb	Sr	Y	Zr	Nb	Mo	Tc	Ru	Rh	Pd	Ag	Cd	In	Sn	Sb	Te	I	Xe
Cs	Ba	La	Hf	Ta	W	Re	Os	Ir	Pt	Au	Hg	Tl	Pb	Bi	Po	At	Rn
Fr	Ra	Ac	Rf	Db	Sg	Bh	Hs	Mt	Ds	Rg	Cn	Uut	Fl	Uup	Lv	Uus	Uuo

Ce	Pr	Nd	Pm	Sm	Eu	Gd	Tb	Dy	Ho	Er	Tm	Yb	Lu
Th	Pa	U	Np	Pu	Am	Cm	Bk	Cf	Es	Fm	Md	No	Lr

17-23 The alkaline-earth metals

17-24 The skeleton of a passenger plane includes magnesium alloys.

used for glow-in-the-dark watches and clocks. (Such instruments now use LEDs or nonradioactive materials for illumination.)

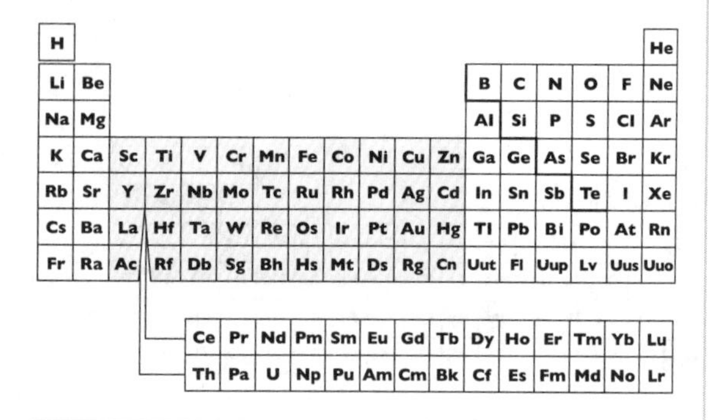

17-25 The transition elements

The Transition Elements (Groups 1B–8B)

Families in the middle of the periodic table, Groups 1B–8B (3–12), are sometimes called *subgroups.* They all have similar chemical and physical properties. Nearly all have either one or two valence electrons. These elements, all metals, are called the **transition elements**.

Very few of these metals are used in their pure forms; they're normally combined in alloys. Sterling silver is an alloy of copper and silver. The gold in jewelry is an alloy of gold, silver, and copper. Most steel is a simple alloy of iron and carbon called carbon steel. Exceptions include pure copper used in electrical wiring, mercury used in electrical switches, and platinum used as a catalyst (see Chapter 19).

The transition elements contain an unfilled electron energy level below the highest energy level. With each higher atomic number within the transition elements, an additional electron is added to this inner, unfilled energy level rather than to the outermost energy level.

The Boron Family (Group 3A)

Group 3A (13) is somewhat unusual because it contains boron (B), which is a metalloid; but the elements aluminum (Al), gallium (Ga), indium (In), and thallium (Tl) are all metals. These elements have three valence electrons.

Aluminum is the most abundant metal in the earth's crust. It is used for a variety of products, from pots and pans to engine blocks. Gallium is a liquid metal when warmed to just above room temperature. It's used in semiconductors and in thermometers requiring a wide temperature range. Indium is used in semiconductors, liquid crystal displays, and alloys with low melting points. It also makes brilliant mirrors when painted on glass. Thallium is very soft and can be cut with a knife. It is toxic, possibly causing cancer. Its use is limited to glass mixtures and alloys with special properties.

H He
Li Be B C N O F Ne
Na Mg Al Si P S Cl Ar
K Ca Sc Ti V Cr Mn Fe Co Ni Cu Zn Ga Ge As Se Br Kr
Rb Sr Y Zr Nb Mo Tc Ru Rh Pd Ag Cd In Sn Sb Te I Xe
Cs Ba La Hf Ta W Re Os Ir Pt Au Hg Tl Pb Bi Po At Rn
Fr Ra Ac Rf Db Sg Bh Hs Mt Ds Rg Cn Uut Fl Uup Lv Uus Uuo
Ce Pr Nd Pm Sm Eu Gd Tb Dy Ho Er Tm Yb Lu
Th Pa U Np Pu Am Cm Bk Cf Es Fm Md No Lr

17-26 The boron family

17-27 Aluminum is almost indispensable in today's society.

The Carbon Family (Group 4A)

Carbon (C), the first element in Group 4A (14), gives the **carbon family** its name. Though carbon is a nonmetal, this group includes the metalloids silicon (Si) and germanium (Ge) and the soft metals tin (Sn) and lead (Pb). All have four valence electrons.

Carbon and silicon form the bases for most known compounds. Carbon is part of almost all *organic compounds*, molecules associated with life. Most rocks of the earth's surface contain compounds of silicon. Germanium is an important metalloid used in semiconductors. Lead is used extensively in lead-acid batteries. It is excellent for shielding against certain kinds of ionizing radiation. Lead-lined aprons are often used to minimize a patient's exposure to radiation during dental x-rays.

17-29 Silicon and germanium are used in integrated circuits, such as this quad-core CPU.

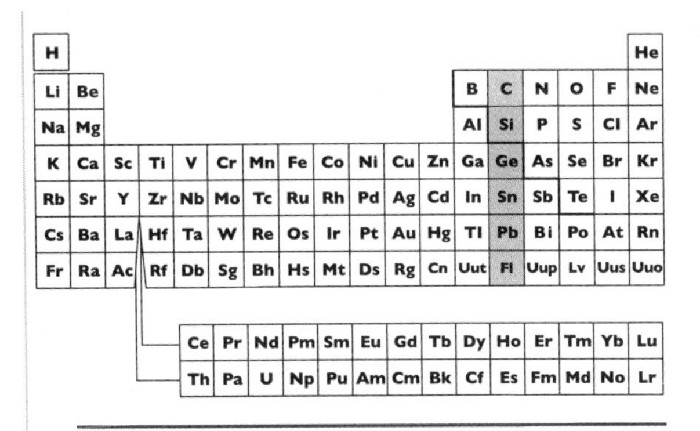

17-28 The carbon family

nitrogen: nitro- (Gk. *nitron*—sodium carbonate) + -gen (Gk. *genēs*—born, forming)

The Nitrogen Family (Group 5A)

Group 5A (15) is called the nitrogen **family** because nitrogen (N) is its first element. Phosphorus (P), arsenic (As), antimony (Sb), and bismuth (Bi) follow. Because these elements can share all five of their valence electrons, the nitrogen family, as a group, can form more chemical bonds per atom than any other family.

Air is nearly 79% diatomic nitrogen (N_2). Both nitrogen and phosphorus are necessary for plant growth. These nonmetals are also found in DNA. Arsenic, a metalloid, forms many poisonous compounds. Antimony is used in semiconducting materials that make effective heat detectors. Bismuth forms alloys with low melting points. Some of its compounds are used in cosmetics.

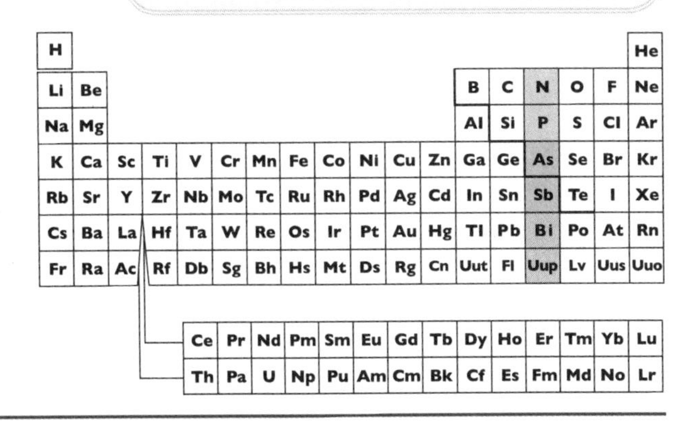

17-30 The nitrogen family

oxygen: oxy- (Gk. *oxus*—acid) + -gen

The Oxygen Family (Group 6A)

Group 6A (16) is called the oxygen **family**. It includes oxygen (O), sulfur (S), selenium (Se), tellurium (Te), and polonium (Po). Each element has six valence electrons.

Oxygen, in its diatomic form (O_2), is essential to most life on Earth. The ozone layer, a triatomic form of oxygen (O_3), protects life on the earth's surface by absorbing UV rays but is otherwise toxic. Sulfur, an important industrial nonmetal, forms many compounds used in manufacturing. Selenium is a nonmetal used in solar cells and semiconductors. Tellurium, though toxic, is useful in many alloys and semiconductors. Polonium is a radioactive metal used as a power source for deep-space probes. It emits alpha particles and is dangerous to handle without proper protective equipment.

H He
Li Be B C N O F Ne
Na Mg Al Si P S Cl Ar
K Ca Sc Ti V Cr Mn Fe Co Ni Cu Zn Ga Ge As Se Br Kr
Rb Sr Y Zr Nb Mo Tc Ru Rh Pd Ag Cd In Sn Sb Te I Xe
Cs Ba La Hf Ta W Re Os Ir Pt Au Hg Tl Pb Bi Po At Rn
Fr Ra Ac Rf Db Sg Bh Hs Mt Ds Rg Cn Uut Fl Uup Lv Uus Uuo
Ce Pr Nd Pm Sm Eu Gd Tb Dy Ho Er Tm Yb Lu
Th Pa U Np Pu Am Cm Bk Cf Es Fm Md No Lr

17-31 The oxygen family

The Halogens (Group 7A)

Elements in the **halogen family** (Group 7A or 17) have seven valence electrons. The halogens are very reactive because they lack only one electron to have a complete valence octet. Group 7A contains fluorine (F), chlorine (Cl), bromine (Br), iodine (I), and astatine (At). With the possible exception of astatine, they exist in their pure form only as diatomic molecules. They are found in many salt compounds, especially those in seawater.

As their atomic numbers increase, the halogens' colors darken and densities increase. Fluorine is a light yellow gas. Chlorine is a greenish yellow gas. Bromine is a dark-reddish liquid. Iodine is a purplish gray solid.

As pure elements, the halogens are biologically hazardous, but they do have many uses. For example, chlorine, a poisonous gas, is dissolved in water to kill bacteria. Iodine is used for its antibacterial properties to clean skin injuries. An iodine solution is often used to sterilize a patient's skin before surgery. Astatine is a radioactive halogen with a half-life of just a few hours. Its properties are not well known.

halogen (HAL uh jun): halo- (Gk. *hals*—salt) + -gen; thus, the "salt formers"

H																	He
Li	Be											B	C	N	O	F	Ne
Na	Mg											Al	Si	P	S	Cl	Ar
K	Ca	Sc	Ti	V	Cr	Mn	Fe	Co	Ni	Cu	Zn	Ga	Ge	As	Se	Br	Kr
Rb	Sr	Y	Zr	Nb	Mo	Tc	Ru	Rh	Pd	Ag	Cd	In	Sn	Sb	Te	I	Xe
Cs	Ba	La	Hf	Ta	W	Re	Os	Ir	Pt	Au	Hg	Tl	Pb	Bi	Po	At	Rn
Fr	Ra	Ac	Rf	Db	Sg	Bh	Hs	Mt	Ds	Rg	Cn	Uut	Fl	Uup	Lv	Uus	Uuo

Ce	Pr	Nd	Pm	Sm	Eu	Gd	Tb	Dy	Ho	Er	Tm	Yb	Lu
Th	Pa	U	Np	Pu	Am	Cm	Bk	Cf	Es	Fm	Md	No	Lr

17-32 The halogens

Astatine has a half-life of only 8.1 hours. Only very tiny amounts of it exist in uranium deposits in nature.

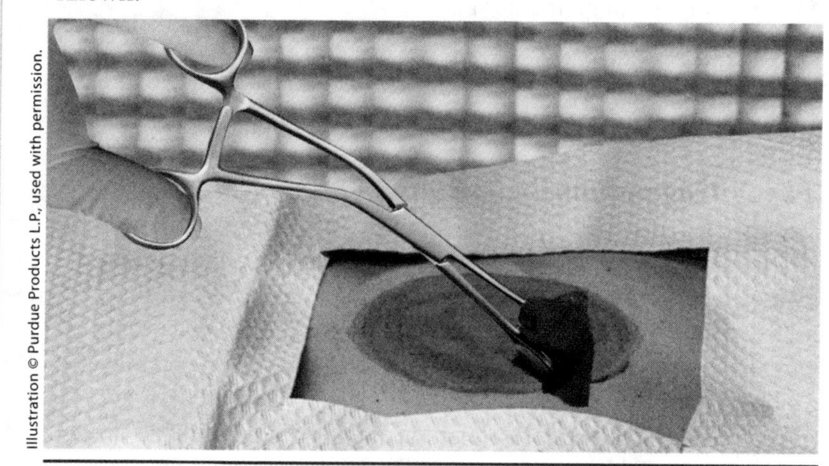

Illustration © Purdue Products L.P., used with permission.

17-33 Surgeons often prepare a surgical site with an iodine solution.

The Noble Gases (Group 8A)

The elements in the last family in the periodic table, Group 8A (18), have eight valence electrons except for helium, which has two. These elements are very stable because they have a complete set of valence electrons. Since they don't readily bond with other atoms, they exist in nature as pure monatomic gases. In fact, they bond with other elements only under high temperatures and pressures. This "snobbish" tendency to avoid joining "lesser" elements has been compared to the behavior of the nobility. This comparison led British scientists to call them the **noble gases**. Helium (He), neon (Ne), argon (Ar), krypton (Kr), xenon (Xe), and radon (Rn) are all present in very small amounts in our air. Argon, making up about 1% of the atmosphere, is the most abundant. Even though helium is the second-most abundant element in the universe (it makes up a large part of stars), there is very little of it or any other noble gas present on Earth.

Noble gases have many uses because of their lack of chemical activity. Helium, a nonflammable gas, is used in balloons and blimps. When liquefied, helium is used for low-temperature cooling applications, such as the magnets in MRI machines and particle accelerators.

H																	He
Li	Be											B	C	N	O	F	Ne
Na	Mg											Al	Si	P	S	Cl	Ar
K	Ca	Sc	Ti	V	Cr	Mn	Fe	Co	Ni	Cu	Zn	Ga	Ge	As	Se	Br	Kr
Rb	Sr	Y	Zr	Nb	Mo	Tc	Ru	Rh	Pd	Ag	Cd	In	Sn	Sb	Te	I	Xe
Cs	Ba	La	Hf	Ta	W	Re	Os	Ir	Pt	Au	Hg	Tl	Pb	Bi	Po	At	Rn
Fr	Ra	Ac	Rf	Db	Sg	Bh	Hs	Mt	Ds	Rg	Cn	Uut	Fl	Uup	Lv	Uus	Uuo

Ce	Pr	Nd	Pm	Sm	Eu	Gd	Tb	Dy	Ho	Er	Tm	Yb	Lu
Th	Pa	U	Np	Pu	Am	Cm	Bk	Cf	Es	Fm	Md	No	Lr

17-34 The noble gases

Helium has one energy level, which can contain only two electrons. Because helium has two electrons, it has a full valence level.

Argon is an inert gas used for special welding processes. Bathing a hot welding arc with argon displaces oxygen, which causes corrosion. Neon, krypton, and xenon are commonly used for special lights. Neon-filled tubes are used in colorful advertising signs. Xenon-filled lamps make very bright headlights.

Since natural gas comes from deep in the earth and the products of radioactive minerals in rocks emit alpha particles, what do you think could be an important source of helium?

Group Number and Valence Electrons

You may have noticed an easy way to determine the number of valence electrons for many elements by using the periodic table. If it has an *A* suffix, the group number equals the number of valence electrons in the neutral atoms of that group's elements. For example, since germanium is in Group 4A, it has four valence electrons. Krypton (Group 8A) has eight.

Elements in the B subgroups (the transition elements) usually have one or two loosely held valence electrons. The elements in the two rows below the periodic table, called the **inner transition elements**, typically have two valence electrons. It is not necessary to memorize which transition elements have one or two valence electrons.

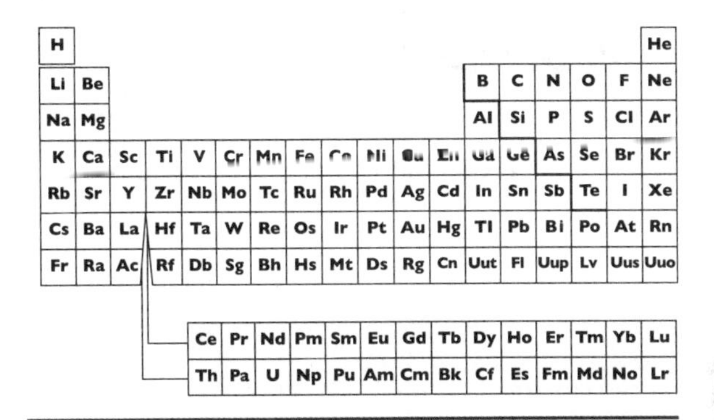

17-35 The inner transition elements

17.15 *Element Periods*

The horizontal rows of the periodic table are called **periods**, or *series*. Elements in the same period do not have similar properties. As atomic numbers increase from left to right across a period, the elements' metallic character decreases and their nonmetallic properties increase. It is the change of electron configuration (number of electrons in each energy level) that causes the properties to change. As you read across a period from left to right, the number of electrons in the outer level increases from one to eight (except for the first period). These trends are discussed further in the next section.

Period numbers are listed on the left side of the table. A period number tells how many main energy levels the elements in that period have in their ground state. Furthermore, the period number also equals the number of the energy level containing the valence electrons. For example, sulfur is in the third period. It therefore has three occupied energy levels, and its valence electrons are in the third main energy level.

You can see that an element's location in the periodic table can tell a lot about its electron structure and properties. The group number tells how many valence electrons an atom of an element has, and its period number tells where they are and gives a general idea of how far from the nucleus they reside.

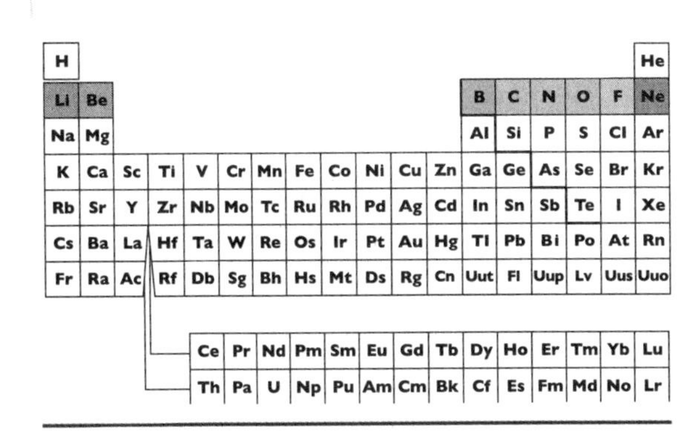

17-36 Period 2 of the periodic table

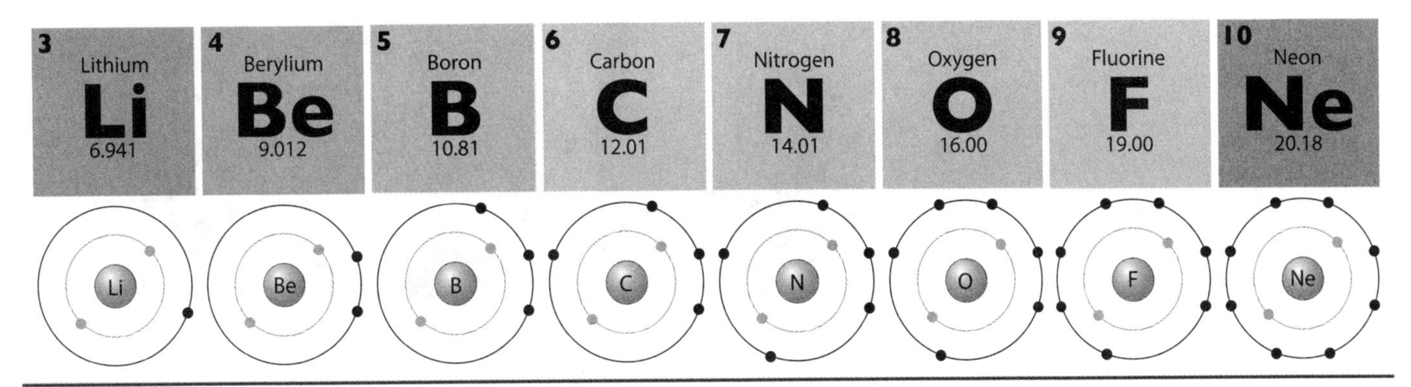

17-37 Period 2 begins with the element lithium. Each succeeding element has one more valence electron.

17.16 Using Element Properties to Solve Problems

17-38 An electric car powered by fuel cells

Common solutions people offer for exhaust pollution are public transportation and carpooling. While these methods help, the average person finds them unappealing because they restrict travel and independence. Also, though they reduce the amount of exhaust in the air, they do not diminish the pollutants in the exhaust. Can we use what we know about the properties of elements to reduce pollution from car emissions?

The best way to reduce pollutants in exhaust is to change vehicle fuel. Using the properties of hydrogen and oxygen, scientists are developing an alternative to gasoline—*fuel cells*, battery-like devices that convert chemical energy to electrical energy. Two hydrogen molecules at one end of the fuel cell split to form four hydrogen cations and release four electrons. These electrons travel through a circuit as electricity, doing electrical work. The hydrogen ions move through a membrane to the outlet of the fuel cell. On the other end of the fuel cell, oxygen splits into two anions as it briefly accepts the used electrons. This action attracts the hydrogen cations to form two molecules of water and release thermal energy in the process.

$$2H_2 + O_2 \longrightarrow 2H_2O + Q$$

This symbolic expression is called a *chemical equation*. The elements on the left combine to produce water and heat on the right. Recall that Q stands for heat.

Several cells combine in a fuel cell stack to produce enough electricity to power a car. The beauty of this technology is that the energy conversion is two to three times as efficient as in gasoline-powered cars, and it doesn't generate any toxic waste products—just steam! Fuel cells also cut down on the noise pollution caused by cars.

So why aren't all cars powered by fuel cells? Think about the fuel in fuel cells. Though oxygen for fuel cells can come directly from the air, the hydrogen must be carried in the vehicle. Safely transporting hydrogen fuel is difficult. Recall that the main engines of the space shuttle were fueled by liquid hydrogen. Just imagine what might happen if a hydrogen fuel tank ruptured in an accident! These and other technological challenges must still be mastered to make fuel cells commonplace.

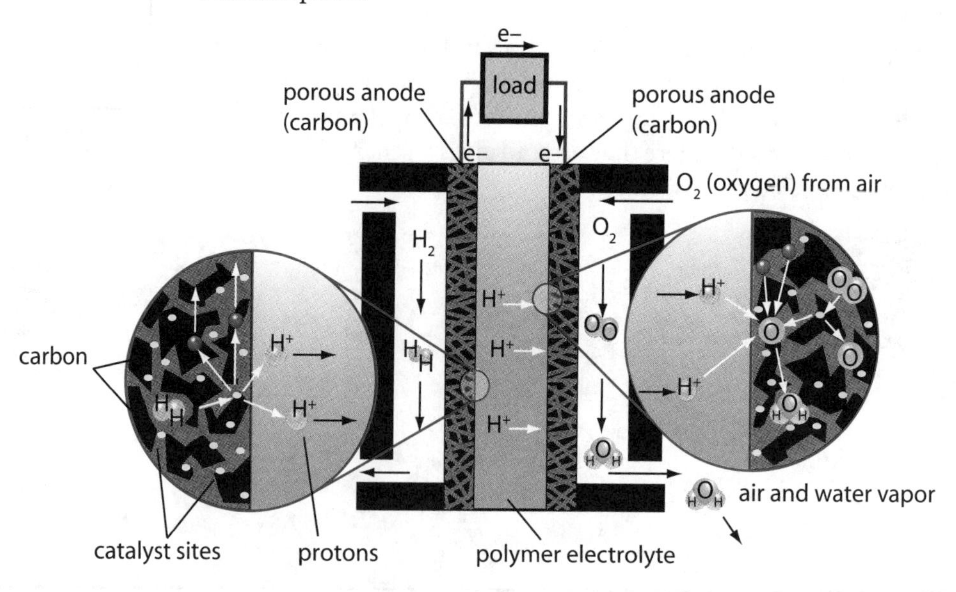

17-39 How a fuel cell works

FACETS OF SCIENCE — UNUSUAL ELEMENTS

Scientists use laws and theories to describe and explain their observations. Although a scientific model may be "set in stone," a few notable exceptions almost always exist. For example, everyone knows that mammals do not lay eggs—that is, nearly all of them don't. The platypus and spiny anteater (echidna) are the lone holdouts, and they do indeed lay eggs.

Elements in the same family share the same characteristics—that is, most of them do. There are a few elements that don't fit the patterns. For example, all metals are solids at room temperature except mercury, which is a liquid. Let's examine some other elements with some quirky, different, or downright bizarre characteristics.

Francium (Fr)

Mendeleev predicted the existence of francium ($Z = 87$), but it was not discovered until 1939 at the Curie Institute in France. Of the first 101 elements, francium is by far the least stable. It is formed by the radioactive decay of actinium ($Z = 89$). Francium is so rare that scientists have estimated that only about 30 g exist in the entire earth's crust at any given time. It decays so quickly that scientists have been unable to measure any quantities or isolate it in its pure form. The most stable isotope is francium-223, which has a half-life of only 22 min. Only minute traces of this isotope have ever been found, mainly in uranium deposits.

Technetium (Tc)

Most radioactive elements are "heavy," meaning that they have fairly large atomic numbers and masses. But technetium ($Z = 43$) is a rarity. Right in the middle of the transition metals, this radioactive element seems out of place. Unlike the other transition metals, it has no stable isotopes. It is not found naturally on the earth but is formed by bombarding molybdenum ($Z = 42$) with hydrogen nuclei. In fact, in 1937 it was the first element to be artificially formed in a laboratory.

technetium (tek NEE she um): (Gk. *technētos*—artificial)

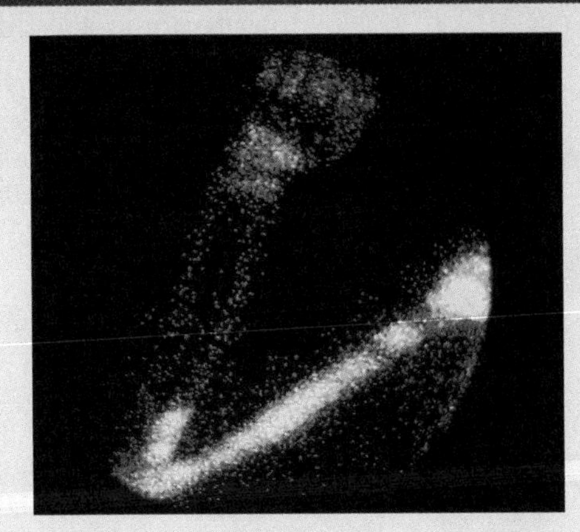

Unlike francium, technetium can be produced in large quantities and has several uses. Physicians inject technetium into a patient's veins, where it concentrates in body tissues. As it decays, it gives off small amounts of gamma radiation, allowing an image similar to an x-ray to be obtained using special equipment (see above).

Manganese (Mn)

Manganese ($Z = 25$), also located in the middle of the transition elements, is an important exception to the general description of metals. Chemically, it is a metal because it has electrons in its outermost energy level that are easily removed. But in its stable form it is a gray-white, brittle substance, too weak for use in engineering. Not ductile, not malleable, and without a metallic luster, it seems an outcast among the transition metals. Manganese behaves like the other transition elements only when alloyed with other metals. Then it gains strength and flexibility. In fact, there are approximately 33 kg of manganese in every metric ton (1000 kg) of special structural steel.

Biologists have also found that many living things need a small amount of manganese in their diets. It is a *trace mineral*—a substance that the body needs to survive. The human body contains only about 0.1 g of manganese.

Hydrogen (H)

With one proton, one electron, and no neutrons, the element hydrogen ($Z = 1$) bears no family resemblance to any group on the periodic table. It is often grouped with the alkali metals, but this colorless, odorless gas is anything but metallic. It exists only as a gas. To be a solid or liquid, it would have to be lowered to a temperature near absolute zero and placed under great pressure. Such conditions may exist on the planet Jupiter.

Hydrogen could theoretically be a metal under the extreme pressures present deep in the atmosphere of Jupiter.

Since the first energy level needs only two electrons to be filled, hydrogen needs to gain only one to be satisfied. This trait would seem to place it as a distant relative of the halogens, but it doesn't have properties resembling them either.

Hydrogen, the lightest element of all, was used in the past to fill blimps; however, its extreme flammability ended that practice. It is probably the most abundant element in the universe. Scientists have estimated that 90% of the atoms in the visible matter of stars, nebulas, and galaxies are hydrogen. Yet on Earth it ranks only tenth in order of abundance and makes up less than 1/1,000,000 of our atmosphere.

There are so many hydrogen compounds that it would be a major challenge to list them all, but the most common hydrogen compound is plain old H_2O—water.

17C Section Review

1. Why do metals have similar properties?
2. How are nonmetals different from metals? Why is this so?
3. What are metalloids? How are they important to modern technology?
4. How can you identify the elements in a family on the periodic table? What do neutral atoms of these elements have in common?
5. What element in the first column of the periodic table is not one of the alkali metals? What does it have in common with them?
6. Identify the family name and the number or range of valence electrons for neutral atoms of each of the following elements using the periodic table on pages 392–93.
   a. astatine (At)
   b. radon (Rn)
   c. tungsten (W)
   d. tin (Sn)
   e. antimony (Sb)
   f. potassium (K)
   g. thallium (Tl)
   h. oxygen (O)
7. Name one use for each of the nine families of elements mentioned in Subsection 17.14. How are these uses an example of fulfilling the Creation Mandate?
8. How is an element's period related to its electron configuration?
9. DS ✪ State two advantages and two disadvantages of using fuel cells for propelling cars with the technology of today.
10. (True or False) Two elements with the same number of valence electrons do not necessarily share the same physical and chemical properties.

17D Objectives

After completing this section, you will be able to

✓ explain how the electron structure of an atom determines its properties.

✓ draw an electron dot structure for a given atom.

✓ explain how conductivity and atomic size generally vary with atomic number from left to right and top to bottom on the periodic table.

17D Periodic Trends

17.17 Electron Structure and the Periodic Table

The number of valence electrons is important in determining the chemical and many physical properties of an element. Properties change with the increase of atomic number, and those changes reveal predictable trends across periods and within families.

Many properties of an atom depend on the distance between its valence electrons and nucleus. The more energy levels an atom has, the larger it is. The larger the atom is, the farther the valence electrons are from the nucleus. Thus, the attraction of the nucleus on valence electrons is weaker in larger atoms. Valence electrons are also shielded from the positive nucleus by the layers of negative electrons occupying inner energy levels. This shielding lessens the attraction between the positive nucleus of a large atom and its negative valence electrons. The more energy levels an atom has, the greater this effect. Since an element's period equals the number of energy levels in its atoms, we can predict how tightly its valence electrons are held.

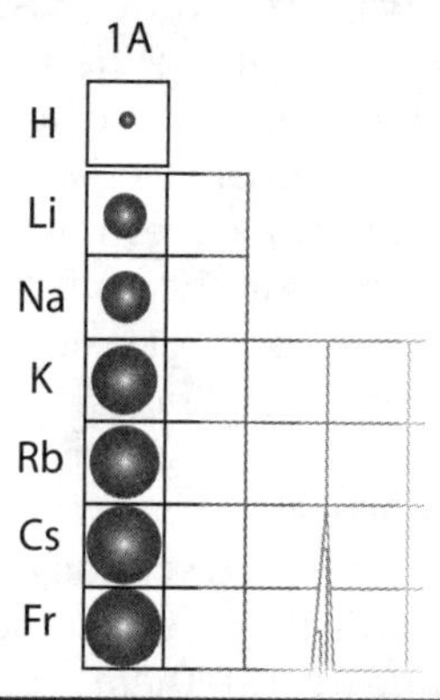

17-40 Relative sizes of the atoms in Group 1A. Atomic size increases with every additional main energy level.

Another factor affecting an element's properties is valence electron arrangement. Electrons naturally pair up in *orbitals*, or sublevels, within an energy level. A pair of electrons in an orbital is more stable than an unpaired electron. Period 1 elements have one orbital in their valence energy level. All others have a maximum of four orbitals in their outermost energy level, so there can be a maximum of only eight valence electrons. The number of occupied orbitals and the number of orbitals containing paired electrons strongly affect the chemical activity of an element.

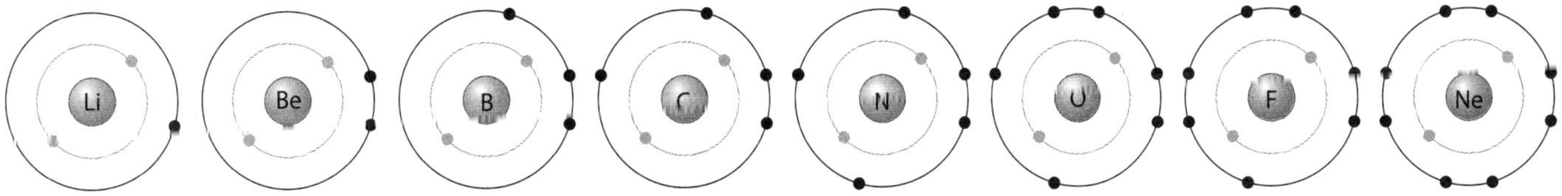

17-41 The arrangement of valence electrons has a lot to do with how reactive an element is. Notice how the valence electrons pair up with increasing atomic number.

To illustrate the arrangement of valence electrons in an atom, chemists created a special system called **electron dot notation**. In this notation, dots representing valence electrons surround an element's symbol. Figure 17-42 shows the standard order in which dots are placed around the symbol. In general, this order reflects the way electrons pair up in orbitals as atomic number increases across a period. An element's group number determines how many dots surround its symbol.

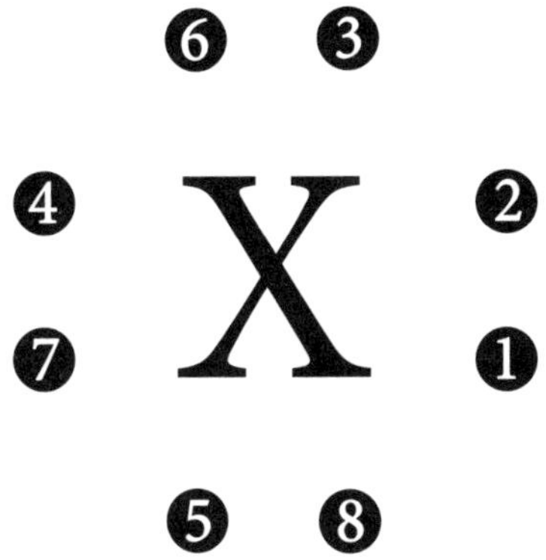

17-42 Standard electron dot notation

17-2 Electron Structure of Series 2 Elements

Group	1A	2A	3A	4A	5A	6A	7A	8A
Bohr model	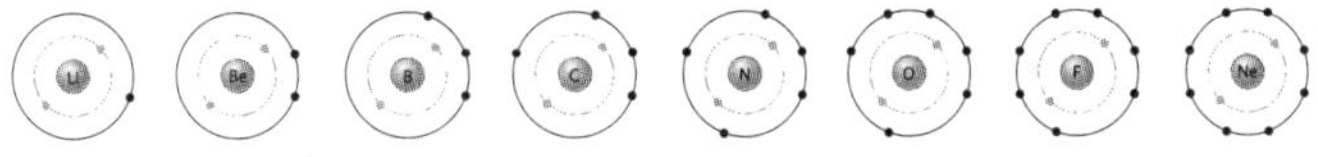							
Valence electrons	1	2	3	4	5	6	7	8
Dot notation	Li•	Be:	Ḃ:	•Ċ:	•Ṇ̇:	•Ö̤:	:F̤̈:	:N̤̈e:

17.18 Summary of Periodic Trends

The periodic table is arranged so that element properties repeat in a periodic way. For example, the metals in Groups 1A and 2A and the transition elements (the B groups) all have one or two valence electrons. These electrons are not strongly held and can be easily removed. This characteristic is responsible for nearly all metallic properties. Continuing to the right within a period, elements have more protons and valence electrons. The stronger nuclear charge holds the valence electrons more tightly. Thus, metallic character decreases across a period from left to right. Conductivity, a distinctive property of metals, drops off dramatically when moving from the metals into the nonmetals.

Predictable changes occur within families also. As you move down a chemical group, each element has an additional energy level. Atomic diameter increases with more energy levels, so atoms generally increase in size toward the bottom of a column. The attraction of the nucleus for valence electrons is weaker in larger atoms. Thus, metallic character, such as electrical conductivity, generally increases down a column of elements (there are many exceptions to this rule). This trend explains why elements at the top of some higher-numbered A groups are nonmetals but those at the bottom are metals. Combining these two trends across periods and down families, we can predict that, *in general*, the more electrically conductive elements are toward the lower left of the periodic table and the lesser conductive are toward the upper right.

Atomic size is another periodic trend. Atomic diameter tends to decrease from left to right across a period because electrons are more strongly attracted by an increasingly positive nucleus. As just mentioned, the largest atoms in a family are toward the bottom of the column because they have more energy levels. Thus, generally the largest atoms are found in the lower left of the periodic table and the smallest in the upper right.

These are just a few properties that follow a periodic pattern with increasing atomic number. There are many others besides those mentioned here. Chapter 18 discusses more periodic properties that predict the kinds of bonds formed by atoms.

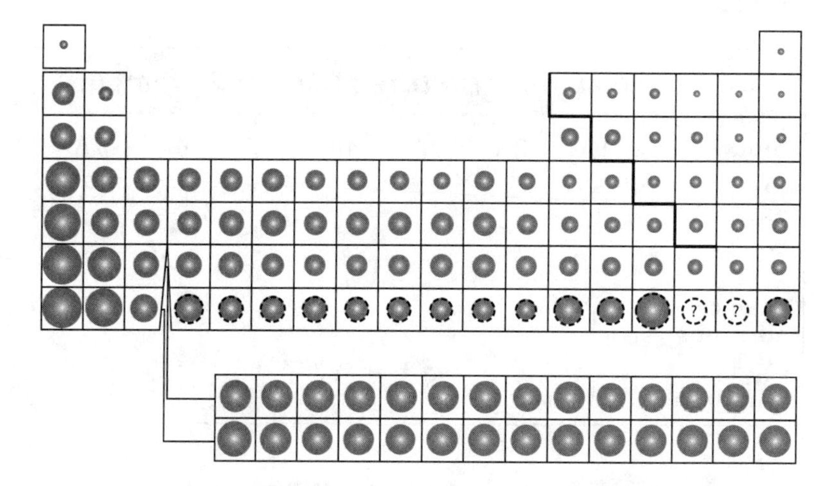

17-43 The general trend of atomic diameters in the periodic table. Dashed outlines indicate atomic sizes predicted but not measured.

17D Section Review

1. What are two important factors that affect valence electrons and contribute to the chemical and physical properties of elements?
2. How does the number of main energy levels in an atom generally affect how tightly its valence electrons are held by the nucleus?
3. What is the purpose of electron dot notation?
4. Place the elements in the following groups in order of *increasing* conductivity according to the trends discussed in this section.
   a. Ba, Be, Mg
   b. Br, Co, Ge
   c. As, Bi, N
   d. Ag, Au, Cu
5. Place the elements in the following groups in order of *increasing* atomic diameter according to the trends discussed in this section.
   a. H, K, Li
   b. Ar, He, Xe
   c. C, Li, Ne
   d. C, Pb, Si
6. Which nonradioactive element would you predict has the largest atoms?
7. (True or False) Since oxygen and fluorine atoms are relatively small and lack only a few valence electrons to complete their octets, their atoms attract and strongly hold electrons.
8. Write electron dot notations for the following elements.
   a. bromine (Br)
   b. helium (He)
   c. nitrogen (N)
   d. barium (Ba)
   e. boron (B)
   f. carbon (C)

Scientifically Speaking

alchemy	383
periodic table of the elements	388
periodic law of the elements	394
metal	394
nonmetal	395
metalloid	396
family	396
alkali metal	397
alkaline-earth metal	397
transition element	398
carbon family	399
nitrogen family	399
oxygen family	399
halogen family	400
noble gas	400
inner transition element	401
period	401
electron dot notation	405

Chapter Review

1. When solid elements are found in their pure form in nature, what is this form called? Give two examples of such elements.
2. What was alchemy's contribution to identifying elements?
3. How did discoveries in the 1700s help define what an element is?
4. Who developed the symbols used today for elements?
5. Why do some elements have three letters in their chemical symbol?
6. Why does the symbol of an element not necessarily stand for the smallest naturally occurring quantity of the element? Give examples in your answer.
7. Why did chemists need to organize the elements into a table?
8. ✪ Suggest a reason why scientists thought that relating element periodicity to octave music scales was not considered scientific.
9. How did Mendeleev organize the elements on his periodic table?
10. Why was Mendeleev's periodic table finally accepted by the scientific community?
11. How does the periodic table tell how many electrons are in a neutral atom?
12. The properties of metals are determined by what aspect of their valence electrons?
13. Why do properties vary so much among the nonmetals?
14. Give an example that shows why the number of valence electrons alone is not enough to predict the general properties of an element.
15. Why are alkali and alkaline-earth metals never found in their pure form in nature?
16. What makes the halogens so reactive?
17. How does the classification of elements change from left to right across a period?
18. DS ✪ Examine the fuel-cell reaction equation on page 402. Why are the symbols for hydrogen and oxygen written with subscripts of 2?
19. When the inner energy levels shield electrons from the attraction of the nucleus, how is atomic size affected?
20. How could electron dot notation help predict an element's chemical activity?
21. What two aspects of an atom's electron structure does its location on the periodic table reveal?

True or False

22. The materials mentioned in the Bible known now as elements were not known as elements when the Bible was written.
23. The modern definition of a chemical element was first stated by Antoine Lavoisier.
24. Only one metal is a liquid at room temperature (25 °C).

25. Dmitri Mendeleev was the first to recognize the periodicity of element properties with increasing atomic mass.
26. There are four types of elements: metals, nonmetals, metalloids, and noble gases.
27. Although hydrogen is in the same group as the alkali metals, it is not a metal.
28. Group 1B and Group 1 are the same family.
29. The alkali metals and alkaline-earth metals are commonly found in native mineral deposits all over the world.
30. Group 5A (the nitrogen family) has very different elements in it, even though they all have the same number of valence electrons.
31. A period contains all the elements that have the same number of valence electrons.

DS ✪ 32. One difficult aspect of fuel-cell technology is safely storing liquid oxygen and hydrogen in a vehicle.

33. Elements with nearly eight valence electrons tend to be chemically active.
34. Electron dot notation shows all the electrons in the atom at a glance.
35. The metallic character of elements in a family tends to increase with atomic number.

✪ 36. Draw the Bohr models and electron dot structures for the following elements.
 a. calcium (Ca)
 b. argon (Ar)
 c. carbon (C)
 d. helium (He)

✪ 37. Copy the table to a separate sheet of paper. Fill in the missing information.

Symbol	Family name	Period number	Valence electrons	Metal (M), nonmetal (N), or semimetal (S)	Electron dot notation
As					
		6	8		
	alkali metal	3			
S					
		2	7		
	noble gases	1			

✪ 38. Research some of the different types of steel, the elements used in them, and their applications.

DS ✪ 39. Research the methods that are being developed to store and sell hydrogen for use in fuel-cell-powered vehicles. Include in your report the places where these ideas are already becoming reality.

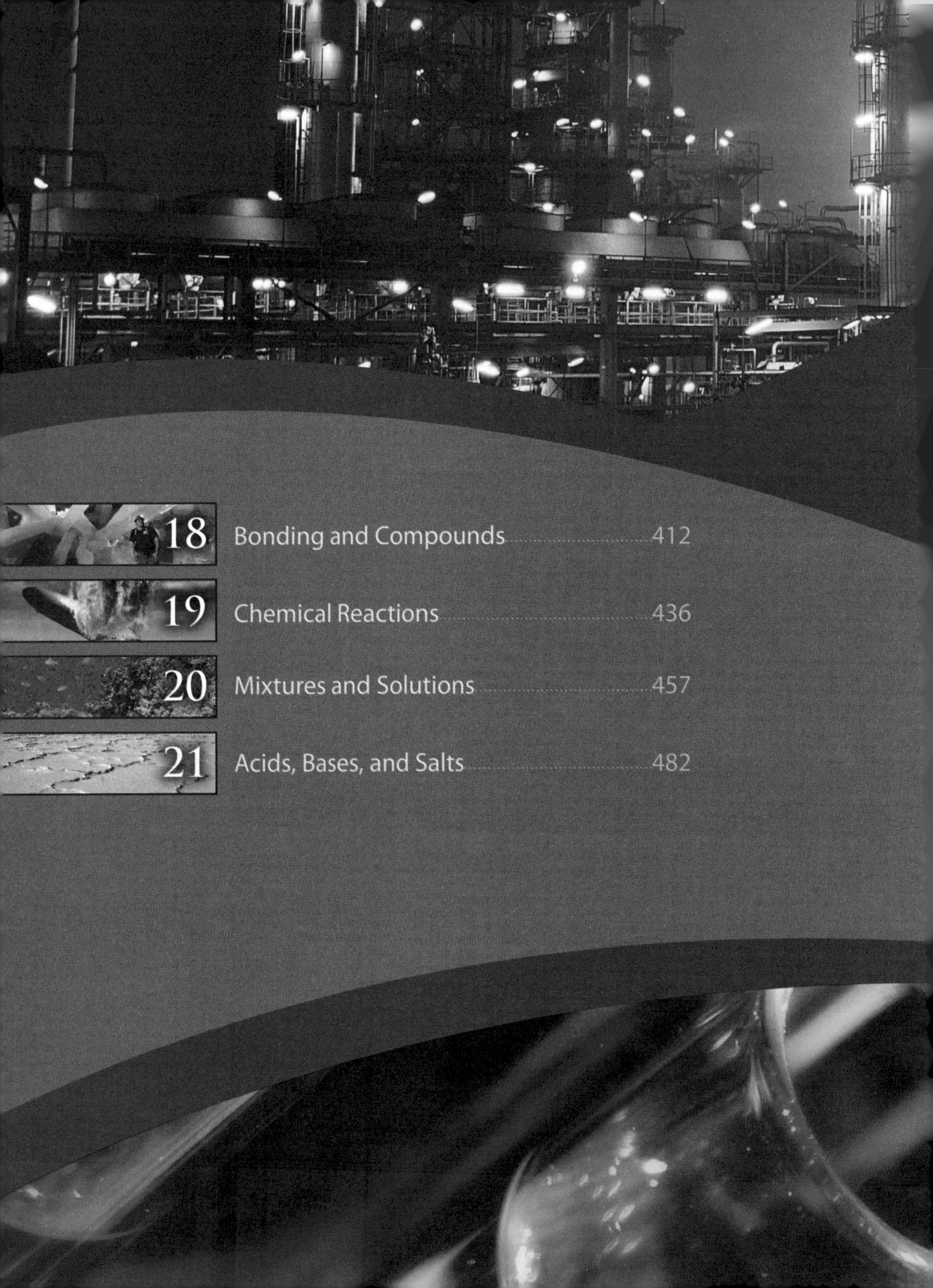

Introduction to Chemistry

If it's matter, it's chemistry.

Chemicals make up everything in our material world. In addition to our bodies and the food we eat, chemicals form the clothes we wear as well as the soap that washes them, the dwellings we live in and all their furnishings, the vehicles we travel in and the fuels that propel them, the air we breathe, our medicines, and even the batteries in our cell phones! Fuel cells, space probes, plastics, ceramics, fertilizers, and skyscrapers are all products of chemistry. Every major area of science deals with chemistry because it is the study of matter and its behavior in the presence of other matter.

In Unit 6 you will be introduced to chemistry. You will learn how atoms bond to form compounds. You will explore how elements and compounds undergo changes to form new substances. You will learn to distinguish between mixtures and solutions and describe the properties of acids and bases. All these topics have practical applications in your daily life.

Unit 6

BONDING AND COMPOUNDS

CHAPTER 18

DOMINION SCIENCE PROBLEM

Pain Relief

A throbbing headache, sore muscles the day after a workout, a bad cold—all involve something quite common to the human experience: pain. Though we aren't fond of pain, it's an important part of the body's alarm system. The skin alone uses about three million pain receptors to warn us when something isn't right. Sometimes pain is merely annoying, but when it's severe or long lasting, it's a signal to see a doctor. But can't we get some relief?

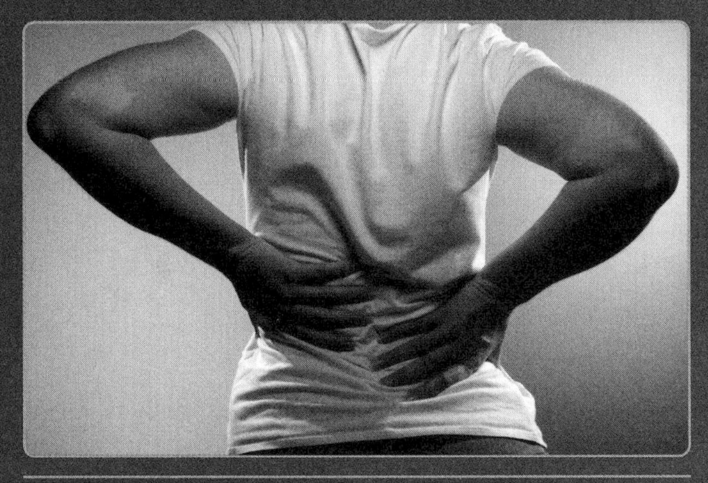

18-1 Pain is a sensation with which we are all too familiar in this fallen world.

18A Principles of Bonding

18.1 Bonding: Holding Atoms Together

This textbook spends much time examining the composition of matter. The basic principles are quite simple. Every physical thing is made of atoms. We know of 118 different types of atoms that form substances we call elements, and they are arranged in a useful diagram called the periodic table. Most can be found in pure form, blended together in mixtures, or chemically bonded together as molecules and compounds. Some bond to each other quickly just by being mixed together. Others refuse to form compounds with each other under almost any condition. Why do atoms vary so much in their bonding behavior?

The arrangement and number of electrons in atoms determine how they bond. For example, aluminum bonds to chlorine to form a white absorbent crystal (aluminum chlorohydrate), the active ingredient in many deodorants. It also bonds with magnesium to form a lightweight alloy used in aircraft called *magnalium*. The electron structures of chlorine and magnesium interact differently with aluminum's electron structure.

Consider graphite and diamond. Pencil lead and the stone in an engagement ring are very different, but they contain substances made of the same element—carbon. Graphite, which looks like soot, is inexpensive, black, soft, and slippery. Diamond, on the other hand, is valuable, transparent, nearly colorless, and the hardest of all natural substances.

Graphite and diamond illustrate that the same element with the same electron structure can often bond in more than one arrangement. In graphite, six carbon atoms link together to form rings. These rings are joined to form sheets. Each carbon atom is bonded to three other carbon atoms. Though the bonding within each sheet is strong, the sheets are not strongly attracted to each other, allowing them to slide over one another. This property makes graphite a slippery lubricant. In diamond, carbon

18A Objectives

After completing this section, you will be able to

- ✓ explain how the octet rule accounts for why atoms bond together.
- ✓ differentiate between electron affinity and electronegativity and describe their general trends on the periodic table.
- ✓ briefly describe covalent, ionic, and metallic bonds.
- ✓ explain how relative electron affinities are important to the kind of bond formed.

iRocks.com photo

18-2 Both graphite and diamond are made of carbon. However, the atoms are bonded together differently in the two substances.

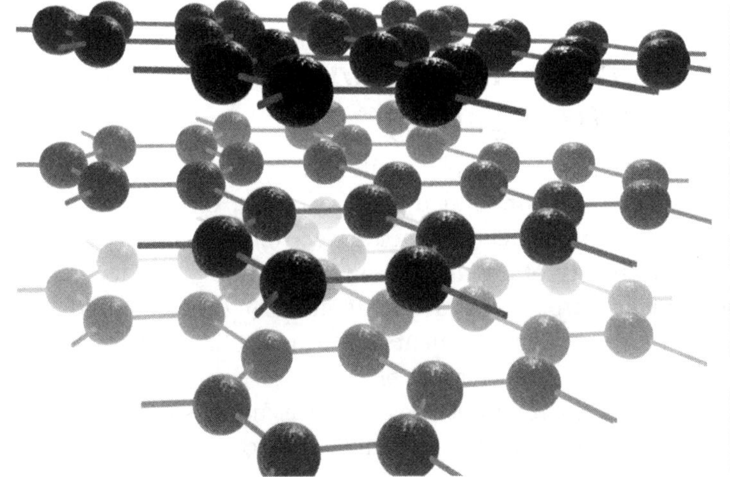

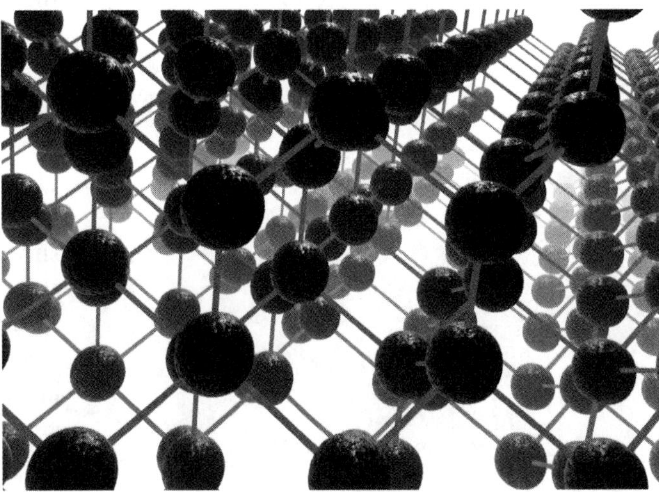

18-3 The structure of graphite (left) and diamond (right)

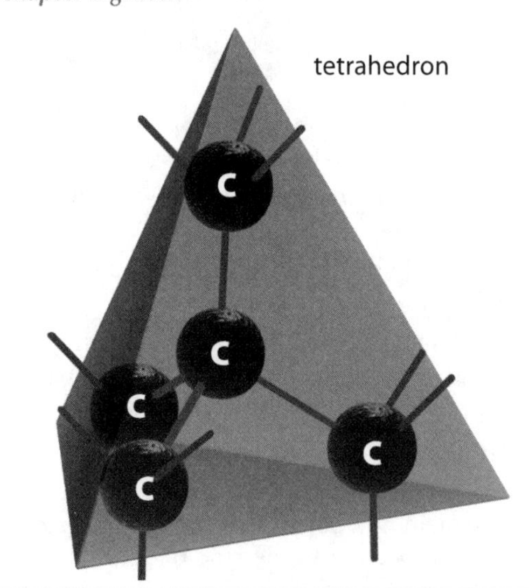

18-4 In diamond, carbon atoms bond together in a repeating pyramid pattern.

A bonded atom's outer energy level isn't always the same thing as its valence energy level. Metals, like lithium, beryllium, and boron, bond by losing their valence electrons, exposing the complete octet of the next lower energy level—the new outer energy level.

affinity: affin- (L. *affinis*—related to, bordering on) + -ity

atoms bond to each other to form a structure in which every carbon atom is surrounded by four other carbon atoms in a pyramid arrangement, not in a sheet like they are in graphite. This repeating structure creates a vast three-dimensional network of carbon atoms that composes the hardest natural material.

These examples show that when atoms bond, the characteristics of their bond(s) affect the properties of the substance they form. *Chemical bonds involve mainly the valence electrons of an atom—* those in the outermost energy level.

18.2 The Octet Rule

The tendency for atoms to form bonds is governed by the second law of thermodynamics. This law states that all natural processes move toward a state of minimum energy. Most atoms are more chemically unstable when they are not bonded to another atom. This instability, the tendency to form chemical bonds, is mainly due to incomplete valence energy levels (see Chapter 17).

Atoms generally are most stable when they have a full eight electrons in their valence energy level. This principle is called the **octet rule** of bonding. Exceptions to this rule are hydrogen and helium, which can have at most two valence electrons, and the next three elements, lithium, beryllium, and boron, which usually lose their valence electrons when bonding. More often the rule is applied in chemistry to determine how atoms bond to other atoms. Extensive experimental evidence shows that an atom can fill an octet in its outer level by bonding with other atoms to form molecules and compounds.

Atoms achieve greater stability through bonding in one of two ways. First, atoms can *share electrons*. The shared electrons can fill the valence levels of all the atoms bonded together at the same time. Second, atoms can *gain or lose electrons* to gain a full octet. Atoms with few valence electrons, which are loosely held, can easily lose them, exposing the full octet of the next lower energy level. On the other hand, if an atom needs a few electrons to complete an octet, it can acquire them from other atoms or its surroundings. Ions formed by losing or gaining electrons are pulled together by their opposite charges. No matter the type of bonding, the end result is the same—atoms are more stable with a completely filled outer energy level.

18.3 Electron Affinity and Electronegativity

How an atom attains a complete outer energy level depends on its electron affinity. An unbonded atom's *electron affinity* is a measure of how well it attracts and holds electrons. Chemists measure the electron affinity of an element by determining how its energy changes when electrons are added to its atoms. Figure 18-5 shows the electron affinities of some elements on the periodic table. A positive affinity means that energy must be added for the element to take electrons. Higher energy indicates that the element is more unstable with extra electrons. A negative affinity means that the element gives

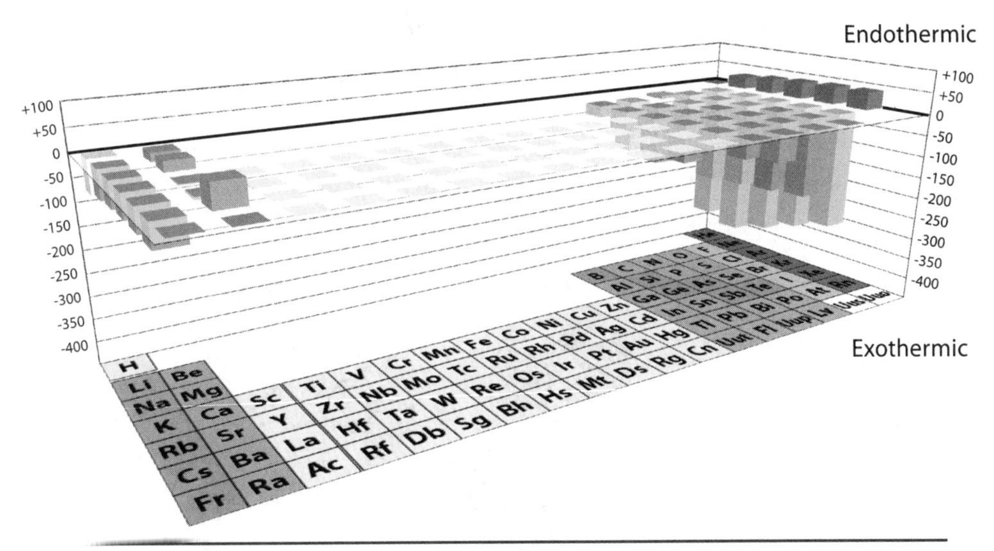

18-5 Electron affinity of the elements. Positive values indicate weak affinities. Large negative values indicate strong affinities. Electron affinity is measured in joules per mole of atoms of an element.

off energy when it gains electrons. A lower energy state is more stable according to the second law of thermodynamics. Atoms with high electron affinities will take or share electrons from other atoms. Atoms with low electron affinities will lose their valence electrons or have a very small part in sharing them when bonding.

Halogens and most nonmetals have the largest electron affinities (the most negative values), while most of the metals have low affinities (small positive or negative values). Noble gases have positive electron affinities because their valence energy levels are already full.

The famous chemist Linus Pauling devised a scale to compare an atom's ability to attract and hold electrons *when bonded to other atoms*. This ability is known as an element's **electronegativity**. Electronegativity varies periodically with atomic number. Atoms with nearly full valence levels have high electronegativities. Atoms whose outer levels are nearly empty have very low electronegativities. Elements within a given group tend to have similar electronegativities. Figure 18-6 compares the electronegativities of the elements on the periodic table.

An atom's electron affinity and electronegativity are related because they are measures of an atom's ability to attract and hold electrons. But they apply to different situations and have different units. Electron affinity applies to *unbonded* atoms and is measured in units of energy gained or lost per mole of atoms. Electronegativity applies to *bonded* atoms and is measured in small positive *electronegativity numbers (EN)*. An atom with a large electronegativity has a large negative electron affinity.

Linus Carl Pauling (1901–94) was one of the most important American chemists and medical researchers of the twentieth century. He won the 1954 Nobel Prize for Chemistry for describing the nature of chemical bonds.

H 2.20																	He
Li 0.98	Be 1.57											B 2.04	C 2.55	N 3.04	O 3.44	F 3.98	Ne
Na 0.93	Mg 1.31											Al 1.61	Si 1.90	P 2.19	S 2.58	Cl 3.16	Ar
K 0.82	Ca 1.00	Sc 1.36	Ti 1.54	V 1.63	Cr 1.66	Mn 1.55	Fe 1.83	Co 1.88	Ni 1.91	Cu 1.90	Zn 1.65	Ga 1.81	Ge 2.01	As 2.18	Se 2.55	Br 2.96	Kr 3.00
Rb 0.82	Sr 0.95	Y 1.22	Zr 1.33	Nb 1.6	Mo 2.16	Tc 1.9	Ru 2.2	Rh 2.28	Pd 2.20	Ag 1.93	Cd 1.69	In 1.78	Sn 1.96	Sb 2.05	Te 2.1	I 2.66	Xe 2.6
Cs 0.79	Ba 0.89	La 1.1	Hf 1.3	Ta 1.5	W 2.36	Re 1.9	Os 2.2	Ir 2.20	Pt 2.28	Au 2.54	Hg 2.00	Tl 1.62	Pb 2.33	Bi 2.02	Po 2.0	At 2.2	Rn
Fr 0.7	Ra 0.9	Ac 1.1	Rf	Db	Sg	Bh	Hs	Mt	Ds	Rg	Cn	Uut	Fl	Uup	Lv	Uus	Uuo

Ce 1.12	Pr 1.13	Nd 1.14	Pm 1.13	Sm 1.17	Eu 1.2	Gd 1.2	Tb 1.1	Dy 1.22	Ho 1.23	Er 1.24	Tm 1.25	Yb 1.1	Lu 1.27
Th 1.3	Pa 1.5	U 1.38	Np 1.36	Pu 1.28	Am 1.13	Cm 1.28	Bk 1.3	Cf 1.3	Es 1.3	Fm 1.3	Md 1.3	No 1.3	Lr

18-6 Electronegativities of the elements. Larger numbers indicate stronger electronegativities.

A mole is the basic SI unit for amount of substance. One mole is nearly 10^{24} particles!

18.4 Types of Chemical Bonds

covalent (ko VAY lunt)

Different types of bonds form between atoms based on their relative electronegativities. If two atoms both have large electronegativities, they can't bond by taking electrons from the other. However, they can share electrons to form a ***covalent** bond*. Covalent bonds occur most often between nonmetals.

18-7 Similar electronegativities—sharing electrons between two atoms

When atoms with very different electronegativities bond, one loses nearly all influence on its electrons and the other gains nearly complete influence over the "lost" electrons. The atom with low electronegativity, which loses electrons, becomes a cation. The atom with high electronegativity, which gains electrons, becomes an anion. These ions attract to form an *ionic bond*. Typically, ionic bonds form between metals and nonmetals.

18-8 Very different electronegativities—taking or giving away electrons

Atoms with weak electronegativities usually have only a few loosely held valence electrons. These atoms bond by sharing their easily lost electrons among many atoms. These mobile electrons are not associated with any specific nucleus. Metal atoms normally share electrons this way to form *metallic bonds* in the solid state.

18-9 Similar electronegativities—sharing electrons among many atoms

The remainder of this chapter provides a look at each of these bond types and the properties they give their compounds.

18A Section Review

1. How are graphite and diamond related? How are they different?
2. Why do atoms form chemical bonds?
3. What is the octet rule? Explain the exceptions to this rule.
4. What are the two main ways that atoms can acquire a full outer energy level?
5. What basic property of atoms determines how easily they can attract and hold electrons? What does it mean when this property has a negative value?
6. What is the difference between electronegativity and electron affinity?
7. Name and briefly describe the three basic bond types.
8. (True or False) The type of chemical bond between atoms often determines many of the chemical and physical properties of an element or compound.

18B Objectives

After completing this section, you will be able to

- ✓ describe a covalent bond, including the kind of element that usually forms covalent bonds.
- ✓ illustrate the pairing of electrons to form covalent bonds using electron dot notation.
- ✓ draw Lewis structures of simple molecules.
- ✓ compare and contrast single, double, and triple covalent bonds.
- ✓ explain what makes a bond polar and how it affects the properties of covalent compounds.
- ✓ explain how bonding and nonbonding electrons affect a molecule's shape.

18B Covalent Bonds

18.5 Forming Covalent Bonds

Suppose two nonmetal atoms with similar, strong electronegativities have the opportunity to form a bond. Both lack a few electrons. The only way for both atoms to attain a full octet is to share valence electrons. Electrons pair up when possible within an atom's energy levels. In a similar way, atoms share electrons as pairs in **covalent bonds**. Normally, each atom contributes one electron to the shared pair, but sometimes it contributes more than one to form multiple electron pairs.

A covalent bond exists between only two atoms. However, an atom can form multiple covalent bonds; that is, it can be covalently bonded to several atoms at the same time. An atom will form as many covalent bonds as it needs to acquire a stable valence electron configuration.

18.6 Illustrating Covalent Bonds

Let's look at how a molecule of chlorine (Cl_2) forms. Both chlorine atoms have the same electronegativity, so neither can completely take electrons from the other. Both atoms have seven valence electrons, so each needs one more electron to complete its octet. When each chlorine atom shares its one unpaired electron with the other atom, both atoms have an octet. The shared pair of electrons is called a *bonding pair.* A single covalent bond consists of one bonding pair of electrons.

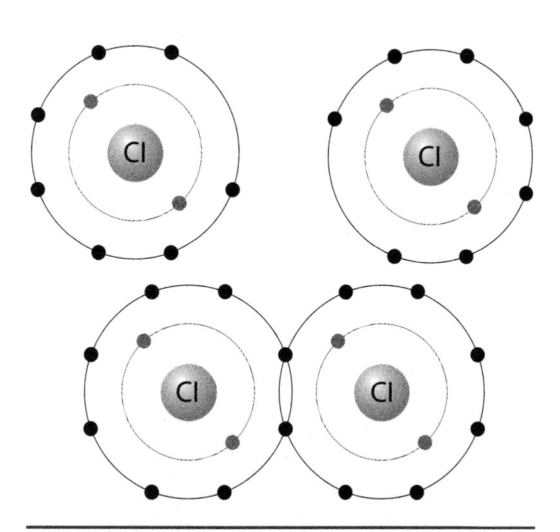

18-10 Chlorine forms a covalent bond by sharing a single electron with another chlorine atom.

Shared electrons spend most of their time between atoms, forming a negative region between the two positive nuclei. The electrostatic attraction between nuclei and electrons is mostly responsible for holding the two atoms together. Using Bohr models, Figure 18-10 shows how shared electrons complete the valence energy level for both atoms of chlorine.

A simpler way to illustrate a covalent bond is *electron dot notation.* You can replace the Bohr models in Figure 18-10 with electron dot symbols for the valence electrons of each chlorine atom. By rotating the location of the unpaired dots around the element symbol, you can place the single dots next to each other to form an electron pair. The pair of dots between the symbols identifies a covalent bond. This molecular notation is called a **Lewis structure**, after Gilbert Lewis. The Lewis structure for the bonding of chlorine is

$$:\ddot{\underset{\cdot\cdot}{Cl}}\cdot + \cdot\ddot{\underset{\cdot\cdot}{Cl}}: \rightarrow :\ddot{\underset{\cdot\cdot}{Cl}}:\ddot{\underset{\cdot\cdot}{Cl}}: \text{ or } Cl-Cl.$$

Covalent bonds may be represented by a pair of dots, but more often a single dash between the bonded atoms is used to distinguish between the electrons involved in bonding and those that are not. When dashes are used in Lewis structures, the nonbonding electrons are usually omitted.

Gilbert Newton Lewis (1875–1946) was an American chemist who contributed to the chemical revolution in America with his detailed research and description of valence electrons and isotopes. He was the first person to describe the covalent bond.

Electrons of one atom are no different from electrons of another. In the electron dot notations shown in the text, however, the electrons are colored differently to more easily show their sources.

18-11 Chlorine helps keep bacteria and algae levels low in swimming pools.

18.7 Bonds in Diatomic Molecules

The chlorine molecule is an example of a *diatomic molecule*—a molecule containing just two atoms. Chlorine is one of the seven naturally occurring diatomic elements. All diatomic elements consist of two of the same kind of nonmetal atom connected by one or more covalent bonds. Hydrogen is another naturally diatomic element. Each hydrogen atom has one valence electron and needs only one more to fill its only energy level. Thus, two hydrogen atoms share their single electrons to form a covalent bond. The Lewis structure for hydrogen may be either

$$\mathrm{H{:}H} \text{ or } \mathrm{H{-}H}.$$

It is optional to show nonbonding electrons in Lewis structures.

Oxygen is a diatomic element as well. It has six valence electrons, so each atom lacks two electrons to make a complete valence octet. If each atom shares two electrons, they can make up the difference. Sharing two pairs of electrons forms a double covalent bond. *Double bonds* are stronger than single bonds. In Lewis structures they are represented by two dashes instead of two pairs of dots. Similarly, covalent *triple bonds* form when each atom must share three electrons to complete its octet. Diatomic nitrogen is one such element.

18-1 Diatomic Bonds

Diatomic element	Structural development	Lewis structure	Covalent bond type
hydrogen (H_2)	H• + •H → H:H	H–H	single bond
oxygen (O_2)	:Ö• + •Ö: → :Ö::Ö:	O=O	double bond
nitrogen (N_2)	:N• + •N: → :N:::N:	N≡N	triple bond

18.8 Bonds in Polyatomic Molecules

The electron dot notation for an element can help determine how many covalent bonds it can form. In general, for nonmetal elements *the number of unpaired dots is the number of covalent bonds the atom can form with other atoms.* These unpaired electrons are called *bonding sites* (see Figure 18-12). Oxygen has two possible bonding sites, nitrogen has three, carbon has four, and fluorine has one.

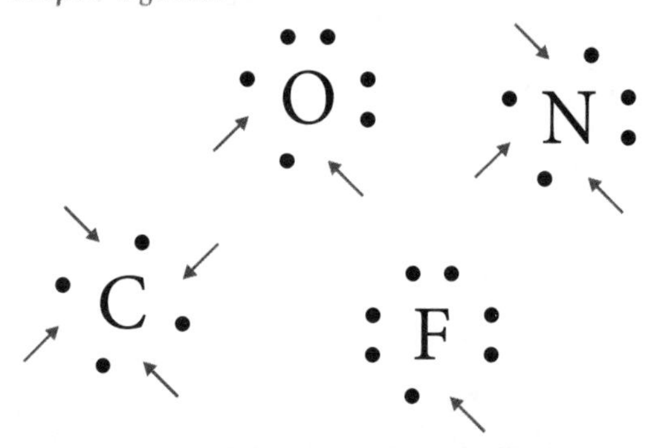

18-12 Unpaired electrons in nonmetals are potential locations for covalent bonds.

·C:

Carbon's standard electron dot symbol is unusual in that one side lacks any electrons. By splitting up its single pair of electrons, a single dot can be placed on each of the four sides. Thus, electron dot notation shows that carbon can form four covalent bonds.

The prefix *poly-* means "many."

Since nonmetals all have similarly large electronegativities, they can be expected to form covalent bonds with each other. There are many examples of covalently bonded molecules such as water (hydrogen and oxygen), ammonia (hydrogen and nitrogen), and methanol (hydrogen, oxygen, and carbon).

Molecules of three or more atoms are called *polyatomic molecules.* A water molecule (H_2O) is made of three atoms, an oxygen atom and two hydrogen atoms. An oxygen atom (Group 6A, or 16) has six valence electrons, two short of an octet. The hydrogen atoms (Group 1A, or 1) have one valence electron apiece. Each needs one more electron to fill its first energy level. When these three atoms bond, oxygen forms a covalent bond with each hydrogen atom to fill the outer energy level of all the atoms.

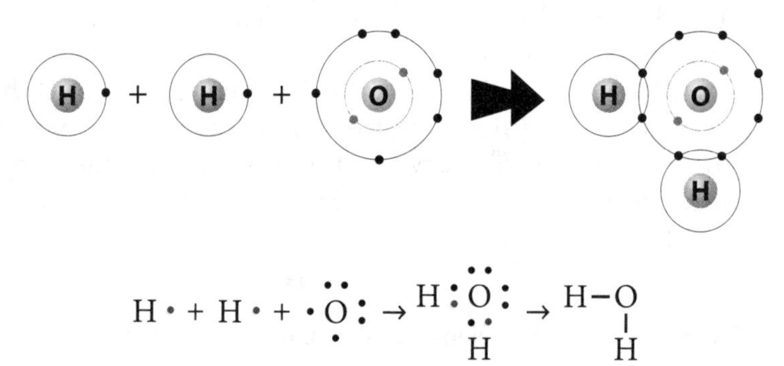

Ammonia is another example. Before nitrogen atoms bond with hydrogen atoms to form ammonia (NH_3), each nitrogen atom (Group 5A, or 15) has five valence electrons. It needs three more to achieve an octet. To do this, it shares each of its three unpaired electrons with three hydrogen atoms. Each hydrogen atom contributes one electron to produce three covalent bonds.

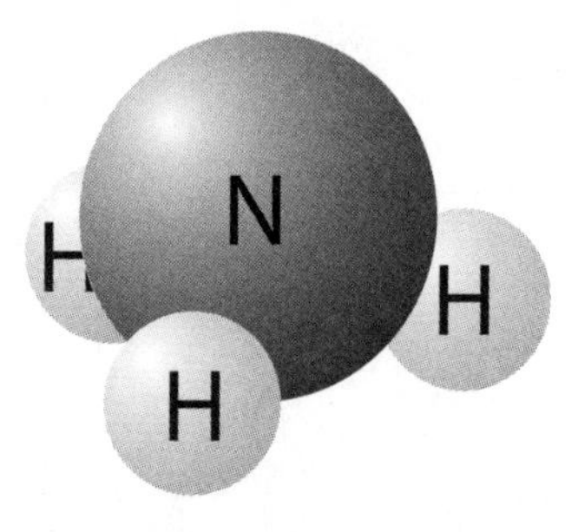

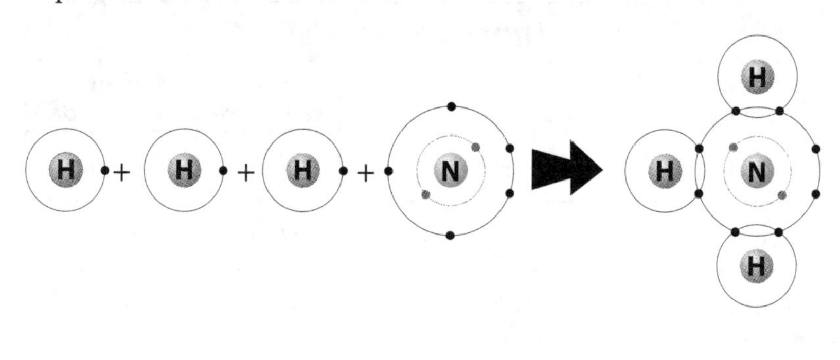

18-13 Ammonia (NH_3) and its compounds are an important part of many fertilizers.

The third example—a molecule of methanol (CH_3OH)—is a bit more complicated. Methanol, used in the production of many important chemicals, is known as an *organic compound* because it contains carbon and at least one hydrogen atom. Since each carbon atom (Group 4A, or 14) has four valence electrons, it needs four more to attain an octet. You already know that oxygen needs two more electrons to achieve an octet. Hydrogen needs one more electron to acquire its full valence energy level.

$$H\cdot + H\cdot + H\cdot + H\cdot + \cdot\overset{\cdot}{\underset{\cdot}{C}}\cdot + \cdot\overset{\cdot\cdot}{\underset{\cdot}{O}}: \rightarrow H:\overset{\overset{H}{\cdot\cdot}}{\underset{\underset{H}{\cdot\cdot}}{C}}:\overset{\cdot\cdot}{\underset{\underset{H}{\cdot\cdot}}{O}}: \rightarrow H-\overset{\overset{H}{|}}{\underset{\underset{H}{|}}{C}}-\underset{\underset{H}{|}}{O}$$

Carbon shares three of its electrons with three hydrogen atoms and one electron with an oxygen atom, forming a total of four covalent bonds. Oxygen forms a second covalent bond when it shares a pair with another hydrogen atom. In this way, all six atoms fill up their valence energy levels—oxygen and carbon with octets and each hydrogen with two.

18.9 Covalent Bond Polarity

In a diatomic element the electronegativities of the two bonded atoms are identical. The difference of their electronegativity numbers (ΔEN) is zero. When a covalent bond's ΔEN is zero, the bonding electrons are shared equally and the bond is characterized as a purely covalent bond.

However, when different nonmetal atoms, such as oxygen and hydrogen, form a covalent bond, their electronegativities can differ greatly. Look at Figure 18-6 on page 415. Oxygen's electronegativity is 3.44, while hydrogen's is 2.20. The nonzero ΔEN of the bond between oxygen and hydrogen means that the electrons are shared unequally between the atoms. They are unequally shared because oxygen has a greater attraction for electrons than hydrogen does. Therefore, the shared electrons are more likely to be found nearer the oxygen. This kind of uneven electron sharing is called a *polar covalent bond.* Diatomic elements form bonds with no polarity called *nonpolar bonds.*

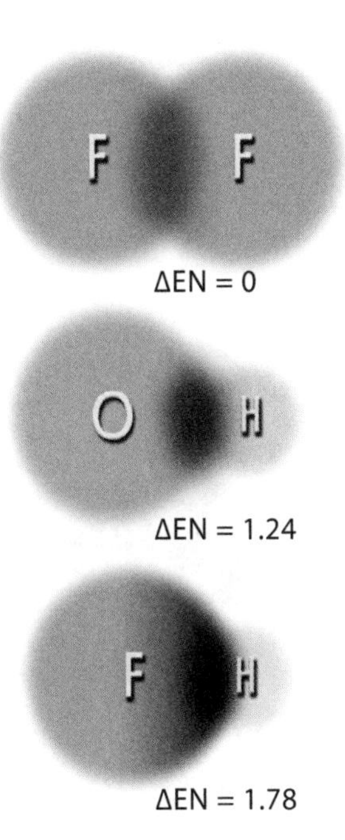

18-14 Covalent bonds become more polar as the ΔEN between the bonded atoms increases.

The atoms at the ends of a polar covalent bond acquire partial electrical charges that act like electrical poles. Depending on the arrangement of polar bonds within it, a molecule itself can have electrical poles, making it a *polar molecule.* The oppositely charged regions on nearby polar molecules can attract, pulling molecules together. These *intermolecular forces* hold together the molecules of many covalent solids, such as ice and sugar. See the box on the next page.

Most nonmetals have similar electronegativities, and thus the bonds between them have low polarity. Hydrogen is an important exception. Because it has and needs only one electron, its low electronegativity is unique. Bonds between hydrogen and other nonmetals with large electronegativities, such as fluorine or oxygen, form highly polar covalent bonds.

Intermolecular Forces

You've learned that covalent bonds hold atoms together in a molecule, but what holds molecules together in solids and liquids? Molecules exert forces on each other. These intermolecular forces hold them together.

There are three types of intermolecular forces: dipole-dipole forces, hydrogen bonds, and London dispersion forces. The type of force depends on which atoms are in the molecule.

Polar molecules can be modeled as three-dimensional objects that have areas of charge on their surfaces. These charged regions are the result of the molecule's arrangement and the electronegativity differences between the molecule's atoms. Areas of opposite charge on different molecules attract with a *dipole-dipole force*. *Hydrogen bonds* are a special type of dipole-dipole force exerted between hydrogen atoms and highly electronegative atoms in other molecules, such as fluorine, oxygen, or nitrogen. Hydrogen can become a strong positively charged pole when covalently bonded to these elements in its own molecule.

Nonpolar molecules cannot attract each other through dipole-dipole forces. But both nonpolar and polar molecules can come together through *London dispersion forces*. Within a molecule the electrons involved in bonding are in constant motion and can at any instant be more concentrated on one side of a molecule than the other. These instantaneous and momentary charge imbalances on the surface of molecules can draw them together through weak electrostatic forces.

These three types of intermolecular forces affect the physical and chemical properties of compounds. The table below compares the bonds between atoms and molecules.

Structural Bond Strength Comparison

Bonds between	Bond type	Increasing strength
ions	ionic bond	↑
metal cations (metals)	metallic bond	
polar molecules containing H-F, H-O, or H-N bonds	hydrogen bond	
polar molecules	dipole-dipole force	
nonpolar molecules	London dispersion force	

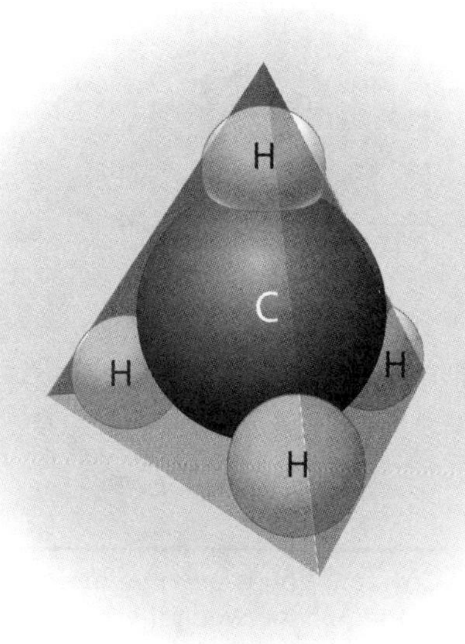

18-15 The methane molecule (CH_4) has a tetrahedral geometry.

18.10 Covalent Bonding and Molecular Structure

Do you remember why diamond is so hard? Each carbon atom in the crystal lattice is covalently bonded to four other carbon atoms. The three-dimensional arrangement of atoms around the central atom forms the points of a *regular tetrahedron*, a four-sided pyramid with faces that are equilateral triangles. The four pairs of valence electrons in the bonds repel each other. The positions at which they are farthest from the other electron pairs form the points of the regular tetrahedron arranged around the nucleus. Methane (CH_4) is another example of how carbon forms a molecule with a tetrahedral shape. This geometric arrangement of valence electron pairs around an atom is not unique to carbon. Rather, it is believed to influence the location of bonding and nonbonding pairs of electrons in most nonmetals.

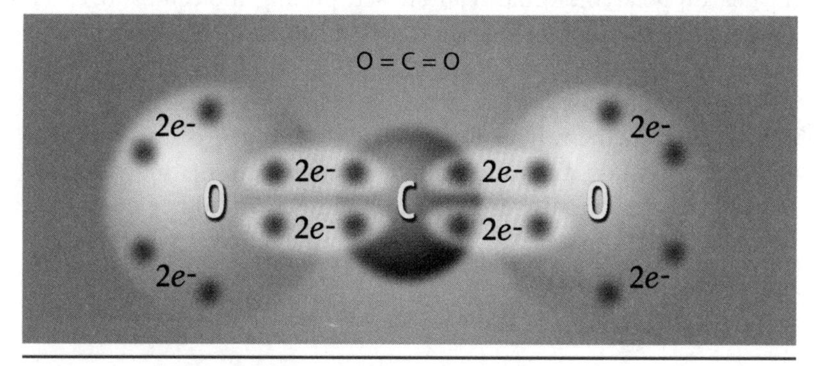

18-16 Carbon dioxide is a linear molecule because of the two double bonds formed with the oxygen atoms.

The number of bonds an atom forms affects the molecule's shape. The atoms in diatomic molecules are always in a line, forming linear molecules. The shapes of polyatomic molecules depend on their types of bonds and how many nonbonding pairs of electrons are associated with the central atom(s). The water molecule, for example, has a bent shape because the central atom, oxygen, has two valence electron pairs that are bonding pairs and two that are not. The two single bonds with hydrogen cannot be arranged in a line around the oxygen atom because the nonbonding pairs repel them. The actual shapes of simple molecules can be suggested by their Lewis structures.

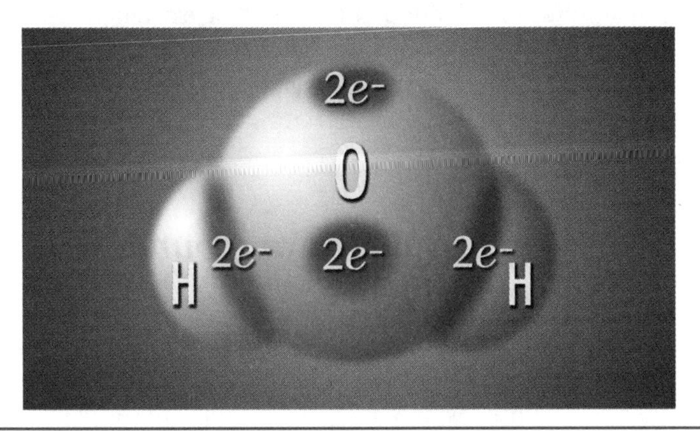

18-17 The locations of the bonding and nonbonding pairs of electrons in a water molecule help determine its shape.

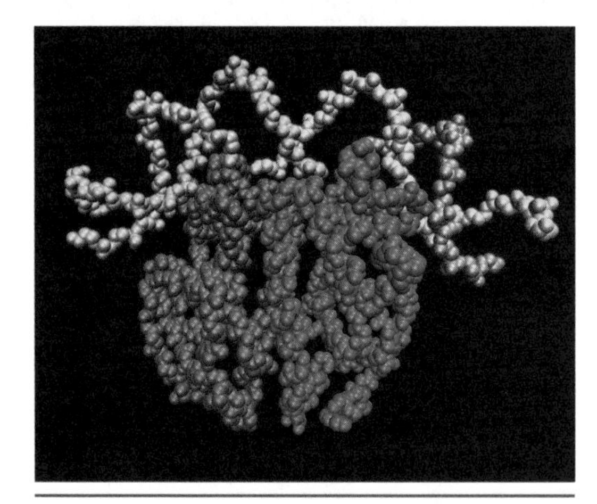

18-18 The specific shape of a highly folded protein molecule has a great influence on its biological function. This protein turns on a specific gene in a cell.

The shapes of molecules are often very important to their functions in chemical reactions. The protein molecules in your body are a good example. They consist of chains of hundreds of atoms folded into intricate shapes. The shape of the individual protein determines how it works in the cell.

18.11 Properties of Covalent Compounds

Covalent compounds consist of covalently bonded molecules. Although there is some attraction between covalent molecules, especially polar molecules, they remain separate, recognizable particles. The attraction between covalent molecules is generally not as strong as the electrostatic forces between the ions in ionic compounds. This weak degree of attraction has a significant effect on the physical properties of covalent compounds.

Almost all covalent compounds have relatively low melting points, and nearly all pure compounds that are gases or liquids at standard conditions are covalent compounds. Covalent gases include carbon dioxide (CO_2), ammonia (NH_3), and methane (CH_4). Examples of covalent liquids are water (H_2O), methanol (CH_3OH), and hydrogen disulfide (H_2S_2).

Some covalent compounds are solids at standard conditions. Many organic molecules are very large molecules called *macromolecules*. They have different properties from "ordinary" covalent molecules. For example, the proteins, fats, and starches that are

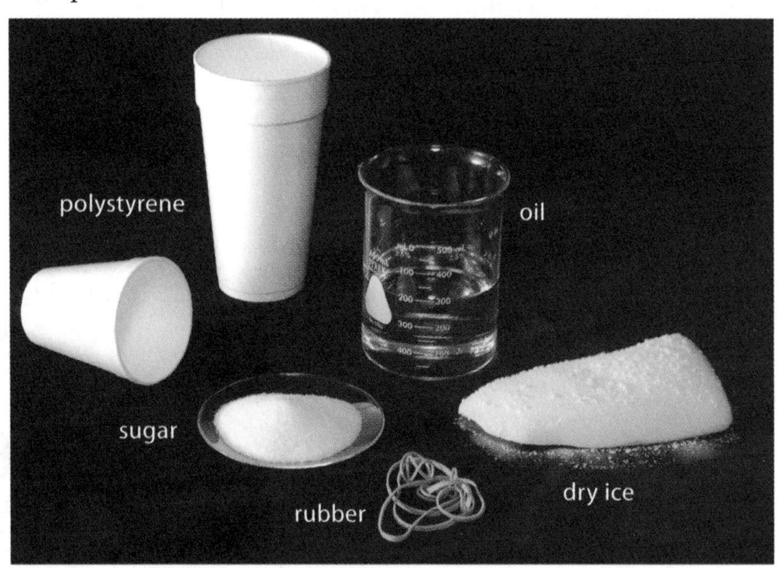

18-19 Covalent compounds are quite varied in their properties.

18-20 The marvel of the human body, God's greatest creation, depends on thousands of covalent compounds.

in food or that are part of your body are solid covalent compounds at room temperature. Most of these compounds begin to melt or decompose at temperatures of less than a few hundred degrees Celsius.

Properties of covalent compounds are quite varied. Some, such as rubber, show great elasticity. Few conduct heat or electricity well, making them excellent insulators. Some covalent compounds can store tremendous energy; explosives are an example of this property.

God's design is evident in the covalent compounds He used to form our bodies. Ionic compounds (see Section 18C) would be brittle and soluble in water (they *do* play a role in the skeleton, though), and metallic solids (see Section 18D) would be extremely heavy and prevent flexible movement. But the covalent compounds that form our body structures are strong, lightweight, and flexible. They don't dissolve easily in water and are perfectly suited for maintaining the life God gave us.

18.12 Using Covalent Compounds to Solve Problems

How can covalent compounds and their properties give us pain relief? When we get a throbbing headache, many of us reach for a bottle of *ibuprofen*, whose formula is $C_{13}H_{18}O_2$. You can see from Figure 18-21 that ibuprofen is a compound in which hydrogen and oxygen atoms are covalently bonded to carbon atoms. It is also a compound whose properties decrease pain.

When a person's body is injured, pain receptors in the injured area relay this information as an electrical impulse to the brain, where it is interpreted as pain. At the site of pain stimulation, the body releases an enzyme that helps produce pain-enhancing hormones called *prostaglandins*. Ibuprofen interferes with the production of the enzyme, blocking the body's ability to produce prostaglandins and decreasing both pain and inflammation.

Besides relieving pain and inflammation, ibuprofen also reduces fevers, and some research indicates that it may even help delay or prevent the onset of Alzheimer's and Parkinson's diseases. Though ibuprofen is incredibly useful, there are some possible adverse side effects. However, if used properly, its beneficial effects far outweigh the hazards.

Pain Relief and Dominion Science

Matthew 8:14–17 and Revelation 21:4 indicate that God desires to relieve humans of their pain. How should this truth affect our view of ibuprofen and other pain relievers?

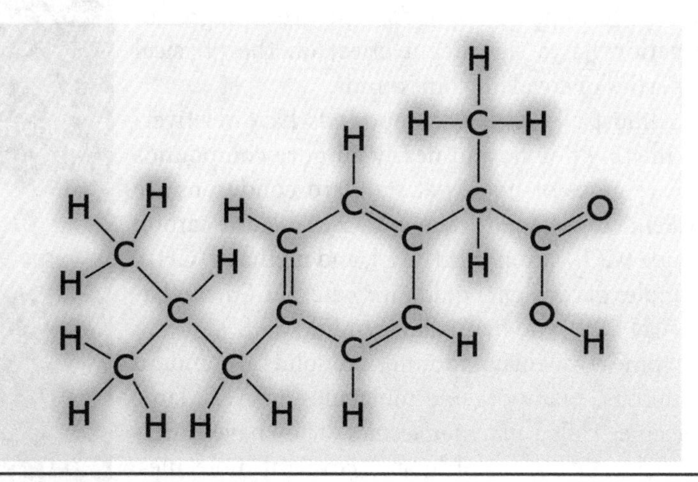

18-21 The ibuprofen molecule

18B Section Review

1. What is a covalent bond?
2. How does a covalent bond hold atoms together?
3. How is electron dot notation useful for predicting the number of covalent bonds a nonmetal atom can form? What do we call the symbols that use electron dot notation to represent molecules?
4. How must the standard electron dot notation of Group 4A (14) elements be modified to show the number of potential covalent bonds they can form?
5. What is a polyatomic molecule? What can you say about the valence electron structure of all atoms in a polyatomic molecule?
6. Which of the following molecules contain polar bonds? How can you tell? Which has the most polar bond? (*Hint*: Calculate the ΔEN for each pair of bonded atoms.)
   a. H – H
   b. H–C–H (with H above and below C)
   c. H–O (with H below O)
   d. I – I
7. What class of elements forms covalent compounds? List three properties of covalent compounds.
8. DS ✪ When you examine the formula for ibuprofen, $C_{13}H_{18}O_2$, why can you state that its atoms are held together with covalent bonds?
9. DS ✪ In some people, ibuprofen can severely irritate the stomach and intestines and damage the liver when taken at high dosages for long periods. How can these problems be avoided?
10. (True or False) When oxygen bonds with hydrogen in a water molecule (H_2O), it forms double covalent bonds.
11. ✪ For each of the following elements, write the electron dot notation. Then determine how many covalent bonds each could form to complete its valence energy level.
   a. sulfur (S)
   b. hydrogen (H)
   c. bromine (Br)
   d. phosphorus (P)
12. ✪ Draw the Lewis structures for the following covalent compounds.
   a. carbon tetrachloride (CCl_4)
   b. hydrogen sulfide (H_2S)
   c. nitrogen trifluoride (NF_3)
   d. hydrogen chloride (HCl)

18C Objectives

After completing this section, you will be able to

- ✓ explain how an ionic bond forms.
- ✓ tell which kinds of elements generally form ionic compounds.
- ✓ use electron dot symbols to describe the net electron transfers in ionic bonds.
- ✓ describe how ions assemble in crystal lattices.
- ✓ identify the structural subdivisions that make up a crystal lattice.
- ✓ describe the general properties of ionic compounds.

18C Ionic Bonds

18.13 Forming Ionic Bonds

What happens when bonds form between atoms with *very* different electronegativities? For example, sodium and chlorine combine to form common table salt, sodium chloride (NaCl). Sodium's electronegativity is 0.93, and chlorine's is 3.16. The difference between the two, 2.23, is much larger than the ΔEN between nonmetal elements found in covalent compounds. Such a large value means that when sodium and chlorine combine, the bonding electrons have a very high probability of being found around the chlorine nucleus and almost no probability of being near the sodium nucleus.

At any given moment, it would appear that the sodium atom has transferred its one valence electron to the chlorine. A complete electron transfer between atoms in a compound is not possible; however, the electron in such a bond can be considered transferred for all practical purposes. The chlorine acquires a net negative charge and the sodium a net positive charge. The strong electrostatic force between the two ions pulls them tightly together to form an **ionic bond**. Substances formed by ionic bonds are called **ionic compounds**.

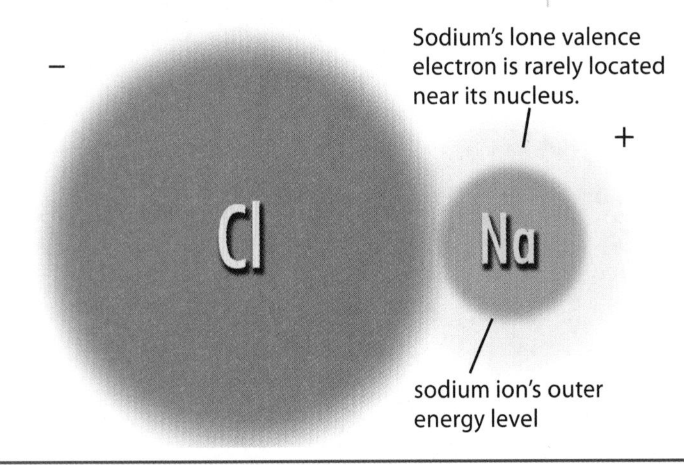

18-22 The transfer of sodium's single valence electron to chlorine is nearly total. There is almost no probability of finding it near the sodium nucleus.

While covalent bonds tend to be somewhat polar due to small values of ΔEN, ionic bonds take bond polarity to the extreme. If pure covalent bonds (ΔEN = 0) are at one end of a spectrum, ionic bonds are at the other. Pure covalent bonds involve equal sharing of electrons, and the bond is nonpolar. In the most ionic bond possible (between cesium and fluorine), there is almost no electron sharing. Remember that an ionic bond never involves a *complete* transfer of electrons because no element has zero electronegativity; in other words, *there is no purely ionic bond.*

How can you tell if a bond between two atoms is ionic or covalent? This is not always an easy question to answer. Chemists determine *bond character* based on the ΔEN of the atoms involved in bonding. In general, ionic bonds form between metals and nonmetals. The greater the ΔEN between the bonded atoms, the stronger the ionic bond between them.

No complete electron transfer occurs when atoms are ionically bonded in a solid compound. However, if the atoms in an ionic compound separate, such as when they dissolve in water or form a plasma, the particles become true ions and the electrons stay with the anions.

18.14 Illustrating Ionic Bonds

Just as for covalent compounds, electron dot notation can illustrate changes in atoms' valence electron structure for ionic compounds. However, it has its limitations. Because of the difference between the types of bonding, covalent compounds can exist as isolated molecules while ionic compounds exist only as immense arrays of ions. For covalent compounds, Lewis structures illustrate distinct molecules, while electron dot notation for ionic compounds shows only the compound's *ratio of ions*. The same ionic bonds actually take place on a vast scale. Ionic bonding electron dot symbols do not represent molecule-like particles.

Let's look at how a strongly electronegative halogen (Group 7A, or 17), such as fluorine, bonds with a weakly electronegative alkali metal (Group 1A, or 1), such as sodium. Fluorine requires one electron to complete its valence octet. The strong electron affinity of fluorine tends to hold on to the bonding electron. Sodium achieves a complete octet when it *loses* its lone valence electron, exposing its next lower complete energy level. This arrangement makes both atoms more stable. Even though the electron transfer is not complete or permanent, electron dot notation can show the probable location of the bonding electron.

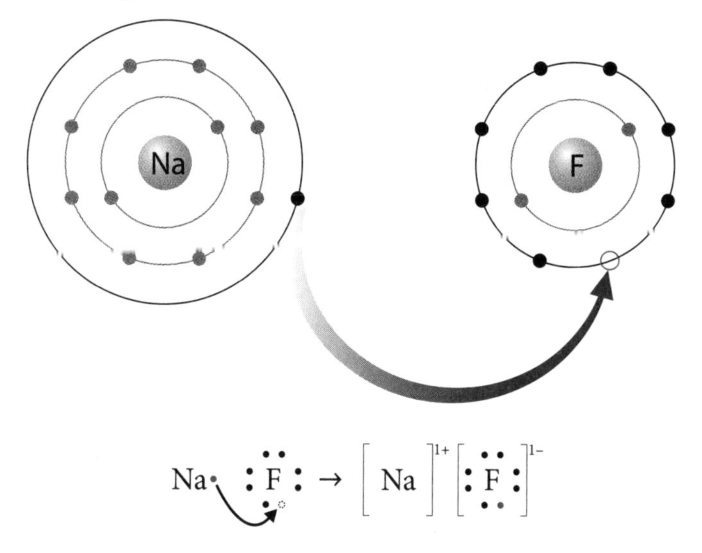

$$\text{Na}\cdot \; :\ddot{\text{F}}: \rightarrow [\text{Na}]^{1+}\,[:\ddot{\underset{\cdot\cdot}{\text{F}}}:]^{1-}$$

When electron dot notation is used to represent an ionic bond, the brackets surrounding each ion show its *original* valence energy level *after* the electron transfer. In this example, sodium forms a cation after losing its valence electron (no electrons are left in its original valence energy level). Fluorine forms an anion and has a complete valence octet (there are now eight electrons in its original valence energy level). The charges of the cation and anion appear as superscripts of the brackets surrounding the element's symbol. The compound sodium fluoride (NaF) can form when a certain number of sodium atoms lose one valence electron to an equal number of fluorine atoms. Thus, only a single transfer needs to be illustrated. The ratio of sodium to fluorine atoms is 1:1.

Both the sign and the magnitude of the charges on the ions must be shown. The number comes *before* the plus or minus sign in ionic charges. The total negative charges of the anions must equal the total positive charges of the cations.

Some ionic bonds involve the transfer of two or even three electrons. For example, when magnesium bonds with sulfur, two electrons transfer. Sulfur has six valence electrons and needs two more to complete its octet. Magnesium has two valence electrons. Its low electronegativity allows the more-electronegative sulfur to take those two electrons. Each ion that forms will then have a full outer energy level.

$$\text{Mg}: \; :\ddot{\text{S}}: \rightarrow [\text{Mg}]^{2+}\,[:\ddot{\underset{\cdot\cdot}{\text{S}}}:]^{2-}$$

Magnesium loses two electrons and has a 2+ charge. Sulfur gains two electrons and has a 2– charge. The two ions are held together by the electrostatic attraction of their opposite charges. The ratio of ions is still 1:1, but this time two electrons were transferred.

Ionic bonds can also involve more than two ions. In magnesium fluoride (MgF_2), two fluorine atoms meet their need for electrons when each takes one electron from a magnesium atom's valence energy level.

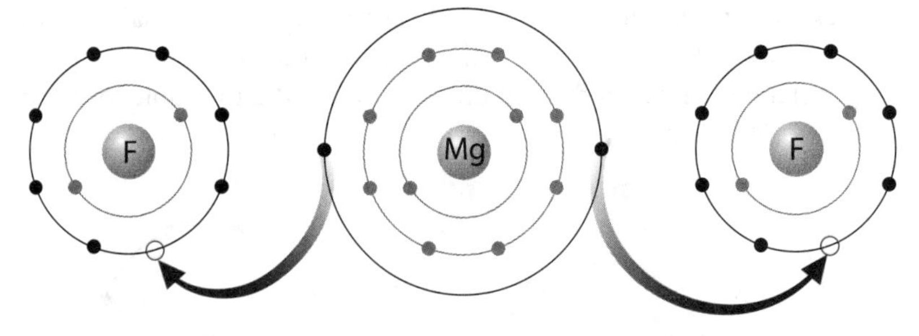

For every 2+ magnesium ion, there are two 1– fluorine ions.

$$Mg\!:\ \ :\ddot{F}:\ \ :\ddot{F}: \rightarrow \left[:\ddot{\underset{..}{F}}:\right]^{1-} \left[Mg\right]^{2+} \left[:\ddot{\underset{..}{F}}:\right]^{1-}$$

In this case, two electrons are transferred, but the ratio of magnesium to fluorine ions is 1:2.

Sometimes more than one metal atom is necessary to provide the electrons for a single nonmetal atom. For example, in sodium sulfide (Na_2S), sulfur acquires the two electrons it needs by taking one electron from two sodium atoms.

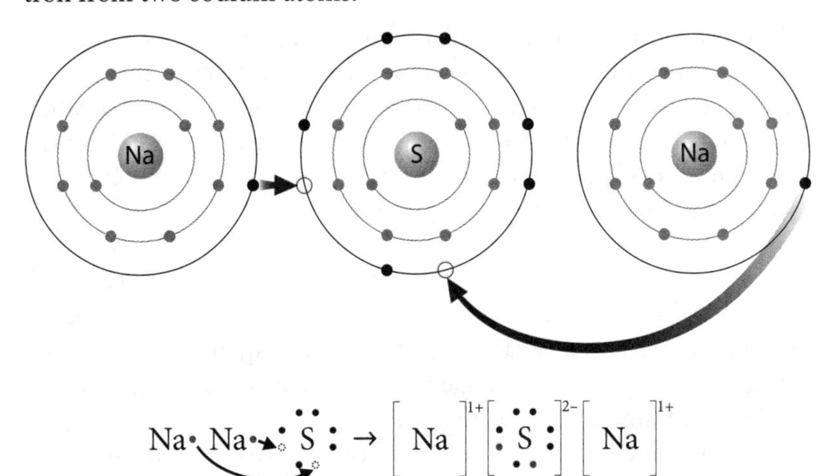

$$Na\cdot\ \ Na\cdot\ \ :\ddot{S}: \rightarrow \left[Na\right]^{1+} \left[:\ddot{\underset{..}{S}}:\right]^{2-} \left[Na\right]^{1+}$$

18.15 Structure of Ionic Compounds

Ionic compounds are composed of innumerable ions. Just think about how many atoms must be present in a grain of salt! When salt crystallizes, the sodium and chlorine ions are attracted to each other to form the solid compound.

The attraction between individual ions doesn't stop with just two ions. Every ion can attract ions of opposite charge. Ions dissolved in a liquid are very mobile. They move toward opposite charges and clump together. An orderly pattern appears as positive and negative ions line up. Eventually, they form an immense three-dimensional pattern called a *crystal lattice*.

Since an ionic compound doesn't form molecules, how do we determine its chemical formula? For example, why is table salt's formula NaCl? Why not $Na_{10}Cl_{10}$ or $Na_{1,300,457}Cl_{1,300,457}$? The smallest ratio of elements in an ionic compound that describes its chemical compo-

18-23 Crystallization of sodium chloride

sition is called its **formula unit**. For example, table salt's formula unit is NaCl because the ratio of sodium ions to chlorine ions is 1:1. The formula unit for barium chloride is $BaCl_2$ because the ratio of barium to chlorine ions is 1:2.

Ionic crystal lattices are composed of repeating structural units. These groupings of ions, called *unit cells*, are used to classify lattices. The shape and size of a unit cell are determined by the relative sizes and numbers of the different ions in the compound. There are only seven different shapes of unit cells that can be repeatedly joined to form an infinite lattice array. The simplest is shaped like a cube. Figure 18-25 shows a cubic unit cell for sodium chloride. Notice that ions on the faces, edges, and corners of the cube are shared with adjacent unit cells. Most unit cells contain several of the compound's formula units. For example, the unit cell for table salt contains four formula units of NaCl.

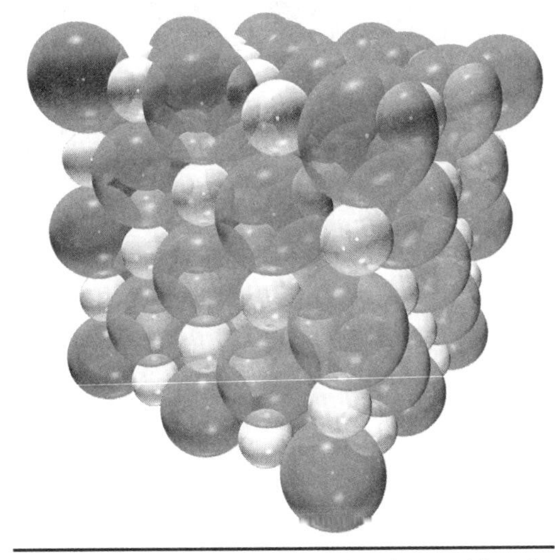

18-24 The crystal lattice of sodium chloride (NaCl)

18.16 Properties of Ionic Compounds

At room temperature most common ionic compounds are solid, brittle, crystalline substances. When ionic compounds solidify slowly, they form distinctive and often beautiful crystals. Many pure minerals in the earth's crust are complex ionic compounds that form crystals, which can be used to identify the mineral.

Most ionic compounds have very high melting points because of the strength of the electrostatic force holding their ions together. For example, table salt has a melting point of 800.7 °C (1473.3 °F) at standard conditions. Similarly, the boiling points of most ionic compounds are very high.

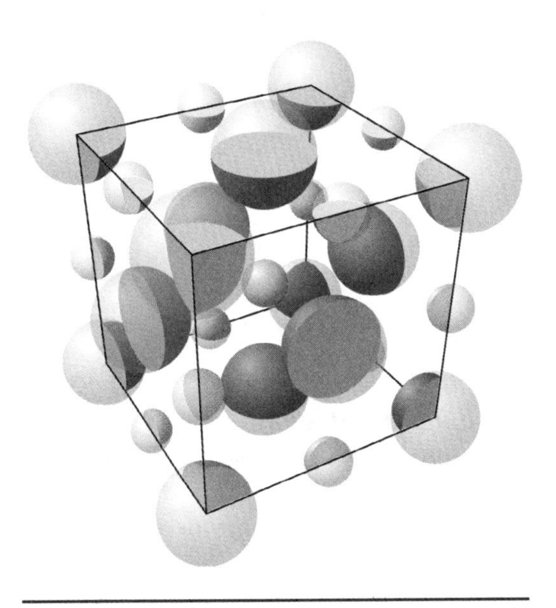

18-25 A unit cell of sodium chloride

Recall from Chapter 9 that most pure materials have characteristic latent heats of fusion (L_f) and vaporization (L_v). These are the amounts of heat that must be absorbed per gram of a substance for it to melt or boil, respectively. Ionic compounds have much lower heats of fusion and vaporization than many covalent compounds do. They are lower because by the time an ionic compound has been heated to near its melting or boiling point, it requires little additional energy to separate the ions. In covalent compounds the changes of state occur at much lower temperatures, but much more energy is required to completely break down intermolecular forces.

Many ionic compounds, like table salt, dissolve in water. When an ionic solid is dissolved or melted, its ions are mobile. This condition makes them an excellent conductor of electricity. An ***electrolyte*** is a substance that produces ions when it dissolves in a liquid. Positive and negative ions conduct current through the liquid. The current is a flow of ions rather than electrons. When electrodes connected to a battery are placed in an *electrolytic solution*, two currents flow. Cations flow in one direction toward the negative electrode (the *cathode*), while anions flow in the other direction toward the positive electrode (the *anode*). Solid salts, however, cannot conduct electricity because their ions and valence electrons are held firmly in place by the ionic bonds.

electrolyte (ih LEK truh LITE)

18C Section Review

1. What makes certain combinations of atoms form ionic bonds rather than other kinds of bonds?
2. Where will the bonding electrons most likely be found in an ionic bond?
3. Why is a purely ionic bond impossible in an ionic compound?
4. Compare the polarity of a pure covalent bond with that of an ionic bond.
5. ✪ What two nonradioactive elements could theoretically form the strongest ionic bond? Give the formula for the compound and explain your answer.
6. What does the electron dot notation for an ionic compound represent? What does it *not* represent?
7. What is true of the sum of all charges in an ionic compound's electron dot notation? Why must this be so?
8. How are particles arranged in an ionic compound?
9. (True or False) The character of a bond—whether ionic or covalent—between two atoms can be viewed as the degree of sharing of the bonding electrons.
10. ✪ Write the electron dot notations for the following ionic compounds. Be sure to include the charges of the ions.
    a. beryllium chloride ($BeCl_2$)
    b. sodium oxide (Na_2O)

18D Metallic Bonds

18D Objectives

After completing this section, you will be able to

✓ describe the simplest model of metallic bonding.

✓ explain how metal atoms are arranged in elements and in compounds (alloys).

✓ explain how metallic bonds account for the properties of metals.

18.17 Metal Bonding Theory

How do metal atoms "stick" together? What kind of bond holds the atoms together in pure aluminum, for example? Each neutral aluminum atom has three valence electrons. To form an ionic bond, it would have to give away all three of these electrons and another aluminum atom would have to accept them. But why would identical atoms become either cations or anions? In addition, the aluminum anions would still not have a complete octet. Can aluminum atoms form covalent bonds with each other? No, there is no apparent way that aluminum atoms could share electrons with each other so that they all end up with stable octets. Chemists developed several models to explain the bonding of metals and their properties. The simplest explanation is called the *electron sea* or *free electron theory.*

Ninety percent of the metals with known electron structures have only one or two valence electrons. As you know, these electrons are loosely held. According to the electron sea model, the valence electrons in a metal atom are shared among all the other bonded atoms in the element or alloy. This differs from sharing electrons between two individual atoms, as in covalent bonding (which cannot explain the behavior of metals).

Metallic bonds involve positive metal ions arranged in a crystal lattice immersed in a "sea" of negative electrons. These valence electrons are free to roam the entire lattice structure. All the nuclei

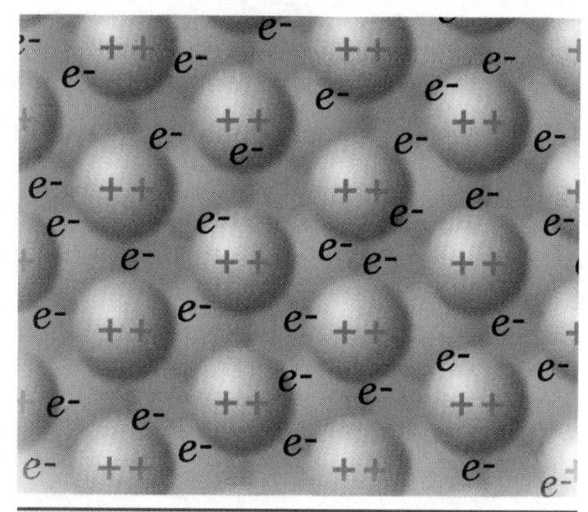

18-26 In metals the cations are surrounded by a "sea" of mobile, freely shared electrons.

share all the former valence electrons. Metallic bonding is a different method of achieving chemical stability *without directly satisfying the octet rule for individual atoms.* No electron dot notation can be used to represent metallic bonds.

Atoms of different metals can bond together just like atoms of the same metal. Compounds of different metals are called *alloys.* While some alloys are true compounds, others are not. They are heterogeneous mixtures at the microscopic level.

18.18 Metallic Lattice Structure

The structure of metal elements and compounds at the particle level is a crystal lattice, similar to ionic compounds. In pure metals, atoms are naturally arranged to take up the smallest amount of space possible as the bonding electrons draw the metal cations together.

The most efficient layout of identical spherical objects, such as metal atoms, is called *hexagonal close-packed (HCP).* In this arrangement, layers of atoms are nestled in the nooks formed by the layers above and below them. For an illustration of how this arrangement works, imagine hexagons of seven atoms (a ring of six surrounding a central atom) in alternating layers separated by triangles of three atoms nestled between them (see Figure 18-27). Figure 18-28 shows particles of carbon in an HCP array.

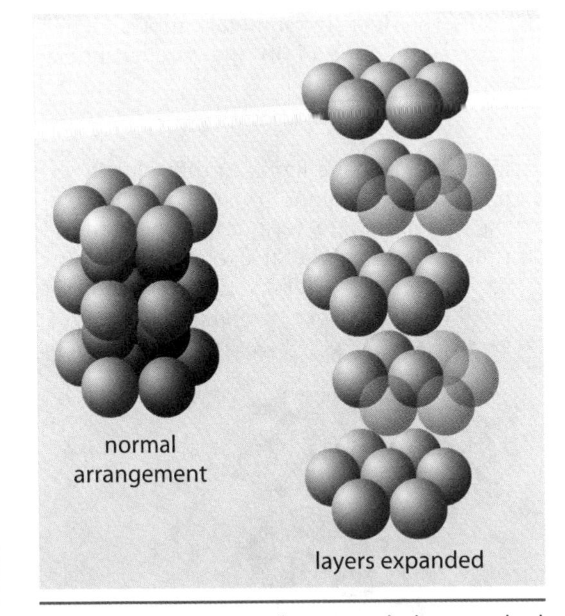

18-27 The common hexagonal close-packed (HCP) arrangement of metal atoms

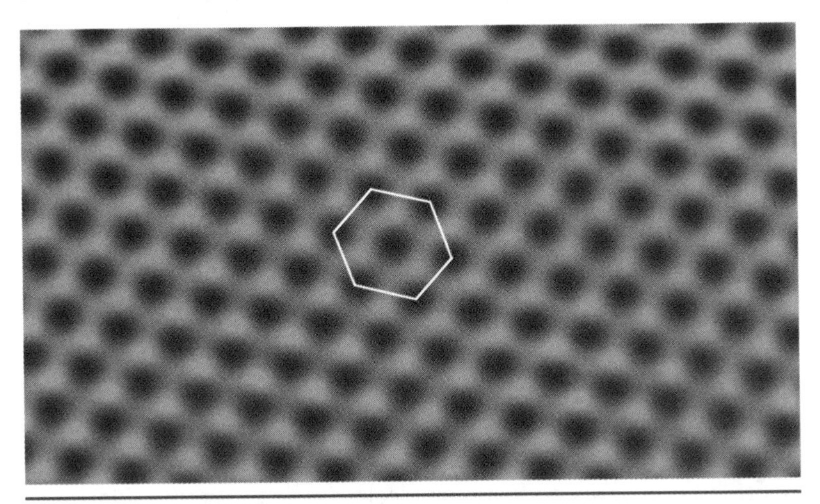

18-28 STM image of particles arranged in an extensive HCP array

Some alloys that are true compounds have a uniform mixture of different atoms scattered throughout the crystal lattice. The different sizes and electron configurations of just a few of these atoms can make a tremendous difference in the properties of the alloy. For instance, when a small percentage of atoms in a piece of iron is replaced by nickel atoms, the metal, an alloy called stainless steel, becomes far less subject to rusting. Other alloys contain a mixture of microcrystals of different metals uniformly mixed throughout the lattice. Although not uniform at the atomic level, such alloys often have beneficial properties such as extreme hardness or corrosion resistance.

18-29 In this iron-carbon alloy, the different constituents exist as microscopic crystals mixed uniformly together.

18-30 The ductility of metals allows them to be pulled into fine wire by special machines.

Most metals have limits to their ductility. If deformed beyond their *elastic limits,* the metals will fracture or fail. This trait is called *metal fatigue.* It explains why you can break the tab off a soda can by bending it back and forth several times.

18-31 The luster of metals is a property of metallic bonding.

18.19 Properties of Metals

Most metals are solids at room temperature. Mercury is the only exception. Most also have the properties of malleability and ductility. *Malleability* is the ability to be hammered or rolled into thin sheets. *Ductility* is the ability to be stretched without breaking, as when metal is drawn into a wire. Although metal atoms are held rigidly in place, they are not immovable. When sufficient force is applied, metal atoms can shift positions without breaking their bonds with adjacent atoms. Thus the shape of metal can change without damaging its underlying structure. Solid ionic materials (and even some covalent solids) are generally brittle. If struck with a sharp blow, they shatter.

Metallic bonding accounts for nearly all electrical and thermal properties unique to metals. Solid metals conduct electricity well because their electrons move easily when an electrical potential is applied. Most covalent and ionic solids do not conduct electricity because their electrons are tightly held in bonds. Free electrons also allow metals to conduct heat very well.

Another distinctive property of metals is their shiny luster. Bonding electrons in metals can have a wide range of energies, since they are not bound to any particular atom. When a photon in an incident light ray strikes them, they immediately emit the photon back as a reflected ray. Since many different wavelengths of light are reflected and the direction of the reflection is orderly due to the orderly crystal array of the atoms, metals appear shiny and mirror-like, especially if they have flat, polished surfaces.

18.20 Summary of Chemical Bonds

Each of the three types of chemical bonds gives its compounds specific properties. So we can predict with some accuracy the inner structure of an element or compound through observation of its properties. Though we can't see covalent bonds in a compound, we can deduce their presence if a substance is flexible or if it is a liquid or a gas at room temperature, does not readily dissolve in water, and doesn't conduct electricity. Though we can't see ionic bonds in a compound, we can deduce their presence if a material has a high melting point, does not conduct electricity or heat in its solid state, but conducts electricity well when dissolved in water. Though we can't see metallic bonds, we can assume that they are present if a material is solid, has a shiny metallic luster, is malleable and ductile, and conducts electricity well.

18-2 Characteristics of Bonding Types

	Covalent	Ionic	Metallic
Valence electrons	shared between two atoms per bond	highly unequal sharing; essentially transferred	free; shared among all atoms in the metal
Electronegativity	elements with similarly high ENs	elements with greatly different ENs	elements with similarly low ENs
Types of elements	between nonmetals	between metals and nonmetals	between metals
Melting points	low	high	high
Solubility in water	generally poor	generally good	generally poor
Conductivity	poor	poor	good

18D Section Review

1. Why does the octet rule not apply to iron atoms when they bond to other iron atoms?
2. Name and describe the model that accounts for metallic bonding.
3. How do similar metal atoms most often arrange themselves in the solid state? What does this arrangement achieve?
4. What property of metallic bonds affects the thermal and electrical conductivity of metals?
5. How does the model for metallic bonding account for the properties of malleability and ductility?
6. Why are metals shiny?
7. How does the electronegativity number (EN) of atoms account for the character of the bond between them?
8. (True or False) Metals are arranged in hexagonal groups of seven atoms called hexagonal close-packed (HCP) units.

Scientifically Speaking

octet rule	414
electronegativity	415
covalent bond	418
Lewis structure	418
covalent compound	423
ionic bond	426
ionic compound	426
formula unit	429
metallic bond	430

Chapter Review

1. What ultimately determines how atoms bond?
2. Which electrons in an atom are generally involved in chemical bonding?
3. Why are bonded atoms generally more chemically stable than unbonded neutral atoms?
4. Why does a positive electron affinity energy indicate a very weak affinity for electrons?
5. In general, how do electron affinity and electronegativity change in the periodic table?
6. How do covalent bonds form? Include how many atoms are usually involved in a covalent bond and compare their electronegativity numbers.
7. What actually holds atoms together in a covalent compound?
8. Why is methane (CH_4) an organic compound?
9. Why are covalent bonds between different elements always polar? When are covalent bonds nonpolar?
10. What determines the shape of simple molecules formed around a central atom?
11. Compounds that are liquids or gases at standard conditions usually are what kind of compounds?
12. What are the beneficial properties of covalent compounds as the main components of living organisms?
13. DS ✪ Ibuprofen functions by attaching itself like a lock to another molecule to prevent the production of prostaglandins. The ibuprofen molecule's shape and atomic arrangement are necessary to fulfill its purpose. How effective would it be after dissolving in the blood if it were an ionic compound, which breaks up into ions, rather than a covalent compound?
14. DS ✪ Explain why discovering ways to relieve pain is an important aspect of dominion science.
15. Explain why the electron transfer in an ionic bond is never complete.
16. When illustrating an ion with electron dot notation, what do the symbols for the ions represent?
17. Why is an ionic compound represented by a formula unit rather than a molecular formula?
18. In a solid metal, what do we call the ions surrounded by a sea of electrons? If some of the electrons are drawn away from one end of a metal bar by an electrical potential, what happens to the atoms there?
19. What evidence do we have that metal atoms are arranged in crystal lattices?
20. ✪ Why is pure sodium metal so soft that it can be cut with a knife? Include in your answer the influence of the strength of the sodium cation charge (that is, the number of protons in its nucleus) compared to the charge of the electron sea surrounding the cations.

True or False

21. Most metals have large electronegativities.
22. All chemical bonds involve at least a tiny probability of sharing electrons.
23. The shared electrons in a covalent bond always come in pairs.
24. A triple covalent bond holds three atoms together.
25. Sodium chloride (NaCl) is a diatomic molecule.
26. A polar bond between chlorine and oxygen is slightly positive at the oxygen end compared to the chlorine end.
27. The water molecule (H_2O) is "bent" because the bonds and nonbonding electron pairs cannot be arranged symmetrically around the oxygen atom.
28. A pure cooking oil is a covalent compound.
29. Ionic bonds form through the *transfer* of electrons from atoms with low electronegativities to atoms with high electronegativities.
30. The electron dot symbol for the calcium cation is

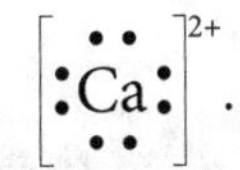

31. A formula unit of an ionic compound contains the smallest ratio of atoms of each element in the compound.
32. A uniform mixture of different metal atoms is a metal compound called an alloy.
33. The kind of chemical bond in a material can often be deduced by the properties of the material.

34. Draw the electron dot symbols for the following elements and ions.
    a. potassium (K)
    b. argon (Ar)
    c. lithium (Li^+)
    d. boron (B)
    e. barium (Ba)
    f. oxygen (O^{2-})
35. Draw the Lewis structures for the following molecules.
    a. hydrogen peroxide (H_2O_2)
    b. bromium oxide (Br_2O)
    c. carbon tetraiodide (CI_4)
    d. sulfur dichloride (SCl_2)

DS 36. A rare genetic disorder can cause a person to experience no pain. For example, researchers studied a Pakistani street performer with this disorder who would stab himself with knives and walk on burning coals with no pain. He later died from injuries after jumping off a roof. Though pain in life probably increased as a result of the Curse following Adam's sin, how can it be considered a good thing?

CHEMICAL REACTIONS

CHAPTER 19

DOMINION SCIENCE PROBLEM

Hard Knocks

Irregular sounds in a car engine usually spell trouble. One such common noise is a knocking sound. What causes it? Is it a problem? A modern car engine takes in air and mixes it with a spray of gasoline from the fuel injector in each cylinder. This liquid gasoline immediately vaporizes. The mix of gasoline vapor and air is compressed by a piston and then ignited with a spark plug. The rapidly burning fuel expands, forcing the piston down. This action turns the crankshaft, which turns the wheels of the car. Sometimes the vapor explodes as the piston compresses it, before the normal spark plug ignition, making the engine *knock* or ping. Knocking is bad for the engine because it can damage the engine's moving parts and make them wear out faster than expected. How can knocking be reduced in car engines?

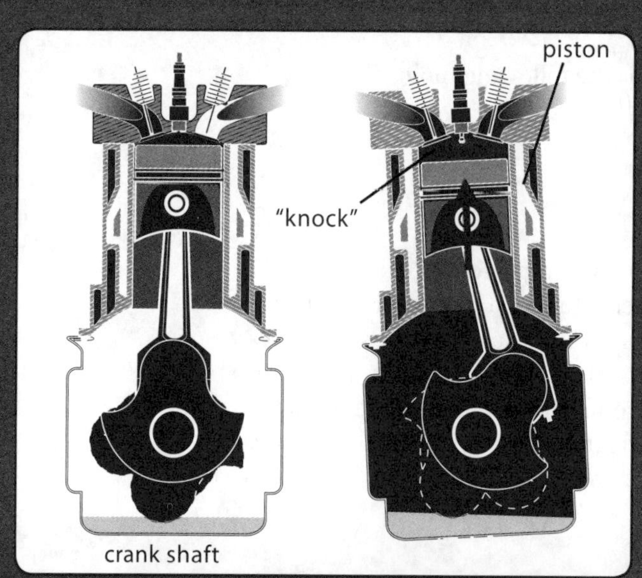

19-1 Normal engine ignition occurs when the fuel is fully compressed (left). Knocking occurs when fuel ignition occurs before the fuel is fully compressed (right).

19A Compounds and Chemical Formulas

19.1 Introduction

What is the most common compound on the surface of the earth? If you've taken an earth science course, you might remember that silicate (SiO_4) compounds make up more than 90% of the earth's crust. But is this true for the earth's *surface*? The compound that is most common on the visible surface of the earth is H_2O—water. The world's oceans cover more than 71% of the earth's surface.

Other compounds are also vital to life on Earth. All animals and plants obtain solar energy either directly or indirectly. Plants receive it directly and store it in sugar glucose ($C_6H_{12}O_6$), which is produced by a compound in green plants called chlorophyll ($C_{55}H_{72}O_5N_4Mg$). Animals obtain solar energy indirectly by eating plants or other animals that have eaten plants. Without chlorophyll, sugar compounds could not be stored in food and complex forms of life could not survive.

There are millions of known compounds and probably an equal number that are yet to be discovered or developed. How do we identify compounds? Where do their names come from? How do they react with other substances? This chapter begins to answer these and many other questions about compounds.

19A Objectives

After completing this section, you will be able to

- ✓ differentiate between empirical and chemical formulas.
- ✓ use the periodic table to state the usual oxidation number for familiar elements.
- ✓ use oxidation numbers to write compound formulas.
- ✓ determine the charge of a polyatomic ion if given its formula.
- ✓ name binary compounds and compounds containing polyatomic ions from their formulas.
- ✓ write correct formulas for binary compounds from their names.

19-2 What is the most common compound on the face of the earth?

19-3 These products contain thousands of useful chemical compounds.

19.2 Chemical Formulas

A **chemical formula** is a shorthand way to tell the identities and the amounts of elements in a compound. For example, a molecule of methane, part of natural gas, is made of one carbon atom bonded to four hydrogen atoms. Its formula is CH_4. Element symbols tell which elements are in a compound. Subscripts tell how many atoms of each element are present in a molecule or formula unit of a compound.

The chemical formula of each compound is unique. For example, every molecule of methane has the same combination of carbon and hydrogen atoms. A molecule with any other combination of carbon and hydrogen would not be methane. This rule is generally true for most simple compounds. However, larger molecules and more complex compounds may have the same numbers and kinds of atoms arranged in different ways.

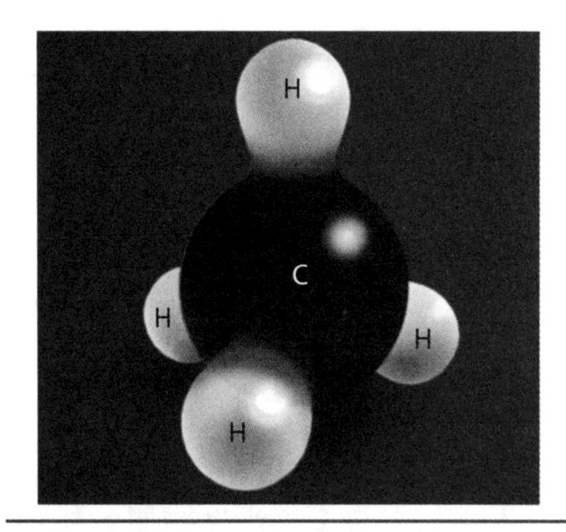

19-4 A molecular model of methane

If no subscript number appears after an element symbol in a chemical formula, a subscript of 1 is understood.

The empirical formula of a compound can be the same as its *molecular formula*. For example, both the empirical and chemical formula for methane is CH_4.

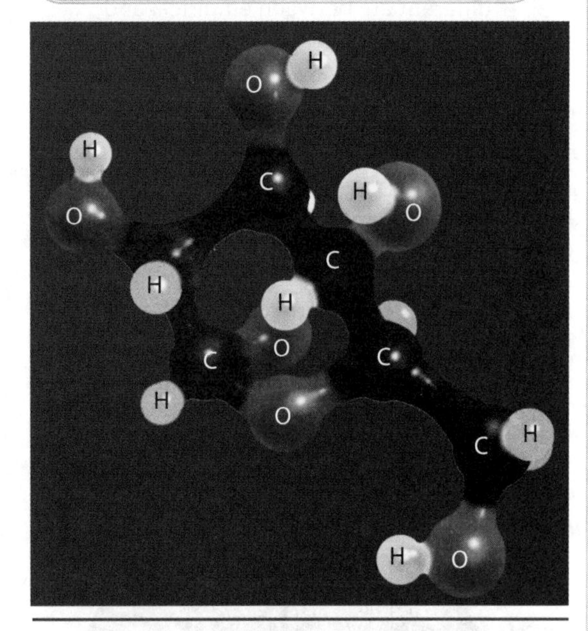

19-5 A molecular formula represents the actual number of atoms in a molecule (a glucose molecule is shown). Its empirical formula (CH_2O) shows only the smallest whole-number ratio of the atoms of the different elements.

Sometimes chemists are not as interested in the exact formula of a compound as in the smallest whole-number ratio of its components. In such cases, they need the compound's *empirical formula*. The empirical formula's subscripts show only the relative proportion of the elements in a compound. In a covalent compound the actual formula will be a whole-number multiple of the subscripts of the empirical formula. For example, the molecular formula for glucose is $C_6H_{12}O_6$; each glucose molecule has six carbon atoms, twelve hydrogen atoms, and six oxygen atoms. However, the empirical formula for glucose is CH_2O. The atoms for the different elements in glucose are in a 1:2:1 ratio. The molecular formula for glucose has six times the number of atoms in its empirical formula. Note that the formula unit of an ionic compound is *always* an empirical formula because it represents the smallest ratio of ions in the compound.

19.3 Oxidation Numbers

How do we know that the formula for table salt is NaCl and not Na_2Cl, $NaCl_2$, or some other ratio of atoms? That is, how do we know that each formula unit has exactly one sodium atom and one chlorine atom?

To answer these questions, chemists developed a method that allows them to predict chemical formulas based on the numbers of electrons involved when atoms bond or become ionized. They determined these electron numbers for each element after studying thousands of compounds through *analytical chemistry*. As a result, chemists have assigned chemical oxidation numbers to each element. An element's **oxidation number (ON)** is the *number of electrons that a bonded atom or ion would have to gain or lose to return to its neutral state.*

Oxidation numbers may be positive or negative. Remember that atoms bond according to the octet rule. Some lose electrons to bond with other atoms and become positively charged. To return to a neutral state, they must gain electrons. Therefore, these atoms are assigned positive oxidation numbers (electrons must be added). Atoms with high electronegativities tend to gain electrons when they bond. To return to their neutral state, they must lose electrons. Thus, they are assigned negative oxidation numbers (electrons must be subtracted). Oxidation numbers are also used to keep track of electrons during chemical reactions.

How are specific numbers assigned? Look at the families of the periodic table. A neutral atom of the alkali metals (Group 1A) has one valence electron. It loses this electron when it bonds, acquiring a charge of 1+. It needs to gain one electron to become neutral again. Each alkali metal is assigned an oxidation number of +1. Similarly, the oxidation number for elements in Group 2A is +2, and it is +3 for those in Group 3A because these metal families readily lose their valence electrons.

The noble gases (Group 8A) have eight valence electrons. According to the octet rule, they don't normally gain, lose, or share electrons to form chemical bonds. Therefore, their oxidation number is 0. Their neighbors, the halogens (Group 7A), however, strongly attract one electron to form bonds. As unbonded free ions, they are negatively charged. Since they must

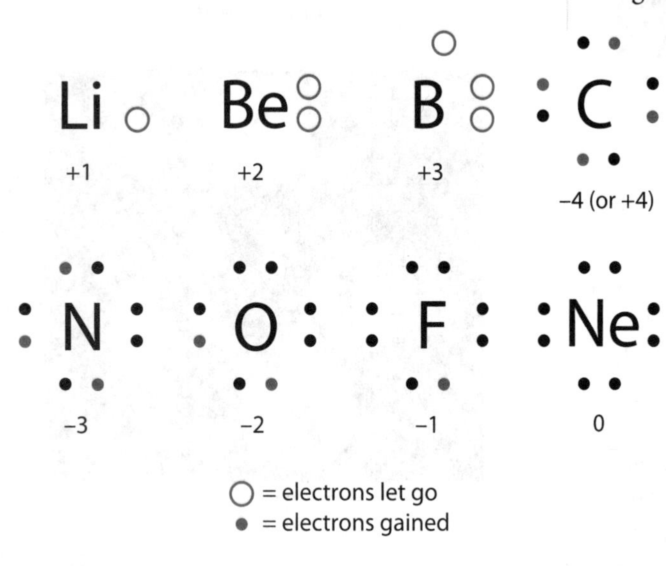

19-6 A *bonded* element's common ON is equal to the number of electrons it must gain or lose to reach the neutral state. The *sign* of the ON is positive if it must gain electrons and negative if it must lose them.

lose this extra electron to become neutral, their oxidation number is −1. Similarly, elements in Group 6A are normally assigned an oxidation number of −2.

Oxidation Number Rules

Oxidation numbers can be assigned using a set of simple rules that work in most situations.

ON Rule 1 (The Free Element Rule): The oxidation number of pure elements is 0. Thus, individual neutral atoms and covalently bonded elements have an oxidation number of 0.

$$\text{Examples: } \overset{0}{Na},\ \overset{0}{Fe},\ \overset{0}{Ar},\ \overset{0}{He},\ \overset{0}{O_2},\ \overset{0}{Cl_2},\ \overset{0}{S_8}$$

In these examples, the element's ON is written above its symbol. This notation is useful for helping you learn the concept of oxidation numbers but it is *not* part of standard chemical symbol notation.

ON Rule 2 (The Ion Rule): The oxidation number of a monatomic ion is equivalent to the charge of the ion. When a chlorine (Cl) atom gains an electron to become an ion, its oxidation number is −1, which shows that the ion must lose one electron to become a neutral atom again.

$$\text{Examples: } \overset{+1}{K^+},\ \overset{+2}{Ag^{2+}},\ \overset{+3}{Fe^{3+}},\ \overset{+4}{Sn^{4+}},\ \overset{-1}{Cl^-}$$

Though the charge of an ion and its ON are not the same thing, they are numerically equal and have the same sign.

ON Rule 3 (The Specific Oxidation Number Rule): Elements in certain families regularly have the same oxidation numbers.

a. *Alkali metals (Group 1A) always have a +1 oxidation number in compounds.*

b. *Alkaline-earth metals (Group 2A) always have a +2 oxidation number in compounds.*

c. *Hydrogen (H) usually has a +1 oxidation number when bonded to another nonmetal.* Most nonmetal elements are more electronegative than hydrogen.

d. *Hydrogen has a −1 oxidation number when bonded to metals.* Metals bonded to hydrogen form compounds called *metal hydrides.*

e. *Oxygen (O) always has a −2 oxidation number except when bonded to fluorine (F).* This exception exists because oxygen is the second-most electronegative element after fluorine.

f. *Halogens (Group 7A) have an oxidation number of −1 when bonded to metals.* Fluorine's oxidation number is always −1 because it is the most electronegative element.

Examples of metal hydrides include lithium hydride (LiH) and sodium hydride (NaH).

Oxidation Number Rule Summary

Rule 1: free atoms/elements; ON = 0
Rule 2: ions; ON = ion's charge
Rule 3a: Group 1A; ON = +1
Rule 3b: Group 2A; ON = +2
Rule 3c: H (with nonmetals); ON = +1
Rule 3d: H (with metals); ON = −1
Rule 3e: O; ON = −2
Rule 3f: Group 7A; ON = −1
Rule 4: compounds; sum of ONs = 0
Rule 5: polyatomic ions; sum of ONs = charge (see Subsection 19.5 on page 441)

ON Rule 4 (The Zero Sum Rule): The sum of the oxidation numbers of all the atoms in a compound must equal zero because compounds are not electrically charged.

$$\overset{+1}{Na^+} + \overset{-1}{Cl^-} \longrightarrow \overset{0}{NaCl}$$

This rule applies to covalent, as well as ionic, compounds. Shared electrons reside closer to the more electronegative atom in a covalent bond. This element is assigned a negative oxidation number based on how many electrons it acquires by sharing. The less electronegative atom has a positive oxidation number, which equals the number of electrons it "donates" to the more electronegative atom. Generally, the element with the largest electronegativity determines the oxidation numbers of the other elements in a compound. The following formula for water illustrates the Zero Sum Rule for a molecular compound:

$$\overset{\overset{0}{2(+1)+(-2)}}{H_2O}$$

Many elements have more than one oxidation number. For example, nitrogen (N; Group 5A) has five valence electrons. It can combine with less electronegative elements and have an oxidation number of −3. In other compounds it donates three, four, or even all five of its valence electrons. In such cases nitrogen has an oxidation number of +3 to +5.

$$\overset{\overset{0}{(-3)+3(+1)}}{N\,H_3} \qquad \overset{\overset{0}{(+3)+3(-1)}}{N\,F_3}$$

19-1 Common Oxidation Numbers of Some Elements

Periods	Group 1A	Group 2A	Typical transition metals			Group 3A	Group 4A	Group 5A	Group 6A	Group 7A	Group 8A
1	H +1 −1										He 0
2	Li +1	Be +2				B +3	C +4 −4 +2	+5 +2 −1 +4 +1 −2 +3 −3	O −2	F −1	Ne 0
3	Na +1	Mg +2				Al +3	Si +4	P +5 +4 −3	S −2	Cl −1	Ar 0
4	K +1	Ca +2	Fe +2 +3	Cu +1 +2	Zn +2			As +5		Br −1	Kr 0
5		Sr +2		Ag +1	Cd +2		Sn +4 +2			I −1 +1	Xe 0
6		Ba +2		Au +1	Hg +1 +2		Pb +2 +4				Rn 0

19.4 *Writing Chemical Formulas*

You can write the formula of a compound if you know the elements and their oxidation numbers. For example, you can use the information in Table 19-1 to write the formula for a compound containing calcium and fluorine.

Step 1 is generally true for *inorganic* compounds. Other rules apply when writing the formulas for compounds of carbon and nitrogen.

Recall that the elements with lower electronegativities are generally toward the left and bottom of the periodic table.

One clue that CaF is not the correct formula is that the sum of the formula's ONs is not zero.

Step 1: Write the symbols of the two elements next to each other. Always write the less electronegative element first.

$$Ca\ F$$

Step 2: Use Table 19-1 to find the oxidation numbers of the two elements. Calcium's (Group 2A) is +2; fluorine's (Group 7A) is −1. Write these numbers above the symbols.

$$\overset{+2}{Ca}\ \overset{-1}{F}$$

Step 3: The quantities of the two elements must be adjusted so that the sum of the oxidation numbers equals zero. For this example, you must add another fluorine atom. The presence of two fluorine atoms is indicated by a subscript.

$$\overset{+2}{Ca}\ \overset{2(-1)}{F_2}$$

The +2 oxidation number of the single calcium atom is now balanced by the two –1 oxidation numbers of the fluorine atoms. The correct compound formula for calcium fluoride is

$$CaF_2.$$

Note that the correct formula for a compound does not contain charges or ONs—only subscripts.

If the oxidation numbers of the two elements are not whole-number multiples of one another, they cannot be balanced by simply adding more atoms of one element. For example, one of the compounds used in safety matches is composed of phosphorus (ON = +5) and sulfur (ON = –2). After Step 2 of the process, you might run into a roadblock.

$$\text{Assign oxidation numbers:} \quad \overset{+5}{P}, \overset{-2}{S}$$

How do you know that the ON of phosphorus is not –3?

The sum of the oxidation numbers will not balance by adding either phosphorous or sulfur. A method involving the *least common multiple (LCM)* of the two oxidation numbers can be used to balance the formula. The LCM is the smallest number that is a whole-number multiple of all the numbers being considered. The LCM of 5 and 2 is 10. The subscript for each element in the compound is found by dividing the LCM by the *magnitude* of its oxidation number. So then the subscript of phosphorus becomes 2 (since 10 ÷ 5 = 2), and the subscript of sulfur becomes 5 (since 10 ÷ 2 = 5). Multiplying phosphorus's oxidation number by 2 and sulfur's by 5 allows the oxidation numbers to add up to zero.

$$2 \times \text{phosphorous ON} + 5 \times \text{sulfur ON} = 0$$

$$2(+5) + 5(-2) = 0$$

$$10 - 10 = 0$$

These numbers are used as subscripts in the formula.

$$\text{Empirical formula: } P_2S_5$$

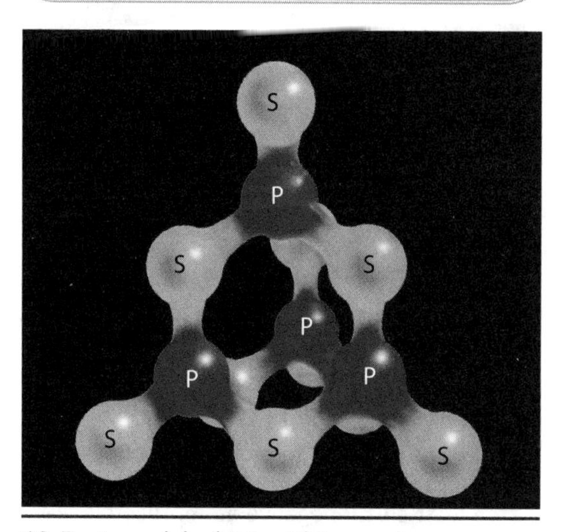

19-7 A model of P_4S_{10}. This covalent compound is named *tetraphosphorus decasulfide*.

Is this empirical formula the same as the molecular formula? Chemical analysis reveals that this form of phosphorus sulfide actually has twice as many atoms in each molecule (see Figure 19-7). Its molecular formula is P_4S_{10}. The sum of all oxidation numbers still equals zero.

19.5 Polyatomic Ions

Many ions are made of two or more covalently bonded atoms that act as a single particle. Such ions are known as molecular or *polyatomic ions*. They can bond with other atomic or polyatomic ions to form ionic compounds. This ability makes them important in many chemical reactions. Table 19-2 lists some common polyatomic ions and their charges. Memorize this table for further use. When writing formulas with polyatomic ions, you can assume that their charges are equivalent to oxidation numbers, even though oxidation numbers don't properly apply to groups of atoms. This leads to the fifth rule of oxidation numbers:

ON Rule 5 (The Polyatomic Ion Charge Rule): The sum of the oxidation numbers of all atoms in a polyatomic ion is equal to its charge.

Polyatomic ions containing strongly electronegative elements will trap electrons or hydrogen ions from their surroundings to complete valence vacancies if they cannot get the needed charges from their own atoms. The simplest polyatomic ion you will work with is the hydroxide ion (OH^-). Its charge (1–) is determined by

19-2 Polyatomic Ions

Name	Formula	Charge	"ON"
ammonium	NH_4^+	1+	+1
acetate	$C_2H_3O_2^-$	1–	–1
bicarbonate	HCO_3^-	1–	–1
carbonate	CO_3^{2-}	2–	–2
hydroxide	OH^-	1–	–1
nitrate	NO_3^-	1–	–1
phosphate	PO_4^{3-}	3–	–3
sulfate	SO_4^{2-}	2–	–2
sulfite	SO_3^{2-}	2–	–2

19-8 Copper (II) sulfate ($CuSO_4$) is a compound of the polyatomic sulfate anion.

19-9 This limestone contains the polyatomic carbonate ion.

In ionic compound formulas the atomic cation (positive ion) is written first because it is the least electronegative atom. This rule includes the polyatomic ammonium ion (NH_4^+).

A polyatomic ion must always be enclosed in parentheses when it has a subscript.

19-10 The mineral halite is nearly pure sodium chloride (NaCl).

the sum of the oxidation numbers of oxygen (−2) and hydrogen (+1). The oxygen atom gets one of the two valence electrons it needs from the hydrogen atom. The remaining electron is taken from another atom in its surroundings, resulting in a net 1− charge.

Example Problem 19-1

Polyatomic Ion Charge

Using the oxidation numbers of its bonded atoms, verify that the charge of the carbonate ion (CO_3^{2-}) is 2−.

Known: common ON of carbon (C): +4
ON of oxygen (O): −2
number of C: 1
number of O: 3

Sum ONs:

C	1(+4)	=	+4
O	3(−2)	=	−6
			−2

Answer: The charge of the carbonate ion is 2−.

Example Problem 19-2

Writing Formulas with Polyatomic Ions

Magnesium hydroxide is a compound in medicines used to help control stomach acid. It is composed of magnesium ions and hydroxide ions. Write the formula for magnesium hydroxide.

Ion symbols: Mg^{2+} OH^-

Oxidation numbers: $\overset{+2}{Mg^{2+}}$, $\overset{-1}{OH^-}$

Balanced ONs: $\overset{+2}{Mg^{2+}}$, $\overset{2(-1)}{2(OH^-)}$

Compound formula: $Mg(OH)_2$

19.6 Naming Compounds

There are millions of compounds. Some have more than one correct name, but the name for any compound must clearly communicate its unique composition. *Binary compounds*, compounds of only two elements, have a first and a last name. Chemists can use several methods to name binary compounds. In this textbook, to make things easier, you will use one for ionic compounds, the other for covalent compounds.

Binary Ionic Compounds

How would you name common table salt (NaCl)? Since Na is a metal and Cl is a nonmetal, NaCl is an ionic compound. Binary ionic compounds have a first and a last name. The metal is named first, then the nonmetal with its ending changed to *-ide*. Na represents sodium, so the first part of the chemical name is *sodium*. Cl represents chlorine, but the last syllable is changed to *-ide*, making the second part *chloride*. Putting both parts together, the correct name for NaCl is *sodium chloride*.

The other halogens are treated the same as chlorine, where *-ine* is changed to *-ide*. Some other nonmetals that appear frequently in binary compounds undergo name changes that are not quite as straightforward. For example, sulfur becomes *sulfide*, oxygen becomes *oxide*, phosphorus becomes *phosphide*, and nitrogen becomes *nitride*. You can see examples of binary ionic compounds and their names in Table 19-3.

Some metals can have more than one oxidation number when forming ionic compounds. To distinguish between these compounds, which are often significantly different, chemists use the Stock naming system. In the Stock system a roman numeral in parentheses comes after the name of the metal, indicating its oxidation number in that compound. For example, the oxidation number of iron in Fe_2O_3 is +3, so the name of this type of rust is iron (III) oxide. Iron (III) oxide results from the corrosion of iron and steel in a humid environment. In another kind of iron oxide, FeO, the oxidation number of iron is +2, so this compound is called iron (II) oxide. This form occurs as a rare natural mineral called wüstite found in meteorites and some iron ores.

19-3 Selected Binary Compounds

Formula	Name
NaF	sodium fluoride
K_2S	potassium sulfide
Ag_2O	silver oxide
Zn_3N_2	zinc nitride
$AlBr_3$	aluminum bromide

The Stock system of naming binary compounds was developed in 1919 by the German chemist Alfred T. Stock (1876–1946).

wüstite (VOOS tite)

19-11 Iron (II) oxide (FeO)

Binary Covalent Compounds

Chemists generally use a different method for naming *binary covalent* compounds. This method is necessary because the same two elements often bond together in different numbers producing compounds with vastly different properties. Binary covalent compounds also have a first and last name. A prefix is placed before one or both of the elements' names to indicate how many atoms of that element are present in the molecule. The prefixes are taken from the Greek words for those numbers. Prefixes let us easily distinguish between the deadly gas carbon *mon*oxide (CO) and carbon *di*oxide (CO_2), the gas in soft drinks. Common prefixes and examples of their uses are listed in Table 19-4.

19-4 Binary Covalent Compound Prefixes

Prefix	Number	Example
mon-, mono-	1	nitrogen monoxide (NO)
di-	2	silicon dioxide (SiO_2)
tri-	3	boron trifluoride (BF_3)
tet-, tetra-	4	carbon tetrachloride (CCl_4)
pent-, penta-	5	diphosphorus pentoxide (P_2O_5)
hex-, hexa-	6	sulfur hexafluoride (SF_6)

Binary covalent compounds use prefixes to indicate the numbers of atoms, while binary ionic compounds do not.

Note that the prefix for one atom, *mon-* or *mono-*, applies only to the second element. It isn't needed when a molecule has only one atom of the first element because the element's name without a prefix stands for a single atom. Other prefixes (*di-*, *tri-*, etc.) are always needed.

19-12 Transferring Mr. Simons from teaching chemistry to teaching cooking was a big mistake.

These naming rules apply mainly to *inorganic* covalent compounds. There are thousands of *organic* compounds, consisting of carbon, hydrogen, and other elements, which use different naming rules to tell them apart.

Sometimes the common name of a compound is used more often than its scientific name. You'll probably never hear water (H_2O) called "dihydrogen monoxide" nor someone refer to ammonia (NH_3) as "nitrogen trihydride." The common name of a compound, however, doesn't give any clues about its chemical formula, so you need to memorize the formulas of compounds known by their common names, such as water and ammonia.

Compounds with Polyatomic Ions

We name compounds that contain polyatomic ions in a way similar to the method used for binary ionic compounds. The first name is the name of the cation (it has a positive oxidation number), which begins the formula. Usually, this is the metal's name since there is only one common polyatomic cation—the ammonium ion, NH_4^+. The second name comes from the name of the anion. If the anion is a polyatomic ion, use the anion's full name (see Table 19-2). If the anion is a single element, an *-ide* ending is used as usual. The Stock system is used for metals with multiple oxidation numbers. Normally, no prefixes are used to indicate the number of each ion since we are not dealing with binary nonmetal compounds.

19-5 *Naming Compounds with Polyatomic Ions*

Formula	Name
NH_4Cl	ammonium chloride
NaOH	sodium hydroxide
K_2CO_3	potassium carbonate
$(NH_4)_3PO_4$	ammonium phosphate
$BaSO_4$	barium sulfate

19A Section Review

1. What information does a chemical formula provide? How may a chemical formula differ from an empirical formula?
2. How do chemists determine the oxidation numbers of bonded elements? What are oxidation numbers used for?
3. Why do some elements have multiple oxidation numbers?
4. What property usually determines which element is written first in a chemical formula?
5. How do you know if you have the correct number of atoms for each element in a chemical formula?
6. How can you determine the sign and value of the charge on a polyatomic ion?
7. How do you indicate the presence of more than one occurrence of a polyatomic ion in an ionic compound's formula unit?
8. What is the general structure of a binary compound's name?
9. When is the Stock system used to name ionic compounds? Explain how it is used.

10. (True or False) In a binary compound the less electronegative element is also the one with the positive oxidation number.

11. Identify the oxidation numbers for the atoms and ions in the following compounds.
    a. CaF_2
    b. NO_2
    c. MgH_2
    d. Li_2S
    e. $FeBr_2$
    f. $NaC_2H_3O_2$

12. For the following elements and ions, use the common oxidation number(s) for each to write the compound formula.
    a. aluminum and chlorine
    b. mercury and the nitrate ion (two possible answers)
    c. carbon and fluorine
    d. oxygen and potassium
    e. the phosphate ion and sodium
    f. sulfur and copper (two possible answers)

13. Write the names for the following compounds.
    a. NaH
    b. S_3Cl_2
    c. FeS
    d. NF_3
    e. $Mg(OH)_2$
    f. $CuCO_3$

19B Chemical Changes

19.7 Evidences for Chemical Changes

A chemical change takes place when substances react to form new substances. Chemical changes are as varied as the rusting of iron, the explosion of dynamite, and the bleaching of dark fabric in sunlight. Photosynthesis and digestion are common chemical changes that are necessary for our survival. Industry uses chemical changes to produce plastics, textile fibers, and many other materials. Many of these chemical changes, or chemical reactions, affect our daily lives.

In a **chemical reaction**, atoms are not destroyed or changed but rearranged into different substances. When the explosive pentaerythritol tetranitrate (PETN; $C_5H_8N_4O_{12}$) detonates, the bonds between the carbon, hydrogen, nitrogen, and oxygen atoms are broken and new bonds form to produce different substances: carbon dioxide (CO_2), water (H_2O), nitrogen (N_2), and oxygen (O_2).

How do we know different substances were formed in the explosion of PETN? Though we cannot watch individual atoms rearrange when substances react, we observe one or more of the following signs that a chemical reaction may have occurred.

1. A solid separates from a liquid mixture of chemicals.
2. Visible bubbles of gas rapidly form or an explosion of burning gases occurs.
3. A substance's color changes; such a change is usually permanent.
4. The temperature of a mixture of chemicals changes.
5. Light, sound, or another type of energy is produced.

19B Objectives

After completing this section, you will be able to

✓ recognize evidence for a chemical reaction.
✓ state the information provided by chemical equations.
✓ describe the structure of chemical equations.
✓ translate word equations into formula equations.
✓ balance a chemical equation.
✓ identify the phases of reactants and products in chemical equations by their special symbols.

19-13 The explosive PETN is often used to quickly demolish old buildings to make way for new construction.

Sometimes physical changes, such as a substance dissolving in water, may produce one or more of these signs. If a chemical reaction has truly taken place, the change cannot be undone by physical processes. Several of the signs of a chemical reaction can be observed in the explosion of PETN. Gases are produced, colors change, and sound, thermal energy, and light are produced. This reaction cannot be reversed by a physical process.

19.8 Chemical Equations

Chemists use a **chemical equation** to describe what happens during a chemical reaction. It represents a chemical reaction in a symbolic way just as a mathematical equation shows equality between different number quantities. Chemical changes in matter can be described by equations because of the *law of the conservation of matter.* The total amount of matter present before a reaction must equal the total amount of matter present after the reaction.

Chemical equations tell what chemicals are present at the beginning of a chemical change and what chemicals are present afterward. They also allow us to account for the numbers of atoms throughout the reaction. The total number of atoms of each element present at the beginning of a reaction must be present after the chemical change is complete.

The following equation describes the reaction between zinc and oxygen to form zinc oxide:

Zinc plus oxygen produces zinc oxide.

Chemical equations written out in this way are called *word equations.* Word equations can be useful, but they do not clearly show how matter is conserved during a chemical reaction. To do this, scientists use *formula equations.* The formula equation for this reaction is

$$Zn + O_2 \longrightarrow ZnO.$$

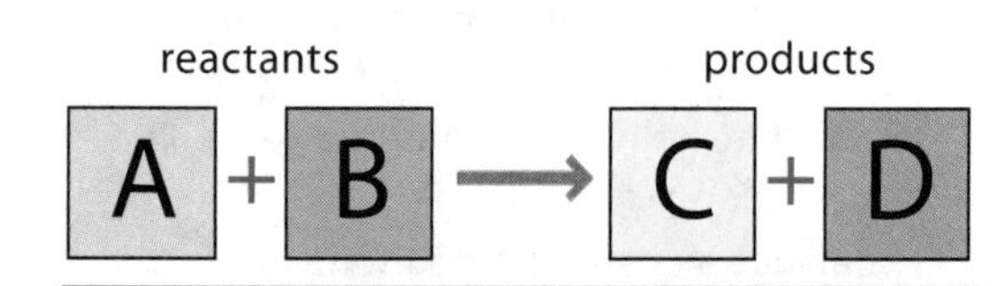

19-14 The structure of a chemical equation. The arrow always points toward the products.

The symbols in a formula equation show how the atoms of each element are both rearranged and conserved during the reaction. The formulas on the left side of an arrow are the **reactants.** They show how atoms are arranged before the reaction takes place. The formulas on the right side of an arrow are the **products.** They show how atoms are arranged after the reaction. The arrow indicates the direction of the reaction.

Notice that the zinc oxide equation does not show the same number of oxygen atoms on both sides. This inequality violates the law of the conservation of matter. A chemical equation must be *balanced* to show that matter is conserved. The numbers and kinds of atoms must be equal on both sides of the equation. To balance an equation, ***coefficients*** are placed in front of the formulas. Coefficients indicate how many atoms, molecules, or formula units of reactants and products are needed to balance the equation. In the example a coefficient of 2 must be placed in front of the zinc and the zinc oxide to show that two atoms of zinc react with an oxygen molecule to form two formula units of the ionic compound zinc oxide.

coefficient (KO uh FISH unt)

$$2Zn + O_2 \longrightarrow 2ZnO$$

Writing Balanced Chemical Equations

You can write chemical equations correctly using three steps. First, write the word equation. Then, write the formula equation. Last, check to see if the law of the conservation of matter holds true, and balance the equation if necessary.

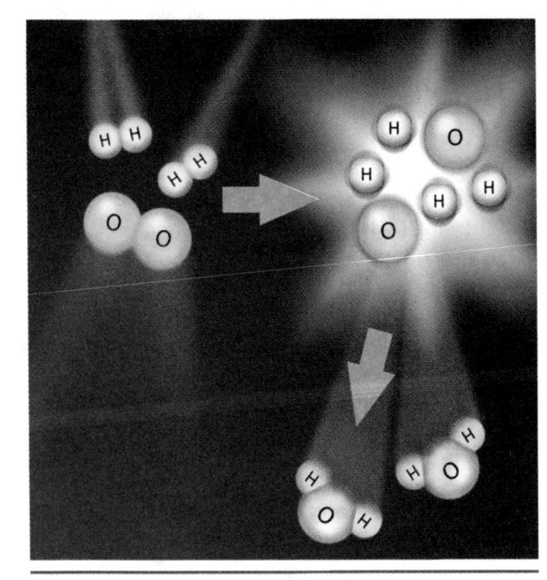

19-15 The formation of water from burning hydrogen in oxygen

Example Problem 19-3

Writing Balanced Equations

Hydrogen and oxygen combine to form water.

Step 1: *Write the word equation.*

Hydrogen plus oxygen produces water.

Step 2: *Write the formula equation.* Hydrogen and oxygen are both diatomic molecules; therefore, their formulas are H_2 and O_2. The correct formula for water is H_2O.

$$H_2 + O_2 \longrightarrow H_2O$$

Step 3: *Balance the formula equation with coefficients, if necessary.* You can immediately see that the oxygen atoms are not balanced. Placing a 2 before the H_2O doubles the number of water molecules and provides two oxygen atoms in the product. This balances the oxygen atoms but unbalances the hydrogen atoms.

$$H_2 + O_2 \longrightarrow \mathbf{2}H_2O$$

Adding a 2 coefficient to the hydrogen molecule in the reactants balances the hydrogen atoms and gives the balanced equation for the reaction.

$$\mathbf{2}H_2 + O_2 \longrightarrow 2H_2O$$

The number of each kind of atom in a formula preceded by a coefficient is found by multiplying the subscript of the element by the coefficient. For example, in $2H_2O$ there are four hydrogen atoms and two oxygen atoms.

Example Problem 19-4

Writing Balanced Equations

Oxygen reacts with iron in a complex process to form reddish brown iron (III) oxide, or rust.

Step 1: *Write the word equation.*

Iron plus oxygen produces iron (III) oxide (rust).

Step 2: *Write the formula equation.* Iron (Fe) is a metal. Oxygen is a diatomic nonmetal (O_2). The formula for iron (III) oxide is Fe_2O_3.

$$Fe + O_2 \longrightarrow Fe_2O_3$$

Step 3: *Balance the formula equation with coefficients, if necessary.* This equation is not balanced. There are two oxygen atoms on the reactant side and three on the product side. Use the least common multiple of the oxygen subscripts 2 and 3, which is 6, to provide six oxygen atoms on both sides. Therefore, a 3 should be placed in front of the O_2 and a 2 in front of the Fe_2O_3 to balance the oxygen atoms.

$$Fe + \mathbf{3}O_2 \longrightarrow \mathbf{2}Fe_2O_3$$

There are four iron atoms on the product side. Placing the coefficient 4 in front of Fe on the left side of the equation balances the iron atoms.

$$\mathbf{4}Fe + 3O_2 \longrightarrow 2Fe_2O_3$$

Never alter a formula to balance an equation. Changing the formula changes the compound.

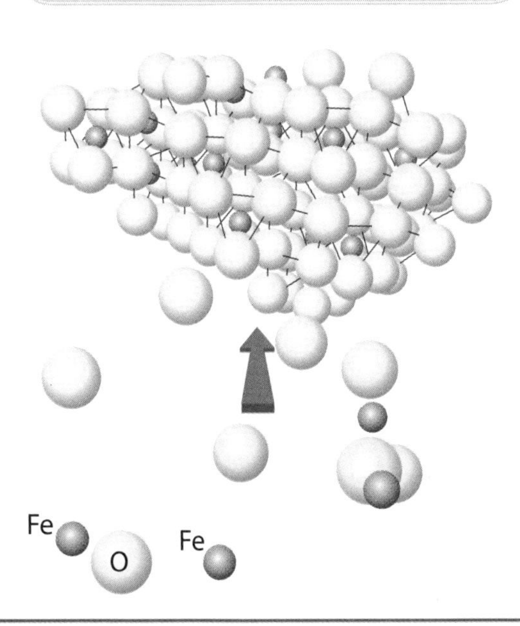

19-16 The net reaction that forms rust

The abbreviation aq stands for the word *aqueous*, which comes from the Latin word for "water."

Even though chemists use a down arrow (↓) to indicate how a precipitate can settle to the bottom of a liquid, the solid often doesn't move from where it forms. The precipitate product may build up right on the solid reactant (see Figure 19-19 on page 451), or even float at the top of a test tube solution (see Figure 19-21 on page 452) where drops of a liquid chemical react with the solution.

Special Symbols in Chemical Equations

Special symbols are sometimes included in chemical equations to show the physical state of reactants or products. If reactants or products are dissolved in water, the symbol (*aq*), including parentheses, appears after their formulas. If the substance is a solid, (*s*) follows the formula; if a liquid, (*l*); and if a gas, (*g*).

Sometimes a solid product from a reaction settles out of solution as a *precipitate*. Such reactions are often called *precipitation reactions*. A down arrow (↓) appears after the solid product. Similarly, if a reaction produces a gaseous product that leaves the reaction site, an up arrow (↑) follows the gaseous product. The arrows are not put in parentheses and may be used instead of the parenthetical symbols given in the previous paragraph. Chemists don't always include these special symbols. They are intended to clarify the conditions that exist during a reaction.

Other symbols indicate special conditions necessary for a particular chemical change. These symbols usually appear above the reaction arrow. The Greek letter *delta* (Δ) above the arrow indicates that the reactant(s) must be heated for the reaction to occur. An element's symbol or a molecular formula above the arrow indicates that the substance is a *catalyst* for the reaction (see the facet on the facing page). A catalyst sometimes merely speeds up a reaction, but sometimes it must be present for the reaction to take place under normal conditions. A catalyst is neither a reactant nor a product and does not undergo a chemical change.

19B Section Review

1. Name five signs that a chemical change may have taken place.
2. What information does a chemical equation provide?
3. What are the products and reactants of a chemical change?
4. What is a balanced chemical equation? Why is an unbalanced chemical equation invalid?
5. How would you write the element chlorine in a chemical equation? the metal silver? Explain your answers.
6. Why do chemists sometimes include special symbols in chemical equations?
7. What do we call the solid product that settles out of a solution following a reaction? How can you indicate this kind of product in a chemical equation?
8. (True or False) In a chemical reaction, both the kinds of atoms and their arrangements change.
9. ✪ Write the *unbalanced* formula equations from the following word equations.
   a. Iron combines with sulfur to produce iron (I) sulfide.
   b. Calcium carbonate breaks down into calcium oxide and carbon dioxide.
   c. Water and carbon dioxide combine to form hydrogen carbonate (carbonic acid).
   d. Carbon monoxide burns in oxygen to produce carbon dioxide.
10. ✪ Balance the following chemical equations.
    a. $Pb + O_2 \longrightarrow PbO$
    b. $FeBr_3 + Cl_2 \longrightarrow FeCl_3 + Br_2$

c. $H_2O_2 \longrightarrow H_2O + O_2$
d. $Na_2CO_3 + Ca(OH)_2 \longrightarrow NaOH + CaCO_3$

✪ 11. Write and balance the complete formula equations for the following word equations.
   a. Barium hydroxide combines with hydrogen chloride to produce barium chloride and water. (*Hint*: Write the formula for water as "hydrogen hydroxide.")
   b. Propane (C_3H_8) burns in oxygen to produce carbon dioxide and water.
   c. Iron reacts with magnesium oxide to produce iron (III) oxide and magnesium metal.
   d. Silicon dioxide combines with pure carbon to produce pure silicon and carbon monoxide.

DS ✪ 12. Balance the following chemical equation, which shows how gasoline (in this case, pure octane) burns in an engine:

$$C_8H_{18} + O_2 \longrightarrow H_2O + CO_2$$

FACETS OF SCIENCE

Reaction Helpers: Catalysts and Enzymes

Catalysts are substances that can change the rate of a chemical reaction without being consumed themselves. They usually speed up the process, but in some instances they slow it down.

The mechanism of how catalysts work is based on the *collision theory* of chemical reactions. This theory is ultimately based on the kinetic-molecular model. The warmer the substance is, the more rapid the motion of its particles. When two particles collide at the right speed and in the right direction, they can be chemically changed.

The speed of the particles determines their kinetic energies. The minimum kinetic energy that the reactant molecules need for the reaction to take place is called the *activation energy*. If the molecules don't have at least the activation energy when they collide, they just bounce apart like billiard balls.

The presence of a catalyst changes how the reactants behave. It provides another way for the reaction to take place. This alternative path usually allows the reaction to proceed many times faster and with a much lower activation energy.

There are many different kinds of catalysts. The largest group of catalysts known, consisting of more than 4000 substances, is the *biocatalysts* called *enzymes*. These compounds are produced by living things and act as catalysts in metabolic processes. One example is *ptyalin* in saliva, which begins the process of digesting the starches in food. This step in digestion is one reason to chew food thoroughly so that the saliva is completely mixed with the food before swallowing. Your tears contain the enzymes *lysozyme*, *lactoferrin*, and *lipocalin*, which all break down bacteria trapped in the tear layer of the eye and protect against eye infections.

To realize the effectiveness of biocatalysts, consider that the ultimate chemical products of the metabolism of food are carbon dioxide and water—the same compounds produced by burning the food! The energy found in food is released at a much lower rate (and lower temperature) through the action of catalysts.

Other catalysts include stain removers containing enzymes that weaken the chemical bonds of stains. The catalytic converters on cars allow harmful and poisonous exhaust gases to be changed into relatively harmless carbon dioxide and nitrogen. It is estimated that catalysts are used at one or more stages of production for 60% of all commercial chemical products. More uses for catalysts are discovered every year.

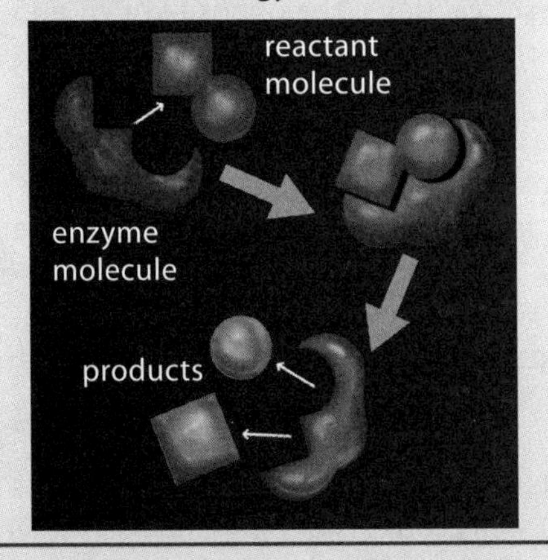

An enzyme catalyst (purple) breaks apart a reactant into simpler products. The enzyme itself is unchanged.

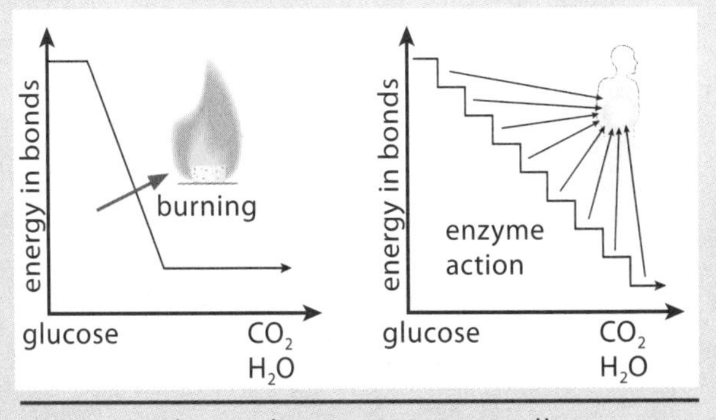

Burning releases the energy in sugar all at once. Digestion, involving enzymes, releases the energy gradually, keeping our bodies warm.

19C Objectives

After completing this section, you will be able to

✓ recognize the four general kinds of chemical reactions.

✓ explain how decomposition reactions usually occur.

✓ describe the difference between endothermic and exothermic reactions.

19-17 Powdered zinc and sulfur combine violently to produce zinc sulfide (ZnS).

Carbonic acid helps give soft drinks their sharp taste. It also forms in raindrops and is responsible for some of the acidity in acid rain.

19C Types of Chemical Reactions

19.9 Classifying Chemical Reactions

Some chemical reactions combine two or more substances into a more complex compound. Others take compounds apart. Some reactions replace one element in a compound with another, while still other reactions trade elements between two different compounds. These distinctions are the key to classifying four general types of common chemical reactions. Knowing the type of chemical reaction and the reactants can allow you to predict the products of a reaction.

19.10 Composition Reactions

Reactions that combine two or more substances into a single, more complex compound are called **composition reactions**. They are sometimes also called *synthesis* or *combination reactions*. They generally take the form

$$X + Y \longrightarrow XY,$$

where X and Y represent two different substances. These substances can be two elements. For instance, zinc and sulfur combine to form zinc sulfide.

$$\mathrm{Zn} + \mathrm{S} \longrightarrow \mathrm{ZnS}$$
$$X + Y \rightarrow XY$$

Composition reactions may also involve compounds that combine to form a more complex compound. For example, soft drinks are carbonated by bubbling carbon dioxide under pressure through water. Some of the carbon dioxide reacts with the water to produce *carbonic acid*.

$$\mathrm{CO_2}\,(g) + \mathrm{H_2O}\,(l) \longrightarrow \mathrm{H_2CO_3}\,(aq)$$
$$X + Y \rightarrow XY$$

The distinguishing feature of composition reactions is the formation of a single product. A composition reaction begins with two or more reactants but always produces one compound.

19.11 Decomposition Reactions

Reactions that break apart or decompose a substance are called **decomposition reactions**. These reactions, the opposite of composition reactions, generally take the form

$$XY \longrightarrow X + Y.$$

Usually, electrical or thermal energy is necessary to decompose substances. One such process is called *electrolysis*. In the case of water, an electrical current can break down its molecules into free hydrogen and oxygen gases.

$$2\mathrm{H_2O}\,(l) \xrightarrow{\text{elec}} 2\mathrm{H_2}\uparrow + \mathrm{O_2}\uparrow$$
$$XY \rightarrow X + Y$$

The main reactions in oil refineries are decomposition reactions. The crude oil pumped from oil wells consists of long-chained molecules, which are heated until they break down into smaller, more useful molecules, such as those in gasoline.

electrolysis (ih lek TRAHL ih sis): electro- (E.—relating to electricity) + -lysis (Gk. *lusis*—breaking apart)

The abbreviation *elec* above the arrow indicates that electrical energy is required to cause the reaction.

The distinguishing feature of decomposition reactions is a single reactant. A decomposition reaction always begins with a single reactant and produces multiple products. Reactions that require the addition of energy to occur are often decomposition reactions.

19.12 Single-Replacement Reactions

In a **single-replacement reaction**, one element in a compound is replaced by another element. These reactions generally take the form

$$XY + Z \longrightarrow ZY + X.$$

A good example of a single-replacement reaction involves solid copper metal (Cu) and a solution of silver nitrate ($AgNO_3$). The silver nitrate solution contains Ag^+ and NO_3^- ions. If a piece of copper is dropped into the solution, copper ions form and dissolve while silver atoms precipitate out of the solution. If the solution were evaporated, silver metal and the ionic compound copper (II) nitrate would remain.

$$\underset{XY}{2AgNO_3\,(aq)} + \underset{Z}{Cu\,(s)} \longrightarrow \underset{ZY}{Cu(NO_3)_2\,(aq)} + \underset{X}{2Ag\downarrow}$$

Note that you need two silver nitrate formula units on the left to balance the two nitrate anions in copper nitrate on the right.

Copper is said to be "more active" than silver; that is, it has a greater tendency than silver to give up its valence electrons. Thus, the silver ions accept the copper atoms' electrons, resulting in copper ions replacing silver ions in the solution.

The distinguishing feature of single-replacement reactions is the presence of both an element and a compound before and after the reaction. The element with the greater activity replaces a similar element of lower activity in a compound.

Another common example of a single-replacement reaction involves acids and metals. Acid compounds contain hydrogen atoms that easily become free ions in solution. The hydrogen ions make water solutions acidic (see Chapter 21). A familiar acid compound is hydrogen chloride gas (HCl), which forms hydrochloric acid when dissolved in water. Many metals react with acids because their atoms are more active than the hydrogen ions in acid solutions. The metal atoms give up their valence electrons to the hydrogen ions, which combine to form free hydrogen gas. The solution fizzes and bubbles as the metal atoms replace the hydrogen ions in solution. The metal sample is "eaten away" in the process. Metallic zinc is readily dissolved by hydrochloric acid in the following reaction:

$$\underset{Z}{Zn\,(s)} + \underset{XY}{2HCl\,(aq)} \longrightarrow \underset{ZY}{ZnCl_2\,(aq)} + \underset{X}{H_2\uparrow}$$

19.13 Double-Replacement Reactions

In **double-replacement reactions**, two compounds swap cations or anions with each other. The general equation for a double-replacement reaction shows how the compounds might trade parts. Remember that ionic compounds are written with the cation first and the anion second.

$$WX + YZ \longrightarrow WZ + YX$$

19-18 The electrolysis of water is a *decomposition reaction*. The inset shows oxygen forming at the positive electrode. In the other tube, twice as much hydrogen is trapped as oxygen, reflecting the ratio of the molecular products ($2H_2 + O_2$).

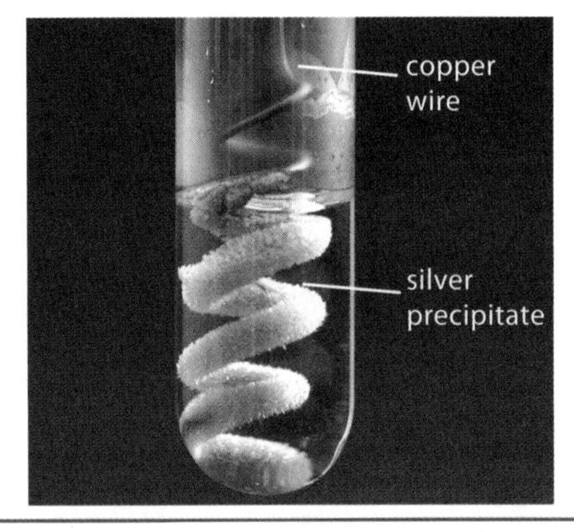

19-19 Metallic silver precipitates when it is replaced in solution by the more active copper in a single-replacement reaction.

19-20 Hydrogen gas is produced when zinc atoms replace hydrogen ions in solution in a single-replacement reaction.

19-21 Silver chloride precipitates when aqueous silver nitrate reacts with aqueous sodium chloride in a double-replacement reaction.

The order in which the products or reactants are written in the equation is not important, just as it makes no difference whether you add 4 to 6 or 6 to 4—the result is still 10.

In one double-replacement reaction, sodium chloride and silver nitrate exchange ions to become sodium nitrate and silver chloride.

$$\begin{array}{ccccccc} NaCl\,(aq) & + & AgNO_3\,(aq) & \longrightarrow & NaNO_3\,(aq) & + & AgCl\downarrow \\ WX & + & YZ & \rightarrow & WZ & + & YX \end{array}$$

The ↓ following the silver chloride indicates that the product is a solid that precipitates from the solution.

The distinguishing feature of double-replacement reactions is the presence of two reactants and two products whose ions exchange places. A precipitate often forms.

19.14 Chemical Thermodynamics

In Subsection 19.7 you learned that temperature changes and the release of energy, such as light or sound, could signal that a chemical change has occurred. Reactions can also be classified by how energy moves into or out of the reaction process—an area of study called *chemical thermodynamics.* For example, natural gas, which is mostly methane (CH_4), provides energy to cook, dry clothes, and heat homes through a chemical reaction known as *combustion,* or burning. As natural gas burns, it combines with oxygen to produce carbon dioxide and water.

$$CH_4 + 2O_2 \longrightarrow CO_2 + 2H_2O + \mathbf{Q}$$

Combustion reactions are useful because they give off large quantities of heat (*Q*). Reactions that produce thermal energy are called **exothermic reactions**. Chemists measure this heat in joules when studying the thermodynamics of a reaction.

exothermic: exo- (Gk. *exō*—out, outside) + -therm- (Gk. *thermē*—heat) + -ic

Another example of an exothermic chemical process is the metabolism of your body. Large amounts of heat are released as your body utilizes the food you eat. These exothermic reactions are used to keep your body at the temperature it should be.

19-22 These exothermic reactions are huge natural gas flares from oil wells in Iraq.

Some reactions occur only if the reactants take energy from their surroundings. Reactions that absorb thermal energy are called **endothermic reactions**. Most decomposition reactions are endothermic in that they require an input of energy to occur. Some examples discussed earlier include the electrolysis of water and the breaking down of crude oil.

In endothermic reactions, heat is considered a reactant. The decomposition of calcium carbonate (limestone) into calcium oxide (quicklime) and carbon dioxide illustrates the symbols of an endothermic reaction:

$$CaCO_3\,(s) \xrightarrow{\Delta} CaO\,(s) + CO_2\uparrow\text{, or}$$

$$CaCO_3 + Q \longrightarrow CaO + CO_2$$

endothermic: endo- (Gk. *endon*—within) + -therm- (Gk. *thermē*—heat) + -ic

19-23 An ice pack becomes cold because of an endothermic reaction. Chemicals are stored in two separate compartments in the package. When mixed, they absorb thermal energy to produce the chemical reaction, making the pack feel cold to the touch.

19.15 Using Compounds to Solve Problems

Do you know what gasoline is made of? It's actually a mixture of many different compounds. Most are hydrocarbons, chains of five to twelve carbons hooked together. When they are burned, the energy in their covalent bonds is released and used to propel a car. The tendency of these compounds to easily ignite contributes to engine knocking, the dominion science problem discussed in this chapter.

The hydrocarbons in gasoline are obtained directly from crude oil or are produced by oil refineries. The resistance of gasoline against early ignition (knocking) due to fuel compression is called its *octane rating*. This rating is determined by comparing the tendency of the gasoline to knock with that of a fuel consisting of 100% octane, an eight-carbon hydrocarbon compound found in gasoline. Octane is the standard because it resists ignition through compression better than any other hydrocarbon. Regular-grade gasoline, rated at 87 octane, has the ability to resist knocking equal to a mixture containing 87% octane. However, crude oil naturally contains only a small percentage of octane, so oil companies must process the oil to increase the octane content. This extra processing is why premium gasoline with octane ratings nearer 100% costs more.

19-24 For many decades of the twentieth century, gasoline containing tetraethyl lead was used to prevent engine knocking. Most countries in the world have stopped using this compound because of health and environmental concerns.

19-25 The minimum octane rating for most of the United States is 87.

Early in the twentieth century, car manufacturers tried to redesign their engines to eliminate knocking but met with little success. The damage this problem caused to engines was costing car owners millions of dollars in car repairs. The oil companies then tackled the

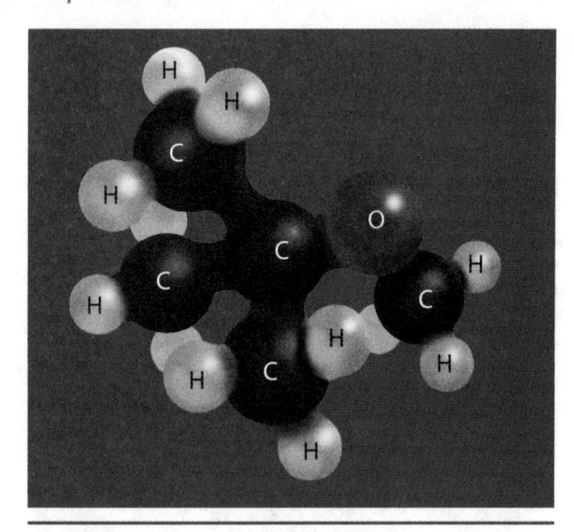

19-26 Methyl tert-butyl ether

problem by mixing in compounds called *additives* to increase the octane rating. This eliminated the explosive burning that caused knocking. These additives are called *antiknock agents*.

Today, one antiknock agent used in gasoline is methyl tert-butyl ether (MTBE), which has the empirical formula $C_5H_{12}O$. It boosts octane ratings in gasoline and adds oxygen to make gasoline burn more completely, helping prevent harmful exhaust emissions. As with most gasoline additives, however, it has been found in alarming amounts in ground water. Little research has been done on the health effects of MTBE. Recently, some people in the United States have promoted ethanol as an environmentally friendly replacement for MTBE. Others, however, cite evidence that ethanol expels substances from gasoline that contribute to smog and cause cancer.

19C Section Review

1. What is the distinguishing feature of each of the following reactions?
   a. composition reaction
   b. decomposition reaction
   c. single-replacement reaction
   d. double-replacement reaction
2. What must usually be added to make a compound decompose?
3. What area of chemistry involves the study of the energy changes in a chemical reaction?
4. ✪ A plant absorbs sunlight to combine carbon dioxide and water to produce sugars. What two reactions can this chemical change be classified as? What two kinds of reactions can occur when you digest food from plants to produce carbon dioxide, water, and heat that warms your body?
5. DS ✪ Is MTBE ($C_5H_{12}O$) a covalent or ionic compound? How can you tell?
6. (True or False) Thermal decomposition of a compound always produces two or more products.
7. ✪ Classify each of the following chemical reactions.
   a. $2Na + 2HOH^* \longrightarrow 2NaOH + H_2\uparrow$
   b. $2Hg + O_2 \longrightarrow 2HgO$
   c. $AgNO_3 + NaCl \longrightarrow AgCl\downarrow + NaNO_3$
   d. $Fe_2O_3 + 2Al \longrightarrow 2Fe + Al_2O_3$
   e. $2KClO_3 \xrightarrow{\Delta} 2KCl + 3O_2\uparrow$
   f. $Mg + 2HCl \longrightarrow MgCl_2 + H_2\uparrow$

*The formula HOH is an alternative way to write H_2O, especially when another compound in the reaction includes the hydroxide ion.

Chapter Review

1. What does a chemical formula tell about a compound?
2. How are positive and negative oxidation numbers assigned to elements in a compound?
3. How can you tell if an empirical formula is correct?
4. How is the charge of a polyatomic ion determined by the oxidation numbers of its elements?
5. What method would you use to name ionic compounds of copper and chlorine in which the copper ions have different oxidation numbers?
6. Identify whether each of the following is a chemical change or a physical change.
   a. grinding a piece of chalk into powder
   b. stirring orange drink mix into water, producing an orange liquid
   c. striking a match, which bursts into flame
   d. growth of a child
   e. tarnishing of a silver teapot
   f. melting of an ice cube
   g. flashing of a firefly
   h. formation of a greenish powdery substance on a copper water pipe
7. Why should you never change the subscripts in the formulas of a chemical equation to balance the equation?
8. How many atoms are represented in this expression?

$$2(NH_4)_3PO_4$$

9. What kind of reaction always has one product? one reactant?
10. How is heat shown as a product in a chemical equation? What kind of reaction produces heat?
11. DS ✪ In the late 1970s gasoline companies switched from cheaper leaded gasoline, which contained the antiknock agent tetraethyl lead (TEL), to more expensive unleaded gasoline. People were concerned about the lead's harmful effects on health and the environment. Gasoline prices went up because higher octane gasoline without TEL was more expensive to produce. Was this the right thing to do? Why or why not?

True or False

12. A single atom of an element in molecular and empirical formulas does not require a subscript.
13. An oxidation number is defined as the charge on a bonded atom.
14. As a general rule, the element with the largest electronegativity in a compound determines the oxidation numbers of the other elements in the compound.
15. The name *disodium oxide* for the compound Na_2O is correct according to the naming rules discussed in this chapter.

Scientifically Speaking

Term	Page
chemical formula	437
oxidation number (ON)	438
Stock system	443
chemical reaction	445
chemical equation	446
reactant	446
product	446
composition reaction	450
decomposition reaction	450
single-replacement reaction	451
double-replacement reaction	451
exothermic reaction	452
endothermic reaction	453

16. When bright yellow sulfur powder and dark gray zinc powder are mixed and ignited, a large quantity of smoke and flame produces a fluffy yellowish white solid. You can conclude that a chemical change took place.
17. The equation $H_2 + O_2 \longrightarrow H_2O$ can be balanced only by adding a subscript 2 to the oxygen in H_2O.
18. In a chemical equation the special symbols (*s*) and ↓ on the product side mean the same thing.
19. A chemical equation with the arrow $\xrightarrow{\text{elec}}$ indicates that energy must be added to make the reaction occur.
20. When a metal is placed in a strong acid solution, the acid converts the metal into a gas that bubbles away.
21. Thermal energy may be considered either a reactant or a product in chemical reactions.

DS ✪ 22. The octane rating of gasoline is *not* a measure of how much energy the gasoline has.

✪ 23. Determine the oxidation numbers of the elements and ions in the following compounds.
   a. FeS
   b. N_2O_3
   c. $Ca(OH)_2$
   d. NaBr
   e. CS_2
   f. Li_2O

✪ 24. For the following elements and ions, use the common oxidation number for each to write the compound formula.
   a. sodium and the bicarbonate ion
   b. iodine and oxygen
   c. hydrogen and the sulfate ion
   d. carbon and bromine

✪ 25. Write the names of the following compounds.
   a. SiO_2
   b. B_2F_4
   c. AuCl
   d. Cu_3PO_4
   e. Si_3N_4
   f. Ba_3P_2

✪ 26. Write and balance the complete formula equations for the following chemical reactions.
   a. Iron metal reacts with aqueous copper (II) sulfate to produce copper metal and aqueous iron (II) sulfate.
   b. Aqueous hydrogen chloride combines with aqueous barium hydroxide to produce aqueous barium chloride and liquid water.
   c. Solid silver oxide decomposes with heat into silver metal and oxygen.

✪ 27. Identify the general type of chemical reaction represented by each of the following chemical equations.
   a. $Fe_2O_3\,(s) + 2Al\,(s) \longrightarrow 2Fe\,(s) + Al_2O_3\,(s)$
   b. $H_2SO_4\,(aq) + 2KOH\,(aq) \longrightarrow 2HOH\,(l) + K_2SO_4\,(aq)$
   c. $N_2\,(g) + 3Mg\,(s) \longrightarrow Mg_3N_2\,(s)$
   d. $2HgO\,(s) \xrightarrow{\Delta} 2Hg\,(l) + O_2\,(g)$

MIXTURES AND SOLUTIONS

CHAPTER 20

DOMINION SCIENCE PROBLEM

When a Solution Is the Problem

All it takes is a hot afternoon in the baking sun to understand one of the most basic human needs—water. Many Middle Eastern countries that border the ocean, such as Kuwait, Qatar, Saudi Arabia, and the United Arab Emirates, are deserts that lack a sufficient supply of natural fresh water. Why can't these countries just draw from the limitless sources of water in the adjacent seas? After all, salt water *does* contain water.

Humans cannot drink seawater or even the slightly salty water from wells near seacoasts. As many sailors adrift at sea have learned, drinking seawater (a salt solution) just makes one thirstier. Salt water dehydrates the body, causing severe upset of the internal organs, delirium, and eventually death. Yet the oceans beckon as the solution to the worldwide need for drinking water in populous coastal countries. How can a salt solution be the answer to the problem of supplying fresh water?

20-1 This raft of Cuban refugees was sighted adrift about 20 mi off the coast of Cuba. Such people often succumb to the lack of fresh water.

20A Heterogeneous Mixtures

20A Objectives

After completing this section, you will be able to

- ✓ differentiate between homogeneous and heterogeneous mixtures.
- ✓ identify the continuous phase and the dispersed phase in a mixture.
- ✓ describe and give examples of the different kinds of heterogeneous mixtures.
- ✓ explain the differences between colloids and suspensions.

20.1 *Introduction*

"Whew! Is it ever hot today!" you exclaim as you finish weeding the flower beds and vegetable garden. You wipe the beads of sweat from your forehead before they can accumulate to form salty trickles that run into your eyes and make them sting. At that moment your thoughtful mom appears with an insulated cup containing your favorite soft drink—ice cold! You wipe the soil off your hands and move to the shade to enjoy a refreshing pause in your yard work.

We live in a world full of mixtures. The story above and the adjacent figure provide examples of mixtures—not only liquids, but solids and gases as well. This chapter answers some important questions about the different types of mixtures.

20-2 How many mixtures can you identify in this image?

All matter can be classified as either a pure substance (element or compound) or a *mixture*. The components of a mixture may be elements, compounds, or mixtures themselves. In general, all mixtures have two common properties.

1. Components of a mixture are not chemically combined. They can be separated by physical means.
2. Mixtures vary in their composition. The ratio of their components can fluctuate.

The sizes of the particles that make up the different parts of the mixtures and how evenly they are mixed determine how the mixtures are classified. If the particles in a mixture are small and uniformly mixed, then the mixture is homogeneous. Another name for a homogeneous mixture is *solution* (homogeneous mixtures are discussed in Section 20B). If the particles in a mixture consist of clumps of atoms or molecules, especially if the clumps are very large, then the mixture is *heterogeneous*. The term *mixture* by itself usually refers to a heterogeneous mixture.

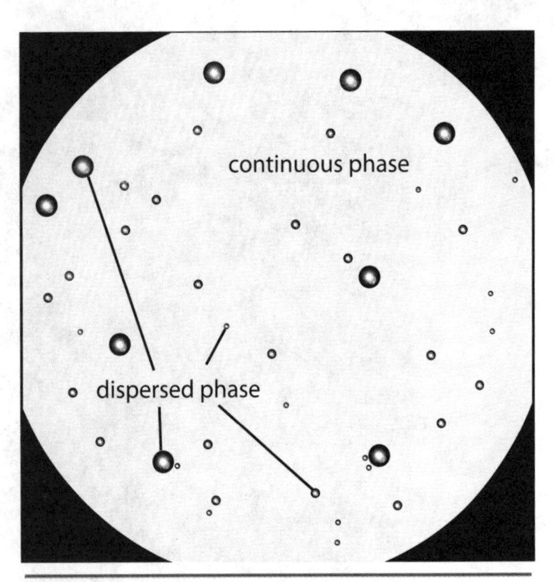

20-3 This microscopic view shows how many heterogeneous mixtures consist of one or more dispersed phases mixed within a continuous phase.

20.2 *Classifying Mixtures by Phase*

Heterogeneous mixtures contain two or more phases, or distinct parts. Sometimes there is one unbroken phase, called the **continuous phase**, in which the other phases are mixed. The other phases are the **dispersed phases**. Scientists categorize mixtures by their phases and how they are mixed.

Emulsions

Mixtures that contain two or more distinct liquid phases are called *emulsions*. The phases in an emulsion cannot be blended into one another—they have the property of **immiscibility**. Because the liquid droplets are extremely small and often highly viscous, the mixture can be very "thick" or "stiff" when they stick to each other. You probably eat several emulsions each day, such as milk, mayonnaise, and butter. Each dispersed phase will usually separate from the continuous phase over time to form a layer. If the dispersed phase has a lower density than the continuous phase, it will form a layer on top of the continuous phase, like cream separating from milk. This process can be slowed but not entirely stopped by *homogenization.* In this process the emulsion is mixed to thoroughly break up and separate dispersed particles. Emulsions are cloudy or opaque because light reflects off the tiny dispersed liquid droplets.

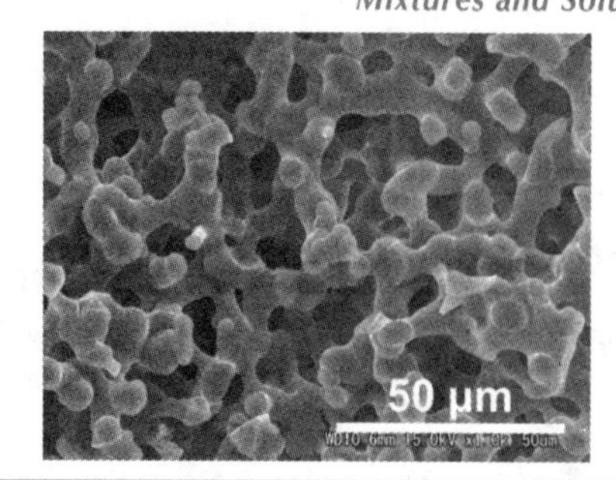

20-4 This photomicrograph shows the structure of a hair gel. The scale bar indicates the sizes of the solid portions of mixture.

Gels and Sols

Gels are two-phase mixtures. They consist of a solid dispersed in a liquid. Although gels are mostly liquid, they have some properties of a solid. The gel particles form an open network that gives the mixture some rigidity. Dessert gelatin is one example of a gel. When the liquid component is removed, some gels can be converted into unusual, extremely lightweight solids called *aerogels* (see the facet on page 194).

If the solid-in-liquid mixture is more like a liquid than a gel, it is called a *sol.* The solid particles in sols do not form an open network of particles. Sols are often thick or viscous liquids. Paints and inks are familiar sols.

Foams

A *foam* forms when a gas is combined with a liquid by whipping so that it makes a mixture of tiny bubbles within the liquid. Depending on the viscosity and other properties of the liquid phase, the foam may last for quite a while. Whipped cream and some shaving creams are examples of foams. Over time, the gas bubbles *coalesce,* or join together, forming larger bubbles that eventually leave the continuous phase.

20-5 Foamy shaving cream is a gas-in-liquid mixture.

Aerosols

Tiny solid particles or liquid drops that are dispersed in a gas form an *aerosol.* Natural and man-made sources of aerosols fill the atmosphere with suspended particles that may take years to leave the air and fall to earth. Particles of dust, salt crystals, water droplets, and chemicals compose atmospheric aerosols like smoke and fog.

Metal Mixtures—Alloys

Most of the known elements are metals (see Chapter 17). While many metals are useful in their pure state, history and research have shown that mixtures of metals—alloys—can have much better strength, ductility, malleability, and corrosion resistance. In addition, they may have many other properties. At the atomic level, metals form by packing atoms close together in crystalline structures (see Section 18D). In alloys, as long as the atoms of different elements are similar in size, the atoms bond to form a fairly regular, undistorted lattice. The resulting metal appears uniform under a microscope and we classify it as a *homogeneous alloy.* These alloys are true metal solutions.

However, if the metal alloy contains a mixture of elements whose atoms are greatly different in size, the resulting metal lattice could warp or have great gaps due to the atomic size differences. Because of this, as the molten mixture solidifies, the different atoms tend to bond with like-sized atoms and exclude larger or smaller atoms from the lattice. These clumps of atoms form distinct crystals of the different elements mixed within the cooling alloy (see Figure 20-7). Microscopically, such metals are *not* uniform and are called *heterogeneous alloys.*

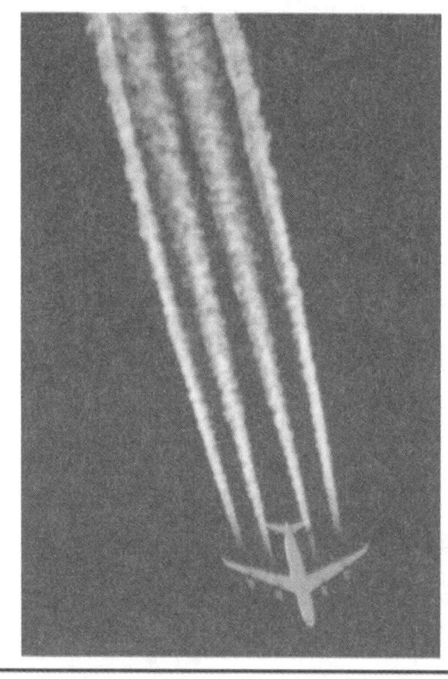

20-6 An airplane contrail forms an aerosol of ice crystals high in the atmosphere.

20-7 A photomicrograph of high carbon steel, a heterogeneous alloy

Amalgams are liquid-in-solid mixtures. They can be heterogeneous or homogeneous. See Subsection 20.6 under "Solid Solutions" for more information about these mixtures.

Heterogeneous alloy crystal size depends on cooling rate and other factors that are often difficult to control during solidification. One way that *metallurgists* (chemists specializing in metals) can control crystal size and distribution of the metal elements is to start the alloy process with a batch of fine powders of the pure elements evenly mixed in the required proportions. They heat the mixture to almost melting and then run it between rollers to crush the particles together. The result is a uniform mixture of microscopic crystals bonded to adjacent crystals. These heterogeneous *powder alloys* are made for specialized purposes, such as strong magnets and fuel plates in nuclear reactors.

20-1 Heterogeneous Mixtures

		Continuous Phase		
		Gas	Liquid	Solid
Dispersed Phase	Gas	does not exist	foams	solid foams (pumice)
	Liquid	aerosols	emulsions	heterogeneous amalgams
	Solid	aerosols	sols and gels	heterogeneous alloys

20.3 Classifying Mixtures by Particle Size

Besides being classified by phase, mixtures can also be classified by the sizes of their particles. Two specific classes of heterogeneous mixtures can be distinguished by particle size: colloidal dispersions and suspensions.

Colloidal Dispersions

Many heterogeneous mixtures often appear to be homogeneous. However, at the microscopic level these mixtures consist of tiny bits of matter or droplets that are uniformly dispersed throughout the mixture. If the particles' largest dimension is between 1 nm and 1 μm, the mixture is called a *colloidal dispersion*. The dispersed particles are called **colloids**. The colloids are too small to settle out of the continuous phase by the force of gravity alone. The thermal vibrations of continuous phase particles keep the dispersed phase particles uniformly suspended. Many gels, sols, emulsions, foams, and aerosols with particles in this size range are classified as colloidal dispersions.

If all the mixture's particles are smaller than 1 nm, then the mixture is considered homogeneous (a solution).

Because dispersed colloid particles are so small, their particles do not readily scatter light. Light from the blue end of the visible spectrum scatters more easily than light from the red end because these wavelengths are closer to the size of colloidal particles. Thus, colloidal dispersions often appear to have a bluish tinge in white light. Skim milk illustrates this property. The sky appears blue for a similar reason. Sunlight photons collide with the molecules of oxygen and nitrogen in the air. The blue wavelengths of light are scattered in all directions, while the red wavelengths go straight through. When looking at the sky away from the sun, we see these scattered light rays. This also explains why sunsets appear reddish. The direct path of the sunlight through the atmosphere to us is so long when the sun is near the horizon that most of the shorter wavelengths have been scattered to the side. By the time we see the sunlight, only the direct red wavelengths are left.

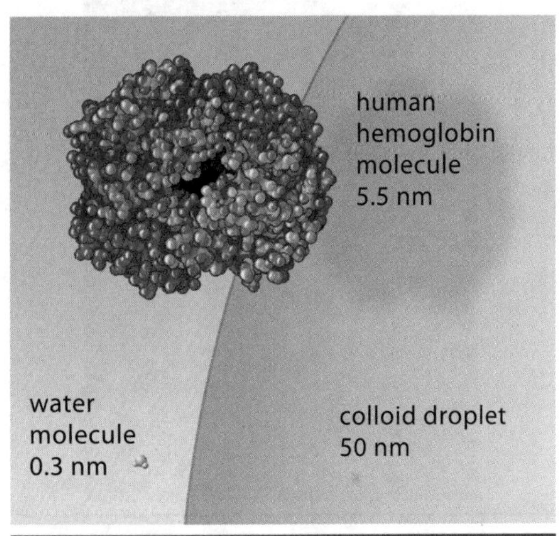

20-8 Size comparisons between a water molecule (0.3 nm), a molecule of hemoglobin (5.5 nm), and a colloidal droplet that is 50 nm in diameter

Suspensions

Sunbeams result from the dispersion of beams of direct sunlight by tiny water droplets and colloidal dust particles suspended in the atmosphere.

Suspensions are fluid mixtures whose dispersed particles, whether solid or liquid, are larger than about 1 μm in size, the maximum size for colloid particles. Particles of this size will eventually settle out of the mixture under the influence of gravity. They will form a layer of *sediment* at the bottom of the fluid if it is left undisturbed. However, if there is even a slight agitation of the fluid, the smallest suspended particles may remain dispersed throughout the continuous medium for a long time. An example of the suspension effect can be seen if you mix a sample of soil in a jar full of water. After you shake the jar, the large pieces of soil settle out immediately, but it may take days with the jar resting motionless on a shelf for the tiny clay particles to finally fall to the bottom of the jar. Your blood is another example of a suspension. The various kinds of cells dispersed in a sample of your blood would eventually settle to the bottom of a stationary test tube if it didn't clot.

Clear colloidal dispersions and suspensions can be distinguished from solutions by the way their dispersed particles scatter light. When a bright beam of light shines through one of these mixtures, the beam can be clearly seen. This phenomenon is called the **Tyndall effect.** It is responsible for visible sunbeams in the clouds and headlight beams in fog. True solutions do not exhibit the Tyndall effect because atoms and molecules are too small to visibly scatter light.

In some suspensions, individual particles are large enough that they can be seen glinting in the liquid or gas. Also, the reflected light from suspensions looks more natural in white light than the reflected light from colloidal dispersions. The particles are large enough to reflect longer wavelengths of light as well as those toward the blue end of the spectrum.

John Tyndall (1820–93) was an Irish scientist who first described the scattering of light in colloidal dispersions and suspensions. He was also a persuasive and popular public speaker who supported Charles Darwin's evolutionary theories. Many people in the United Kingdom and the United States turned from believing in the authority of Scripture because of his influence.

20A Section Review

1. What kinds of substances may form mixtures? What two properties do all mixtures have?
2. Name and describe the two phases of a heterogeneous mixture in which one phase is mixed in another.
3. How do metallurgists avoid problems in alloys of elements having atoms of very different sizes?
4. Name six kinds of heterogeneous mixtures and briefly describe their composition. Give an example of each.
5. What is the main difference between a solution, a colloidal dispersion, and a suspension? Give the general limits that define these mixtures.
6. What happens to light when it passes through a colloidal dispersion or suspension? What is this phenomenon called?
7. What is often visibly different about a suspension compared to a similar colloidal dispersion?
8. (True or False) The main difference between colloidal dispersions and suspensions is the size of the particles.

God and Scientific Discovery

Can an atheistic scientist who believes in Darwinism make discoveries that are useful as dominion science?

20B Objectives

After completing this section, you will be able to

- ✓ list the properties that characterize solutions and identify which ones are shared with other mixtures.
- ✓ identify the kinds of solutions possible.
- ✓ describe the properties of water that make it a good solvent.
- ✓ explain how solute particles dissolve.
- ✓ differentiate between dissociation and ionization.
- ✓ define solubility and explain what factors affect it.
- ✓ compare unsaturated, saturated, and supersaturated solutions.
- ✓ identify conditions that can affect the rate of solution.

solute (SAHL yoot)

A membrane is a very thin sheet of material. Biological membranes in cells can be only a few molecules thick.

20B Homogeneous Mixtures: Solutions

20.4 Defining Solutions

Solutions are homogeneous, or uniform, mixtures of pure substances that consist of a single phase. The word *solution* is related to the word *dissolve*. Both words come from the Latin word *solvere*, meaning "to loosen." In its simplest form, a solution is one substance dissolved in another. The substance that is dissolved is called the solute. The substance that does the dissolving is called the **solvent**. There are some cases in which it is difficult to determine which substance is being dissolved and which is doing the dissolving. For example, automobile antifreeze is a solution of half water and half ethylene glycol, both liquids. The question may be posed, which liquid is dissolving which? Thus, a more complete definition states that *the solute is the substance in a solution that is present in the smaller amount, while the solvent is the substance present in the larger amount.*

20.5 Properties of Solutions

Suppose you dissolved some table salt in water. Even if you examined the salt water under a powerful microscope, it would appear homogeneous. You would not be able to filter out the salt particles from the water with standard laboratory filter papers (although some membranes can filter certain molecules and ions). The salt (the solute) in the salt water would not settle out (even after a long period of time), assuming that no water (the solvent) evaporated from the mixture. If you boiled the saltwater solution or simply allowed it to stand for some time so that the water evaporated, only the salt would be left. You might also notice that the saltwater mixture was clear. And, unless you added quite a bit of salt to the water initially, you would find that more could be added and still dissolve.

The characteristics discussed above apply to all true solutions. They can be summarized as follows:

- Solutions are homogeneous.
- Solutes cannot be filtered out of a solution.
- Solutes do not settle out of a solution.
- True fluid solutions are transparent and do not display the Tyndall effect.

Solutions also share the following properties with all mixtures:

- Components of a solution are not chemically combined.
- The compositions of solutions can vary.

The main difference between solutions and other mixtures is particle size. Solute and solvent particles are much smaller than 1 nm, the lower limit of the colloid's size.

20-9 Suspensions and colloidal dispersions (left) produce the Tyndall effect, while true solutions (right) do not.

20.6 Types of Solutions

When you hear the term *solution*, you probably think of liquids. Liquid solutions are the most common type, but solutions can occur in solids, liquids, and gases. Solutions are normally categorized by the solvent's state of matter.

Liquid Solutions

The solvent in liquid solutions is a liquid. The solute may be a solid, a liquid, or a gas. In brine—a saltwater solution—the solute is the solid sodium chloride and the solvent is water. A solution of salt water is classified as a solid-in-liquid solution. In vinegar the solute consists of a liquid (acetic acid) dissolved in water—a liquid-in-liquid solution. Carbonated beverages contain a gaseous solute (carbon dioxide) dissolved in flavored water—a gas-in-liquid solution.

Liquid solutions are not necessarily just mixtures of a single solvent and a single solute. Many liquid solutions have numerous solutes. The complex mixture that makes up our blood is a liquid solution that also contains suspended cells and macromolecules. Our bodies depend on substances like proteins, minerals, sugars, and even gases that are dissolved in the watery plasma of our circulatory system.

The solutions of the waters in lakes, rivers, and oceans are extremely complicated mixtures of solids and gases, not to mention the living and nonliving matter suspended in them. Liquid solutions can contain solvents other than water. Many liquid solutions form with organic solvents, such as alcohol and acetone. Fingernail polish removers and perfumes are solutions of such organic solvents. Gasoline is also a common organic solution.

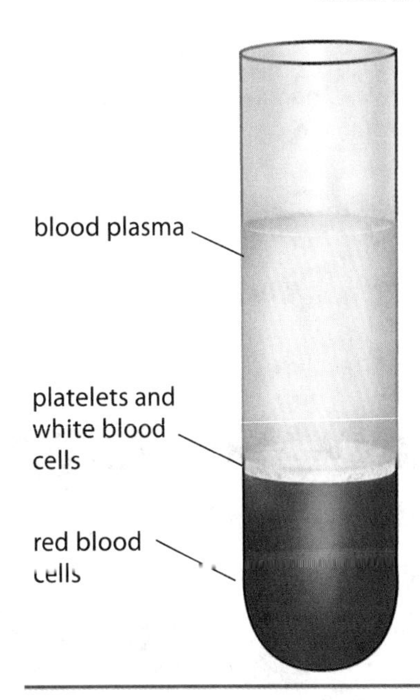

20-10 Whole human blood consists of a complex solution of chemicals in the clear liquid blood plasma (at the top of the tube) and a suspension of blood cells and large protein molecules (in the bottom of the tube).

Solid Solutions

Although it may seem unlikely, it is quite possible to form solid solutions. Gases, liquids, and other solids can dissolve in solid solvents. Hydrogen gas and some noble gases will dissolve into metals such as platinum (Pt) and palladium (Pd). When hydrogen gas is produced industrially, it is purified by passing through the walls of palladium cylinders. This metal allows hydrogen to pass through but stops all other gases at the metal's surface. Helium (He) and argon (Ar) gases, the products of radioactive decay, are known to pass through the crystal lattices of minerals in rocks. Under these conditions, they form gas-in-solid solutions.

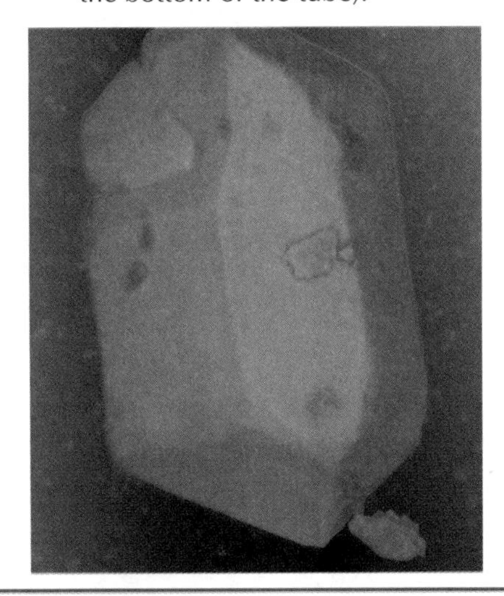

20-11 Creationary geologists use the rate of diffusion of helium gas dissolved in tiny crystals of zircon as evidence for a young earth.

20-2 Common Alloys

Name	Components	Typical applications
brass	copper, zinc	plumbing fixtures, musical instruments
bronze	copper, tin	ball bearings, seawater valves, gears
dental amalgam	silver, mercury, copper, zinc	dental fillings
gold (18 karat)	gold, silver, copper, nickel	jewelry
solder (plumbing)	silver, copper, zinc	connecting metal drinking water pipes
sterling silver	silver, copper	flatware, jewelry
Wood's metal	bismuth, lead, tin, cadmium	automatic fire sprinklers

amalgam (uh MAL gum): (possibly from Gk. *malagma*—soft mass)

Certain liquids will dissolve in solids. For instance, mercury easily dissolves in gold and silver, forming an ***amalgam***. Dentists sometimes use a silver amalgam to fill cavities in teeth.

Most true solid-in-solid solutions are alloys—metals or other solid elements dissolved in metals. Homogeneous alloys are generally formed by uniformly mixing them while they are molten and then allowing the mixture to cool. Brass is an alloy made by mixing molten copper and zinc, and bronze is an alloy of copper and tin. These metal solutions are useful because the properties of the alloy often differ greatly from those of the individual metals.

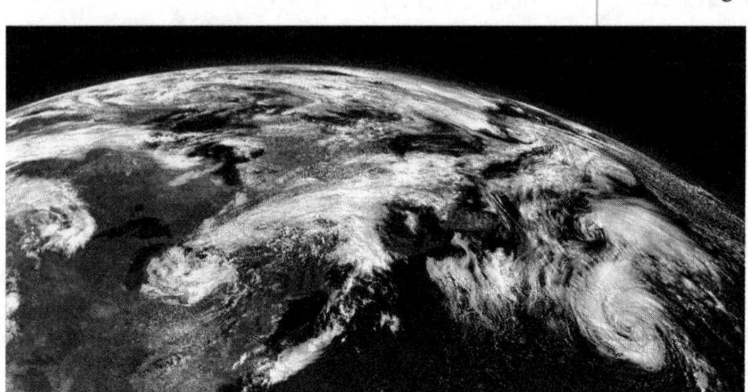

20-12 The earth's atmosphere is a solution of oxygen and other gases in nitrogen.

Gaseous Solutions

In a gaseous solution, both the solvent and solutes must be gases. Suspended liquid or solid solute particles would settle out or form a colloidal dispersion because they would exceed the maximum particle size limit for true solutions.

The most common example of a gaseous solution is air—composed of approximately 78% nitrogen, 21% oxygen, and 1% other gases. Since nitrogen is the major component, air could be considered a solution of oxygen and other gases (the solutes) dissolved in nitrogen (the solvent).

Since true liquid-in-gas and solid-in-gas solutions are not possible, only seven types of solutions involving one solute can exist. These are summarized in Table 20-3.

Solutions with more than one solute can be very complex. Seawater, for example, is a complicated solution containing hundreds of different solids, liquids, and gases.

20-3 Solutions

		Solvents		
		Gas	**Liquid**	**Solid**
Solutes	**Gas**	gas-in-gas (O_2 in N_2, air)	gas-in-liquid (soft drink)	gas-in-solid (H_2 in Pd)
	Liquid	does not exist	liquid-in-liquid (vinegar)	liquid-in-solid (amalgam)
	Solid	does not exist	solid-in-liquid (salt water)	solid-in-solid (alloy)

20.7 Water—The "Universal Solvent"

Of all the substances that God created, water comes closest to being a universal solvent. It is the most common solvent in liquid solutions. In fact, it dissolves so many materials that totally pure water is extremely rare, if not impossible to find.

Water is an effective solvent because of its molecular structure. As you have learned in earlier chapters, a water molecule consists of two hydrogen atoms covalently bonded to a central oxygen atom. Because the oxygen-hydrogen bonds are so polar, partial electrical charges occur at opposite ends of the bonds. Oxygen is strongly electronegative. It acquires a slightly negative charge, and the hydrogen atoms become slightly positive.

If the water molecule were linear, the equal positive charges at both ends of the molecule would result in a molecule without electrical poles. However, the nonbonding pairs of electrons in the oxygen atom's outer energy level repel the bonding pairs, forcing the molecule into a bent shape. If you draw lines between the centers of the hydrogen atoms and the oxygen atom, they form an obtuse angle. This arrangement places the positive hydrogen atoms toward one side of the molecule. The negative oxygen atom is exposed on the other side (see Figure 20-13). Thus, the water molecule is electrically polar. The negative end will attract positive ions and the positive regions on other polar molecules. The positive end will attract negative ions and the negative regions on nearby polar molecules. These attractions play a key role in the dissolving action of water.

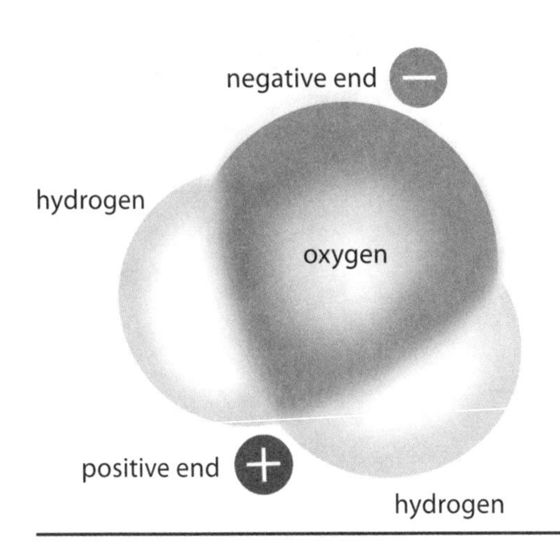

20-13 The water molecule's structure makes water an excellent solvent due to its polarity.

20.8 The Solution Process

Consider the dissolving of an ionic solid, such as sodium chloride, in water. Since ionic solids consist of networks of positive and negative ions, several things must happen for them to dissolve. According to the kinetic-molecular theory, solvent water molecules are highly mobile and are continuously and forcefully bumping against the surface of the salt crystal. As they collide with the sodium and chloride ions, the polar ends of the water molecules attract the ions, weakening the pull of their neighbors. With each collision, ions are pulled and wedged from the crystal in a process called **dissociation**. Eventually, the water molecules surround and isolate the ions from the crystal. This process is called *hydration*. The solute particle is said to be *hydrated*. For solvents other than water, the surrounding of a solute particle by solvent molecules is called *solvation*.

The molecules of solid crystalline covalent compounds, such as table sugar, are also separated by the solvating action of solvents like water and ethyl alcohol. However, there is no special term like *dissociation* to describe this process for molecular compounds. When some covalent compounds are hydrated, they break apart to form ions in a process called *ionization*. For example, when the covalent gas hydrogen chloride (HCl) is bubbled through water, its molecules ionize to form hydrogen (H^+) and chloride (Cl^-) ions. The solution is known as hydrochloric acid.

20.9 Solubility

A substance is *soluble* when it can be dissolved by a solvent. Substances that are soluble in water are usually polar. The rule that governs the solubility of substances is *like dissolves like*. The more similar the chemical structures of solutes are to a solvent, the more soluble they will be in that solvent. Alcohols, such as ethanol (ethyl alcohol), dissolve well in water because both have a hydroxide (OH^-) group as part of their structure (see Figure 20-14). (A water molecule can be considered hydrogen bonded to a hydroxide group.) The hydroxide provides a very polar end to the ethanol molecule, which is strongly attracted to the highly polar water molecules. This similarity allows them to attract one another and mix completely. *Liquids that can freely mix in any proportion are said to be miscible.* As with any solution, miscible liquids form a single phase.

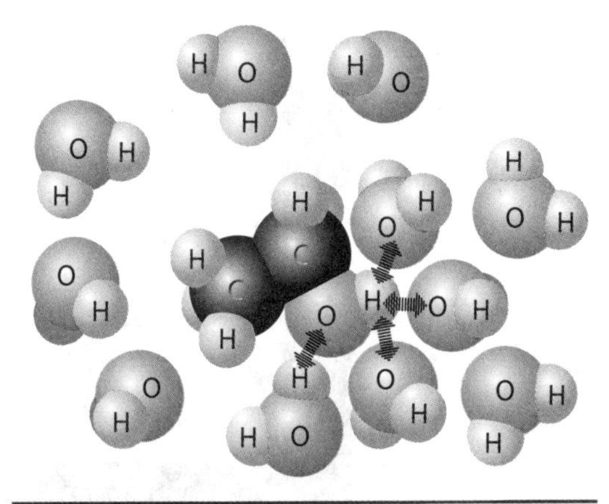

20-14 Water and ethanol are completely miscible because of their similar structures.

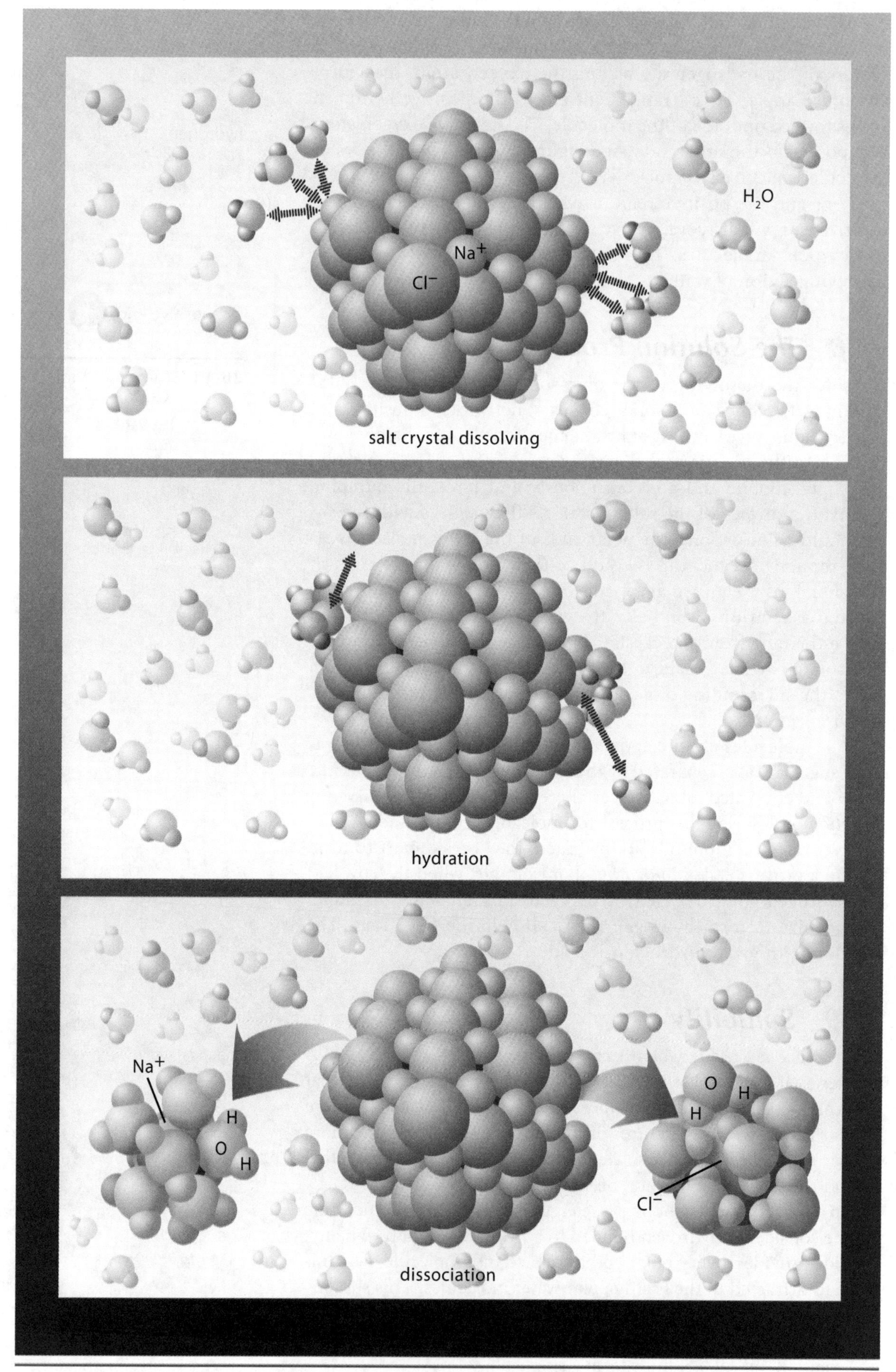

20-15 The water solution process includes removing the solute particle from the solid state, surrounding it with water molecules (hydration), and removing it from the solute's location.

Water's ability to dissolve polar or ionic substances is astounding. For example, 100 mL of water at 100 °C can dissolve 1024 g of ammonium nitrate, a common fertilizer—a solute mass more than 10 times the mass of the solvent! **Solubility** is the maximum amount of solute that will dissolve in a given amount of solvent at a given temperature. For aqueous solutions, solubility is usually expressed in grams of solute per 100 mL of water. Table 20-4 lists solubilities of some common substances in water near room temperature.

20-4 Solubility of Common Substances

Name	Formula	Solubility*
calcium carbonate (chalk)	$CaCO_3$	0.0015
carbon dioxide	CO_2	0.178
ethyl alcohol	C_2H_5OH	no limit
sodium bicarbonate (baking soda)	$NaHCO_3$	8.7
sodium chloride	$NaCl$	35.9
sucrose (sugar)	$C_{12}H_{22}O_{11}$	201.9

*grams solute/100 mL of H_2O at 20 °C and 1 atm

Very few solutes are completely insoluble in water. Though its ions resist hydration, even calcium carbonate has a very small solubility.

Though water is called the universal solvent, some substances do not mix with water. Solutes that cannot mix in certain solvents are said to be *insoluble*. Insoluble liquids are *immiscible*. Water-insoluble substances usually consist of nonpolar molecules. When mixed with water, they form different phases. A familiar combination of immiscible liquids is Italian salad dressing. The nonpolar oil molecules cannot be hydrated by the polar water molecules because there are no charged sites on the surface of the oil molecules. When the mixture is shaken vigorously, it appears for a few moments that the oil and water have mixed. The liquid appears milky and uniform. But soon a layer of oil begins to form on top of the water. Eventually, the two liquids are distinctly separate layers.

Miscibility applies only to liquid-in-liquid solutions.

Nonpolar substances include thousands of organic compounds, such as oils and greases. Though water cannot dissolve these substances, *organic solvents*, such as acetone and mineral spirits, can. Organic solvent molecules are nonpolar or have a region that is nonpolar. They work by surrounding solute molecules through thermal motion to dissolve them. Since the intermolecular forces between nonpolar molecules are so weak (see the box on page 422), the solvent is able to dissolve the solute fairly easily.

FACETS OF SCIENCE

How to Get Squeaky Clean

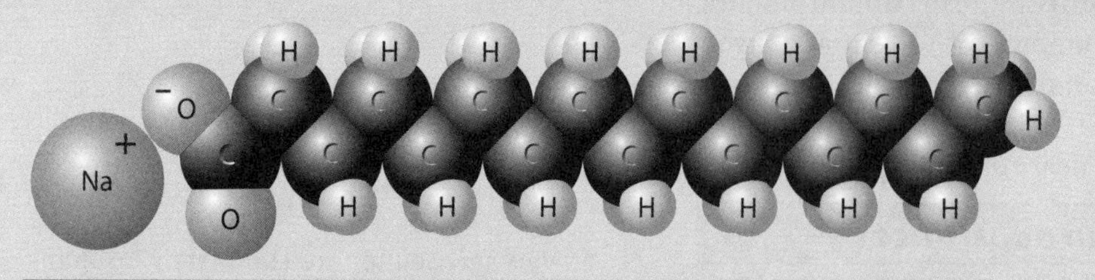

A typical soap molecule. The long chain of carbon atoms is nonpolar. The negatively charged oxygen provides a polar end that attracts water molecules.

Why do we use soaps to get things clean? The answer lies in a basic principle of solutions: *like dissolves like.* Water is a polar molecule. Dirt and greasy oils are usually nonpolar molecules. They will have nothing to do with water. Since water cannot dissolve them, these oil and dirt particles would remain on our clothing and us if it were not for a go-between that can dissolve them and yet is soluble in water. Soap is the perfect middleman for the job!

Look at the structure of a typical soap (detergent) molecule. The molecule itself resembles both polar and nonpolar molecules. It is slightly related to a fat molecule (oils are fats). The part of the molecule that is like oils and dirt is the long chain of carbon and hydrogen atoms. This nonpolar end is attracted to the dirt and oils on your skin and clothes. The opposite end has a pair of ions, sodium and oxygen. When soap is dissolved in water, the sodium ions dissociate, leaving a negatively charged site that attracts water molecules. Thus, the soap molecule has a polar end that is soluble in water. Once the nonpolar end grabs the dirt and oils, an emulsion forms. All we have to do is rinse the soap molecules away with the water!

Traditional soaps are produced by combining lye (sodium hydroxide) with animal, vegetable, or artificial fats, a process called *saponification.*

The first soaps were probably produced as early as 2000 BC, but no one was really interested in them. In fact, the grand baths that the Romans built probably lacked soap! Even the great Cleopatra of Egypt was "soapless." She is said to have used very fine sand for its abrasive properties. It wasn't until the thirteenth century that soap making became an industry in Europe. In the early history of the United States, soap making in the home was as common as spring cleaning or crop harvesting. Common household items provided the materials. Lye was obtained from ashes, and fat came from butchered animals. Animal fat and ashes were saved so that most homes had a fairly large quantity of both for making soap.

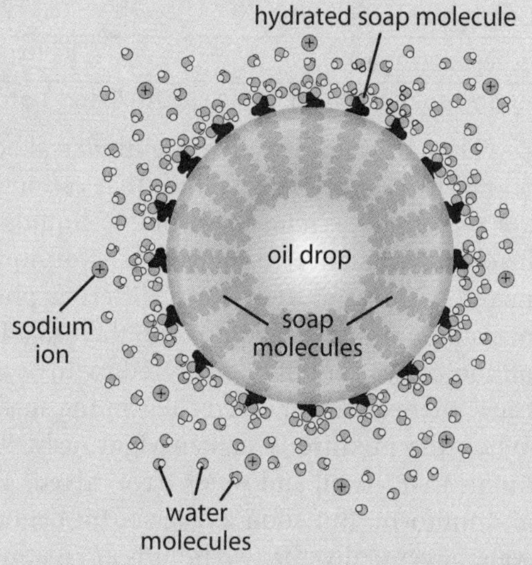

An oil drop in water, surrounded by soap molecules and emulsified

20.10 Factors Affecting Solubility

Temperature

At 10 °C, 100 mL of water can dissolve 118 g of ammonium nitrate, while 100 mL of near-boiling water can dissolve 1024 g of the compound. Temperature obviously affects solubility. Though there are some exceptions, most solids become more soluble in liquids as temperature increases. The greater kinetic energy of the solvent molecules in a hot liquid makes the job of breaking up the solute particles much easier.

A temperature increase also affects solid and liquid solutes by increasing the vibration of their particles and weakening the bonds between them. This condition results in less force being needed to separate the solute particles. Solvent particles also spread farther apart, providing more room for solute particles to occupy. These three effects account for increased solubility with increased temperature.

Figure 20-16 shows the relationship between temperature and solubility in water for several ionic substances. Note that two of them are only slightly affected by changes in temperature, while the solubility of the others increases dramatically as the temperature rises. Generally, liquid solutes follow the same trend as solid solutes in liquid solvents.

When gases are dissolved in liquids, an increase in temperature has the opposite effect—gas solubility decreases with temperature. After a bottle or can of cold carbonated drink is opened, most of the carbon dioxide stays in the solution. But if you open a bottle or can of warm soft drink, you could have a mess on your hands—literally! When the soft drink's temperature rises, the solubility of the carbon dioxide in water decreases; the gas is forced to remain in the solution only because the bottle is sealed and vapor pressure keeps the gas dissolved. Once the seal is broken, the gas escapes rapidly from the warm solution. It forces the liquid out of the container and leaves you with a flat-tasting soft drink.

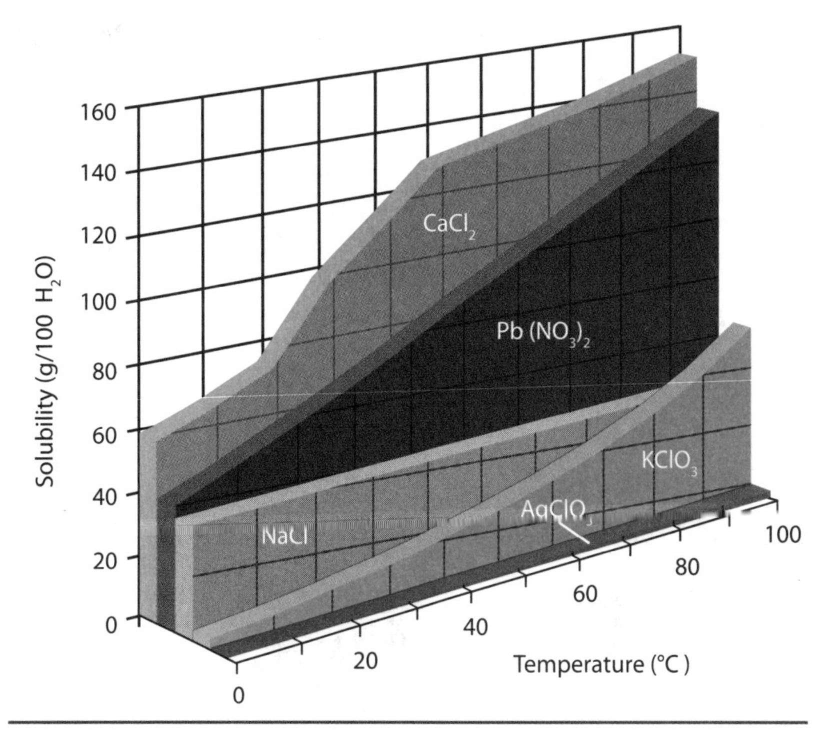

20-16 Comparison of solubilities in relation to temperature for several substances

A gas's solubility decreases with a rise in temperature because the kinetic energies of its molecules dramatically increase with temperature. You can observe this by allowing a glass of cold tap water to warm to room temperature. The small bubbles you see forming on the inside of the glass are a result of air dissolved in the water coming out of the solution. Different gases have different solubilities. For example, carbon dioxide, with a relatively low solubility in water at room temperature, escapes the solution quickly when its molecules gain the kinetic energy that they need. Oxygen has an even lower water solubility. Because more oxygen dissolves in cold water than in warm water, very active fish, such as trout, can thrive in only cold streams. The colder water contains more of the oxygen that they need. Shallow, still water warmed by the sun may cause them to suffocate.

Common atmospheric gases, such as oxygen, nitrogen, and carbon dioxide, consist of nonpolar molecules. Water has little attraction for them, so their solubilities are very low.

20-17 These trout thrive in stream rapids, where turbulent cold water dissolves a lot of oxygen.

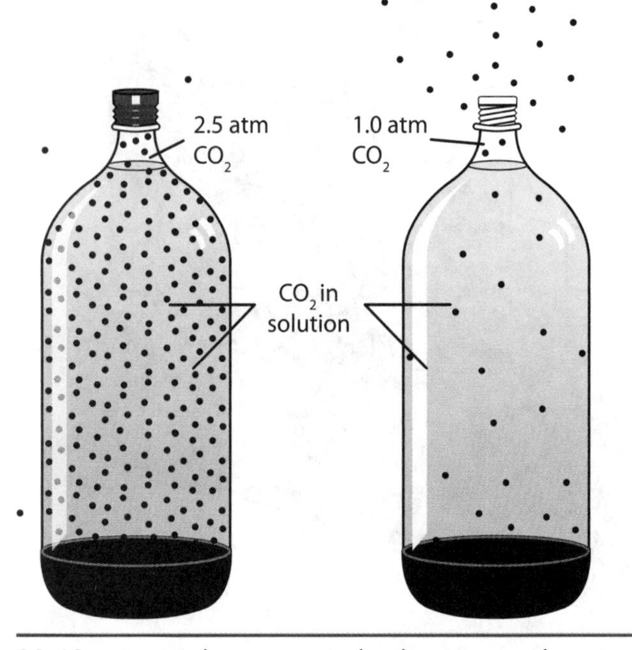

20-18 Henry's law accounts for the amount of gas in a solution in relation to the gas's pressure above the solution.

Pressure

A gas's pressure exerted on a gas-in-liquid solution affects the gas's solubility. In 1801 William Henry described this relationship in **Henry's law**. It states that *the greater the pressure of a gas on a liquid, the greater the mass of the gas that will remain dissolved at any given temperature.* Carbon dioxide gas remains in a soft drink while it is sealed under pressure, but as soon as the pressure is released, the gas begins escaping to reduce the mass of gas dissolved in solution.

20.11 Saturation

Some substances are infinitely soluble in each other. In other words, the substances may be mixed in each other in any ratio. Either substance can be the solvent or the solute in the solution. For example, ethyl alcohol is listed as having no limit of solubility in water (see Table 20-4 on page 467).

Most substances do have a limit to their solubility in a given solvent based on the characteristics of both the solute and the solvent. Some substances, such as the fertilizer ammonium nitrate, dissociate very easily in water. But only so much ammonium nitrate can dissolve. At 80 °C, 630 g can dissolve in 100 mL of water. If more ammonium nitrate is added than can be dissolved at this temperature, it settles to the bottom of the container. Such a solution is said to have reached the point of **saturation**. Excess gaseous and liquid solutes will form a separate phase in *saturated* solutions as well. Solutions that contain less solute than their solubility at a specific temperature are *unsaturated.*

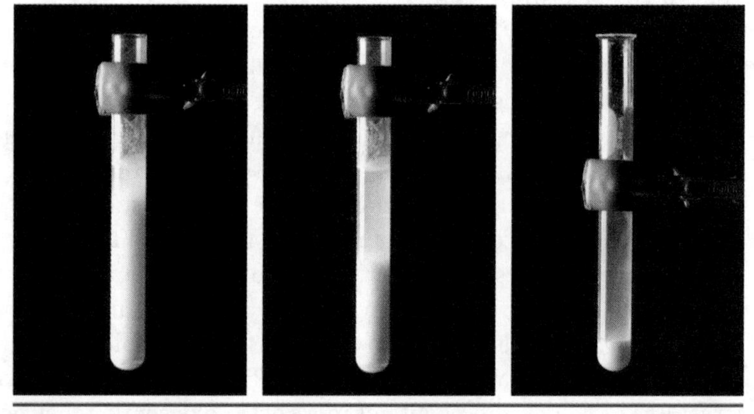

20-19 Substances that are nearly insoluble, such as calcium carbonate, form a saturated solution very quickly.

The solubility of a solute usually depends on the temperature of the solution. Solubility generally increases with solution temperature. When a solution that is saturated at a higher temperature is cooled slowly, it can sometimes become a *supersaturated* solution. A supersaturated solution has more dissolved solute than a saturated solution should at a specific temperature. A supersaturated solution is quite unstable. If it has been cooled many degrees below its normal saturation point, even a small disturbance will cause the excess solute to precipitate, or rapidly settle out of the

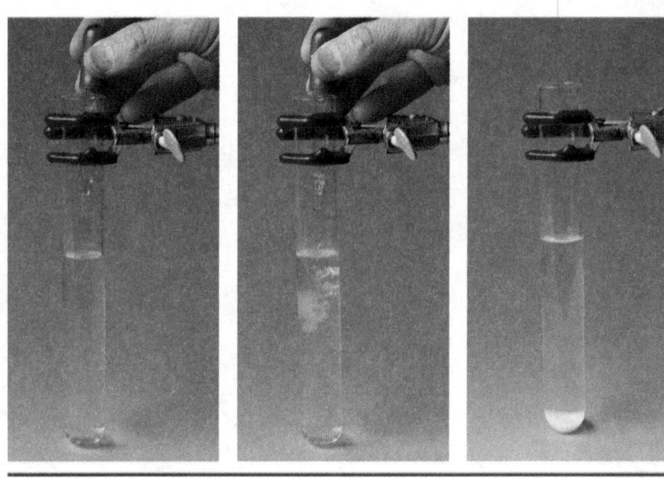

20-20 When disturbed, unstable supersaturated solutions can instantly form crystals as the excess solute precipitates. In this demonstration, a drop of the original unsaturated solution is released into a cool, supersaturated solution, causing precipitation of the solute.

solution. The remaining solution will be saturated for the existing temperature.

Supersaturated solutions are often used to make pure crystals of compounds. A solution of a compound is mixed to saturation at a high temperature. A seed crystal is placed in the solution. Then the solution is slowly and evenly cooled to create a supersaturated condition. The excess solute particles bond to the seed crystal and it grows larger. Very large pure crystals can be manufactured this way.

Gases in a solution can also be supersaturated. If a new bottle of a carbonated drink is carefully uncapped, excess carbon dioxide in solution can be made to rapidly enter the gaseous phase by striking the bottom of the bottle sharply on a table.

20.12 Rate of Solution

How *much* solute dissolves in a solvent is different from how *fast* it dissolves. Solubility involves how much solute will dissolve in a solvent. On the other hand, the rate of solution involves how fast a solute dissolves in a solvent. Three primary factors affect this rate.

Stirring

If you add sugar to your drink, you probably instinctively stir it. If a solid solute is added to an undisturbed solvent, solvated solute particles remain near the solute and are more likely to work their way back to the crystal and recrystallize. Eventually, the rate of particles entering the solution equals the rate of particles leaving the solution to become part of the solid again. At this point, the solid stops visibly dissolving. Stirring moves solvated particles away from the crystal so that they cannot rejoin the solid. Stirring also brings more solvent molecules in contact with the solute.

20-21 Stirring granulated solutes permits faster dissolving, especially in cold liquids.

Temperature

Since solvent molecules move faster at a higher temperature, more of them collide with solute particles in a given amount of time, speeding up the solution process. For example, sweetened tea is often prepared by dissolving sugar in tea while it is hot. Sugar added to hot tea rapidly dissolves, even with little stirring. When sugar is added to cool tea, it takes much longer to completely dissolve.

Surface Area

The third factor affecting the rate of solution is the surface area of the solute exposed to the solvent. For example, a sugar cube dissolves more slowly than an equal mass of granulated sugar at the same temperature. Why? Although the mass of the two sugar samples is the same, thousands of tiny granules have more surface area than does one large cube. The solution process can occur only at the surface of a solid solute. Therefore, greater surface area means that there can be more collisions of the solvent molecules with the solute in a given time; more rapid solvation results. Thus, grinding or crushing the solute increases the rate of solution.

20-22 Assume that the size of a sugar grain is 0.25 mm across. The surface area of a mass of granular sugar would be approximately 40 times the surface area of a 1 cm^3 sugar cube of equal mass.

20B Section Review

1. How can you tell the difference between solutes and solvents in a solution when they are in the same state?
2. What are the properties of solutions? Which of these are shared with *some* heterogeneous mixtures?
3. What gas solutions are not possible? Why?
4. Why is water such an effective solvent?
5. Describe the three steps in the solution process.
6. How is dissociation different from ionization?
7. What kinds of substances does the term *miscible* refer to? What is true about the possible proportions of completely miscible substances in solutions?
8. How does the pressure of a gas above a solvent affect the amount of gas that dissolves in the solvent? What is this principle called?
9. How can an unsaturated solution become supersaturated without any solute being added or any solvent being removed?
10. Which parts of the solution process are enhanced by stirring? by heating? by crushing or grinding?
11. (True or False) Solutions differ from most colloidal dispersions in that their solute particles are uniformly dispersed.

20C Objectives

After completing this section, you will be able to

✓ compare the terms *concentration*, *concentrated*, and *dilute*.

✓ describe two ways to measure the concentration of a solution.

✓ define the mole and state its value.

✓ state and describe three colligative properties.

✓ explain the processes of osmosis and reverse osmosis.

20C Solution Concentration

20.13 Measuring Concentration

Have you ever been served a glass of fruit punch that was too weak? Or perhaps you've been served lemonade that was too strong. Generally speaking, we could say that the fruit punch was too *dilute* and the lemonade was too *concentrated*. You may use these terms frequently, but do you understand what they mean? What determines whether the substance is dilute or concentrated? The terms *dilute* and *concentrated* are informal ways to compare the amount of solute in solutions. If a solution has a lot of solute, it is concentrated. If it has little solute, it is dilute. There is no way to quantify these terms unless a solution's concentration can be measured. **Concentration** is the amount of solute in a certain amount of solvent or solution. There are many ways to express concentration.

Percentage by Mass

One way to measure solution concentration is called *mass-percent* or **percentage by mass.** The word *percent* literally means "per hundred." Percentage by mass, then, is the mass of solute per 100 g of the solution. Since the mass of water is approximately 1 g/mL, it is simple to make a solution with a certain percentage by mass. Just determine the required amount of solute needed in 100 g of solution and subtract its mass from 100 g to find the amount of water it must be dissolved in. If more or less solution is needed, just use proportional amounts of solute and water.

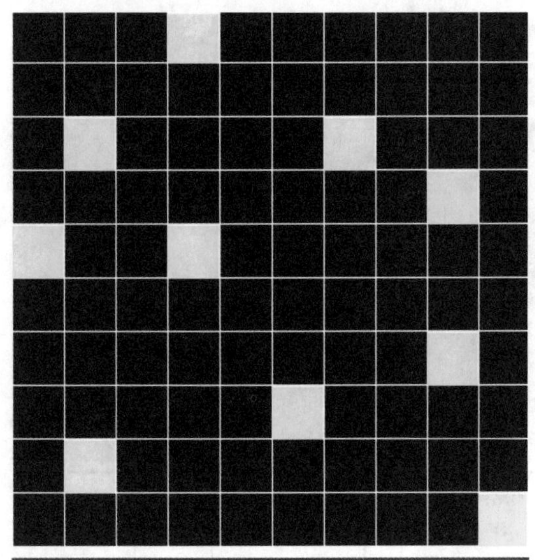

20-23 Illustration of percentage by mass. Each block represents a mass unit. There are 100 blocks. This solution has a concentration of 10% solute (yellow) by mass.

Hospital patients often receive a solution of sodium chloride that has the same salinity as bodily fluids. This solution is called *physiological saline*. The concentration of sodium chloride in the saline solution is 0.9%. To prepare the solution, you would dissolve 0.9 g of sodium chloride in 99.1 g (99.1 mL) of water (100.0 g – 0.9 g). This would give 100.0 g of solution. You could also dissolve 0.45 g in 49.55 g of water if you wanted only 50.00 g of solution.

Molarity

Solution concentration, as defined earlier, has to do with the number of solute particles in solution. For most scientific purposes, this is the most accurate way to measure concentration. The number of particles (atoms, molecules, or ions) or formula units present in a quantity of matter is measured with the SI unit called the *mole (mol)*. Because atoms and molecules are so tiny, it takes a lot of them to make up a measurable amount of matter. The value of a mole is about 6.02×10^{23} particles!

The numerical value of the mole, 6.02×10^{23}, is called *Avogadro's number*, after the Italian lawyer Amedeo Avogadro (1776–1856), who first suggested that there could be a molar number of particles.

The mass of 1 mol of any element or compound, in grams, has a value equal to its atomic or formula mass measured in atomic mass units (u). For elements, this value is the atomic mass shown on the periodic table. For example, the atomic mass of sodium is 22.99 u. One mole of sodium contains 6.02×10^{23} atoms and has a mass of 22.99 g. The mass of 1 mol (6.02×10^{23} atoms) of chlorine is 35.45 g. One mole of sodium chloride (6.02×10^{23} formula units of NaCl) has a mass of 58.44 g (22.99 g + 35.45 g).

Chemists can measure solution concentration in moles of solute per liter of solution, or *molarity (M)*. A 1 L solution containing 58.44 g, or 1 mol, of NaCl dissolved in water forms a one-molar (1 *M*) solution of sodium chloride. Knowing the concentration of solute particles in a solution measured in molarity helps chemists measure a variety of other properties of solutions.

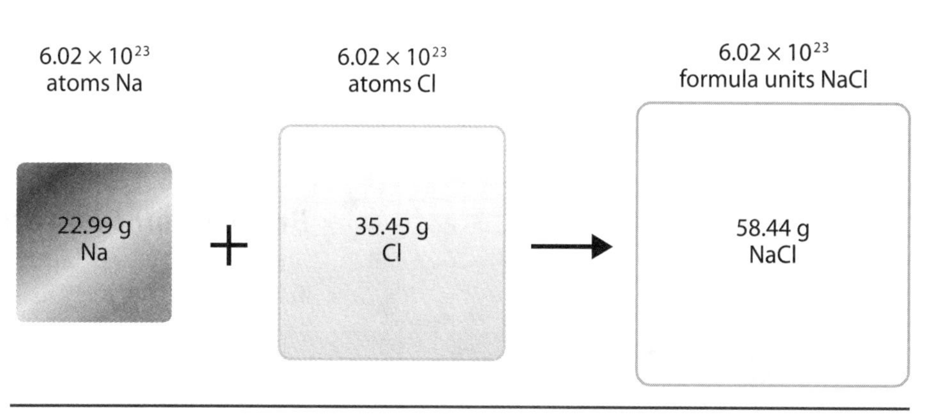

20-24 How a mole of a compound is determined in a laboratory

20.14 Colligative Properties

People who live in places where winters are long and snowy are familiar with using salt on roads and sidewalks to melt accumulated ice. Why does this work? The answer lies in understanding a whole class of properties related to the *concentration effects* of solvents. These properties result from the *number* of solute particles in a solution and are not due to the properties of the solute itself. Chemists call these effects **colligative properties**.

colligative (KAHL ih GAY tiv)

Freezing-Point Depression

In winter when temperatures hover around freezing for long periods of time, snow melts at the middle of the day and then refreezes as the sun sets. Ice always has a very thin layer of liquid water a few molecules thick covering its surface unless its temperature is extremely cold. When salt crystals are spread on ice, the water molecules begin dissociating the salt ions. These ions interfere with the refreezing process, making it much more difficult for water molecules to reenter the ice crystal lattice. The molecules in the liquid state begin knocking more water molecules out of the ice lattice. These molecules are then free to dissociate more salt ions. An increasing concentration of salt ions in the solution keeps water molecules in the liquid state.

When salt is applied to ice at a certain temperature, the solid ice melts because of the presence of salt ions. To refreeze, it has to be chilled to a colder temperature. Thus, salt water has a lower freezing point than pure water—its freezing point has been *depressed.* **Freezing-point depression** is a concentration effect that is directly proportional to the number of particles, either ions or molecules, in solution. This colligative property is not caused by the kind of solute particle involved. Similar concentrations of other water-soluble solutes produce the same effect. Water's freezing point is lowered 1.86 °C for every mole of solute particles per liter of solvent.

Road salt is commonly calcium chloride ($CaCl_2$) rather than sodium chloride (NaCl). Notice that $CaCl_2$ produces three ion particles for every two ions from NaCl in solution. Therefore, its effect on freezing-point depression is greater.

Freezing-point depression is involved when making homemade ice cream in a hand-cranked or motorized ice-cream freezer. Very cold crushed ice is used to chill the ice-cream mix. However, the ice pieces alone make poor thermal contact with the ice-cream mix can. Rock salt is added to the ice to lower its melting point, producing a cold ice-water slurry. This improves heat transfer out of the ice-cream mix. The brine slurry surrounding the can increases the surface area of the heat "sink" in contact with the mix can, and it is colder than just a plain ice-water bath. So adding the salt doesn't lower the temperature of the ice; it just makes the process of removing heat from the ice-cream mix more efficient. When the solid ice cream forms, it is colder than it would be if only plain ice were used.

Boiling-Point Elevation

Just as solute particles interfere with the freezing process, they also affect vaporization. Solute particles are part of a solution because they have an attraction for the solvent particles. At the solvent's boiling point, its particles must gain enough kinetic energy to escape into the gaseous phase. When solute particles are present, they hold back the solvent particles. The solvent particles must gain greater kinetic energy from their neighboring particles to leave the liquid phase. Thus, the solvent boils at a higher temperature when a solute is present. This concentration effect is called **boiling-point elevation**. Boiling-point elevation is directly proportional to the concentration of the solute particles.

Antifreeze used in cars is an important application of both boiling-point elevation and freezing-point depression. Adding an equal volume of antifreeze to the water in the radiator produces a mixture with a higher boiling point and a lower freezing point than pure water. The engine's coolant mixture is in little danger of boiling over in hot summer driving. More importantly, it will not freeze and expand, which could damage the engine, when exposed to freezing weather.

Osmotic Pressure

Osmosis is the movement of solvent from a high concentration to a low concentration through a *semipermeable membrane,* a thin sheet of material that allows solvent particles to pass through but prevents the passage of solute particles. Osmosis is similar to *diffusion* (see Chapter 2), but the membrane barrier restricts the diffusion of the solute.

Remember that diffusion is the movement of a substance from an area of high concentration to an area of low concentration. Osmosis is easily understood if you consider the concentration of the solvent in the solution. In a 5% aqueous solution of sodium chloride, the concentration of the solvent (water) is 95%, whereas the pure solvent has a water concentration of 100%. In this process the concentration of the solvent on the solution side of the membrane is lower than on the side with the pure solvent (see Figure 20-25). Solvent particles will diffuse through the membrane from the pure solvent side to increase the relative concentration of the solvent in the solution. Because of the structure of the membrane, the solute particles are prevented from diffusing in the other direction.

Scientists who first studied osmosis designed their apparatus using tall cylinders cross-connected at the bottom with the semipermeable membrane inserted between them. They noticed that the liquid level on the side with the pure solvent dropped while the level on the side with the solution rose. Eventually the levels stabilized. The difference in heights could be equated to a difference in liquid pressure between the columns. Thus, the driving force for osmosis came to be known as *osmotic pressure*. Osmotic pressure is directly proportional to the concentration of the solute and the absolute temperature of the solution.

Osmosis is vital to the health of living cells. Cells contain hundreds of solutes. If the cells didn't control the osmosis of water, they could swell so much that they would burst or dehydrate and shrink. The cellular membrane is a complex semipermeable membrane. It allows the transport of vital nutrients, cellular products needed by the rest of the body, and waste materials. Water and some other substances simply diffuse according to their concentrations on either side. Most other substances are "pumped" across the membrane by special complexes of macromolecules. The cell is designed so that it can control the osmotic pressure while carrying out its function within the organism.

Osmotic pressure is defined as the pressure that must be applied to the solution side of the semipermeable membrane to prevent the net diffusion of the pure solvent through the membrane into the solution.

The osmotic pressure of seawater, which contains about 35 g of dissolved solute per kilogram, is approximately 27 atm!

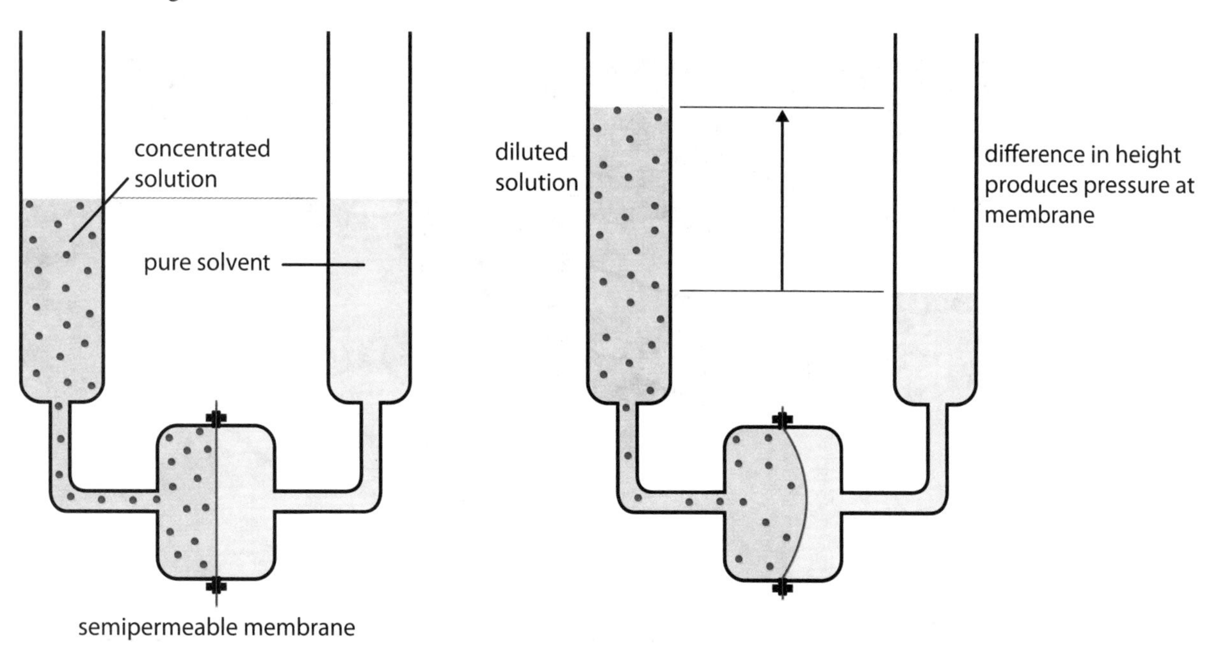

20-25 Osmotic pressure is produced by the flow of solvent into a concentrated solution separated by a semipermeable membrane.

20.15 Using a Salt Solution to Solve Problems

How do we obtain drinking water from seawater? Getting the salt out (*desalination*) is not an easy proposition. Because seawater is a solution, the salt does not settle out by the action of gravity, nor can it be filtered out by normal fiber filters. One very inefficient way involves boiling seawater, trapping the steam, and condensing it. This simple distillation method requires immense amounts of energy (remember water's heat of vaporization?) and some means of cooling the steam to condense it. In addition, the boiling water can trap saltwater droplets in the steam, so the condensed water is not always pure.

Recall from the previous subsection that semipermeable membranes *can* trap hydrated salt ions. If there were a way to force the water solvent in a salt solution through a semipermeable membrane *against* the normal osmotic pressure, then water pure enough to drink could be obtained on the other side of the membrane. This is exactly what happens in a process called *reverse-osmosis* desalination.

In reverse-osmosis desalination plants, seawater is first filtered to remove solid suspended matter. Then it is pressurized to greater than 40 atm (600 lb/in.^{2}). The water molecules pass through the tiny pores of the membrane, leaving very salty brine behind. Potable (drinkable) water is collected on the low-pressure side of the membrane. Some processing may be required to purify the discharged water, since bacteria and contaminants can be forced through the membrane. Reverse osmosis requires far less energy than simple boiling and condensation; however, it is still fairly expensive because it requires electricity and an industrial plant to produce the quantities of water necessary to support a community. Still, there are more reverse-osmosis plants in the world than any other kind of desalination facility.

20-26 Desalination of seawater can provide much-needed drinking water in arid locations near the ocean.

20C Section Review

1. How is the solubility of a solute related to its concentration in general?
2. When mixing a solution to attain a certain concentration measured in percentage by mass, how do you determine the amount of solvent needed?
3. What is a mole of a substance? When is the mole unit necessary in science?
4. What unit of concentration is related to the number of moles? How is this concentration measured?
5. What is a colligative property? What is it proportional to? What does it *not* depend on?
6. Does a colligative property affect the solvent's properties or the solute's?
7. How does the presence of a solute affect the boiling point of a solvent? the freezing point?
8. Compare and contrast diffusion and osmosis.
9. How do cells control the concentration of solutes inside their cellular membranes?
10. (True or False) Adding a solute to a liquid will both increase the temperature at which it boils and lower the temperature at which it freezes.

DS ✪ 11. Why is pressure required to produce reverse osmosis?

✪ 12. What is the percentage by mass of a solution containing 18.0 g of sodium hydroxide and 82.0 g of water?

FACETS OF SCIENCE

Desalting the Sea

No doubt you've heard stories of sailors in lifeboats drifting aimlessly on the high seas waiting for rescue. They may have survived the sinking of the ship, but if they don't get drinking water, they are as good as dead men. The cruel irony is that the billions and billions of gallons of seawater that surround them are unfit for human consumption. Drinking seawater doesn't quench thirst—it intensifies it. Perhaps the greater irony is that nearly three-fourths of the earth's surface is covered with water, yet there are vast areas of desert where finding even small amounts of water is an exercise in futility. Now more than ever, scientists are seeking ways to make the oceans a profitable source of fresh and clean water.

Desalination is not an easy task. Remember that salt water is a solution. The salt does not separate out of still water, nor can it be filtered out by ordinary means. One of the earliest methods (and still one of the most effective) is *flash distillation*. Water is pumped from the ocean into a series of coils where it is heated to 100 °C or higher, which is above the normal boiling point of water. (The water may be pressurized to raise the boiling point.) The pressure is then released, and the water immediately flashes to steam, leaving behind salt and other impurities in the remaining brine. The water vapor is condensed on pipes carrying colder seawater into the unit. It is then collected and piped to neighborhoods, businesses, and farms or is stored in large tanks. Flash distillation accounts for more than 80% of the industrial distilled water produced around the world.

Another widely used method is called *reverse osmosis (RO)*, which uses a *semipermeable membrane* to separate the filtered seawater from the pure-water product. Under normal conditions the pure water would flow through the membrane into the salt water, diluting it in a process called *osmosis*. But when the process is reversed, pure water actually flows out of the seawater solution, leaving the salt behind. Many atmospheres of pressure are applied to the saltwater side. The water solvent flow is reversed so that the salt is left behind and fresh water flows through the membrane, hence the term *reverse osmosis*.

A modern RO filter is made up of alternating layers of fibrous materials rolled up into a spiral cylinder. The high-pressure seawater layer contains a fine wire mesh that provides structural reinforcement for the cylinder. The other layer is sealed inside semipermeable membranes and receives the fresh water that passes through the membranes. The pure-water product spirals inward through the fibrous material to the perforated tube at the center, which delivers the water for further processing. The concentrated brine flows out the end of the filter and is discharged back to the sea.

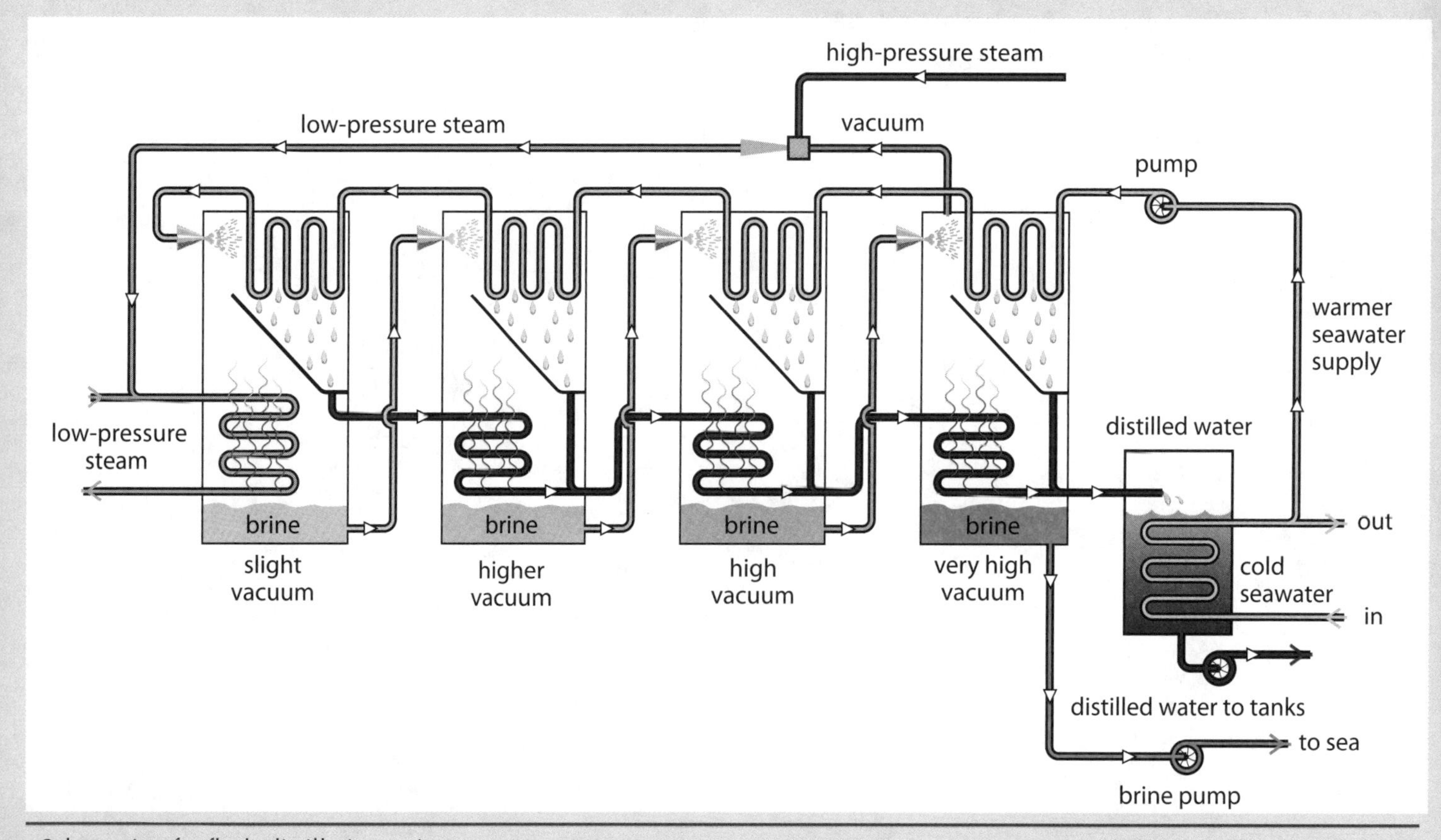

Schematic of a flash-distillation unit

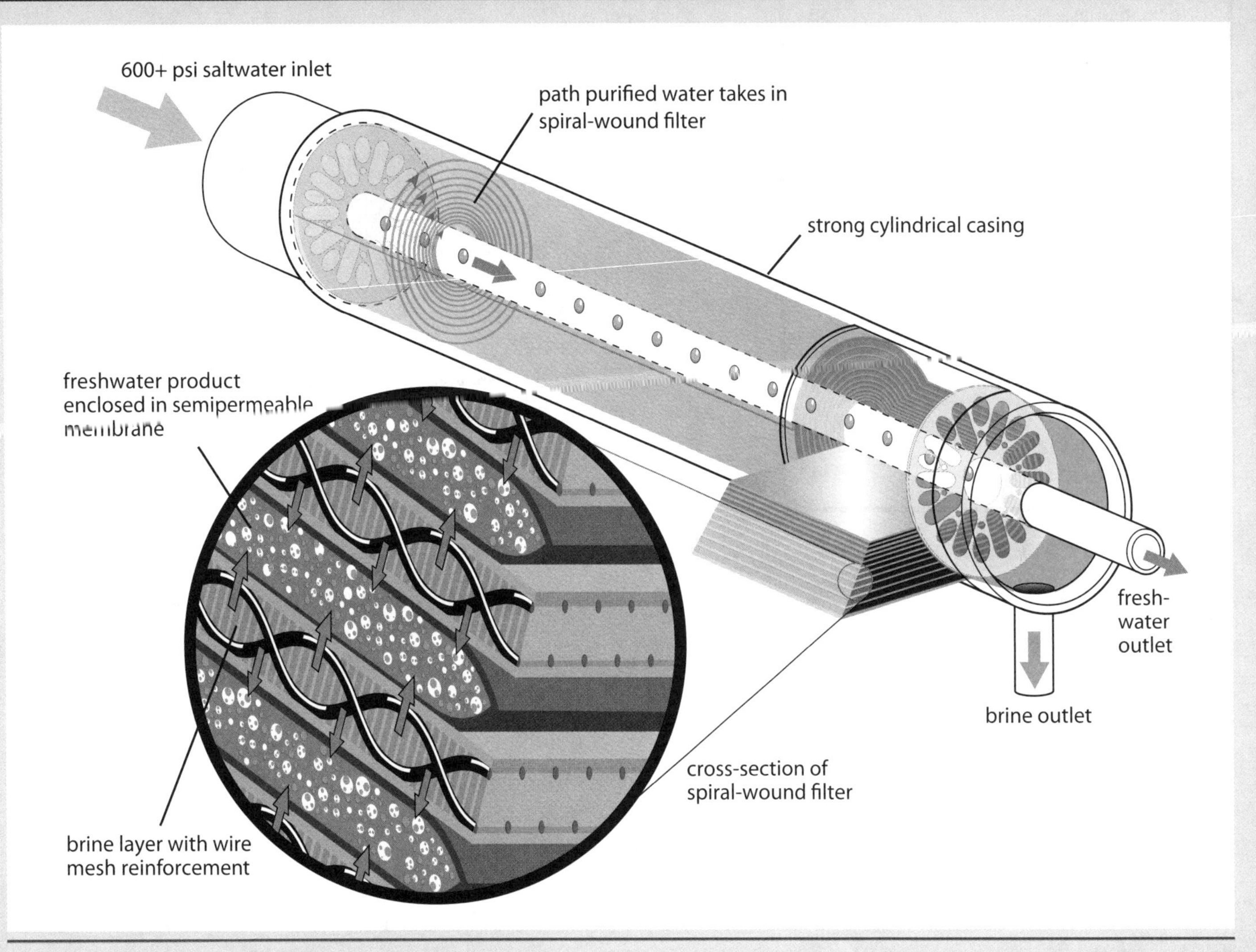

A modern spiral-wound RO cell

As the world's population grows, the availability of fresh water will become more crucial. Even though the demand is great, the vast majority of the world still gets its potable water from freshwater sources—lakes, reservoirs, rivers, and ground water. Only recently have desalination efforts become profitable, widespread enterprises. One hundred twenty countries now have some type of desalination program for salt or brackish water. Most desalination plants are located in the dry regions of the Middle East and North Africa, but they can also be found on many islands, the Florida Keys, and along the California coast.

Desalination does have its drawbacks. Of every gallon of seawater that is pumped into a desalination plant, less than half becomes potable water. The brine left over from the process is often dumped back into the ocean, a practice that concerns some environmentalists. It takes electricity, manpower, and specialized equipment to run these plants, and they must be located close to the ocean—prime real estate that homeowners are reluctant to give up. For customers, desalinated water is more expensive than ground water, but the prices continue to go down as the technology for reverse osmosis becomes more advanced. Besides, if fresh water had to be shipped in from faraway sources, the cost would be even greater.

Since humans began repopulating the earth after the Flood, man has settled in areas where fresh water was plentiful. Look on a map sometime and see how many cities are located along lakes and rivers. Even a small oasis in the desert is a welcome area because of the precious water that flows there. But as the populations of coastal and desert areas continue to grow, it has become clear that in some places on the earth, converting seawater to fresh water is no longer a novelty but a necessity.

Scientifically Speaking

Chapter Review

1. How is a mixture determined to be heterogeneous or homogeneous?
2. What are the phases of matter that form gels? sols? How are gels and sols different?
3. What property of the liquids in an emulsion would seem to work against their forming any kind of mixture?
4. How can you tell that a rock is a heterogeneous mixture?
5. What property do colloidal dispersions and solutions share that other kinds of mixtures do not?
6. What is one property of skim milk indicating that it is a colloidal dispersion?
7. How might you visually tell the difference between a suspension and a colloidal dispersion?
8. What can happen when metals with very different-sized atoms are mixed in heterogeneous alloys?
9. Which part of a solution is the solvent? How is this information used to classify the solution?
10. When you mix a solid and a liquid, what are two indications that a solution was formed and not a new compound?
11. What are solutions of different metals called? Why are they often preferred to pure metals?
12. Why is it not possible to have a solid-in-gas solution?
13. Why is the water molecule electrically polar?
14. Why is solvation of a solute particle an important step in dissolving the solute? What do we call the solution process for ionic solutes?
15. What kind of solvent could dissolve a nonpolar solute like sulfur? Explain your answer.
16. Explain the difference between the terms *soluble* and *solubility.*
17. Explain why the solubility of gases normally decreases with rising temperature.
18. How can you tell when a solid-in-liquid solution is saturated?
19. Where does the solution process take place on a piece of a solid solute? How can you increase the rate of solution in light of this fact?
20. What instrument could be used to measure a mole of solid solute particles? Why can this instrument be used?
21. Which solution would have a lower boiling point, one with 1.0 mol of solute particles or one with 2.0 mol? Explain.
22. DS ✪ Some conservationists are disturbed by the large amount of brine that is pumped back into the ocean by desalination plants. Why could it be a problem? What could be done to reduce the effects of returning concentrated brine to the sea?

True or False

23. Seawater is a mixture.
24. A pure quartz crystal made of silicon dioxide (SiO_2) is a mixture.
25. A mixture is a chemical combination of two or more substances.
26. A rigid but lightweight solid can be formed from a gel.
27. Aerosols may be either colloidal dispersions or suspensions.
28. Some alloys are heterogeneous mixtures of different metals.
29. Filtered seawater will display the Tyndall effect.
30. To be a solution, a mixture must be uniformly mixed and all particle sizes must be less than 1 nm.
31. Water can dissolve many substances because of the arrangement of atoms and kind of bonds in its molecules.
32. A solid solution can have a liquid solute.
33. The solvation of an ionic substance is called *ionization.*
34. A pond near sea level will have more oxygen dissolved in it than one near a high mountaintop at the same temperature.
35. A solution is saturated only under specific conditions.
36. A solution's concentration is a relative measure of how strong or weak the solution is.
37. A 0.5 *M* solution will have a higher freezing point than a 0.75 *M* solution of the same substances.

✪ 38. You must prepare a 15% by mass solution of sodium hydroxide in water. If you intend to have 100.0 g of solution, how much sodium hydroxide and water do you need? How much do you need for 75.0 g of solution?

DS ✪ 39. How could the construction of a desalination plant near an arid or swampy coastline transform the usefulness of the region to people who live in the area?

ACIDS, BASES, AND SALTS

CHAPTER 21

DOMINION SCIENCE PROBLEM

Bad Gut Feelings

At Thanksgiving and Christmas we all like to eat—and eat and eat! The stomach contains a strong acid, hydrochloric acid, which helps digest our food. Overeating, eating spicy or greasy foods, or catching a virus can cause unpleasant feelings called *indigestion*. Heartburn, stomach cramps, and nausea can come from the production of excess hydrochloric acid, which also can sometimes produce stomach ulcers. While these problems are usually due to occasional eating indiscretions, some people's stomachs naturally make extra stomach acid all the time, resulting in chronic discomfort. How can acid indigestion be treated?

21-1 Acid indigestion—an unpleasant consequence of overeating

21A Acids and Bases

21.1 *Describing Acids and Bases*

Have you ever wondered why foods taste the way they do? For example, grapes contain *tartaric acid*, citrus fruits contain *citric acid* and *ascorbic acid* (vitamin C), carbonated beverages contain *carbonic acid* and sometimes *phosphoric acid*, apples contain *malic acid*, and many vegetables also contain ascorbic acid. Bitter foods, like unsweetened chocolate, raw almonds, and certain kinds of herbs, often contain *ammonium hydroxide*, *potassium hydroxide*, *sodium hydroxide*, or other bases. The tastes of these foods are directly related to the compounds that they contain.

Even in ancient times, people recognized acids and bases by how they tasted or acted. In fact, the word *acid* comes from the Latin word *acidus*, meaning "sour." Acids were known to react with metals to produce hydrogen gas and were used as food preservatives. Acids turn litmus paper—a paper containing a substance from a lichen—red. Bases were known to taste bitter and feel slippery. They were used to prepare soaps; they also turn litmus paper blue. But how can acids and bases be defined in a chemical sense?

21A Objectives

After completing this section, you will be able to

- ✓ define acids and bases using both the Arrhenius and Brønsted-Lowry models.
- ✓ describe the process of water ionization that occurs in pure water and all water solutions.
- ✓ describe the structure of molecular acids.
- ✓ give examples of conjugate acid-base pairs.
- ✓ explain the relationship between conjugate acids and bases.
- ✓ list the common properties of most water-soluble acids and bases.

21-2 You can tell foods that contain acids and bases by their taste.

21.2 *Ionization of Water*

Chapter 20 discusses some interesting properties of water, such as its ability to dissolve many substances. Unlike most substances, it also has a tendency to become less dense and float when it freezes. Nearly all of water's special properties are due to its strongly polar molecule. Most acid and base compounds that you encounter from day to day are dissolved in water solutions because of the way they interact with water. Before we discuss the definitions of acids and bases, you need to understand another property of water—water ionization.

When substances are dissolved in water, they become hydrated, or surrounded by water molecules. If the substance is a soluble ionic compound, it dissociates into ions that become hydrated. If the substance is a soluble covalent compound, the action of the water molecules simply separates the molecules. For some covalent compounds the action of the water is so strong that it can break the covalent bonds to form ions, which are then hydrated. Recall that this process is called *ionization*, a term which applies only to molecular compounds.

Ionization occurs even in pure water. In this case, water itself is the "victim" of its own strong hydrating action. In a very small percentage of the molecules in pure water, one water molecule will attract a hydrogen ion from another water molecule. This action can occur because the hydrogen-oxygen bond is so polar that the bonding electrons spend most of their time around the oxygen atom. Thus, the hydrogen atoms can be pulled away with some ease when another water molecule bumps into it. This produces two new ions, called the **hydronium ion** (H_3O^+) and the **hydroxide ion** (OH^-). The following equation shows the reaction:

$$2H_2O \longrightarrow H_3O^+ + OH^-$$

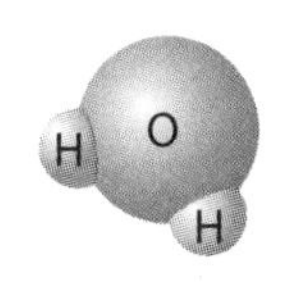

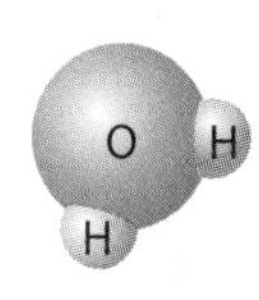

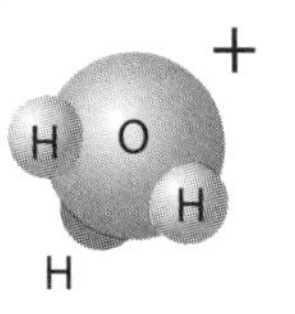

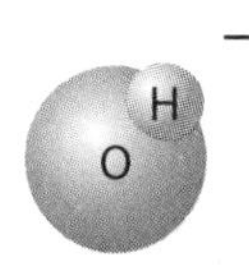

The number of hydronium and hydroxide ions is fairly constant in pure water, though individual molecules change constantly. The opposite reaction occurs just as easily.

$$H_3O^+ + OH^- \longrightarrow 2H_2O$$

The number of hydronium ions, 10^{-7} mol, works out to a mere 60 million billion ions per liter of water!

At room temperature only 10^{-7} mol of hydronium ions and hydroxide ions exist in a liter of pure water at any given moment.

This constant but continuously changing process is an example of a *dynamic equilibrium*. Dynamic equilibria are common in many physical and chemical processes, especially in solutions. A crowd of people Christmas shopping in a large department store is an illustration of a dynamic equilibrium. The number of shoppers may remain about the same over time, but the individuals who actually make up the crowd change constantly. The dynamic equilibrium of water ionization contributed to the first definition of acids and bases.

21-3 The number of people present at any given time in Grand Central Terminal may be about the same as at other times, but the individuals are constantly changing—a dynamic equilibrium.

21.3 The Arrhenius Definition

When speaking of ionizable protons in acids, remember that they are hydrogen ions (hydrogen nuclei), *not* protons taken from the nucleus of a larger atom.

In 1883 the brilliant Swedish chemist Svante Arrhenius published an important paper on the existence of ions in water solutions and their role in many chemical processes. He was the first to give simple and straightforward definitions for acids and bases.

Arrhenius defined an **acid** as any substance that produces hydrogen ions in a water solution. Hydrogen ions (H^+) are bare protons pulled from molecular compounds by the action of water molecules, just as in water ionization. Bare protons are highly unstable and instantaneously bond to water molecules to form hydronium ions. So an Arrhenius acid is any substance that forms hydronium ions in a water solution.

An Arrhenius acid is responsible for some painful insect bites. For example, the wood ant sprays formic acid when it bites, producing a stinging sensation. When formic acid ($HCHO_2$) dissolves in water, it ionizes into a *formate ion* (CHO_2^-) and a hydrogen ion

21-4 A wood ant *(Formica rufa)*

(H^+), which immediately attaches to a water molecule to form a hydronium ion.

$$HCHO_2(aq) \longrightarrow CHO_2^-(aq) + H^+(aq)$$

$$H^+(aq) + H_2O \longrightarrow H_3O^+$$

Use of the symbol (*aq*) indicates that the ions are dissolved in a water solution.

Hydrogen ions that can be pulled from molecules in an aqueous solution are called *ionizable hydrogen atoms.*

Similarly, an Arrhenius **base** produces hydroxide ions in a water solution. For example, when sodium hydroxide dissolves in water, it produces sodium ions and hydroxide ions.

$$NaOH\ (aq) \longrightarrow Na^+(aq) + OH^-(aq)$$

This model of acids and bases, called the **Arrhenius model**, produced a flurry of new chemical discoveries in the last decades of the 1800s and the beginning of the 1900s. While it helped advance science during this period, it did not explain every kind of reaction involving acids or bases. The Arrhenius model assumes that

FACETS OF SCIENCE

Svante Arrhenius—Father of Physical Chemistry

Svante August Arrhenius (1859–1927)

Many famous and influential people in history discovered their calling and made their most important contributions later in life. Others got an early start and were leaders for much of their lives. Svante August Arrhenius was definitely the latter kind.

Arrhenius was born in February 1859 to a Swedish land surveyor near Uppsala, Sweden. At age three he wanted someone to teach him to read, but everybody told him he was too young. "Wait a few years," he was told. So he taught himself to read. He also taught himself addition, subtraction, multiplication, and division by watching his father as he calculated the data in his survey reports. He started school in the fifth grade at age eight. When Arrhenius graduated from the equivalent of high school, he was at the head of his class and younger than any of the other students.

At the University of Uppsala, he was dissatisfied with the quality of instruction in physics and chemistry, so he transferred to the Physical Institute of the Swedish Academy of Sciences in Stockholm, where he studied the conductivities of electrolytes—solutions that conduct electricity.

Arrhenius found that salts, acids, and bases, when mixed with a solvent to form an electrolyte, were able to conduct electricity because the compounds dissociated into positive and negative ions, strange particles that carried charges. (This was before the discovery of the electron, a time when scientists could not readily distinguish between atoms and molecules and when atoms were believed to be indivisible.) His brilliant and thorough 150-page doctoral dissertation on this subject had to be approved by the professors at Uppsala. Whether they couldn't understand it or it was just too revolutionary for them, they awarded it the lowest passing grade possible. Nineteen years later (1903), the scientific community recognized its value and Arrhenius was awarded the Nobel Prize in Chemistry. In 1905 he was appointed rector of the new Nobel Institute for Physical Research in Stockholm.

As an extension of his ionic theory, Arrhenius developed new definitions of acids and bases. He proposed that acids are substances that produce hydrogen ions in water solutions and that bases are substances that produce hydroxide ions. His theory accounted for weak and strong acids and bases by their percentage of ionization or dissociation. He also could explain the rate of chemical activity, or chemical affinity, by temperature and the concentration of ions.

Arrhenius's brilliant work in electrochemistry has been acknowledged as one of the great contributions to modern physical chemistry, even though modifications have been necessary due to more recent discoveries. He also made contributions to many other fields, such as biology, geology, astronomy, cosmology, and astrophysics. He wrote many books and articles, both for the highly technically trained and for the nonscientific reader.

acids and bases exist only in water solutions and that hydronium and hydroxide ions are always formed in solutions of acids or bases. However, reactions of acids or bases can occur between gases, in solutions in which the solvent is not water, and in situations where hydronium or hydroxide ions are not produced.

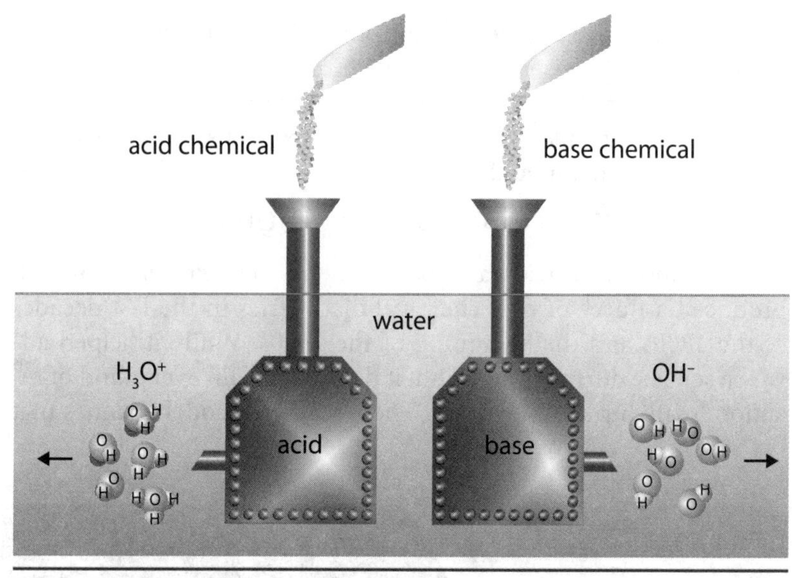

21-5 The Arrhenius model of acids and bases

21.4 The Brønsted-Lowry Definition

Two chemists in the 1920s, Johannes Brønsted and Thomas Lowry, independently discovered that acids were *proton donors.* Similarly, they showed that bases were *proton acceptors.* These statements define the **Brønsted-Lowry model** of acids and bases. There are four important principles of the Brønsted-Lowry definition:

1. An acid is any compound that is a proton (H^+) donor. This means that any substance with ionizable hydrogen atoms can be a Brønsted-Lowry acid, not just those that produce hydronium ions.
2. A base is any compound that accepts protons, not just those that produce hydroxide ions (which are also proton acceptors).
3. Acids and bases do not have to be dissolved in aqueous solutions.
4. Some compounds can behave as either acids or bases depending on the other reactants.

Hydrochloric acid is a common example of a Brønsted-Lowry acid (as well as an Arrhenius acid). Besides its role as the main gastric acid in your stomach, it is also the chemical called *muriatic acid,* a strong cleaning agent sold in building supply stores. The acid solution is commercially formed by dissolving the gas hydrogen chloride (HCl) in water.

Johannes Nicolaus Brønsted (1879–1947) was a Danish physical chemist who is best known for his work with acid and base reactions and developing the proton-donor definition of acids.

Thomas Martin Lowry (1874–1936) was an English physical chemist who studied the optical properties of solutions, which led to his defining acids and bases at the same time as Brønsted.

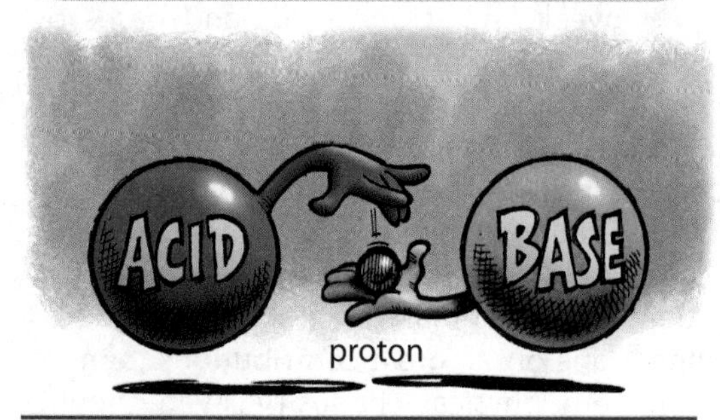

21-6 The Brønsted-Lowry model of acids and bases

In this textbook, you will study the reactions of acids and bases as defined by the Brønsted-Lowry model, even though most of the examples involve water solutions and conform to the Arrhenius model.

$$H_2O\ (l) + HCl\ (g) \longrightarrow H_3O^+(aq) + Cl^-(aq)$$

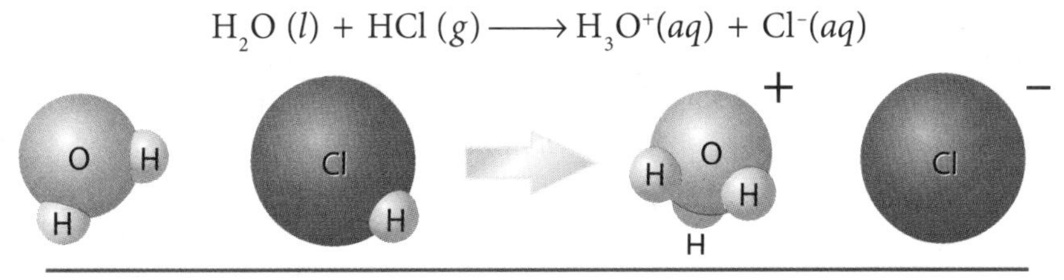

21-7 Ionization of hydrogen chloride in water to form hydrochloric acid

Hydrochloric acid has many industrial uses, such as cleaning metals and brick, refining sugar, and producing corn syrup.

An example of a Brønsted-Lowry base is ammonia (NH_3). When ammonia gas molecules dissolve in water, some of them remove a proton from water molecules, resulting in NH_4^+ and OH^- ions. That is why the ammonia solution is often referred to as ammonium hydroxide.

$$NH_3(g) + H_2O\ (l) \longrightarrow NH_4^+(aq) + OH^-(aq)$$

$$\begin{matrix} & \\ H:\ddot{\underset{..}{N}}:H \\ H \end{matrix} + \begin{matrix} H:\ddot{\underset{..}{O}}: \\ H \end{matrix} \rightarrow \left[\begin{matrix} H \\ H:\ddot{\underset{..}{N}}:H \\ H \end{matrix}\right]^+ + \left[H:\ddot{\underset{..}{O}}:\right]^-$$

The reaction between ammonia and water is possible because of the ammonia molecule's structure. A nonbonding pair of electrons on the nitrogen atom in ammonia attracts a proton and forms a weak bond with it. Thus, ammonia is a base according to the Brønsted-Lowry definition.

Although water was needed for the previous examples of acids and bases, water is *not* necessary for compounds to act as Brønsted-Lowry acids and bases. Let's examine the reaction of hydrogen chloride gas (HCl) and ammonia gas (NH_3). When these two gases mingle in the air, a hydrogen chloride molecule donates its hydrogen atom as an ion to a strongly polar ammonia molecule. After accepting the proton, the positively charged ammonium ion (NH_4^+) immediately bonds with the leftover negative chloride ion, forming the ionic compound ammonium chloride (NH_4Cl), a white powdery solid. The chemical equation for this reaction is

$$HCl\ (g) + NH_3(g) \longrightarrow NH_4^+ + Cl^- \longrightarrow NH_4Cl\ (s)$$

or simply

$$HCl\ (g) + NH_3(g) \longrightarrow NH_4Cl\ (s).$$

Note that this reaction takes place in air, not in water. The Arrhenius model of acids and bases does not account for this reaction.

21.5 Polyprotic Acids

Two factors determine a compound's ability to act as an acid: the number of ionizable hydrogen atoms in its molecule (or its ion) and how tightly they are bonded. If an acid molecule has only one ionizable hydrogen atom (a single ionizable proton), it is a *monoprotic acid.* Hydrogen chloride is a monoprotic acid. If a molecule contains two or more ionizable hydrogen atoms, the acid is a

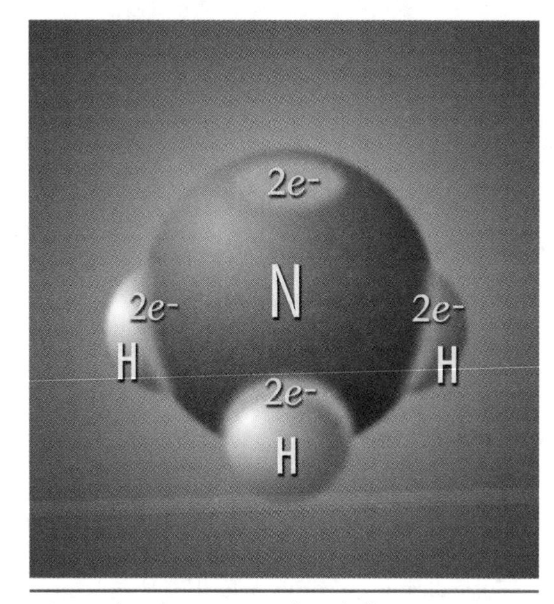

21-8 The ammonia molecule has a nonbonding pair of electrons that produces a negative region on the surface of the molecule, making it very polar.

21-9 An acid-base reaction without water, producing solid ammonium chloride (the white cloud above the bottles)

Dihydrogen phosphate ($H_2PO_4^-$) is both a Brønsted-Lowry acid and base. It can accept a hydrogen ion to form phosphoric acid (H_3PO_4) or lose a hydrogen ion to form hydrogen phosphate (HPO_4^{2-}).

polyprotic acid. For example, sulfuric acid (H_2SO_4) has two ionizable hydrogen atoms. It is a *diprotic* acid. Phosphoric acid (H_3PO_4) has three, making it a *triprotic* acid. Examine the reactions below that show how phosphoric acid loses its ionizable hydrogen atoms.

$$H_3PO_4(aq) + H_2O\ (l) \longrightarrow H_2PO_4^-(aq) + H_3O^+(aq)$$

$$H_2PO_4^-(aq) + H_2O\ (l) \longrightarrow HPO_4^{2-}(aq) + H_3O^+(aq)$$

$$HPO_4^{2-}(aq) + H_2O\ (l) \longrightarrow PO_4^{3-}(aq) + H_3O^+(aq)$$

It becomes more difficult to remove a hydrogen ion with each ionization, because a negative ion would rather gain a positive hydrogen ion than lose another one.

It is tempting to assume that you can tell the number of ionizable hydrogen atoms from the number of hydrogens in the molecular formula, but this is generally not the case. For instance, in Table 21-1 you see the formula for ascorbic acid, $H_2C_6H_6O_6$. Although its formula contains eight hydrogen atoms, no more than two hydrogen ions can be donated. This fact is indicated by the first two hydrogen atoms in the formula.

In this textbook, ionizable hydrogen atoms are written at the beginning of an acid's formula. However, some textbooks present complex formulas in alphabetical order. For example, ascorbic acid ($H_2C_6H_6O_6$) can be written $C_6H_8O_6$. Chemists often must know something of a compound's molecular structure to determine how many ionizable hydrogen atoms it has.

21-1 Common Acids and Bases

Source	Name	Formula
vinegar	acetic acid	$HC_2H_3O_2$
vegetables, vitamin C	ascorbic acid	$H_2C_6H_6O_6$
soft drinks	carbonic acid	H_2CO_3
citrus fruit	citric acid	$H_3C_6H_5O_7$
insect acid	formic acid	$HCHO_2$
sour milk, sore muscles	lactic acid	$H_2C_3H_4O_3$
rhubarb	oxalic acid	$H_2C_2O_4$
household ammonia	ammonium hydroxide	NH_4OH
milk of magnesia	magnesium hydroxide	$Mg(OH)_2$
lye	sodium hydroxide	NaOH

21.6 Conjugate Acids and Bases

If you examine acid-base reactions, you may notice that when a molecule is ionized to form a hydrogen ion and an anion, the anion is capable of accepting a hydrogen ion again. In each of these cases, the anion is a Brønsted-Lowry base. Chemists have given these ions a special name—conjugate bases. A **conjugate base** is the ion that remains after a molecular compound has donated a hydrogen ion. Every acid compound therefore becomes its own conjugate base after donating a proton.

conjugate (KON juh GIT): con- (L. *con*—together) + -jug- (L. *jugare*—to join) + -ate; closely related to one another

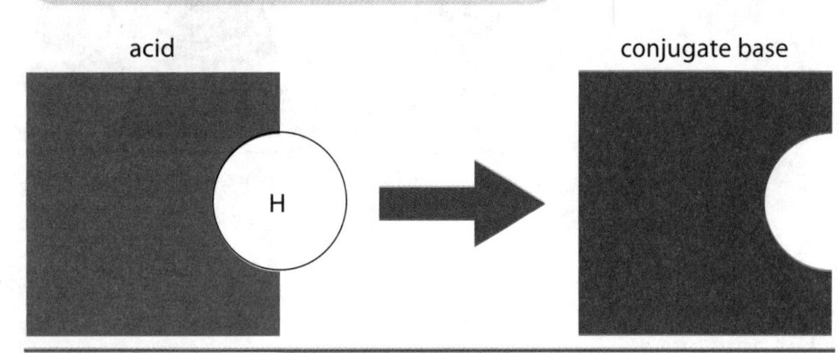

21-10 An acid and its conjugate base

Let's begin with water ionization. When water self-ionizes, it produces a hydronium ion and a hydroxide ion. The water molecule that donates the hydrogen ion is the acid molecule; the remaining hydroxide anion is its conjugate base. Similarly, the water molecule that accepts the hydrogen ion is the

FACETS OF SCIENCE: The King of Chemicals

Sulfuric acid is used in such large quantities that it is shipped in railroad tank cars.

Chemical companies in the United States produce more sulfuric acid than any other single chemical. Production of sulfuric acid in America began in Philadelphia 31 years after the signing of the Declaration of Independence. By the time of the Civil War, sulfuric acid had become one of the most important chemical products of our nation.

Today this acid is used in almost every industrial process. In fact, economists consider sulfuric acid so essential to industry that they sometimes measure the economic condition of a country by how much sulfuric acid it uses. Generally, when a nation's usage of sulfuric acid drops, its whole economy is headed for a downturn.

You can easily understand why sulfuric acid is often referred to as the king of chemicals. Its properties make it extremely useful. A dense, oily substance with a high boiling point, concentrated sulfuric acid is highly corrosive and can eat through many metals in a matter of minutes! Most importantly, sulfuric acid can react with numerous other chemicals to produce thousands of useful products. More than 60% of the sulfuric acid used worldwide goes into the manufacture of phosphate fertilizers and phosphate detergents. The second largest consumer of sulfuric acid is the chemical industry. Chemists use sulfuric acid to produce paints, dyes, plastics, fibers, and a vast array of other chemical products. Another major use is the decomposition of mineral ores to assist in the extraction of industrial metals.

base, and the resulting hydronium ion is the **conjugate acid** of water because it can now donate a proton.

$$\underset{\text{acid}}{H_2O\,(l)} + \underset{\text{base}}{H_2O\,(l)} \longrightarrow \underset{\text{conjugate base}}{OH^-(aq)} + \underset{\text{conjugate acid}}{H_3O^+(aq)}$$

In a monoprotic inorganic acid such as hydrochloric acid, the acid molecule is HCl. Its conjugate base is the chloride ion (Cl^-) because the chloride ion can now receive a hydrogen ion. Polyprotic acids can produce a variety of acids and bases when they ionize. Sulfuric acid (H_2SO_4) ionizes to produce a hydrogen ion and its conjugate base, the hydrogen sulfate anion (HSO_4^-). But the hydrogen sulfate ion can donate a hydrogen ion, so it is itself an acid particle. It ionizes to form a hydrogen ion and its conjugate base, the sulfate anion (SO_4^{2-}). Table 21-3 (see page 496) lists an assortment of acids and their conjugate bases.

21.7 Properties of Acids and Bases

Acids

The properties of acids and bases in this subsection apply to mainly *water-soluble* acid compounds.

Acids have many characteristic properties, some of which have long been used to identify them. One example previously mentioned is the sourness of foods containing acids. The sour or tart taste of foods is usually due to the presence of hydronium ions. *You should never taste any chemical in a laboratory to see if it is an acid!*

One of the first chemical properties of acids to be discovered was the ability of acid solutions to conduct electricity. Before Arrhenius produced his first major research paper, he discovered that solutions of certain substances could carry an electrical current. He called these substances *electrolytes*. He concluded that electrolytes conduct electricity because of the ions that they produce in a solution. All acids and bases are electrolytes, as are dissolved salts, the other class of compounds discussed in this chapter. He classified these three groups of compounds according to the type of ions each produces in a solution. The ability of most acids and bases to form ions in an aqueous solution is very important. It means that electrical charges can be stored in and transported through a solution. A car can start only because aqueous acids are electrolytes. The sulfuric acid solution in a car battery produces and conducts the electrical current needed to start the car's engine.

One of the best-known properties of an acid is its ability to corrode metals. Again, the ionizable hydrogen atoms of acids play the leading role in the reaction of acids with metals. Acidic corrosion is a single-replacement reaction. When a metal is placed into an acid solution, the metal atoms become ions in the solution as their electrons are taken from them by the hydrogen ions. The hydrogen atoms form molecules and leave the solution as a gas. For example, zinc metal and hydrochloric acid react to form hydrogen gas. The zinc atom loses two electrons to become Zn^{2+} and achieve a full octet. Having a stronger attraction for electrons than zinc, the protons held by hydronium ions each pull one electron from the zinc to become atoms. Two of these hydrogen atoms instantly join to form a gaseous hydrogen molecule (H_2). As the zinc atoms become ions, they leave the metal lattice and enter the solution. The metal appears to be eaten away.

$$2HCl\,(aq) + Zn\,(s) \longrightarrow ZnCl_2(aq) + H_2(g)$$

21-11 Hydrochloric acid readily reacts with zinc metal, liberating hydrogen gas.

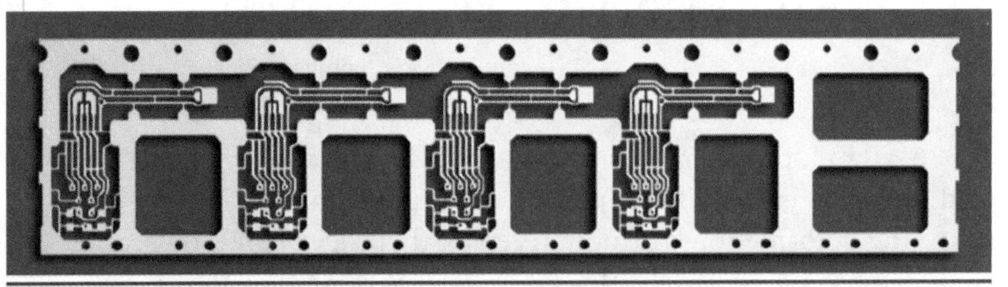

21-12 Acids are often used to dissolve unwanted metal from delicate circuit patterns and model parts in a process called *photoetching*.

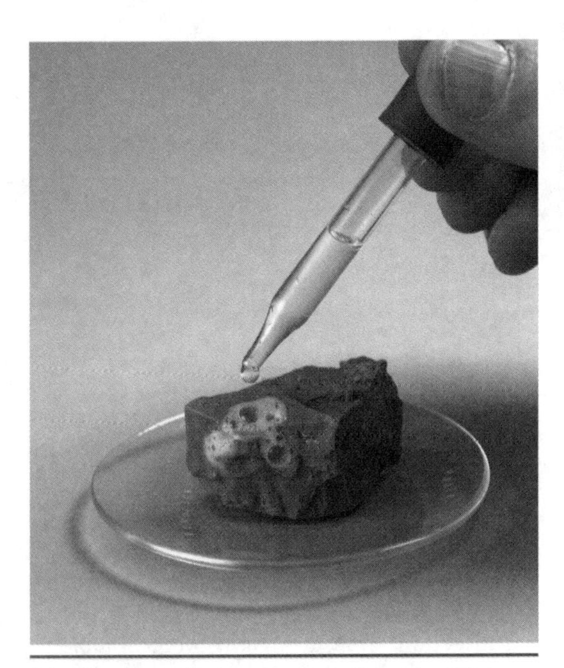

21-13 The effervescent reaction of acids is used to confirm the presence of carbonates, such as limestone and marble, in rock samples. If calcium carbonate is present, the rock will bubble when hydrochloric acid is applied.

$CaCO_3 + 2HCl \rightarrow CaCl_2 + H_2O + CO_2$

Another useful chemical property of acids is that they react with carbonates and bicarbonates to form carbon dioxide gas. An example is the reaction between acetic acid ($HC_2H_3O_2$) in vinegar and sodium bicarbonate (baking soda). When combined, they produce violent bubbling, called *effervescence*.

$$HC_2H_3O_2(aq) + NaHCO_3 \longrightarrow NaC_2H_3O_2(aq) + H_2O\,(l) + CO_2(g)$$

Bases

If you've ever tasted soap, you know that bases have a bitter taste. Dark unsweetened chocolate is required to contain a certain amount of a base to be marketed as such. Bases also have a slippery feel because they react with the oils of your skin. The same caution regarding acids applies here too: *never touch or taste a chemical to see if it is a base unless you are instructed to do so by your teacher or unless the chemical's label states that it is safe to do so!*

Bases have the ability to emulsify organic materials, such as fats and proteins. For this reason, strong bases are used as oven cleaners and for unclogging sink drains. Many strong bases can cause severe burns and if taken internally can cause serious injury or death. You should always wear protective equipment, such as rubber gloves and goggles, when using strong solutions containing bases.

Acids and bases can cancel out each other's chemical action in a reaction called *neutralization*. When equal numbers of hydronium and hydroxide ions are combined, the resulting solution is neither acidic nor basic—it is *neutral*. Acids can neutralize bases, and bases can neutralize acids. For water solutions, the hydronium ion (H_3O^+) and the hydroxide ion (OH^-) combine to form water molecules as described in Subsection 21.2. This reaction is the heart of the neutralization process, which is covered more fully in Section 21B.

21A Section Review

1. Describe how water molecules can hydrate various substances.
2. Describe the two ionization reactions that occur in pure water. What do we call the condition that keeps the concentrations of ions at approximately constant levels?
3. What are hydrogen ions, and where do they come from? How do they exist in a water solution?
4. What are the major differences between the Arrhenius and Brønsted-Lowry models of acids and bases? Could one be considered a special case of the other? Explain.
5. Does a reaction involving an acid and a base have to take place in a water environment? Justify your answer.
6. ✪ Which of the following could be a base in a water solution?
   a. Br^-
   b. NH_4^+
   c. H_2O
   d. NO_3^-
   e. Ca^{2+}
   f. $C_2H_3O_2^-$
7. DS ✪ A popular commercial antacid contains both $Al(OH)_3$ and $Mg(OH)_2$. Are these compounds Arrhenius bases, Brønsted-Lowry bases, or both? Explain your answer.
8. DS ✪ Classify stomach acid (HCl) on the basis of the number of ionizable hydrogens it has.
9. State the general properties of water-soluble acids discussed in this section.
10. Explain why it is never a good idea to use the sense of taste in a laboratory to identify an acid or a base.
11. (True or False) The water molecule can be either an acid or a base.

21B Objectives

After completing this section, you will be able to

- ✓ explain the fundamental chemical reaction in a neutralization reaction.
- ✓ identify the cation of a base and the anion of an acid.
- ✓ predict the salt compound formed by the neutralization of an acid and a base.
- ✓ discuss how antacids function to ease acid indigestion symptoms.

Salts are compounds that consist of any two oppositely charged ions (excepting the hydrogen and hydroxide ions).

salt = base cation + acid anion

Some neutralization reactions produce other substances in addition to a salt, such as hydrogen or carbon dioxide gas.

The aqueous H^+ ion represents a hydronium ion.

$$H^+ (aq) = H_3O^+$$

21B Salts

21.8 Neutralization Reactions

Table salt (sodium chloride) is only one of a large class of compounds known as **salts**. Salts are ionic compounds produced from an acid-base reaction, or **neutralization reaction**. Neutralization reactions are double-displacement reactions that generally produce a salt and water. You are probably familiar with several salts in the form of seasonings that are used in food.

Salts are composed of a cation from a base and an anion from an acid. For example, sodium chloride contains a sodium cation (Na^+) and a chloride anion (Cl^-). The sodium cations come from a sodium base, such as sodium hydroxide. The chloride anions come from a chloride acid, such as hydrochloric acid. Look at the neutralization reaction that forms this salt:

$$HCl\ (aq) + NaOH\ (aq) \longrightarrow NaCl\ (aq) + H_2O\ (l)$$

Since these compounds completely ionize or dissociate in water, this equation could be written

$$H^+(aq) + Cl^-(aq) + Na^+(aq) + OH^-(aq) \longrightarrow$$
$$Na^+(aq) + Cl^-(aq) + H_2O\ (l).$$

The double-displacement reaction produces water, which mingles with the water in the solution, and soluble table salt, which can be removed by evaporation.

You learned in Section 21A that when a molecular acid compound dissolves in water, it ionizes to produce positive hydronium ions and hydrated negative ions. For example, in the case of nitric acid, the ions in the solution are hydronium ions and nitrate anions. Nitric acid is produced by the following reactions:

$$4NO_2(g) + O_2(g) + 2H_2O\ (l) \longrightarrow 4HNO_3(aq)$$

$$HNO_3(aq) \longrightarrow H^+(aq) + NO_3^-(aq)$$

Similarly, when a base compound that is a metal hydroxide dissolves in water, it dissociates into positive metal ions and negative hydroxide ions. For the strong base potassium hydroxide (KOH), the ions formed are potassium cations and hydroxide anions.

$$KOH\ (s) \longrightarrow K^+(aq) + OH^-(aq)$$

If nitric acid and potassium hydroxide solutions are mixed together, all four ions meet in the same solution. The excess hydroxide ions immediately pull the extra protons from the hydronium ions to form water. They neutralize each other.

$$K^+(aq) + NO_3^-(aq) + OH^-(aq) + H_3O^+(aq) \longrightarrow K^+(aq) + NO_3^-(aq) + 2H_2O\ (l)$$

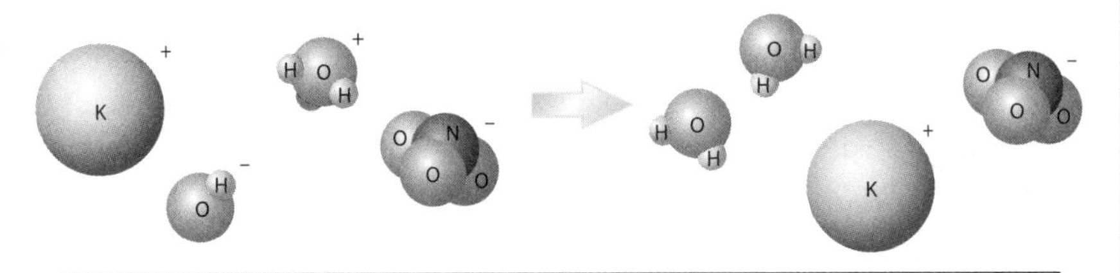

21-14 This reaction takes place between only the hydroxide and hydronium ions. The potassium and nitrate ions in Figure 21-14 are like spectators at a game—they do not participate. Thus, such ions are called *spectator ions*.

The cations and anions remain in the solution unless they can react to form a compound with a low solubility in water (see Chapter 20). In that case a precipitate forms and settles to the bottom.

To isolate the soluble salt produced in an aqueous neutralization reaction, the water must be evaporated from the solution. As the water evaporates, these ions *associate* (join together) to form an ionic compound—the salt. In the above example the potassium cations would combine with the nitrate anions to form solid potassium nitrate (KNO_3) when the water is driven off as a vapor.

$$K^+(aq) + NO_3^-(aq) \longrightarrow KNO_3(s) + H_2O\ (g)$$

21.9 Variety of Salts

Since there are many different acids and bases, thousands of different salts can be produced from them. Theoretically, any acid could react with any base to produce a unique salt. Table 21-2 lists three acids and three bases. Notice the nine different salts that can be produced from their neutralization reactions with one another. The composition of salts and their properties depend on the acids and bases used to produce them.

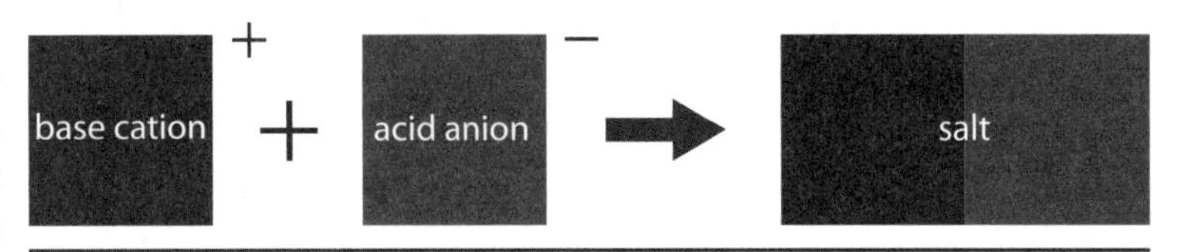

21-15 Acid-base neutralization

21-2 Neutralization Reaction Salts

		Bases		
		NaOH (sodium hydroxide)	**KOH (potassium hydroxide)**	**$Mg(OH)_2$ (magnesium hydroxide)**
Acids	**HCl (hydrochloric)**	NaCl (sodium chloride)	KCl (potassium chloride)	$MgCl_2$ (magnesium chloride)
	HNO_3 (nitric)	$NaNO_3$ (sodium nitrate)	KNO_3 (potassium nitrate)	$Mg(NO_3)_2$ (magnesium nitrate)
	H_2SO_4 (sulfuric)	Na_2SO_4 (sodium sulfate)	K_2SO_4 (potassium sulfate)	$MgSO_4$ (magnesium sulfate)

21.10 Using Neutralization Reactions to Solve Problems

How can the symptoms associated with too much gastric acid be treated? One way is to take an antacid, a base compound used to relieve the symptoms of indigestion. Common bases found in antacids are calcium carbonate, sodium bicarbonate, magnesium hydroxide, and aluminum hydroxide. These compounds work to neutralize stomach acid. For example, one brand containing aluminum hydroxide ($Al(OH)_3$) neutralizes hydrochloric acid in the following reaction:

$$Al(OH)_3(s) + 3HCl\ (aq) \longrightarrow AlCl_3(aq) + 3H_2O\ (l)$$

Several shorthand notations are used in the antacid equation:

$HCl\ (aq) = H_3O^+ + Cl^-\ (aq)$

$AlCl_3\ (aq) = Al^{3+}\ (aq) + 3Cl^-\ (aq)$

You can see that the product of this reaction is aluminum chloride salt and water.

Many people find great relief through these kinds of products. But, as with any medicine, antacids must be taken in limited doses. The equation above shows that aluminum ions are released as a by-product of the neutralization reaction. High concentrations of some metal ions are not healthy for the human body.

Treatment of chronic (that is, continuous or repeated) acid indigestion is crucial since excess stomach acid can damage the esophagus in cases where the muscular valve that shuts off the esophagus at the top of the stomach leaks. This damage can lead to a very deadly kind of cancer in the esophagus. Doctors generally treat chronic acid indigestion with a class of drugs known as *proton pump inhibitors*.

Normally, an enzyme known as a proton pump transports H^+ ions formed from water molecules within the cells that line the stomach into tiny ducts in the stomach wall. There they combine with Cl^- ions to form hydrochloric acid (HCl). Proton pump inhibitors block the enzyme so that it can't carry the H^+ ions into the ducts. Controlling the H^+ ions here stops or greatly reduces the production of hydrochloric acid, preventing chronic acid indigestion.

21-16 Many commercial antacids contain a base that neutralizes stomach acid.

21B Section Review

1. What class of compounds are salts? In view of the topics in this chapter, where can the ions of a salt come from?
2. In water solutions, what occurs in every neutralization reaction?
3. How do you know when an insoluble salt is formed by an acid-base neutralization?
4. How can you recover a soluble salt produced by an acid-base neutralization reaction?
5. DS Some antacids contain magnesium hydroxide ($Mg(OH)_2$). Write a balanced chemical equation showing how it neutralizes hydrochloric acid (HCl), using the pattern in Subsection 21.10.
6. DS Some antacid tablets contain sodium bicarbonate ($NaHCO_3$), or baking soda. If you combine an acid with baking soda, it fizzes. Examine the reaction below:

   $$NaHCO_3(s) + HCl\ (aq) \longrightarrow NaCl\ (aq) + CO_2(g) + H_2O\ (l)$$

   What is the "fizz" produced in this reaction?
7. (True or False) Salts produced by a neutralization reaction are a combination of the base's anion and the acid's cation.

21C Acidity and Alkalinity

21.11 Acid Strength

The sulfuric acid in a car battery can corrode painted surfaces and burn holes in clothing. Yet we touch and eat other acids every day. The aspirin that you take for a headache is an acid, the sharp flavor of soft drinks is caused by an acid, and several of the vitamins that you take or that are found in foods are actually acids. What makes some acids strong, corrosive, and dangerous, and other acids weak and harmless?

Acids are classified as strong or weak according to their ability to donate hydrogen ions. This ability is directly related to the acid molecule's *degree of ionization*. In water solutions, the more hydrogen ions donated, the more hydronium ions produced and the stronger the acid. Thus, *strong acids ionize to a large extent*, even 100%, and *weak acids ionize only slightly*.

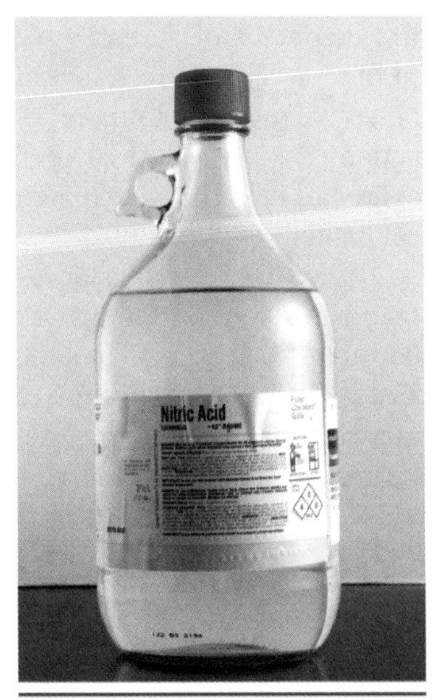

21-17 Nitric acid is a strong acid because 100% of its molecules ionize in a dilute water solution.

The number of ionizable hydrogen atoms in the molecular formula does not necessarily indicate whether or not an acid will be strong. If it did, you might expect that acids with two or more ionizable hydrogen atoms would be stronger acids than those with only one. This is not usually the case. For example, hydrochloric acid (HCl), a monoprotic acid, can be virtually 100% ionized and is therefore a very strong acid. Sulfuric acid (H_2SO_4), a diprotic acid, is a strong acid because, in a water solution, 100% of its formula units lack one hydrogen atom due to ionization. However, the remaining hydrogen sulfate ions (HSO_4^-) do not completely ionize.

An acid's strength is determined by the percentage of its *first available hydrogen atoms* that ionize. For example, less than 100% of the first available hydrogen atoms in phosphoric acid (H_3PO_4) ionize. It is, therefore, only a moderately strong acid, even though it is a triprotic acid. Citric acid ($H_3C_6H_5O_7$), also triprotic, is a very weak acid because it ionizes only slightly. Table 21-3 (page 496) lists a selection of acids by their relative strengths.

Another observation that you can make based on Table 21-3 is that acids with the same number of ionizable hydrogen atoms do not necessarily have the same strength. You can see that the monoprotic acids range from very strong ($HClO_4$) to very weak (HCN). The stronger an acid molecule "holds" its ionizable hydrogen atoms, the weaker its strength as an acid.

21C Objectives

After completing this section, you will be able to

- ✓ differentiate between an acid's strength and its concentration in a solution.
- ✓ discuss how the molecular structure of acids can influence their strength.
- ✓ relate a base's strength to its concentration in a solution.
- ✓ state the properties that make for strong bases.
- ✓ compare the strengths of acids and their conjugate bases and vice versa.
- ✓ describe the pH scale and its basis.
- ✓ discuss the action of pH indicators and give several examples.
- ✓ discuss the general principle behind the operation of a pH meter.

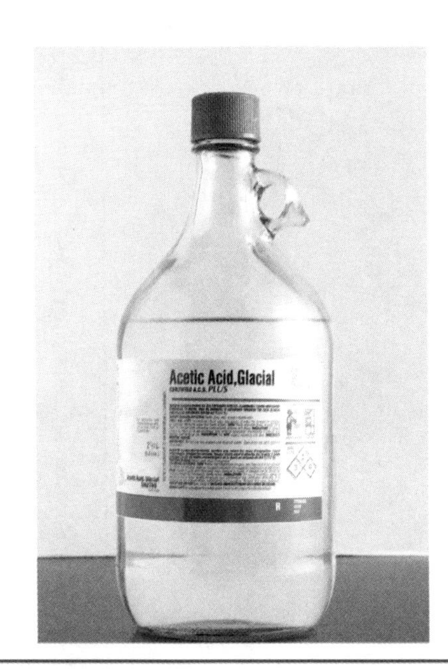

21-18 Acetic acid is a weak acid because only about 1% of its molecules ionize in a dilute water solution.

21.12 Base Strength

Bases have varying strengths just as acids do. Some bases are very damaging to clothing and tissue, and others are used in cosmetics or cleaning solutions and need no special safeguards. Strong bases have a strong attraction for hydrogen ions. Sometimes it is useful to think of them as the conjugate bases of weak acids. They also include ionic compounds that dissociate completely to produce many hydroxide ions or other anions that are strong proton acceptors. Weak bases are the conjugate bases of strong acids and thus have a small attraction for hydrogen ions. Weak bases also include ionic hydroxide compounds that do not dissociate easily and so produce only a few hydroxide ions (proton acceptors) in a solution (see Table 21-3). A strong base can be very dangerous to living tissues because of its proton-attracting action. Loosely held hydrogen atoms in proteins and other molecules can be removed, changing their function or destroying them completely.

What kinds of substances act as strong bases? Scientists have found that ions of the alkali metals and the alkaline-earth metals (Groups 1A and 2A) combine with hydroxide ions to form strong bases. These compounds can release large amounts of hydroxide ions in a solution. Compounds such as sodium hydroxide (NaOH), potassium hydroxide (KOH), and calcium hydroxide ($Ca(OH)_2$) readily

Though calcium hydroxide is not very soluble in water, the formula units that do dissolve dissociate completely.

21-3 Strengths of Conjugate Acids and Bases

	Acid		Base		
Strong acids	perchloric acid	$HClO_4$	perchlorate ion	ClO_4^-	Weak bases
	sulfuric acid	H_2SO_4	hydrogen sulfate ion	HSO_4^-	
	hydriodic acid	HI	iodide ion	I^-	
	hydrobromic acid	HBr	bromide ion	Br^-	
	hydrochloric acid	HCl	chloride ion	Cl^-	
	nitric acid	HNO_3	nitrate ion	NO_3^-	
Increasing acid strength →	hydronium ion	H_3O^+	water	H_2O	←Increasing base strength
	hydrogen sulfate ion	HSO_4^-	sulfate ion	SO_4^{2-}	
	phosphoric acid	H_3PO_4	dihydrogen phosphate ion	$H_2PO_4^-$	
	hydrofluoric acid	HF	fluoride ion	F^-	
	nitrous acid	HNO_2	nitrite ion	NO_2^-	
	acetic acid	$HC_2H_3O_2$	acetate ion	$C_2H_3O_2^-$	
	carbonic acid	H_2CO_3	bicarbonate ion	HCO_3^-	
	hydrosulfuric acid	H_2S	hydrogen sulfide ion	HS^-	
	ammonium ion	NH_4^+	ammonia	NH_3	
	hydrogen cyanide	HCN	cyanide ion	CN^-	
Weak acids	water	H_2O	hydroxide ion	OH^-	Strong bases
	ammonia	NH_3	amide ion	NH_2^-	
	hydrogen	H_2	hydride ion	H^-	

dissociate into metal ions and hydroxide ions. The following equation shows how the strong base NaOH dissociates into solvated ions when dissolved in water. To simplify the equation, the water involved in the hydration is represented by the symbol H_2O over the arrow.

$$NaOH\ (s) \xrightarrow{H_2O} Na^+(aq) + OH^-(aq)$$

The bond between hydroxide ions and these active metals can easily be broken by water molecules, making these compounds strong bases.

As stated previously, the strength or weakness of a base is measured by the number of proton acceptors it produces in an aqueous solution. Ammonia is a weak base because it weakly ionizes water molecules to produce relatively few hydroxide ions. Because it is weak, it is reasonably safe to use in cleaning solutions. Some metal hydroxides, such as aluminum hydroxide ($Al(OH)_3$), are very weak bases because they are so insoluble in water that very few hydroxide ions dissociate into the solution.

Chapter 20 discusses how concentration is a measure of the amount of solute that is dissolved in a solution. Sometimes people use the terms *strong* or *weak* to refer to a concentrated or dilute solution, respectively. However, the *strength* of an acid in solution—the percentage of molecules ionized in solution—must not be confused with the acid's *concentration*. On the other hand, the concentration of a base compound in its solution is directly related to its ability to be dissociated into ions and thus its strength as a base. It is possible to have a concentrated solution of a weak acid and a dilute solution of a strong acid because the solubility of an acid compound is not related to its strength as an acid. However, a concentrated solution of a base implies a strong base. A dilute base solution may result from a weak base with low solubility or a small amount of a strong base dissolved in the solution. Because of the possibility of confusion, it is best to refer to solution concentrations as dilute or concentrated rather than as weak or strong.

21-19 Lye (sodium hydroxide) is the main component of oven cleaners and products that unclog drains. Very nasty stuff!

> The strength of a metal hydroxide base is directly related to its solubility in water. Soluble metal hydroxides, like all soluble ionic compounds, must dissociate into their component ions to dissolve. Those that dissolve completely are strong bases. Those that have limited solubility are weaker.

21.13 The pH Scale

Since the strength of an acid compound and its solution concentration can vary widely, how is the *acidity* of a solution measured? Acidity is directly related to the concentration of the hydronium ions in the solution. Because the strongest concentration of hydronium ions is a hundred trillion (10^{14}) times as strong as the weakest concentration, chemists developed the **pH scale** to make measuring and reporting acidity easier. The term *pH* comes from the phrase *power* (or *potential*) *of hydrogen*. The value of pH ranges from less than 0 to around 14. Values less than 7 are *acidic* and values greater than 7 are *basic*, or *alkaline*. A pH equal to 7 is *neutral*. Pure water has a pH of 7. When comparing the acidity of two solutions, the one with the *lower* pH is more acidic and the one with the *higher* pH is more alkaline.

A solution with 10^{-4} mol/L of hydronium ions is ten times as concentrated as a solution with 10^{-5} mol/L.

The hydronium ion concentration is expressed as a power of 10, such as 10^{-5} mol/L. The pH is the negative of the exponent of that concentration. A difference of 1 pH unit equals a difference of one power of 10, or a tenfold difference in the hydronium ion concentration. For example, if one solution has a pH of 5 and another a pH of 4, the one with a pH of 4 is ten times as acidic—its hydronium ion concentration is ten times as great as the one that has a pH of 5.

Because of the way in which pH is defined mathematically, the larger the concentration of hydronium ions, the smaller the value of pH and the more acidic the solution. Conversely, the smaller the concentration of hydronium ions, the larger the value of pH and the more basic the solution. This relationship is most easily remembered by associating *lower pH* values with *higher acidity.* As the concentration of hydronium ions decreases, the concentration of hydroxide ions increases. At a pH of 7, when the solution is neutral, the concentrations of hydronium and hydroxide ions are equal. You can see a comparison of the pH values and the associated hydronium ion concentrations for a number of common substances in Figure 21-20.

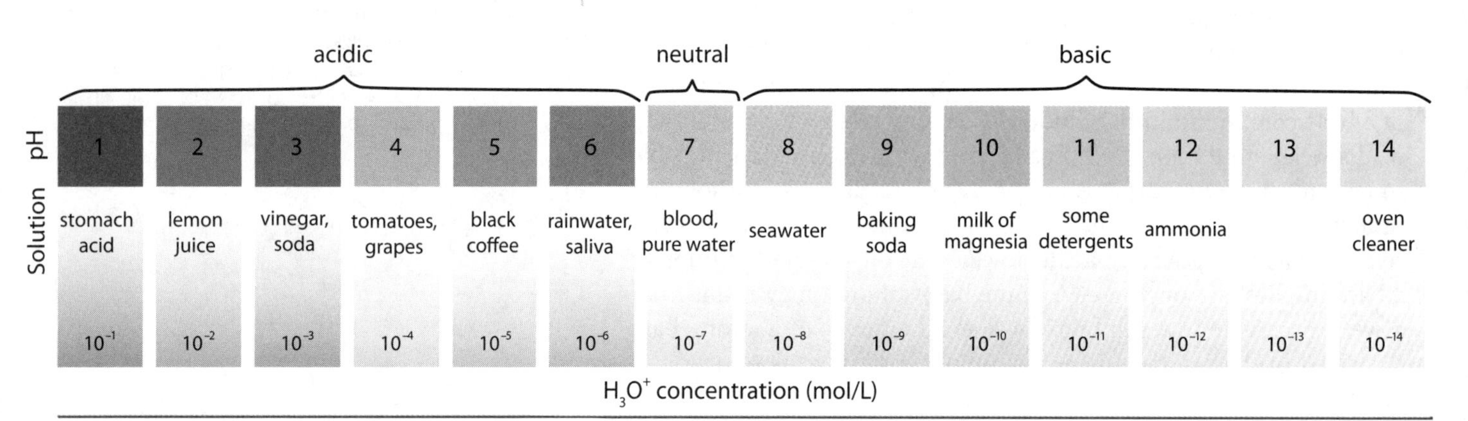

21-20 The color span in this figure is based on the pH response of pigments in the leaves of red cabbage. Many natural and artificial substances change color in response to changes in pH.

As you can see from Figure 21-20, solutions of soaps (and some detergents) are basic. They can effectively remove oils and grease from skin, clothing, and hair. However, when used too frequently on hair, soaps tend to remove too much of the oil, leaving it dull, brittle, and "lifeless." This is not surprising since strongly alkaline drain cleaners, which have large pH values, actually dissolve hair in drainpipes! Thus, shampoos that are only slightly acidic—usually marketed as *nonalkaline*—are better for hair because they allow it to retain some of its natural oils and shine.

Math Help: Logarithms

You have learned that the equilibrium concentration of H_3O^+ in plain water is 1×10^{-7} mole per liter. Written out, that is 0.000 000 1 mol/L. In a strongly acidic solution, there can be 1×10^{-1} mol/L, and in a strongly basic solution, there might be only 1×10^{-14} mol/L of H_3O^+. Using such small numbers, or even scientific notation, when you just want to talk about acidity is quite cumbersome. That is why scientists often use a *logarithmic scale* for quantities that can change over many powers of 10 (see the Math Help box on page 50).

A *logarithm* is just an exponent of a base number to which the base is raised to equal the desired number in standard notation. *Common logarithms* use a base of 10, but base 2 or other numbers are also used in science.

Let's look at an example. The logarithm of 1000 is 3 in base 10 because $10^3 = 1000$. We write this expression like this:

$$3 = \log_{10}(1000)$$

and we say "3 is the log of 1000, base 10." This process seems pretty simple for whole-number logarithms, but what about logarithm values between the whole numbers?

Logarithms can be broken down into two parts: one evenly divisible by 10 and a decimal part. A number with a common logarithm of 1.000 is exactly $10^{(1 + 0.000)}$, or just 10. But a logarithm of 1.699 is for a decimal number equal to $10^{(1 + 0.699)}$. This number lies somewhere between 10^1 (or 10) and 10^2 (or 100). You can interpret the decimal part of the logarithm from the scale on the right. It shows the relationship of decimal numbers and their logarithms between two whole powers of 10. Notice that 10 raised to the 0.699 fraction of a logarithm equals about 5. According to the laws of exponents, $10^{(1 + 0.699)} = (10^1)(10^{0.699})$, and so $10^{1.699} = (10)(5) = 50$. If the logarithm were 3.699, the decimal value of $10^{3.699}$ would be (1000)(5), or 5000.

What about negative logarithms? Recall that negative exponents simply mean the reciprocal of a number raised to a power. So the decimal value of $10^{-1.699}$ would be $1/10^{1.699}$, or $1/50 = 0.2$. You can experiment with logarithms if you have a scientific calculator with a LOG key.

So, how do logarithms apply to the pH scale? The acidity of a water solution depends on the concentration of hydronium ions (H_3O^+) in the solution. Concentration of these ions is measured in moles per liter of solution. Chemists use a special symbol for solution molar concentration: [*(substance)*]. The brackets mean "moles per liter" of whatever substance is shown between them. So we symbolize the molar concentration of hydronium ions as $[H_3O^+]$.

In logarithmic notation, the molar concentrations of the three solutions given in the first paragraph would be –7 for the neutral; –1 for the acidic; and –14 for the basic solution. To make the system easier to use, scientists have simply defined a pH value as the negative common logarithm of the H_3O^+ molar concentration, or

$$pH = -\log_{10}[H_3O^+].$$

A moderately acidic solution of vinegar might have a pH of 2.2, which works out to $10^{-2.2}$ mole of H_3O^+ ions per liter. If needed, this exponential quantity can be rewritten in standard notation (0.006 309 mol/L), or in scientific notation (6.309×10^{-3} mol/L).

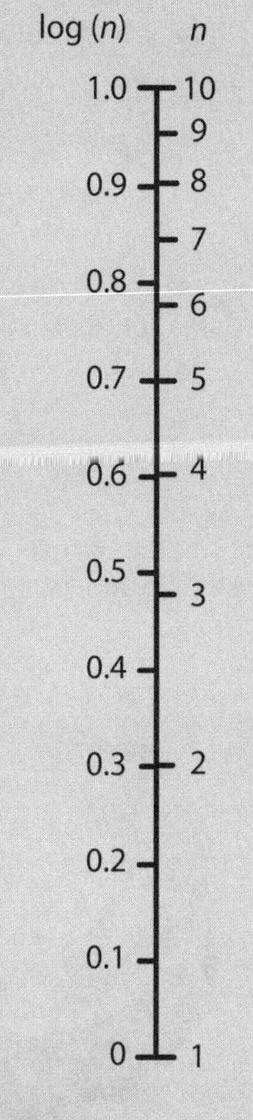

21.14 Measuring Solution pH

A solution's pH can be an important property. The pH of human blood, for example, must be maintained within the narrow range of 7.35–7.45. A blood pH value outside this range can be a symptom of a disease. Therefore, measuring solution pH is an important task of medical science.

pH Indicators

Some substances, called *pH indicators*, change color in the presence of acids and bases. This color change is directly related to the acid's or base's pH. One of the oldest-known common organic indicator compounds is *litmus*. It is obtained from several lichens. After it is extracted from the lichen and treated, litmus is a purplish chemical that turns red in an acidic solution and blue in a basic solution. Both red and blue litmus papers are available for testing whether a solution is acidic or basic. Red litmus paper will change to blue if the solution is basic, and blue litmus paper will change to red if the solution is acidic. A simple way to remember the action of litmus is the expression,

AciD turns litmus reD; Base turns litmus Blue.

21-21 A lichen (*Roccella fuciformis*) that contains the litmus pigment

Litmus is red at a pH below 4.5, gradually changes to blue for values between 4.5 and 8.3, and is blue for a pH above 8.3.

ACID
RED

21-22 An easy way to remember the action of the litmus indicator

phenolphthalein (FEE nawl THAY LEEN)

Besides litmus, another common indicator in the laboratory is ***phenolphthalein***. This substance is a white powder that is insoluble in water but dissolves in alcohol. It is colorless when added to a neutral or acidic solution but turns a deep pink when added to an alkaline solution with a pH above 8.2.

anthocyanin (AN tho SYE uh nin)

There are many other indicators that can be used to determine whether the solution of a substance is acidic or basic. Many plants contain substances called ***anthocyanins***, which change color with pH. For example, the flowers of hydrangeas can be many different colors depending on the acidity of their soil.

By knowing the color ranges of several indicators, a good estimate can be made of a solution's pH. Manufacturers have even developed indicator papers and solutions that combine the properties of multiple indicators so that pH values can be measured to the nearest whole pH unit or better. These are known as *universal indicators*. The color they produce in a solution is compared to the color of a calibrated scale similar to Figure 21-20 but with finer gradations.

21-23 A pH indicator test kit

pH Meters

Indicators depend on a person's ability to compare a color to a standard or to recognize a color change. While simple and useful for measurements that don't require high precision, this method is ultimately influenced by human perception. A much more objective technique for most scientific research uses an instrument known as a pH meter. A *pH meter* determines the concentration of the hydronium ions indirectly by measuring the voltage they produce between two points and comparing the voltage to a standard. This instrument uses a *pH probe*, a sensor that responds only to hydronium ions. The meter also contains a reference probe and usually a temperature sensor or adjustment control since the measurement of pH is temperature dependent. The voltage produced by the probe is directly proportional to pH. The slope of the pH probe's output is determined by temperature (see Figure 21-25). The pH meter converts the voltage and temperature information into a digital display of pH. These meters are *standardized* against solutions of known pH before measurements are taken to assure accuracy.

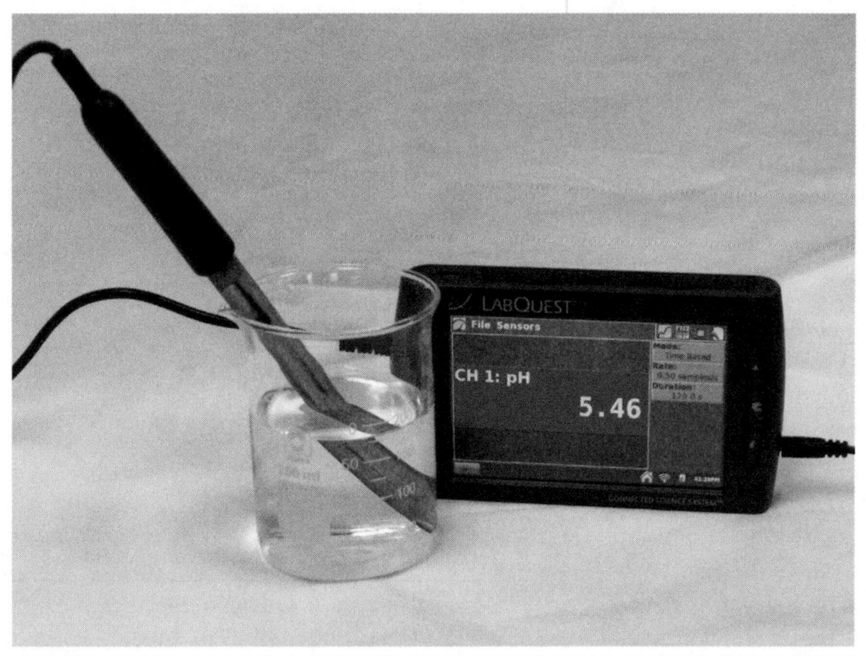

21-24 A pH meter and probe

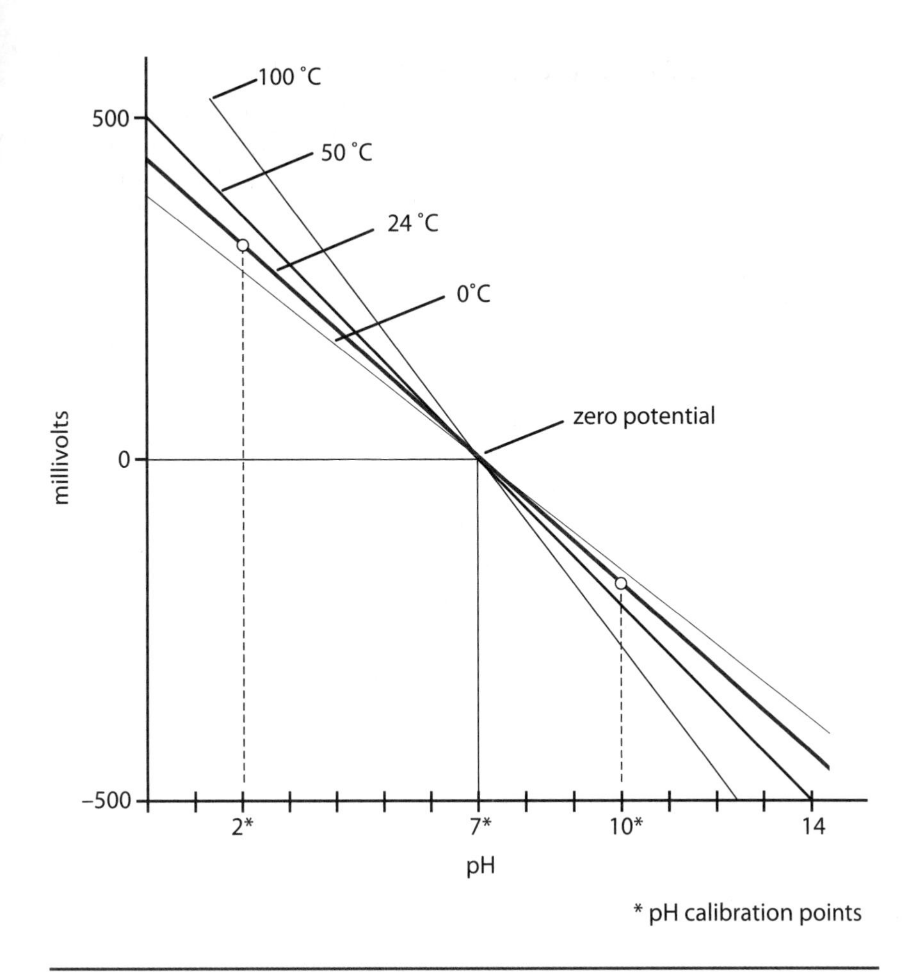

21-25 Standardization of a pH meter adjusts the zero point of the probe, and temperature compensation adjusts the slope of the probe's output signal.

Looking Back and the View Ahead

Throughout this course, we have tried to reveal to you important aspects of God's physical creation in which we live. Understanding the nature of matter, forces, energy, and the changes that matter can undergo is essential for becoming an adult who can make right decisions as a Christian and as a citizen of your nation. Even more importantly, we have presented a view of the many ways that science both glorifies the Creator (Matt. 22:37) and helps us to serve our fellow bearers of God's image as Jesus Christ told us to (Matt. 22:39). Almost limitless opportunities exist for us to use science to make this fallen world a better place.

We, the authors of this textbook, encourage you to prayerfully consider *now* what vocation God would have you pursue in the future. Not everyone can or should work in science, but if you are called to seek a profession in science (and it is a calling, just like any other area of Christian service), you must begin *now* to prepare for it. In school, take every math and science course offered. Seek opportunities to volunteer or work with technology and science during vacations. Devote your spare time to investigating things in nature or in technology. Borrow, buy, or build good quality instruments such as the following:

- a microscope
- a telescope
- data-logging equipment and associated probes
- a camera
- basic measuring equipment used in labs
- mechanical drawing equipment
- a multimeter

Take drawing or photography classes to help you learn how to observe, record, and portray nature accurately. Follow your own interests using these instruments and education to learn more about your world.

Participate in science or math competitions. Build a robot or refurbish an old car. Take things apart to see how they work and then put them back together again. Build things of your own design. Raise plants, animals, or even a microscopic zoo. Keep a freshwater or reef aquarium. Practice keeping a journal of your observations and research. Learn to find information, not only on the Internet but also in libraries with real books and scientific journals. Above all, read, read, read. Critically think about things that you learn, and compare new ideas (and old) with the kind of biblical worldview principles that you have learned in this class. As we noted in Chapter 1, preparing for a lifetime of work in science isn't easy, and the world and nature will work against any Christian who seeks to do God's will. But we assure you that you will lead a satisfying and meaningful life if you seek His will first. We are praying for you, as we do for all students who use our textbooks.

21C Section Review

1. What is a strong acid? Why is an acid's strength not related to its concentration?
2. Refer to Table 21-3 on page 496. How many acid compounds actually exist in an aqueous solution of sulfuric acid (H_2SO_4)? Name them or write their formulas. Also, state their relative strengths.
3. Do diprotic acids always yield twice the number of hydronium ions as monoprotic acids? Explain your answer.
4. In what two ways may various compounds act as strong bases? Classify each according to the acid-base model with which it can be associated.
5. Why is it better not to use the terms *weak* and *strong* when describing the concentration of an acid or base solution? What are the preferred terms?
6. DS ✪ Using the information in Table 21-3, which do you think would be a more effective antacid, sodium bicarbonate ($NaHCO_3$) or magnesium hydroxide ($Mg(OH)_2$), given the same concentration of the compounds?
7. How is the acidity or alkalinity of a solution measured? What is the significance of the numbers in this scale?
8. What is the relative difference in hydronium ion concentration between a solution with a pH of 9 and one with a pH of 8?
9. What are two different ways to measure the pH of a solution? Briefly describe each method.
10. (True or False) A solution with lower pH is less acidic.

Scientifically Speaking

hydronium ion	483
hydroxide ion	483
acid	484
base	485
Arrhenius model	485
Brønsted-Lowry model	486
conjugate base	488
conjugate acid	489
salt	492
neutralization reaction	492
pH scale	497

Chapter Review

1. What compounds produce bitter tastes in foods? Which produce sour tastes?
2. Does a sample of pure water contain only water molecules (H_2O)? Explain your answer.
3. What do chemists call a condition in which the number of particles is relatively constant over time but the identity of the individual particles is continuously changing?
4. Classify the following substances as an acid, a base, or both. Also identify the acid-base model that applies to each. Both definitions may apply.
   a. HCl
   b. H_2O
   c. HSO_4^-
   d. OH^-
   e. NaOH
   f. H_3O^+
5. How do hydrogen ions exist in an aqueous solution? Why is this the case?
6. From what you know about acids and bases, are all Brønsted-Lowry bases anions? Explain your answer.
7. Name the kind of acid compound according to the number of ionizable hydrogen atoms each has. Example: H_2SO_4, diprotic.
   a. HNO_3
   b. H_3BO_3
   c. HF
   d. H_2CO_3
   e. $HC_2H_3O_2$
   f. HCN
8. According to the Brønsted-Lowry model, what remains behind when an acid compound transfers away an ionizable proton (a hydrogen ion)?
9. Why does an acid make an excellent electrolyte?
10. What substance is often liberated when a metal reacts with an acid?
11. Describe a neutralization reaction.
12. What must be done to obtain the solid salt resulting from an acid-base reaction in an aqueous solution? Describe an acid-base reaction where the salt is produced directly, without being dissolved in a solution.
13. Besides water and a salt, what other compounds may be formed in a neutralization reaction?
14. DS ✪ What kind of chemical substance is an antacid? What is one potential hazard of taking too many doses or too large a dose of an antacid?
15. What is the main difference between the same concentration of a strong molecular acid and a weak molecular acid in solution?
16. How do strong bases damage biological tissues?
17. What are the two factors that determine the concentration of hydronium (or hydroxide) ions in an acid (or base) solution?
18. What quantity is pH a measure of?
19. Describe how a pH indicator works.

True or False

20. The hydronium ion (H_3O^+) is just one of many acid ions.
21. Water-soluble acids seem to be mostly molecular compounds.

DS ✪ 22. People with severe acid indigestion suffer from an excess of sulfuric acid in their stomachs.

23. The ammonium ion (NH_4^+) is the conjugate acid of the ammonia molecule (NH_3).
24. Sodium hydroxide *ionizes* in water to form a strong base.
25. Only certain acids will neutralize a given base.
26. The corrosion of a metal by an acid is a kind of neutralization reaction.

DS ✪ 27. An effective antacid reacts with hydronium ions in the stomach.

28. All monoprotic acids are strong acids.
29. With few exceptions, the strongest bases are ionic compounds that contain one or more hydroxide ions in their formula unit.
30. In solutions with a pH higher than 7, there are fewer hydronium ions than there are hydroxide ions.
31. A pH meter is normally more precise than other pH indicators for most scientific work.

✪ 32. Complete the following chemical equations by writing the correct formulas (including coefficients, if applicable). Use your own paper.

a. $H_2O + \_\_\_\_\_\_ \longrightarrow H_3O^+ + OH^-$

b. $2HF\ (aq) + Ba(OH)_2(aq) \longrightarrow \_\_\_\_\_\_ (aq) + 2H_2O$

c. $Zn\ (s) + 2H_3O^+ + 2NO_3(aq) \longrightarrow$
$Zn(NO_3)_2(aq) + 2H_2O\ (l) + \_\_\_\_\_\_ (g)$

d. $CaCO_3 + \_\_\_\_\_\_ \longrightarrow$
$CaCl_2(aq) + H_2O\ (l) + \_\_\_\_\_\_ (g)$

✪ 33. The formula for pH is

$$pH = -\log[H_3O^+],$$

where "$[H_3O^+]$" is the concentration of the hydronium ion in moles per liter and "log( )" is the common logarithm (in base 10) of whatever is in the parentheses following the log symbol. Why is there a negative sign in front of the right-hand term?

APPENDIX A
UNDERSTANDING WORDS IN SCIENCE

You may find science a little intimidating because of all the long words scientists use. Some of these words may be unfamiliar to you. However, you can often understand these words by breaking them down into simple parts that *do* have meaning to you. When you see a difficult scientific word, look at the parts on this table to help you understand it. Many of these parts may come at the beginning, end, or middle of the term, depending on its meaning.

Example: *magnetohydrodynamics*

Magneto deals with magnetism or magnetic energy.

Hydro deals with water or liquids in general.

Dynamics deals with the application of forces and use of energy.

Therefore, *magnetohydrodynamics* has something to do with magnetic forces within a fluid.

a-, an-	without, not
ab-, abs-	away from
ac-	toward
acous-	hear
aer-	in or of the atmosphere
alter-	change
amal-	soft
amphi-	on both sides
anti-	opposed to
audio-	hear
auto-	from or by self
bar-	pressure
bi-	two
bio-	life
calc-	calcium
calor-	heat
cent-	hundred
chem-	chemistry
chrom-	color
co-, com-, con-	together
cupr-	copper
de-	loss, removal
deci-	tenth
deka-	ten
di-	two
-duce, -duct-	to lead
dyna-	power
eco-	house, abode
elect-	electricity
endo-	inside
epi-	upon, on top of
equa-, equi-	same
exo-, exter-	outside
extra-	beyond
fil-	a thread
fissi-	split, divide
flam-	fire
fund-	the bottom
fusi-	melt, pour
gastr-	the stomach
geo-	earth
glob-	a ball, a globe
-gnosis	knowledge
-grade	step, walk, slope
-graph	record
grav-	heavy
gyr-, gyro-	round, turning
halo-	salt
helio-	sun
hemi-	half
hetero-	different
hexa-	six
homo-	same
hydro-	water, fluid
hyper-	excess
hypo-	less than, below
icon-	image
inter-	between
intra-	within
-ism	belief, process of
iso-	equal
kine-	moving
lith	stone
-logy	study of
luna-	moon
macro-	large, long
magneto-	magnetic force
mal-	bad
mega-	great, large
meteor-	in or of the atmosphere
-meter	measure
micro-	small, millionth
milli-	thousandth
-mit	to send
modus	measure
mono-	one
nona-	nine
nucle-	core, inner part
oct-, octa-	eight
opti-	the eye, vision
organ-	living, carbon bearing
ortho-	regular, perpendicular
-osis	condition of, disease
ox-, oxy-	oxygen
pan-	all
pend-	hanging
penta-	five
peri-	about, around
phon-, phono-	sound, voice
photo-	light
poly-	many
post-	after
pre-	before
pulmo-	lung
pyro-	fire
quad-	four
radi-	spoke, ray
retro-	backward, reverse
sal-	salt
scient-	knowledge
-scope	seeing, visual
-sect	cut
selen-	moon
semi-	half
septa-	seven
sex-	six
sol-	sun
sono-	sound
spec-	see, look at
stella-	star
sub-	underneath
super-	above, greater than
syn-, sys-	with, together
tele-	far, remote
telo-	end, complete
tetra-	four
therm-	heat, temperature
trans-	across
tri-	three
uni-	one, single
vacu-	empty
vitre-	glass, glass-like
volu-	bulk, amount, quantity

APPENDIX B
BASE AND DERIVED UNITS OF THE SI

Base Units			
Dimension	**Name**	**Symbol**	**Definition**
length	meter	m	The meter is the length of the path traveled by light in vacuum during a time interval of 1/299,792,458 of a second.
mass	kilogram	kg	The kilogram is the unit of mass; it is equal to the mass of the international prototype of the kilogram.
time	second	s	The second is the duration of 9,192,631,770 periods of the radiation corresponding to the transition between the two hyperfine levels of the ground state of the cesium 133 atom at rest at 0 K.
electrical current	ampere	A	The ampere is that constant current which, if maintained in two straight parallel conductors of infinite length, of negligible circular cross-section, and placed 1 meter apart in vacuum, would produce between these conductors a force equal to 2×10^{-7} newton per meter of length.
temperature	kelvin	K	The kelvin, unit of thermodynamic temperature, is the fraction 1/273.16 of the thermodynamic temperature of the triple point of water having a specific isotopic composition.
amount of substance	mole	mol	1. The mole is the amount of substance of a system which contains as many elementary entities as there are atoms in 0.012 kilogram of unbound carbon 12 atoms in their ground state; its symbol is "mol." 2. When the mole is used, the elementary entities must be specified and may be atoms, molecules, ions, electrons, other particles, or specified groups of such particles.
luminous intensity	candela	cd	The candela is the luminous intensity, in a given direction, of a source that emits monochromatic radiation of frequency 540×10^{12} hertz and that has a radiant intensity in that direction of 1/683 watt per steradian.

plane angle	radian	rad	The radian is the plane angle between two radii of a circle that cuts off on the circumference an arc equal in length to the radius. A radian is equal to approximately 57.30°.
solid angle	steradian	sr	The steradian is the solid angle that, having its vertex in the center of a sphere, cuts off an area of the surface of the sphere equal to the radius of the sphere squared. The central angle of a steradian is approximately 65.54°.

The radian and steradian are angular units included for completeness. They are not used in this textbook. All other unit definitions are a contribution of the National Institute of Standards and Technology and come directly from the NIST website.

continued

Derived Units			
Dimension	**Name**	**Symbol**	**Definition**
electrical charge	coulomb	C	A·s
electrical potential difference	volt	V	$\frac{J}{C} = \frac{kg\cdot m^2}{A\cdot s^3}$
electrical resistance	ohm	Ω	$\frac{V}{A} = \frac{kg\cdot m^2}{A^2\cdot s^3}$
energy, work, heat	joule	J	$N\cdot m = kg\cdot m^2\cdot s^{-2}$
force	newton	N	$kg\cdot m/s^2$
frequency	hertz	Hz	$\frac{1}{s} = s^{-1}$
power	watt	W	$\frac{J}{s} = \frac{kg\cdot m^2}{s^3}$
pressure	pascal	Pa	$\frac{N}{m^2} = \frac{kg}{m\cdot s^2}$
temperature	degree (Celsius)	°C	$T - 273.15°$

Experimentally Derived Units

particle mass	atomic mass unit	u	The atomic mass unit is equal to 1⁄12 of the mass of an atom of the nuclide ^{12}C; 1 u = $1.660\,538\,92 \times 10^{-27}$ kg.
unit distance	astronomical unit	AU	The average distance between the earth and the sun; defined as exactly 149,597,870,700 m in 2012 by the IAU
interstellar distance	parsec	pc	The interstellar distance corresponding to an observational parallax of 1 arcsecond generated across a baseline of 1 AU; 1 pc equals 3.09×10^{13} km (1.92×10^{13} mi) or 3.26 ly

Other Units Used with the SI

pressure	bar	bar	1 bar = 10^5 Pa
level of power	decibel	dB	1 dB = $10 \log(P/P_0)$, where P is measured power and P_0 is the reference power.

Appendix C
Metric Prefixes

Metric Prefixes					
Prefix	**Origin/Meaning**	**Symbol**	**Factor**	**Example**	**Application**
exa-	Greek/six (i.e., 1000^6)	E-	$\times 10^{18}$	exahertz (EHz)	frequency of a gamma ray
peta-	Greek/five (i.e., 1000^5)	P-	$\times 10^{15}$	petabyte	Google's annual data volume
tera-	Greek/"monster"	T-	$\times 10^{12}$	terawatt (TW)	power generation in the United States
giga-	Greek/"giant"	G-	$\times 10^{9}$	gigabyte (GB)	hard-drive storage capacity
mega-	Greek/"great"	M-	$\times 10^{6}$	megajoule (MJ)	work done by a bulldozer
kilo-	Greek/thousand	k-	$\times 10^{3}$	kilometer (km)	terrestrial distances
hecto-	Greek/hundred	h-	$\times 10^{2}$	—	—
deka-	Greek/ten	da-	$\times 10^{1}$	—	—
(base)	—	—	$\times 10^{0}$	—	—
deci-	Latin/tenth	d-	$\times 10^{-1}$	decibel (dB)	sound loudness
centi-	Latin/hundred(th)	c-	$\times 10^{-2}$	centimeter (cm)	distances in a laboratory
milli-	Latin/one thousand(th)	m-	$\times 10^{-3}$	millivolt (mV)	EKG signal from heart at skin
micro-	Greek/"small"	u-	$\times 10^{-6}$	micropascal (uPa)	sound wave pressures
nano-	Greek/"dwarf"	n-	$\times 10^{-9}$	nanometer (nm)	size of atoms
pico-	Italian/"small"	p-	$\times 10^{-12}$	picometer (pm)	long gamma-ray wavelength
femto-	Danish/fifteen	f-	$\times 10^{-15}$	femtometer (fm)	short gamma-ray wavelength
atto-	Danish/eighteen	a-	$\times 10^{-18}$	attosecond (as)	period required to image an orbiting electron

APPENDIX D
UNIT CONVERSIONS

Each row shows the conversion factors for the unit in the left column. For example, to convert centimeters to inches, you would multiply by 0.3937.

Length	cm	m	km	in.	ft	mi
1 cm =	1	10^{-2}	10^{-5}	0.3937	3.281×10^{-2}	6.214×10^{-6}
1 m =	100	1	10^{-3}	39.37	3.281	6.214×10^{-4}
1 km =	10^5	1000	1	3.937×10^4	3281	0.6214
1 in. =	2.540	2.54×10^{-2}	2.540×10^{-5}	1	8.333×10^{-2}	1.578×10^{-5}
1 ft =	30.48	0.3048	3.048×10^{-4}	12	1	1.894×10^{-4}
1 mi =	1.609×10^5	1609	1.609	6.336×10^4	5280	1

1 angstrom (Å) = 1.000×10^{-10} m
1 nautical mile (nm) = 1852 m ≈ 1.151 mi ≈ 6076 ft
1 ly = 9.461×10^{15} m
1 astronomical unit (AU) = 1.496×10^{11} m

Mass	g	kg	u	oz	lbm	t (short)
1 g =	1	0.001	6.022×10^{23}	3.527×10^{-2}	2.205×10^{-3}	1.102×10^{-6}
1 kg =	1000	1	6.022×10^{26}	35.27	2.205	1.102×10^{-3}
1 u =	1.661×10^{-24}	1.661×10^{-27}	1	5.857×10^{-26}	3.661×10^{-27}	1.830×10^{-30}
1 oz =	28.35	2.835×10^{-2}	1.707×10^{25}	1	6.250×10^{-2}	3.125×10^{-5}
1 lbm =	453.6	0.4536	2.732×10^{26}	16	1	5.0×10^{-4}
1 t (short) =	9.072×10^5	907.2	5.463×10^{29}	3.2×10^4	2000	1

Time	s	min	h	d	y
1 s =	1	1.667×10^{-2}	2.778×10^{-4}	1.157×10^{-5}	3.169×10^{-8}
1 min =	60	1	1.667×10^{-2}	6.944×10^{-4}	1.901×10^{-6}
1 h =	3600	60	1	4.167×10^{-2}	1.141×10^{-4}
1 d =	8.640×10^4	1440	24	1	2.738×10^{-3}
1 y =	3.156×10^7	5.259×10^5	8.766×10^3	365.24	1

Pressure	atm	in. Hg	torr	Pa	mb	lbf/in.^{2}
1 atm =	1	29.92	760.0	1.013×10^5	1013	14.70
1 in. Hg =	3.342×10^{-2}	1	25.40	3386	33.86	0.4911
1 torr =	1.316×10^{-3}	3.937×10^{-2}	1	133.3	1.333	1.934×10^{-2}
1 Pa =	9.869×10^{-6}	2.953×10^{-4}	7.501×10^{-3}	1	0.01	1.450×10^{-4}
1 mb =	9.869×10^{-4}	2.953×10^{-2}	0.7501	100	1	1.450×10^{-2}
1 lbf/in.^{2} =	6.805×10^{-2}	2.036	51.71	6895	68.95	1

Volume	cm^3	m^3	L	in.^{3}	ft^3
1 cm^3 =	1	1.0×10^{-6}	1.000×10^{-3}	6.102×10^{-2}	3.531×10^{-5}
1 m^3 =	10^6	1	1000	6.102×10^4	35.31
1 L =	1000	1.000×10^{-3}	1	61.02	3.531×10^{-2}
1 in.^{3} =	16.39	1.639×10^{-5}	1.639×10^{-2}	1	5.787×10^{-4}
1 ft^3 =	2.832×10^4	2.832×10^{-2}	28.32	1728	1

1 gal = 4 qt = 8 pt = 128 fl. oz = 231 in.^{3}

Appendix E
Math Helps

This appendix provides you the necessary helps for solving the many kinds of physical science problems found in this textbook. It's not intended to be all-inclusive. We assume that you understand or are learning the methods for solving single-variable algebra equations, including the order of arithmetic operations, and rearranging equations to solve for an unknown quantity.

Math Help Boxes

Math Help boxes provide detailed information on a useful math concept that apply to the associated science topic in the textbook. Math Help boxes cover the following topics:

Converting a Measured Quantity from One Unit to Another

You can learn to convert a measurement from one kind of unit to another in Subsection 3.5 on pages 51–55.

Mathematics and Measurement

When measured quantities appear in equations as terms or factors, you follow the same arithmetic principles that you have learned for solving any equations. However, acting as a scientist, you have to think about some extra things before arriving at a solution to a problem.

Measured quantities include *units*. We treat units, such as meters (m), seconds (s) and meters per second (m/s), as factors of the measurements. For example, 35 m/s means $35 \times \frac{m}{s}$. So any arithmetic operation must apply to both the number and the unit factors in a measurement.

Measured quantities also have *precision* (see Subsection 3.8). Precision is the exactness of a measurement. Scientists spend a lot of effort to ensure that their data is no more and no less exact than their instruments can read. Scientists tell other scientists the precision of their measurements by the number of *significant digits* they include in their reported data. We discuss significant digits in Subsections 3.10–3.11.

On the following pages you will find some rules for doing basic arithmetic operations using measurements.

You can't add measurements with different units any more than you can add apples and oranges!

Adding and Subtracting Physical Science Data

Math Rule 1: Added or subtracted data must have the same units. Even if the kinds of data are the same, if their units are not the same, you are adding apples and oranges. For example, you can't add the lengths 3.1 m and 45 cm together without converting one of the measurements to the other unit.

Math Rule 2: The result of adding or subtracting data can't be more precise than the least precise data used in the sum or difference. After adding or subtracting, round the result to the decimal place of the estimated digit in the *least* precise data. Review rounding rules in the Math Help box below.

When adding measured data in a column, you can quickly identify the location of the estimated digit in the sum. Place a vertical dashed line to the right of the estimated digit with the largest place value (the estimated place value farthest to the left). The sum must be rounded to the place value to the left of this line, in this case, 3.5 m.

Example Problem 1

Adding Data: Length

Add 3.1 m and 45 cm.

Unit Conversion: $45\,\text{cm} \times \frac{1\,\text{m}}{100\,\text{cm}} = 0.45\,\text{m}$

Add data in column:

$$\begin{array}{r} 3.\underline{1}\;\;\text{m} \\ +0.4\underline{5}\,\text{m} \\ \hline 3.\underline{5}\underline{5}\,\text{m} \end{array} \approx 3.6\,\text{m}$$

The estimated digits in each piece of data and the sum are underlined. The sum contains two underlined digits. These digits result from a sum, including an estimated digit in that column. The sum must be rounded to the estimated digit having the largest place value—0.1 m.

Math Help: Rounding Rules

We round numbers all the time. Sometimes it's convenient to round the number of students in your school to the nearest 10 students, the number of residents in your city to the nearest 1000 people, the distance around the earth to the nearest 1000 miles, and so on.

Rounding is useful for several reasons. It helps us grasp large numbers when the specific number is not needed but the approximate size is. Rounding also allows us to report measured scientific data correctly.

For the purposes of measuring scientific data and performing calculations, scientists have agreed on certain rounding rules. These ensure that calculation solutions are properly rounded to show the needed precision of numerical quantities. We suggest that you use the following rules for problems in this textbook to imitate closely those used by scientists.

1. Identify the place value that you are going to round to. (This will be determined by other rules that you will learn.) This is the rounded place value. We underline the rounded place value in example problems in this textbook.

   Example: The following length must be rounded to the nearest 0.1 m.

   $12.\underline{3}4$ m

2. If the digit to the right of the rounded place value is 0–4, the digit in the rounded place value remains unchanged ("round down"). If the digit to the right is 5–9, the digit in the rounded place value is increased by 1 ("round up").

   Examples:
   $56.\underline{7}39\,\text{mL} \Rightarrow 56.7\,\text{mL}$
   $39.\underline{4}7\,\text{mm} \Rightarrow 39.5\,\text{mm}$
   $74.\underline{6}07\,\text{s} \Rightarrow 74.6\,\text{s}$

3. If the rounded place value is to the *right* of the decimal point, all digits to the *right* of the rounded place value will be dropped after rounding.

   Examples:
   $105.\underline{6}39\,\text{g} \Rightarrow 105.6\,\text{g}$
   $32.\underline{9}5\,\text{cm} \Rightarrow 33.0\,\text{cm}$

   If the rounded place value is to the *left* of the decimal point (or assumed decimal point), all digits between the rounded place value and the decimal point become zeros and you leave out the decimal point after rounding.

   Examples:
   $1\underline{6}94\,\text{m} \Rightarrow 1700\,\text{m}$
   $\underline{1}9.6\,°\text{C} \Rightarrow 20\,°\text{C}$

4. You may have noticed in the examples above that rounding to a particular place value may affect other digits to the left of the rounding place value. This occurs when a 9 is rounded up. The 9 becomes 0 and 1 is added to the digit to the left.

Multiplying and Dividing Physical Science Data

Math Rule 3: The result of a multiplication or division can't have more SDs than the data with the fewest SDs used in the calculation. After multiplying or dividing, round the results to the same number of significant digits as the data with the fewest SDs.

You CAN divide apples by oranges!

EXAMPLE PROBLEM 2

Dividing Data: Density

Calculate the density of a sample of quartz. Its mass is 27.55 g and its volume is 10.4 cm³. Round to the correct number of significant digits.

Given data: mass = 27.55 g (4 SDs)
volume = 10.4 cm^3 (3 SDs)

Calculation:

$$\text{density} = \frac{\text{mass}}{\text{volume}} \quad \frac{27.55\ \text{g}}{10.\underline{4}\ \text{cm}^3}$$

$= 2.6\underline{4}9\,038\,462\ \text{g/cm}^3$ (3 SDs allowed)

$\approx 2.6\underline{4}9\ \text{g/cm}^3$ (drop all but one extra digit)

$\approx 2.65\ \text{g/cm}^3$ (round as usual)

Since 10.4 cm^3 has only 3 SDs, only 3 SDs are allowed in the quotient. Notice the extra digit carried in the result before rounding. All other digits are dropped (not rounded) except for that one extra digit, which is then used to determine how the least significant digit is rounded in the final answer.

Any measured data can be multiplied or divided together, as long as the resulting combination of units represents a meaningful and useful quantity.

Math Rule 4: The result of a multiplication or division between a measured quantity and a pure number has the same number of decimal places, or the same precision, as the measured quantity used in the calculation.

Examples:

a. 7 × 2.35 cm = 16.45 cm (*not* 16.5 cm)

b. 2.63 cm ÷ 5 = 0.53 cm (*not* 0.526 cm)

Notice that the number of SDs in the measured data do *not* determine the number of SDs in the result when multiplying or dividing by pure numbers.

Math Rule 4 ensures that you preserve the precision of the original measurement, even if you change the number by multiplying or dividing by a pure number.

For example, you shouldn't expect that you could improve the precision of the measurement of a circle's radius with a ruler by dividing its diameter by 2.

Proportionalities in Physical Science

As you study physical science, you will discover that many measurable quantities change or vary in a reliable, predictable way with other quantities. For example, the gravitational potential energy of a ball increases with its height above the ground. Double its height and you double its potential energy. Similarly, if you double your running speed around a cross-country track, you halve the time it takes to run the course. These kinds of relationships are called *proportionalities*. Study the following Math Help box to better understand these relationships.

Math Help: Proportionality

A minute has 60 seconds, right? How many seconds are in two minutes? Correct, 120 seconds. The number of seconds is *directly proportional* to the number of minutes. The word *proportional* means "having a constant ratio." In the case of seconds and minutes, the ratio is always 60 s:1 min or 1 min:60 s—they are directly proportional.

If two numbers, *A* and *B*, are directly proportional, then they can be set *equal* to each other by including a *proportionality constant* factor—*k*. The equation is

$$A = k \times B,$$

where

$$\frac{A}{B} = k$$

and *k* is a constant for all ratios of these two quantities. For our example, *A* is time in minutes and *B* is time in seconds. The proportionality constant *k* is equal to1 min/60 s, as below:

$$A = (1 \text{ min}/60 \text{ s}) \times B$$

You'll work with many quantities that are directly proportional. In this textbook, direct proportions that you will study include mechanical potential energy, momentum, acceleration, mechanical advantage, and many others.

You'll also find that other pairs of quantities are *inversely proportional*. If the value of one increases, the other decreases in proportion. The equations for these proportions look like this:

$$A = k \div B \text{ or } A = k \times \frac{1}{B}$$

If you multiply both sides of the equation by *B*, you get

$$AB = k.$$

You can see immediately that if *k* is a constant, then as *A* increases, *B* must decrease in the same proportion to maintain the equality.

Examples of inverse proportions are pressure and volume of a confined gas (Boyle's law), Ohm's electrical power law, frequency and period, and wavelength and frequency of a wave in a uniform medium.

Lastly, there are other kinds of proportionalities that are neither entirely direct nor inverse relationships. For example, the kinetic energy of an object increases much faster than expected by just doubling its speed. Double a baseball's speed, and it has four times as much kinetic energy as at the slower speed. Similarly, the force of gravity and the intensity of light both decrease much faster with distance from their sources than would be the case with a simple inverse proportion. Double the distance from a light's source and the light's intensity is only one-fourth the value at the closer distance. These kinds of proportionalities include an extra math operation in the direct and inverse relationships. Scientists give the latter case a special name—the *inverse square law*. You will study these and other phenomena where this law applies in this textbook.

Appendix F
Commonly Used Abbreviations and Symbols

Unit Abbreviations					
Unit	**Abbreviation**	**Dimension**	**Unit**	**Abbreviation**	**Dimension**
ampere	A	current	light-year	ly	length/distance (interstellar)
atmosphere	atm	pressure	liter	L	volume
atomic mass unit	u	mass	meter	m	length/distance
bar	b	pressure	millibar	mb	pressure
candela	cd	light intensity	minute	min	time; 60 s
coulomb	C	charge	molarity	M	solution concentration
day	d	time; 24 h	mole	mol	quantity (particles)
decibel	dB	relative power	nautical mile	NM	distance
degree	°	temperature or angle	newton	N	force
degree (Celsius)	°C	temperature	ohm	Ω	electrical resistance
degree (Fahrenheit)	°F	temperature	pascal	Pa	pressure
foot	ft	length/distance (US)	percent	%	ratio (per hundred)
gallon	gal	volume (US)	pound (force)	lbf	force (US)
gram	g	mass	pound (mass)	lbm	mass (US)
hertz	Hz, s^{-1}	frequency	revolutions per minute	rpm	rate of rotation
hour	h	time; 60 min	second	s	time
inch	in.	length (US)	volt	V	electrical potential
joule	J	energy or work	watt	W	power
kelvin	K	temperature	week	wk	time; 7 d
kilogram	kg	mass	yard	yd	length/distance (US)
kilowatt-hour	kWh	energy	year	y	time; 365.24 d

Other Abbreviations

Abbreviation	Term
×	magnification power
☢	radiation/radioactive property (element)
α, $^{4}_{2}He$	alpha particle
AC	alternating current
AFCI	arc-fault circuit interrupter
AM	amplitude modulation
AMA	actual mechanical advantage
(*aq*)	chemical state; aqueous
AUV	autonomous underwater vehicle
β, $^{0}_{-1}e$	beta particle
BEA	Bureau d'Enquêtes et d'Analyses
BIPM	Bureau International des Poids et Mesures
CAT/CT	computed axial tomography
CCD	charge-coupled device
CMYK	cyan-magenta-yellow-black color model
ΔEN	difference in electronegativity numbers
DC	direct current
DNA	deoxyribonucleic acid
DTV	digital television
e	electron
EM	electromagnetic
EN	electronegativity number
EPE	elastic potential energy
EVLA	Enhanced Very Large Array (radio telescope)
FM	frequency modulation
γ	gamma ray
(*g*)	chemical state; gas
GFCI	ground-fault circuit interrupter
GPE	gravitational potential energy
GPS	Global Positioning System
HCP	hexagonal close-packed
HDTV	high-definition television
HSV	hue-saturation-value color model
IMA	ideal mechanical advantage
IR	infrared
ISS	International Space Station
ITER	(formerly) International Thermonuclear Experimental Reactor
IUPAC	International Union of Pure and Applied Chemistry
KE	kinetic energy
(*l*)	chemical state; liquid
LCM	least common multiple
LED	light-emitting diode
MA	mechanical advantage
MRI	magnetic resonance imaging
n	neutron
NIST	National Institute of Standards and Technology
ON	oxidation number
p	proton
PE	potential energy
pH	acidity/alkalinity (power of hydrogen)
QSO	quasi-stellar object
RFID	radio-frequency identification
ROV	remotely operated vehicle
RREA	Relativistic Runaway Electron Avalanche
(*s*)	chemical state; solid
SD	significant digit
s.g.	specific gravity
SI	Système International d'Unités
STM	scanning tunneling electron microscope
STP	standard temperature and pressure
TOH	threshold of hearing
TPI	threads per inch (mechanical)
UV	ultraviolet

Formula Symbols and Quantities

Symbol	Quantity	Symbol	Quantity
A	area	n	index of refraction
a or $\mathbf{a}$	acceleration	N_{in}	number of turns, input
A	atomic mass	N_{out}	number of turns, output
C	heat capacity	π	pi
c	speed of light	P	power
c_{sp}	specific heat capacity	P	pressure
d	distance	p or $\mathbf{p}$	linear momentum
$\mathbf{d}$	displacement	Δp	momentum, change of
f	frequency	Q	charge
F or $\mathbf{F}$	force	Q	heat
f or $\mathbf{f}$	friction	ρ	density
F_e	effort force	r	radial distance
F_g or $\mathbf{F}_g$	gravitational force	R	electrical resistance
F_r	resistance force	S	entropy
g or $\mathbf{g}$	gravitational acceleration	T	Kelvin temperature
G	universal gravitation constant	T	period
h	height	t_C or t	Celsius temperature
h	Planck's constant	t_F or t	Fahrenheit temperature
H_I	image height	Δt	temperature change
H_O	object height	Δt	time interval
I	electrical current	V	electrical potential
k	constant (unspecified)	v	speed
λ	wavelength	$\mathbf{v}$	velocity
ℓ	length/lever arm	V_{in}	input voltage
ℓ_e	effort arm	V_{out}	output voltage
ℓ_r	resistance arm	V	volume
L_f	latent heat of fusion	w	weight
L_v	latent heat of vaporization	W	mechanical work
m	mass	Z	atomic number

Note: Bold symbols represent the vector form of each quantity.

APPENDIX G
PERIODIC TABLE OF THE ELEMENTS

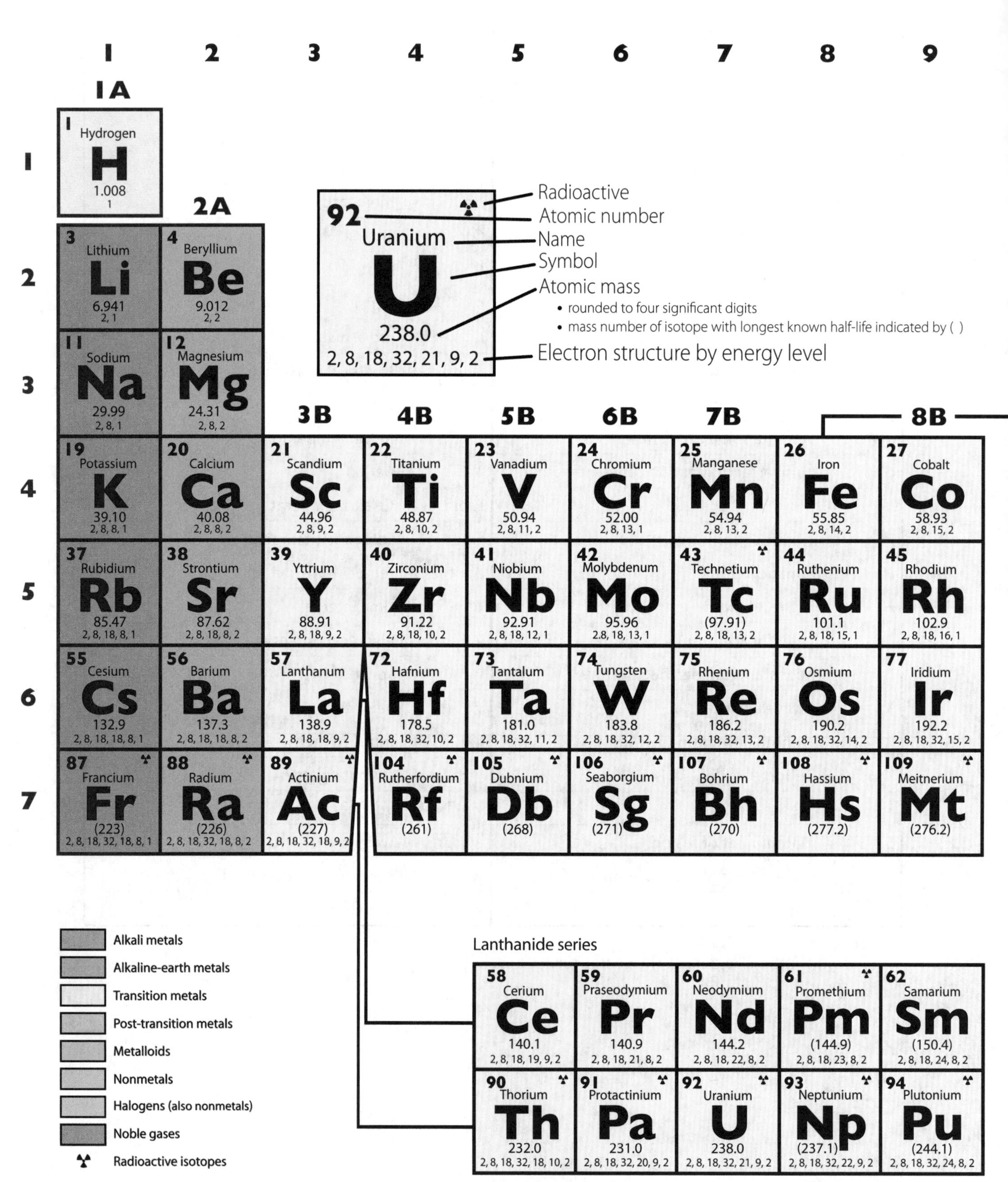

10	11 1B	12 2B	13 3A	14 4A	15 5A	16 6A	17 7A	18 8A
								2 Helium **He** 4.003 2
			5 Boron **B** 10.81 2, 3	6 Carbon **C** 12.01 2, 4	7 Nitrogen **N** 14.01 2, 5	8 Oxygen **O** 16.00 2, 6	9 Fluorine **F** 19.00 2, 7	10 Neon **Ne** 20.18 2, 8
			13 Aluminum **Al** 26.98 2, 8, 3	14 Silicon **Si** 28.09 2, 8, 4	15 Phosphorus **P** 30.97 2, 8, 5	16 Sulfur **S** 32.07 2, 8, 6	17 Chlorine **Cl** 35.45 2, 8, 7	18 Argon **Ar** 39.95 2, 8, 8
28 Nickel **Ni** 58.69 2, 8, 16, 2	29 Copper **Cu** 63.55 2, 8, 18, 1	30 Zinc **Zn** 65.38 2, 8, 18, 2	31 Gallium **Ga** 69.72 2, 8, 18, 3	32 Germanium **Ge** 72.63 2, 8, 18, 4	33 Arsenic **As** 74.92 2, 8, 18, 5	34 Selenium **Se** 78.96 2, 8, 18, 6	35 Bromine **Br** 79.90 2, 8, 18, 7	36 Krypton **Kr** 83.80 2, 8, 18, 8
46 Palladium **Pd** 106.4 2, 8, 18, 18	47 Silver **Ag** 107.9 2, 8, 18, 18, 1	48 Cadmium **Cd** 112.4 2, 8, 18, 18, 2	49 Indium **In** 114.8 2, 8, 18, 18, 3	50 Tin **Sn** 118.7 2, 8, 18, 18, 4	51 Antimony **Sb** 121.8 2, 8, 18, 18, 5	52 Tellurium **Te** 127.6 2, 8, 18, 18, 6	53 Iodine **I** 126.9 2, 8, 18, 18, 7	54 Xenon **Xe** 131.3 2, 8, 18, 18, 8
78 Platinum **Pt** 195.1 2, 8, 18, 32, 17, 1	79 Gold **Au** 197.0 2, 8, 18, 32, 18, 1	80 Mercury **Hg** 200.6 2, 8, 18, 32, 18, 2	81 Thallium **Tl** 204.4 2, 8, 18, 32, 18, 3	82 Lead **Pb** 207.2 2, 8, 18, 32, 18, 4	83 Bismuth **Bi** 209.0 2, 8, 18, 32, 18, 5	84 Polonium **Po** (209) 2, 8, 18, 32, 18, 6	85 Astatine **At** (210.0) 2, 8, 18, 32, 18, 7	86 Radon **Rn** (222.0) 2, 8, 18, 32, 18, 8
110 Darmstadtium **Ds** (281.2)	111 Roentgenium **Rg** (280.2)	112 Copernicium **Cn** (285.2)	113 Ununtrium **Uut** (284.2)	114 Flerovium **Fl** (289.2)	115 Ununpentium **Uup** (288.2)	116 Livermorium **Lv** (293)	117 Ununseptium **Uus** (294)	118 Ununoctium **Uuo** (294)

The names given to elements 113, 115, 117, and 118 represent the Latin and Greek names for their arabic numbers.

63 Europium **Eu** 152.0 2, 8, 18, 25, 8, 2	64 Gadolinium **Gd** 157.3 2, 8, 18, 25, 9, 2	65 Terbium **Tb** 158.9 2, 8, 18, 27, 8, 2	66 Dysprosium **Dy** 162.5 2, 8, 18, 28, 8, 2	67 Holmium **Ho** 164.9 2, 8, 18, 29, 8, 2	68 Erbium **Er** 167.3 2, 8, 18, 30, 8, 2	69 Thulium **Tm** 168.9 2, 8, 18, 31, 8, 2	70 Ytterbium **Yb** 173.0 2, 8, 18, 32, 8, 2	71 Lutetium **Lu** 175.0 2, 8, 18, 32, 9, 2
95 Americium **Am** (243.1) 2, 8, 18, 32, 25, 8, 2	96 Curium **Cm** (247.1) 2, 8, 18, 32, 25, 9, 2	97 Berkelium **Bk** (247) 2, 8, 18, 32, 27, 8, 2	98 Californium **Cf** (251.1) 2, 8, 18, 32, 28, 8, 2	99 Einsteinium **Es** (252.1) 2, 8, 18, 32, 29, 8, 2	100 Fermium **Fm** (257.1) 2, 8, 18, 32, 30, 8, 2	101 Mendelevium **Md** (258.1) 2, 8, 18, 32, 31, 8, 2	102 Nobelium **No** (259.1) 2, 8, 18, 32, 32, 8, 2	103 Lawrencium **Lr** (262) 2, 8, 18, 32, 32, 9, 2

Appendix H
Worldview, Apologetics, and Evidence

From the first chapter of this textbook, we have tried to show the importance and usefulness of physical science. When viewing the world through the lens of the Christian Bible, we see science as an important tool that helps us to become wise stewards of God's world. But many people today use science to discredit the Bible and even to question the existence of God! To many Christians, science seems to be a great evil that seeks to defeat our faith in God and His Word.

In this appendix, we will briefly look at science, what it can and cannot do, and ways that we can use and misuse science in defense of our faith.

What's a Worldview? Why Is It Important?

Every person who has ever lived had and has a worldview. *You* have a worldview. Your worldview is probably the most important aspect of your character. It determines how you interpret all sensory and intellectual information that you receive from the world around you—what you see, hear, read, and touch. Your worldview defines your expectations about how things should be, your goals in life, what you believe and don't believe, and even what's right and wrong.

Every person's worldview is different, though worldviews can be similar in many ways. We develop our worldview starting very early in life. Certain parts come from our parents and interactions with brothers and sisters (for example, obedience, family loyalty, love, sharing, personal ownership). Some parts come from early school (view of authority figures, public behavior, non-family loyalty, group interaction). As we grow older and become aware of the greater world, listen to news reports, and begin to think about growing up, we develop definite ideas about how people and organizations act. We learn about government, wars, the history of the earth, and human history. We understand that people are expected to work as adults. Most importantly, our religious training, from whatever source, defines for us what is true and what is valuable. This training shapes every part of our lives and gives us our sense of right and wrong.

As we mature, we learn to reason, using our minds to address questions and problems that we face in life. Most people are not aware that our reasoning is built on many basic assumptions that come from our worldview. These basic assumptions are called *presuppositions*, which is just a formal word that means "assumed beforehand." They are assumptions that we believe to be true whether we are aware of them or not. You likely assume that parents should take care of their children, that teachers and textbooks don't knowingly provide false information, that people should be properly dressed in public, and that if you hold a job, that you should be paid for your work. These are examples of presuppositions.

The Christian Worldview

In 2 Corinthians 5:17, the Bible tells us that when a person becomes a Christian, he becomes a *very* different person. A drastically different worldview framework accompanies salvation. A new believer begins to build on that framework by reading the Bible with understanding. He receives knowledge and spiritual guidance from preachers and teachers, and the Holy Spirit helps him to understand the great truths of Scripture. A growing Christian who is maturing in faith learns how to interpret the various kinds of passages found in the Bible. Some books are mainly historical, such as Genesis and the book of Acts. Some reveal spiritual truths through poetic imagery, such as Psalms and Proverbs. Some are sources of Christian wisdom, doctrine, standards of behavior, and God's plan for future history, such as the prophetical books and the New Testament epistles. A Spirit-filled understanding of the Bible supplies us with presuppositions, such as:

- God is the preeminent being in the universe—the Creator, the source of all truth, our Judge, and our Savior.
- The earth, the universe, and life were created supernaturally, only about 7000 years ago.
- Humans are a special form of life, created separately from animals in the image of God.
- Disasters, evil, and death in the world originated with Adam's sin and from God's curse on his sin as just punishment.
- The earth's surface and climate were fundamentally changed during and after a global flood about 5500 years ago.
- Distinct people groups and their languages resulted from the dispersion of Noah's descendants following the confusion of languages at Babel several hundred years after the Flood.
- The Bible is the source of absolute standards of human behavior that can't be set aside by human laws.
- God established the institutions of the family, human government, and the church.
- This life is not the only period of human existence. An eternal, physical existence awaits everyone—everlasting, blessed life with God or eternal, conscious torment apart from Him—so Christians must be active in this life leading others to eternal life through Jesus Christ.

A Christian worldview is a central part of what a Christian is. It's not an add-on or something that can be ignored when convenient. A Christ-centered person cannot *not* express his Christian worldview.

All other possible worldviews differ from a Christian worldview. They are based on different cultural influences, educational principles, religions, and philosophies. It shouldn't seem strange that there are overlapping agreements among worldviews. We are, after all, made in God's image and bear His law written on our hearts (Rom. 2:15). But it also shouldn't surprise us when we realize that all other worldviews are, to some degree, contrary or even hostile to the Christian worldview. When someone disagrees with a Christian who holds to a Christian worldview about an important issue of life, it usually occurs along worldview lines.

In school, you will see these disagreements most clearly in certain branches of science, especially the earth sciences and biology. The age of the earth and the idea of a global flood are the most controversial issues in the earth sciences, because they directly influence the interpretation of basic scientific evidence. Secular biology presumes that nearly every feature of living things is the result of random mutations and natural selection (evolution) over hundreds of millions of years. The Christian worldview, on the other hand, presumes that everything was specially created over a six-day period of time. There is no way to avoid worldview conflicts in these areas since these presuppositions are incompatible.

Christian Apologetics

Differences in worldviews can dissolve into heated arguments. But should Christians involve themselves in such disputes? The Apostle Peter reminds us that we should always be ready to give a reason for our faith with respect and gentleness (1 Pet. 3:15). The formal word for the act of giving an account or explanation of the Christian faith is *apologetics* (from the Greek word for "speaking in defense").

Christians have often used apologetics:

- to explain to others the reasons for their beliefs and practices, as the apostles did through their epistles and when standing before various authority figures;
- to try to persuade a non-Christian to become a Christian;
- to try to persuade another professing Christian who holds to a non-biblical position to change his mind; and
- to encourage believers that their beliefs are reasonable.

Like any kind of effective argumentation, apologetics relies on persuasively presenting credible evidence in a logical presentation. Christian apologetics draws mainly from:

- biblical evidence that comes from properly interpreted scriptural passages;
- non-biblical evidence (historical, scientific, etc.);
- a well-constructed argument that follows accepted rules of logic and avoids logical fallacies; and
- a respectful and loving appeal to understand, to change one's belief, or to be strengthened in one's faith.

The Limits of Evidence as an Apologetic

A key part of convincing someone that your position is correct is presenting evidence to support your position. But this is where worldviews come into the picture. *All* evidence is filtered and interpreted through one's worldview. With the exception of evidence from the Bible (which Christians hold to be perfect and without error), no other kind of evidence is truly objective or certain, not even scientific evidence!

Each person's worldview will guide his interpretation of the significance, or even the validity, of the evidence. And no one is absolutely logical or flawless in his reasoning. Our fallen minds can misinterpret, misunderstand, or draw illogical conclusions from evidence. Normally, if evidence doesn't fit into our view of things or makes us uncomfortable at some deep level in our minds, we either reject it or explain it away.

When using non-biblical evidence to present arguments for what we believe, we must be careful about its quality. Some non-biblical evidence that we might have used at one time is no longer considered valid. For example, many creationary scientists in the 1970s argued for a young creation because of the lack of a deep dust layer on the moon. Today, we realize that this evidence isn't very useful because our understanding of how lunar dust accumulates has changed.

On the other hand, high-quality non-biblical evidence can be effective in getting unbelievers to question their presuppositions. For instance, geologists can describe a single stratum of sedimentary rock at least 1000 miles across covering portions of the central United States. It is distinctive because it's mostly sandstone and shows evidence of deposition under rapidly flowing water (cross-bedding). Most geologists interpret this feature as sediments gradually deposited by a large system of meandering rivers during a vast period of time, but the evidence better fits deposition by a broad, fast-moving flood all at once in a short period of time.

Some disagreements resulting from differences between Christian and secular worldviews don't have clear non-biblical answers. The "starlight and time" problem in cosmology is an excellent example of this. If the universe is only 7000 years old, then how can we see light from objects that are more than 13 billion light-years away if the speed of light is unchanging? And even disciplines like physics and chemistry, which are generally considered more objective and straightforward than cosmology or biology, don't produce truth, but simply workable models of the physical world. The evidence they supply is not worldview neutral anyway.

When trying to change the heart of an unbeliever through apologetics, the most concrete, impressive evidence will not work on its own (Luke 16:27–31). Only the Holy Spirit through the presentation of Scripture convinces and persuades (Rom. 10:8–17). Piling up non-biblical evidence will save no one, and the unbeliever can make his own pile of evidence to counter yours. The relative heights of the piles cannot determine truth!

Even a believer who bases some of his convictions on non-biblical evidence is on shaky ground. A Christian's life ultimately must rest on faith, not sight (2 Cor. 5:7). Belief based on non-biblical evidence or improperly interpreted biblical evidence is unstable because the view of such evidence changes with time. For example, since the late 1800s, many Christians believed that the Canopy theory was true. But recent atmospheric modeling and evaluation of the theory by qualified creationary atmospheric scientists have concluded that a vapor canopy could not possibly have been the source for the rains during the Flood, and would have made life on Earth impossible before the Flood. If a person bases his faith on evidences such as these, he could easily become disillusioned and perhaps bitter if the evidence is later shown to be invalid. This is particularly true for professing believers who have no root of faith (see Mark 4:16–17). Only sound biblical evidence obtained from proper *hermeneutics* (methods of interpreting the Bible) and accompanied by the ministry of the Holy Spirit will convince and persuade.

A Proper View of Evidence as an Apologetic

God is able to use all sorts of means to convince an unbeliever. And evidence and logical argument are two of those potential means—means Paul himself employed. He pointed to eyewitnesses (1 Cor. 15:5–6), and he used reason and persuasion (Acts 19:8–9).

But Paul knew, as we should, that no amount of evidence, on its own, is sufficient to convert someone. Only God's powerful Holy Spirit can do that, because man's ultimate problem is not in his brain or in his science textbook. Humanity's shared problem is that we are all born loving all sorts of things other than God.

Even when you do appeal to evidence, don't let your debate stay there. Use evidence as a shovel to dig deeper, to question the other person's presuppositions. The moral law is a good way to do that. Secular scientists have recently claimed that evolution can account for rise of love, justice, and self-sacrifice as important emotions that help survival of the species. But how do these work with the underlying evolutionary assumption that natural selection acts through untold millennia of ape-eat-ape survival of the fittest? This contradiction isn't easily explained, but the evolutionist will come up with one because he looks around him and sees that love, justice, and self-sacrifice exist. They *must* be useful for survival.

But this is a non-answer. This is retreating to one's presupposition that naturalistic, evolutionary explanations are the only ones permitted. There is and can be no "evidence" that this evolutionary explanation is accurate. Any evidence for it only counts as evidence given the presuppositions of an evolutionary worldview.

So (graciously) push back: "Can you really live as if love and justice aren't *right*—just useful for preserving the species? How can a worldview based on natural selection and random chance account for our shared human conviction that some things are right and others wrong?" This would mean that an individual or a government can decide on their own that love of one's neighbor or justice in some circumstances doesn't help survival of the human species.

A Christian worldview accounts for love and justice because their source is an absolute, personal, triune God. Love exists because it is part of God's nature as a three-person being to love—the Father loved the Son, the Son the Father, the Spirit the Son, from eternity past.

All of life is a journey of discovery. You're always learning and growing (or you should be!). You may grow to find that some of your understanding of Scripture was wrong or, perhaps, incomplete. This is part of life in a fallen world. Even dedicated Christians are still finite. They can't assume that they understand the Bible perfectly. But you can, with the Spirit's help, obey what you do understand and seek to refine that obedience and understanding over time.

When building up a believer's faith and tearing down strongholds of unbelief (2 Cor. 10:4–5), begin with careful, consistent interpretations of Scripture. Then show how careful, consistent interpretations of scientific or cultural evidence hold to these biblical truths. Remember, however, that presuppositions play a role in everyone's evaluation of the evidence. You can't test the big bang *or* a 7000-year-old earth with the scientific method. So biblical presuppositions must interpret all the evidence you encounter. Science and human reasoning must never be your starting point; the fear of the Lord is the beginning of knowledge.

GLOSSARY

Glossary entries in **bold** text are definitions of vocabulary terms listed at the end of each chapter. Entries in ***bold italic*** text are important scientific concepts and terms that will help you better understand the subject. Cross-references listed at the end of entries link to other terms in the Glossary.

A

absolute pressure **(155)** A pressure reading referenced to a vacuum. In contrast, see *gauge pressure.*

absolute temperature (T) **(140)** The temperature of a system referenced to absolute zero. The scale's degree is the kelvin (K). See also *Kelvin scale*. See *temperature (t); absolute zero; degree (°).*

absolute zero **(187)** The lowest theoretical temperature; defined as 0 K (about −273.15 °C or −459.67 °F). Physicists believe that all motion of matter ceases at this point. Laboratory temperatures have been attained to within billionths of a degree of this point. See *temperature (t).*

accelerated reference frame **(77)** A reference frame that is not stationary or moving at a constant velocity. As a result, systems do not appear to obey Newton's laws of motion without the invention of imaginary forces or *pseudoforces*. A *rotational reference frame* is a common example. See *frame of reference; laws of motion; rotational motion.*

acceleration (*a*) (84) The rate of change of the velocity or speed of a system during a time interval. It may be a vector or scalar quantity. See *interval (Δt); velocity (**v**); speed (v); vector; scalar.*

acceptable value **(56)** The value of a measureable dimension that represents its "actual value" for the purposes that the measurement is intended. It is impossible to determine the "actual value" for any given measurement. Also called the *accepted value.* See *dimension; measurement.*

accuracy (57) An assessment of the measurement error. An indication of how close a measurement is to its acceptable value. A smaller error means a more accurate measurement. See *measurement; error; acceptable value.*

acid (484) According to the Arrhenius model, any substance that produces hydronium ions (H_3O^+) in an aqueous solution. According to the Brønsted-Lowry model, any substance that donates hydrogen ions (protons, or H^+). Other definitions of acids exist. See *Arrhenius model; aqueous; Brønsted-Lowry model; hydronium ion.*

acidic **(497)** Having a pH value less than 7. See *pH scale.*

acidity **(497)** A measure of the concentration of the hydronium ions in solution. Units are in pH for concentrations greater than 1×10^{-7} mol/L. See *concentration; hydronium ion; pH.*

acoustic energy (115) The transmission of energy through matter by the periodic longitudinal motion of particles (mechanical waves). See *longitudinal wave; periodic motion.*

acoustic spectrum (280) The continuum of all possible sound wave frequencies from infrasonic to ultrasonic. See *continuum; sound; frequency (f); infrasonic wave; ultrasonic wave.*

acoustic synthesizer **(290)** An electronic device capable of artificially producing most instrumental sounds. See *acoustics; sound.*

acoustics (282) The scientific study of sounds and how they are generated, transmitted, and heard. Acoustics also refers to how a room or building affects sounds. See *sound.*

action-reaction principle **(97)** See *law of action-reaction.*

activation energy **(449)** The minimum energy required for a chemical reaction to take place. Exothermic reactions typically have low activation energies; endothermic reactions have much higher. See *chemical reaction; exothermic reaction; endothermic reaction.*

additive primary color (333) One of the three primary hues sensed by the human visual system (red, green, or blue) that when mixed in various ratios can produce all other colors of the visible spectrum. See *primary hue; visual system; color.*

aerodynamics **(163)** The physics of objects, such as airfoils and aircraft, moving through gases. See *airfoil; gas.*

aerogel **(194)** A low density solid material that begins as a gel before the liquid phase is removed. Aerogels are exceptionally good insulators and have great promise in many applications. See *gel; phase (mixture); insulator.*

aerosol **(459)** A heterogeneous mixture consisting of tiny solid particles or liquid drops dispersed in a gas. See *heterogeneous mixture; dispersed phase.*

airfoil **(166)** A streamlined shape designed to produce lift as it moves through the air or as air moves past it. An equivalent shape designed to work in liquids is a *hydrofoil*. See *lift.*

alchemy (**383**) An ancient form of chemistry that also included philosophy, religion, observational science, art, mysticism, and ritual. It was often associated with occult practices and efforts to change base materials into gold or discover elixirs of youth. Practitioners were called *alchemists*.

alkali metal (**397**) An element in Group 1A of the periodic table; a strongly reactive metal with an oxidation number of +1. See *element*; *group*; *periodic table of the elements*; *reactive*; *oxidation number (ON)*.

alkaline (**497**) Having a pH value greater than 7; basic. See *pH scale*.

alkaline-earth metal (**397**) An element in Group 2A of the periodic table; a very reactive metal having an oxidation number of +2. See *element*; *group*; *periodic table of the elements*; *reactive*; *oxidation number (ON)*.

alkalinity (**497**) A measure of the concentration of the hydronium ions in solution. Units are in pH for concentrations less than 1×10^{-7} mol/L. See *concentration*; *hydronium ion*; *pH*.

alloy (**397**) A solid mixture of two or more elements, at least one of which is a metal. Alloys can be homogeneous (a single phase) or heterogeneous (multiple phases). See *metal (chemistry)*; *homogeneous mixture*; *heterogeneous mixture*; *phase (mixture)*.

alpha (α) particle (**371**) A high-mass positive particle emitted during nuclear decay consisting of two protons and two neutrons. It is a completely ionized helium nucleus. See *nuclear decay*; *proton*; *neutron*; *ion*.

alpha decay (**372**) A nuclear decay resulting in the emission of an alpha particle. It results in a decrease of the mass number of the remaining nucleus by 4 and the atomic number by 2. See *nuclear decay*; *alpha (α) particle*; *mass number (A)*; *atomic number (Z)*.

alternating current (AC) (**237**) An electrical current that periodically changes direction. In the United States, the frequency for a complete AC cycle is 60 Hz. Most of the countries in the world use 50 Hz AC.

amalgam (**464**) A solution of two metals, one of which is often liquid mercury. Amalgams are usually very malleable and easy to work. See *solution*; *malleable*.

amorphous solid (**31**) A solid substance whose particles are arranged in a random, disorganized way. Wax and common forms of glass are amorphous solids.

ampere **(A)** (**219**) The SI base unit of electrical current. The amp is often defined as 1 A = 1 C/s for practical purposes. See Appendix B for the formal SI definition. See also *electrical current (I)*.

amplification (**290**) The mechanical or electronic process of making a sound louder. See *process*; *loudness*; *mechanical amplification*; *electronic amplification*.

amplitude (**256**) The maximum displacement from the rest position of a system experiencing periodic motion. It is half the distance between the extreme positions of the system. See *periodic motion*; *rest position*.

amplitude modulation **(AM)** (**315**) A method of transmitting radio communication by varying (or *modulating*) the amplitude of a constant-frequency radio wave. Contrast with *frequency modulation (FM)*. See *amplitude*; *frequency (f)*; *radio wave*.

analytical chemistry (**438**) The area of chemistry that deals with determining the composition of substances and developing methods to do so. See *composition*.

aneroid barometer (**156**) A mechanical instrument for measuring atmospheric pressure that consists of a sealed flexible can that expands and contracts with changes in air pressure, moving a pointer to indicate pressure on a scale. See *instrument*; *pressure (P)*.

angle of incidence (**337**) The angle between the incident ray and an imaginary line normal to the reflective surface at the point of incidence. See *incident ray*; *normal*; *point of incidence*.

angle of reflection (**337**) The angle between the reflected ray and an imaginary line normal to the reflective surface at the point of incidence. See *reflected ray*; *normal*; *point of incidence*.

angle of refraction (**341**) The angle between the refracted ray and an imaginary line normal to the boundary between refractive media at the point of incidence. See *refracted ray*; *normal*; *medium*; *point of incidence*.

angular momentum (**121**) The momentum of a rotating system. It is proportional to the speed of rotation and a factor relating to the arrangement of the mass of the system around the axis of rotation. Compare with *linear momentum*. See *momentum (p)*; *system*; *rotational motion*; *axis*.

anion (**27**) A charged atomic or molecular particle with a net negative charge. See *atom*; *molecule*.

anode (**429**) In an electrical circuit that includes an electrolytic solution, the anode is the electrode toward which dissolved anions move. In electronics, a positively charged electrode. Compare with *cathode*. See *electrical circuit*; *electrolytic solution*; *anion*.

antenna (**314**) In physics, any device designed to radiate or receive electromagnetic energy in the radio-frequency spectrum. See *electromagnetic (EM) energy*; *radio-frequency spectrum.*

anthocyanin **(500)** A natural plant pigment that changes color in response to pH. The anthocyanin in red cabbage is a common natural pH indicator. See *pigment*; *pH*; *pH indicator.*

antiparticle **(363)** An elementary particle that has the same mass and basic characteristics as its corresponding "normal" particle. Other key properties, such as charge and spin, are opposite. Particles and their antiparticles annihilate each other when they meet. See *elementary particle.*

aqueous **(448)** Having to do with water solutions. See *solution.*

arc-fault circuit interrupter **(AFCI) (224)** An overcurrent protection device that opens a switch when it detects the pulsing, high-current conditions associated with arcing. It often includes a microprocessor-controlled sensor and can be reset after the fault is corrected. See *overcurrent protection*; *arcing.*

Archimedes's principle (158) The buoyant force exerted by a fluid on an immersed object is equal to the weight of the fluid that the object displaces. See *buoyant force*; *fluid.*

arcing **(224)** (ARK ing) The passage of electrical current through a cloud of ionized gases between electrical conductors. Often melts the conductors and spatters hot, molten metal. If the current is not turned off, arcing can result in a fire as nearby combustible material ignites. See *electrical current (I)*; *ionization*; *gas*; *conductor.*

Arrhenius model (485) One of the first useful models of acids and bases in which acid compounds form hydronium ions (H_3O^+) and bases form hydroxide ions (OH^-) in aqueous solutions. Compare with *Brønsted-Lowry model.* See *acid*; *base*; *hydronium ion*; *hydroxide ion.*

association **(493)** In chemistry, the reorganization of dissolved ions from solution into a solid ionic crystal lattice. Compare with *dissociation.* See *ion*; *solution*; *crystal lattice.*

astrology **(100)** An occult practice of attempting to understand the influences of celestial objects on events on Earth and predicting the future on the basis of the positions of the sun, moon, and planets in the sky. While it was the basis for modern astronomy, God repeatedly spoke of the ineffectiveness of astrological predictions in the Old Testament (cf. Isa. 47:13).

atmospheric pressure **(30)** Air pressure at a location on the earth's surface or in its atmosphere. As a standard condition, refers to average atmospheric pressure at sea level, or about 1013 mb or 14.7 lb/in.^{2} See *pressure (P)*; *standard conditions.*

atom **(26)** The basic particle of matter from which all other matter is constructed. It consists of protons, electrons, and (usually) neutrons.

atomic bomb **(376)** A powerful weapon based on an uncontrolled nuclear fission chain reaction. Also called a *fission bomb*, *A-bomb*, or *nuclear bomb.* The first and only atomic bombs ever used in combat ended the Pacific War in World War II. See *nuclear fission*; *chain reaction.*

atomic clock **(273)** A high-precision electronic timekeeping device that uses some aspect of the vibrations of atoms to subdivide time into extremely small intervals. Errors are typically given in microseconds per year. See *precision*; *vibration*; *time (t)*; *interval (Δt).*

atomic mass (366) The mass of an atom or an element expressed in atomic mass units; for an element, the weighted average of all naturally occurring isotopes of the element. See *atomic mass unit (u)*; *isotope.*

atomic mass unit **(u) (366)** A relative unit of mass used for measuring the mass of particles of matter in physics and chemistry. It is equal to 1/12 the mass of a carbon atom with six protons and six neutrons (carbon-12). The term is abbreviated "amu" but the dimensional unit is "u," which comes from the official name *unified atomic mass unit.*

atomic number (*Z*) (364) The number of protons in the nucleus of an atom, which determines the element identity of the atom. See *proton*; *nucleus*; *element.*

atomism **(22)** See *particle theory.*

autonomous undersea vehicle **(AUV) (14)** A small submersible unmanned vehicle carrying various kinds of sensors, including sonar, lights, and cameras, that can be programmed to follow preset tracks and depths for underwater research. AUVs are not connected in any way with surface vessels or shore facilities. Scientists recover data by periodic radio transmissions from the AUV or by recovery of the vehicle itself.

average acceleration **(a_{avg}) (84)** The average rate of change of motion over a time interval when acceleration is not constant. Since system speed, direction, or both may change during the interval, determining average acceleration can be complicated unless direction is held constant. See *motion*; *interval (Δt)*; *acceleration (a)*; *system*; *speed (v).*

average speed **(v_{avg}) (81)** The average rate of motion of a system during a time interval that accounts for variations in speed. It is a positive scalar quantity. See *speed (v)*; *interval (Δt).*

Avogadro's number **(473)** The number of particles in a mole of matter. Named after Amedeo Avogadro, who first suggested the molar concept. Equal to about 6.02×10^{23} particles. See *particle*; *mole (mol).*

axis **(135)** An imaginary line around which a system rotates or orbits. The center of rotational motion. Compare with *coordinate axis.* See *system*; *rotational motion.*

axle **(140)** The support for a wheel around which it rotates. The rotational axis of a wheel and axle system. See *rotational motion*; *axis*; *wheel and axle*; *system.*

B

balance **(64)** A simple instrument that compares a known mass to an unknown mass using a lever arm. When balanced, the weight of the known mass equals the weight, and thus the mass, of the object being weighed. Also called a *mass balance.* See *instrument*; *mass (m)*; *torque arm (ℓ)*; *weight (w).*

balanced force **(94)** Any force that in combination with other forces acting on the same system produces a zero net force on the system. See *force (F)*; *net force.*

ball lightning **(212)** A rare form of lightning that appears as a small glowing ball of plasma, moving relatively slowly above the ground or through open or closed doors and windows. A small percentage end with a loud bang. Several hypotheses have been offered to explain them. See *lightning*; *plasma.*

base **(485)** According to the Arrhenius model, any substance that produces hydroxide ions (OH^-) in an aqueous solution. According to the Brønsted-Lowry model, any substance that accepts hydrogen ions (protons, or H^+). Other definitions of bases exist. See *Arrhenius model*; *hydroxide ion*; *aqueous*; *Brønsted-Lowry model.*

basic **(497)** See *alkaline.*

battery **(217)** A source of electrical potential consisting of one or more *voltaic cells* (electrochemical cells) connected in series. See *electrochemical cell.*

beam **(336)** A continuous stream of light photons moving in the same direction. See *photon.*

beat **(271)** A large-amplitude, slowly varying wave that is the result of interference between two continuous waves having slightly different wavelengths. It is especially noticeable in acoustic wave interference. See *amplitude*; *interference*; *wavelength (λ)*; *acoustic energy.*

Bernoulli's principle **(165)** States that the total energy (represented by kinetic energy, potential energy, and pressure) for a confined ideal fluid flowing through a pipe is conserved at all locations within the pipe. See *energy*; *kinetic energy (KE)*; *potential energy (PE)*; *ideal fluid*; *law of conservation of energy.*

beta (β) particle **(371)** An electron that is emitted during a nuclear decay when a neutron changes into a proton. See *nuclear decay*; *electron*; *proton*; *neutron.*

beta decay **(372)** A nuclear decay resulting in the emission of a beta particle. The remaining nucleus has the same mass number as before the decay, but its atomic number is reduced by 1. See *nuclear decay*; *beta (β) particle*; *mass number (A)*; *atomic number (Z).*

bias **(7)** An inclination one has about an idea after thinking about its pros and cons. Also a built-in error in an instrument. See *systematic error.*

binary compound **(442)** A compound composed of only two different elements. See *compound*; *element.*

binary system **(245)** A two-digit number system consisting of 0s and 1s. It is the basis for all computer programming languages and digital data.

binding energy **(117)** The nuclear energy required to bind together the protons and neutrons within the nucleus of an atom. See *nuclear energy*; *proton*; *neutron*; *nucleus.*

bioluminescence **(331)** An organic form of cold light produced by animals and plants. Compare with *chemiluminescence.* See *cold light.*

block and tackle **(143)** An arrangement of fixed and movable pulleys connected by ropes or cables that provides a mechanical advantage to do mechanical work. Also called a *pulley gang.* See *pulley*; *mechanical advantage (MA).*

blue jet **(212)** Blue plasma discharges that project upward from the tops of thunderheads that are associated with heavy hail. These are not necessarily triggered by lightning discharges.

Bohr model **(367)** See *planetary model.*

boiling **(42)** Rapid vaporization that occurs when a liquid's vapor pressure exceeds atmospheric pressure and the static pressure in the liquid, causing it to form vapor bubbles. Boiling may occur anywhere in the liquid. See *vaporization*; *vapor pressure.*

boiling point **(42)** The temperature and pressure at which boiling occurs. See *temperature (t)*; *pressure (P)*; *boiling.*

boiling-point elevation **(474)** The number of degrees the boiling point of a solvent is increased per mole of dissolved solute particles. It is a colligative property of the solvent. See *degree (°)*; *boiling point*; *mole (mol)*; *solute*; *colligative properties*; *solvent.*

bond **(24)** See *chemical bond.*

bond character **(426)** The type of bond predicted by the difference in electronegativity numbers (ΔEN) between the atoms making the chemical bond. Large ΔENs between metal and nonmetal elements generally produce ionic bonds; zero and small to medium ΔENs between nonmetals result in covalent bonds; very small ΔENs between metal atoms generally produce metallic bonds. See *chemical bond; electronegativity; ionic bond; covalent bond; metallic bond.*

boundary layer effect **(167)** The region of no-flow to low-flow near a surface as a fluid flows by the surface. Fluid particle speed rapidly decreases from full-flow speed to near zero next to the surface. The effect depends on particle temperature, density, and viscosity.

Bourdon tube **(156)** The C-shaped tube that forms the main pressure-sensing element of a mechanical pressure gauge. As pressure increases, the tube tends to unbend, actuating a mechanical linkage to a needle or other kind of display system to indicate pressure.

Boyle's law **(170)** The volume of a fixed quantity of a confined gas is inversely proportional to its pressure when its temperature is held constant. See *volume (V); pressure (P); temperature (t).*

brass **(384)** A family of alloys made from copper and zinc. In some versions of the Bible, the term could refer either to pure copper or to bronze. See *alloy; bronze.*

Brønsted-Lowry model **(486)** A model of acids and bases (more complete than the Arrhenius model) that defines acidity and alkalinity on the basis of hydrogen ion transfer. Acids *donate* hydrogen ions (H^+, or protons), and bases *accept* hydrogen ions. Compare with *Arrhenius model.* See *acid; base; acidity; alkalinity; ion.*

bronze **(384)** A family of alloys made from copper and tin, sometimes with other trace elements. See *alloy.*

Brownian motion **(26)** The microscopic, random jostling of suspended matter due to the collisions of innumerable gas or liquid particles in which the matter is suspended.

brush (electrical) **(239)** In a DC machine, one of the many curved blocks of conductive material that ride on a rotor commutator to connect the rotor coils to the external electrical circuit. See *direct current; rotor; commutator.*

buoyancy **(158)** The property describing an object's tendency to float when immersed in a fluid. *Positively buoyant* objects float because their density is less than that of the fluid; *neutrally buoyant* objects neither rise nor sink because their densities precisely equal that of the fluid; *negatively buoyant* objects sink because they are denser than the fluid. See *fluid; density (ρ).*

buoyant force **(158)** The lifting force exerted by a fluid against gravity on an object immersed in the fluid. See *gravity; fluid.*

C

calibration **(500)** The process of adjusting an instrument to its design accuracy. This is done by comparing its readings of a known standard to the acceptable value(s) of a prime standard. Also means *standardization.* See *process; accuracy; acceptable value; standard.*

caloric theory **(179)** A model of heat as a material fluid that flows from hot to cold objects. When scientists discovered that it could not explain certain observations, they replaced it with the kinetic-molecular model of thermal energy and heat. See *kinetic-molecular theory; heat (Q); thermal energy.*

calorimeter **(195)** A chamber insulated from its surroundings that is used to measure heat transfer between systems contained within it. See *heat (Q).*

candela **(cd)** **(328)** The SI base unit of light intensity. See Appendix B for a formal definition. See also *intensity (light).*

capacitor **(213)** A modern charge-storage device used in electrical and electronic circuits, consisting of two or more conductive plates or sheets separated by an insulator.

carbon-14 dating **(375)** A radiometric dating technique that attempts to assign an age to ancient organic materials on the basis of the measured half-life of the radioactive isotope carbon-14 and an assumed amount of the isotope in the original material. See *radiometric dating technique; organic; half-life; isotope.*

carbon family **(399)** Group 4A of the periodic table. The elements have variable chemical activity and usually have an oxidation number of +4 or −4, depending on the other elements they bond with. See *element; family; group; periodic table of the elements; oxidation number (ON).*

Carnot cycle **(181)** The ideal relationship between the changing temperature and pressure of the working fluid within a heat engine as it is doing work. Developed by the French engineer Sadi Carnot in 1824, it defines the maximum efficiency that a heat engine can have as it operates between two temperatures. See *temperature (t); pressure (P); heat engine.*

catalyst **(449)** A substance that can change the rate of a chemical reaction without participating in the reaction. Typically, catalysts lower the activation energy of the reaction. Biological catalysts are called *enzymes*. See *chemical reaction*; *activation energy*.

cathode **(429)** In an electrical circuit that includes an electrolytic solution, the cathode is the electrode toward which dissolved cations move. Compare with *anode*. In electronics, a negatively charged electrode. See *electrical circuit*; *electrolytic solution*; *cation*.

cathode ray tube **(CRT) (215)** A hollow, vacuum-filled and sealed glass device containing electrical plates with opposite charges at high voltages. Three "electron guns" project electrons toward the positively charged screen. Magnetic coils deflect the beams to scan the screen horizontally and vertically to form an image on a phosphorescent coating. The CRT formed the basis for older TV screens and computer monitors. Almost all such devices have been replaced by flat-screen LCD or plasma screens.

cation **(27)** A charged atomic or molecular particle with a net positive charge. See *atom*; *molecule*; *particle*.

Celsius scale (186) A temperature scale with fiducial points at the freezing point (0 °C) and the boiling point (100 °C) of pure water at 1 atm of pressure. See *temperature (t)*; *fiducial point*; *freezing point*; *boiling point*.

center of mass **(260)** The geometric point within a system where we can assume the system's entire mass is located. A valuable and common concept in mechanics. See *system*; *mechanics*.

chain reaction **(376)** A nuclear fission process in which neutrons produced by fission trigger more fissions in a continuous way. Uncontrolled chain reactions are the basis for atomic bombs and nuclear reactor accidents. Controlled chain reactions are the basis for nuclear power plants. See *nuclear fission*; *process*; *atomic bomb*; *nuclear reactor*.

charge (electrical) **(205)** A concentration of at least one but typically many fundamental electrical charges. Represents a distinct source for the electrostatic force. See *fundamental electrical charge (e)*; *electrostatic force*.

charge coupled device **(CCD) (317)** A light-sensitive semiconductor surface used in cameras in place of photographic film. It transforms the pattern of light received on its surface into pixels identified by binary code that can be used by digital processors in cameras to create photographs. See *semiconductor*; *pixel*; *binary system*.

Charles's law (172) The volume of a fixed quantity of a confined gas is directly proportional to its absolute temperature when its pressure is held constant. See *volume (V)*; *gas*; *absolute temperature (T)*; *pressure (P)*.

chemical bond **(24)** Any permanent attraction between atoms resulting from the sharing of valence electrons to a greater or lesser extent. Chemical bonds cannot be broken by simple physical changes. See *valence electron*; *physical change*; *covalent bond*; *ionic bond*; *metallic bond*.

chemical change **(36)** See *chemical reaction*.

chemical energy (116) The potential energy stored in the chemical bonds between atoms that is released or absorbed during chemical reactions. See *potential energy (PE)*; *chemical bond*; *chemical reaction*.

chemical equation (446) An equation that symbolically represents the identity and amounts of reactants and products that take part in a chemical reaction. A *balanced* chemical equation enforces the law of conservation of matter—it has the same number and kinds of atoms on both the reactant and product sides. See *reactant*; *product*; *chemical reaction*; *law of conservation of matter*.

chemical formula (437) A shorthand way to represent the identity and the relative amounts of elements in a compound. See *element*; *compound*.

chemical property **(36)** A property of a substance that describes how its chemical identity changes in the presence of another substance or under certain conditions. See *property*.

chemical reaction (445) Any change in a substance that alters its composition. See *composition*.

chemical symbol **(386)** A unique symbol used to represent an element, consisting of one or two letters derived from the element's current or Latin name. See *element*.

chemical thermodynamics **(452)** The study of how energy moves and changes forms during chemical reactions. See *energy*; *chemical reaction*.

chemiluminescence **(331)** An artificial source of cold light. Compare with *bioluminescence*. See *cold light*.

chemistry **(15)** The study of the structure, composition, and properties of matter, and how matter acts in the presence of other matter. See *composition*; *matter*.

circuit breaker **(224)** An overcurrent protection device that opens a switch when circuit current exceeds a certain value. It can be reset after the fault is corrected. See *overcurrent protection*; *switch*; *fuse*.

Coandă effect (166) The tendency of a fluid flowing past a curved surface to follow the surface. An important contributor to lift in an airfoil. See *fluid*; *lift*; *phenomenon*; *airfoil*.

coefficient **(446)** A whole-number multiplier appearing in front of a chemical formula in a balanced chemical equation. It indicates the proportion of each reactant or product needed to ensure the conservation of matter during a chemical reaction. When no coefficient appears in front of a chemical formula, a coefficient of 1 is assumed. See *chemical formula*; *chemical equation*; *reactant*; *product*; *law of conservation of matter*; *chemical reaction.*

coherent light (330) In-phase, monochromatic light waves from a single source. See *in phase*; *monochromatic*; *laser.*

cold light (330) Visible light produced by chemical reactions at temperatures far below those required for incandescence. See *incandescence.*

colligative properties (473) A physical property of a solvent that depends on the concentration of the solute particles and not the kind of solute. Common colligative properties are freezing-point depression, boiling-point elevation, and osmotic pressure. Also called *concentration effects.* See *property*; *solvent*; *solute*; *freezing point depression*; *boiling point elevation*; *osmotic pressure.*

collision theory **(449)** A particle model of chemical reactions. The probability of two substances reacting is related to their orientation and energies when the particles collide. See *chemical reaction.*

colloid (460) A heterogeneous fluid mixture in which the particles of the dispersed phase are between 1 nm and 1 µm in diameter. *Colloidal dispersions* exhibit the Tyndall effect. See *heterogeneous mixture*; *dispersed phase*; *Tyndall effect.*

color **(326)** A property of light frequencies lying within the visible portion of the electromagnetic spectrum as perceived by the human visual system. We visually differentiate frequencies as differences in color. See *property*; *electromagnetic (EM) spectrum*; *visual system.*

color space **(336)** All possible colors that can be produced by the combination of a specific set of primary colors—the color system. See *color*; *primary hue*; *color system.*

color system (336) The primary colors that, when combined, create a color space for a particular purpose. The red-green-blue (RGB) color system produces all the colors displayed on an RGB monitor, for instance. See *color*; *color space.*

combustion **(452)** An exothermic reaction that involves burning, where a substance combines with oxygen to form various products and heat. See *exothermic reaction*; *chemical reaction*; *product*; *heat (Q).*

common logarithm **(499)** A logarithm of a number in base 10. See *logarithm.*

commutator **(239)** A device that converts the alternating current inside a DC generator to a DC output, or converts a DC supply to an alternating current inside a DC motor. It consists of segments of metal mounted on the rotor shaft that are connected to the electromagnet coils mounted on the rotor of the machine. Brushes conduct electricity to or from the rotating commutator. See *alternating current (AC)*; *direct current (DC)*; *generator (electrical)*; *motor*; *electromagnet*; *rotor*; *brush (electrical).*

complementary color **(334)** A color produced by the mixing of any pair of additive or subtractive primary colors. See *color*; *additive primary color*; *subtractive primary color.*

composite solid **(32)** A solid material made of two or more substances having different particle structures. Examples include wood, bone, and carbon-fiber resin. See *solid.*

composition **(34)** The identity of the chemical or physical substances that make up an object. The chemical composition of a pure substance is identified by its chemical formula. A mixture's composition is identified by the list of substances and their proportions. See *pure substance*; *chemical formula*; *mixture.*

composition reaction (450) A chemical reaction yielding a single product from several reactants. Also called a *synthesis reaction* or a *combination reaction.* See *chemical reaction*; *product*; *reactant.*

compound **(28)** A pure substance consisting of atoms of two or more elements bonded together in a fixed ratio. See *pure substance*; *atom*; *element.*

compressibility **(31)** The property of a substance that indicates how readily its particles can be pushed closer together, reducing the volume and increasing the density of the substance. See *property*; *volume (V)*; *density (ρ).*

compression (278) In longitudinal waves, regions of maximum particle density. A compression corresponds to the crest of a transverse wave. See *longitudinal wave*; *crest*; *transverse wave.*

computed axial tomography **(CT) (313)** A medical imaging technology that uses x-rays to obtain cross-sectional images or "slices" of the body, which can then be assembled by computer to produce a three-dimensional image. Also called CAT or CT scans. See *x-ray*; *medical imaging.*

concave mirror (339) A slightly bowl-shaped mirror with the reflective surface on the interior of the dish or depression. The mirror surface bulges away from the object being reflected. See *mirror.*

concentration (472) The measured amount of solute in a given amount of solvent. It may be measured in units of percentage by mass, specific gravity, moles per liter, or other units. See *solute*; *solvent*; *percentage by mass*; *specific gravity (s.g.)*; *mole (mol)*.

concentration effects (473) See *colligative properties.*

conceptual model (9) An explanation or description of a phenomenon in the form of a mathematical equation, a computer program, or a graph. See *model.*

condensation (43) The change from a vapor to the liquid state when the vapor cools below its dew point temperature. See *vapor*; *liquid*; *dew point.*

conduction (190) One of three methods of heat transfer from a hotter to a cooler object. Conduction involves direct contact. See *heat (Q).*

conductor (191) A material through which heat and electrical current easily flow. Good conductors are usually materials that contain mobile electrons, such as most metals. Compare with *insulator.* See *heat (Q)*; *electrical current (I)*; *metal (chemistry).*

cone cell (335) The light receptor cell in the retina of the human eye responsible for color perception. There are three kinds of cone cells, sensitive to red, green, and blue light. They have the appearance of elongated cones containing many cellular plates through which light must pass. Compare with *rod cell.*

conjugate acid (489) The proton donor ion produced after a Brønsted-Lowry base accepts a proton. See *ion*; *Brønsted-Lowry model*; *base*; *ionizable hydrogen.*

conjugate base (488) The proton acceptor ion produced after a Brønsted-Lowry acid donates a proton. See *ion*; *Brønsted-Lowry model*; *acid*; *ionizable hydrogen.*

conservation law (120) One of several natural laws of science affirming the observation that many quantities such as energy and matter, as well as others, cannot be created or destroyed, only converted from one form to another. See *law of conservation of energy*; *law of conservation of matter*; *first law of thermodynamics.*

conservation of momentum (122) A fundamental natural conservation law. In a system of colliding objects, the sum of their momentums before the collision is equal to the sum of their momentums afterward if no external forces act on the objects. See *conservation law*; *momentum (p)*; *external force.*

constructive interference (271) The interaction of two overlapping wave crests or troughs that produces a new wave when added together. The resulting wave amplitude is much larger than those of the original waves. It is most significant when the waves are in phase. Compare with *destructive interference.* See *crest*; *trough*; *amplitude*; *wave phase.*

contact force (92) A force that acts between systems only when one system touches another. See *force (F)*; *system.*

continuous phase (458) An unbroken phase that makes up most of a heterogeneous mixture in which other phases are dispersed. See *phase (mixture)*; *heterogeneous mixture*; *dispersed phase.*

continuous spectrum (326) A complete visual spectrum with no frequency gaps, emitted only by an ideally luminous object. See *electromagnetic (EM) spectrum*; *frequency (f)*; *luminous.*

continuum (78) Any unbroken sequence of values in which the difference between adjacent values is nearly indistinguishable, but the ends or limits of the sequence are greatly different. The set of real numbers is an example of a continuum. Continuums may be one-, two-, or three-dimensional.

controlled experiment (15) An artificial situation created by a scientist to investigate the cause-and-effect relationship of a phenomenon with other factors under the control of the scientist. See *phenomenon.*

convection (191) One of three methods of heat transfer from hotter to cooler locations. Convection involves the flow of a fluid within a gravitational field, as warmer, less dense fluid is displaced upward by descending, cooler fluid. See *heat (Q)*; *fluid*; *gravitational field*; *density (ρ).*

convection current (191) The flow of matter in a fluid experiencing convection. With a continuous heat input, the flow follows a cyclical path. See *matter*; *convection*; *cycle.*

conventional current (216) The flow of positive charges through a conductor or electrolytic solution toward a more negatively charged region. This flow is opposite to the flow of electrons in a wire. See *conductor*; *electrolytic solution*; *charge (electrical)*; *electrolyte.*

converging lens (344) A lens that is thicker at its optical center than at its edges. It focuses light rays to a point. Also known as a *convex lens.* See *lens.*

conversion factor (52) Any factor (multiplier) we use to convert from one unit of measure to another. It is a fraction—equal to 1—that consists of the appropriate ratio of the two units. See *unit.*

convex mirror (340) A slightly outward-bulging mirrored surface. The surface bulges toward the object being reflected. See *mirror.*

cooling (114) The process of removing thermal energy from a system so that its temperature decreases. See *process*; *thermal energy*; *temperature (t).*

coordinate axis (78) An imaginary line that contains an *origin* as a zero point of position reference and is marked off in distance units. A system's position is a coordinate on the axis. The plural of axis is *axes*. Compare with *axis*. See *position*.

core-envelope model **(356)** An atomic model developed by John Dalton, who suggested that each element has its own kind of indivisible atom and that atomic mass differences are due mainly to the varying thickness of their caloric theory "heat envelopes." See *caloric theory*.

Coriolis effect **(77)** The deflection of the path of a moving system over the surface of the earth. This deflection results from the system's inertia and the change of the direction of the earth's surface rotation in relation to the system's path as it moves over the earth. See *system*; *inertia*; *rotational motion*.

corpuscle **(357)** In physics, especially before the 1900s, a general term for any unidentified minute particle that obeyed the laws of physics. The term applied to particles eventually known as photons, electrons, ions, atoms, and so on.

cosmic ray **(212)** An extremely high energy particle originating mainly from outside the solar system, apparently from supernovas in our galaxy. Nearly all cosmic rays are bare nuclei and most of these are protons (hydrogen nuclei). A very small fraction are lone electrons. See *proton*; *electron*.

coulomb **(C)** **(206)** The SI derived unit of electrical charge. In SI base units, 1 C = 1 A·s. The coulomb is equivalent to the charge of 6.24×10^{18} electrons or protons. See *SI base unit*.

covalent bond **(418)** A chemical bond in which two atoms, usually nonmetals, share a pair of electrons, or a *bonding pair*. The atoms of many nonmetal elements can be covalently bonded with more than one atom at the same time. See *nonmetal*; *chemical bond*.

covalent compound **(423)** A compound consisting of covalently bonded nonmetal atoms. Also called a *molecular compound*, since a covalent compound consists of molecules. See *compound*; *covalent bond*; *nonmetal*; *molecule*.

Creation Mandate **(4)** God's command given to mankind in Genesis 1:28 to exercise dominion over the world by wisely using the resources He has placed here.

crest **(265)** Compared to the rest position in a waveform, it is the highest point of a wave (as in water waves) or the location of greatest particle density (as in sound waves). See *rest position*; *waveform*; *wave*.

critical angle of incidence **(342)** The angle of incidence that produces an angle of refraction of 90° for a light ray traveling from a medium with a higher index of refraction to one with a lower index; that is, the refracted light ray is parallel to the boundary surface between the media. See *angle of incidence*; *angle of refraction*; *medium*; *index of refraction (n)*.

crystal lattice **(428)** An extensive three-dimensional structure of atoms or ions built up by repeating subunits that may be unit cells, as in ionic compounds, or individual atoms or molecules, as in metals or covalent solids. Crystal lattices are responsible for the distinctive crystalline shapes of minerals. See *unit cell*; *ionic compound*; *covalent compound*; *mineral*; *molecule*.

crystalline solid **(31)** A solid substance whose particles are arranged in a crystal lattice. Diamond is a crystalline solid. See *particle*; *crystal lattice*.

cubit **(54)** The ancient and probably original base unit of length. The *common cubit* was based on the length of the forearm of a prominent person from elbow to fingertip—approximately 17.7 in. or 45 cm. The *royal cubit*, used for constructing government structures, temples, and possibly Noah's Ark, was about 20.4 in. (51.8 cm), or even a few inches longer.

Curie temperature **(233)** The temperature above which a permanent magnet loses its ferromagnetic properties. See *permanent magnet*; *ferromagnetism*.

current See *electrical current (I)* or *convection current*.

current electricity **(215)** Electrical phenomena and applications that depend on flowing electric charges (current). See *static electricity*; *electrical current (I)*.

cycle **(256)** One complete back-and-forth motion or rotation in periodic motion, where the system has returned to its starting point. See *periodic motion*.

D

damping **(259)** Reducing the amplitude of periodic motion through friction or other resistance. See *periodic motion*; *amplitude*.

data **(6)** Plural form of the Latin word *datum*, but usually used for both the singular and plural forms. Information collected through observation and often used as the basis for scientific models. See *model*.

deceleration **(85)** The slowing of a system during a time interval. We usually consider it to be a negative acceleration. See *interval* (Δt); *acceleration (a)*.

decibel **(dB)** **(281)** In acoustics, a unit of sound intensity. More generally, a unit that is the ratio of the measured power of a quantity to a reference power, expressed on a logarithmic scale. See *intensity (sound)*; *power (P)*; *logarithmic scale*.

decomposition reaction **(450)** A chemical reaction yielding two or more products from a single reactant. It is usually an endothermic reaction. See *chemical reaction*; *product*; *reactant*; *endothermic reaction*.

deferent **(10)** In the geocentric system, the main circular orbit of the moon, sun, or a planet around the earth. See *rotational motion*; *geocentric model*.

degree **(°)** **(184)** A dimensional unit of temperature that assigns a value of "hotness" or "coldness" to a system. Its magnitude depends on which temperature scale is being used. See *dimension*; *unit*; *temperature (t)*.

density (ρ) **(67)** The mass of a substance or object contained within a stated volumetric unit. For example, the density of water at room temperature is about 1.00 g/cm^3. See *mass (m)*; *volume (V)*.

deposition **(43)** The change of state of a vapor directly to a solid, bypassing the liquid state. This occurs when the vapor is in contact with a surface whose temperature is below its freezing point. See *vapor*; *liquid*; *freezing point*.

desalination **(476)** Removing salt from brine solutions or from seawater.

destructive interference **(271)** The interaction of two overlapping waves so that the trough of one subtracts from the crest of the other. The resulting wave amplitude is much smaller than those of the original waves. It is most significant when the waves are out of phase. Compare with *constructive interference*. See *crest*; *trough*; *amplitude*; *wave phase*.

dew point **(43)** The temperature at which condensation occurs for a given percentage of water vapor by mass in the air (i.e., *relative humidity*). See *condensation*; *percentage by mass*.

diamagnetism **(232)** The natural tendency of all matter to weaken a magnetic field, especially in the absence of other kinds of magnetism. It generates a magnetic field that opposes the external magnetic field that causes it. See *magnetic field*.

diaphragm **(287)** The large dome-shaped muscle that separates the interior region of the body containing the lungs and heart from the rest of the internal organs. Its main function is to control breathing, acting with the muscles between the ribs.

diatomic element **(387)** An element that occurs naturally as a molecule of two identical atoms. Only seven nonmetal elements are naturally diatomic. See *element*; *molecule*; *nonmetal*.

diffraction **(272)** The bending of waves passing around the edges of objects or after passing through openings or gratings. See *wave*.

diffuse reflection **(337)** The most common type of reflection in which photons reflect off an uneven surface in all directions. An image of the light source cannot be formed from a diffuse reflection. See *reflection*; *photon*.

diffusion **(25)** The process of spreading out and mixing due to particle motion. See *process*; *particle*; *motion*.

dimension **(48)** Any measurable property of the physical universe. Examples include length, mass, time, volume, and temperature. See *property*.

dipole **(208)** An object or particle with two electrical or magnetic poles of opposite nature.

direct current (DC) **(216)** Electrical current that flows in only one direction.

dispersed phase **(458)** A phase that is a minority in a heterogeneous mixture and that is dispersed or suspended throughout the continuous phase. See *heterogeneous mixture*; *continuous phase*; *suspension*.

dispersion **(344)** The angular separation of the various electromagnetic energy frequencies in a beam of light by refraction, as with a prism. The dispersion angle is greater for higher frequencies. See *electromagnetic (EM) energy*; *frequency (f)*; *beam*; *refraction*; *prism*.

displacement (d) **(80)** A quantity that describes the net distance and direction of motion. It is always a vector quantity and is represented graphically by a vector arrow whose tail is at the starting point of motion and whose tip is at the ending point. See *distance (d)*; *motion*; *vector*.

dissociation **(465)** The breaking apart of the ions in a solid ionic compound by the action of the solvent to form a solution. Compare with *association*. See *ion*; *ionic compound*; *solvent*; *solution*.

distance (*d*) **(80)** A positive scalar quantity that is the total linear dimension traveled by a moving object during a time interval. It also may be the magnitude of displacement. See *scalar*; *dimension*; *interval* (Δt); *magnitude*; *displacement* (**d**).

distance principle **(134)** When using a simple machine, the distance through which the effort force acts equals the distance the load moves multiplied by the MA of the machine. This is a consequence of the law of the conservation of energy, which states that in any process, the work input must equal work output. See *effort force* (F_e); *simple machine*; *load*; *mechanical advantage (MA)*; *process*.

distillation **(478)** Purifying contaminated or salt water by first boiling it, and then condensing the steam back to the liquid state as drinkable water. Compare with *reverse osmosis (RO)*. Also refers to any method used in chemistry for purifying or concentrating a substance using a similar process. See *boiling*; *condensation*; *process*.

diverging lens **(344)** A lens that is thicker at its edges than at its optical center. It causes an incident beam to spread or diverge as it passes through the lens. Also known as a *concave lens.* See *lens.*

domain (magnetic) **(231)** A microscopic region of a metal where most of the magnetic dipoles (atoms) are oriented in a single direction, giving the region a distinct magnetic orientation. See *magnetic dipole.*

dominion **(4)** In a biblical sense, the human work of caring for the world and using it for the benefit of others.

dominion science **(11)** Operational science used as a tool to obey the Creation Mandate (Gen. 1:28) for God's glory and in love for our fellow human beings. See *operational science; Creation Mandate.*

Doppler effect **(273)** The apparent difference in observed frequency of wave phenomenon compared to the base frequency of a wave source due to the relative motion between the observer and the source. See *wave; frequency (f).*

double bond **(419)** Two covalent bonds between two atoms. See *covalent bond.*

double-replacement reaction **(451)** A chemical reaction in which two reactant ionic compounds swap cations when forming the two product compounds. One of the products often becomes a precipitate. See *chemical reaction*; *reactant*; *ionic compound*; *cation*; *product*; *precipitate.*

dowsing **(261)** A mystical practice intended to reveal hidden knowledge, such as dowsing for groundwater, ores, or even fugitives. Methods usually involve some kind of material object, such as a pendulum or specially shaped stick, to aid the dowser. No scientific basis exists for dowsing and it should be avoided by believers.

drag **(105)** A form of friction exerted by a fluid on an object moving through the fluid or as the fluid flows around the object. See *friction*; *fluid.*

dry cell battery **(217)** A battery that contains a thick paste electrolyte. Not rechargeable. Commonly used for portable DC electrical equipment such as flashlights and portable radios. See *battery*; *electrochemical cell*; *electrolyte*; *direct current (DC).*

ductile **(395)** Able to be easily drawn into wire or otherwise deformed by stretching without breaking. It is a common property (called *ductility*) of most metals. See *property*; *metal (chemistry).*

dynamic equilibrium **(42)** A condition of a system where two opposite processes occur at the same rate. This results in the amounts of all substances involved in the processes being the same over time. See *process*; *system.*

dynamics **(91)** The study of forces and how forces interact with matter, including their effects on an object's motion. Newton's three laws of motion form the basis of dynamics. See *force (F)*; *matter*; *motion*; *laws of motion.*

E

echo **(283)** The reflected sound heard by a listener or an acoustic receiver from a separate source of sound. Compare with *reverberation*; *echolocation.* See *reflection*; *sound*; *acoustics*; *receiver.*

echolocation **(291)** The biological or artificial use of acoustic echoes to detect distance and direction of objects of interest. See *acoustics*; *echo.*

ecliptic **(77)** The apparent path of the sun across the background of stars. Astronomers use the term *ecliptic plane* to describe the plane that passes through the combination of this imaginary line, the earth, and the sun. It is a useful frame of reference for locating objects in space with respect to the earth-sun system. See *frame of reference*; *space.*

effervescence **(490)** The vigorous production of a gas that forms bubbles; often from a chemical reaction involving an aqueous solution. See *chemical reaction*; *aqueous.*

efficiency **(119)** For a machine or process, the ratio of energy or work produced to the energy or work that was put into the machine or process. It is a measure of the effectiveness of converting energy from one form to another. See *process*; *energy*; *mechanical work (W).*

effort arm **(ℓ_e)** **(137)** The torque arm of the effort force in a lever system. See *effort force (F_e)*; *lever*; *torque arm.*

effort force **(F_e)** **(133)** The force exerted on a simple machine to do work on a load. Also called the *effort.* See *simple machine*; *load.*

elastic collision **(122)** A collision between two objects in which the momentums and kinetic energies of the colliding objects are conserved. See *conservation of momentum*; *law of conservation of energy.*

elastic force **(113)** A variable force exerted by an elastic object (e.g., a spring or a rubber band) that increases in proportion to the object's deformation from its relaxed state. The source of elastic potential energy. See *elastic potential energy (EPE).*

elastic potential energy **(EPE)** **(113)** The potential energy of a system due to an *elastic force* acting on it and its distance from a zero reference position.

electrical circuit **(216)** A complete path for an electrical current. It includes a current source, such as a battery or a generator, a conductor, an electrical load, and a conductor by which the current returns to the current source. See *electrical current (I).*

electrical conductor **(211)** See *conductor.*

electrical current (I) **(215)** A continuous flow of electrical charges. Measured using an *ammeter.*

electrical discharge **(208)** The loss of static charge on an object as the surroundings supply or absorb charges to restore a neutral condition on the object.

electrical energy **(115)** The ability to do work through the action of the electromagnetic force on or by electrical charges. See *electromagnetic force.*

electrical ground **(213)** A connection between an electrical device or appliance and the earth. This path allows the earth, which is considered to be an infinite source or sink for electrons, to maintain the electrical neutrality of the connected device. For electrical safety, the ground path ensures that a faulty piece of electrical equipment discharges safely to ground rather than through the user.

electrical induction **(208)** The creation of a charged region on a neutral object when it is exposed to an electrical field.

electrical insulator **(211)** See *insulator.*

electrical load **(216)** Any device that purposely converts electrical energy to another form of energy in an electrical circuit. See *electrical circuit.*

electrical potential **(219)** See *voltage.*

electrical potential energy **(216)** The energy change or work that can be done by charges moving between two points of different voltages.

electrical resistance (*R*) **(219)** The property of all electrical circuit elements that impedes the flow of current to some extent. Electrical resistance is measured in ohms using an *ohmmeter.* See *property; ohm (Ω).*

electrical utility **(238)** A facility whose primary purpose is to generate electricity and deliver it to customers throughout a geographic area. It includes the power generation facility itself, the system of transmission towers, wires, substations, and customer connections, and the business administrative functions.

electric field **(207)** The region of influence surrounding a source of electrical charge in which an electrostatic force is exerted. It is modeled by lines of force. See *charge (electrical); electrostatic force.*

electrochemical cell **(217)** A device containing a combination of an electrolytic medium and two conductors of different materials that creates an electrical potential between the conductors. When connected to an outside circuit, current will flow between the conductors and through the cell until the chemicals within the cell are exhausted. See *electrolyte; conductor; electric current (I).*

electrolysis **(450)** The decomposition of a chemical compound into two or more substances by applying an electrical current. See *decomposition reaction.*

electrolyte **(429)** A substance that dissociates or ionizes in water to produce an electrolytic solution. See *dissociation; ionization; conduction; electrolytic solution.*

electrolytic solution **(429)** A water solution of an ionic or ionized compound that can conduct an electrical current. Salt and acid solutions are good electrolytic solutions. See *solution; ionic compound; ionization; electrical current (I); salt; acid.*

electromagnet **(236)** A solenoid that contains a ferromagnetic core. The core greatly increases the magnetic field strength of the solenoid. See *ferromagnetic material; solenoid.*

electromagnetic (EM) energy **(116)** The combined action of electrical and magnetic energies in the form of wavelike, radiant energy. (Electricity and magnetism are different forms of electromagnetic or EM energy.) Also called *radiant energy.* See *energy.*

electromagnetic (EM) spectrum **(300)** The entire frequency continuum of electromagnetic waves. See *frequency (f); continuum; electromagnetic wave.*

electromagnetic force **(92)** The attractive or repulsive force produced by static and moving charges in atoms. It is the second-strongest fundamental force. Physicists have shown that it is the source of both the electrical and magnetic field forces. See *fundamental force; magnetic field; electric field; field force.*

electromagnetic induction **(236)** Electricity generated from a changing magnetic field or a magnetic field generated by an electrical current through a conductor. See *magnetic field; electrical current (I).*

electromagnetic wave **(300)** The wave model for the transmission of electromagnetic energy. Useful for studying the wave properties of light and other forms of electromagnetic energy. See *wave; model; electromagnetic (EM) energy.*

electron **(26)** An elementary particle with unusual properties that carries a single negative *fundamental charge.* Within an atom, it occupies the spherical volume of space around the nucleus. The electrons and their arrangement within an atom determine its chemical and many of its physical properties. See *elementary particle; fundamental electrical charge (e); nucleus; chemical property; physical property.*

electron affinity **(414)** A measure of how well an unbonded atom can attract and hold an electron. Energy must be added to atoms of elements with low affinities to get them to accept an electron (less stable); energy is released from atoms of elements with high affinities that take an electron (more stable). See *electron*; *electronegativity*.

electron diffusion theory **(215)** A model of electrical current flow that views electrons as flowing by diffusion through the conductor from areas of high electron concentration to those of low concentration.

electron dot notation **(405)** A chemical symbol of an element that displays the number and basic arrangement of the valence electrons of an atom in its ground state. See *chemical symbol*; *valence electron*; *ground state*.

electronegativity **(415)** A measure of an element's ability to attract and hold electrons *when bonded to other atoms*, designated by a small decimal number called the *electronegativity number (EN)*. Compare with *electron affinity*. See *electron*; *chemical bond*.

electronic amplification **(290)** An electronic device that makes a sound louder by converting it to an electronic signal, increasing the signal's energy, and then converting the signal back to sound energy. See *amplification*.

electronic scale **(64)** A laboratory instrument that measures weight (or mass) by converting the pull of gravity on an object resting on the instrument's sample plate into an electronic signal that is processed by a built-in computer circuit. The instrument displays the weight/mass on a digital display with the associated units. See *instrument*; *weight (w)*; *mass (m)*; *gravity*; *force (F)*.

electron sea theory **(430)** See *free electron theory*.

electroscope **(210)** An instrument that can detect the presence of an electrical charge but not its polarity or strength.

electrostatic force **(205)** The field force exerted by electrical charges. It may be repulsive or attractive depending on the kinds of charges interacting. Also called *electrical force*. See *field force*.

electrostatic precipitator **(214)** A device in a smoke exhaust or ventilation system that electrostatically charges smoke particles, and then pulls them from the air as they pass through oppositely charged plates. See *charge (electrical)*.

element **(28)** A pure substance that contains only one kind of atom. See *atom*.

elementary particle **(24)** One of the hundreds of basic particles that make up an atom. Protons, neutrons, and electrons are the most well known. Protons and neutrons are themselves made of other elementary particles. See *particle*; *atom*; *proton*; *neutron*; *electron*.

empirical formula **(438)** A chemical formula that shows the smallest whole-number ratio of elements in a compound. An ionic compound's formula unit is an empirical formula. Compare with *molecular formula*. See *chemical formula*; *ionic compound*.

emulsion **(458)** A heterogeneous mixture that contains two or more immiscible liquid phases. See *heterogeneous mixture*; *immiscibility*; *phase (mixture)*.

endoscope **(343)** A long, flexible fiber optic bundle that includes a light and lens, used for internal medical examinations of the gastrointestinal tract by *endoscopy*. It is usually inserted through natural openings of the body. See *fiber optics*.

endothermic reaction **(453)** A chemical reaction that must absorb heat from its surroundings to occur. See *chemical reaction*; *heat (Q)*.

energy **(110)** The ability to do work. It is a positive scalar quantity measured in joules (J). *See mechanical work (W)*; *scalar*.

energy level **(367)** A three-dimensional region within an atom occupied by groups of electrons with similar energies. Energy levels are arranged concentrically around the nucleus. They can be viewed as similar to the layers of an onion in order of increasing energies but their actual structure is far more complex. Each level has an allowable number of orbitals and electrons. See *electron*; *nucleus*; *orbital*.

entropy (S) **(199)** The measure of a system's randomness or disorder. Greater entropy implies that matter is more spread out in a random arrangement or that energy is dispersed and in a less useful form.

epicycle **(10)** In the geocentric system, a small circular orbit of a celestial object superimposed on a deferent to account for the apparent backward motion by certain planets at certain points in their orbits around the earth. See *geocentric model*; *deferent*.

equant point **(10)** In the geocentric system, a point not centered on the earth from which the center of an object's epicycle appears to orbit at a constant speed. Every object has its own equant point, and the point itself revolves around the earth. See *geocentric model*; *epicycle*.

error **(56)** The arithmetic difference between a measured value and the acceptable value of the dimension. Every measurement includes some error. See *acceptable value; measurement*.

evaporation **(43)** Relatively slow vaporization that occurs when a liquid's temperature is below its boiling point but above its freezing point. Evaporation occurs only at the surface of the liquid. See *vaporization*; *boiling point*; *freezing point*; *liquid*.

exothermic reaction **(452)** A chemical reaction that liberates heat to its surroundings as it occurs. See *chemical reaction*; *heat (Q)*.

external force **(94)** Any force exerted on a system by something outside of a system's boundaries in its surroundings. See *force (F)*; *system*; *surroundings*.

extrapolate **(374)** To make an estimation of the value of data outside or beyond existing data on the basis of the observed trend of the data. The method can be useful but depends on the assumption that the relationship holds true to the point of extrapolation, which may or may not be valid. See *data*.

F

Fahrenheit scale (185) A temperature scale with fiducial points at the freezing point (32 °F) and the boiling point (212 °F) of pure water at 1 atm of pressure. See *fiducial point*; *freezing point*; *boiling point*.

Fall, the **(3)** The transition of the world from a condition of innocence and goodness into a condition of sinfulness and suffering, because of the disobedience of Adam and Eve in the Garden of Eden. The Fall has left every human a sinner. It has also corrupted the entire physical universe.

family (396) Elements having the same valence electron structure and often similar chemical properties. Elements in a family reside in a column on the periodic table. Families are named by the element at the top of the column. Compare with *group*. See *element*; *valence electron*; *periodic table of the elements*.

ferrimagnetism (232) Property exhibited by materials that are mixtures of two magnetic elements or compounds. Magnetism results from the net difference of opposed magnetic dipole particles. Ferrimagnetism is the basis of natural magnets, such as lodestones. See *property*; *magnetic dipole*.

ferromagnetic material **(231)** A material that naturally exhibits ferromagnetism, like iron or neodymium. See *ferromagnetism*.

ferromagnetism (231) The phenomenon observed in materials that are highly permeable to magnetic lines of force because their magnetic domains align with the field, reinforcing it. Aligned magnetic domains can produce permanent magnetism. See *lines of force*; *domain (magnetic)*; *permanent magnet*.

fiber optics (343) Technology using bundles of long, fine, transparent glass fibers that transmit light along their lengths by total internal reflection. Besides high speed communications, fiber optics are used in *endoscopy*, which utilizes an *endoscope*, in *laparoscopy* for examining and doing surgery inside inaccessible spaces in the body, and for inspecting the internal areas of complex engines using a *borescope*. See *total internal reflection*; *endoscope*.

fiducial point **(185)** Fixed, precise, and easily reproducible values in a dimension used to calibrate a measuring scale. For example, the fiducial points of the Celsius and the Fahrenheit temperature scales are the boiling and freezing points of pure water at 1 atm of pressure. See *boiling point*; *freezing point*; *pressure (P)*.

field **(93)** A volume of space in which every point can have a different value of a dimension of interest. A field can be a *scalar field*, such as one that describes temperature at every point in an air mass, or it can be a *vector field*, such as a field of magnetic forces surrounding a magnet. See *space*; *dimension*; *scalar*; *vector*.

field force (93) A field in which a force vector exists at every point in the field. Isaac Newton originally called field forces *action-at-a-distance forces* to distinguish them from contact forces. See *vector*; *field*; *contact force*.

field strength **(207)** The density of the lines of force passing through a unit area within a force field. See *field force*; *lines of force*.

first-class lever **(137)** A lever system in which the fulcrum is between the resistance and the effort. Its MA can be greater than 1, equal to 1, or less than 1. See *lever*; *system*; *fulcrum*; *resistance force* (F_r); *effort force* (F_e).

first law of thermodynamics (120) Energy and matter cannot be created or destroyed, only changed in form. Compare with *law of conservation of energy*; *conservation of matter*. See *energy*; *matter*.

flammable **(36)** The property of any substance that burns rapidly or easily catches fire. (*Inflammable* means the same thing but should be avoided, since some people think that it means *not* flammable.) Compare with *nonflammable*. See *property*.

fluid **(32)** A substance that can flow. Can be a liquid or a gas. See *liquid*; *gas*.

fluid mechanics (152) The study of how fluids flow and how forces and energy are transmitted through fluids. Subdivided into *hydrostatics* and *hydrodynamics*. See *fluid*; *force (F)*; *energy*.

fluid pressure (153) A property of all fluids in which pressure is exerted equally in all directions at any point in the fluid. See *property*; *pressure (P)*.

fluorescence (330) The emission of visible light by a substance that is continuously exposed to high-frequency electromagnetic energy. See *electromagnetic (EM) energy*.

foam **(459)** A heterogeneous mixture in which a gaseous phase consisting of tiny bubbles is dispersed in a liquid or solid (e.g., pumice). See *heterogeneous mixture*; *dispersed phase.*

focal length **(339)** The distance from the optical center of a lens or mirror to its principal focus. See *principal focus.*

focal point **(339)** The point on the principal optical axis at which incident light rays are focused by a concave mirror or a converging lens. The focal point varies depending on the distance between the object and the mirror or lens forming the image. For parallel incident light rays (when the object is at an infinite distance), the focal point and principal focus are the same point. See *principal optical axis*; *incident*; *concave mirror*; *converging lens*; *principal focus.*

force (*F*) **(91)** A push or pull on a system. It may be a vector or scalar quantity. See *system*; *vector*; *scalar.*

force diagram **(91)** An illustration that includes a system and all the force vectors acting on the system. See *system*; *force (F).*

forensics **(12)** A kind of scientific investigation that attempts to reconstruct the scenario of a crime committed in the past. It involves aspects of both operational and historical science. Forensics usually cannot establish with total certainty what *actually* occurred but rather what *most likely happened.* See *operational science*; *historical science.*

formula **(65)** A mathematical equation that relates measureable physical quantities using symbols. When solved, the combined units as well as the values on both sides of the equation must be the same. See *unit.*

formula unit **(429)** The smallest ratio of elements in an ionic compound that describes its composition. It is also the molecular formula of a covalent compound. See *ionic compound*; *composition*; *molecule; formula*; *covalent compound.*

Foucault pendulum **(262)** A pendulum first constructed in France in 1851 by Jean Foucault to demonstrate that the earth rotates on its axis; any similar pendulum consisting of a large pendulum mass suspended by a long wire from a rigid attachment point. See *pendulum.*

frame of reference **(76)** The geometric space containing the point of reference and coordinate axes from which a person observes or measures position and movement. Also called *reference frame.* See *space*; *point of reference; coordinate axis*; *position*; *motion.*

free electron theory **(430)** The accepted model of bonding in metals. Each metal atom freely shares its few valence electrons with all the other atoms in the crystal lattice contributing to a negative *electron sea* that holds the cations together. See *metallic bond*; *valence electron*; *crystal lattice*; *cation.*

free fall **(102)** The condition of an object accelerated by the force of gravity alone with no other external forces acting on it. True free fall can occur only in a vacuum. See *acceleration (a)*; *gravitational force*; *external force.*

free surface **(32)** The upper, mobile surface formed by a liquid in a less-than-full container situated in a gravitational field. In free fall, a quantity of liquid will form a spherical free surface unless it completely fills a container. See *liquid*; *gravitational field*; *free fall.*

freezing **(40)** The change of state from a liquid to a solid that usually occurs when a substance cools to its freezing point. See *liquid*; *solid*; *freezing point.*

freezing point **(10)** The temperature at which freezing of a substance occurs at 1 atm. For a pure substance, the same temperature as its melting point. See *freezing*; *pure substance*; *temperature (t)*; *melting point.*

freezing-point depression **(474)** The number of degrees the freezing point of a solvent is reduced per mole of dissolved solute particles. It is a colligative property of the solvent. See *degree (°)*; *freezing point*; *mole (mol)*; *solute*; *colligative properties*; *solvent.*

frequency (*f*) **(257)** The number of oscillations or other periodic changes completed per unit of time. It is measured in hertz (Hz) or cycles per second (s^{-1}) in the SI. See *oscillation*; *periodic motion*; *cycle.*

frequency modulation **(FM)** **(315)** A method of transmitting radio communication by varying (or modulating) the frequency of a constant-amplitude radio wave. Contrast with *amplitude modulation (AM).* See *frequency (f)*; *amplitude*; *radio wave.*

friction **(98)** A contact force that opposes the movement of objects past each other. It may be a vector or scalar quantity. Kinds of friction include *kinetic friction*, *static friction*, *rolling friction*, and *fluid friction.* See *contact force*; *vector*; *scalar.*

fuel cell **(402)** An electrical energy source that produces electricity from the chemical reaction of a fuel with oxygen. Most practical fuel cell designs use hydrogen as the fuel but other kinds of fuels are possible.

fulcrum **(135)** The point about which a lever pivots or rotates. See *lever*; *rotational motion.*

fullerene **(387)** Large net-like cylinders or spheres of bonded carbon atoms. Their structures are similar to those of geodesic domes popularized by scientist-philosopher Buckminster Fuller in the second half of the twentieth century. See *chemical bond.*

fundamental (wave) **(282)** The lowest-frequency sound wave in a complex mixture of related harmonic tones. See *frequency (f)*; *sound*; *harmonic (wave).*

fundamental electrical charge (e) **(26)** The magnitude of the charge on an electron or proton. Its value is equal to about 1.602×10^{-19} coulomb and it may be positive or negative. All measureable electrical charges are whole-number multiples of the fundamental charge. Also called the *elementary charge.* See *charge (electrical).*

fundamental force **(92)** One of the four principal forces or interactions that hold matter together and govern the exertions of all contact and field forces in the universe. See *contact force*; *field force.*

fuse **(224)** An overcurrent protection device that melts to open a circuit when current exceeds a certain value. Must be replaced to restore the circuit after the fault is corrected. See *overcurrent protection*; *circuit breaker.*

fusion bomb **(376)** An extremely powerful weapon based on nuclear fusion. Also called a *hydrogen bomb, H-bomb,* or *thermonuclear bomb.* None has ever been used in combat and only six nations have successfully tested one. See *nuclear fusion.*

fusion energy **(377)** A commercial energy source based on nuclear fusion. A practical fusion energy reactor, called the *ITER,* is currently under development but is many years away from operation because of the difficult technical problems involved. See *nuclear fusion*; *ITER.*

G

gamma decay **(371)** A form of nuclear decay resulting in the emission of a high-energy photon. There is no change in the atom except for a decrease in the total energy of its nucleus. See *photon*; *nuclear decay.*

gamma ray **(312)** A photon in the highest frequency band of the electromagnetic spectrum, with frequencies extending upward from 30 EHz. Gamma rays are produced by nuclear changes. They support nonmedical internal imaging, cancer treatment, and high-energy astronomical observation. See *photon*; *electromagnetic (EM) spectrum*; *nuclear change.*

gas **(32)** The state of a substance in which its particles are far apart and have large kinetic energies. A gas has no fixed volume or shape, is highly compressible, and is able to flow. See *particle*; *kinetic energy (KE)*; *volume (V)*; *compressibility.*

gas pressure **(33)** The pressure a gas exerts on its container or on an object immersed in it. See *gas*; *pressure (P).*

gauge **(156)** A mechanical device, usually connected to a fluid system, designed to indicate gas or liquid properties. See *fluid.*

gauge pressure **(156)** A pressure reading referenced to atmospheric pressure, typically displayed by most mechanical pressure gauges. Compare with *absolute pressure.* See *atmospheric pressure (P).*

gear **(141)** A wheel with teeth on its perimeter that mesh with similar teeth of other gears to do work while producing rotational motion. The effort gear is the gear that supplies the force and energy to do work. The resistance gear, and whatever it is attached to, is the resistance or load of the gear system. See *rotation*; *force (F)*; *mechanical work (W)*; *effort force* (F_e); *resistance force* (F_r).

gel **(459)** A two-phase heterogeneous mixture in which dispersed solid particles form an open network throughout a liquid continuous phase, giving some rigidity to the mixture. A gel has some of the properties of a solid but is mostly liquid. Compare with *sol.* See *heterogeneous mixture*; *dispersed phase*; *continuous phase.*

general theory of relativity **(305)** A generalized model of the three-dimensional universe that accounts for the effects of gravity on time, energy, and matter. First proposed by Albert Einstein in 1915. See *gravity*; *time (t)*; *energy*; *matter.*

generator (electrical) **(236)** A mechanical device that produces electricity by the relative motion between electrical conductors and a magnetic field. See *electromagnetic force*; *magnetic field.*

geocentric model **(10)** A model of the universe that places the unmoving earth at its center, with all planets, the sun, and other celestial objects moving about the earth in complex paths. The stellar field is at the inner surface of a rotating sphere located at a large but finite distance from the earth. Compare with the *heliocentric model.*

geographic North Pole **(233)** The point representing the intersection of the earth's axis with its surface in the Northern Hemisphere. All lines of longitude pass through this point. Also called *true north.* Compare with *magnetic North Pole.* See *axis.*

geomagnetic field **(234)** The magnetic field of the earth. Several theories for the origin of the field have been suggested, including the *geodynamic theory,* where convection currents in the earth's liquid core create the field, and the *original creation theory* favored by many young-earth creationists. See *convection current.*

God's image **(4)** The likeness that every person has that reflects God's characteristics in many significant ways, such as intelligence, will, and emotion. It is God's image in us that makes us different from all other creatures.

graduated scale **(48)** A series of markings on an instrument related to the dimension to be measured by the instrument. See *instrument*; *dimension*.

gravitational acceleration **(102)** The acceleration of an object due to gravity. At the earth's surface, its average magnitude is about 9.8 m/s^2, regardless of the object's mass. It may be a vector or scalar quantity. See *acceleration (a)*; *gravity*; *vector*; *scalar*.

gravitational field **(93)** The region around an object within which its gravitational force is exerted. See *gravitational force*.

gravitational force **(92)** The force exerted by all matter on other matter. It is the weakest of all fundamental forces. It may be a vector or scalar quantity. See *gravity*; *vector*; *scalar*; *fundamental force*.

gravitational lensing **(305)** In astronomy, the observed bending of light from a distant celestial object by the intense gravity of an object closer to the observer on Earth. A phenomenon predicted by Einstein's general theory of relativity. See *gravity*; *general theory of relativity*.

gravitational potential energy **(GPE) (112)** The potential energy of a system due to its weight and its height above a zero reference position. See *gravity*; *potential energy (PE)*; *weight (w)*; *zero reference position*.

gravity **(112)** The attraction of the earth, sun, or any other large celestial object for any object near it. A general term for the gravitational force. See *gravitational force*.

ground-fault circuit interrupter **(GFCI) (225)** A protective electrical device that very rapidly opens a circuit when it senses that an abnormal path for the current to ground exists. It is specifically designed to protect human life rather than property and can be reset when the fault is corrected. See *electrical current (I)*; *electrical ground*.

ground state **(367)** The condition where a neutral atom's electrons are at their lowest possible energy state. The standard condition for comparing the electron structures of different elements. See *electron*.

group **(396)** A column of elements in the main part of the periodic table, identified by a group letter/number combination or just a number in the IUPAC form of the periodic table. Compare with *family*. See *element*; *periodic table of the elements; IUPAC*.

H

half-life **(374)** The length of time required for half of the atoms of a radioactive isotope in a sample to experience a certain kind of nuclear decay. Half-life is a measurable, unique property of all radioactive isotopes. See *radioactive*; *isotope*; *nuclear decay*; *property*.

halogen family (400) Group 7A of the periodic table. Very chemically reactive elements that usually have an oxidation number of −1. See *group*; *periodic table of the elements*; *reactive*; *element*; *family*; *oxidation number (ON)*.

hard drive **(245)** A common magnetic digital data storage medium used in most personal computers today. Consists of multiple disk-shaped glass platters coated with a magnetic film mounted on a motorized drive shaft. Read-write heads access and write data through electromagnetic induction as the disks spin. Data is formatted in cylinders, tracks, and sectors so that it can be quickly found by the computer's disk operating system. See *electromagnetic induction*.

harmonic (wave) **(282)** A mechanical wave in a complex mixture of related waves whose frequency is a whole-number multiple of the fundamental wave's frequency. Essentially the same as an *overtone*. See *mechanical wave*; *fundamental (wave)*; *frequency (f)*.

heat (Q) (195) A quantity of thermal energy that flows from one system to another, resulting in a change of temperature or state. It is measured in joules (J). As a verb, it means to add thermal energy to a system. See *thermal energy*; *system*.

heat capacity (C) (194) The amount of thermal energy that an entire object must gain or lose to change its temperature by 1 °C. Heat capacity depends on the mass and thermal properties of the substance(s) in the object. Compare with *specific heat* (c_{sp}). See *thermal energy*; *temperature (t)*.

heat engine **(181)** A device that converts heat into mechanical work or vice versa. Heat engines include steam, internal combustion, and jet engines, as well as refrigeration and heat pump units. See *heat (Q)*; *mechanical work (W)*.

heating **(114)** The process of adding thermal energy to a system so that its temperature increases. See *process*; *thermal energy*; *system*; *temperature (t)*.

heliocentric model **(10)** The modern model of the solar system that places the sun at the gravitational center of all revolving planets, including the earth, and other objects not orbiting around planets. The moon is a satellite of the earth. The earth itself rotates around its own axis. All visible stellar objects are part of the Milky Way galaxy, which is itself part of a vast universe. Compare with *geocentric model*.

Henry's law (470) The greater the pressure of a gas on a liquid, the greater the mass of the gas that can dissolve in the liquid at a given temperature. See *gas pressure*; *mass (m)*; *liquid*.

hertz **(Hz) (238)** The SI unit of frequency, sometimes represented by s^{-1} (1 Hz = 1 cycle/s). See *SI*; *unit.*

heterogeneous mixture **(30)** A nonuniform mixture that contains two or more distinct phases, usually of different kinds of matter. See *phase (mixture).*

hexagonal close-packed **(HCP) (431)** Describes an arrangement that maximizes the density of spherical particles in a given volume. A metallic crystal lattice consists of interlocking hexagons in layers that are nested in gaps between particles in adjacent layers. See *metal (chemistry)*; *crystal lattice.*

historical science (11) The application of scientific principles to create models of events and processes that occurred in the unobservable past, made on the basis of observable evidence that exists today. Also called *origins science.* Compare with *operational science.* See *model*; *process.*

homogeneous mixture **(29)** A uniform mixture of particles of different substances that form a single phase. Also called a *solution.* See *phase (mixture).*

homogenization **(459)** The process of finely mixing two or more immiscible materials to such an extent that they appear and act as a homogeneous mixture, at least for a brief period of time. See *process*; *immiscibility*; *homogeneous mixture.*

hue **(334)** The base color that is modified by saturation and value. Examples include red, blue, green, cyan, magenta, or yellow. See *color*; *saturation (color)*; *value.*

hydration **(465)** The act of surrounding solute particles by water molecules during the solution process. Called *solvation* if the solvent is something other than water. See *solute*; *solution process.*

hydraulic machine **(162)** A machine filled with a special liquid (*hydraulic fluid*) that uses Pascal's principle to do work by converting a small force exerted on a small diameter piston (the *effort piston*) to a large force exerted by a large diameter piston (the *load piston*). See *Pascal's principle*; *mechanical work (W)*; *force (F).*

hydraulics (162) The area of physics that deals with the transfer of forces and doing work by confined fluids according to Pascal's principle. See *force (F)*; *mechanical work (W)*; *Pascal's principle.*

hydrodynamics **(163)** The physics of stationary and flowing liquids. Pertains to ships, submarines, propellers, docks, and any other submerged objects, devices, and instruments. See *liquid*; *fluid mechanics.*

hydrometer **(159)** An instrument for measuring the specific gravity of a liquid. It functions by floating in the liquid. The specific gravity is read on a graduated scale where the liquid surface cuts across the scale. See *specific gravity*; *graduated scale.*

hydronium ion (483) The cation formed by the self-ionization of water; H_3O^+. See *cation*; *ionization.*

hydrophone **(292)** A waterproof microphone used passively to detect sounds under water.

hydrostatic pressure (155) Pressure at a point within a volume of water (or any liquid) based only on the depth of the point within the liquid.

hydroxide ion (483) The anion formed by the self-ionization of water; OH^-. See *anion*; *ionization.*

hypothesis (14) A temporary, testable explanation of a phenomenon that stimulates and guides further scientific investigation.

I

ideal fluid **(165)** A flowing fluid that has the following properties:

- At every fixed point in the fluid, the fluid's speed as it flows through the point is constant.
- It flows smoothly and without turbulence.
- It is incompressible; its density is constant at all points.

No real fluid actually has these properties, though many, such as water, come close enough that we can develop workable models of fluid flow. See *fluid*; *property*; *model.*

illuminated **(327)** The condition of receiving and reflecting light from another source. Compare with *luminous.*

illumination **(329)** A measure of the amount of light received at a surface; depends on the intensity of the light source and the distance of the surface from the source. See *intensity (light).*

immiscibility (458) Not able to dissolve in another liquid. Used of liquids. The liquid phases of an emulsion are immiscible. See *phase (mixture)*; *emulsion.*

incandescence (329) Light produced by materials that are heated until they glow. Incandescent sources radiate in the infrared, visible, and ultraviolet light bands of the electromagnetic spectrum. See *electromagnetic (EM) spectrum*; *luminous.*

incident **(337)** Description of light or another kind of energy illuminating or falling on a surface. Contrast with *transmitted.*

incident ray (337) A light ray approaching a reflective surface or a boundary between different refractive media. See *ray*; *reflection*; *refraction*; *medium.*

inclined plane (145) In its most basic form, a two-dimensional surface whose opposite ends are at different heights. It forms the basis for wedges, stairways, ramps, and screws. See *wedge*; *screw*.

index color **(326)** One of the seven familiar colors that identify important regions of the visual color spectrum: red, orange, yellow, green, blue, indigo, and violet. See *electromagnetic (EM) spectrum*.

index of refraction (*n*) (341) The ratio of the speed of light in a vacuum to the speed of light in a given medium; a measure of a medium's optical density. See *speed of light (c)*; *medium*; *optical density*.

indigestion **(482)** Any symptom associated with the stomach, esophagus, or upper intestinal tract, that causes discomfort, such as burning or nausea, along with other symptoms of discomfort. Often caused by excess stomach acid. See *acid*.

Industrial Revolution **(128)** The major cultural transition from a farm-centered society with handmade goods to a factory-based, industrialized society. Beginning in the late 1700s in Great Britain, it quickly spread to Western Europe and America through the mid-1800s. Energy sources shifted from wood and water to coal and steam. Steam engines revolutionized land and water transportation.

inelastic collision **(123)** A collision between two objects in which the deformation is so severe that they stick together. While total momentum is conserved, total kinetic energy is greatly reduced by conversion into the energy required to deform the objects. See *conservation of momentum*; *law of conservation of energy*.

inert **(401)** Having almost no chemical reactivity except under extreme conditions. An important property of the noble gases. See *reactive*; *property; noble gas*.

inertia (95) The tendency of any kind of matter to resist change in its motion. Mass is a measure of inertia. See *motion*; *mass (m)*.

inertial reference frame **(77)** A reference frame within which systems tend to move in straight lines or in smooth, predictable curves. Their motions are affected by forces originating from *within* the frame of reference. Motions and change in motions obey Newton's laws of motions without the need to invent forces that seem to exist but have no real source. See *motion*; *frame of reference*; *laws of motion*; *force (F)*.

inflammable **(36)** See *flammable*.

infrared (IR) (308) The electromagnetic spectrum extending from 300 GHz to about 384 THz, produced by the thermal motion of atoms and molecules. Infrared supports remote thermal imaging, "night vision," digital communications, and remote control of modern electronic technology. See *electromagnetic (EM) spectrum*.

infrasonic wave (293) A sound with a frequency below the range audible to humans (typically less than 20 Hz). The study of such waves is *infrasonics*. See *acoustic spectrum*; *frequency (f)*.

inner ear **(289)** The parts of the ear that assist with balance and convert the mechanical vibrations of sound waves into nerve impulses that are sent to the brain. Includes the *cochlea*, the *auditory nerve*, and the *semicircular canals*.

inner transition element (401) Any member of the two rows of metal elements usually placed below the periodic table, which typically have an oxidation number of +2. They are called the Lanthanide series and the Actinide series. The chemical and physical properties of these elements are very uniform. See *element*; *metal (chemistry)*; *periodic table of the elements*; *oxidation number (ON)*; *period (chemistry)*.

inorganic See *organic*.

inorganic compound **(29)** A compound that either does not contain the element carbon or is generally not produced *only* by living organisms. Examples include calcium carbonate and carbon dioxide. See *compound*; *element*; *organic*.

in phase **(271)** The condition where the waveforms of two or more wave phenomena coincide, crest-to-crest and trough-to-trough. This results in constructive interference producing waves with amplitudes greater than that of any individual wave. Compare with *out of phase*. See *wave phase*; *waveform*; *crest*; *trough*; *constructive interference*; *amplitude*.

insoluble **(467)** Describes a solute that is unable to dissolve in a certain solvent. Liquids that are insoluble in another liquid are *immiscible*. See *solvent*.

instantaneous speed **(81)** The speed of a system at any moment. See *speed (v)*; *system*.

instrument **(48)** Any artificial device made for the purpose of refining, extending, or substituting for the human senses when measuring. See *measurement*.

insulator **(193)** A material that does not easily conduct thermal energy or electricity. Insulators are poor conductors with tightly bound valence electrons. Compare with *conductor*. See *thermal energy*; *electrical energy*.

intensity (light) **(328)** The rate at which a light source radiates energy. It is measured in candelas. Also called *brightness*. See *candela (cd)*.

intensity (sound) (281) The measure of a sound wave's power (its rate of energy transmission). It is proportional to the square of the wave's amplitude. See *sound*; *power (P)*; *amplitude*.

interference **(271)** The interaction of two or more overlapping waves, resulting in a net waveform with characteristics different from those of the original waves. See *waveform.*

intermolecular force **(422)** Any force acting between molecules resulting from permanent or momentary charge differences between regions on the surfaces of the molecules. The most important are *hydrogen bonds, dipole-dipole forces,* and *London dispersion forces.*

internal energy **(182)** The sum of the kinetic and potential energies of particles together with the energies needed to assemble the particles into structured matter. See *potential energy (PE); kinetic energy (KE); particle; matter.*

interval (Δ*t*) **(78)** A span of time during which we observe a phenomenon. We calculate an interval by subtracting the initial time from the final time. It is always a positive number. Compare with *temperature difference (Δt).* See *time (t); phenomenon.*

inverse square law **(329)** A mathematical relationship in which the magnitude of one quantity varies with the inverse of the square of another (usually the distance between two points where the first quantity is measured). It applies to many areas of physics. See also *gravitational force; illumination.*

ion **(27)** An atom or molecule that has gained or lost electrons, thus producing a charge imbalance between the number of protons and electrons in the particle. See *atom; molecule; electron; charge (electrical); proton.*

ionic bond **(426)** A chemical bond between oppositely charged ions. The bonded atoms have very different electronegativities (e.g., metals and nonmetals). The electrostatic attraction of the ions is due to the nearly complete transfer of one or more electrons to the anion(s). See *chemical bond; ion; electronegativity; electrostatic force; anion; metal (chemistry); nonmetal.*

ionic compound **(426)** A solid compound consisting of ionically bonded metal and nonmetal ions. The ions of a solid compound are arranged in an immense crystal lattice consisting of repeating structures called unit cells. See *compound; ionic bond; ion; crystal lattice; unit cell; metal (chemistry); nonmetal.*

ionizable hydrogen **(485)** Any hydrogen atom in a covalent compound molecule that can be easily hydrated in an aqueous solution, resulting in the ionization of the compound. See *covalent compound; hydration; aqueous; solution; ionization.*

ionization **(465)** The breaking apart of the molecules of a covalent compound into ions by the action of the solvent to form a solution. Compare with *dissociation.* See *molecule; covalent compound; ion; solvent; solution.*

ionizing radiation **(313)** Any type of nuclear particle or electromagnetic energy that can produce molecular or atomic ions when it interacts with matter. This kind of radiation can be very damaging to living biological tissues. See *nuclear radiation; electromagnetic (EM) energy; ion.*

isotope **(364)** Any atom of an element that has a different number of neutrons compared to other atoms of the same element. All elements have two or more isotopes. See *element; neutron.*

isotopic notation **(365)** A special symbol that distinguishes between different isotopes of elements. It is written in the form $^{\text{mass number}}_{\text{atomic number}}\text{X}$, where X is the element's symbol. See *isotope; mass number (A); atomic number (Z).*

ITER **(378)** The name of the experimental fusion energy reactor project being built near Cadarache in southern France. Originally, it stood for *International Thermonuclear Experimental Reactor,* but the name was officially shortened to just *ITER,* which in Latin means "the way" (to a future of peaceful use of nuclear energy). See *fusion energy.*

IUPAC **(396)** The International Union of Pure and Applied Chemistry, which is the organization that sets the standards for the practice of chemistry around the world.

J

joule **(J)** **(110)** The SI unit for energy and work (1 J = 1 kg·m²/s² = 1 N·m). See *energy; mechanical work (W).*

K

Kelvin scale **(187)** Also called the *absolute temperature scale,* whose theoretical zero point is absolute zero. Its single fiducial point is the triple point of pure water (273.16 K). One kelvin is the same size unit as one degree Celsius. See also *absolute temperature scale (T).* See *absolute zero; fiducial point; triple point; Celsius scale.*

kilowatt-hour **(kWh)** **(221)** The unit of electrical energy used by utilities to bill for electricity.

kinematics **(80)** The science of describing *how* things move. It involves the measurements and calculations of position, time, velocity, acceleration, and displacement within a reference frame. See *position;*

time (t); velocity (v); acceleration (a); displacement (d); frame of reference.

kinetic energy (KE) **(111)** The energy of motion that depends only on the system's mass and speed. A form of mechanical energy. See *motion; system; mass (m); speed (v); mechanical energy.*

kinetic-molecular theory **(26)** The concept that tiny particles in constant, random motion make up all matter. See *particle; motion; matter.*

L

larynx **(286)** A structure located at the top of the trachea (windpipe) that contains the vocal cords and is the source of the human voice. Also called the *voice box,* it forms the Adam's apple in the throat. It is protected by the epiglottis when swallowing. See *vocal cords.*

laser **(331)** An abbreviation for *l*ight *a*mplification by *s*timulated *e*mission of *r*adiation. A method of producing monochromatic, coherent light that can be tightly focused and very intense. See *monochromatic; coherent light.*

latent heat of fusion (L_f) **(197)** The amount of thermal energy absorbed per gram as a pure solid melts (fuses) at its melting point. The same amount of heat per gram must be released to freeze the substance. See *thermal energy; melting; melting point; freezing.*

latent heat of vaporization (L_v) **(197)** The amount of thermal energy absorbed per gram as a pure liquid vaporizes. The same amount of heat per gram must be released to condense the vapor to a liquid. See *thermal energy; liquid; vaporization; heat (Q); condensation.*

lattice **(29)** See *crystal lattice.*

law **(11)** A model, often expressed as a mathematical equation, that describes phenomena under certain conditions. A law makes no attempt to explain or account for the phenomena. See *model; phenomenon.*

law of accelerated motion **(96)** Newton's second law of motion. The acceleration of a system is directly proportional to the net force acting on the system and is inversely proportional to the system's mass. See *system; net force; acceleration (a); mass (m).*

law of action-reaction **(97)** Newton's third law of motion. For every external force exerted on a system by its surroundings, the system exerts an equal but opposite force on its surroundings. Also called the *action-reaction principle.* See *external force; system; surroundings.*

law of charges **(206)** Like electrical charges repel and unlike charges attract. See *charge (electrical).*

law of conservation of energy **(120)** A fundamental natural conservation law. The total amount of energy entering a process equals the sum of all forms of energy that exist at the end of the process. See *conservation law; energy; first law of thermodynamics; process.*

law of conservation of matter **(37)** A fundamental natural conservation law. Matter can neither be created nor destroyed, but only changed from one form to another. This means that matter must be conserved in any physical change or chemical reaction it can undergo. See *conservation law; matter; physical change; chemical reaction.*

law of definite proportions **(356)** The masses of chemical substances combine in definite, characteristic integer ratios when forming compounds. It was first stated in the 1790s by the French chemist Joseph Proust. See *compound.*

law of inertia **(95)** Newton's first law of motion. Objects at rest remain at rest and objects in motion continue in a straight line at a constant speed unless acted on by a net external force. See *motion; speed (v); net force; external force.*

law of magnetism **(231)** Opposite magnetic poles attract and like magnetic poles repel. See *magnetic pole.*

law of octaves **(389)** A principle stated by chemist John Newlands in 1866 that the properties of the 49 known elements repeated every eighth element, as in a musical octave. See *element; octave.*

law of reflection **(337)** The angle of an incident ray equals the angle of the reflected ray. Both angles are measured in relation to the normal at the point of incidence. See *incident ray; reflected ray; normal; point of incidence.*

law of torques **(136)** In a lever system in rotational equilibrium, the effort torque equals the resistance torque. See *lever; system; rotational equilibrium; torque.*

law of universal gravitation **(99)** The force of gravity between two objects is directly proportional to the products of their masses and inversely proportional to the square of the distance between their centers of mass. Isaac Newton developed this fundamental law of physics. See *gravitational force; mass (m); distance (d).*

laws of motion **(95)** Generally refers to Newton's three laws of motion, which define the science of dynamics and apply to all areas of mechanics. These are the *law of inertia*, the *law of accelerated motion*, and the *law of action-reaction*. See *dynamics; mechanics.*

laws of planetary motion **(100)** Kepler's laws. Planets move in elliptical paths around the sun. Radial lines between the sun and a planet sweep out equal areas in equal times. The square of the orbital period of a planet is proportional to the cube of its average distance from the sun. Applies to any object orbiting a gravitational center. Developed by Johannes Kepler early in the seventeenth century.

least common multiple **(LCM) (441)** The smallest integer that can be divided without remainder by a given group of two or more integers.

lens (344) A curved disk of transparent material that refracts light so that the light converges or diverges. See *refraction*; *converging lens*; *diverging lens*.

lever (135) A simple machine that can be modeled as a rigid bar that pivots about a fulcrum to exert an effort torque against a resistance torque. See *fulcrum*; *torque*.

lever arm **(137)** See *torque arm*.

Lewis structure (418) A method of showing the covalent bonds and arrangement of a molecule using electron dot notation. See *covalent bond*; *electron dot notation*.

Leyden jar (213) An early charge-storage device that consisted of a glass jar lined and coated on the outside with lead. It used electrical induction and grounding to greatly increase its charge storage capacity. It is an ancestor to the modern capacitor. See *electrical ground*; *capacitor*.

lift **(166)** The supporting force on an airfoil or hydrofoil created as it moves through a fluid. See *force (F)*; *airfoil*.

light-emitting diode **(LED) (327)** A semiconductor device that emits a specific frequency range of light when it conducts electricity. A *diode* is an electrical component that will pass a current in one direction, but not the other. See *semiconductor*.

light-year (ly) **(328)** The distance that light travels in one year in a vacuum, approximately 9.6 trillion km (6 trillion mi); the unit of distance (not time) best applied to objects far outside our solar system. See *distance (d)*; *speed of light (c)*; *time (t)*.

lightning **(212)** A huge electrical discharge between a weather-related cloud and the ground or from cloud to cloud. Similar discharges occur within volcanic eruption clouds and can be created artificially on a smaller scale in laboratories. Astronomers have observed lightning on other planets.

lightning rod **(211)** A metal rod attached to the highest point of a building. It is designed to conduct a lightning discharge safely through cables to the ground, thus protecting the building from damage. See *lightning*; *conductor*.

linear momentum **(121)** The momentum of a system moving in a straight line and that is directly proportional to its mass and speed ($p = mv$). Compare with *angular momentum*. See *momentum (p)*; *system*; *mass (m)*; *speed (v)*.

lines of force **(207)** Imaginary lines used to model electric and magnetic fields. Their density and direction represent the strength and direction of the field force. See *model*; *field force*.

line spectrum **(327)** A display produced by a spectrograph consisting of lines representing discrete frequencies or wavelengths in an electromagnetic emission spectrum. If the lines are colors against a dark background, it is a *bright-line spectrum*; if black lines against a continuous spectrum, it is a *dark-line spectrum*. See *spectrograph*; *electromagnetic (EM) spectrum*; *continuous spectrum*.

liquid (32) The state of a substance in which its particles are close together but mobile. A liquid has a definite volume but no fixed shape. It has low compressibility, can flow, and has a free surface in a less-than-full container. See *state*; *particle*; *volume (V)*; *compressibility*; *free surface*.

lithium ion battery **(217)** Refers to a rechargeable dry cell battery technology that uses a special electrolyte paste containing mobile lithium ions. Commonly used to power portable electrical devices. Lithium ion batteries operate longer and at higher current rates than most other battery types. There is also a lithium-based, non-rechargeable battery type. See *dry cell battery*; *electrolyte*.

litmus **(499)** One of a family of organic pH indicators that are extracted from various species of lichens. It typically turns red in acidic solutions and blue in alkaline ones. See *organic*; *pH indicator*; *acidic*; *solution*; *alkaline*.

load **(133)** The system on which work is done by a machine. The resistance force acting on a simple machine. Also called *resistance*. See *simple machine*; *resistance force* (F_r).

logarithm **(499)** An exponent of a given base number (e.g., 2 or 10). It represents a standard number equal to the base raised to that exponent. Scientists typically use logarithms to represent quantities that can vary over extremely large ranges of values, such as pH and sound intensity, on a *logarithmic scale*. See *pH*; *intensity (sound)*.

logarithmic scale **(281)** A scale that expresses numbers as powers of ten or some other base number. It allows for values of widely varying magnitudes, such as sound intensity, to be represented by a small range of numbers. See *magnitude.*

longitudinal wave **(268)** A wave in which the oscillations are parallel to the direction of wave travel. See *wave; oscillation.*

loudness **(281)** The response of the ear and brain to the intensity of a sound. Intensity is measurable but loudness is not. It is dependent on the listener's perception of intensity. See *sound*; *intensity (sound).*

luminous **(326)** The condition of producing light, usually by incandescence or some other form of light production. Compare with *illuminated.* See *incandescence.*

luster **(395)** The quality of light reflected by a substance. The various lusters are described by referring to common materials having similar reflective properties. A material's luster may be metallic, pearly, glassy, etc.

M

macromolecule **(423)** A very large molecule; a molecule with many atoms. Biological macromolecules can contain hundreds or thousands of atoms (e.g., DNA). See *molecule.*

maglev **(247)** An emerging transportation technology that relies on magnetic levitation and in some cases propulsion. Several nations are researching and building maglev trains as a form of high-speed, economical transportation.

magnet **(229)** Any object that possesses or can generate a magnetic field. See *magnetic field.*

magnetic core memory **(245)** An early magnetic digital data storage medium consisting of tiny ferrite rings nested in a matrix of fine wires. *Binary bits* were set by passing current through wires that would magnetize a core in one direction ("0") or the other ("1"). See *binary system.*

magnetic declination **(233)** At a given location, the difference in direction, measured in degrees, to the magnetic North Pole and to the geographic North Pole. If the magnetic pole lies to the right of true north by compass, declination is east; if to the left, declination is west. To convert a magnetic compass reading to a true direction, you add west declination to and subtract east declination from the compass direction. See *geographic North Pole*; *magnetic North Pole.*

magnetic dipole **(230)** Any object or particle with normal magnetic properties, i.e., having two magnetic poles. All magnets are magnetic dipoles. See *magnetic pole.*

magnetic energy **(116)** The ability of a magnetic field to do work on magnetic objects and moving electrical charges. See *magnetic field*; *mechanical work (W)*; *charge (electrical).*

magnetic field **(230)** The region of influence surrounding a magnet where it exerts a magnetic force, modeled by lines of force. See *magnet*; *field force.*

magnetic North Pole **(233)** The point in the Northern Hemisphere where the earth's magnetic field lines are most concentrated and nearly vertical as they pass into the earth's interior. It is also the south magnetic pole of the earth's geomagnetic field. Compare with *geographic North Pole.* See *geomagnetic field.*

magnetic pole **(229)** One of two areas of concentrated magnetic field lines on a magnet's surface. The field direction for each pole is the opposite of the other. See *magnetic field*; *lines of force.*

magnetic resonance imaging **(MRI)** **(316)** A form of medical imaging that surrounds a patient with strong oscillating fields of magnetic energy. Under its influence, the molecules in the patient's tissues produce radio waves that are detected and converted into a three-dimensional image of the patient's internal tissues. See *medical imaging*; *oscillation*; *magnetic field*; *radio wave.*

magnification power (×) **(346)** The ratio of the image size produced by an optical instrument to the object's size when viewed without the instrument. In practice, it is the ratio of the angular size of the image to the angular size of the object. For example, if the object appears to be 1° high to the unaided eye but appears to be 10° high through a telescope, then the telescope has a magnification power of 10×.

magnitude **(81)** The size of any quantity expressed as a positive number in appropriate units. The length or size of a vector is its magnitude. See *vector.*

malleable **(395)** Able to be easily deformed by hammering or cold rolling. It is an important property (called *malleability*) of most metals. See *property*; *metal (chemistry).*

mass (*m*) **(64)** The measure of the inertia of matter in an object or within the boundaries of a physical system. See *inertia; matter*; *system.*

mass defect **(117)** The difference in mass between a nucleus and the masses of the uncombined protons and neutrons that make up its nucleus. The difference in mass is converted to the energy required to hold the nucleus together. Compare with *binding energy.* See *mass (m)*; *nucleus*; *proton*; *neutron*; *energy.*

mass energy (**117**) The energy equivalent to matter itself, according to the equation in Einstein's special theory of relativity, $E = mc^2$. See *special theory of relativity.*

mass number **(*A*)** (**364**) A whole number representing the sum of the protons and neutrons in the nucleus of an atom. It uniquely identifies isotopes of an element. See *proton*; *neutron*; *nucleus*; *isotope.*

mass-percent (**472**) See *percentage by mass.*

matter (**21**) One of the fundamental concepts of creation. Anything that occupies space and has mass. This is an operational definition—a statement of conditions that something must meet in order to be considered matter. We cannot define matter in simpler terms. See *space*; *mass (m).*

measurement (**48**) The comparison of a dimension of an object or substance to an appropriate standard, usually an instrument with a graduated scale. See *dimension*; *standard*; *instrument*; *graduated scale.*

mechanical advantage (MA) (**133**) A measure of the reduction in effort to do a certain amount of work when using a simple machine. MA has no units. *Actual mechanical advantage (AMA)* is the ratio of the actual resistance force, including friction, to the effort force; *ideal mechanical advantage (IMA)* is the ratio of just the resistance force to the effort force without considering system friction. See *simple machine*; *effort force* (F_e); *resistance force* (F_r); *friction.*

mechanical amplification (**290**) A mechanical device, such as a megaphone, that focuses and directs sound waves in a particular direction, resulting in an increase in loudness. See *amplification*; *loudness.*

mechanical energy (**112**) The energy resulting from the motion of matter or its position relative to an external force acting on it, i.e., *kinetic energy (KE)* and *potential energy (PE).* See *motion*; *position*; *external force.*

mechanical equilibrium (**96**) The condition in which all forces acting on a system are balanced as indicated by the system's lack of acceleration. See *force (F)*; *system*; *balanced force*; *acceleration (a).*

mechanical equivalent of heat (**180**) The direct, equal relationship between mechanical work and thermal energy. It is known that 4.18 N·m of mechanical work is equivalent to the heat necessary to raise 1 g of water 1 °C. In the SI, this quantity is known as the joule (J). See *mechanical work (W)*; *thermal energy*; *heat (Q)*; *joule (J).*

mechanical force (**93**) The proper name for a contact force. These forces include *compression*, *tension*, *shear*, and *torsional forces.* See *contact force.*

mechanical tolerance (**62**) The allowable variation in a manufactured dimension of a part that assures that the part will fit and work properly when assembled with the other parts in a man-made product. Usually given as a ± value from the intended dimension. See *dimension.*

mechanical wave (**264**) A wave that travels through matter by the periodic motion of its particles. See *wave; periodic motion.*

mechanical work (*W*) (**129**) The energy transferred to a system by an external force when it acts on the system to move it. A scalar quantity measured in joules (J) in the SI. For positive work, the force vector acts in the same direction that the system moves; for negative work, the force vector acts in the direction *opposite* to system motion. Zero work is done on a system by a force if the system does not move or if the applied force is perpendicular to the direction of motion. See *external force*; *system*; *vector*; *scalar.*

mechanics (**75**) The scientific study of forces and motion, which consists of kinematics, dynamics, and statics. See *force (F)*; *motion*; *kinematics*; *dynamics*; *statics.*

medical imaging (**316**) Any method of obtaining images of the interior of a medical patient without the need for major surgery. Familiar techniques include x-ray, ultrasound, MRI, and CT images. See *x-ray*; *ultrasound*; *magnetic resonance imaging (MRI)*; *computed axial tomography (CT).*

medium (**265**) The matter through which a wave travels. A medium is required for all mechanical waves but impedes electromagnetic waves. See *wave*; *mechanical wave; electromagnetic wave.*

melting (**40**) The change of state from a solid to a liquid that occurs when a substance's temperature rises to its melting point. See *state*; *solid*; *liquid*; *temperature (t)*; *melting point.*

melting point (**40**) The temperature at which a solid turns into a liquid at 1 atm. For a pure substance, the same temperature as its freezing point. See *temperature (t)*; *solid*; *liquid*; *pure substance*; *freezing point.*

membrane (**462**) A thin, sheet-like material that forms a flexible barrier between two media or environments. See *medium.*

meniscus lens (**346**) A lens that has one convex side and one concave side. Preferred for glasses because it provides a larger area of corrected vision. Can act as a converging or diverging lens, depending on its shape. See *converging lens*; *diverging lens.*

mercury barometer (**155**) An instrument that measures atmospheric pressure. It consists of a mercury-filled glass tube, sealed at one end and open at the other.

The mercury column is supported by air pressure acting on a reservoir of mercury in which the tube's open end rests. The height of the column is proportional to air pressure, which is read on a graduated scale in inches or units of pressure. See *instrument*; *atmospheric pressure*; *graduated scale*.

metal (chemistry) (394) An element that is typically dense, solid, ductile, malleable, highly conductive, and chemically reactive, especially in the presence of nonmetal elements. Metals have few valence electrons and comprise almost three-fourths of all elements. See *ductile*; *malleable*; *conductor*; *reactive*; *nonmetal*; *valence electron*.

metallic bond (430) A chemical bond characteristic of metals in which the atoms freely share their few valence electrons with all the other metal atoms. See *metal (chemistry)*; *chemical bond*; *valence electron*.

metalloid (396) An element that is neither a metal nor a nonmetal that can display some characteristics of either under certain conditions. Also called a *semimetal*. See *element*; *metal (chemistry)*; *nonmetal*.

***metric instrument* (58)** Any instrument with a graduated scale divided with 10 marks between significant subdivisions (a *decimal* scale). See *instrument*; *graduated scale*.

metric system (49) A system of measurement in which all units of a given dimension are related to each other by powers of 10. The main metric system used today is the SI. See *measurement*; *dimension*; *SI*.

***microscope* (347)** An optical instrument that magnifies extremely tiny objects not normally visible to the unaided eye. Basic microscopes include an objective lens, which forms the image of the object, and an eyepiece lens, which magnifies and focuses the image for viewing. Modern microscopes have objectives and eyepieces containing multiple lens elements.

microwave (307) A subset of the radio-frequency band of the electromagnetic spectrum extending from 100 MHz to 300 GHz. Microwaves support line-of-sight communications, high-speed data links, and high-accuracy radar. See *electromagnetic (EM) spectrum*; *radio wave*.

***middle ear* (288)** The parts of the ear that amplify and prepare sound waves for conversion into nerve impulses in the inner ear. Includes the *eardrum*, the *hammer*, *anvil*, and *stirrup* bones, and the *oval window* of the cochlea. See *inner ear*.

***mineral* (31)** A naturally occurring, inorganic, crystalline solid that has a definite chemical composition. See *inorganic compound*; *crystalline solid*; *composition*.

***mirage* (343)** An optical phenomenon that alters the appearance of distant objects as light from the objects is refracted through atmospheric layers of different optical densities. See *refraction*; *optical density*.

***mirror* (338)** A surface that reflects nearly all incident light rays as a specular reflection. See *incident ray*; *specular reflection*.

***mirror image* (338)** The image reflected by a plane mirror. The image is upright; the left side is reflected on the left and the right side is reflected on the right; often incorrectly interpreted to be reversed left-to-right. See *plane mirror*; *reflection*.

***miscibility* (467)** The solubility of one liquid in another liquid. Used only with liquids. Liquids that do not mix are *immiscible*. See *solubility*.

mixture (29) A nonchemical combination of two or more substances.

model (8) A workable explanation or description of a natural or artificial phenomenon. Models may be physical, virtual, or conceptual. Also, the act of creating a model. See *workability*; *phenomenon*.

***molarity* (*M*) (473)** Solution concentration measured in units of moles of solute per liter of solution (mol/L). Scientists prefer the phrase *amount-of-substance concentration* of the solute, or simply the *concentration* of the solute. *Molarity* is becoming obsolete in the scientific community. See *solution*; *concentration*; *mole (mol)*.

***mole* (mol) (473)** The SI base unit for the quantity of matter in a substance, especially for numbering particles of a pure substance. A mole is equal to Avogadro's number of particles, or to the mass in grams of that number of particles. See *SI base unit*; *particle*; *Avogadro's number*; *pure substance*.

***molecular formula* (438)** The chemical formula of a molecular compound. Molecular formulas show the actual numbers of atoms bonded in a formula unit of the compound. Compare with *empirical formula*. See *chemical formula*; *covalent compound*; *formula unit*.

***molecule* (26)** A distinct particle formed when two or more atoms covalently bond together. See *particle*; *atom*; *covalent bond*.

***moment* (136)** See *torque*.

momentum (*p*) (121) A property of a moving system that is proportional to its speed and mass. Isaac Newton called it the *quantity of motion*. See *property*; *speed (v)*; *mass (m)*.

***monatomic element* (387)** An element that occurs naturally as a single atom. Only the six natural noble gases occur as monatomic elements. See *element*; *noble gas*.

monochromatic **(327)** The property of having a single color. See *property*; *color.*

monopole **(230)** A hypothetical object or particle that has only a single magnetic pole. Such an object has never been observed. See *magnetic pole.*

monoprotic acid **(487)** An acid that can donate only one hydrogen ion. See *acid*; *ionizable hydrogen.*

motion **(80)** A change of position during a time interval. See *position*; *interval* (Δt).

motor **(241)** A rotating machine that converts electrical current into mechanical motion. It may be an AC or DC motor, depending on the intended kind of electricity supplying it. See *alternating current (AC)*; *direct current (DC).*

N

native mineral **(383)** Any solid element that occurs in its pure form in nature, such as native copper, gold, or sulfur. See *element.*

natural frequency **(259)** The frequency at which an oscillating object experiences resonance. See *frequency (f)*; *oscillation*; *resonance.*

natural philosopher **(75)** An early worker in science who might have studied all aspects of the physical world. When biology, physics, and chemistry became separate areas of science, the name went out of use.

nebular hypothesis **(301)** A hypothesis for the secular, natural origin of stellar systems, and specifically our solar system; first proposed in 1734 by Emanuel Swedenborg and developed by Pierre Simon de Laplace in 1796. See *hypothesis.*

net force **(94)** The single zero or nonzero force acting on a system that is the sum of *all* external forces acting on the system. If the external forces are balanced, then the net force is zero. See also *balanced force; unbalanced force.* See *system; external force.*

neutral **(497)** Used to describe a chemical solution that has a pH at or near 7. See *solution*; *pH.*

neutralization reaction (492) A reaction between an acid and a base. In aqueous solutions, the products are a salt and water. See *acid*; *base*; *aqueous*; *solution*; *salt.*

neutron **(26)** A neutral elementary particle located in the nucleus of an atom (a *nucleon*). It has about the same mass as a proton, and its purpose seems to be to stabilize the nucleus. The number of neutrons can vary within the atoms of a given element. See *elementary particle*; *nucleon*; *proton*; *nucleus*; *atom*; *element.*

newton **(N) (91)** The SI unit of force. In SI base units, $1 \text{ N} = 1 \text{ kg·m/s}^2$. See *SI base unit*; *force (F).*

nitrogen family (399) Group 5A of the periodic table. The chemical activity of this group is highly variable. Several elements have as many as eight different oxidation numbers. See *element*; *family*; *group*; *periodic table of the elements*; *oxidation number (ON).*

noble gas (400) An element in Group 8A of the periodic table. The noble gas elements are essentially inert, monatomic gases with an oxidation number of 0. See *element*; *group*; *periodic table of the elements*; *inert*; *monatomic element*; *oxidation number (ON).*

node **(271)** A point in a standing wave that experiences no displacement. Nodes are spaced at half-wavelengths apart. See *standing wave; wavelength* (λ).

nonflammable **(36)** The property of a substance that does *not* burn rapidly or easily catch fire. Compare with *flammable.* See *property.*

non-ionizing radiation **(371)** Lower-energy forms of mainly electromagnetic energy that do not cause atomic or molecular ionizations when absorbed by organic or inorganic materials. It includes low-energy UV light, visible light, infrared, microwave, and radio-frequency radiation. Compare with *ionizing radiation.* See *electromagnetic (EM) energy*; *radio-frequency spectrum*; *radiation*; *ion.*

nonmetal (395) An element that typically has four or more valence electrons and does *not* exhibit the general properties of metals. A nonmetal may be a gas, a liquid, or a dull, brittle solid at standard conditions. They are generally poor conductors, and their chemical activity is highly variable, from very reactive to inert. See *element*; *valence electrons*; *metal (chemistry)*; *standard conditions*; *conductor*; *reactive*; *inert.*

nonpolar bond **(421)** A bond between two atoms of the same nonmetal element. Electrons are evenly shared, so no partial charges exist at the ends of the bond.

normal **(337)** An imaginary line that is perpendicular to a surface at a point of incidence. Angles of incidence, reflection, and refraction are measured relative to normal lines. See *point of incidence*; *angle of incidence*; *angle of reflection*; *angle of refraction.*

normal science **(10)** The day-to-day activities of scientists who are working to refine the accuracy and precision of scientific knowledge and extend the boundaries of the scientific paradigm within which they work. See *accuracy*; *precision*; *scientific paradigm.*

nuclear bombardment reaction **(375)** A nuclear change that occurs when a nucleus is struck by a high-energy particle or another nucleus. See *nucleus.*

nuclear change **(38)** The change in the energy or composition of an atom's nucleus when it emits or absorbs a particle or ray. See *composition*; *atom*; *nucleus*; *nuclear radiation*.

nuclear chemistry **(370)** The study of changes that occur in atomic nuclei. See *nucleus*.

nuclear decay **(371)** Any spontaneous change in the particle makeup and/or total energy of the nucleus of an unstable atom. Nuclear decays occur when alpha particles are ejected, neutrons decay, or gamma rays are emitted. See *nucleus*; *neutron*; *alpha (α) particle*; *gamma ray*.

nuclear energy **(117)** The potential energy, stored in an atom's nucleus, that is released or absorbed when an atom experiences nuclear fission or fusion. See *potential energy (PE)*; *nucleus*; *nuclear fission*; *nuclear fusion*.

nuclear fission **(376)** The splitting of a large nucleus into two smaller nuclei and several free neutrons. Nuclear fission most commonly occurs from artificial neutron bombardment in nuclear power reactors. See *nucleus*; *nuclear bombardment reaction*; *nuclear reactor*.

nuclear fusion **(376)** The violent fusing of two small nuclei together to form a larger one under extremely energetic conditions. Nuclear fusion produces far more energy per event than a single nuclear fission event. See *nucleus*; *nuclear fission*.

nuclear model **(359)** An atomic model developed by Ernest Rutherford, who theorized that every atom has an extremely tiny, positively charged nucleus at its center, where most of the atom's mass is concentrated, and that the empty space around this nucleus contains the electrons. See *electron*; *nucleus*.

nuclear physics **(363)** The study of the particles, their arrangement, structure, and energy, and the changes that can occur within an atom's nucleus. See *nucleus*.

nuclear radiation **(370)** The rays and particles emitted during nuclear decay. See *ray*; *particle*; *nuclear decay*.

nuclear reactor **(376)** A complex device that controls a nuclear fission chain reaction to produce thermal energy used for generating steam in electrical power plants or ship propulsion. See *nuclear fission*; *chain reaction*.

nucleon **(362)** A proton or neutron bound in the nucleus of an atom. See *nucleus*; *atom*; *proton*; *neutron*.

nucleus **(26)** The center of mass of an atom. It contains protons and usually neutrons, which make up nearly all the mass of the atom.

nuclide **(365)** An isotope of an element. The term is usually used when discussing isotopes of different elements. See *isotope*; *element*.

O

octave **(389)** In music, a pattern of ascending or descending notes that repeats every eighth note.

octet rule **(414)** Atoms are generally most stable when they have a full set of electrons in their outer energy level. Except for elements with $Z = 1$ to 5, this is usually eight electrons. See *electron*; *energy level*; *atomic number (Z)*.

ohm **(Ω)** **(220)** The SI derived unit of electrical resistance ($1\ \Omega = 1\ \text{V/A}$). See *electrical resistance (R)*.

Ohm's law **(220)** An electrical law relating to DC circuits that states the relationship between *voltage (V)*, *current (I)*, and *resistance (R)* ($V = I \times R$). See *voltage (V)*; *electrical current (I)*; *electrical resistance (R)*.

operational science **(11)** Scientific work that involves observing and testing present-day phenomena. Compare with *historical science*. See *normal science*; *phenomenon; dominion science*.

optical density **(341)** A measure of a transparent material's ability to impede (slow down) the movement of light. See also *index of refraction (n)*.

orbital **(361)** A three-dimensional region surrounding an atom's nucleus within which a pair of electrons is most likely to be found. The shape and size of an orbital is determined by the electron's quantized energy. See *quantized*; *electron*; *nucleus*.

ore **(383)** A rocky earth material that is mined as a source for a useful metal.

organic **(375)** Refers to anything that is or was living, or is a substance uniquely produced by organisms. Compare with *organic compound*.

organic compound **(29)** A compound that always contains carbon and usually hydrogen. Oxygen, nitrogen, and phosphorus are other elements often included in organic compounds found in living organisms.

oscillation **(255)** In this textbook, visible, relatively large or slow repetitive motions. Clock pendulums and wind-blown leaves on trees oscillate. Compare with *vibration*. See *repetitive motion; periodic motion*.

osmosis **(474)** The diffusion of a solvent through a semipermeable membrane from a region of higher concentration of the solute to one of lower concentration. See *solvent; semipermeable membrane*; *concentration*; *solute*.

osmotic pressure **(475)** The minimum pressure that must be applied to the solution side of a semipermeable membrane to prevent the net diffusion of the pure solvent through the membrane into the solution. See *fluid pressure*; *solution*; *semipermeable membrane*; *diffusion*; *solvent*.

outer ear **(287)** The parts of the ear that catch and direct sounds into the middle ear. Includes the external ear structure and the ear canal. See *middle ear*.

out of phase **(271)** The condition where the waveforms of two or more wave phenomena align crest-to-trough. This results in destructive interference, producing waves with lower amplitudes than that of any individual wave. Compare with *in phase*. See *wave phase*; *waveform*; *crest*; *trough*; *destructive interference*; *amplitude*.

overcurrent protection **(224)** A protective electrical device designed to open the circuit when excessive current flows due to a fault in the circuit. See *electrical current (I)*; *short circuit*.

overtone **(282)** See *harmonic (wave)*.

oxidation number (ON) (438) The number of electrons that a bonded atom or ion must gain or lose to return to its neutral state. Oxidation numbers are used mainly for writing the formulas of inorganic compounds. See *inorganic compound*.

oxygen family (399) Group 6A of the periodic table. Most elements in this chemically reactive family have an oxidation number of –2. See *group*; *periodic table of the elements*; *reactive*; *element*; *family*; *oxidation number (ON)*.

ozone **(387)** A form of oxygen molecule containing three covalently bonded oxygen atoms. Ozone occurs naturally in the stratosphere and is an air pollutant near the ground. See *molecule*; *covalent bond*.

P

paradigm **(10)** See *scientific paradigm*.

parallel circuit (223) An electrical circuit or portion of a circuit with multiple current paths so that the supplied current must split up to flow through each load in the circuit. See *electrical current (I)*; *load*.

paramagnetism (232) Property exhibited by materials that are only slightly attracted to magnets. Permeability to a magnetic field varies with temperature. Paramagnetic materials do not become permanent magnets. See *property*; *permanent magnet*.

partially elastic collision **(123)** A collision between two objects in which momentum is conserved but some of their kinetic energies is lost to other forms of energy during the collision. The objects are slightly or permanently deformed but rebound from the collision. See *conservation of momentum*; *kinetic energy (KE)*; *law of conservation of energy*.

partial pressure **(169)** In a mixture of gases, the contribution of the pressure of a given gas to the total pressure of the sample. The total pressure of the gas sample is equal to the sums of the partial pressures of the individual gases. See *pressure (P)*.

particle **(21)** In physics, an extremely tiny bit of matter that can be treated as a physical system and which obeys physical principles. Particles can be molecules, atoms, or elementary particles like electrons or protons. In some areas of study, such as ray optics, even electromagnetic photons are considered to be particles. See *system*; *molecule*; *atom*; *elementary particle*; *ray optics*; *photon*.

particle theory **(22)** The concept that all matter is made of exceedingly small particles. Also called *atomism*. The particle theory of matter is one of the most fundamental concepts supporting models of the universe. See *particle*.

pascal **(Pa) (154)** The SI derived unit of pressure (1 Pa = 1 N/m^2). See *SI*; *unit*; *pressure (P)*.

Pascal's principle (161) States that changes of pressure on the surface of a confined fluid are exerted equally throughout the fluid and at all points on the fluid's container. See *pressure (P)*; *fluid*.

pendulum (260) A mass attached to the end of an arm suspended from a pivot point that is free to swing back and forth under the influence of gravity. See *center of mass*; *pendulum arm*.

pendulum arm **(260)** The distance between a pendulum's pivot point and the center of the pendulum's mass. See *pendulum*; *center of mass*.

percentage by mass (472) Solution concentration found by taking the ratio of the mass of the solute to the mass of the solution and expressing it as a percentage. Also called *mass-percent*. See *solution*; *concentration*; *solute*.

period (chemistry) (401) A horizontal row in the periodic table of the elements. Also called a *series*. See *periodic table of the elements*.

period (*T*) (256) The time interval required to complete one cycle of a system experiencing periodic motion. It is measured in seconds (s) in the SI. See *interval* (Δt); *cycle*; *periodic motion*.

periodicity **(389)** In chemistry, the idea that many properties of elements appear to repeat in regular or periodic ways in relation to some basic dimension, such as atomic mass or atomic number. See *element*; *atomic mass*; *atomic number (Z)*.

periodic law of the elements (**394**) The properties of the elements vary with their atomic numbers in a periodic way. See *element*; *periodicity*; *atomic number (Z)*.

periodic motion (**255**) The back-and-forth or cyclical repetitive motion of a system having a regular period. See *repetitive motion*; *period*.

periodic table of the elements (**388**) A table of the chemical elements that is arranged to display their periodic properties in relation to their atomic numbers. This name usually refers to the Mendeleyev/Moseley form of the periodic table, although other formats exist. See *element*; *atomic number (Z)*; *periodicity*.

permanent magnet (**232**) Any magnet in which its magnetic domains remain more or less aligned without the influence of an external magnetic field. Ferromagnetic materials can become permanent magnets. See *magnetic domain*; *ferromagnetic material*.

pH (**497**) The acidity or alkalinity of a solution. See *acidity*; *alkalinity*; *pH scale*; *solution*.

pH indicator (**499**) Any chemical compound, often an organic one, that represents the solution's pH by turning a specific color in a solution. See *compound*; *organic compound*; *solution*; *pH*.

pH meter (**500**) An instrument designed to measure the pH of a solution, using the electrical potential of the hydronium ions in the solution. Its most important component is the *pH probe*, which is a special sensor that responds to the presence of hydronium ions. See *pH*; *voltage*; *hydronium ion*; *solution*.

pH scale (**497**) A scale that indicates the acidity or alkalinity of a solution and that is based on the logarithm of the hydronium ion concentration in moles per liter. See *acidity*; *alkalinity*; *logarithm*; *hydronium ion*; *concentration*.

phase (mixture) (**30**) Matter that has distinctly different properties from other matter that it is mixed with. One of the physical states of a substance in a mixture of its different states. A separate part of a heterogeneous mixture. See *heterogenous mixture*.

phase (wave) (**271**) See *wave phase*.

phase change (**39**) Any change of physical state, e.g., from solid to liquid or liquid to gas vapor.

phenomenon (**8**) An observable or measureable object, process, or property. Science studies phenomena. See *process*; *property*.

phosphorescence (**330**) The continuing emission of visible light by a substance after having been exposed to high-frequency electromagnetic energy. See *electromagnetic (EM) energy*.

photoetching (**490**) A chemical process of producing fine metal objects and parts using an acid solution. The object pattern, made out of an acid-resistant chemical, is first printed on a thin sheet of metal, often using a photocopier-like machine. The metal is then placed in the acid solution and the unprotected portions of the sheet are dissolved. Used mainly for electronic circuit board manufacturing, model parts, and fine jewelry. See *process*; *acid*.

photometer (**329**) An optical or electronic instrument that measures the illumination by light at the instrument. See *illumination*.

photon (**304**) A packet of electromagnetic energy modeled as a particle rather than as a wave. The photon is one of the fundamental elementary particles of the standard model of matter. See *electromagnetic (EM) energy*; *elementary particle*; *standard model*.

photovoltaic* (PV) *cell (**218**) A semiconductor device that converts light energy into an electrical potential. Also called a *solar cell*. See *semiconductor*; *electrical potential energy*.

physical change (**35**) Any change that does not alter the composition of a substance or its nuclear properties (e.g., melting and crushing). See *composition*; *property*.

physical model (**9**) A model of any object, real or imaginary, that can be handled and examined to understand the relationships of its parts to the whole. See *model*.

physical property (**34**) Any property of matter that can be observed or measured without altering its chemical composition. See *property*; *composition*.

physical science (**3**) A subset of science that deals primarily with physics, chemistry, and the earth sciences. While physical science principles apply to life processes, living organisms are not its primary focus. See *process*.

physics (**15**) The scientific study of matter, energy, and work, and how they are related and can be used. See *matter*; *energy*; *mechanical work (W)*.

pigment (**334**) A chemical compound whose main purpose is to supply color to matter. It works by absorbing and reflecting specific frequencies of incident or transmitted light. See *compound*; *color*; *incident*; *transmitted*.

pitch (fastener) (**148**) The distance between two adjacent threads on metric screws and bolts, measured in millimeters. A smaller pitch implies greater holding power (greater mechanical advantage). See *metric system*; *mechanical advantage (MA)*.

pitch (sound) **(280)** The property of an audible sound related to its frequency. Higher frequencies have higher pitches. Frequency is measurable but pitch is not; it is dependent on the listener's perception. See *property*; *sound*; *frequency (f)*; *dimension*.

pixel **(317)** The smallest definable element of a digital image that can be represented in the binary system. It is a combination of the words *picture* and *element*. See *binary system*.

plane mirror **(338)** A flat mirror. See *mirror*.

planetary model **(360)** An atomic model resembling a solar system, developed by Niels Bohr, who suggested that electrons orbit at specific distances from the nucleus according to their energies. Also called the *Bohr model*. See *electron*; *nucleus*.

plasma **(21)** A gas-like substance, formed at very high temperatures, that consists of high-energy ions.

plum pudding model **(357)** An atomic model developed by J. J. Thomson that views an atom as a sphere of positive charge in which negative electrons are embedded. The model was the first to suggest that the atom was divisible into smaller parts. See *electron*.

pneumatics **(163)** The physics of stationary and flowing gases in confined piping systems.

point of incidence **(337)** The point at which the incident light ray strikes the reflective surface or boundary between two refractive media. It is the same as the point of reflection and refraction. See *incident ray*; *reflection*; *refraction*; *medium*.

point of reference **(76)** The location within a reference frame from which an observer measures all positions and motions. See *position*; *motion*; *frame of reference*.

polar **(421)** Used to describe the property of having concentrations of positive and negative electrical charges or *poles*. Especially used when describing covalent bonds and molecules with such charge concentrations. See *property; charge (electrical)*; *covalent bond*; *molecule*.

polar covalent bond **(421)** A covalent bond between the atoms of two different elements, especially those having greatly different electronegativity numbers (EN). The bond tends to be polar, that is, more negative toward the element with the higher electronegativity. See *covalent bond*; *electronegativity*; *polar*.

polyatomic ion **(441)** An ion made of two or more covalently bonded atoms that acts as a single charged particle. See *ion; covalent bond*.

polyatomic molecule **(420)** A molecule with more than two atoms. Used to distinguish larger molecules from simple diatomic (two-atom) molecules. See *molecule*; *atom*.

polycrystalline solid **(31)** A substance composed of innumerable microscopic crystals compacted into a solid material. See *composition*; *crystalline solid*; *solid*.

polymer **(31)** A solid material resulting from the creation of long chains of identical molecular units bonded together (*polymerized*). See *solid*; *molecule*; *chemical bond*.

polyprotic acid **(488)** An acid that can donate two or more hydrogen ions. A diprotic acid can donate two ions. Triprotic acids can donate three ions. See *acid*; *ionizable hydrogen*.

position **(80)** A location identified by one or more coordinates, depending on the number of spatial dimensions defined by coordinate axes. Linear positions require one position coordinate; positions on a plane require two; positions in space require three. See *spatial*; *dimension*; *coordinate axis*; *space*.

positivism **(10)** A philosophy of science that claims that all matters of fact can be understood only by experience through the senses, and by human logic and mathematical treatment. One of its presuppositions is that no supernatural being or event has influenced the natural world. See *presupposition*.

potential energy (PE) **(111)** The energy of a system due to its position or condition. A form of mechanical energy. See *position*; *mechanical energy*.

power (*P*) **(131)** The rate of doing work, measured in watts (W) in the SI. See *mechanical work (W)*; *SI*.

precipitate **(448)** An insoluble solid produced by a chemical reaction. In a water medium, it separates from the solution as a solid phase and often settles out. A precipitate also forms when the concentration of a solid solute exceeds its solubility limit. See *solid*; *chemical reaction*; *solution*; *phase (mixture)*; *concentration*; *solubility*; *insoluble*.

precision **(57)** An assessment of the exactness of a measurement. A more precise measurement has more known digits than a less precise measurement of the same quantity. The fineness of an instrument's scale markings determines the maximum precision of a measurement. See *measurement*; *instrument*; *graduated scale*.

prejudice **(7)** An inclination one holds about an idea without thinking about its reasonableness.

pressure (*P*) **(152)** The force exerted perpendicularly on a unit of area. Units of pressure include N/m^2 (or pascals) and $lbf/in.^2$ See *force*; *unit*; *pascal (Pa)*.

presupposition **(6)** An assumption that a person makes about the world that helps him to investigate the world and understand it from a certain point of view. The essential presupposition of a Christian worldview is that the Bible is the complete and true revelation of God. From the Bible we acquire the additional presuppositions that God created the world, that the creation is in a fallen condition, and that God will redeem this fallen world to Himself through Jesus Christ.

primary hue **(333)** A single color of a small set of distinct colors that can be mixed to produce the other colors in a color system. See *color*; *color system*.

prime mover **(236)** A mechanical device that rotates an electrical generator. This concept includes various kinds of turbines and external and internal combustion engines. See *generator (electrical)*; *turbine*; *heat engine*.

prime standard **(57)** See *standard*.

principal focus **(339)** The point at which light rays (initially parallel to the principal axis) converge when focused by a concave mirror or converging lens; the focal point of parallel incident light rays. See *principal optical axis*; *concave mirror*; *converging lens*; *focal point*.

principal optical axis **(339)** An imaginary line normal to the optical center of a concave or convex mirror or lens. In ray optics, the incident and reflected/refracted rays are compared to the optical axis of the mirror or lens. See *normal*; *concave mirror*; *converging lens*; *ray optics*.

prism **(344)** A wedge-shaped, transparent optical component that can disperse light into a spectrum of colors. Prisms also act as nearly perfect mirrors (by total internal reflection) for redirecting light in optical instruments such as binoculars and telescope erecting eyepieces. See *dispersion*; *mirror*; *total internal reflection*.

process **(7)** A sequence or series of actions or changes that result in something (an effect). Every natural and artificial process known has a cause and results in an effect.

product **(446)** A substance that is produced in a chemical reaction. It normally appears on the right side of a chemical equation. See *chemical reaction*; *chemical equation*.

property **(34)** An identifiable characteristic of a system, object, or idea. Measurable properties are associated with dimensional units. See *system*; *dimension*; *unit*.

proton **(26)** An elementary particle located in the nucleus of an atom (a *nucleon*) that carries a single positive *fundamental charge*. Its mass is about 1836 times that of an electron. The number of protons in an atom determines its identity as an element. See *elementary particle*; *nucleon*; *fundamental electrical charge (e)*; *electron*; *element*.

proton pump inhibitor **(494)** A medicine that is prescribed by doctors to combat excess acid production in patients' stomachs. See *acid*; *indigestion*.

pulley **(142)** A wheel and axle system with a groove around the perimeter of the wheel in which a rope, cable, or belt moves with the wheel as it rotates. An application of the lever principle. A single fixed pulley simply changes the direction of effort force and has a MA of 1. A single moveable pulley splits the resistance force between a fixed support and the effort force, so it has a MA of 2. See *wheel*; *axle*; *system*; *lever*; *effort force* (F_e); *resistance force* (F_r).

pulsar **(322)** A supernova remnant that collapsed into a very compact state of matter that is detectable mainly by radio telescopes. Astronomers believe its rapid spin accounts for the pulsing character of its signal. See *radio-frequency spectrum*.

pulse **(266)** A single waveform or a very short sequence of waves. See *waveform*.

pure substance **(28)** A substance that contains only a single element or compound. See *element*; *compound*.

Q

qualitative data **(48)** Scientific observations resulting from detailed word or other non-numerical descriptions (e.g., hand drawings, descriptions of color, texture, or sound). Compare with *quantitative data*. See *data*.

quality **(281)** The listener's perception of the mixture of fundamental and harmonic frequencies in a complex sound, such as that from a musical instrument. Also called *timbre (TAM bur)*. See *frequency (f)*; *fundamental (wave)*; *harmonic (wave)*.

quanta **(304)** Plural of quantum; a distinct amount of electromagnetic energy absorbed or released by electrons as they change energy states within an atom. Einstein believed quanta were equivalent to photons. See *electromagnetic (EM) energy*; *photon*.

quantitative data **(48)** Numerical scientific observations that result from measurements with instruments or from a method relating to counting (surveys, statistics, etc.). Usually consists of a number and a unit. Compare with *qualitative data*. See *data*; *measurement*; *unit*.

quantized **(360)** The condition where a measurable quantity can take on only certain allowable values. These could be multiples of some basic value (such as the fundamental electrical charge) or the solutions to special kinds of equations (quantum mechanics), for example. See *fundamental electrical charge (e)*; *quantum mechanics*.

quantum mechanics **(363)** The area of physics that deals with modeling the atomic and subatomic aspects of the physical universe. The quantum model of the atom is one result of quantum mechanics. See *quantum model.*

quantum model (360) The current atomic model developed during the twentieth century with the establishment of quantum mechanics. In this model, the tiny, dense atomic nucleus is surrounded by a "cloud" of electrons occupying three-dimensional orbitals organized by energy levels. See *quantum mechanics*; *orbital*; *energy level*; *electron*; *nucleus.*

quasar **(322)** A word that stands for *quasi-stellar.* An unusual celestial object that is extremely bright and observed in the distant regions of the universe. Quasars, or QSOs, are observable in the visible and radio-frequency portions of the electromagnetic spectrum. See *visible light*; *radio-frequency spectrum*; *electromagnetic (EM) spectrum.*

R

radar (315) Acronym for *ra*dio *d*etection *a*nd *r*anging; a technology that uses echolocation by radio-frequency waves to determine distance, direction, speed, or shape of an object. See *radio-frequency spectrum*; *echolocation*; *radio wave.*

radiant energy (116) See *electromagnetic (EM) energy.*

radiation (192) One of three methods of heat transfer from hotter to cooler locations involving radiant energy. The process of energy or particles moving directly away from a source. It also refers to the nuclear particles or electromagnetic waves themselves that are in such motion. See *process*; *energy*; *particle*; *heat (Q)*; *nuclear radiation*; *electromagnetic wave.*

radioactive **(370)** Having the property of emitting nuclear radiation. Compare with *radioactivity.* See *property*; *nuclear radiation.*

radioactivity **(370)** The emission of rays and particles during nuclear decay. Physicists measure it as the *rate* of such emissions. Compare with *nuclear radiation.* See *ray*; *particle*; *nuclear decay.*

radio astronomy (321) The study of celestial objects by observing the radio waves that they emit and the study of near-earth objects using active radar pulses. See *radio wave.*

radio-frequency identification (RFID) (319) Technology that uses a radio-frequency transmitter to activate a "tag" containing an antenna, receiver, and microprocessor. When activated, the tag replies by transmitting a unique ID code. It is used for inventory control, identity and security access control, and automatic debit systems, such as at highway toll booths. See *radio-frequency spectrum*; *antenna*; *transmitter*; *receiver.*

radio-frequency spectrum **(314)** That portion of the electromagnetic spectrum that includes all radio waves useful for communications by radio waves. See *electromagnetic (EM) spectrum*; *radio wave.*

radio-frequency technology **(314)** Any electronic device or method that works in or uses the radio-frequency portion of the electromagnetic spectrum. See *radio-frequency spectrum*; *electromagnetic (EM) spectrum.*

radio navigation (320) Any method of navigation using radio signals. Important systems include global navigation satellite systems, such as GPS and LORAN-C. See *radio wave.*

radio wave (306) All electromagnetic waves with frequencies below 300 GHz. They are produced by the acceleration of electrons. Radio waves support television signals, radio signals, and radio astronomy. See *electromagnetic (EM) energy*; *radio-frequency spectrum.*

radioisotope **(372)** An isotope of an element that has a known probability of experiencing nuclear decay. Also called a *radioactive isotope.* See *isotope*; *nuclear decay*; *radioactive.*

radiometric dating technique **(374)** Any method of determining the age of an object by measuring the current radioactivity of the sample. The scientist computes the age of the sample by applying certain assumptions about the decay history of the material's isotopes and the amounts of radioactive materials originally in the sample. It is the principal way secular scientists support an unimaginably old age for the earth and for the universe. See *radioactivity*; *isotope*; *radioactive.*

radiotherapy **(313)** The use of high-energy nuclear or electromagnetic radiation to treat cancers. See *nuclear energy.*

random error **(58)** An error in measurement that results during the correct and normal usage of a given instrument. It can result from slight differences in reading by the user, minor fluctuations within the instrument, or random environmental influences on the act of measuring. See *measurement*; *instrument.*

rarefaction (278) In longitudinal waves, regions of minimum particle density. A rarefaction corresponds to the trough of a transverse wave. See *longitudinal wave*; *trough*; *transverse wave.*

ray (336) Mathematically, a line segment with an arrowhead showing direction. In ray optics, a line that represents the path and direction of a photon of light when viewed as a particle. See *ray optics*; *photon.*

ray optics **(336)** The mathematical analysis of optical phenomena, particularly reflection and refraction, by treating light as rays. See *reflection*; *refraction*; *ray*.

reactant **(446)** A substance that is present before a chemical reaction and takes part in it. It normally appears on the left side of a chemical equation. See *chemical reaction*; *chemical equation*.

reactive **(397)** A characteristic of a substance described as easily undergoing chemical changes in the presence of other substances. Compare with *inert*. See *chemical change*.

real image **(338)** An image formed when light rays from an object converge after reflecting off a concave mirror or passing through a converging lens. A real image exists apart from any visual system perceiving it. It is upside down compared to the object that it represents, is reversed, and can be projected onto a screen. Compare with *virtual image*. See *concave mirror*; *converging lens*; *visual system*.

receiver **(307)** Any electronic or acoustic device whose purpose is to collect a specific kind of energy and then convert it into a form that can be understood by humans. Televisions and radios are common types of receivers.

Redemption **(5)** God's work of restoring the world to its original glory and purpose through the work of Jesus Christ. Redemption includes the future restoration of the earth as well as the resurrection of all believers. It also includes, at the present time, the rescuing of believers' minds and behavior from sin.

reflected ray **(337)** A light ray moving away from the point of incidence (reflection) at a reflective surface or the boundary between different refractive media. See *ray*; *reflection*; *refraction*; *medium*.

reflection **(337)** The change of direction of a wave or moving particle at the boundary between two media as it returns to the original medium. See *wave*; *particle*; *medium*.

refracted ray **(342)** A light ray moving into the second medium away from the point of incidence at the boundary between different refractive media. See *ray*; *medium*; *point of incidence*; *refraction*.

refracting telescope **(347)** An optical instrument that magnifies distant objects using an objective lens for gathering light and forming an image and an eyepiece lens that magnifies and focuses the image for viewing. *Binoculars* are basically two small, portable refracting telescopes built into a single instrument to provide stereoscopic viewing.

refraction **(269)** The change of a wave's speed as it moves through a boundary between different media. If its angle of incidence is other than 90°, the refracted wave bends from its initial path. See *medium*; *angle of incidence*.

Relativistic Runaway Electron Avalanche **(RREA)** **(212)** A model of lightning formation that suggests that lightning originates from some disturbance high in the atmosphere, possibly from cosmic rays striking atmospheric molecules. This event initiates an avalanche of electrons. This "electron snowball effect" was predicted to produce x-rays and gamma rays, which were later observed. See *lightning*; *cosmic ray*; *x-ray*; *gamma ray*.

relay **(246)** A remotely operated electrical switch activated by an electromagnetic device, usually a solenoid. See *solenoid*; *electromagnet*.

remotely operated vehicle **(ROV)** **(14)** An unmanned underwater robot controlled by operators located in surface ships through long power and control cables. ROVs carry lights, cameras, and strong mechanical manipulators, and are typically used for deep-water, heavy-duty, manual work.

renewable energy **(177)** Any energy source that is naturally replenished, such as hydroelectric, wind, solar, or biofuels. See *energy*.

repeatability **(58)** An assessment of the random errors associated with a series of measurements. Results showing small random errors around an average value for a given measurement have good repeatability. See *random error*; *measurement*; *instrument*.

repetitive motion **(255)** Any kind of motion that more or less repeats itself following a similar path each time. See also *vibration*; *oscillation*; *periodic motion*.

resistance (electrical) **(219)** See *electrical resistance (R)*.

resistance (mechanical) **(133)** See *load*.

resistance arm **(ℓ_r)** **(137)** The torque arm of the load or resistance force in a lever system. See *torque arm*; *load*; *resistance force (F_r)*; *lever*; *system*.

resistance force **(F_r)** **(133)** The force exerted on a simple machine by a load. See *force (F)*; *simple machine*; *load*.

resistor **(219)** An electrical circuit component whose intended purpose is to impede current flow by a specific amount or to create a potential difference of a specific magnitude between two points in an electrical circuit. See *electrical resistance (R)*; *electrical circuit*.

resonance **(259)** The condition of a vibrating or oscillating system when its amplitude increases due to reinforcement by energy added at its natural frequency. It is the transfer of energy from one vibrating or oscillating object to another with the same natural frequency. See *vibration*; *oscillation*; *amplitude*; *natural frequency*.

restoring force **(258)** The force that acts on an oscillating system to return it to its rest position. See *oscillation; rest position.*

rest position **(256)** The position of an oscillating system exactly halfway between the maximum displacements of its oscillations. See *oscillation.*

reverberation **(283)** Multiple echoes of a sound as it reflects off surfaces in an enclosed space. Low reverberation can give a sense of largeness or richness to music or voice. Large reverberations can make words and notes unintelligible. See *echo*; *sound*; *reflection.*

reverse osmosis **(RO) (476)** The process of removing solutes from a solvent by applying a pressure greater than the osmotic pressure of the solvent across a semipermeable membrane. This forces the solvent from the solution side to flow "backward" through the membrane, thus purifying it. Commonly used instead of distillation for producing potable water. See *process*; *fluid pressure*; *osmotic pressure*; *solvent*; *semipermeable membrane*; *distillation.*

right-hand rule of magnetism (235) A method for determining the direction of the magnetic field surrounding a current-carrying conductor. It is indicated by the direction that the fingers wrap around the conductor if it is grasped with the right hand with the thumb pointing in the direction of the conventional current flow. See *magnetic field*; *conventional current.*

rod cell **(335)** The light receptor cell in the retina of the human eye responsible mainly for sensing changes in illumination, low light level reception, and motion detection. There are about 20 times as many rods as cone cells in the human eye. They have a long rod- or cylinder-like shape. Compare with *cone cell.* See *illumination.*

rotational equilibrium (135) The condition in which the sum of all the torques on a system is zero—the system will either rotate at a constant speed or not begin to rotate at all. The rotating system equivalent of Newton's first law for linear motion. Compare with *law of inertia.*

rotational motion **(135)** A general description of circular motion around an axis of rotation. If the axis passes through the system, the motion is called *spin.* If the axis *does not* pass through the system, the motion is called *orbital motion.* See *axis*; *system.*

rotor **(237)** The part of an electrical motor or generator that rotates. Depending on the application, it may contain magnets or electromagnets, or the coils in which a current is induced. See *generator (electrical)*; *motor*; *electromagnet*; *electromagnetic induction*; *stator.*

S

salt (492) In general, any ionic compound of a base cation and an acid anion. More commonly, a compound consisting of a halogen and a metal element. See *ionic compound*; *base*; *cation*; *acid*; *anion*; *halogen family*; *metal (chemistry)*; *element.*

saponification **(468)** The conversion of a fatty substance into soap when heated with a strong base, such as lye (sodium hydroxide, NaOH). See *base.*

saturated **(470)** See *saturation (color)*; *saturation (solution).*

saturation (color) **(334)** The apparent intensity of a color. The saturation range varies from no color to the maximum brightness of color. See *color*; *intensity (light)*; *hue*; *value.*

saturation (solution) (470) A comparison of the concentration of a solute in a solution to its concentration at its solubility limit. A solution is said to be *saturated* at the saturation limit of the solute. See *solution*; *solute*; *solubility.*

scalar **(79)** Any measurable quantity that can be completely described by a single piece of information, usually a number. A scalar can be positive, negative, or zero. We represent scalar quantities by italic symbols in scientific formulas. See *measurement*; *formula.*

science (8) The collection of observations, inferences, and models produced through a systematic study of nature for the purpose of enabling humans to exercise good and wise dominion over God's world. Also, the systematic methods that produce the observations, inferences, and models. See *model*; *dominion science.*

scientific method **(14)** See *scientific process.*

scientific notation (61) A convenient way to express very large or small numbers. We write the notation in the form $M \times 10^n$, where M is a number greater than or equal to 1 and less than 10, and n is a positive or negative integer.

scientific paradigm **(10)** The accepted body of knowledge, theories, hypotheses, and experimental approaches to answering questions of science. It dictates the presuppositions used in a branch of science, and all workers in the science agree to certain rules set by the paradigm. See *theory*; *hypothesis*; *presupposition.*

scientific process **(14)** Sometimes called the *scientific method*, it is a logical sequence of steps acceptable to the scientific paradigm that ensures that the models and conclusions that result from the work are valid and useful for supporting further scientific work. See *process*; *scientific paradigm*; *model.*

scientific revolution **(10)** Any change in a scientific paradigm that results in a new and fundamentally different way of looking at phenomena. See *scientific paradigm*; *phenomenon*.

screw **(147)** A fastener that consists of a long, thin wedge wrapped around a shaft in a spiral or *helix*, called a *screw thread*. Screws exert large forces to hold objects together. *Bolts* are similar to screws except that they either exert their fastening force by use of a separate *nut* threaded onto the shaft of the fastener or are inserted into a pre-threaded hole in one part to be fastened to the other. See *force (F)*; *mechanical advantage (MA)*.

second-class lever **(138)** A lever system in which the resistance is between the fulcrum and the effort. Its MA is always greater than 1. See *lever*; *system*; *fulcrum*; *effort force* (F_e); *mechanical advantage (MA)*.

second law of thermodynamics **(199)** Every natural process moves toward a condition of lowest usable energy and highest entropy. See *process*; *energy*; *entropy (S)*.

secular **(7)** A description of someone or of an idea that is not influenced by religious principles.

seismoscope **(261)** A device that uses seismic vibrations to indicate the occurrence and direction to an earthquake. See *vibration*.

semiconductor **(211)** An artificial crystalline solid composed of metalloid elements with trace amounts of other elements used as the basis for microprocessors and other kinds of integrated circuit electronics. It allows limited electron flow so that it can act as either a conductor or an insulator depending on the polarity and the magnitude of the imposed electrical potential across it. See *crystalline solid*; *metalloid*; *element*; *conductor*; *insulator*; *voltage*.

semimetal **(396)** See *metalloid*.

semipermeable membrane **(474)** A membrane that selectively allows solvent particles to diffuse through but blocks the passage of some or all of the solute particles. See *membrane*; *solvent*; *diffusion*; *solute*.

series **(401)** See *period (chemistry)*.

series circuit **(223)** An electrical circuit or portion of a circuit with a single current flow path through a connected set of electrical loads. See *electrical current (I)*; *load*.

shadow zone **(272)** The region behind an obstacle in the path of waves where the diffracted wave energy is diminished or absent. It is most noticeable when the obstacle is much larger than the wavelength of the waves. See *diffraction; wavelength* (λ).

sheet lightning **(212)** A broad brightening of a cloud caused by lightning hidden within it or by reflection of an unobserved lighting bolt. See *lightning*.

short circuit **(223)** A location (usually a fault) in an electrical circuit where the majority of the current bypasses a load to take a low-resistance path back to the current's source. See *electrical resistance (R)*; *electrical current (I)*; *load*.

SI **(50)** The global scientific metric system. The abbreviation stands for the French name *Système International d'Unités*. See *metric system*.

SI base unit **(50)** One of seven dimensional units carefully defined with reference to standard conditions (or an object in the case of the kilogram) that can be recreated in any properly equipped laboratory. See Appendix B. See *dimension*; *unit*; *standard conditions*.

SI derived unit **(50)** Any metric unit that can be expressed as a combination of two or more SI base units. Some derived units have their own names and symbols, such as the newton (1 N = 1 kg·m/s^2) and liter (1 L = 1 dm^3). Others are simply combinations of base units, as for momentum (kg·m/s). Compare with *SI base unit*.

SI unit prefix **(51)** A prefix to an SI base or derived unit name that indicates the power-of-10 factor (multiplier) to be used with the unit. SI prefixes allow us to use convenient numbers with the units for the dimension that we are measuring. See *SI base unit*; *unit*; *dimension*.

significant digit (SD) **(59)** A digit in a measurement used to communicate the precision of the measurement. The significant digits in a measurement are all the digits known from the instrument scale plus one estimated digit determined by the user. See *measurement*; *precision*; *instrument*; *graduated scale*.

simple machine **(132)** A basic mechanical device that either reduces effort when doing a certain amount of work or increases the magnitude of motion for a given input motion. See *effort force* (F_e); *mechanical work (W)*; *motion*.

single-replacement reaction **(451)** A chemical reaction in which one element in a reactant compound is replaced by another reactant element, forming a new compound product. The replaced element becomes one of the products and often forms a precipitate or a gas. See *chemical reaction*; *reactant*; *product*; *precipitate*.

sol **(459)** A two-phase heterogeneous mixture in which the dispersed solid phase does not form an open network in the liquid continuous phase as in a gel. Its particles may still have some attraction for each other, producing a viscous liquid. Compare with *gel*. See *heterogeneous mixture*; *dispersed phase*; *continuous phase*; *viscosity*.

solar thermal* (ST) *power plant **(182)** A power plant that generates electricity by boiling water or other fluids to a superheated vapor using concentrated solar energy. The working fluid then immediately drives turbine generators or is stored underground for use at night. See *fluid*; *energy*.

solenoid **(236)** An electromagnetic device consisting of a cylindrical coil of many wraps of wire. When carrying a current, a strong magnetic field forms within the coil of wire. See *magnetic field*; *electromagnet*.

solid **(31)** The state of a substance in which its particles occupy fixed positions. A solid is rigid, or nearly so, and has a definite volume and shape, and low compressibility. See *state*; *particle*; *volume (V)*; *compressibility*.

***solid state disk* (SSD)** **(245)** A common non-magnetic data storage medium for computers consisting of semiconductor modules that mimic the operation of hard drives but with no moving parts. See *hard drive*.

solubility **(467)** The maximum amount of solute that can be dissolved in a given amount of a particular solvent at a specified temperature. A characteristic property of a solute in the solvent. See *solute*; *solvent*; *property*.

soluble **(465)** Describes a substance that is able to dissolve in a certain solvent. See *solvent*.

solute **(462)** The substance dissolved in the solvent in a solution; the minority component in a solution. See *solvent*; *solution*.

solution **(462)** A homogeneous mixture of two or more pure substances. See *homogeneous mixture*; *pure substance*.

solution process **(465)** The process at the particle level that a solute goes through as it is dissolved by the solvent. See *process*; *solute*; *solvent*.

solvation **(465)** See *hydration*.

solvent **(462)** The substance that does the dissolving in a solution; the majority component of a solution. See *solution*.

sonar **(291)** An acronym for *so*und *na*vigation and *r*anging. An artificial device that either listens passively or uses active echolocation in bodies of water for navigation, research, or military purposes. Can also refer to methods using such a device. Also, an informal name for any natural use of echolocation, such as with bats. See *echolocation*.

sonogram **(296)** An image generated by ultrasound technology. See *ultrasound*.

sound **(278)** The form of wave-transmitted energy that is detectable by the ear. More generally, energy transmitted through a medium by mechanical waves. Also called *acoustic energy*. See *medium*; *mechanical wave*.

sound pressure level **(281)** The maximum pressure produced by a sound wave. For normal sounds, it is measured in fractions of a pascal (Pa). See *pressure (P)*; *sound*; *pascal (Pa)*.

space **(65)** One of the fundamental concepts of creation. It is any region described by three dimensions, such as length, width, and height (although there are other ways to describe space). Measureable spaces are called volumes and are measured in volume units. See *dimension*; *volume (V)*; *unit*.

spatial **(86)** Relating to any property that has significance in three-dimensional space. For example, an athlete must have good spatial awareness, or a flying aircraft's GPS position consists of three spatial coordinates—latitude, longitude, and altitude. See *property*.

special theory of relativity **(117)** Albert Einstein's revolutionary theory of space, time, and matter. It assumes that the only universal constant when considering motion is the speed of light (*c*). Time and mass of a moving system are dependent on the speed of the system within the observer's frame of reference. Mass and energy are interchangeable as expressed by the equation $E = mc^2$. See *space*; *time (t)*; *matter*; *motion*; *system*; *speed (v)*; *frame of reference*.

specific gravity (s.g.) **(159)** The ratio of a substance's density to water's density; a unitless quantity numerically equal to the density of the substance. Also called *relative density*. See *density* (ρ).

specific heat (c_{sp}) **(195)** The amount of thermal energy 1 g of a substance must gain or lose to change its temperature 1 °C. Properly called *specific heat capacity*. See *thermal energy*; *temperature (t)*; *heat capacity (C)*.

spectator ion **(493)** Any ion in an aqueous solution that does not take part in a chemical reaction between other substances in the solution. They are "spectators" of the action. See *ion*; *aqueous*; *solution*; *chemical reaction*.

spectrograph **(327)** A photograph, printout, or other display of a spectrum produced by a spectroscope. See *electromagnetic (EM) spectrum*; *spectroscope*.

spectroscope **(327)** An optical device that disperses a source of light into a continuous or line spectrum so that the spectrum can be observed, or recorded as in a spectrograph. See *dispersion*; *continuous spectrum*; *line spectrum*; *spectrograph*.

specular reflection **(337)** Reflection of photons off a microscopically smooth surface in the same direction. An image of the light source can be formed from specular reflection. See *reflection*; *photon*.

speed (*v*) **(81)** The rate of motion of a system. As with any rate, we express it as a change of position (a distance) with time. It may also be the magnitude of velocity. It is always a scalar quantity. See *motion*; *system*; *position*; *time (t)*; *velocity (**v**)*; *scalar*.

speed of light (*c*) **(302)** The speed of any electromagnetic wave in a vacuum, approximately 3.00×10^8 m/s. See *speed (v)*; *electromagnetic (EM) energy*.

speed of sound **(279)** The rate at which a sound wave travels through a medium. Measured in meters per second (m/s). Under standard conditions, the speed of sound in air is 331.4 m/s. See *sound*; *medium*; *standard conditions*; *speed (v)*.

spring scale **(64)** A laboratory instrument that measures weight (or mass) by balancing the pull of the earth's gravity on an object being weighed against the force exerted by a stretched spring inside the instrument. See *instrument*; *weight (w)*; *mass (m)*; *gravity*; *force (F)*.

sprite **(212)** Upper atmosphere discharges of glowing balls of "cold" plasma associated with lightning discharges in large thunderstorm systems. See *plasma*; *lightning*.

standard **(48)** A known quantity that everyone agrees will be used for comparison when measuring. A standard may be an artificial object (e.g., the *standard kilogram*) or a reference point known to be accurate on the instrument in use. It may also be a separate instrument that has been accurately calibrated with which you can compare the instrument being used. See *calibration*.

standard conditions **(30)** A set of specified environmental or laboratory conditions that allow scientists to define SI base units or to compare various properties of matter in a uniform and meaningful way. See *standard temperature and pressure (STP)*; *SI base unit*; *property*; *matter*.

standard kilogram **(49)** A platinum-iridium cylinder that represents the base SI mass unit—the kilogram, stored at the International Bureau of Weights and Measures (BIPM) at Sèvres, France. Properly called the *International Prototype Kilogram (IPK)*, it is the only SI unit based on a manufactured standard.

standard model **(363)** The accepted model of matter that accounts for all known elementary particles and all fundamental forces except for the gravitational force carrier. See *elementary particle*; *fundamental force*; *gravitational force*.

***standard temperature and pressure* (STP)** **(30)** Standard conditions for comparing the properties of different substances. Usually 1 atm pressure and 0 °C. See *standard conditions*; *property*; *atmospheric pressure*.

standardization **(500)** See *calibration*.

standing wave **(271)** A waveform that appears to be stationary, but rises and falls between nodes. It results when two identical waves moving in opposite directions interfere as they pass through each other. See *waveform*; *node*; *interference*.

state **(30)** A physical form of matter determined by the arrangement and energy of its particles. The three most common states of matter are solid, liquid, and gas.

static electricity **(205)** Any electrical phenomenon relating to essentially stationary electrical charges and the forces that they exert. See *charge (electrical)*; *electrostatic force*.

statics **(75)** The description of how stationary objects or systems exert or react to external forces. See *system*; *external force*.

stator **(237)** The stationary part of an electrical motor or generator. Depending on the application, it may contain the magnets or coils that surround the rotor. See *generator (electrical)*; *motor*; *electromagnetic induction*; *rotor*.

Stock system **(443)** A method of naming different ionic compounds made of the same elements. Many metals can have different oxidation numbers. A Roman numeral in parentheses following the metal's name represents the value of the metal's oxidation number for that compound. See *oxidation number (ON)*; *ionic compound*.

strong nuclear force **(92)** One of the four fundamental forces. The attractive force that holds protons and neutrons together in nuclei. It is the strongest of all fundamental forces. At distances much shorter than the size of a proton, it strongly repels. See *fundamental force*; *proton*; *neutron*.

subatomic particle **(26)** See *elementary particle*.

subgroup **(398)** In the periodic table, the columns of elements lying in the transition element section of the table (IUPAC groups 3–12). Subgroups are designated by a "B" in the older group designation method. The elements in each subgroup tend to have similar properties because of the similar electron arrangements in their valence and innermost energy levels. See *group*; *periodic table of the elements*; *transition element*; *valence electron*; *energy level*.

sublimation **(43)** The change of state directly from a solid to a vapor, bypassing the liquid state, at temperatures below the melting point of the substance. See *state*; *solid*; *vapor*; *liquid*; *temperature (t)*; *melting point*.

subtractive primary color (**334**) One of several distinct colors of pigment (e.g., cyan, magenta, yellow) that, when combined in various ratios, produce all colors in the associated color system by the absorption of incident or transmitted light. See *pigment*; *color*; *color system*; *incident*; *transmitted*.

superconducting magnet (**246**) An electromagnet made of a material that has zero resistance to current flow, usually at temperatures very close to absolute zero. See *electrical resistance (R)*; *absolute zero*.

supersaturated (**470**) Describes a solution in which the solute concentration exceeds its normal solubility limit under existing conditions. This is a highly unstable condition that can result in rapid precipitation of the solute until its concentration stabilizes at its solubility limit. See *solution*; *solute*; *concentration*; *precipitate*.

surroundings (**76**) Everything outside a system's boundary. See *system*.

susceptible object (**93**) As defined in this textbook, any object that has properties that respond to the influence of a particular kind of field force. For example, iron objects are susceptible to the force exerted by a magnetic field. See *property*; *field force*.

suspension (**461**) A heterogeneous fluid mixture in which the solid particles of the dispersed phase are larger than 1 μm. Over time, gravity will cause these particles to settle out of the mixture. See *heterogeneous mixture*; *dispersed phase*.

switch (**217**) A device that opens or closes a break in an electrical circuit to control the flow of current. See also *electrical circuit*; *circuit breaker*; *fuse*.

system (**76**) A distinct part of the universe, from elementary particles to galaxies, that we may want to study or measure. It is separated from its surroundings by an actual or imaginary boundary. See *elementary particle*; *surroundings*.

systematic error (**58**) An error that results from a bias within an instrument or the method used to make a measurement. It results in a consistent offset of the measured value from the acceptable value of the dimension being measured. See *error*; *bias*; *instrument*; *measurement*; *acceptable value; dimension*.

T

television (**317**) Technology used to transmit, receive, and display moving images with synchronized sound using formatted radio-frequency signals. Can also refer to the display device itself. See *radio-frequency technology*.

temperature (*t*) (**184**) A measure of the average kinetic energy of the particles in a substance; the hotness or coldness of an object, measured in degrees. See *kinetic energy (KE)*; *particle*; *degree (°)*.

temperature difference (**Δt**) (**181**) For a single system, the difference of the final temperature and the initial temperature ($t_f - t_i$) during a time interval. When considering two systems, or a system and its surroundings, it is the difference of temperature between the temperatures of the two points being compared. See *system*; *interval (Δt)*; *surroundings*.

terahertz band (**309**) A subset of the IR band extending from 300 GHz to 3 THz. It is a relatively new focus of study. It supports medical imaging, low-energy astronomy, and other emerging technologies. See *infrared (IR)*; *medical imaging*.

terminal velocity (**105**) The maximum possible speed that an object in free fall can achieve in a fluid, such as the atmosphere. It occurs when fluid drag balances the force of gravity. See *speed (v)*; *free fall*; *fluid*; *drag*; *gravitational force*.

test charge (**207**) A theoretical concept to identify the direction of electrical field lines of force. The ideal test charge is a positive point charge so small that it cannot affect the electrical lines of force around it. See *lines of force*.

theory (**11**) A scientific model that explains a related set of phenomena according to a certain paradigm. This definition is valid even though the theory may not be ultimately true or even plausible. Truth and plausibility are established by the relationship of the paradigm to the absolute truths found in the Bible. See *model*; *phenomenon*; *scientific paradigm*.

thermal energy (**114**) The average sum of the kinetic energies of all the particles in an object. Directly proportional to the temperature of the object. Compare with *heat (Q)*. See *kinetic energy (KE)*; *particle*; *temperature (t)*.

thermal equilibrium (**190**) The condition of a system that is at the same temperature as its surroundings so that there is no net flow of thermal energy into or out of the system. See *system*; *temperature (t)*; *thermal energy*.

thermal expansion (**187**) A thermal property of most materials in which length and/or volume increase in proportion with increasing temperature. See *thermal energy*; *property*; *length*; *volume (V)*; *temperature (t)*.

thermodynamics (**179**) The study of thermal energy, heat, and their useful applications. See *thermal energy*; *heat (Q)*.

thermometer (**184**) An instrument that uses a thermometric property to measure and display temperature. See *instrument*; *thermometric property*; *temperature (t)*.

thermometric property **(184)** Any property of matter that predictably varies in proportion to changes in temperature. See *property*; *matter*; *temperature (t)*.

thermoscope **(185)** Invented by Galileo, one of the earliest temperature-measuring instruments. It relied on the change of volume of a confined sample of air with temperature. Its accuracy was affected by changes in air pressure. See *temperature (t)*; *instrument*; *accuracy*; *atmospheric pressure*.

third-class lever **(138)** A lever system in which the effort is between the fulcrum and the resistance. Its MA is always less than 1. Its main purpose is to magnify motion of the load for a given amount of effort force movement. See *lever*; *system*; *effort force* (F_e); *fulcrum*; *resistance force* (F_r); *mechanical advantage (MA)*; *motion*.

thread count **(TPI) (147)** The number of threads per inch counted along the shaft of a screw. Greater TPI implies greater holding power (mechanical advantage). See *screw*; *mechanical advantage (MA)*.

threshold of hearing **(TOH) (281)** The lowest sound intensity detectable by an average healthy person—typically 0 dB. The threshold of hearing declines with age. See *intensity (sound)*; *decibel (dB)*.

threshold of pain **(281)** The loudness of sound that can first cause pain and damage to a person's hearing. Occurs around 120 dB. See *loudness*; *decibel (dB)*.

time **(*t*) (78)** One of the fundamental concepts of creation. A nonphysical continuum that orders the sequence of events, processes, and other phenomena. See *continuum*; *process*; *phenomenon*.

torque (136) A twisting action produced by a force exerted perpendicularly to a torque arm. In the US, *torque* is used by physicists for both the tendency to cause rotation or to actually cause or change rotational motion. US mechanical engineers use the term *moment* to describe a force that creates a tendency to twist or bend. See *force (F)*; *torque arm*; *rotational motion*.

torque arm **(ℓ) (136)** The distance along a lever between the point that a force is applied and the lever fulcrum. Also called *lever arm*. See *torque*; *lever*; *force (F)*; *fulcrum*.

total internal reflection (342) Reflection of a light ray approaching the boundary between two refractive media from within the medium with the higher index of refraction. Total internal reflection occurs when the incident angle exceeds the critical angle for the two media. See *reflection*; *index of refraction (n)*; *medium*; *angle of incidence*; *critical angle of incidence*.

transducer **(292)** A device that both transmits and receives pulsed sound waves. An underwater version functions as a key part of an active sonar system. They are also used in ultrasonic medical devices. See *sonar*; *sonogram*; *ultrasound*.

transformer (242) An electrical device that uses electromagnetic induction to raise or lower the available voltage to the voltage required by an AC circuit. Consists of an input coil and an output coil of wire wound on a single ferromagnetic core, usually shaped like a hollow square or a donut. *Step-up transformers* raise output AC voltage compared to input voltage; *step-down transformers* lower output voltage. Transformers can also isolate the conductors of one AC circuit from the conductors of another without changing voltage (*isolation transformer*). See *electromagnetic induction*.

transition element (398) A metal element found in the B-subgroups (IUPAC groups 3–12) of the periodic table. See *element*; *metal (chemistry)*; *subgroup*; *periodic table of the elements*.

transmitted **(334)** Description of light or other kind of energy that is sent or that passes through a medium or space. Light is transmitted or passes through transparent materials. Radio waves are transmitted through space. Contrast with *incident*.

transmitter **(314)** A device that generates electromagnetic waves to be radiated from an antenna. In radio-frequency communications, the transmitter is supplied with a signal containing information that it converts to radio waves in the antenna. See *electromagnetic wave*; *antenna*; *radio-frequency spectrum*; *radio wave*.

transuranic element **(391)** Any element with an atomic number greater than $Z = 92$ (i.e., uranium). Nearly all nuclides of these elements are artificial. See *element*; *atomic number (Z)*; *nuclide*.

transverse wave (267) A wave in which the oscillations are at right angles to the direction of wave travel. See *wave*; *oscillation*.

trilateration **(321)** A method of finding one's position by determining distances to at least three known locations or landmarks. This method is used by the GPS. See *radio navigation*.

triple bond **(419)** Three covalent bonds between two atoms. See *covalent bond*.

triple point **(187)** The pressure and temperature conditions at which the solid, liquid, and gaseous phases of a substance simultaneously exist in a stable condition. The triple point of water is 0.01 °C and 611.73 Pa. See *pressure (P)*; *temperature (t)*; *solid*; *liquid*; *gas*; *phase (mixture)*; *pascal (Pa)*.

trough **(265)** Compared to the rest position in a waveform, it is the lowest point of a wave (as in water waves) or the location of least particle density (as in sound waves). See *rest position*; *waveform*; *wave.*

tuning fork **(278)** A device made from a U-shaped metal bar with an attached handle. When struck, tuning forks generate nearly pure musical tones with insignificant harmonics. See *harmonic (wave).*

turbine **(238)** A mechanical engine that converts fluid motion into rotary motion as the fluid passes through propeller-like blades. Common types include *steam*, *water*, *wind*, and *gas turbines*. Often used as a prime mover for electrical generators. See *fluid*; *prime mover*; *generator (electrical).*

Tyndall effect (461) The scattering of light off dispersed or suspended particles in a heterogeneous fluid mixture; the definitive test that differentiates between a colloidal dispersion and a true solution. See *dispersed phase*; *suspension*; *colloid*; *solution.*

U

ultrasonic wave (294) A sound with a frequency above the range audible to humans (typically greater than 20 kHz). See *acoustic spectrum*; *frequency (f).*

ultrasound **(296)** Related technologies designed to produce images through the use of ultrasonic waves. Applications include both medical and industrial imaging. See *ultrasonic wave*; *sonogram.*

ultraviolet (UV) (310) The electromagnetic spectrum extending from about 789 THz to 30 PHz, produced by the emission of photons during large changes of electron energy levels within atoms. Ultraviolet light aids in antibacterial sterilization, adhesive fixatives, and moderately high-energy imaging of the universe. It also causes sunburn and some skin cancers. Sometimes referred to as *UV rays* or *UV radiation*. See *electromagnetic (EM) spectrum.*

unbalanced force **(94)** Any external force that in combination with other forces acting on the same system results in a nonzero net force and produces a change in motion of the system. See *external force*; *system*; *net force*; *motion.*

uniform acceleration **(104)** A constant acceleration, such as that of an object in free fall. See *acceleration (a)*; *free fall.*

unit **(48)** For measuring purposes, the subdivision or portion of a dimension assigned a value of 1. For example, a unit of length is 1 foot or 1 meter. See *measurement*; *dimension.*

unit cell **(429)** A repeated structural unit from which a crystal lattice is assembled. It is classified by the arrangement of atoms that it contains. A unit cell usually contains multiples of the formula unit of a chemical compound. See *crystal lattice*; *formula unit*; *compound.*

universal indicator **(500)** A pH indicator that is responsive to the entire pH range. Usually is a combination of several compounds that can cover portions of the pH scale range. See *pH indicator*; *compound*; *pH scale.*

unsaturated **(470)** Describes a solution in which the solute concentration is below its solubility limit (additional solute can dissolve under the existing conditions). See *solution*; *solute*; *concentration*; *saturation (solution).*

urban heat island effect **(109)** The tendency of large concentrations of buildings, roads, and parking lots to absorb solar energy and re-radiate it as thermal energy, raising the average urban temperatures many degrees above the temperatures of the city's surroundings. See *thermal energy*; *temperature (t).*

V

valence electron (368) An electron in the highest or outermost energy level of a neutral atom. Unpaired valence electrons are usually involved in chemical bonding. See *electron*; *energy level*; *chemical bond.*

value **(334)** The darkness or lightness of a color. Value varies from black to maximum lightness available in a color system. See *color*; *color system*; *hue*; *saturation (color).*

vapor **(41)** The gaseous phase of a substance. A true vapor of a colorless substance cannot be observed, since it is a gas. See *gas*; *phase (mixture).*

vaporization (41) Any process in which particles of a liquid enter the gaseous phase. See also *evaporation*; *boiling*; *sublimation*. See *process*; *particle*; *liquid*; *gas*; *phase (mixture).*

vapor pressure **(42)** The gas pressure exerted on the surface of a liquid by its vapor in a closed container when the gas and liquid are in equilibrium. A measurable property of the *liquid* at standard conditions. See *gas pressure*; *liquid*; *vapor*; *property*; *standard conditions.*

vector **(79)** A measurable quantity that requires *two* pieces of information to fully describe it—a positive scalar value (its magnitude) and a direction. In special *vector equations*, we show vectors in bold characters. When all the vectors in a problem are parallel to the same coordinate axis, we can represent them in a formula using their italic scalar symbols for simplicity. See *scalar*; *coordinate axis*; *formula.*

velocity (v) (82) The rate of displacement of a system. It is always a vector quantity that points in the same direction as the displacement vector of the system. See *displacement (**d**)*; *vector*; *system*.

venturi **(166)** A specially designed constriction in a pipe, used to measure fluid flow rate by comparing the differences in fluid pressure before and within the constriction that occur according to Bernoulli's principle. See *fluid pressure*; *Bernoulli's principle*.

vibration **(255)** In this textbook, very small and rapid repetitive motions. Atoms and tuning forks vibrate. Vibrations tend to be a complex combination of several kinds of periodic motions. Compare with *oscillation*. See *repetitive motion*; *periodic motion*.

virtual image **(338)** An image perceived by a visual system that does not exist where it appears to be. It is visible when the visual system processes diverging light rays reflecting off a mirror or passing through a diverging lens. A virtual image is upright and cannot be projected onto a screen. Compare with *real image*. See *visual system*; *mirror*; *diverging lens*.

virtual model **(9)** A model, usually in 3D, of a real or imaginary object or environment that is presented on some form of digital display to give a sense of its properties. See *model*; *property*.

viscosity **(32)** A measure of a fluid's resistance to flow. Can be thought of as the "thickness" of a liquid or gas. See *fluid*.

visible light **(310)** The portion of the electromagnetic spectrum that human eyes can detect, extending from approximately 384 THz at the red end of the spectrum to 789 THz at the violet end. Visible light is produced by the emission of photons as electrons drop from higher energies to lower energies within atoms. Sight and optical technologies use this band of frequencies. See *electromagnetic (EM) spectrum*.

visual system **(326)** An organic or artificial system that can receive, transmit, interpret or display images. In humans, it consists of the eye, the retina, the optic nerves, and the regions of the brain associated with interpretation of these nerve signals. Artificial visual systems include any that can form and display a photographic image, such as cameras, television, telescopes, and microscopes.

vocal cords **(286)** The two flat pieces of tissue contained within the larynx that vibrate and produce many of the sounds of the human voice. See *larynx*; *vibration*.

volt **(V)** **(219)** The amount of work required to move a unit charge between two points in a circuit or field; the SI derived unit for potential difference (1 V = 1 J/C). See *voltage*.

voltage **(*V*)** **(219)** The electrical potential difference between two locations, usually but not necessarily within an electrical circuit. Measured using a *voltmeter*. See *electrical potential energy*.

volume **(*V*)** **(65)** The space enclosed or occupied by an object or within the boundaries of a physical system. Scientists derive volumetric units from the cube of units of length. For example, a liter (L) is equal to a cubic decimeter (dm^3). See *system*; *unit*.

voxel **(316)** The smallest element of a 3D MRI image, just as a pixel is the smallest element of a digital 2D picture. Voxel is a combination of *volume* and *pixel*. See *magnetic resonance imaging (MRI)*; *volume (V)*.

W

watt **(W)** **(131)** The SI unit of power (1 W = 1 J/s). See *SI*; *unit*; *power (P)*.

wave **(264)** An oscillation in matter or a model for periodically changing, mutually interacting electromagnetic fields. Waves transfer energy from one place to another. See *oscillation*; *mechanical wave*; *electromagnetic wave*.

wave base **(265)** In a water wave, the depth where the circular motion of water particles ceases. This depth is about two wavelengths below the surface in deep-water waves. See *wavelength (λ)*.

waveform **(265)** The shape of a wave that is displayed as a displacement from its rest position over time. Periodic waves show a regular pattern of crests and troughs. Waveforms from non-periodic sources, such as a person's voice, can be very complex. See *wave*; *rest position*; *periodic motion*; *crest*; *trough*.

wave height **(265)** The total displacement of the wave medium between the crest and the trough of a wave. It is twice the wave amplitude. Most commonly used to describe water waves. See *medium*; *crest*; *trough*; *amplitude*.

wavelength (λ) **(265)** The distance between corresponding points on adjacent waveforms, such as crest-to-crest or trough-to-trough. See *waveform*; *crest*; *trough*.

wave phase **(271)** The comparison of two interfering waveforms. When the crest of one wave aligns with the crest of the other, the waves are *in phase*; when the crest of one aligns with the trough of the other, the waves are *out of phase*. See *interference*; *waveform*; *crest*; *trough*.

weak nuclear force **(92)** One of the four fundamental forces. A force exerted between elementary particles of matter inside protons and neutrons. It is the third-strongest fundamental force. It is involved in some forms of nuclear decay. See *fundamental force*; *elementary particle*; *proton*; *neutron*; *nuclear change*.

wedge **(147)** A three-dimensional inclined plane used to exert a force in order to spread a material apart as it is forced into the material. See *inclined plane*.

weight (*w*) **(64)** The measure of the earth's gravity acting on the matter in an object. Weight is a force measured in newtons (N). See *gravity*; *force*.

wet cell battery **(217)** A set of electrochemical cells that uses a liquid electrolyte. Usually used as an electrical power source in land or water vehicles or in stationary applications. *Gel cell batteries* with gelatinous electrolytes are similar, but allow more motion of the battery without spilling. See *electrochemical cell*; *electrolyte*; *gel*.

wheel and axle **(140)** Any rotational mechanical system having unrestricted range of motion. The wheel and axle is a modification of a lever system. The axle is the axis on which the wheel rotates and is functionally the same as a fulcrum in a lever system. See *rotational motion*; *lever*; *system*; *axis*; *fulcrum*.

work **(110)** See *mechanical work (W)*.

workability **(8)** The usefulness of something for a particular purpose. The workability of a scientific model is its most important property. See *model*; *property*.

worldview **(7)** The perspective from which a person sees or interprets all matters of life. Compare with *scientific paradigm*.

X-Y-Z

x-ray **(311)** A photon in the portion of the electromagnetic spectrum extending from 30 PHz to 30 EHz, although there is some overlap with gamma rays at the upper end. X-rays are produced by high-energy collisions of electrons with atoms. They aid medical and nonmedical internal imaging, cancer treatments, and astronomical studies of high-energy phenomena. See *electromagnetic (EM) spectrum*; *gamma ray*; *medical imaging*.

x-ray diffraction **(305)** An analytical technique for studying the fine structures of matter by creating a diffraction pattern resulting from the interaction of x-rays with crystal lattices. See *diffraction*; *x-ray*; *crystal lattice*.

zero reference position **(112)** When computing the mechanical potential energy of a system, the position or height from which the system's position is measured. See *potential energy (PE)*; *system*; *position*.

Index

Bold face page numbers denote illustrations.
Italicized page numbers denote margin notes.

A

B

D

F

M

N

O

P

T

U

V

Photograph Credits

Key: (t) top; (c) center; (b) bottom; (l) left; (r) right; (i) inset; (bg) background

Cover

© iStockphoto.com/virtualphoto

Front Matter

viit The Asahi Shimbun/Getty Images; **vii**b "PS20andPS10" by Koza1983/Wikimedia Commons/CC BY 3.0; **viii**t USDA Photo by Scott Bauer; **viii**b "Diamagnetic graphite levitation" by Splarka/Wikimedia Commons/Public Domain; **ix**l © iStockphoto.com/Andrei Tchernov; **ix**r © PHILIPPE PLAILLY/PHOTO RESEARCHERS, INC.

Unit 1

xii–1t NASA; **xii–1**b © iStockphoto.com/Jack Thornton; **xii, 2**t Gary S. Settles/Photo Researchers, Inc; **xii, 20**t © 2007 JupiterImages Corporation/Hemera Technologies; **xii, 47**t Dr. Jeremy Burgess/Science Source; **2**b "PKIERZKOWSKI 070328 FGZCP CDG" by Pawel Kierzkowski/Wikimedia Commons/CC SA 3.0, GNU FDL 1.2; **3**t Gonzalo Fuentes/Reuters; **3**b © iStockphoto.com/Yuri_Arcurs; **4** © ClickPop - Fotolia.com; **5**t LatinContent/Getty Images; **5**b AP Photo/Brazil's Air Force; **6, 13, 14**t BEA; **7** AP Photo/Science; **8, 14**b, **15**t, **15**c, **15**b BEA/ECPAD; **9**t © G00b | Dreamstime.com; **9**c NASA/JPL/UCLA; **9**b © iStockphoto.com/Andrei Tchernov; **10** Public Domain; **11**t NASA/MSFC; **11**b STR/Stringer/AFP/Getty Images; **12** Peter Dazeley/Photographer's Choice/Getty Images; **16**t David SPRAGUE/Stringer/AFP/Getty Images; **16**b Copyright 2012 EUMETSAT; **17** © Lightpoet | Dreamstime.com; **20**b AP Photo/Michael A. Mariant; **21**t © iStockphoto.com/kevin miller; **21**b Courtesy of Dr. Robert Wolkow, National Research Council of Canada; **22** © PHILIPPE PLAILLY/PHOTO RESEARCHERS, INC.; **23** "Aristotle Altemps Inv8575"/Jastrow/Wikimedia Commons/Public Domain; **25**l, **25**c, **25**r, **29**l, **31**l, **31**c, **32, 33, 48, 49, 52, 53, 57**t, **57**b, **59, 64**t, **67**bl, br BJU Photo Services; **29**c © iStockphoto.com/Bradley Mason; **29**r © iStockphoto.com/Shaun Lowe; **30**t Copyright Dr. Rich Busch; **30**b Aaron Dickey; **31**r © iStockphoto.com/Jonathan Ling; **35**l E R DEGGINGER/Photo Researchers/Getty Images; **35**c © iStockphoto.com/Valeriy Gontar; **35**r Courtesy of Baker Hughes; **37**t © Sinclair Stammers/Photo Researchers, Inc.; **37**b Photo is courtesy of DuPont; **43**t Top Photo Group/Thinkstock; **43**b Vladimir Shulevsky/age fotostock/Getty Images; **47**b NASA/JPL-Caltech; **50** Copyright Robert Rathe/NIST; **51** Alex Wild/Visuals Unlimited, Inc./Getty Images; **58** © iStockphoto.com/james forte; **61** Hemera/Thinkstock; **62**t Sarah Ensminger; **62**c, b Courtesy of Andrew Tool and Machining; **63** PhotoDisc, Inc.; **64**b De Agostini/Getty Images; **67**t © iStockphoto.com/Pierre-Emmanuel Turcotte

Unit 2

72–73t © iStockphoto.com/Leonardo Fav; **72–73**b, **146** PhotoDisc, Inc.; **72, 74**t © iStockphoto.com/Sergii Tsololo; **72, 90**t Nancie Battaglia/Sports Illustrated/Getty Images; **72, 109**t Courtesy of Los Alamos National Laboratory; **72, 128**t Photography by Deryc Sands. UK Parliament copyright; **72, 151**t USCG photo by Petty Officer Kevin Martin; **72, 177**t USDA Forest Service Photo; **74**b Photo by Victor Keppler/George Eastman House/Getty Images; **75**l Courtesy of Cedar Point; **75**r Ivan Petrovich Keler-Viliandi/The Bridgeman Art Library/Getty Images; **77**t NASA; **77**c Photograph by Alexander C. Treneff; **77**b Rubberball/JupiterImages; **82** John Berry/Getty Images Sport/Getty Images; **83**t Segway Inc.; **83**b "JRW-500 V2 inHimeji" by Rsa/Wikimedia Commons/GNU FDL 1.2; **84** © Shiningcolors | Dreamstime.com; **85**t U.S. Navy Photo by Jonathan Snyder; **85**b Mike Hewitt/Allsport/Getty Images; **86**t Hal Pierce, Laboratory for Atmospheres, NASA Goddard Space Flight Center; **86**b Photo courtesy of First Technology Safety Systems, Plymouth, MI; **90**bl, **116**b © 2006 JupiterImages Corporation; **90**br © iStockphoto.com/Adrian Dracup; **92** NASA/JPL/Space Science Institute; **95** "Newton-Principia (1687), title, p. 5, color"/Piero/Wikimedia Commons/Public Domain; **96**bg, **102** NASA; **98** iStockphoto/Thinkstock; **99** U.S. Navy photo by Aaron Burden; **100** Imagno/Hulton Fine Art Collection/Getty Images; **103** DeBeers; **105**l Cpl. Joel Abshier, USMC; **105**r Courtesy of US Army; **106** Everett Collection/SuperStock; **109**bl © iStockphoto.com/josemoraes; **109**br Heat Island Group at Lawrence Berkeley National Laboratory, Berkeley, California, USA; **110**l © iStockphoto.com/Ryan Lindsay; **110**r © Skyscan/Photo Researchers, Inc.; **111**t "Queen Nefertiti Rock in Arches NP" by Daniel Mayer/Wikimedia Commons/CC-BY-SA 2.5; **111**b © iStockphoto.com/John Muchow; **112** Cynthia Johnson/Time Life Pictures/Getty Images; **113**t R. Terrance Egolf; **113**b SSGT Bob Simons; **114**t MCT/McClatchy-Tribune/Getty Images; **114**b Digital Vision; **115**t The Verdin Company; **115**bl © iStockphoto.com/Scott Lomenzo; **115**br Flagstaffoto; **116**t © iStockphoto.com; **116**c Cusp/SuperStock; **117** Courtesy of Energy Northwest; **119** Cesar Lucas Abreu/age fotostock/Getty Images; **121, 123**t, **153**t, b, **166**l, c, **190** BJU Photo Services; **122**t © iStockphoto.com/Bob Coyle; **122**b © Ktree | Dreamstime.com; **123**c Courtesy of SC Highway Patrol; **123**b Aaron Dickey; **125, 130, 138**l, **162**t, **178**r © 2007 JupiterImages Corporation; **128**b "James Pollard - The Louth-London Royal Mail Travelling by Train from Peterborough East, Northamptonshire - Google Art Project" by James Pollard/Wikimedia Commons/Public Domain; **129** U.S. Navy photo by Michael R. McCormick; **131**t © iStockphoto.com/Philip Lange; **131**b U.S. Navy photo by Kristopher Wilson; **132** Smith of Derby, UK. Regulator covered by patent; **135, 162**b, **172**b, **173** Susan Perry; **138**r GUSTOIMAGES/Science Photo Library/Getty Images; **142** Fuller Manual Transmission provided by Eaton Corporation; **143** © iStockphoto.com/Sergey Chushkin; **148** Copyright © Lakeside Equipment; **151**b AP Photo/Kyodo News, File; **152**l US Geological Survey, photo by Dale Blank; **152**r © iStockphoto.com/Glen Jones; **161** "Octopus3" by albert kok/Wikimedia Commons/GNU FDL 1.2,

CC-BY-SA 3.0; **166**r Photo courtesy of Englert LeafGuard; **167** Paul Bowen/Science Faction/Getty Images; **172**t 2007 Copyright, © University Corporation for Atmospheric Research; **177**bl "Turbiny wiatrowe w Szwecji, 2011 ubt" by Tomasz Sienicki/Wikimedia Commons/GNU FDL 1.2, CC BY 3.0; **177**br © Clouston/Shutterstock; **178**l Stock Montage/Archive Photos/Getty Images; **179** "Count Rumford"/Corel Photo CD, MASTERS I/Wikimedia Commoms/Public Domain; **180**tl © Sheila Terry/Photo Researchers, Inc.; **180**r, bl SSPL via Getty Images; **181**l Hulton Archive/Stringer/Getty Images; **181**r © J-L CHARMET/PHOTO RESEARCHERS, INC.; **182**t © Peter Menzel/Science Source; **182**b "PS20andPS10" by Koza1983/ Wikimedia Commons/CC BY 3.0; **183** Science and Society/ SuperStock; **184** © iStockphoto.com; **185**l © Scala/Art Resource, NY; **185**r AP Photo/Gdansk University of Technology Press Office; **186** "Anders-Celsius" by Olof Arenius/Wikimedia Commons/Public Domain; **187** © iStockphoto.com/Jeremy Edwards; **188**t © Jonathan Wilson | Dreamstime.com; **188**b © ROBIN SCAGELL/SCIENCE PHOTO LIBRARY; **189** Dollie Harvey/University of Connecticut; **192**t Royal Swedish Academy of Sciences; **192**b © 2005 Derek Ramsey/Wikimedia; **193**t © iStophoto.com/Sergey Kashkin; **193**b Pella Windows and Doors; **194**tl, tr JPL/NASA; **194**b © ITAR-TASS Photo Agency/Alamy; **196** Photo courtesy of Superbolt®, Inc. www.superbolt.com; **199**l "Plagiomnium affine laminazellen" by Kristian Peters--Fabelfroh/Wikimedia Commons/GNU DFL 1.2, CC-BY-SA 3.0; **199**r © iStockphoto.com/Monika Lassaud

Unit 3

202–3t, **245**l, i Don Congdon; **202–3**b Wayne R Bilenduke/ Photographer's Choice/Getty Images; **202**, **204**t © SNL/DOE/ Photo Researchers, Inc.; **202**, **228**t CORDELIA MOLLOY/ Science Photo Library/Getty Images; **204**b, **205**c, **221**br, **228**br © 2007 JupiterImages Corporation; **205**t, **209**, **210**b Charles D. Winters/Photo Researchers/Getty Images; **205**b, **223**, **231**l, **235**l, r, **236**l Aaron Dickey; **206**l, r, **211**, **231**r Susan Perry; **210**tl, tr, **213**, **217**, **222**, **236**r BJU Photo Services; **212**r "BigRed-Sprite" by Eastview/Wikimedia Commons/Public Domain; **212**l Photo Researchers/Getty Images; **215** NASA; **218**t Rich LaSalle/The Image Bank/Getty Images; **218**b GI-PhotoStock/Photo Researchers/Getty Images; **219** © Andrew Lambert Photography/Photo Researchers, Inc.; **220**t Science and Society/SuperStock; **220**b © iStockphoto.com/Nicolae Popovici; **221**t Jim Zimmerman; **221**bl © iStockphoto.com/ Norman Morin; **224**t Courtesy of Greenville Fire Department; **224**cr, **243** Coy Moore; **224**cl © iStockphoto.com/Edward Todd; **224**bl, br Photo courtesy of Schneider Electric, 2007; **225** Joyce Landis; **228**bl Stock Connection/SuperStock; **230** Stock Connection/SuperStock; **233** "Diamagnetic graphite levitation" by Splarka/Wikimedia Commons/Public Domain; **238**t "Steam Turbine" by TDC/Wikimedia Commons/GNU FDL 1.2, CC-BY-SA 3.0; **238**b Photo courtesy of DTL Hydro; **239**t © iStockphoto.com/James Graham; **239**b © iStockphoto .com/jim pruitt; **241**t, **244**tl Department of Energy; **241**b Franz Von Riedel; **244**tr © iStockphoto.com/David Woods; **244**b ©Alex Bartel/Photo Researchers, Inc.; **245**r © iStockphoto .com/Janne Ahvo; **246** Fred Ullrich, Fermilab Visual Media Services; **247**r Ross Engineering Corporation, 540 Westchester Drive, Campbell, CA 95008; **247**l "A maglev train coming out, Pudong International Airport, Shanghai" by Alex Needham/ Wikimedia Commons/Public Domain; **249**l Stock Montage/ Archive Photos/Getty Images; **249**r Smithsonian Institute, Neg. #74-249

Unit 4

252–53t David Eppstein; **252–53**b © iStockphoto.com/Slawomir Jastrzebski; **252**, **254**t © iStockphoto.com/Ivan Dinev; **252**, **277**t © Brandon Cole Marine Photography/Alamy; **252**, **299**t Image courtesy of NRAO/AUI and Image courtesy of Stephen White, University of Maryland, and of NRAO/AUI/ NSO/NASA/NASA/Goddard/SDO AIA Team/NOAA/NASA/ Lockheed Martin/COMPTEL team, University of New Hampshire; **252**, **325**t "Full double rainbow" by Niklas Tyrefors/ Wikimedia Commons/CC-By SA 3.0; **254**b Rare Books and Special Collections, University of Sydney Library; **255**l Jacques Descloitres, MODIS Rapid Response Team, NASA/GSFC; **255**c © iStockphoto.com/Alexander Sakhatovsky; **255**r NASA and The Hubble Heritage Team (STScI/AURA); **256**l, r, **279**t, **342**bl BJU Photo Services; **257**, **290**, **291**t, **293**, **326**, **336** © 2007 JupiterImages Corporation; **258** © iStockphoto.com/José Carlos Pires; **259** © fotolia.com/Danny Hooks; **260** Ted Kinsman/ Science Source; **261**t Museum of New Zealand Te Papa Tongarewa; **261**b Christian Huygens/The Bridgeman Art Library/ Getty Images; **262** "Péndulo de Foucault (M. Ciències Valencia) 02" by Manuel M. Vicente/Wikimedia Commons/CC BY 2.0; **265** Pacific Stock - Design Pics/SuperStock; **266**t U.S. Navy photo by Ensign John Gay; **266**c AP Photo/Goran Stenberg; **266**b Aaron Dickey; **277**b Andrea Booher/FEMA; **278**t © Will & Deni McIntyre/Photo Researchers, Inc.; **278**b © Smileus | Dreamstime.com; **279**b Edgar Fahs Smith Collection, University of Pennsylvania Library; **282** Architect of the Capitol; **283**t Photo by David McNew/Getty Images; **283**b, **284**t Susan Perry; **284**b © iStockphoto.com/Andrea Leone; **286** Courtesy of Dr. Parsons; **288**t © CNRI/Photo Researchers, Inc.; **288**b © Dave Roberts/Photo Researchers, Inc.; **289** This research, Validation/Dissemination of Virtual Temporal Bone Dissection, is supported by a grant from the National Institute on Deafness and Other Communication Disorders, of the National Institutes of Health, 1R01 DC06458-01A1; conducted by Nikolai Svakhin, Don Stredney, David Ebert; **291**b © Sheila Terry/Photo Researchers, Inc.; **294** Courtesy Birring NDE Center, Houston; **296**l © iStockphoto.com/Kenneth C. Zirkel; **296**c Courtesy of Heather Stanley; **296**r Amdent; **299**b age fotostock/SuperStock; **301** Pantheon/SuperStock; **305**t © SPL/Photo Researchers, Inc.; **305**b NASA, Andrew Fruchter and the ERO Team [Sylvia Baggett (STScI), Richard Hook (ST-ECF), Zoltan Levay (STScI)] (STScI); **307** Image courtesy of NRAO/AUI and J. M. Uson; **308** © iStockphoto.com/Mark Jensen; **309**tl, tc, tr Bob Jones University; **309**b © iStockphoto .com; **311**l Courtesy of Merck & Co., Inc.; **311**r © Tom Grill/ Corbis; **312**tr © Jack Hollingsworth/Corbis/Media Bakery; **312**l NASA/CXC/STScI/U.North Carolina/G.Cecil; **312**br Newco, Inc.; **313**t DOE; **313**b Elekta, Inc.; **315** Harbour Light of the Windwards; **316**t © iStockphoto.com/Brad Wieland; **316**b Courtesy of FONAR Corporation; **317** Copyright Bettmann/Corbis/AP Images; **318**tl iStockphoto/Thinkstock; **318**r JupiterImages/Pixland/Thinkstock; **318**bl Hemera/Thinkstock; **319**t © iStockphoto.com/Peter Elvidge; **319**br Photo courtesy of VDOT; **319**bc Scott Olson/Getty Images; **319**bl Photo courtesy of The Kennedy Group, Inc.; **320**t Photo courtesy of Furuno USA, Inc.; **320**b ProMark3 RTK GPS system, ©2007 Magellan Navigation Inc.; **321**t Tom Brown; **321**b, **322**r Image courtesy of NRAO/AUI; **322**l Image courtesy of NRAO/AUI and Chandra: NASA/CXC/SAO/B.Gaensler et al.; **325**b National Library of Medicine; **327** Michael Stegina/Adam Block/NOAO/AURA/NSF; **328** NASA, ESA, M. Robberto (Space Telescope Science Institute/ESA) and the Hubble Space Telescope Orion Treasury Project Team; **329** © iStockphoto.com/Daniel Albiez; **330**tl © Mark A. Schneider/Photo Researchers, Inc.; **330**bl © iStockphoto.com/Robert Kyllo; **330**r Photo provided by AmeriGlo, Atlanta, GA; **331**t U.S. Navy; **331**b Sarah Ensminger; **332** © Fotolia/O.M.; **334** Don Congdon; **335** © Eye of Science/Photo Researchers, Inc.; **337**t © iStockphoto.com/Paiwei Wei; **337**b PhotoDisc, Inc.; **338**r © iStockphoto.com/Dragan Trifunovic; **338**l Photo Researchers, Inc.; **340** © iStockphoto.com/Dorit Jordan Dotan; **342**t Tammy Fisher; **342**br Pekka Parviainen; **343** © Phanie/Photo Researchers, Inc.; **344** Adam Hart-Davis; **348** picture-alliance/dpa

Unit 5

352–53t Science Source; **352–53**b © Dirk Wiersma/Photo Researchers, Inc.; **352**, **354**t, **361** © iStockphoto.com/Andrei Tchernov; **352**, **382**t iRocks.com photo; **354**b © iStockphoto.com/Scott Leman; **355** © Dr Tim Evans/Photo Researchers, Inc.; **356** Pantheon/SuperStock; **357** The Cavendish Laboratory; **358** Universal Images Group/Getty Images; **359**tr Elliott & Fry/Stringer/Hulton Archive/Getty Images; **359**br © Princeton University/AIP/Photo Researchers, Inc.; **359**l Alex Maclean/Getty Images; **362**l © iStockphoto.com/Olaru Radian-Alexandru; **362**c © Warren Price | Dreamstime.com; **362**r Photo Researchers/Getty Images; **364**l © University Corporation for Atmospheric Research, photo by Terry Hock; **364**r © Martyn F. Chillmaid/Photo Researchers, Inc.; **370** "Portrait of Antoine-Henri Becquerel" by Paul Nadar/Wikimedia Commons/Public Domain; **373** "InsideSmoke-Detector" by MD111/Wikimedia Commons/CC-BY-SA 2.0; **374** Lloyd Homer/GNS Science; **376** Library of Congress; **377** National Nuclear Security Admin/Nevada Site Office; **378** © ITER Organization, http://www.iter.org; **382**b © 2007 Insadco Photography and World of Stock; **383** An Alchemist in his Workshop (oil on canvas), Teniers, David the Younger (1610–90)/Musee de Tesse, Le Mans, France/Giraudon/The Bridgeman Art Library; **384** De Agostini Picture Library/Getty Images; **385**t "Portrait of Antoine-Laurent Lavoisier and his wife" by Jacques-Louis David/Wikimedia Commons/Public Domain; **385**c William L. Krewson/bibleplaces.com; **385**b BJU Photo Services; **387** SCOTT CAMAZINE/Photo Researchers/Getty Images; **390**t Edgar Fahs Smith Collection, University of Pennsylvania Library; **390**b © SPL/Photo Researchers, Inc.; **391**l Photo: A. Zschau, GSI; **391**c Photo Courtesy of National Nuclear Security Administration/Nevada Site Office; **391**r Sabina Fletcher and Thomas Tegge/Lawrence Livermore National Laboratory; **394** National Bureau of Standards Archives, courtesy AIP Emilio Segre Visual Archives, W. F. Meggers Collection; **397**t, **402** Charles D. Winters/Photo Researchers/Getty Images; **397**b copyright © Boeing; **398**, **403**bl BJU Photo Services; **399** © 2007 Advanced Micro Devices, Inc.; **400** Illustration © Purdue Products L.P., used with permission.; **403**tl © SIMON FRASER/MEDICAL PHYSICS, RVI, NEWCASTLE UPON-TYNE/Photo Researchers; **403**r NASA/ESA/GSFC

Unit 6

410–11t Digital Vision; **410–11**b MedioImages/Getty Images; **410**, **412**t © Javier Trueba/MSF/Photo Researchers, Inc.; **410**, **436** National Archives; **410**, **457**t PhotoDisc, Inc.; **410**, **482** © iStockphoto.com/Georg Hafner; **412**b © iStockphoto.com/Pali Rao; **413**r iRocks.com photo; **413**l "Graphite-233436" by Rob Lavinsky, iRocks.com/Wikimedia Commons/CC-BY-SA 3.0; **419** © iStockphoto.com/Adivin; **420** Lynn Betts/USDA NRCS; **423**t © Dr Tim Evans/Photo Researchers, Inc.; **423**b Bob Jones University; **424** Banana Stock/© 2007 JupiterImages, Inc.; **429** © Russell Kightley; **431**t Courtesy of Advanced Surface Microscopy, Inc.; **431**b © Astrid & Hanns-Frieder Michler/Photo Researchers, Inc.; **432**t Ken-tron Mfg., Inc., Owensboro, KY; **432**b © 2007 JupiterImages, Inc.; **437** USDA; **442**t Photo Researchers/Getty Images; **442**c © iStockphoto.com/malerapaso; **442**b Charles D. Winters/Science Source; **443** "Iron(II) oxide" by Fuzik/Wikimedia Commons/Public Domain; **445** Jim Redyke of Dykon Explosive Demolition Corp., Tulsa, OK; **450**, **451**c, **459**c, **462**, **470** all, **471**t, **483**, **490**tl, bl, **495**t, b BJU Photo Services; **451**t, **487** © Charles D. Winters/Photo Researchers, Inc.; **451**b Charles D. Winters/Photo Researchers/Getty Images; **452**t © Martyn F. Chillmaid/Photo Researchers, Inc.; **452**b AFP/Getty Images; **453**tr Carolyn A McKeone/Photo Researchers/Getty Images; **453**l © iStockphoto.com/Karin Lau; **453**br Bryan Mullennix/Iconica/Getty Images; **457**b AP Photo/Jeffrey Boan; **459**t Courtesy of Drs. Kazuyoshi Kanamori and Kazuki Nakanishi; **459**b © iStockphoto.com/Eric Naud; **460** © Astrid & Hanns-Frieder Michler/Photo Researchers, Inc.; **461** Hemera/Thinkstock; **463** Mark H. Armitage, M.S., Ed.S., Creation Research Society; **464** Image created by Reto Stockli with the help of Alan Nelson, under the leadership of Fritz Hasler/NASA; **469** Reinhard Dirscherl/WaterFrame/Getty Images; **471**b Sarah Ensminger; **476** courtesy of VID Desalination; **484**t Leland Bobbe/Stockbyte/Getty Images; **484**b "A Formica rufa sideview" by Richard Bartz/Wikimedia Commons/CC-BY-SA 2.5; **485** Time & Life Pictures/Getty Images; **489** © iStockphoto.com/Ken Cameron; **490**r Photofabrication Engineering, Inc.; **494** © iStockphoto.com/Lisa Thornberg; **497** Claire-Sprayway, Inc.; **499** Mike Sutcliffe; **500**t © iStockphoto.com/Sabine Kappe; **500**b Don Congdon

Appendix

520 Andrew Naprienko